INDEX OF APPLICATIONS

(Continued in back of book)

Mathematics

All Around

Second Edition

Thomas L. Pirnot
Kutztown University of Pennsylvania

PEARSON

Addison
Wesley

Boston San Francisco New York
London Toronto Sydney Tokyo Singapore Madrid
Mexico City Munich Paris Cape Town Hong Kong Montreal

Publisher:	**Greg Tobin**
Senior Acquisitions Editor:	**Anne Kelly**
Project Editor:	**Rachel S. Reeve**
Associate Project Editor:	**Joanne Ha**
Assistant Editor:	**Jaime Bailey**
Editorial Intern:	**Marcia Emerson**
Managing Editor:	**Karen Guardino**
Senior Production Supervisor:	**Peggy McMahon**
Marketing Manager:	**Becky Anderson**
Marketing Coordinator:	**Julia Coen**
Associate Media Producer:	**Sara Anderson**
Software Development:	**Mary Dougherty, Gail Light**
Text and Cover Designer:	**Barbara T. Atkinson**
Cover Photo:	**© Ed Bock/Corbis**
Senior Prepress Supervisor:	**Caroline Fell**
Composition and Illustration:	**Progressive Information Technologies**
Production Services:	**Progressive Publishing Alternatives**

For permission to use copyrighted material, grateful acknowledgment is made to the copyright holders on page 905 in the back of the book, which is hereby made part of this copyright page.

Library of Congress Cataloging-in-Publication Data
Pirnot, Thomas L.
    Mathematics all around/Thomas L. Pirnot. — 2nd ed.
      p. cm.
    Includes index.
    ISBN 0-201-79511-6
    1. Mathematics. I. Title.

QA39.3.P57 2003
510--dc21

2002032603

ISBN 0-201-79511-6

2 3 4 5 6 7 8 9 10—QWT—06050403

*To my wife Ann whose loving support and encouragement
have been the wind beneath my wings.*

Contents

Preface

I have written *Mathematics All Around* based on the belief that there are three things a student must focus on in order to learn and remember mathematics—understanding, understanding, understanding! Instead of simply presenting the students with an equation or method and asking them to repeat the procedure, I explain the thinking behind the subject so they have a better grasp of the material and an easier time with the work. With this approach, we can discuss topics usually considered "too difficult" for liberal arts students such as fuzzy logic and dynamical systems. As a result, students end up with an understanding of and a positive attitude toward many different and often challenging mathematical topics.

I, and many other instructors who have used the first edition, have found that students respond enthusiastically to the realistic, modern applications presented in this text. I believe that the students who use this book will become more educated consumers of the vast amount of technical and mathematical information that they encounter daily. I also want to present to them some of the most interesting and exciting ideas of mathematics at a level accessible to all students.

Some of the topics from this text that students have enjoyed are:

- Using **fuzzy logic** in complex decision making

- Using **dynamical systems** to explore antibiotics in the bloodstream and the twentieth-century arms race

- Organizing large projects by using **directed graphs**

- Understanding the role mathematics plays in the politics of **voting and apportionment**

- Using mathematics to help them make better **financial decisions** involving loans, credit cards, mortgages, and annuities

- Understanding how **counting and probability** theory can explain lotteries, gambling, and the genetics of inherited diseases

- Using linear, quadratic, exponential, and logistic equations for **modeling** realistic situations such as the ups and downs of presidential popularity polls and the growth of epidemics while appreciating the appropriateness, strengths, and most importantly, the weaknesses of these models

- Seeing how **geometry** is used to solve practical everyday problems and understanding how **fractal geometry** can pattern the way blood vessels and air passageways are arranged in the human body and how it can be used to create realistic, computer-generated, natural scenes

- Understanding how to use **statistics** to understand and compare data and also learning how to write a warranty that will almost always run out before the product breaks down.

As with the first edition, I have written the second edition to provide students majoring in the liberal arts, the social sciences, education, business, and other nonscientific areas an understanding and appreciation of mathematics and its many fascinating applications. This book is particularly appropriate for students who need to satisfy a one- or two-course requirement in mathematics for graduation or to transfer to another institution.

In this second edition, I have responded to the advice of reviewers and previous users by expanding the content (as described below), refining the presentations of some material to make it easier to teach and learn, and enlarging the exercise sets to allow students more opportunity to practice their newly learned skills. Additionally, I have incorporated their suggestions regarding communicating mathematics and using technology throughout the book via the exercises and Web site.

New Content

I have added three chapters to this text:

- *Numeration Systems* (Chapter 4)—Includes a discussion of modular arithmetic and secret codes
- *Number Theory and the Real Number System* (Chapter 5)—Includes sections on exponents and scientific notation and sequences
- *Modeling with Systems of Linear Equations and Inequalities* (Chapter 7)—Includes a section on linear programming and its applications

Additionally, I have added sections to the following chapters:

- Chapter 1, *Set Theory,* now contains a new section on estimation (1.2).
- Chapter 2, *Logic,* has a new section on inductive and deductive reasoning (2.1).
- Chapter 6, *Algebraic Models,* has a section dealing with proportion and variation (6.5).
- Chapter 8, *Geometry,* now contains a section on the metric system and dimensional analysis (8.5).
- Chapter 9, *Apportionment,* has been streamlined and I have added a new section discussing Jefferson's, Adams', and Webster's apportionment methods (9.4).
- Chapter 14, *Descriptive Statistics,* now includes a brief discussion of gathering reliable data (14.1).

We have deleted the chapter on matrices, which included a section on matrix representations of geometrical transformations and computer graphics. However, those who still wish to use this material can download it in its entirety from the book's Web site, *www.aw.com/pirnot.*

Pedagogical Features

The many strong pedagogical features of Mathematics All Around *will make this text a useful tool for the student.*

Motivation with Emphasis on Applications

Applications throughout the text motivate the discussion of the mathematics and increase the student's interest in the material. In addition, each chapter opener emphasizes realistic problem situations and gives a broad overview of the chapter material. Toward the end of this brief section, I frequently present problems to motivate the mathematics we will be discussing. After developing the necessary mathematical tools, I return to solve the problem, reinforcing the value of the concepts studied in the chapter.

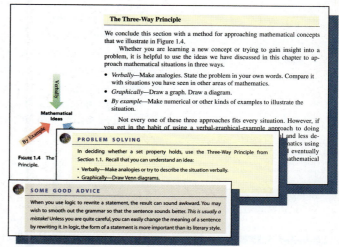

CHAPTER **3**

Graph Theory:*

The

branch offices wishes to minimize

re airport, air traffic controllers
o get passengers to their destina-

te the tribal alliances among

works administrator in Buffalo,
to locate the most efficient routes

congressional subcommittee
who exerts the greatest influence.

nrelated, each of these people is
ects to each other. The sales man-
between cities in terms of the cost
uffalo administrator has to know
nnected by streets. The political
he committee members are

page 140

3.2 THE TRAVELING SALESPERSON PROBLEM

Some problems in mathematics are so simple to state that a child can understand them, yet the greatest mathematicians in the world cannot solve them. This is the case with a famous and difficult problem in graph theory called **the traveling salesperson problem (TSP)**. The TSP gets its name from the problem of determining the most efficient way for a salesperson to schedule a trip to a series of cities and then return home.

For an example of a TSP, suppose that Danielle, who is regional sales manager for a publishing company, lives in Philadelphia and must make visits next week to branch offices in New York City, Cleveland, Atlanta, and Memphis (see Figure 3.21). In order to determine which would be her cheapest trip, she has obtained prices of flights between each pair of

FIGURE 3.21 Cities Danielle must visit.

page 159

Problem Solving, expanded

Problem solving is one of the main themes of this text. Section 1.1 discusses strategies and principles that help the student understand and attack problems more effectively. In this section I introduce the Three-Way Principle, which encourages students to approach mathematical ideas graphically, verbally, and by constructing examples. I return to these discussions frequently, via the **Problem Solving** boxes, thereby making problem solving an integrated component of each chapter. Remarks titled **Some Good Advice** point out common mistakes, provide further advice on problem solving, and make connections between different areas of mathematics. Some Good Advice includes learning not to use *equal* and *equivalent* interchangeably or learning that in logic, the form of a statement is more important than its literary style. I have increased the number of **Problem Solving** and **Some Good Advice** boxes in this edition.

The Three-Way Principle

We conclude this section with a method for approaching mathematical concepts that we illustrate in Figure 1.4.

Whether you are learning a new concept or trying to gain insight into a problem, it is helpful to use the ideas we have discussed in this chapter to approach mathematical situations in three ways.

- *Verbally*—Make analogies. State the problem in your own words. Compare it with situations you have seen in other areas of mathematics.
- *Graphically*—Draw a graph. Draw a diagram.
- *By example*—Make numerical or other kinds of examples to illustrate the situation.

Not every one of these three approaches fits every situation. However, if you get in the habit of using a verbal-graphical-example approach to doing ... and less de... ...atics using ... eventually ... athematical

Verbally

Mathematical Ideas

By Example

FIGURE 1.4 The Principle.

PROBLEM SOLVING

In deciding whether a set property holds, use the Three-Way Principle from Section 1.1. Recall that you can understand an idea:

- Verbally—Make analogies or try to describe the situation verbally.
- Graphically—Draw Venn diagrams.

SOME GOOD ADVICE

When you use logic to rewrite a statement, the result can sound awkward. You may wish to smooth out the grammar so that the sentence sounds better. *This is usually a mistake!* Unless you are quite careful, you can easily change the meaning of a sentence by rewriting it. In logic, the form of a statement is more important than its literary style.

pages 12, 45, 96

Format

The format makes this text easy to use and makes learning mathematics easier for the student. I introduce ideas conversationally and then illustrate them with clear, simple examples. Summary boxes help students locate information when doing exercises. After reinforcing highlighted points with more examples, students test their understanding by taking a **Quiz Yourself**. More detailed examples occur later in a section after the ideas have been clearly discussed.

page 523

Quiz Yourself, expanded

Each section contains numerous short quizzes called "Quiz Yourself," which students can use to check their understanding of the material immediately preceding the quiz. These quizzes can be used as a break in the flow of the lecture material and to encourage student participation. The number of **Quiz Yourself** exercises has been increased.

page 449

Highlights, expanded

Each chapter contains **Highlight** boxes that discuss the history and applications of the topics being presented. These highlights help students understand the material that they are learning in a broader context. There are more highlights, particularly those titled **Historical Highlights.** A listing of the historical figures mentioned in the text is located in the inside of the front cover and includes individuals such as Aristotle, George Polya, Sophie Germain, and Georg Cantor.

Using Technology, expanded

Also in the **Highlight** boxes, I encourage students to take advantage of available technology. I explain how tools such as spreadsheets, graphing calculators, and computer algebra systems can be used effectively to solve problems. Although I encourage the students to use these technologies, I have purposely not made technology a requirement for this book. Throughout the text, I make frequent references to the technology section of the Web site mentioned earlier. There the instructor and student can download tutorials and examples using spreadsheets, graphing calculators, and computer algebra systems.

page 509

Topic Statements

Topic statements divide each section into subsections to provide students with a useful outline of the chapter's material as well as providing a clear idea of the concept being discussed.

Exercises, expanded

In general, the exercise sets begin with straightforward problems that closely follow the examples and **Quiz Yourself** questions. In addition to these problems, other exercises require the student to think critically about extensions of ideas. Many exercises are application oriented, using real data. More challenging and open-ended exercises are placed in a section labeled **Further Exercises.**

Many of the exercise sets have been greatly expanded. We also have labeled numerous exercises **Communicating Mathematics,** which ask the student to write about the mathematics he or she is studying. We have also indicated a number of exercises (their numbers are circled in red) that can be used as **Group Exercises.**

Summary and Chapter Test

Each chapter has a summary followed by a chapter test. Chapter test questions are cross-referenced to the sections that cover the points being tested. This enables a student who is not comfortable with a particular concept to return to the section covering that material.

Of Further Interest

Brief sections called **Of Further Interest** introduce subjects that go beyond the main chapter material but are of interest for students. The section "The Annual Percentage Rate" in the consumer mathematics chapter illustrates this intent. Because the **Of Further Interest** section appears at the end of the chapter, it can be treated as a bonus topic or easily omitted at the instructor's discretion.

132   Chapter 2   Logic: The Study of What's True or False or Somewhere in Between

Of Further Interest: FUZZY LOGIC

To say "It is warm today" may not mean the same thing on the first day of March as it does in the middle of August. Statements we make in everyday life have shades of meaning and, in that sense, differ from the statements that we have been studying in symbolic logic. Although it may be warm today, perhaps it is not as warm as it was two weeks ago. To simply say "It is warm today" ignores exactly how warm it is. The strict condition that statements must be either true or false makes symbolic logic unsuitable for representing many real-life situations.

Statements in fuzzy logic have truth values between 0 and 1.

In order to apply the techniques of logic more widely, mathematicians developed *fuzzy logic.* This term might seem contradictory, but, as you will see shortly, it is a perfectly acceptable, well-developed area of mathematics. Once we define the meaning of fuzzy statements, and explain how fuzzy connectives behave, we can work with them as we did for other (nonfuzzy*) statements.

> **DEFINITIONS**
>
> In **fuzzy logic**, a **statement** is a declarative sentence that has an associated **truth value** between 0 and 1 inclusive.

EXAMPLE 1   **The Truth Values of Statements in Fuzzy Logic**

Here are some examples of statements in fuzzy logic.
a) "I like peach ice cream," with a truth value of 0.9.
b) "Tiger Woods is a great golfer," with a truth value of 0.92.
c) "Maine is a large state," with a truth value of 0.45.
d) "*Gone with the Wind* is a great movie," with a truth value of 0.73.
e) "You find mathematics interesting," with a truth value of 0.7.

The way we assigned truth values in Example 1 is somewhat arbitrary. You may want to assign a truth value of 0.85 to the statement "*Gone with the Wind* is a great movie" because you feel that it is a greater movie than we do. *This is perfectly OK and is really the point of fuzzy*

_____
* Some use the term crisp statements to describe the statements we studied earlier.

Supplements

SUPPLEMENTS FOR THE INSTRUCTOR

Instructor's Edition
ISBN 0-201-79512-4

This version of the text includes answers to all of the problems in addition to the material found in the student edition.

Instructor's Solutions Manual
Heidi Howard, Florida Community College – Jacksonville
ISBN 0-321-10900-7

This manual contains detailed, worked-out solutions to all the exercises in the text.

Instructor's Testing Manual
ISBN 0-321-10901-5

The Testing Manual includes three alternative tests per chapter. These items may be used as actual tests or as references for creating tests.

TestGen with QuizMaster
ISBN 0-201-92038-7

TestGen enables instructors to build, edit, print, and administer tests using a computerized bank of questions developed to cover all the objectives of the text. TestGen is algorithmically based so that multiple yet equal versions of the same question or test can be generated at the click of a button. Instructors can also modify test bank questions or add new questions by using the built-in question editor, which allows users to create graphs, import graphics, insert math notation, and insert variable numbers or text. Tests can be printed or administered online via the Web or other network. Many questions in TestGen can be expressed in a short-answer or multiple-choice form, giving instructors greater flexibility in their test preparation. TestGen comes packaged with QuizMaster, which allows students to take tests on a local area network. The software is available on a dual-platform Windows/Macintosh CD-ROM.

MathXL

MathXL is a Web-based program that provides online diagnostic testing, homework, and tutorial help using algorithmically generated exercises correlated closely to the textbook. Students can take chapter tests, receive individualized study plans based on their test results, work unlimited practice problems and receive tutorial instruction for areas in which they need improvement, and take further tests to gauge their progress. With MathXL, instructors can assign and customize preloaded tests and homework assignments or create their own tests and homework and track all student results and practice work in the online gradebook. MathXL is free when the access code is bundled with a new Addison Wesley text.

MyMathLab

MyMathLab is a complete, online course for Addison Wesley mathematics textbooks that provides interactive, multimedia instruction correlated to the textbook content. MyMathLab is easily customizable to suit the needs of students and instructors and provides a comprehensive and efficient online course-management system that allows for diagnosis, assessment, and tracking of students' progress.

MyMathLab features

- Chapter and section folders in the online course mirror the textbooks' Table of Contents and contain a wide range of multimedia instruction, including video lectures, tutorial software and electronic supplements.

- The actual pages of the textbook are loaded into MyMathLab, and as students work through a section of the online text, they can link to multimedia resources—such as video and audio clips, tutorial exercises, and interactive animations—that are correlated directly to the examples and exercises in the text.

- Hyperlinks take you directly to online testing, diagnosis, tutorials, and tracking in MathXL—Addison Wesley's tutorial and testing system for mathematics and statistics.

- Instructors can create, copy, edit, assign, and track all tests and homework for their course as well as track students' results and practice work.

- With push-button ease, instructors can remove, hide, or annotate Addison Wesley preloaded content, add their own course documents, or change the order in which material is presented.

- Using the communication tools found in MyMathLab, instructors can hold online office hours, host a discussion board, create communication groups within their class, send e-mail, and maintain a course calendar.

- Print supplements are available online, side by side with their textbooks.

- All of the material found at *www.aw.com/pirnot* is also available through MyMathLab.

For more information, visit our Web site at *www.mymathlab.com* or contact your Addison Wesley sales representative for a live demonstration.

Videotapes
ISBN 0-321-10902-3
Videotapes correlated to each important topic are available to departments. Contact your Addison Wesley Sales Consultant.

SUPPLEMENTS FOR THE STUDENT

Student's Solutions Manual
Heidi Howard, Florida Community College – Jacksonville
ISBN 0-321-10904-X
This manual contains detailed worked-out solutions to all the odd-numbered section exercises and to all Chapter Test exercises. Students will find this manual very helpful.

Addison Wesley Math Tutor Center
The Addison Wesley Math Tutor Center is staffed by qualified math and statistics instructors who provide students with tutoring on examples and odd-numbered exercises from the textbook. Tutoring is available via toll-free telephone, fax, email, or the Internet, and White Board technology allows tutors and students to actually see the problems worked while they "talk" in real time over the Internet during tutoring sessions.

An access card is required. For more information, go to *www.aw.com/tutorcenter.*

Digital Video Tutor

ISBN 0-321-10903-1

The Digital Video Tutor provides our videos on CD-ROM, making it easy and convenient for students to watch video segments from a computer at home or on campus. This complete video set is ideal for distance learning or extra instruction.

Web site

http://www.aw.com/pirnot

The Web site for the book, *www.aw.com/pirnot,* contains extensive bibliographies, Web site references for each chapter and additional resources. In addition it contains chapters 7, *Data,* and 13, *Matrices,* from the first edition. As mentioned previously, the book's site also contains numerous graphing calculator, spreadsheet, and computer algebra system examples and tutorials for those who wish to incorporate technology into their courses.

Acknowledgments

Again, I would like to acknowledge the wonderfully talented team at Addison Wesley who produced this edition. I thank my senior acquisitions editor Anne Kelly and my project editor Rachel S. Reeve for their sound advice and generous encouragement throughout this project. I appreciate the work of Peggy McMahon who supervised the production of the text and also thank Barbara Atkinson, senior designer, for her beautiful design.

I also would like to thank Heidi Howard, who wrote the solutions manuals, corrected my errors, and suggested many improvements for the exercises in this edition. I appreciate the care that Kelly Carey and Cathy Ferrer took in checking for accuracy in the text and the exercise answers. Jaime Bailey and Joanne Ha contributed in so many ways in producing this second edition. Emily Hui updated the data and examples from the first edition and I thank her for her careful work. Our marketing manager Becky Anderson and Julia Coen have helped greatly in spreading the message of *Mathematics All Around.* Sara Anderson has produced an excellent Web site as well as the fine material for the MyMath-Lab supplement that accompanies this book.

I also appreciate Greg Tobin's continuing confidence and support for this project.

I would particularly like to thank my wife Ann and children, Matt, Tony, Joanna, and Mike who supported me and encouraged me throughout this project.

Finally, I would like to thank the reviewers and the previous users who helped me to reshape *Mathematics All Around* into a better text. Again, I have listened carefully to you and this edition would not be what it is today without your thoughtful advice and your many constructive suggestions. The following is a list of reviewers of the first and second editions (reviewers of the second edition are marked with an "*").

*Gisela Acosta	Valencia Community College
*Randall Allbritton	Daytona Beach Community College
*Seth Armstrong	Southern Utah University
James Arnold	University of Wisconsin
*Carmen Q. Artino	The College of Saint Rose
John Atkinson	Missouri Western State College
René Barrientos	Miami Dade Community College–Kendall
Thomas Bartlow	Villanova University
*Linda Barton	Ball State University

Kathleen Bavelas	Manchester Community Technical College
*Kari Beaty	Midlands Technical College — Airport
*David Behrman	Somertset Community College
Terence R. Blows	Northern Arizona University
*Warren S. Butler	Daytona Beach Community College
*David Cochener	Austin Peay State University
Linda F. Crabtree	Metropolitan Community College
*Debra Curtis	Bloomfield College
Walter Czarnec	Framingham State College
Greg Dietrich	Florida Community College at Jacksonville
Margaret M. Donlan	University of Delaware
John Emert	Ball State University
*Catherine Ferrer	Valencia Community College
Tom Foley	Glendale Community College
*Olivia Garcia	University of Brownsville
*John Gaudio	Waubonsee Community College
John Gimbel	University of Alaska – Fairbanks
*Gail Gonyo	Adirondack Community College
Donald R. Goral	Northern Virginia Community College
Glen Granzow	Idaho State University
Tom Greenwood	Bakersfield College
*Joshua Guest	University of Mississippi
*Ward Heilman	Bridgewater State College
*Nyeita Irish-Schult	St. Petersburg Junior College
*Stacey Jones	Benedict College
*Anne Jowsey	Niagara County Community College
Kathryn Lavelle	Westchester Community College
Ana Leon	Lexington Community College
Maita Levine	University of Cincinnati
Rhonda MacLeod	Florida State University
Rafael Marino	Nassau Community College
Karen L. McLaren	University of Maryland at College Park
*Kenneth Myers	Bloomfield College
Bill Naegele	South Suburban College
Nancy H. Olson	Johnson County Community College
John L. Orr	University of Nebraska – Lincoln
Joanne Peeples	El Paso Community College
Scott Reed	College of Lake County
*Nancy Ressler	Oakton Community College
*Nolan Rice	College of Southern Idaho
Len Ruth	Sinclair Community College
*Gene Schlereth	University of Tennessee — Chattanooga
Daniel Seabold	Hofstra University
Mary Lee Seitz	Erie Community College – City Campus
Kara Shavo	Mercer County Community College
Marguerite Smith	Merced College
Jamal Tartir	Catholic University
*Jamal Tartir	Youngstown State University
*Beverly Taylor	Valencia Community College
*William Thistleton	State University of New York — Utica/Rome
Mark Tom	College of the Sequoias
Robert R. Urbanski	Middlesex County College
Sheela L. Whelan	Westchester Community College
*Carol S. White	Flagler College

John L. Wisthoff	Anne Arundel Community College
Robert Woods	Broome Community College
* Rob Wylie	Carl Albert State University
Marvin Zeman	Southern Illinois University

I would welcome comments and suggestions from the users of this book. If you wish to contact me, my address is:

Thomas Pirnot
Department of Mathematics and Computer Science
Kutztown University
Kutztown, PA 19530
e-mail: *pirnot@kutztown.edu*

Set Theory:*
Using Mathematics to Classify Objects

How would you describe yourself? Would your reply be like one of these?

> *"I am a Puerto Rican. I come from Philadelphia."*†
> *"I am an 18-year-old white male and am friendly."*
> *"I am a musician, I work part-time, and I have a good sense of humor."*
> *"I am a single mother and part-time college student. I am athletic."*

These individuals have specified characteristics, talents, and interests that classify them as members of larger, more recognizable groups. People classify and categorize the world around them in order to make sense of their environment and their relationships to others. We hear about Generations X and Y and the baby boomers—age groupings that have special needs and aspirations. Stores and libraries organize books and recordings so that we can find items quickly and efficiently. The Yellow Pages lists businesses by category to help us locate products and services.

Mathematicians are no different from others in their need to collect objects into meaningful groups. However, the process they use is more precise than the informal groupings we use in everyday life. In this chapter we will introduce the methods that mathematicians use to describe relationships among sets and to calculate with sets to solve problems.

Set theory has many practical applications. If you have ever used an Internet search engine to scour millions of Web sites, you have

---

* You can find further resources on problem solving and set theory at www.aw.com/pirnot.
† These answers are a sample of responses from the students in one of my classes.

benefited from set theory principles. Businesses use databases built from set theory to organize vast amounts of data. For example, in marketing a luxury sports car, a car manufacturer may be interested in the buying habits of single males who earn over $70,000 per year. When we search the Internet or query a database, we are selecting items from a huge pool of possible candidates. As you learn about set theory in this chapter, you will understand these applications and learn to solve problems that require the organization of large amounts of data.

1.1   PROBLEM SOLVING

The techniques and principles that we describe in this section can help you solve many personal and professional problems. These may include deciding whether to buy or lease a car, estimating how much to save for your retirement, interpreting statistics on your job, or organizing a large project efficiently.

You will become more skillful in problem solving if you tackle the problems in each chapter. By solving these problems, you will not only master the mathematical ideas, but you will find that you will be able to solve similar problems more quickly. Real-life problems can be more complex than textbook problems, and solving them can be challenging. A goal of this text is to teach you problem-solving strategies that will remain with you long after you have read this book.

You cannot rush the problem-solving process. It takes patience and experience to move smoothly from one step in a problem to another. Just as a student driver who has logged only a few hours behind the wheel is not as skilled as a more experienced driver, you should not expect to improve your problem-solving ability without practice. By working diligently—and even struggling with problems—you will grow and become a stronger problem solver.

Problem solving requires preparation, incubation, illumination, and verification.*

- *Preparation.* Suppose you were hired to design an ad campaign for a new line of exercise equipment. You would have to learn about the features of the equipment and identify the intended customers for it. You would try out the equipment, talk to exercise physiologists about its benefits, and study the market research on the target audience. Only after you had thoroughly prepared yourself could you design an effective campaign.

  Solving problems in mathematics is similar. In order to solve a problem, you must know the underlying mathematical concepts. If it is a problem in graph theory, you must know the terminology, notation, and properties used in graph theory. Do not try to absorb information too quickly or attempt to design

---

* This advice on problem solving was put forth by C. Ann Oxrieder and Janet P. Ray in *Your Number's Up* (Addison-Wesley, 1982).

a solution before you have laid the proper groundwork. *Be active* while you prepare to solve a problem. Can you draw a picture or diagram to represent the problem? Can you give examples of some conditions in the problem? Can you rephrase the problem in your own words?

- *Incubation.* Solving a difficult problem takes time. After you have studied the problem and are sure that you understand it, if you feel that you are not getting closer to a solution, set it aside for a while. Many experts on creativity (including mathematical creativity) recommend turning over problems to our subconscious mind.* Perhaps by turning your attention to another subject, getting some exercise, or relaxing, you will have a better idea of how to solve the problem when you return to it. You will be more relaxed about problem solving if you allow sufficient incubation time.

- *Illumination.* As you work for a solution, often you obtain an insight that unlocks the door to your problem. You may notice that the situation parallels one that you've seen before. Perhaps you suddenly see the problem from a different point of view that makes the solution clear. Once you've experienced this insight, you can set up and carry out the solution quickly.

- *Verification.* Once you have found your solution, pause and think about it. Does it make sense? If the problem is about tourists on a trip, there cannot be $19\frac{3}{7}$ travelers. If you have invested \$1,000 for one year, you would probably not earn \$351.68 in interest.

  One way to verify your solution is to see if it fits all of the conditions of the problem. If your solution isn't reasonable, then look for the source of your error. Maybe you have misunderstood an important condition; maybe you have made a simple computational error; maybe the theory you are using doesn't apply to this particular situation. Retrace your steps and reexamine what you have done. Often you will locate your mistake and then find the correct answer.

Problem solving relies on several basic strategies.

Problem solving is more of an art than a science. We will now suggest some useful strategies; however, just as we cannot list a set of rules describing how to write a novel, we cannot specify a series of steps that will enable you to solve every problem. Artists, composers, and writers make creative decisions as to how to use their tools, and so you also must be creative in using your mathematical tools.

Mathematics is not as rigid as you may believe from your past experiences. It is important to use the strategies in this section to keep your focus on understanding concepts rather then memorizing formulas. If you do this, you may be surprised to find that a given problem can be solved in several different ways.

---

* Jacques Hadamard, in his book *An Essay on the Psychology of Invention in the Mathematical Field* (Dover, 1954), gives some interesting insights as to how real mathematicians use the subconscious mind to solve problems.

STRATEGY: Draw Pictures

Problems usually contain several conditions that must be satisfied. You will find it very useful to draw pictures in order to understand these conditions before trying to solve the problem.

EXAMPLE 1   Visualizing a Condition in a Word Problem

Suppose that a problem has the following condition:

> "A pharmacist is adding 10 liters of a 20% alcohol solution to a beaker of pure alcohol to get a mixture which is 25% alcohol."

Draw a diagram to help visualize the situation.

SOLUTION: The drawing in Figure 1.1 is appropriate.

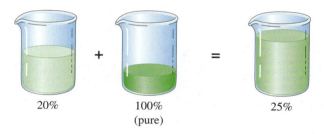

20%   100%   25%
             (pure)

FIGURE 1.1   Illustrating a mixture problem.

To understand the situation better, we might ask the following questions. Which container contains the most liquid? Do you expect to add a lot or a little of the alcohol to bring the 20% mixture to a concentration of 25%? Why? What is the relationship between the amount of liquid in the three containers?

We are not going to solve this problem now. The key idea here is to emphasize how examining an appropriate diagram can increase your understanding of a problem. ◉

STRATEGY: Choose Good Names for Unknowns

It is wise to name the objects in a problem so that you can remember their meaning easily.

EXAMPLE 2   Choosing Good Names for Objects in a Problem

Choose meaningful names for the objects mentioned in each of the following situations.

a) One group of dance students is taking swing; another group is taking Latin American dance.

b) Four people—Frasier, Niles, Daphne, and Roz—are attending a broadcasting awards dinner.

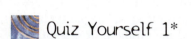

Quiz Yourself 1*

Over lunch, four architects discuss preliminary plans for a new civic center. Each shakes hands with all the others. Draw a picture to illustrate this condition. *Hint:* Use points to represent the four people and use lines joining the points to represent the handshakes.

Quiz Yourself 2

Choose meaningful names for the objects mentioned in each of the following situations.

a) Two amounts are invested— one at a high interest rate, the other at a low rate.

b) We have a number of pennies, dimes, and quarters.

c) A mixture contains quantities of peanuts and cashews.

d) A red-hot steel beam from a rolling mill is cut into two pieces. One piece is 12 feet longer than the other.

SOLUTION: a) The swing students could be called S and the Latin American dance students L.

b) We could label the people using their initials: F, N, D, and R.

c) Good names for amounts of peanuts and cashews would be p and c rather than, say, x and y.

d) We could call the small length s and the large length $s + 12$.

Casey	Sitka	Jena
Fail	Fail	Fail
Pass	Fail	Fail
Fail	Pass	Fail
?	?	?
Pass	Pass	Fail
Pass	Fail	Pass
?	?	?
Pass	Pass	Pass

TABLE 1.1   Systematic listing of outcomes.

STRATEGY: Be Systematic

If you approach a situation in an organized, systematic way, frequently you will gain insight into the problem.

EXAMPLE 3   Systematically Listing Outcomes

Three Seeing Eye guide dogs—Casey, Sitka, and Jena—are taking a pass/fail certification test. In how many different ways can the final scores be assigned?

SOLUTION: In listing these possibilities, we see that there are four cases:

none pass *or* one passes *or* two pass *or* all three pass.

We organize these possibilities in Table 1.1.

STRATEGY: Look for Patterns

If you can recognize a pattern in a situation you are studying, you can often use it to answer questions about that situation.

EXAMPLE 4   Finding Patterns in Pascal's Triangle

You will encounter the pattern that we show in Figure 1.2, called *Pascal's triangle*, in later chapters.

Quiz Yourself 3

Complete Table 1.1.

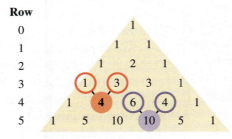

FIGURE 1.2   Pascal's triangle.

Notice how each number is the sum of the two numbers immediately above it, which are a little to the right and a little to the left. Suppose we want to find the total of all the numbers that will be in the ninth row of this diagram.

SOLUTION: (It will be convenient when we discuss this diagram in later chapters to begin numbering rows with 0 instead of 1.)

Notice that in the zeroth row the total is 1; in the first row the total is 2; in the second row the total is 4; and in the third row it is 8. We continue this pattern in Table 1.2.

Quiz Yourself 4

a) List the numbers in the sixth and seventh rows of Pascal's triangle.

b) What are the first two numbers in the 100th row of Pascal's triangle?

Row	0	1	2	3	4	5	6	7	8	9
Total	1	2	4	8	16	32	64	128	256	512

TABLE 1.2   The sum of the numbers in each row of Pascal's triangle.

We now easily see that the desired total is 512.

STRATEGY: Try a Simpler Version of the Problem

You can begin to understand a complex problem by solving some scaled-down versions of the problem. Once you recognize a pattern in the way you are solving the simpler problems, then you can carry over this insight to attack the full-blown problem.

EXAMPLE 5   Counting Combinations of Investment Options

Assume that you have money to invest and your financial advisor suggests that you consider the following possible investments: stocks, bonds, mutual funds, real estate, certificates of deposit, collectibles, and precious metals. How many different ways can you invest your money using these options if you can choose from none to all seven of these options?

SOLUTION: Instead of considering all seven options that you may or may not choose, what if there were only one option? Or two? Or three? Could you answer the question then? Let's consider these simpler cases.

What if we had only one option—stocks? Then there would be only two possibilities—you could choose to invest in stocks or choose not to invest in stocks. What if you had two options, stocks and bonds? Then there would be four possibilities—you could invest in neither, invest in stocks only, invest in bonds only, or invest in both.

We'll next consider three options, but we will organize our findings in Table 1.3. We abbreviate stocks by s, bonds by b, and mutual funds by m.

Comparing the first and last columns, we see a pattern—each time we add one option, the number of possibilities doubles. Can you see why this is so? Assuming that this is the case, we realize that there will be 128 ways to invest using various combinations of the seven investment options.

Quiz Yourself 5

List the different combinations of options if you have four investment options—$s, b, m,$ and r (for real estate). Try to be systematic as you go from the eight combinations you already have for $s, b,$ and m to the sixteen for $s, b, m,$ and r.

Number of Options	Options	Possibilities for Combinations of Options	Number of Combinations of Options
1	*s*	none, *s*	2
2	*s, b*	none, *s, b,* *s + b*	4
3	*s, b, m*	none, *s, b, m,* *s + b, s + m, b + m,* *s + b + m*	8
4			?
5			?
6			?
7			?

TABLE 1.3   Counting combinations of investment options.

STRATEGY: Guessing Is OK

One of the difficulties in solving word problems is that you can be afraid to say something that may be wrong and consequently sit staring at a problem writing nothing until you have the full-blown solution. Making guesses, even incorrect guesses, is not a bad way to begin. It may give you some understanding of the problem. Once you make a guess, evaluate it to see how close you are to meeting all the conditions of the problem.

EXAMPLE 6   **Solving a Word Problem by Guessing**

The curator of the Museum of Natural Science has acquired three mechanical dinosaurs for a new exhibit. The combined weight of the dinosaurs is 50 tons. If the weight of the heaviest dinosaur, Apatasaurus, is seven times that of the smallest, a duckbill, and the weight of the middle dinosaur, Diplodocus, is 14 tons more than the smallest, then what are the weights of each dinosaur?

SOLUTION: In a later chapter we will discuss how to solve problems algebraically, but for now all we want to do is to practice making educated guesses and evaluating them until we arrive at an answer. We will denote the dinosaurs by *s, m,* and *l* (*small, medium,* and *large*).

Trial and error may not be an effective way to solve a more complex problem; however, it still can be useful in giving you an insight as to how you would use algebra to attack the problem.

Guesses for s, m, l	Evaluation of Guess	
	Good Points	**Weak Points**
10, 20, 20	Total is 50.	The weights are not different.
10, 15, 20	Weights are different.	The total is not 50.
10, 12, 28	Total is 50; weights are different.	The heaviest is not seven times the smallest; the middle one is not 14 more than the smallest.
2, 16, 32	Total is 50; weights are different; middle one is 14 more than smallest.	The largest is not seven times the smallest.
4, 18, 28	All conditions are satisfied; we have a solution.	

You may not believe that the guessing approach to problem solving that we suggested in Example 6 is doing mathematics. However, if you are making intelligent guesses and systematically refining them, you probably have some solid, intuitive, underlying logical reasons for what you are doing. And, as a result of this thinking, you are doing legitimate mathematics. The problem with guessing is that it can be very inefficient and, if the answer is complex, you probably will not be able to find it without doing some algebra.

Quiz Yourself 6

We have a collection of pennies, nickels, and quarters. There are twice as many nickels as pennies, and there are two fewer quarters than pennies. If the total amount is $6.70, how many of each coin do we have? We begin a table of guesses that should enable you to solve the problem by improving your guesses until you have a solution. *Hint:* Notice that because the total is $6.70, certain numbers are not appropriate for the number of pennies.

P	N	Q	Total	Comment
0	0	−2		With 0 pennies we can't have two fewer quarters
5	10	3	$1.30	Total is wrong
?	?	?	?	?

STRATEGY: Convert a New Problem to an Older One

An effective technique in solving a new problem is to try to connect it with a problem you have solved earlier. It is often possible to rewrite a condition so that the problem becomes exactly like one you have seen before.

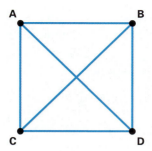

FIGURE 1.3   Diagram representing four architects shaking hands.

EXAMPLE 7   **Converting a New Problem to One You've Solved Before**

Eight branch offices of a bank are to be connected by secure communications links so that any office can send confidential information to any other office. How many links are needed?

SOLUTION: First, recognize that this problem is essentially the same problem as the four architects shaking hands in Quiz Yourself 1. In that problem, we drew the diagram shown in Figure 1.3.

The six lines joining the points in Figure 1.3 represent the different handshakes. With this in mind, you could draw a diagram with eight points joined in all possible ways by lines representing the communication links. Counting the lines in that diagram would give you the number of links.

However, there is another way to approach this problem. Knowing that six links are needed for four offices, we might ask how many would be needed for five offices, and then six, seven, and eight? We convert the problem of linking five offices to the problem of linking four that was solved previously. In order to link the fifth office to the others, we need four more links. These four plus the earlier six means that ten links are needed for five offices.

Next consider the sixth office. It needs five new links. These five plus the earlier ten gives us fifteen links. It now becomes clear what to do. The seventh office requires six more links and the eighth office requires seven more. We can say this in a word equation.

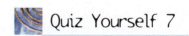

Quiz Yourself 7

In Example 7, how many communications links would be needed for ten offices?

Total number of links = six for the first four offices +
four for the fifth office +
five for the sixth office +
six for the seventh office +
seven for the eighth office = 28 links.

Remembering several fundamental principles helps us with problem solving.

In this section we discuss some basic mathematical principles that we will refer to frequently throughout this text.

The Always Principle

When we say a statement is true in mathematics, we are saying that statement is true *100 percent of the time*. One of the great strengths of mathematics is that we do not deal with statements that are "sometimes true" or "usually true."

EXAMPLE 8   **An Algebraic Statement That Is Always True**

Find two examples to illustrate that the statement

$$(x + y)^2 = x^2 + 2xy + y^2$$

is true.

SOLUTION: In algebra, we prove that the statement $(x + y)^2 = x^2 + 2xy + y^2$ is true for all numbers x and y. This means that any number we substitute for x and y should make the statement true.

a) Suppose that $x = 2$ and $y = 3$. Then the given statement becomes

$$(2 + 3)^2 = 2^2 + 2 \cdot 2 \cdot 3 + 3^2, \text{ or } 25 = 4 + 12 + 9.$$

b) If we let $x = 5$ and $y = -8$, then the statement becomes

$$(5 + (-8))^2 = 5^2 + 2 \cdot (5) \cdot (-8) + (-8)^2, \text{ or } (-3)^2 = 25 + (-80) + 64. \quad \textcircled{\scriptsize ⊚}$$

In a similar way, we accept a mathematical argument only if it holds up under every set of conditions. If we can think of even a single situation in which the argument fails, then the argument is not acceptable.

The Counterexample Principle

An example that shows that a mathematical statement fails to be true is called a *counterexample*. Keep in mind that if you wish to use a mathematical property and someone can find a counterexample, then the property you are trying to use is not allowable. A hundred examples in which a statement is true do not prove it to be *always* true, yet a single example in which a statement fails makes it a false statement. Be careful to understand that when we say a statement is false, we are not saying that it is always false. We are only saying that the statement is *not always true*. That is, we can find at least one instance in which it is false.

EXAMPLE 9   Counterexamples to False Algebra Statements

Although the following two statements are false, those who are not secure in working with algebra often believe them to be true.

a) $(x + y)^2 = x^2 + y^2$ b) $\sqrt{x + y} = \sqrt{x} + \sqrt{y}$

Provide a counterexample for each statement.

SOLUTION: a) Assume that $x = 3$ and $y = 4$. Then

$$(x + y)^2 = (3 + 4)^2 = 7^2 = 49,$$

which is not equal to

$$x^2 + y^2 = 3^2 + 4^2 = 9 + 16 = 25.$$

b) Let $x = 16$ and $y = 9$. Then,

$$\sqrt{x + y} = \sqrt{16 + 9} = \sqrt{25} = 5.$$

However,

$$\sqrt{x} + \sqrt{y} = \sqrt{16} + \sqrt{9} = 4 + 3 = 7. \quad \textcircled{\scriptsize ⊚}$$

 Quiz Yourself 8

Find a counterexample to the following statement:

$$\frac{a + b}{a + c} = \frac{b}{c}.$$

The Order Principle

When you read mathematical notation, pay careful attention to the *order* in which the operations must be performed. The order in which we do things in mathematics is as important as it is in everyday life. When getting dressed in the morning, it makes a difference whether you first put on your socks and then your shoes, or first put on your shoes and then your socks! Although the difference may not seem as dramatic, reversing the order of mathematical operations can also give unacceptable results. Note that we are not saying that it is *always* wrong to reverse the order of mathematical operations; we are saying that if you reverse the order of operations, you *may* accidentally change the meaning of your calculations.

EXAMPLE 10   Reversing the Order of Operations Can Change the Result of Mathematical Computations

Explain how Example 9 illustrates the order principle.

 Quiz Yourself 9

Consider the equation
$\sqrt{x + y} = \sqrt{x} + \sqrt{y}$.

a) On the left side of the equation we are performing two operations. What are these operations and what is their order?

b) What is the order of the operations performed on the right side of the equation?

c) Summarize what is wrong with this equation, as we did in Example 10(a).

SOLUTION: In Example 9, the problem with statements (a) and (b) is that in each case we carelessly changed the order in which we performed the operations.

a) In the equation $(x + y)^2 = x^2 + y^2$, the left side of the equation tells us to *first* add x and y and *then* square the resulting sum. The right side of the equation tells us to *first* square x and y separately and *then* add these two squares. To say this more succinctly, we can say that to "first add and then square" is not the same as to "first square and then add."

b) The problem here is similar. Test your understanding of the order principle by solving Quiz Yourself 9.

The Splitting-Hairs Principle

You should "split hairs" when reading mathematical terminology. If two terms are similar but sound slightly different, they usually do not mean exactly the same thing. In everyday English, we may use the words "equal" and "equivalent" interchangeably; however, in mathematics they do not mean the same thing. The same is true for notation. When you encounter different-looking notation or terminology, work hard to get a clear idea of exactly what the difference is. Representing your ideas precisely is part of good problem solving.

 Quiz Yourself 10

In your own words, explain the differences you see in each pair of symbols.

a) $\supset$ and $\supseteq$

b) $\{\emptyset\}$ and $\emptyset$

c) $\cup$ and $\vee$

d) $\{0\}$ and 0.

EXAMPLE 11   Identifying Differences in Notation

Notice the slight differences in the following pairs of symbols.

a) $<$ and $\leq$. There is an extra line under the $<$ symbol on the right.

b) $\cap$ and $\wedge$. The symbol on the left is rounded; the symbol on the right is pointed.

c) $\in$ and $\subset$. The symbol on the left has an extra line.

d) $\emptyset$ and 0. The symbol on the right is the number 0, the symbol on the left is something else, not a number.

The Analogies Principle

Much of the formal terminology that we use in mathematics sounds like words that we use in everyday life. This is not a coincidence. Whenever you can associate ideas from real life with mathematical concepts, you will better understand the meaning behind the mathematics you are learning.

EXAMPLE 12   **Relating Mathematical Terms to Everyday Language**

In Table 1.4, the left column contains formal mathematical terminology. The right column contains ordinary English words that will help you remember the mathematical concepts.

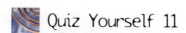

Quiz Yourself 11

Which English words are you reminded of by the following mathematical terminology?
a) intersection
b) simultaneous (as in simultaneous equations)

Mathematical Concept	Related English Ideas
Union	Labor union, marriage union, united
Complement	Complete
Equivalent	Equivalent (in some way the same)
Slope	Ski slope, slope of a roof

TABLE 1.4   Mathematical terms related to words in everyday language.

The Three-Way Principle

We conclude this section with a method for approaching mathematical concepts that we illustrate in Figure 1.4.

Whether you are learning a new concept or trying to gain insight into a problem, it is helpful to use the ideas we have discussed in this chapter to approach mathematical situations in three ways.

- *Verbally*—Make analogies. State the problem in your own words. Compare it with situations you have seen in other areas of mathematics.
- *Graphically*—Draw a graph. Draw a diagram.
- *By example*—Make numerical or other kinds of examples to illustrate the situation.

Not every one of these three approaches fits every situation. However, if you get in the habit of using a verbal-graphical-example approach to doing mathematics, you will find that mathematics is more meaningful and less dependent on rote memorization. If you practice approaching mathematics using the strategies and principles that we have discussed, you will find eventually that you are more comfortable and more successful in your mathematical studies.

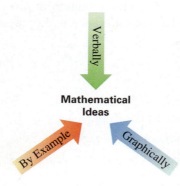

FIGURE 1.4   The Three-Way Principle.

Historical Highlight: *George Polya*

We could argue that the Hungarian mathematician George Polya (1887–1985) is the father of problem solving as we teach it in so many of today's mathematics text books (including this one). As a youth, Polya at first decided to study law but, because of his dislike for memorization, he found it tedious and eventually chose a career in mathematics. While working at one of his early jobs as a mathematics tutor, he began to develop a problem-solving method that led him to write a book called *How to Solve It.** This book, which has been published in almost twenty languages and has sold over one million copies, discusses many of the approaches to problem solving that you have been learning in this section.

Polya loved to solve problems, and it seemed that he could find them everywhere. One day he and his wife-to-be, Stella, were walking randomly in a garden in Switzerland. On their walk, George and Stella encountered another young couple six times and Polya wondered what was the likelihood that on the same walk they would meet the same couple so many times. His attempts to answer this question eventually led him to publish mathematical research papers on the topic of the random walk problem.

In the 1940s, due to their concern about the increasing influence of Nazism in Europe, George and Stella emigrated to the United States. Polya eventually accepted a position at Stanford University in California where he conducted research and worked on problem solving until he was well into his nineties.

Exercises 1.1

In Exercises 1–4, draw a picture to illustrate each situation. You are not being asked to solve the problems—just draw a picture.

1. Five liters of a 10% sugar solution are mixed with pure water to get a 5% solution.

2. Six people clink each other's glasses in a toast.

3. A triangle has two equal sides that are longer than its third side.

4. One craft guild has no members in common with another guild.

In Exercises 5–10, choose names that would be meaningful for each item. Again, you are not being asked to solve the problem—just make a recommendation for some good names.

5. A wooden covered bridge is eight times as long as it is wide.

6. There are three groups of dogs at a dog show. Some are competing in obedience, some in agility, and some in tracking. (Dogs can compete in more than one category.)

7. Monica, Chandler, Rachel, and Joey are planning a party for Ross.

8. Two pathways in the brain cross.

9. A person makes two investments—one in collectibles, the other in precious metals.

10. An Olympic swimmer wishes to include calcium and protein in her diet.

---

* See Exercise 63 at the end of this section.

In Exercises 11–14, list the items mentioned. Try to organize your list in a systematic way.

11. List the different combinations of heads and tails that can occur when a penny and a nickel are flipped.

12. Using the numbers 1, 2, and 3, form as many ordered pairs of numbers as you can. For example, (2, 3) is one such pair; (3, 2) is a different one. You are allowed to repeat numbers, so (1, 1) is an allowable pair.

13. Choose two recording artists to host the Grammys. Your choices are Mariah Carey, Will Smith, Marilyn Manson, and Shania Twain. (Represent the artists by the first letter of their last name. One pair would be Carey and Smith, which you would represent by *CS*. Note that the pair *CS* is the same as *SC*.)

14. In the following diagram, travel from "Begin" to "End." List the different routes you could take.

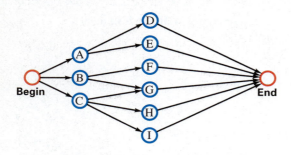

In Exercises 15–18, continue the pattern for five more items in the list. (There may be more than one way to continue the pattern. We will provide only one solution.)

15. 7, 14, 21, 28, . . .

16. 1, 3, 9, 27, . . .

17. *ab, ac, ad, ae, bc, bd, be,* . . .

18. (1, 1), (1, 2), (1, 3), (1, 4), (1, 5), (1, 6), (2, 1), (2, 2), (2, 3), . . .

In Exercises 19–24, we state a problem. Instead of trying to solve that problem, state a simpler problem and solve it instead.

19. Ten people are being honored for their work in reducing pollution. In how many ways can we line up these people for a picture?

20. If you guess at ten true-false questions, how many different ways can you fill in the ten answers?

21. Using all the letters of the alphabet, how many two-letter codes can we form if we are allowed to use the same letter twice? For example, *ah, yy, yo,* and *bg* are all allowable pairs. (*Note: bg* is different from *gb*.)

22. A family has seven children. If we list the genders of the children (for example, *bbggbgb*, where *b* is boy and *g* is girl), how many different lists are possible?

23. An electric-blue sports car comes with seven options: air conditioning, CD player, air bags, antilock brakes, sunroof, extended warranty, and leather seats. You may buy the car with any combination of the options (including none). How many different choices do you have?

24. In printing your résumé, you have 10 different colors of paper to use and 20 different font styles. How many different ways can you print your résumé?

In Exercises 25–34, do not try to solve each problem algebraically. Instead, make a guess that satisfies one or more conditions of the problem and then evaluate your guess, as we did in Example 6. Keep adjusting your guesses until you have a solution that fits all the conditions of the problem.

25. The local historical society wishes to preserve two buildings. The total age of the buildings is 321 years. If one building is twice as old as the other, what are the ages of the two buildings?

26. To celebrate a diner's 40th anniversary, the owners have made a 40-foot-long sandwich. The sandwich is cut into three unequal pieces. The longest piece is three times as long as the middle piece, and the shortest piece is 5 feet shorter than the middle piece. What are the lengths of the three pieces?

27. Janine worked 15 hours last week. One job as a clerk in a sporting goods store paid her $5.25 per hour, while her job giving piano lessons paid $12 per hour. If she earned $119.25 between the two jobs, how many hours did she work at each job?

28. Vince worked 18 hours last week. Part of the time he worked in a fast-food restaurant and part of the time he worked tutoring a high school student in mathematics. He was paid $4.75 per hour in the restaurant and $15 per hour tutoring. If he earned $126.50 total, how many hours did he work at each job?

29. Last season, Emmit carried the football 29 times more than Barry and 51 times less than Ricky. If the three running backs carried the ball 742 times, how many times did each carry the ball?

30. In 1998, Mo Vaughn and Eric Davis scored 188 runs. If Mo Vaughn scored 26 runs more than Eric Davis, how many runs did each score?

31. Heather has divided $8,000 between two investments—one paying 8%, the other 6%. If the return on her investment is $550, how much does she have in each investment?

32. Carlos has $9,000 in two mutual funds. One fund pays 11% interest and the other fund pays 8%. If his income from the two funds last year was $936, how much did he invest in each fund?

33. The administration at Center City Community College has formed a 26-person planning committee. There are five times as many administrators as there are students and five more faculty members than students. How many students are there on the committee?

34. Last week Minxia made 55 phone calls to gather support for the reelection of her local state representative. She contacted three times as many senior citizens as she did young adults. The number of middle-aged adults was one-half the number of senior citizens. How many senior citizens did she contact?

In Exercises 35–44, determine whether each statement is true or false. If a statement is true, give two examples to illustrate it. If it is false, give a single counterexample.

35. Months of the year have 31 days.

36. All past presidents of the United States are deceased.

37. $\dfrac{a}{b} + \dfrac{c}{d} = \dfrac{a + c}{b + d}$

38. If $a < b$, then $a + c < b + c$.

39. If A is the father of B and B is the father of C, then A is the father of C.

40. If X is acquainted with Y and Y is acquainted with Z, then X is acquainted with Z.

41. If $a < b$, then $a^2 < b^2$.

42. $x + y - z = x + z - y$

43. If the lengths of the sides of a square are doubled, then the area of the square is also doubled.

44. If the price on a VCR is increased by 10% and then later reduced by 10%, the price will be the same as the original price.

In Exercises 45–48, decide whether the two sequences of operations will give the same results.

45. Squaring a number, then adding five to it; adding five to a number and then squaring that sum.

46. Squaring two numbers x and y, then subtracting the results; subtracting x and y, then squaring the difference.

47. Adding two numbers x and y, then dividing the result by 3; dividing x and y by 3, then adding the results.

48. Multiplying two numbers x and y by 5, then adding the results; adding x and y, then multiplying the result by 5.

In Exercises 49–54, explain the differences you see in each pair of symbols.

49. 5 and {5}

50. A and A'

51. U and u

52. (1, 2) and [1, 2]

53. {1, 2} and (1, 2)

54. (2, 3) and (3, 2)

Communicating Mathematics   *In later chapters we will use the following mathematical terminology. Of what common English usage do these terms remind you? Explain.*

55. path, circuit, bridge, directed

56. dimension, reflection, translation, transform

57. intercept, best fit, compounding

58. median, deviation, central tendency, correlation

Further Exercises

59. Communicating Mathematics Explain the role of preparation, incubation, illumination, and verification in the problem-solving process.

60. Communicating Mathematics People often ignore the preparation and verification stages of problem solving. Why do you think that this is so? What are the dangers of ignoring these steps?

61. Communicating Mathematics Explain in your own words the Three-Way Principle.

62. Communicating Mathematics Does using the Three-Way Principle in learning mathematics seem different from the ways you have approached learning mathematics in the past? Explain.

63. Try to find a book in your library called *How to Solve It* by George Polya. How is Polya's advice similar to the advice you were given in this section?

64. Write a report on Jacques Hadamard's book on creativity in mathematics (see footnote on p. 3).

1.2 ESTIMATION

In this section we will discuss estimation, which is an important part of effective problem solving. You can use estimation to check if your answer to a problem is reasonable. Also, if you are in a situation where you do not need an exact answer, you can make an estimate to get a quick, approximate answer.

Rounding is an effective way to make an estimate.

We will now examine how to use rounding to do estimation. In rounding, we consider the digit to the right of the digit being rounded.

If that digit to the right is five or more, we round up, otherwise, we round down.

For example, if we wish to round 12,435 to the nearest hundred, we look at the digit in the tens place, which is 3. Because 3 is less than 5, we round 12,435 down to 12,400. On the other hand in rounding 7,864 to the nearest hundred, because the tens digit is 6, we round 7,864 up to get 7,900.

Example 1 illustrates a familiar situation in which rounding is useful.

 Quiz Yourself 12*

Round each of the following numbers to the nearest thousand:

a) 635,421   b) 723,562

EXAMPLE 1   Estimating a Grocery Bill

On your way home from work, you stopped at the market to pick up the items listed in Table 1.5. You also would like to buy a half gallon of ice cream for $3.59, but you remember that you only have $20 in your wallet and you don't want to be caught in the checkout line without enough money. Use rounding to the nearest ten cents to:

a) Estimate the total cost of your purchases.

b) Decide if it is safe to put the ice cream in your cart.

---

* Quiz Yourself answers begin on page 849.

Item	Cost ($)	Cost Rounded to Nearest Ten Cents
Cereal	4.29	4.30
Milk	2.41	2.40
Bread	1.89	1.90
Lunch meat	3.36	3.40
Pickles	2.47	2.50
Dish-washing liquid	2.87	2.90
	Total: $17.29	Total: $17.40

TABLE 1.5   Estimating the cost of grocery items.

SOLUTION: a) You can add the rounded prices mentally to get $17.40.

b) You decide that the half gallon is probably too expensive, and decide to buy a quart of ice cream instead.

Notice in Example 1 that if you had rounded to the nearest dollar, your estimate would have been $16 and you would have taken the half gallon which would have put you over $20.

SOME GOOD ADVICE

In Example 1, you do not need a calculator to add the estimated prices. For example, to add 4.30 and 2.40, first add the dollars to get $2 + 4 = 6$. Then add the cents to get $30 + 40 = 70$ cents, so the total is 6.70. You can easily add in the other numbers one at a time to get the total. Doing this kind of mental arithmetic will make you a stronger mathematics student.

Using compatible numbers simplifies approximate calculations.

Another way to make a quick estimate that is slightly different than rounding is to use compatible numbers. In using compatible numbers, instead of using the given numbers in a problem, we substitute other numbers that are easier to work with. For example, instead of dividing 298 by 14, we might divide 300 by 15 to get 20. Or, instead of multiplying 11 times 73, we multiply 10 times 73 to get 730.

EXAMPLE 2   **Using Compatible Numbers to Estimate a Bill**

Marc is buying 19 blank tapes at $2.89 each. Use compatible numbers to estimate the cost of the tapes.

SOLUTION: We replace $2.89 by $3 and 19 by 20 to get $20 \times 3 = 60$. Because we have replaced the $2.89 and 19 by two larger numbers, we know that Marc's actual bill will be somewhat less than $60. ◎

In Example 3 we use compatible numbers to estimate a quotient.

EXAMPLE 3   Estimating a Population

In 2001, the world's population was approximately 6,157 million (6.157 billion) and the population of North America was 486 million. Estimate what fractional part of the world's population lives in North America?

SOLUTION: We replace 6,157 by 6,000 and 486 by 500 to get $\frac{500}{6,000} = \frac{1}{12}$, so, roughly one-twelfth of the world's population lives in North America. ◎

Quiz Yourself 13

Use compatible numbers to do the following calculations.

a) 26×403 b) $263 \div 49$

■

Estimation can help interpret graphical data.

Often you can use estimation to summarize information that is given to you graphically.

EXAMPLE 4   Estimating Time to Download Computer Files

Marianne is a graphics designer and wants to download some files to her home computer. The size of the files is given in Figure 1.5.

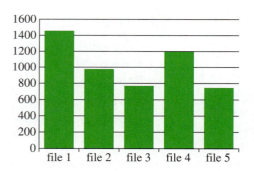

FIGURE 1.5   Size of Marianne's files in kilobytes.

a) Estimate the total size of the files that she is downloading. One kilobyte is roughly one thousand bytes.

b) If she has a 56 K modem (that is, a modem that transfers 56,000 bits per second), how long will it take for her to transfer all of the files? Use the fact that one byte has 8 bits.

SOLUTION: a) Although we cannot tell the exact size of the files, we can estimate their size from the graph in Figure 1.5. Also, because we want to make sure to allow enough time for downloading, we would rather overestimate than under-

Historical Highlight: *An Amazing Estimate**

The ancient Greek mathematician Eratosthenes (276–194 BC) devised a clever way to estimate the Earth's circumference. He believed that the cities Alexandria and Syene (today called Aswan) in Egypt were on the same great circle of the Earth. He also knew the distance between the cities because it had been measured by a surveyor called a *bematistes*, who was trained to walk in equal steps. The bematistes, using a unit of measurement called a stadium, which was 516.73 feet, found the cities to be 5,000 stadia apart.

Using simple but clever geometry calculations, which we explain in Example 3 of Section 8.1, Eratosthenes determined the distance from Alexandria to Syene to be equal to 1/50 of the Earth's circumference. He then used this information to estimate the Earth's circumference to be 24,662 miles, which is just 245 miles less than its true value!

File	Estimated Size
1	1,450
2	1,000
3	800
4	1,200
5	800
Total	5,250

TABLE 1.6 Estimated size of Marianne's files.

estimate their size. File 1 clearly is not 1,500 kilobytes, so we will estimate its size as 1,450. We give the estimated sizes of all the files in Table 1.6.

From Table 1.6, we see that Marianne is planning to download roughly 5,250 kilobytes.

b) Her modem can transfer 56,000 bits per second, or, if we divide by 8, we get 7,000 bytes per second. This means that she can download about 7 kilobytes per second. If we divide 5,250 by 7 we get 750. So, it will take 750 seconds, or 750/60 = 12.5 minutes to download the files.

SOME GOOD ADVICE

When making an estimate, it is a good idea to "estimate your estimate." By that, we mean you should have a rough idea of whether your estimate is too large or too small. In Example 4, we overestimated the time needed to download Marianne's files because we didn't want her to run out of time while downloading.

In Examples 5 and 6, we show you some other ways to make estimates.

EXAMPLE 5   Estimating Distances from a Map

In the Great Alaska Challenge, boats will race from the starting point at Prudhoe Bay (A), to the finish line at Kodiak (F), (see Figure 1.6 on page 20).

a) Estimate the length of the race course. Assume that at any point in the race, boats will take the shortest path.

b) If the boat *The Inyuk* travels at 35 miles per hour, how long will it take to complete the course?

---

* This highlight is based on material from David M. Burton, *The History of Mathematics; An Introduction* (McGraw-Hill, 1999) pp. 179–180.

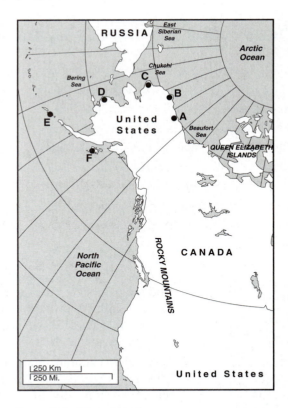

FIGURE 1.6   Map of Alaska.

SOLUTION: a) We have marked points A, B, C, and so on, on the map to approximate the total length of the race. If we measure the total length of these distances on the map, we get $2\frac{1}{4}$ inches. According to the scale* of the map, one inch equals 286 miles; thus the total length of the race course is approximately 644 miles.

b) If *The Inyuk* travels at 35 miles per hour, then the race will take $\frac{644}{35} = 18.4$ hours, or about 18 hours and 24 minutes.

After the Million Mom's March on May 14, 2000 (Mother's Day), the news media estimated that several hundred thousand people had gathered to protest gun violence in Washington D.C. Of course, no one would have actually been able to count all of the people who were there. However, Example 6 illustrates a technique that you can use for making this type of estimate.

EXAMPLE 6   **Estimating the Size of a Large Number of Objects**

Estimate the number of M&Ms that are visible in Figure 1.7.

SOLUTION: In Figure 1.7, we have divided the photograph into 16 equally sized rectangles. It appears that each rectangle has approximately the same

FIGURE 1.7   Estimating number of M&Ms.

---
* The scale shows that $\frac{7}{8}$ inch = 250 miles, so one inch = $\frac{8}{7} \times 250 \approx 286$ miles.

number of candies. We see that the rectangle we have highlighted contains 11 candies (we will count any candy that is partially visible). Therefore, it is reasonable to estimate that there are $11 \times 16 = 176$ candies in the photograph.

Exercises 1.2

Round each of the following numbers to the nearest hundred.

1. 37,264

2. 6,835

3. 8,327

4. 14,872

Estimate the answers to the following problems. Use either rounding or compatible numbers. In some cases, your answers may differ from ours depending on what method you use.

5. $21.6 + 38.93 + 191 + 42.5$

6. $17.18 + 27.2 + 10.31 + 87.6$

7. $34.6 - 15.3$

8. $107.28 - 68.49$

9. 4.75×16.3

10. 1028×14.8

11. $17.4/3.31$

12. $2068/72.4$

13. 0.091×785

14. 0.0008×4026.3

15. $8.7\% \times 1024$

16. $18\% \times 683$

Communicating Mathematics   *Estimate each of the following answers. Explain how you made your estimate and, where possible, state whether your estimate is larger or smaller than the exact answer. We will provide one possible answer and yours may differ from ours depending on how you make your estimate.*

17. Training for cross country. Mike is training for the cross country team. He will run 3.7 miles per day, five days a week. How many miles will he run during the next six weeks?

18. Purchasing supplies. Julia is purchasing 18 packs of pencils at $0.94 per pack. What is her bill?

19. Travel time. Celine is averaging 47.5 miles per hour on her trip to Memphis. If she still has 325 miles to travel and it is now 1:00 PM, about what time should she arrive?

20. The price of gasoline. Ramon spent $15.85 last month for 13.5 gallons of gasoline. What was the price per gallon?

21. Calculating a tip. The Jacksons went out to dinner and their bill was $82.45. If they wish to leave a 15% tip, how much should they leave?

22. Sharing apartment expenses. Bob, Stu, and Carl share an apartment. Their total utilities bill for last month was $76.38. What is each person's share of the bill?

23. Buying plants. Emily is buying plants for spring planting. She bought 3 packs of petunias for $2.95 per pack, 4 packs of vincas for $1.39 per flat, and a package of potting soil for $2.79. What is her total bill?

24. Buying a computer. Shandra bought a new computer for $1,389. If the sales tax is 6%, what is her total bill?

25. Capacity of an elevator. An elevator has a capacity of 2,300 pounds. Alicia and her fifth grade class of 21 students wish to crowd onto the elevator. Do you think it is safe? Explain

26. Capacity of an elevator. Would it be safe for eight linemen for the NFL Washington Redskins to crowd onto the elevator mentioned in Exercise 25? Explain.

27. Calculating taxes. Dan's taxable income is $37,840. If the state income tax rate is 2.4% and the county wage tax is 1.1%, what is the total that he will pay for those two taxes?

28. Estimating a tire warranty. Aaron bought new tires that are guaranteed for 40,000 miles. If he drives 11.5 miles each way to work five days a week and then an additional 14 miles each weekend, how many weeks will it be before his warranty expires?

29. Calculating tax deductions. Mary Rose works at home as a financial consultant. She wants to claim some of her yearly expenses as tax deductions. Her average monthly basic phone bill is $13.75, electricity is $68.45 per month, and water and sewer is $12.80 per month. If she can claim 1/7 of these expenses as deductions, how much will her annual deduction be?

30. Calculating tax deductions. Ben bought a new car for $19,880 and pays a 2.5% sales tax. He can deduct this amount on his federal income tax form. If he is in the 21% bracket, how much will this deduction save him in taxes?

Use the following graph, which shows the relationship between amount of education and average earnings in 1999 according to the United States Bureau of the Census (The New York Times Almanac, 2002), to estimate answers for Exercises 31–34. We will give the exact answers in the back of the text.

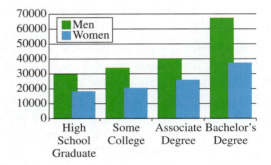

31. What was the average earnings for men with a bachelor's degree?

32. How much higher was the average earnings for men with an associate degree than for women with an associate degree?

33. How much higher was the average earnings for women with a bachelor's degree than for women with a high school diploma?

34. In what category was the gap between the earnings for men and women the largest? What was the difference?

The following graph shows the change in the use of electronic devices from 1995 to 2000 (The World Almanac, 2002). Use this graph to estimate answers for Exercises 35–38. We will give the exact answers in the back of the text.

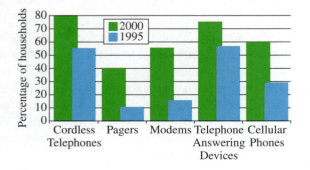

35. What percentage of households had cellular phones in 1995?

36. What percentage of households had modems in 2000?

37. What device showed the largest percentage gain? By how much?

38. What device showed the smallest percentage gain? By how much?

The following pie chart shows revenues for the federal government in billions of dollars for the fiscal year 2002. Use this graph to estimate answers to Exercises 39–42. The total revenue is $2,192 billion.

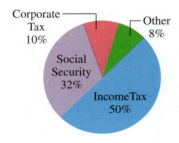

39. How much revenue is derived from Social Security taxes?

40. How much revenue is not derived from income taxes?

41. How much revenue is derived from corporate taxes?

42. How much revenue is derived from Social Security taxes and income taxes combined?

The following pie chart shows a distribution of immigration into the United States in 1998. Use this graph to estimate the answers to Exercises 43–46. There were 705,361 immigrants to the United States in 1998. (Source: The New York Times Almanac, 2002)

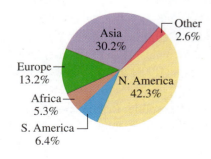

43. How many people immigrated to the United States from Asia?

44. How many people immigrated to the United States from Europe?

45. If 130,661 people immigrated from Mexico, what percentage was that of the total number of immigrants?

46. If 41,034 people immigrated from China, what percentage was that of the total number of immigrants?

Further Exercises

47. Buying fertilizer. The Martinez's yard is 96 feet by 169 feet and a diagram of their yard is given in Figure 1.8. The dimensions of the house, driveway, and garden are also given. The rest of the yard is grass. Mr. Martinez wants to fertilize his lawn. A bag of fertilizer covers 5,000 square feet. Estimate how many bags of fertilizer he will need. Explain how you made your estimate.

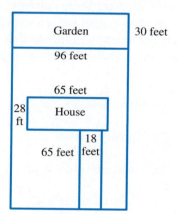

FIGURE 1.8 The Martinez's yard.

48. Purchasing paint. The Wilson's are painting their living room, which is 18.5 feet long and 11 feet wide. The room is 7.75 feet high. If they wish to put two coats of paint on the walls and a gallon of paint will cover 200 square feet, how many gallons of paint should they buy? Explain how you arrived at your estimate.

* **49.** Communicating Mathematics Give several situations in real life where you find estimation to be useful.

50. Estimating the Earth's circumference. Use a map of Egypt to estimate the distance between Alexandria and Aswan. Then multiply your estimate by 50. When Eratosthenes made his estimate of the Earth's circumference, he also added an additional 2,000 stadia to his estimate. Do this also. You need to know that there are 5,280 feet in a mile to find the extra miles in your estimate. How does your estimate compare with Eratosthenes' estimate?

1.3 THE LANGUAGE OF SETS

Mathematicians group items together that have a common characteristic so that they can treat that collection as a single object. In mathematical terms, a collection of objects is called a **set** and the individual objects in this collection are called the **elements** or **members** of the set.

We generally use capital letters to name sets and lowercase letters to denote elements in a set. For example, we may label the set of Olympic gold medal winners by the letter G and denote elements of that set of people by x, a, p, or d. It is important to choose a name for a set to help us remember what it is. We could call a group of Republicans X and a group of Democrats Y, but it is easier to work with these sets if we name the set of Republicans R and the Democrats D.

---

* Exercise numbers circled in red can be used as group exercises.

We represent sets by listing elements or by using set-builder notation.

We often write a set by listing its elements within braces. For example, consider the set of seasons of the year to be the set S. Then we may write $S = \{$spring, summer, fall, winter$\}$. Although it may be possible to list all the elements of a set, it is sometimes inconvenient to do so. If B were the set of all natural numbers* from 1 to 1,000 inclusive, listing all its elements would be cumbersome. Instead, we could write $B = \{1, 2, 3, \ldots, 1000\}$. The first few elements of B are written in order to establish a pattern. The dots, called an ellipsis, indicate that the list continues in the same manner up to the last number in the set, which is 1,000. We would write the set W, consisting of all whole numbers, in a similar fashion as $W = \{0, 1, 2, 3, \ldots\}$. Since we do not put a number after the ellipsis, this means that the list does not end.

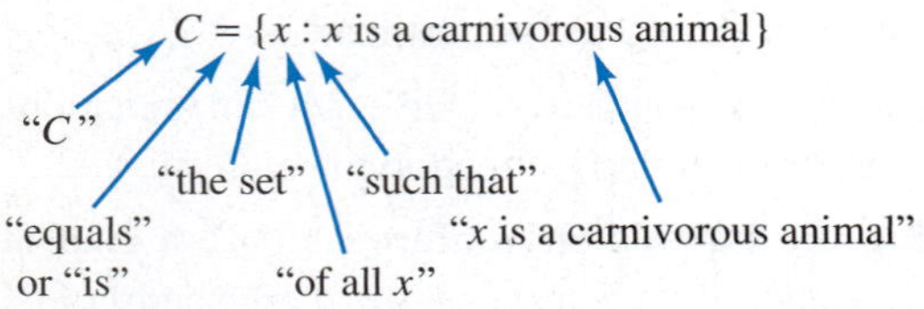

If the elements of a set all share some common characteristics that are satisfied by no other object, then we can use **set-builder notation** to represent the set. For example, suppose that C is the set of all carnivorous (meat-eating) animals. Using set-builder notation, we write

$$C = \{x : x \text{ is a carnivorous animal}\}.$$

As shown in Figure 1.9, we read this equation as:

"C is the set of all x such that x is a carnivorous animal."

Clearly, *lion* is one of the elements of C, but *lamb* is not.

We can write a set represented by the listing method by using set-builder notation, and vice versa.

$C = \{x : x \text{ is a carnivorous animal}\}$

"C"
"equals" or "is"
"the set"
"of all x"
"such that"
"x is a carnivorous animal"

FIGURE 1.9 Reading set-builder notation.

Quiz Yourself 14†

Express each set using an alternative method.
a) $A = \{y : y$ is a day of the week$\}$
b) $B = \{1, 2, 3, \ldots, 60\}$

SOME GOOD ADVICE

Using terminology and symbols properly will improve your work in mathematics. Notice that the set equation in Figure 1.9 sounds like a grammatically correct English sentence when you read it out loud.

EXAMPLE 1 Using Set Notation

Use an alternative method to write each set.
a) $T = \{A^-, A^+, B^-, B^+, AB^-, AB^+, O^-, O^+\}$
b) $B = \{y : y$ is a color of the American flag$\}$
c) $A = \{a : a$ is a counting number less than 20 and is evenly divisible by 3$\}$

---

* We describe some common sets of numbers on page 25.
† Quiz Yourself answers begin on page 849.

Highlight: *Sets of Numbers Commonly Used in Mathematics*

$N = \{x : x \text{ is a natural number}\} = \{1, 2, 3, \ldots\}$

$W = \{x : x \text{ is a whole number}\} = \{0, 1, 2, 3, \ldots\}$

$I = \{x : x \text{ is an integer}\} = \{\ldots -2, -1, 0, 1, 2, 3, \ldots\}$

$Q = \{x : x \text{ is a rational number}\} = \{x : x \text{ is of the form } \frac{a}{b}, \text{ where } a, b \text{ are integers and } b \neq 0\}$

$R = \{x : x \text{ is a real number}\} = \{x : x \text{ has a decimal expansion}\}$

The set N is also sometimes called the set of counting numbers and denoted by C.

SOLUTION: a) We can use set-builder notation to write

$$T = \{x : x \text{ is a blood type}\}.$$

b) B can be written using the listing method as $B = \{\text{red, white, blue}\}$.

c) We can write $A = \{3, 6, 9, 12, 15, 18\}$.

Sets must be well defined.

A set is **well defined** if we are able to tell whether any particular object is an element of that set. Example 2 illustrates this idea.

EXAMPLE 2   Determining Whether a Set Is Well Defined

Which sets are well defined?

a) $A = \{x : x \text{ is a winner of an Academy Award}\}$

b) $T = \{x : x \text{ is tall}\}$

SOLUTION: a) This set is well defined, because we can always determine whether a person belongs to A. Academy Award winners such as Denzel Washington, Jodie Foster, and Jack Nicholson would be members of A. Others such as Hillary Clinton, Bill Cosby, and Barney would not be members of A because they have never won an Oscar.

b) Whether or not a person belongs to this set is a matter of how we interpret *tall*, therefore T is not well defined. Can you think of one situation in which a person who is six feet tall would be considered tall and a different situation in which that same person would be considered short?   ◎

Quiz Yourself 15

Which sets are well defined?

a) $\{x : x \text{ is a mountain over } 10{,}000 \text{ feet high}\}$

b) $\{y : y \text{ is a large number}\}$

In using set-builder notation, it is possible to have a condition that no object satisfies. For example, the set

$$M = \{m : m \text{ is in your math class and is also a U.S. Supreme Court justice}\}$$

has no elements.

DEFINITION

The set that contains no elements is called the **empty set** or **null set**. This set is labeled by the symbol ∅.

The set *M*, therefore, is the empty set. We can write this as $M = \emptyset$. Another set that we frequently use is called the universal set.

DEFINITION

The **universal set** is the set of all elements under consideration in a given discussion. We often denote the universal set by the capital letter *U*.

For example, in a certain problem we may wish to use only the counting numbers from 1 to 10. For this discussion, the universal set would be $U = \{1, 2, 3, \ldots, 10\}$. In another situation we might consider only the female consumers living in the United States. In this case the universal set would be $U = \{x : x$ is a female consumer living in the United States$\}$.

The element symbol expresses that an object is a member of a set.

We use the symbol $\in$ to stand for the phrase *is an element of*. Although $\in$ looks somewhat like the letter *e*, these two symbols are *not the same* and should not be confused. The notation "$3 \in A$" expresses that 3 is an element of the set *A*. If 3 is not an element of set *A*, we write "$3 \notin A$." The Italian mathematician Giuseppe Peano introduced the element symbol in 1889. It is a variation of the Greek letter *epsilon* and is an abbreviation of the Greek word $\epsilon\sigma\tau\iota$, which means "is."*

EXAMPLE 3   Using Set Element Notation

Replace the symbol # in each statement by either $\in$ or $\notin$.

a) 3 # {2, 3, 4, 5}   b) {5} # {2, 3, 4, 5}

c) Bill Gates # $\{x : x$ is a billionaire$\}$

d) jogging # $\{y : y$ is an aerobic exercise$\}$

e) the ace of hearts # $\{f : f$ is a face card in a standard 52-card deck$\}$

SOLUTION: a) $3 \in \{2, 3, 4, 5\}$

b) $\{5\} \notin \{2, 3, 4, 5\}$. Notice that 5 is not the same as {5}

c) Bill Gates $\in \{x : x$ is a billionaire$\}$

d) jogging $\in \{y : y$ is an aerobic exercise$\}$

e) the ace of hearts $\notin \{f : f$ is a face card in a standard 52-card deck$\}$.

 Quiz Yourself 16

Decide whether each statement is true or false.

a) $3 \in \{x : x$ is an odd counting number$\}$

b) $2 \notin \emptyset$

c) skim milk $\notin \{x : x$ is a food containing calcium$\}$

---

* Ralph P. Grimaldi, *Discrete and Combinatorial Mathematics* (Addison-Wesley, 1994).

Historical Highlight: *Set Theory and the Foundations of Mathematics*

Just as cracks in a bridge or the wing of an airplane concern the ordinary citizen, cracks in the structure of mathematics trouble mathematicians. In the late nineteenth century, several contradictions, called *paradoxes*, arose in mathematics, which caused concern and vigorous debate among mathematicians. As a result, scholars who carefully examined the structure of mathematics and scrutinized its logical princi-

ples realized that mathematics had not been built as carefully as previously had been thought. For example, although people had been working with fundamental ideas such as numbers for thousands of years, these basic ideas had never been defined carefully.

Mathematicians used the precision of set theory to define the underlying ideas of mathematics, and thus helped to put mathematics on a firmer foundation.

The cardinal number of a set indicates its size.

In solving set theory problems, we often need to know the number of elements in a set.

DEFINITIONS

The number of elements in set A is called the **cardinal number** of set A and is denoted $n(A)$. A set is **finite** if its cardinal number is a whole number. An **infinite** set is one that is not finite.

EXAMPLE 4   Finding the Cardinal Number of a Set

State whether each set is finite or infinite. If it is finite, state its cardinal number using $n(A)$ notation.

a) $P = \{x : x$ is a planet in our solar system$\}$

b) $N = \{1, 2, 3, \ldots\}$

c) $A = \{y : y$ is a person living in the United States who is not a citizen$\}$

d) $\varnothing$

e) $X = \{\{1, 2, 3\}, \{1, 4, 5\}, \{3\}\}$

Quiz Yourself 17

Find the cardinal number of each set.

a) $\{2, 4, \ldots, 20\}$

b) $\{\{1, 2\}, \{1, 3, 4\}\}$

c) $\{s : s$ is one of the United States$\}$

SOLUTION: a) There are nine planets, therefore, this is a finite set. $n(P) = 9$.

b) The set of counting numbers is an infinite set.

c) There are a finite number of people living in the United States who are not citizens; however, we probably do not know $n(A)$.

d) The empty set has no elements, so it is a finite set. Thus $n(\varnothing) = 0$.

e) The set X contains three objects: the set $\{1, 2, 3\}$, the set $\{1, 4, 5\}$, and the set $\{3\}$. Therefore, $n(X) = 3$.

Exercises 1.3

In Exercises 1–12, use set notation to list all the elements of each set.

1. The natural numbers from 10 to 15 inclusive

2. The letters of the alphabet that follow *e* and come before *k*

3. {17, 18, . . . , 25}

4. The integers between −5 and 5, not inclusive

5. The natural numbers less than 30 that are divisible by 4

6. {*y* : *y* is an odd natural number between 6 and 20}

7. The days of the week

8. {*x* : *x* is one of the 50 U.S. states that begins with the letter "A," "B," or "C"}

9. {*y* : *y* is a natural number less than zero}

10. {11, 13, 15, . . . , 25}

11. The set of U.S. presidents who served after Jimmy Carter but before George W. Bush

12. {*x* : *x* is one of the 50 U.S. states that borders no other state}

In Exercises 13–24, use an alternative method to express each set. There may be several acceptable answers.

13. {3, 6, 9, 12}

14. {*t* : *t* is a letter in the word *member*}

15. {*m, u, s, i, c*}

16. {*t* : *t* is a digit in the number 173,268}

17. {*z* : *z* is a negative integer}

18. {*y* : *y* is a month of the year}

19. {*y* : *y* is a letter of the word *set* and also a letter of the word *ring*}

20. {Aries, Taurus, Gemini, Cancer, . . . , Aquarius, Pisces}

21. {*y* : *y* is a natural number greater than 100}

22. {2, 4, 6, 8, 10, . . .}

23. {2, 4, 6, 8, 10, . . . , 100}

24. {*x* : *x* is a natural number that is evenly divisible by three}

In Exercises 25–28, use this table of general education electives to describe each set in an alternative way.

25. {History012, History223, English010, English220, Anthropology111}

26. {English010, English220, Psychology200}

27. {*x* : *x* satisfies a world culture requirement}

28. {*x* : *x* does not satisfy a cultural diversity requirement}

	Humanities	Writing	World Culture	Cultural Diversity
History012	yes	yes	yes	no
History223	yes	yes	yes	yes
English010	yes	yes	no	no
English220	yes	yes	no	no
Psychology200	no	yes	no	no
Geography115	no	no	yes	yes
Anthropology111	yes	no	yes	yes

In Exercises 29–36, determine whether each set is well defined.

29. {*x* : *x* lives in Michigan}

30. {2, 4, 6, 8, 10, . . .}

31. {*y* : *y* has an interesting job}

32. {*t* : *t* has traveled much}

33. {*x* : *x* is a ferocious animal}

34. {*y* : *y* is a mammal}

35. {1, −3, 5, −7, 9, −11, . . .}

36. {*y* : *y* is an easy telephone number to remember}

In Exercises 37–48, replace each # with either ∈ or ∉ to express a true statement.

37. 3 # {2, 4, 6, 8}

38. 3 # {*x* : *x* is a whole number}

39. Franklin Roosevelt # {x : x is a past president of the United States}

40. Albert Einstein # {y : y is a living American poet}

41. IBM # {w : w is a manufacturer of computers}

42. Tiger Woods # {a : a is a professional ice skater}

43. 5 # {x : x is a rational number}

44. −5 # {y : y is a real number}

45. 0 # ∅

46. ∅ # {0}

47. Florida # {x : x is a state south of Pennsylvania}

48. {Florida} # {x : x is a state east of the Mississippi}

Find n(A) for each of the following sets A.

49. {1, 3, 5, 7, . . . , 11}

50. {3, 4, 5, . . . , 13}

51. {x : x is a living American president born before 1900}

52. {x : x is one of the continental United States}

53. {x : x is a letter in the word Mississippi}

54. {x : x is a vowel in the alphabet}

55. {{1, 2}, {1, 2, 3}}

56. {{1}, ∅, 0, {0}}

Describe each of the following sets as either finite or infinite.

57. {1, 4, 7, . . . , 16}

58. {1, 4, 7, 10, 13, 16, . . .}

59. {x : x is a natural number greater than 58}

60. {x : x is a word written by Shakespeare}

61. {y : y is the number of people who have walked on the moon}

62. {y : y is a number between 4 and 5}

63. {x : x is an element of the empty set}

64. {x : x is one of the ways we can combine different letters of the alphabet}

In Exercises 65–72, find an element of set A that is not an element of set B. There are many correct answers.

65. A = {y : y is a number between 4 and 10}
B = {y : y is a natural number between 4 and 10}

66. A = {y : y is a member of the human race}
B = {y : y is a citizen of the United States}

67. A = {y : y is manufacturer of electronic products}
B = {y : y is a company based in the United States}

68. A = {y : y is an animal}
B = {y : y is covered with fur}

69. A = {y : y is a world political leader}
B = {y : y is European}

70. A = {y : y is a car manufacturer}
B = {y : y is a company based in the United States}

71. A = {y : y is a day of the week}
B = {y : y is a weekday}

72. A = {y : y is a state whose name begins with the letter "A," "B," or "C"}
B = {y : y is a state whose name begins with the letter "A" or "B"}

Further Exercises

(73.**) **Sets of famous people**. Consider the universal set

U = {Michael Jordan, Babe Ruth, Beethoven, Bach, Leonardo da Vinci, Ernest Hemingway, Bart Simpson, Hillary Clinton, Napoleon, Troy Aikman, Rembrandt, Winston Churchill, Julius Caesar, Shakespeare, Harry Truman, Madonna, Barney}.

From U we can choose a set of elements that all share some common characteristic and can give this set using

---

both the listing method and set-builder notation—for example, {Shakespeare, Winston Churchill}, which can be written as $\{y : y$ is British$\}$. Find as many such sets as you can and give them using both the listing method and set-builder notation.

74. **Sets of common objects.** Consider the universal set

$U = \{$apple, TV set, hat, radio, fish, sofa, washing machine, shoe, dog, automobile, potato chip, toenail clipper, bread, banana, vacuum cleaner, hammer, bed, pizza$\}$.

From U we can choose a set of elements that all share some common characteristic and can give this set using both the listing method and set-builder notation—for example, {TV set, radio, washing machine, vacuum cleaner} = $\{x : x$ is an electrical appliance$\}$. Find as many such sets as you can and give them using both the listing method and set-builder notation.

We mentioned earlier that serious paradoxes arose in mathematics. We will define a paradox as a statement that contradicts itself, or a statement that can be proved to be both true and false at the same time. Use this definition in thinking about Exercises 75–77.

75. **A paradox.** A small town has only one barber, who is male. Furthermore, this barber shaves those men, and only those men, who do not shave themselves. Who shaves the barber? One of two things can happen—either the barber shaves himself or he does not. First assume that the barber shaves himself. What must you conclude? Now assume that the barber does not shave himself. What must you conclude?

76. Communicating Mathematics Consider the statement, "This sentence is false." Is this sentence true or false? First assume that the sentence is true. What must you conclude? Now assume that the sentence is false. What must you conclude?

77. Communicating Mathematics Let A be the set $\{1, 2, 3\}$. Clearly $A \notin A$. Now think of the set of all such sets that are not elements of themselves and call that set S. That is, $S = \{X : X$ is a set and $X \notin X\}$. Now we ask the question, "Is $S \in S$?" First assume that $S \in S$. What must you conclude? Now assume that $S \notin S$. What must you conclude?

1.4 COMPARING SETS

One benefit of organizing objects into sets is that by doing so we can compare them similarly to the way we compare other mathematical objects, such as numbers in algebra and shapes in geometry.

Equal sets contain the same members.

One of the most fundamental things we need to know about two sets is when do we consider them to be the same.

> **DEFINITION**
>
> Two sets A and B are **equal** if they have exactly the same members. In this case we write $A = B$. If A and B are not equal, we write $A \neq B$.

This definition says that for sets A and B to be equal, every element of A must also be a member of B and every element of B must also be in A.

Quiz Yourself 18*

Determine whether each statement is true or false.

a) {Socrates, Shakespeare, Beethoven} = {Shakespeare, Beethoven, Socrates}

b) {tiger, gray whale, giant panda} = {y : y is an endangered species}

EXAMPLE 1   Set Equality

Which pairs of sets are equal?

a) $\{A^-, A^+, B^-, B^+, AB^-, AB^+, O^-, O^+\}$ and $\{A^-, B^-, AB^-, O^-, A^+, B^+, AB^+, O^+\}$

b) $A = \{x : x$ is a citizen of the United States$\}$ and $B = \{y : y$ was born in the United States$\}$

SOLUTION: a) Notice that the left-hand set and the right-hand set contain exactly the same elements. The order of the elements is not important, therefore the two sets are equal.

b) Because A contains elements that are not in B, these sets are not equal.

One set is a subset of another if all of its elements are found in the other set.

Another way we compare sets is to determine whether one set is part of another set.

In order to show that $A \subseteq B$, we must show that every element of A also occurs as an element of B. To show that A is not a subset of B, all we have to do is find one element of A that is not in B.

---

* Quiz Yourself answers begin on page 849.

EXAMPLE 2   Identifying Subsets

Determine whether either set is a subset of the other.

a) $A = \{1, 2, 3\}$ and $B = \{1, 2, 3, 4\}$

b) $L = \{x : x$ lives in Los Angeles$\}$ and $C = \{y : y$ lives in California$\}$

SOLUTION: a) Every member of A is also in B, so we can say $A \subseteq B$. Since there is an element of B that is not in A, we write $B \not\subseteq A$.

b) $L \subseteq C$. However, C is not a subset of L, since there are people who live in California who do not live in Los Angeles.

If A is any set, then $A \subseteq A$ because clearly each element of A is an element of A. Also, the empty set is a subset of every set. For example, $\varnothing \subseteq \{1, 2, 3\}$. Even though this sounds strange, it is true that every element of the empty set is also an element of $\{1, 2, 3\}$. According to the Counterexample Principle from Section 1.1, to show that this statement is false, *you must find a counterexample*. That is, you must find an element of $\varnothing$ that is not in $\{1, 2, 3\}$. Since this is impossible, our statement is true.

Venn diagrams represent set relationships graphically.

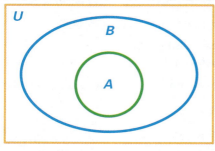

FIGURE 1.10   A Venn diagram illustrating that A is a subset of B.

Drawings called **Venn diagrams** are used to visualize relationships among sets. Figure 1.10 is a Venn diagram that illustrates A is a subset of B.

We represent the universal set by the rectangular region, labeled U. The region labeled A is completely contained in region B, indicating that all of A's elements are also in B.

DEFINITION

The set A is a **proper subset** of the set B if $A \subseteq B$ but $A \neq B$. We write this as $A \subset B$. If A is not a proper subset of B, then we write $A \not\subset B$.

From this definition, we see that $\{2, 4, 6, \ldots\} \subseteq \{1, 2, 3, \ldots\}$. Also $\{2, 4, 6, \ldots\} \subset \{1, 2, 3, \ldots\}$ because $\{1, 2, 3, \ldots\}$ contains elements that are not members of $\{2, 4, 6, \ldots\}$.

SOME GOOD ADVICE

Distinguish between the notation for subset and proper subset. The lower half of the subset symbol in the notation "$A \subseteq B$" looks somewhat like an equal sign because it is *possible* that sets A and B are equal. Of course, the sets do not have to be equal. When we write $A \subset B$, the lower line is missing to remind us that the sets *cannot* be equal.

EXAMPLE 3 Identifying Subsets

Consider the following table of information regarding Olympic athletes.

	Event	Medal	Country
Gina Gogean	Balance beam	Bronze	Romania
Carl Lewis	Long jump	Gold	United States
Marie-Jose Perec	200-meter sprint	Gold	France
Shannon Miller	Balance beam	Gold	United States
Simona Amanar	All-around (gymnastics)	Silver	Romania
Joe Greene	Long jump	Bronze	United States
Simona Amanar	Floor exercise	Silver	Romania
Michael Johnson	200-meter sprint	Gold	United States
Lilia Podkopayeva	Floor exercise	Gold	Ukraine

Marie-Jose Perec

Assume that this set of athletes is the universal set and define the following sets:

T = the set of 200-meter medal winners
G = the set of gold medal winners
A = the set of U.S. athletes
L = the set of long jump medalists

Which statements are true?

a) $L \subseteq A$ b) $L \subset A$ c) $G \subseteq A$

SOLUTION: a) This is true because every long jump medalist is also a U.S. athlete.

b) This is true. We already know that L is a subset of A. Notice that A contains elements such as Shannon Miller that are not members of L, which makes L a proper subset of A.

c) This is false. Set G has elements such as Marie-Jose Perec that are not elements of A.

We sometimes want to find all subsets of a given set, such as $\{1, 2, 3, 4, 5\}$. When asked to do this you may respond:

$$\{1, 2\}, \{3, 4\}, \{5\}, \{1, 2, 3, 4, 5\}, \{1\}, \{1, 2, 3\}, \text{ and so on.}$$

The trouble with this random approach is that after identifying about eleven or twelve subsets, it becomes harder and harder to think of new ones, and you might stop without creating a complete list. As we pointed out in Section 1.1, it is important to attack a problem systematically. We illustrate an organized solution to this type of problem in Example 4.

EXAMPLE 4 Finding All Subsets of a Set Systematically

Find all subsets of the set $\{1, 2, 3, 4\}$.

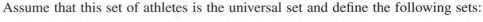

Quiz Yourself 19

Decide whether each statement is true or false.

a) $\{2,4,6,8, \ldots\} \subseteq \{1,2,3,4, \ldots\}$

Use the table in Example 3 to answer (b).

b) $T \subseteq A$

SOLUTION: We can organize this problem by considering subsets according to their size, going from 0 to 4. This method is illustrated in the following table.

Size of Subset	Subsets of This Size	Number of Subsets of This Size
0	$\emptyset$	1
1	{1}, {2}, {3}, {4}	4
2	{1, 2}, {1, 3}, {1, 4}, {2, 3}, {2, 4}, {3, 4}	6
3	{1, 2, 3}, {1, 2, 4}, {1, 3, 4}, {2, 3, 4}	4
4	{1, 2, 3, 4}	1
		Total = 16

Quiz Yourself 20

Find all subsets of the set {1, 2, 3}.

If we can generalize the solution to one problem, we can often use this knowledge to solve other, related problems. Let us try to generalize the solution that we found in Example 4. In order to discover a pattern, we look at further examples having different-sized sets S.

S	All Subsets of S	Number of Subsets of S
$\emptyset$	$\emptyset$	1
{1}	$\emptyset$, {1}	2
{1, 2}	$\emptyset$, {1}, {2}, {1, 2}	4
{1, 2, 3}	See Quiz Yourself 20	8
{1, 2, 3, 4}	See Example 4	16

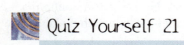

Quiz Yourself 21

a) If a set has five elements, how many subsets will it have?

b) How many subsets can be formed using letters of the alphabet?

We see that each time we add an element to S, the number of subsets of S doubles. The pattern $1 = 2^0$, $2 = 2^1$, $4 = 2^2$, $8 = 2^3$, and $16 = 2^4$ gives us the general relationship that we seek.

> A set that has k elements has 2^k subsets.

Equivalent sets have the same number of elements.

Another relationship that can occur between two sets is that the elements of one set may match up with the elements in the other set.

Historical Highlight: *Georg Cantor and Set Theory*

In 1879, the German mathematician Georg Cantor wrote a series of papers on set theory that began one of the fiercest battles in the history of mathematics. Using ingenious methods, he was able to prove remarkable results about infinite sets. For example, he proved that infinite sets can be of different sizes—that is, there are different infinite cardinal numbers. In fact, he proved the amazing fact that there are an infinite number of different infinite cardinal numbers.

Because Cantor's results were so revolutionary, they were not accepted by many of his contemporaries. His most devastating critic was his former teacher, Leopold Kronecker, who concluded that these new methods were "a dangerous type of mathematical insanity." The distinguished French mathematician Henri Poincaré proclaimed, "Later generations will regard [Cantor's set theory] as a disease from which one has recovered."

To make matters worse, toward the end of the nineteenth century, mathematicians found contradictions in Cantor's set theory. These "paradoxes," as they were called, prompted two of the foremost mathematicians of the time, David Hilbert of Germany and Bertrand Russell of England, to modify the theory to eliminate the paradoxes. Unimpressed, the Dutch mathematician L. Brouwer wrote, "A false theory which is not stopped by a contradiction is nonetheless false."

Although the battle still continues, most mathematicians have accepted Cantor's results. Today set theory is recognized as a convenient language for expressing mathematical ideas and provides a foundation from which virtually all mathematics can be developed. It is perhaps, in the words of Hilbert, "the most admirable fruit of the mathematical mind and indeed one of the highest achievements of man's intellectual processes."

DEFINITION

Sets A and B are **equivalent**, or in **one-to-one correspondence**, if $n(A) = n(B)$. Another way of saying this is that two sets are equivalent if they have the same number of elements.*

The sets $\{1, 2, 3\}$ and $\{4, 5, 6\}$ are equivalent because they have the same number of elements. In Example 3, the set of gold medal winners was {Lewis, Perec, Miller, Johnson, Podkopayeva}, which is not equivalent to the set of U.S. athletes, {Lewis, Miller, Greene, Johnson}.

SOME GOOD ADVICE

The words *equal* and *equivalent* sound similar but do not have the same meanings. *Be careful not to use them interchangeably.*

---

* We have been discussing the cardinal numbers of finite sets. We will discuss one-to-one correspondences between infinite sets in the Of Further Interest section.

Exercises 1.4

In Exercises 1–10, decide whether each pair of sets is equal. Justify your answer.

1. $\{1, 3, 5, 7, 9\}$ and $\{1, 5, 9, 3, 7\}$

2. $\{a, e, i, o, u\}$ and $\{a, b, \ldots, u\}$

3. $\{x : x$ is a counting number between 5 and 19 inclusive$\}$ and $\{y : y$ is a rational number between 5 and 19 inclusive$\}$

4. $\{x : x$ is a counting number between 3 and 10 inclusive$\}$ and $\{y : y$ is a whole number between 3 and 10 inclusive$\}$

5. $\{1, 3, 5, \ldots, 99\}$ and $\{x : x$ is an odd counting number between 0 and 100$\}$

6. $\{3, 6, 9, 12, 15\}$ and $\{x : x$ is a counting number that is a multiple of 3$\}$

7. $\{z : z$ is a month of the year$\}$ and $\{$January, February, March, . . . , December$\}$

8. $\{y : y$ is a weekday$\}$ and $\{$Monday, Tuesday, . . . , Friday$\}$

9. $\varnothing$ and $\{x : x$ is a living American born before 1800$\}$

10. $\varnothing$ and $\{\varnothing\}$

In Exercises 11–20, decide whether each statement is true or false. Justify your answer.

11. $\{1, 3, 5, 9\} \subseteq \{1, 2, 3, \ldots, 9\}$

12. $\{1, 3, 5, 9\} \subset \{1, 2, 3, \ldots, 9\}$

13. $\{1, 3, 5, 9\} \subset \{1, 3, 5, 9\}$

14. $\{1, 3, 5, 9\} \subseteq \{1, 3, 5, 9\}$

15. $\{x : x$ is a letter in the word *happy*$\} \subseteq \{y : y$ is a letter in the word *happiness*$\}$

16. $\{t : t$ is a letter in the word *Ruth*$\} \subset \{z : z$ is a letter in the word *truth*$\}$

17. $\varnothing \subseteq \{1, 3, 5\}$

18. $\varnothing \subset \varnothing$

19. $\{\varnothing\} \subseteq \{0\}$

20. $\{0\} \subseteq \varnothing$

In Exercises 21–30, decide whether each pair of sets is equivalent. Justify your answer.

21. $\{1, 2, 3, 4, 5\}$ and $\{a, e, i, o, u\}$

22. $\{2, 4, 6, 8, 10, 12\}$ and $\{2, 3, 4, \ldots, 12\}$

23. $\{x : x$ is a letter in the word *song*$\}$ and $\{x : x$ is a letter in the word *songs*$\}$

24. $\{x : x$ is a letter in the word *tenacity*$\}$ and $\{x : x$ is a letter in the word *resolve*$\}$

25. $\varnothing$ and $\{\varnothing\}$

26. $\{\varnothing\}$ and $\{0\}$

27. $\{1, 3, 5, 7, \ldots, 15\}$ and $\{4, 6, 8, 10, \ldots, 18\}$

28. $\{a, b, c, d, e, \ldots, z\}$ and $\{3, 4, 5, 6, \ldots, 26\}$

29. $\{x : x$ is a day in the year 2003$\}$ and $\{y : y$ is a day in the year 2004$\}$

30. $\{x : x$ was in the starting lineup of the Baltimore Ravens in the 2001 Super Bowl$\}$ and $\{x : x$ was in the starting lineup of the New York Giants in the 2001 Super Bowl$\}$

31. List all of the two element subsets of the set $\{1, 2, 3\}$.

32. List all of the two element subsets of the set $\{1, 2, 3, 4\}$.

33. List all of the three element subsets of the set $\{1, 2, 3, 4\}$.

34. List all of the three element subsets of the set $\{1, 2, 3, 4, 5\}$.

35. Communicating Mathematics Explain the difference between equal and equivalent sets.

36. Communicating Mathematics How would you help a friend remember the meaning of the notation $n(A)$?

37. Communicating Mathematics What is the difference between the meaning of the notation $\subseteq$ and $\subset$?

38. Communicating Mathematics If $A \subseteq B$, can we conclude that $A \subset B$? Explain. If $A \subset B$, can we conclude that $A \subseteq B$? Explain.

Use the following table to answer Exercises 39–42.

	Major	Class Rank	GPA	Activities
Allen	Music	Freshman	1.9	Drama
Belinda	Art	Senior	2.8	Newspaper
Carmen	English	Freshman	3.1	Baseball
Dana	History	Freshman	2.9	Drama
Elston	Art	Senior	2.8	Band
Frank	Sociology	Sophomore	3.1	Football
Gina	Chemistry	Junior	2.6	Newspaper
Hector	Physics	Freshman	2.2	Band
Ivana	English	Junior	3.5	Basketball
James	English	Sophomore	2.9	Newspaper

In Exercises 39–42, consider the following sets: U (upperclassmen), L (lowerclassmen), S (science majors), V (GPA above 3.0), A (art majors), T (athletes), and D (involved in drama).

39. Find a set that is equal to V.

40. Find a set that is equivalent to S, but not equal to S.

41. Find a set whose cardinal number is the largest of all the sets.

42. Find a set whose cardinal number is the smallest of all the sets.

In Exercises 43–44, recall that in Section 1.2 we defined the following sets:

$N = \{x : x \text{ is a natural number}\}$
$W = \{x : x \text{ is a whole number}\}$
$I = \{x : x \text{ is an integer}\}$
$Q = \{x : x \text{ is a rational number}\}$
$R = \{x : x \text{ is a real number}\}$

43. Which sets are subsets of each other? For example, we can say $N \subseteq W$, and $N \subseteq N$.

44. Which sets are proper subsets of each other? For example, we can say $N \subset R$ and $N \subset Q$.

45. If set A has 5 elements, how many subsets does A have? How many of these subsets are proper?

46. If set A has 7 elements, how many subsets does A have? How many of these subsets are proper?

On an exam, Pete said that the set B has 25 subsets. Use this information to answer Exercises 47–50.

47. Communicating Mathematics What is wrong with Pete's answer?

48. Communicating Mathematics What mistake do you think Pete made?

49. Communicating Mathematics How many elements do you think that B had? Why?

50. Communicating Mathematics How many subsets do you think that B had? Why?

51. Communicating Mathematics How does the following diagram explain why the set $A = \{1, 2\}$ has four subsets? Draw a similar diagram for $B = \{1, 2, 3\}$.

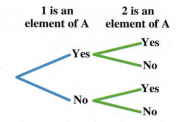

52. Communicating Mathematics If a set has three elements, then it has 8 subsets. Use a tree diagram as in Exercise 51 to give an explanation as to why you would then expect a four-element set to have 16 elements.

Further Exercises

In Exercises 53–60, recall that in Section 1.1 we introduced the following arrangement of numbers, which is called Pascal's triangle.

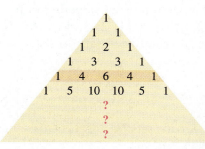

Notice that the fourth line* of this triangle contains the numbers 1, 4, 6, 4, 1, which, as we saw in Example 4, are

precisely the counts of the number of subsets of a four-element set with 0, 1, 2, 3, and 4 elements, respectively.

53. With this observation in mind, how do you interpret the fifth line of Pascal's triangle?

54. How do you interpret the sixth line of Pascal's triangle?

55. Suppose that a newspaper editor has nine reporters and wishes to send three of them to a convention. In how many ways can this be done? (*Hint:* Use Pascal's triangle as you did in Exercises 53 and 54.)

56. The cheerleading team has ten members and wishes to send four to a workshop. In how many ways can this choice be made? (*Hint:* Use Pascal's triangle as you did in Exercises 53 and 54.)

57. Add the numbers across each row of Pascal's triangle. Do the results look familiar?

_____
* We start counting these lines with 0, not 1.

58. Notice that the arrangement of numbers in each row is symmetric. See how the fours are balanced in the fourth row, the fives are balanced in the fifth row, and so on. Can you give a set theory explanation to account for this?

59. A committee consists of Senators Allen, Brown, Cianci, Devlin, and Eastman. Each senator's vote has been weighted so that Allen's and Cianci's votes will be counted twice, Brown's will be counted three times, Devlin's four times, and Eastman's once. In order for the committee to pass a given bill, a total voting weight of 9 or more is required. How many different subsets of this committee have the property that the sum of the voting weights of its members is 9 or more?

60. Let us suppose that Colombia, Brazil, Ecuador, Venezuela, and Mexico are meeting to plan future strategy for controlling world coffee. Each country's vote has been weighted according to their current coffee-producing ability, so that Brazil's vote will count as four votes, Colombia's and Venezuela's votes will each count as three, Ecuador's vote will count as two, and Mexico's vote will count only once. In order for any policy to be accepted by this group, it has been agreed that a total voting weight of ten or more is needed. How many different subsets of these five countries have a total voting weight of ten or more?

We mentioned that the subset notation, $\subseteq$, and the notation for "less than or equal to," $\leq$, appear to be similar. In Exercises 61–62, for each property of $\leq$, state the corresponding property of $\subseteq$. Next, convince yourself that the newly stated property is indeed a valid set theory property.

61. If $a \leq b$ and $b \leq c$, then $a \leq c$.

62. If $a \leq b$ and $b \leq a$, then $a = b$.

1.5   SET OPERATIONS

When you use numerical operations such as addition, subtraction, and so on, you are combining numbers to get other numbers. For instance, adding 2 and 3 you get 5, dividing 12 by 4 you obtain 3. Similarly, we use set operations in various ways to form other sets.

We form the union of sets by joining sets together.

In forming a set union (see below), we join sets together to form a larger set—for example, $\{1, 3, 4, 5\} \cup \{2, 4, 6\} = \{1, 2, 3, 4, 5, 6\}$. Notice that although 4 is a member of both sets, it is not necessary to list 4 twice in the union.

DEFINITION

The **union** of sets A and B, written $A \cup B$, is the set of elements that are members of either A or B (or both). Using set-builder notation,

$$A \cup B = \{x : x \text{ is a member of } A \text{ or } x \text{ is a member of } B\}.$$

The union of more than two sets is the set of all elements belonging to at least one of the sets.

EXAMPLE 1   Finding the Union of Sets

The following table compares several popular fitness activities.

Activity	Calories Burned per Hour*	Requires Special Location	Requires Special Equipment	Can Be Done Alone	Can Be Done in All Kinds of Weather
Ballroom dancing (ba)	300	Yes	No	No	Yes
Bench stepping (be)	610	No	Yes	Yes	Yes
Bicycling (bi)	415	No	Yes	Yes	No
Calisthenics (ca)	300	No	No	Yes	Yes
Hiking (hi)	400	No	No	Yes	No
Jogging (jo)	655	No	No	Yes	No
Stair-climbing machine (st)	680	Yes	Yes	Yes	Yes
Tennis (te)	425	Yes	Yes	No	No

These activities form a universal set,

$$U = \{ba, be, bi, ca, hi, jo, st, te\},$$

where each element is represented by a two-letter abbreviation.
Let's define the following sets:

M = the set of activities that burn more than 650 calories per hour
W = the set of activities that can be done in all kinds of weather
E = the set of activities that need special equipment
L = the set of activities that must be done in a special location

a) Find $M \cup W$.   b) Find $E \cup L$.

SOLUTION: a) Since $M = \{jo, st\}$ and $W = \{ba, be, ca, st\}$, then

$$M \cup W = \{ba, be, ca, jo, st\}.$$

b) We see that $E = \{be, bi, st, te\}$ and $L = \{ba, st, te\}$. Therefore,

$$E \cup L = \{ba, be, bi, st, te\}.$$

---

* "Your Permanent Weight-Loss Workbook," *Prevention Magazine* (January 1995).

It is good to practice reading the statements in Example 1 aloud to become comfortable with set notation and terminology.

■────────────────────

The intersection of sets is the set of elements they have in common.

Another set operation is intersection.

> **DEFINITIONS**
>
> The **intersection** of sets A and B, written $A \cap B$, is the set of elements common to both A and B. Using set-builder notation,
>
> $$A \cap B = \{x : x \text{ is a member of } A \text{ and } x \text{ is a member of } B\}.$$
>
> The intersection of more than two sets is the set of elements that belong to each of the sets. If $A \cap B = \varnothing$, then we say that A and B are **disjoint**.

EXAMPLE 2   Finding the Intersection of Sets

Recall the sets $M = \{jo, st\}$, $W = \{ba, be, ca, st\}$, $E = \{be, bi, st, te\}$, and $L = \{ba, st, te\}$ that we introduced in Example 1.

a) Find $M \cap W$.   b) Find $E \cap L$.

SOLUTION: a) $M \cap W$ is the set of elements common to both M and W. Therefore, $M \cap W = \{jo, st\} \cap \{ba, be, ca, st\} = \{st\}$.

b) Similarly, $E \cap L = \{be, bi, st, te\} \cap \{ba, st, te\} = \{st, te\}$. ◎

It often helps us understand a set theory problem better if we represent the sets in the problem graphically. Figure 1.11 shows how to visualize the union and

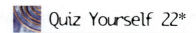

Quiz Yourself 22*

Let $M = \{x : x$ is a letter in the word *mathematics*} and let $B = \{y : y$ is a letter in the word *beauty*}. Find $M \cup B$ and $M \cap B$.

intersection of two sets. We labeled the regions in the Venn diagrams in order to be able to refer to them more easily. Set A consists of regions r_2 and r_3 and B consists of regions r_3 and r_4. In Figure 1.11(a) $A \cup B$ is the shaded region consisting of regions r_2, r_3, and r_4. The shaded region in Figure 1.11(b) represents $A \cap B$ which consists of region r_3.

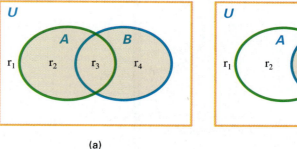

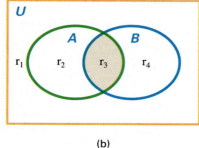

(a)   (b)

FIGURE 1.11   (a) A Venn diagram of $A \cup B$.   (b) A Venn diagram of $A \cap B$.

If sets A and B are disjoint, we draw the Venn diagram so that A and B do not overlap.

The elements not in a set form its complement.

The third set operation that we consider is complementation.

DEFINITION

If A is a subset of the universal set U, the **complement** of A is the set of elements of U that are *not* elements of A. This set is denoted by A'. Using set-builder notation,

$$A' = \{x : x \in U \text{ but } x \notin A\}.$$

EXAMPLE 3   Finding the Complement of Sets

Find the complement of each set. We have stated a universal set for each set.

a) $U = \{1, 2, 3, \ldots, 10\}$ and $A = \{1, 3, 5, 7, 9\}$.

b) U is the set of people living in the United States and S is the set of people living in the United States who smoke.

c) U is the set of cards in a standard 52-card deck and F is the set of face cards.

---

* Quiz Yourself answers begin on page 849.

SOLUTION: a) A' is the set of elements in U that are not in A, so $A' = \{2, 4, 6, 8, 10\}$.

b) S' is the set of nonsmokers living in the United States.

c) F' is the set of nonface cards.

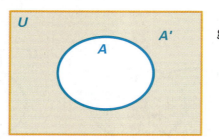

The complement of A is the shaded region in Figure 1.12.

FIGURE 1.12 Venn diagram of the complement of A.

To form a set difference, begin with one set and remove all elements that appear in a second set.

The last set operation that we discuss in this section is set difference.

> **DEFINITION**
>
> The **difference** of sets B and A is the set of elements that are in B but not in A. This set is denoted by $B - A$. Using set-builder notation,
>
> $$B - A = \{x : x \text{ is a member of } B \text{ and } x \text{ is not a member of } A\}.$$

EXAMPLE 4   Finding the Difference of Sets

a) Find $\{3, 6, 9, 12\} - \{x : x \text{ is an odd integer}\}$.

b)   Recall the sets $M = \{jo, st\}$ and $W = \{ba, be, ca, st\}$ from Example 1. Find $M - W$.

SOLUTION: a) We start with $\{3, 6, 9, 12\}$ and remove all of the odd integers to get $\{6, 12\}$.

b)   The set $M - W = \{jo, st\} - \{ba, be, ca, st\} = \{jo\}$. This is the set of activities that burn more than 650 calories per hour but cannot be done in all kinds of weather. Notice that $M - W$ is not the same as $W - M$, which is $\{ba, be, ca\}$.

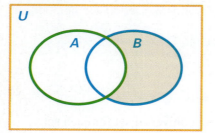

FIGURE 1.13 Venn diagram of $B - A$.

The difference of sets B and A is shaded in Figure 1.13.

 Quiz Yourself 23

a) Let $U = \{a, c, e, g, i, k, m, o, q\}$ and $C = \{x : x \text{ is a letter in the word } game\}$. Find C'.

b) Let $A = \{1, 3, 4, 5, 6\}$ and $B = \{2, 3, 4, 6, 7, 8\}$. Find $B - A$ and $A - B$.

> ### PROBLEM SOLVING
>
> You can remember set theory terminology more easily if you relate it to words that occur in everyday English. For example, consider the following associations:
>
> • Union—labor unions, marriage union, United States, unify, and so on. Each word conveys the notion of *joining things together* to make something larger.
> • Intersection—intersection of streets, intersection of lines, and so on. These terms describe overlapping things, or *the region common to both*.
> • Complement—pieces of clothing complement each other, two colors complement each other, the complement of a military unit, and so on. In each case we see the notion of *completing something*.

Set operations must be performed in correct order.

Just as we must perform arithmetic operations in a certain order, set notation specifies the order in which we perform set operations.

EXAMPLE 5   Order of Set Operations

Let $U = \{1, 2, 3, \ldots, 10\}$, $E = \{x : x \text{ is even}\}$, $B = \{1, 3, 4, 5, 8\}$, and $A = \{1, 2, 4, 7, 8\}$. Find $(A \cup B)' \cap (E' \cup A)$.

SOLUTION: The notation indicates that some of the operations in this expression must be done before others. For example, in considering $(A \cup B)'$, the parentheses tell us to form the union before taking the complement. In the expression $(E' \cup A)$, we first must find E' before calculating the union. The following is one way to do these calculations.

$$\overset{\substack{1 \qquad 2\ 5 \quad 3\ 4}}{(A \cup B)' \cap (E' \cup A)}$$

We follow these steps in order:

1. $(A \cup B) = \{1, 2, 3, 4, 5, 7, 8\}$
2. $(A \cup B)' = \{6, 9, 10\}$
3. $E' = \{1, 3, 5, 7, 9\}$
4. $(E' \cup A) = \{1, 2, 3, 4, 5, 7, 8, 9\}$
5. $(A \cup B)' \cap (E' \cup A) = \{9\}$

Sometimes notation leads us into thinking that something is true when it is not. For example, in algebra we know that $(x + y)^2 \neq x^2 + y^2$ even though it

Historical Highlight: *DeMorgan and Boole*

Just as biologists classify animals and flowers according to their various characteristics, mathematicians classify mathematical systems according to the relations and operations that are present in the systems. Set theory is an example of a system called a *Boolean algebra*, which is named in honor of the British mathematician George Boole (1815–1864). The son of a shopkeeper, Boole was not able to attend the schools of the more privileged students. As a result, he taught himself Latin, Greek, and also mathematics. His book *Investigation of*

the Laws of Thought, published in 1854, is one of the classics in the history of mathematics.

Boole, together with Augustus DeMorgan (1806–1871) and others, formed what some call the "British school" of mathematics. In the nineteenth century, they established a method of doing logical computations in a manner similar to the way in which we perform algebraic computations. The work of Boole and DeMorgan formed the mathematical basis for all of the computer-based devices that are so prevalent today.

looks like it might be so. Similarly, in set theory if we do not think carefully about the statement $(A \cup B)' = A' \cup B'$, it appears as though it could be true. Example 6 shows that you cannot change the order in which you calculate unions and complements; that is, $(A \cup B)' \neq A' \cup B'$.

EXAMPLE 6   **Order of Calculating Unions and Complements**

Let $U = \{1, 2, 3, 4, 5\}$, $A = \{1, 3, 5\}$, and $B = \{1, 2, 3\}$.
a) Find $(A \cup B)'$.   b) Find $A' \cup B'$.

SOLUTION: a) This notation tells us to *first* find the union and *then* take the complement. Hence, $(A \cup B)' = \{1, 2, 3, 5\}' = \{4\}$.

b) Here we first take complements and then find the union. This gives us

$$A' \cup B' = \{2, 4\} \cup \{4, 5\} = \{2, 4, 5\}$$

Thus we see that $(A \cup B)' \neq A' \cup B'$.

If you look carefully at Example 6, you will see that $(A \cup B)' = A' \cap B'$. This is an example of one of DeMorgan's Laws in set theory. Quiz Yourself 24 illustrates the second of DeMorgan's laws.

Quiz Yourself 24

Let $A = \{1, 2, 5, 7, 8, 9\}$ and $B = \{2, 3, 5, 6, 7\}$ be subsets of the universal set $U = \{1, 2, 3, \ldots, 10\}$. Find the following.
a) $(A \cap B)'$ b) A'
c) B' d) $A' \cup B'$

DeMorgan's Laws for Set Theory

If A and B are sets, then $(A \cup B)' = A' \cap B'$ and $(A \cap B)' = A' \cup B'$.

Viewing set theory as a mathematical system, we can ask whether set operations satisfy some of the same properties that are present in number systems. For example, we know that multiplication is distributive over addition—for all integers a, b, and c, $a \times (b + c) = (a \times b) + (a \times c)$. Example 7 investigates whether intersection is distributive over union.

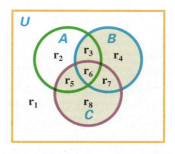

FIGURE 1.14 Venn diagram of $B \cup C$.

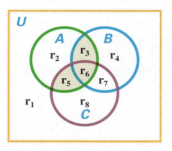

FIGURE 1.15 Venn diagram of $A \cap (B \cup C)$.

EXAMPLE 7   Intersection Distributes over Union

Let A, B, and C be sets in a universal set U. Does intersection distribute over union? That is, does

$$A \cap (B \cup C) = (A \cap B) \cup (A \cap C)?$$

SOLUTION: We will answer this question by drawing a Venn diagram for the set on the left-hand side of the equation and another for the set on the right-hand side.

In shading the left-hand set, the parentheses tell us that we must first take the union before considering the intersection. In Figure 1.14, we have shaded $B \cup C$, which consists of regions r_3, r_4, r_5, r_6, r_7, and r_8.

When we intersect $B \cup C$ with set A, we see that of the shaded regions in Figure 1.14, only regions r_3, r_6, and r_5 are also in A. This gives us the Venn diagram of $A \cap (B \cup C)$ shown in Figure 1.15.

Next, we look at the right-hand side of our equation, $(A \cap B) \cup (A \cap C)$. The parentheses tell us to first consider $(A \cap B)$ and $(A \cap C)$ and then take their union. We shade these sets in Figure 1.16.

Therefore, $(A \cap B) \cup (A \cap C)$ is made up of regions r_3, r_6, and r_5, which is the same set we shaded in Figure 1.15. Because these two Venn diagrams are the same, this shows that $A \cap (B \cup C) = (A \cap B) \cup (A \cap C)$.

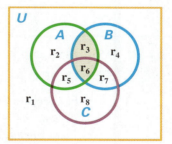 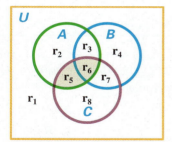

FIGURE 1.16   Venn diagrams of $A \cap B$ and $A \cap C$.

PROBLEM SOLVING

In deciding whether a set property holds, use the Three-Way Principle from Section 1.1. Recall that you can understand an idea:

- Verbally—Make analogies or try to describe the situation verbally.
- Graphically—Draw Venn diagrams.
- By examples—Use sets of numbers.

Often it is necessary to count the elements of a union of sets. Example 8 shows a common mistake we can make when doing this.

EXAMPLE 8   **Incorrectly Counting the Elements in a Set Union**

A manufacturer of camping equipment has purchased two mailing lists and intends to send catalogs to all people on the list. The first list, C, is the list of 2,785 members of a conservation club; the second list, M, is the list of 6,782 subscribers to a fishing and hunting magazine. The head of advertising for the manufacturer has placed an order with a printer for $2{,}785 + 6{,}782 = 9{,}567$ copies of the catalog. What is wrong with doing this?

SOLUTION: If we are careless in our thinking, we might be tempted to say that $n(C \cup M) = n(C) + n(M)$. However, it is possible that some names are on both lists. Let us say, for example, that there are 831 names that are common to both lists. Then we would have counted these people as part of set C and also as part of set M. Therefore to get the correct count, we have to subtract 831 from the total. That is, we would compute

$$n(C \cup M) = n(C) + n((M) - n(C \cap M) = 2{,}785 + 6{,}782 - 831 = 8{,}736. \quad \circledcirc$$

Example 8 illustrates that we should use the following property when counting the elements in the union of two sets.

> If A and B are sets, then $n(A \cup B) = n(A) + n(B) - n(A \cap B)$.

Exercises 1.5

In Exercises 1–12, let $U = \{1, 2, 3, \ldots, 10\}$, $A = \{1, 3, 5, 7, 9\}$, $B = \{1, 2, 3, 4, 5, 6\}$, and $C = \{2, 4, 6, 7, 8\}$. Perform the indicated operations.

1. $A \cap B$

2. $A \cup B$

3. $B \cup C$

4. $B \cap C$

5. $A \cup \varnothing$

6. $A \cap \varnothing$

7. $A \cup U$

8. $A \cap U$

9. $A \cap (B \cup C)$

10. $A' \cap (B \cup C')$

11. $(A - B) \cap (A - C)$

12. $A - (B \cup C)$

In order to increase its readership, a computer magazine conducted a survey of people who have recently purchased a new computer and identified the following groups:

$E = \{x : x$ *will use the computer for education$\}$, $B = \{x : x$ will use the computer for business$\}$, $H = \{y : y$ will use the computer for home management$\}$.*

Use this information to describe verbally each set in Exercises 13–18.

13. $E \cap B$

14. $B \cap H$

15. $E \cap B \cap H$

16. $E \cup B \cup H$

17. $(E \cap B) - H$

18. $E \cap (B \cup H)$

In Exercises 19–24, consider the universal set $U = \{$apple, TV, hat, radio, fish, sofa, automobile, potato chip, bread, banana, hammer, pizza$\}$.

Let $M = \{x : x$ is human-made$\}$, $E = \{y : y$ is edible$\}$, $G = \{t : t$ grows on a plant$\}$. Find each set.

19. $M \cap E$

20. $M - E$

21. $E - M$

22. E'

23. $M' \cap G'$

24. $G \cap (M' \cup E)$

Highlight: *Using a Computer Algebra System to Perform Set Computations*

A *computer algebra system* (CAS), is a program that manipulates mathematical expressions symbolically. The following is a sample of a CAS session that verifies one of DeMorgan's Laws. The "greater than" sign (>) means the program is waiting for a command. The boldface typing that follows the command is the CAS's response.

(Begin by telling the CAS to define sets U, A, B.)

> $U := \{1, 2, 3, 4, 5, 6, 7, 8, 9, 10\}$;
$$U := \{1, 2, 3, 4, 5, 6, 7, 8, 9, 10\}$$
> $A := \{1, 2, 5, 7, 8\}; B := \{1, 4, 5, 8\}$;
$$A := \{1, 2, 5, 7, 8\} \qquad B := \{1, 4, 5, 8\}$$

(First calculate $(A \cup B)'$.)

> U minus (A union B);
$$\{3, 6, 9, 10\}$$

(Next calculate $A' \cap B'$.)

> (U minus A) intersect (U minus B);
$$\{3, 6, 9, 10\}$$

This example shows how easily computer algebra systems handle nonnumerical computations. Therefore, in your mathematical studies you should emphasize concepts and problem solving rather than mechanical computations that can be done quickly and cheaply by a computer.

In the following table, we have listed various features that a person might consider when buying a new car.

	Cost	Size	Warranty (years)	Safety Rating	Anti-Theft Package
a	$10,800	Subcompact	3	Good	Yes
b	$12,500	Midrange	2	Good	No
c	$10,300	Subcompact	3	Moderate	Yes
d	$14,500	Compact	3	Poor	No
e	$10,200	Subcompact	2	Poor	No
f	$11,700	Compact	3	Poor	No
g	$13,000	Compact	4	Moderate	Yes
h	$14,700	Midrange	4	Good	No

In the universal set $U = \{a, b, c, \ldots, h\}$, define subsets of cars with the following characteristics:

P = *price is above $12,000*
C = *is compact*
G = *has good safety rating*
A = *has anti-theft package*
W = *warranty is at least three years*

In Exercises 25–32, first describe each set in words and then find the set.

25. $P \cap C$ **26.** $A \cup G$

27. $W \cap G'$ **28.** $G - A$

29. $P \cap (G \cup W)$ **30.** $G' \cap C'$

31. $P - (G \cup A)$ **32.** $P' - (G \cup C)$

In the following table, we have given nutritional information for the following items available at a fast-food restaurant:* Monster (m), Monster with Cheese (mc), Bacon Cheeseburger (bc), Hamburger (h), Cheeseburger (c), Fish Sandwich (fs), and Ham Cheesy (hc).

% of Minimum Daily Allowance (MDA)			
	Protein	Vitamin A	Vitamin B1	Calcium
m	42	14	25	8
mc	49	21	25	21
bc	49	8	20	17
h	23	3	16	4
c	27	7	16	10
fs	29	—	18	5
hc	37	15	58	19

Let P = {x : x is an item that provides at least 30% of the MDA for protein}, C = {x : x is an item that provides at least 10% of the MDA for calcium}, B = {x : x is an item that provides at least 25% of the MDA for vitamin B1}, A = {x : x is an item that provides at least 15% of the MDA for vitamin A}

In Exercises 33–36, find each set.

33. $P \cap (B \cup A)$

34. $(P \cup C) \cap (B \cup A)$

35. $P \cup C \cup B$

36. $P \cap C \cap B$

In Exercises 37–44, represent each set using a Venn diagram.

37. $A - (B \cup C)$

38. $A \cap (B - C)$

39. $(A \cap B) - C$

40. $(A \cup B) - C$

41. $A \cup (B - C)$

42. $A \cup (B \cup C)$

43. $(A \cup (B \cup C))'$

44. $(A \cap (B \cap C))'$

In Exercises 45–52, describe the shaded region using set theory notation.

45.

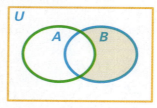

46.

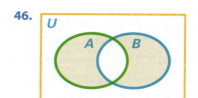

47.

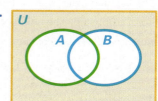

48.

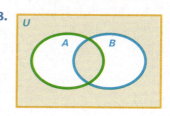

49.

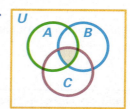

50.

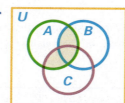

51.

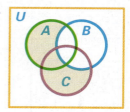

52.

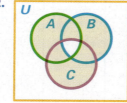

In Exercises 53–56, assume $A \subseteq B$. Express each set in a simpler way.

53. $A \cap B$

54. $A - B$

55. $A \cup B$

56. $A' \cap B'$

We have indicated the number of elements in each region of the Venn diagram to the right. In Exercises 57–64, find the following.

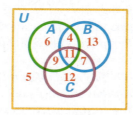

---

57. $n(A)$ **58.** $n(A \cup B)$

59. $n(C')$ **60.** $n(A - C)$

61. $n(A \cap C)$ **62.** $n(A \cap B \cap C)$

63. $n((A \cup B) \cap C)$ **64.** $n((A \cup C) - (B \cup C))$

Communicating Mathematics   *In Exercises 65–78, decide whether each statement is always true. Explain your answer by considering appropriate examples or by drawing a Venn diagram. If you think that a statement is not always true, provide a counterexample. Assume that all sets are finite.*

* **65.** If $A \subseteq B$, then $n(A) < n(B)$.

66. If $A \subset B$, then $n(A) < n(B)$.

67. If $A - B = \varnothing$, then $A \subseteq B$.

68. If $A \subseteq B$, then $A = B$.

69. If $A \subseteq B$ and $B \subseteq A$, then $A = B$.

70. If $n(A) = n(B)$ and $A \subseteq B$, then $A = B$.

71. $n(A \cap B) = n(A) - n(B)$.

72. If $A \cup B = A \cap B$, then $A = B$.

73. $X \subseteq (X \cup Y)$

74. $(X \cup Y) \subseteq Y$

75. $n(X) \leq n(X \cup Y)$

76. $n(X - Y) = n(X) - n(Y)$

77. If $X \subseteq Y$, then $n(Y - X) = n(Y) - n(X)$.

78. If $X \subseteq Y$, then $n(Y - X) = n(Y) - n(X \cap Y)$.

Further Exercises

In comparing number systems with set theory, it is useful to notice that union is in some ways similar to addition and that intersection is in some ways similar to multiplication. In Exercises 79–82, for each familiar property of the real number system,
a. State the corresponding property for set theory.
b. Determine whether the property stated in (a) is true.

79. $a \cdot b = b \cdot a$

80. $a + b = b + a$

81. $(a \cdot b) \cdot c = a \cdot (b \cdot c)$

82. $(a + b) + c = a + (b + c)$

83. We know that multiplication is distributive over addition, and we have shown in Example 7 that intersection distributes over union. Show that addition does not distribute over multiplication by providing an appropriate counterexample.

84. Does union distribute over intersection? Use Venn diagrams to confirm your answer.

In Exercises 85–86, consider the following. When using an Internet search engine, the instructions for the search are given by specifying keywords. For example, you can ask for a search to locate sites that discuss music by speci-fying the keyword "music." If you wish to make a more complex search, you can join keywords using "OR" or "AND." If you wish to search for sites discussing music or art, you can specify "music OR art." If you specify "music AND alternative," the search engine identifies sites dealing with both "music" and "alternative," that is, alternative music.

85. When you use AND, what set operation are you asking the search engine to perform? What about OR?

86. If we specify "music AND American OR British," it is not clear how the search engine would group the words specified. Using parentheses, we can group as follows:

(music AND American) OR British

or

music AND (American OR British).

Are both of these requests the same? What does this have to do with set theory? Discuss. If you have access to a search engine, see whether you can learn how it handles compound requests for searches.

---

* Exercise numbers circled in red can be used as group exercises.

1.6   SURVEY PROBLEMS

Every ten years, the U.S. government hires thousands of workers and sends out millions of questionnaires to gather data about the makeup of the nation's population. In order to design legislation and programs that better fit the country's needs, the nation's leaders need to know which groups (sets) in the country are growing and which are shrinking. For example, in order to write legislation giving tax relief to single-parent heads of households who are attending institutions of higher education, legislators would want to know the number of elements in this set.

Like the government, private companies collect information on their customers, products, and operations. If you were to buy a new personal computer, no doubt the manufacturer would ask you to fill out a survey asking such things as, "Will you use this computer to access the Internet? Will you use this computer for home entertainment?" The company will organize this information in a database and apply the principles of set theory to extract information that it can use to design new products and target new markets.

In this section, we will show how to use set theory to solve survey problems, but first we need to sharpen our skills in working with Venn diagrams.

■

Regions of a Venn diagram can have many different names.

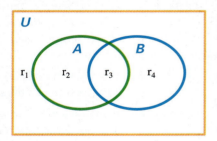

FIGURE 1.17   A Venn diagram with two sets.

In everyday life, it is common for a person to be known by many names. For example, a woman might be known at work by Ms. Smith, at home by Mom or Honey, by her sister as Sis, by her Internet provider as ASmith123, and by the IRS as 178-34-7886. Similarly, the same set can have many different names. For example, consider the four regions shown in the Venn diagram shown in Figure 1.17.

We can use set notation to name these regions. For example, region r_4 is $B - A$, and region r_1 is $(A \cup B)'$. These names are not unique. Using one of DeMorgan's Laws, region r_1 could also be called $A' \cap B'$.

There are numerous relationships among the various sets in this diagram. Because $A \cup B$ is the union of regions r_2, r_3, and r_4, we can write

$$A \cup B = (A - B) \cup (A \cap B) \cup (B - A).$$

Also, if we remove region r_4 from $A \cup B$, we would get set A. We can express this symbolically by writing

$$A = (A \cup B) - (B - A).$$

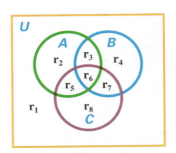

FIGURE 1.18   A Venn diagram with three sets.

Quiz Yourself 25*

Use Figure 1.17 to answer each question.

a) What is a set name for region 8?

b) Express $B - C$ as the union of two sets.

The point is to show you that we can represent the same set in different ways, just as we often express numbers in different ways. Depending on what calculations we are doing, we may write 6 one time as $3 + 3$ and another time as $5 + 1$. These ideas carry over to Venn diagrams with more than two sets.

EXAMPLE 1   Naming Regions of a Three-Set Venn Diagram

Use Figure 1.18 to answer each question.

a) Name the set we get by combining regions r_6 and r_7.

b) What is a set name for region r_2?

c) Express $A - B$ as the union of two sets.

SOLUTION: a) Regions r_6 and r_7 are precisely the regions common to B and C; therefore, this set is $B \cap C$.

b) Region r_2 is clearly *part* of set A; however, elements in region r_2 are not elements of B, nor are they elements of C. Therefore, one name for this set is $A \cap B' \cap C'$.

Notice that another way you can view this set is to start with set A (regions r_2, r_3, r_5, r_6) and remove those regions that are in $B \cup C$, namely, r_3, r_5, and r_6. Thus we also can name region r_2 as $A - (B \cup C)$.

c) $A - B$ consists of regions r_2 and r_5, which have set names $A \cap B' \cap C'$ and $A \cap B' \cap C$. We can write $A - B = (A \cap B' \cap C') \cup (A \cap B' \cap C)$.

Venn diagrams can organize information in a survey problem.

When we collect data and organize it into sets, we then usually want to analyze the information to answer questions about those sets. These types of problems are often called **survey problems**. Example 2 is an example of a survey problem.

EXAMPLE 2   A Survey Problem Involving Energy Policies

A clean-air advocate polled 100 people to determine which energy policies they support. Partial results from the poll are as follows:

a) 12 people favor the increased construction of nuclear plants only.

b) 20 people recommend both solar energy research and increased nuclear plant construction.

c) 22 people favor both tax credits for oil companies and increased nuclear plant construction.

d) 14 people would like to see all three areas pursued.

From this information, determine the total number of people who favored increased nuclear plant construction.

SOLUTION: We represent the information using sets:

$$T = \{x : x \text{ favors tax credits for oil companies}\}$$
$$N = \{x : x \text{ favors increased nuclear plant construction}\}$$
$$S = \{x : x \text{ favors federal aid for solar energy research}\}$$

Then conditions (a)–(d) can be rewritten:

a) $n(N \cap T' \cap S') = 12$,

b) $n(S \cap N) = 20$,

c) $n(T \cap N) = 22$,

d) $n(T \cap N \cap S) = 14$.

It is helpful to display this information in a Venn diagram. For example, we can see in Figure 1.19 the fact that $n(T \cap N \cap S) = 14$ and $n(N \cap T' \cap S') = 12$.

In Figure 1.19, we have also shaded $S \cap N$. From condition (b), we see that there are 20 people in this set. Since 14 of these are already accounted for in the subset $T \cap N \cap S$, there must be 6 people in the set $S \cap N \cap T'$. Similarly, we can use the fact that $n(T \cap N) = 22$ to account for 8 more in set N. Adding these numbers, we see that N contains $8 + 14 + 6 + 12 = 40$ people. ◎

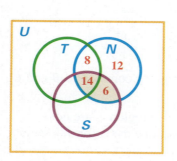

FIGURE 1.19   The shaded region is $S \cap N$.

Example 3 analyzes a survey problem that requires a Venn diagram with three sets.

EXAMPLE 3   A Survey of TV Viewer Preferences

A television network conducted a market survey to determine the evening viewing preferences of people in the 18–25 age bracket. The following information was obtained.

a) 3 prefer a comedy early on weekdays.

b) 14 want to watch TV early on weekdays.

c) 21 want to see comedies early.

d) 8 want comedies on weekdays.

e) 31 want to watch TV on weekdays.

f) 36 wish to watch TV early.

g) 40 wish to see comedies.

h) 13 prefer late, weekend shows that are not comedies.

From this information, determine how many people do not wish to see comedies and how many prefer to watch TV on the weekend.

SOLUTION: The set of people surveyed is a universal set containing three subsets.

$$W = \text{those who prefer to watch TV on weekdays}$$
$$E = \text{those who desire early programming}$$
$$C = \text{those who want to see comedy}$$

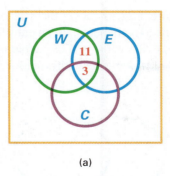
(a)

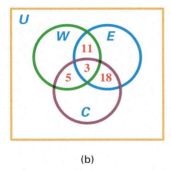
(b)

Figure 1.20   Counting elements in $W \cap E$, $W \cap C$, and $C \cap E$.

We draw these sets in Figure 1.20. By condition (a) we know that there are 3 people in $W \cap E \cap C$. Since condition (b) states there are 14 people in $W \cap E$, and we have counted 3 of them, there must be 11 more in this region. We have recorded this information in Figure 1.20(a).

By condition (d), 8 people wish to see weekday comedies, so we can deduce that 5 have not yet been counted in $W \cap C$. Similarly, from condition (c), we see that 18 people have not been counted in $C \cap E$. This is shown in Figure 1.20(b).

Condition (e) states that 31 people desire weekday programming, but we have already counted 19. Therefore, there are 12 more in set W. Similarly there are 4 more people to be counted in set E and 14 more in set C. Condition (h) tells us that there are 13 people outside sets W, E, and C. We summarize this information in Figure 1.21.

With this information, we can answer the original questions. The people wishing to see noncomedies fall outside C, so they total $12 + 11 + 4 + 13 = 40$. Those who prefer weekend programming are outside W. This count is $14 + 18 + 4 + 13 = 49$. ◎

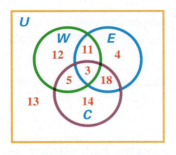

Figure 1.21   The number of elements in each region of the Venn diagram.

A survey problem may contain contradictory information.

It is possible to list a set of conditions on sets that cannot all hold at the same time.

EXAMPLE 4   Inconsistent Survey Data

Suppose that an Internet magazine states the following information concerning which Web browsers its readers use:

a) 316 use WebMagic.

b) 478 use iBrowse.

c) 104 use both WebMagic and iBrowse.

d) 567 use only one of the Web browsers.

Find the inconsistency in these data.

SOLUTION: We begin by drawing a Venn diagram in Figure 1.22(a). Let W be the set of WebMagic users and let I be the set of those who use iBrowse.

Figure 1.22(a) shows the 104 people who use both Web browsers. Because W contains 316 elements, this means that $316 - 104 = 212$ elements are in $W - I$. Similarly, $478 - 104 = 374$ people use only iBrowse, so $I - W$ contains 374 elements. We show this new information in Figure 1.22(b).

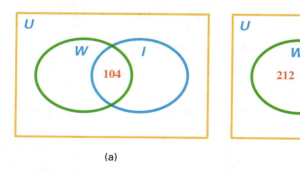

(a)   (b)

FIGURE 1.22   (a) 104 use both WebMagic and iBrowse. (b) 212 use only WebMagic; 374 use only iBrowse.

According to Figure 1.22(a), $212 + 374 = 586$ people use only one of the two Web browsers. This directly contradicts condition (d). Therefore, these data are inconsistent.

Information may be organized in a computer's database in table form. We analyze such a situation in Example 5.

EXAMPLE 5

A media company that produces exercise videos wants to find whether its customers prefer to view its products on videotape, CD-ROM, or DVD. The company has extracted the following table of information from its customer database.

	Videotape (V)	CD-ROM (C)	DVD (D)	Total
Under 41 (Y)oung	20	15	9	44
41 to 55 (M)iddle-aged	44	34	8	86
Over 55 (S)enior	31	14	5	50
Total	95	63	22	180

a) Find the number of elements in $M \cup C$.

b) Find the number of elements in V'.

SOLUTION: a) The number of elements in $M \cup C$ is the number of people who are middle-aged or prefer CD-ROM. Thus,

$$n(M \cup C) = n(M) + n(C) - n(M \cap C) = 86 + 63 - 34 = 115.$$

b) V' is the set of people who do not prefer videotape, so

$$n(V') = 63 + 22 = 85.$$

 Quiz Yourself 26

Use the table in Example 5 to find $n(Y \cup D)$.

Exercises 1.6

In Exercises 1–6, determine which numbered regions make up the indicated set.

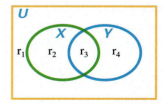

1. X **2.** Y

3. $X \cap Y$ **4.** $X \cup Y$

5. $X - Y$ **6.** $Y - X$

In Exercises 7–18, describe each set by referring to the numbered regions, as in Example 1.

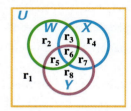

7. W **8.** Y

9. $W \cap Y$ **10.** $W \cup X$

11. $X - W$ **12.** $Y - X$

13. $W - (X \cup Y)$ **14.** $Y - (X \cap W)$

15. $X \cap Y \cap W'$ **16.** $X' \cap Y \cap W'$

17. $X \cap Y \cap W$ **18.** $X \cup Y \cup W$

In Exercises 19–22, find, if possible, the number of elements in sets A, B, and C using the given information. If there is an inconsistency in the information, state where it occurs.

19. $n(A \cap B) = 5$, $\quad n(A \cap B \cap C) = 2$, $\quad n(B \cap C) = 6$, $n(B - A) = 10$, $\quad n(B \cup C) = 23$, $\quad n(A \cap C) = 7$, $n(A \cup B \cup C) = 31$.

20. $n(B - C) = 4$, $\quad n(C - B) = 9$, $\quad n(A \cap B \cap C) = 3$, $n(B \cup C) = 22$, $\quad n(A - C) = 7$, $\quad n(A \cap B) = 7$, $n(A \cap C) = 5$.

21. $A \cap C = \varnothing$, $n(A \cap B) = 3$, $n(C - B) = 2$, $n(B - C) = 7$, $n(B \cup C) = 16$, $n(A \cup B) = 16$.

22. $A \subset B$, $A \cap C = \varnothing$, $n(C - A) = 8$, $n(A \cup C) = 12$, $n(C - B) = 3$, $n(B \cup C) = 17$.

23. Automobile accidents. An investigation of a number of automobile accidents revealed the following information:

a) 18 accidents involved alcohol and excessive speed.

b) 26 accidents involved alcohol.

c) 12 accidents involved excessive speed but not alcohol.

d) 21 of the accidents involved neither alcohol nor excessive speed.

How many accidents were investigated?

24. Perceptions of students. The political science department at a university surveyed a number of politicians regarding their perceptions of college students, and gathered the following information:

a) 15 saw students as hardworking and economically advantaged.

b) 25 saw students as hardworking.

c) 11 believed students were economically advantaged but not hardworking.

d) 16 felt that students were neither hardworking nor economically advantaged.

How many surveyed felt that students were economically advantaged?

25. There are 82 people collecting signatures to protest the destruction of the rain forests. If there are 47 males and 28 teenagers, 13 of whom are girls, then how many females are collecting signatures?

26. There are 95 students who have applied for a scholarship. If there are 41 men and 36 minorities, 19 of whom are women, how many women applied for the scholarship?

27. Survey of vacationers. A survey is taken of 100 people who vacationed at a dude ranch. The following information was obtained:

a) 17 took horseback-riding lessons, attended the Saturday night barbecue, and purchased a tour guide.

b) 28 attended the Saturday night barbecue and purchased a tour guide.

c) 24 took horseback-riding lessons and purchased a tour guide.

d) 42 took horseback-riding lessons but didn't attend the barbecue.

e) 86 took horseback-riding lessons or purchased a tour guide.

f) 14 only purchased a tour guide.

g) 14 did none of these three things.

How many attended the barbecue? How many purchased a tour guide?

28. Repeat Exercise 27, given that

a) 19 took horseback-riding lessons, attended the Saturday night barbecue, and purchased a tour guide.

b) 34 attended the Saturday night barbecue and purchased a tour guide.

c) 30 took horseback-riding lessons and purchased a tour guide.

d) 33 took horseback-riding lessons but didn't attend the barbecue.

e) 86 took horseback-riding lessons or purchased a tour guide.

f) 8 only purchased a tour guide.

g) 3 did none of these three things.

How many attended the barbecue or purchased the tour guide? How many did not purchase a tour guide?

29. Media survey. A survey of young adults was taken to determine which of the various media they use regularly to obtain the news. The following information was obtained:

a) Of the 36 people who read the newspaper, 13 use only the newspaper to learn the news.

b) Of the 48 people who listen to the radio, 11 use only the radio to learn the news.

c) 23 people use all three to learn the news.

How many people use both the radio and television regularly to learn the news?

30. Mass transit survey. A survey was made of 200 city residents to study their use of mass transit facilities. According to the survey:

a) 83 did not use mass transit.

b) 68 used the bus.

c) 44 used only the subway.

d) 28 used both the bus and subway.

e) 59 used the train.

Explain how you can use this information to deduce that some residents must use both the bus and train.

31. MusiChan.com surveyed a group of subscribers regarding which online music channels they use on a regular basis. The following information summarizes their answers:

7 listened to Rap, Heavy Metal, and Alternative Rock

10 listened to Rap and Heavy Metal

13 listened to Heavy Metal and Alternative Rock

12 listened to Rap and Alternative Rock

17 listened to Rap

24 listened to Heavy Metal

22 listened to Alternative Rock

9 listened to none of these three channels

a) How many people were surveyed?

b) How many people listened to either Rap or Alternative Rock?

c) How many listened to Heavy Metal only?

32. *Personal Fitness Magazine* surveyed a group of young adults regarding their exercise programs and the following results were obtained:

3 were using weight training, Tae Bo, and Pilates to improve their fitness

5 were using weight training and Tae Bo

12 were using Tae Bo and Pilates

8 were using weight training and Pilates

15 were using weight training only

30 were using Tae Bo

17 were using Pilates but not Tae Bo

14 were using something other than these three types of workouts

a) How many people were surveyed?

b) How many people were using only Tae Bo?

c) How many people were using Pilates but not weight training?

33. The Dean of Academic Services surveyed a group of students about which support services they were using to help them improve their academic performance and found the following results:

5 were using office hours, tutoring, and online study groups to improve their grades

16 were using office hours and tutoring

28 were using tutoring

14 were using tutoring and online study groups

8 were using office hours and online study groups but not tutoring

23 were using office hours but not tutoring

18 were using only online study groups

37 were using none of these services

a) How many students were surveyed?

b) How many students were using only office hours?

c) How many students were using online study groups?

34. A group of young adults who were asked which issues would be important during the next decade gave the following results:

13 believed that nuclear war, terrorism, and environmental concerns would be important

43 believed that nuclear war would be important

17 believed that nuclear war and terrorism would be important

23 believed that nuclear war or terrorism but not the environment would be important

28 believed that nuclear war and the environment would be important

18 believed that terrorism but not nuclear war would be important

7 believed that only the environment would be a serious issue

6 believed that none of these issues would be important

a) How many people were surveyed?

b) How many people believed that the environment would be an important issue?

c) How many people believed that only terrorism would be an important issue?

In Exercises 35–38, use the following table from Example 5 to find the cardinal number of each set.

	Videotape (V)	CD-ROM (C)	DVD (D)	Total
Under 41 (Y)oung	20	15	9	44
41 to 55 (M)iddle-aged	44	34	8	86
Over 55 (S)enior	31	14	5	50
Total	95	63	22	180

35. $V \cup D$

36. $S \cap C$

37. $V - (M \cup S)$

38. $Y \cap (V \cup C)$

Further Exercises

(39.) Communicating Mathematics   Make up a survey question whose conditions will satisfy the given Venn diagram.

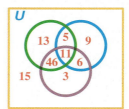

40. Survey of aerobic activities. Draw a Venn diagram that summarizes numerically the information in the following table regarding sets C (burns more than 600 calories per hour), A (can be done alone), and W (can be done in any kind of weather).

Activity	Calories Burned per Hour†	Can Be Done Alone	Can Be Done in Any Kind of Weather
Ballroom dancing (ba)	300	No	Yes
Bench stepping (be)	610	Yes	Yes
Bicycling (bi)	415	Yes	No
Calisthenics (ca)	300	Yes	Yes
Hiking (hi)	400	Yes	No
Jogging (jo)	655	Yes	No
Rowing machine (ro)	655	Yes	Yes
Tennis (te)	425	No	No

* Exercise numbers circled in red can be used as group exercises.
† "Your Permanent Weight-Loss Workbook," *Prevention Magazine* (January 1995).

The following table gives a partial summary of grounds for divorce according to the 1999 World Almanac. Use these data to answer Exercises 41–42.

State	Adultery (AD)	Mental or Physical Cruelty (C)	Desertion (D)	Alcoholism (AL)
Arkansas	Yes	Yes	No	Yes
California	No	No	No	No
Delaware	Yes	Yes	Yes	Yes
Florida	No	No	No	No
Illinois	Yes	Yes	Yes	Yes
New Jersey	Yes	Yes	Yes	Yes
New York	Yes	Yes	Yes	No
Oregon	No	No	No	No
Pennsylvania	Yes	Yes	Yes	No
Texas	Yes	Yes	Yes	No

41. Draw a Venn diagram that summarizes numerically the grounds-for-divorce data regarding adultery, cruelty, and desertion.

42. Draw a Venn diagram that summarizes numerically the grounds-for-divorce data regarding alcoholism, cruelty, and desertion.

(Note: The diagrams that we ask for in Exercises 43 and 44 do exist; if you cannot construct these diagrams yourself, you may make this into a research project to find and present these diagrams to your class.)

43. Draw a Venn diagram showing all possible 16 regions for four sets, *A*, *B*, *C*, and *D*.

44. Draw a Venn diagram showing all possible 32 regions for five sets, *A*, *B*, *C*, *D*, and *E*.

45. Communicating Mathematics   Take a survey of your classmates and gather information to write a survey problem similar to Exercises 31–34. Write the survey problem and have your classmates solve it. Try to be creative in the ways that you give the clues.

46. Communicating Mathematics   Go to an almanac, on the Internet, or some other source. Pick an area of interest and gather information to write a survey problem similar to Exercises 31–34. Write the survey problem and have your classmates solve it. Try to be creative in the ways that you give the clues.

CHAPTER SUMMARY

SECTION 1.1

steps in problem solving
Problem solving involves preparation, incubation, illumination, and verification.

problem-solving strategies
The following strategies are useful in solving problems.
Draw pictures.
Choose good names for unknowns.
Be systematic.
Look for patterns.
Try a simpler version of the problem.
Guessing is OK.
Convert a new problem to an older one.

problem-solving principles
The following principles are helpful in problem solving:
The Always Principle
The Counterexample Principle
The Order Principle
The Splitting-Hairs Principle
The Analogies Principle
The Three-Way Principle

SECTION 1.2

rounding
In rounding, we consider the digit to the right of the digit being rounded. If that digit to the right is five or more, we round up; otherwise, we round down.

compatible numbers
In using compatible numbers, instead of using the given numbers in a problem, we substitute other numbers that are easier to work with.

SECTION 1.3

set, element
A collection of objects is called a set, and the individual objects in this collection are called the elements of the set.

set-builder notation
Set-builder notation is a method of describing a set whose elements all share some common characteristic that is satisfied by no other object.

well defined
A set is well defined if we are able to tell whether or not any particular object is an element of the set.

empty set, null set
The set that has no elements is called the empty set or null set.

universal set
The set U of all elements under consideration in a given discussion is called the universal set.

cardinal number
The number of elements in set A is called the cardinal number of set A and is denoted $n(A)$.

finite, infinite
A set is finite if its cardinal number is a whole number. An infinite set is one that is not finite.

SECTION 1.4

equal sets
Two sets A and B are equal if they have exactly the same members. In this case we write $A = B$. If they are not equal, we write $A \neq B$.

subset

Set A is a subset of set B if every element of A is also an element of B. We write $A \subseteq B$. If A is not a subset of B, we write $A \nsubseteq B$.

Venn diagram

A Venn diagram is a method of visualizing set relationships using rectangles, circles, and other geometric shapes.

proper subset

Set A is a proper subset of set B if $A \subseteq B$ but $A \neq B$. This relationship is written $A \subset B$.

number of subsets of a set

A k-element set has 2^k subsets.

equivalent sets, one-to-one correspondence

A and B are equivalent, or in one-to-one correspondence, if $n(A) = n(B)$.

SECTION 1.5   **union**

The union of sets A and B, written $A \cup B$, is the set of elements that are elements of either A or B (or both).

intersection

The intersection of sets A and B, written $A \cap B$, is the set of elements that are common to both A and B.

disjoint

If $A \cap B = \varnothing$, then A and B are disjoint.

complement

The complement of A, written A', is the set of elements of the universal set that are not elements of A.

difference

The difference of sets B and A, written $B - A$, is the set of elements that are in B but not in A.

DeMorgan's Laws

These laws state that for any sets A and B,

$$(A \cup B)' = A' \cap B' \quad \text{and} \quad (A \cap B)' = A' \cup B'.$$

counting elements in the union of sets

If A and B are sets, then $n(A \cup B) = n(A) + n(B) - n(A \cap B)$.

SECTION 1.6   **solving survey problems**

By drawing Venn diagrams and carefully counting the elements in the various regions of the diagram, we can solve survey problems.

CHAPTER TEST

SECTION 1.1

1. Use the letters *a, b, c,* and *d* to form as many pairs as you can. The order in which you list the letters is important, and you are not allowed to use a letter twice in a pair.

2. At a restaurant, you have 8 appetizers, 20 entrées, and 10 desserts. How many different meals can you choose if you select one appetizer, one entrée, and one dessert? Do not solve this problem, but state a simpler version and solve it instead.

3. Hector worked 20 hours last week. Part of the time he worked as a stockperson for $5.65 per hour. The rest of the time he worked as a ski instructor for $8 per hour. If he earned $141.20, how many hours did he work at each job? Solve this problem by making and evaluating guesses until you find the answer.

4. Is the following statement true or false?

$$\frac{a}{b} + \frac{c}{d} = \frac{a+c}{b+d}$$

If it is true, give two examples. If it is false, give a counterexample.

5. Explain the Three-Way Principle.

SECTION 1.2

6. Round each of the following numbers to the nearest thousand.

a) 46,358

b) 27,541

7. Use compatible numbers to estimate the answers to the following problems. Your answers may differ from ours.

a) $209.35 - 61.19$ b) 5.85×15.64

8. Juana is averaging 52.4 miles per hour on her trip to Miami. If she still has 156 miles to travel and it is now 4:00 PM, about what time should she arrive?

9. Use the following graph to estimate a) the average earnings for women with a bachelor's degree, b) the difference in earnings between men and women who are high school graduates with no further education.

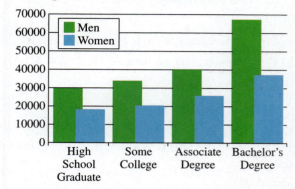

SECTION 1.3

10. Use an alternative method to express each set.

a) $\{1, 3, 5, 7, 9\}$

b) $\{t : t$ is a letter in the word *happiness*$\}$

c) $\{a, b, c \ . \ . \ . \ , z\}$

d) $\{y : y$ is a living person who fought in the American Revolutionary War$\}$

SECTION 1.4

11. Decide whether each pair of sets is equal. Justify your answer.

a) $\{1, 3, 6, 7, 8\}$ and $\{1, 6, 8, 3, 7\}$

b) $\{1, 3, 5, \ . \ . \ . \ , 49\}$ and $\{x : x$ is a counting number between 0 and 50$\}$

c) $\{a, a, b, b, c, c, d, d, \ . \ . \ . \ , z, z\}$ and $\{x : x$ is a letter in the alphabet$\}$

d) $\varnothing$ and $\{\varnothing\}$

12. Decide whether each statement is true or false. Justify your answer.

a) $\{2, 4, 6, 8\} \subseteq \{1, 2, 3, \ . \ . \ . \ , 9\}$

b) $\{2, 4, 6, 8\} \subset \{8, 6, 4, 2\}$

c) $\{x : x$ is a letter in the word *merry*$\} \subseteq \{y : y$ is a letter in the word *marry*$\}$

d) $\varnothing \subseteq \{1, 3, 5\}$

e) $\varnothing \subset \varnothing$

13. Which pairs of sets are equivalent?

    a) $\{10, 20, 30, 40, 50\}$ and $\{a, e, i, o, u\}$

    b) $\{x : x$ is a letter in the word *flower*$\}$ and $\{x : x$ is a letter in the word *tulips*$\}$

    c) $\{\varnothing\}$ and $\{0\}$

14. a) List all the subsets of $\{a, b, c\}$.

    b) How many subsets are there of $\{1, 2, 3, 4, 5, 6, 7, 8, 9, 10\}$?

SECTION 1.5

15. Let $U = \{1, 2, 3, \ldots, 8\}$, $A = \{1, 3, 5, 7\}$, $B = \{1, 3, 4, 6, 7, 8\}$, and $C = \{2, 3, 4, 5, 6, 7\}$. Find the following:

    a) $A \cap B$

    b) $B \cup C$

    c) C'

    d) $C - A$

16. Let $U = \{1, 2, 3, \ldots, 8\}$, $A = \{1, 3, 5, 7\}$, $B = \{1, 3, 4, 6, 7, 8\}$, and $C = \{2, 3, 4, 5, 6, 7\}$.

    a) Find $(B' \cup C) \cap A$.

    b) Find $(A - B) \cup (A - C)$.

17. Represent each set using a Venn diagram.

    a) $A \cap B$

    b) $B \cup C$

    c) $A' \cap B'$

    d) $A - (B \cup C)$

18. Represent $(A \cap B)'$ in an alternative way by using DeMorgan's Laws.

19. a) List three algebraic properties satisfied by the set operation of intersection.

    b) List three algebraic properties satisfied by the set operation of union.

    c) List an algebraic property that relates union and intersection.

SECTION 1.6

20. Use the following information to answer the given questions.

$$n(A \cap B) = 4, n(A \cap B \cap C) = 1, n(B \cap C) = 8,$$
$$n(B - A) = 9, n(B \cup C) = 23,$$
$$n(A \cap C) = 6, n(A \cup B \cup C) = 35, n(B') = 32.$$

    a) How many elements are in $C - B$?

    b) How many elements are in A'?

21. A survey is taken of 100 health club members regarding their use of the club's exercise facilities. The following information was obtained:

    a) 16 use the gym, pool, and weight room.

    b) 29 use both the gym and pool.

    c) 23 use both the gym and weight room.

    d) 40 use the gym.

    e) 68 use either the gym or pool.

    f) 14 use only the pool.

    g) 18 use none of the three facilities.

How many use the pool? How many use only the weight room?

Of Further Interest: INFINITE SETS

In the late nineteenth century, the great German mathematician Georg Cantor* made the remarkable discovery that infinite sets can have different sizes. In this section, we will explain some of Cantor's revolutionary ideas; but first we will use the notion of one-to-one correspondence to make the notion of cardinal number more precise.

> **DEFINITION**
>
> Two sets A and B are in **one-to-one correspondence** if we can pair the elements of the two sets so that every element of A is paired with exactly one element of B and every element of B is paired with exactly one element of A.

For example, the set $P = $ {Reagan, Clinton, Carter} is in one-to-one correspondence with the set $V = $ {Bush, Gore, Mondale}, because we can pair each president with his vice president as follows:

$$
\begin{array}{ccc}
\text{Reagan} & \text{Clinton} & \text{Carter} \\
\updownarrow & \updownarrow & \updownarrow \\
\text{Bush} & \text{Gore} & \text{Mondale}
\end{array}
$$

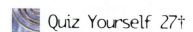 **Quiz Yourself 27†**

Show another one-to-one correspondence between *P* and *V*.

Of course, there are several other ways of pairing the elements of *P* with the elements of *V* in a one-to-one correspondence, and we will investigate this further in the exercises.

We can now say that a set *A* has cardinal number *n*, or is finite, if we can place *A* in a one-to-one correspondence with the set {1, 2, 3, . . . , *n*}. For example, if $A = \{x : x \text{ is a letter of the alphabet}\}$, then $n(A) = 26$ because we have the following one-to-one correspondence between *A* and {1, 2, 3, . . . , 26}:

$$
\begin{array}{cccc}
a & b & c \cdots & z \\
\updownarrow & \updownarrow & \updownarrow & \updownarrow \\
1 & 2 & 3 \cdots & 26
\end{array}
$$

Notice that we cannot place a finite set in a one-to-one correspondence with one of its proper subsets. Cantor used this simple observation to define infinite sets.

> **DEFINITION**
>
> A set is **infinite** if we can place it in a one-to-one correspondence with a proper subset of itself.

---

* See the Historical Highlight in Section 1.4.
† Quiz Yourself answers begin on page 849.

EXAMPLE 1 The Set of Natural Numbers Is Infinite

Show that the set of natural numbers is infinite.

SOLUTION: Let $N = \{1, 2, 3, \ldots\}$ and $E = \{2, 4, 6, \ldots\}$ then the following is a one-to-one correspondence between N and its proper subset E.

$$
\begin{array}{ccccc}
1 & 2 & 3 & \cdots & n & \cdots \\
\updownarrow & \updownarrow & \updownarrow & & \updownarrow \\
2 & 4 & 6 & \cdots & 2n & \cdots
\end{array}
$$

Quiz Yourself 28

Find another one-to-one correspondence between N and one of its proper subsets. Be sure to include the general pattern for the correspondence.

Instead of simply saying that the set of natural numbers is infinite, we can now be more precise.

DEFINITION

Cantor called the cardinal number for the set of natural numbers $\aleph_0$. The symbol $\aleph$, aleph, is the first letter of the Hebrew alphabet. We pronounce $\aleph_0$ as "Al-luf null" or "Al-luf naught." Therefore, we can write $n(N) = \aleph_0$. Any set that is either finite or that is in a one-to-one correspondence with N, and therefore has cardinal number $\aleph_0$, is called a **countable set.**

EXAMPLE 2 The Set of Integers Is Countable

Show that I, the set of integers, can be put in a one-to-one correspondence with the set of natural numbers.

SOLUTION: The following matching shows the correspondence. (In the correspondence, we are assuming that n represents a positive integer.)

$$
\begin{array}{ccccccccccc}
0 & 1 & -1 & 2 & -2 & 3 & -3 & \cdots & n & & -n & \cdots \\
\updownarrow & \updownarrow & \updownarrow & \updownarrow & \updownarrow & \updownarrow & \updownarrow & & \updownarrow & & \updownarrow \\
1 & 2 & 3 & 4 & 5 & 6 & 7 & \cdots & 2n & & 2n+1 & \cdots
\end{array}
$$

Therefore, $n(I) = \aleph_0$, and I is a countable set.

When we work with infinite sets, we find some surprising results. It would seem that there are so many more rational numbers than there are integers, that we might expect that the cardinal number of the rational numbers is larger than $\aleph_0$. As Example 3 shows, this is not the case.

EXAMPLE 3 The Set of Rational Numbers Is Countable

Show that the set of rational numbers is countable.

SOLUTION: First consider the positive rational numbers as we list them in Figure 1.23. The first row of the arrangement has all positive rational numbers with denominator one. The second row has all denominators two, the third row has denominators of three, and so on. We then trace through the arrangement following the line in Figure 1.23, skipping over numbers that we have encountered earlier.

The first number we encounter is 1, then 2, then 1/2, and then 1/3. We will skip 2/2 because we have encountered 1 earlier. So the next numbers are 3, 4, 3/2, 2/3, 1/4, and 1/5. We skip 2/4, 3/3, 4/2 (why?), and then list 5.

1/1 2/1 3/1 4/1 5/1 . . .
1/2 2/2 3/2 4/2 5/2 . . .
1/3 2/3 3/3 4/3 5/3 . . .
1/4 2/4 3/4 4/4 5/4 . . .
1/5 2/5 3/5 4/5 5/5 . . .

FIGURE 1.23 Listing the positive rational numbers.

When we look at the positive rational numbers in this way, we see that there is a first number, then a second, then a third, and so on, and we are listing each number exactly once. Therefore, there is a one-to-one correspondence between N and the positive rational numbers.

To show that the entire set of rational numbers is countable, we would argue as we did in Example 2 to account for zero and the negative rationals; however, we will not go into the details of how to do it. Therefore, the set of rational numbers has the same cardinal number as the natural numbers, which is $\aleph_0$.

So far, the infinite sets that you have studied have all been countable. We will next show you a set that has a cardinal number greater than $\aleph_0$. Because the argument that we will show you is a little sophisticated, we will look at a simpler version of our reasoning before going on to the real thing.

Imagine that you were dealing with a person who had no knowledge of numbers beyond 20. If you were in a classroom containing 29 students and 35 seats, how might you convince him that there were more seats than students? You could ask each student to take a seat and then argue that because there were seats that were not occupied, there were therefore more seats than students. We will use this same type of reasoning in Example 4.

EXAMPLE 4   A Cardinal Number Greater Than $\aleph_0$

We will reproduce Cantor's argument that the set of real numbers between 0 and 1 has cardinal number greater than $\aleph_0$. We begin by assuming that we *can* put the set of numbers between 0 and 1 in a one-to-one correspondence with the natural numbers and show that no matter how we try, there will always be some number that we could not have listed.

Although we would not actually know what the listing would be, for the sake of argument, let us assume that we had listed all of the numbers between 0 and 1 as follows:

$$1 \leftrightarrow 0.6348291347 \ldots$$
$$2 \leftrightarrow 0.2373261008 \ldots$$
$$3 \leftrightarrow 0.4821063391 \ldots$$
$$4 \leftrightarrow 0.6824537128 \ldots$$
$$5 \leftrightarrow 0.4657189233 \ldots$$

. .
. .
. .

Although we have assumed that all numbers between 0 and 1 are listed, we will now show you how to construct a number x between 0 and 1 that is not on this list. We want x to be different from the first number on the list, so we will begin the decimal expansion of x with a digit other than a 6 in the tenths place, say $x = 0.5 \ldots$. Because we don't want x to equal the second number on the list, we make the hundredths place not equal to 3, say 4, so far, $x = 0.54 \ldots$

Historical Highlight: *The Continuum Hypothesis*

Perhaps you believe that mathematics is pretty much "cut and dried" in the sense that if you know certain rules and formulas agreed upon by all mathematicians, then that is all that there is to mathematics. Nothing could be farther from the truth. If you were to go more deeply into mathematics, you would find that at certain critical spots, some mathematicians go one way, while others go in a completely opposite direction.* This is certainly the case with Cantor's set theory. Although Cantor proved what many considered to be remarkable results, others rejected his methods completely.

Among Cantor's discoveries was that there is an infinite number of infinite cardinal numbers. That is, there is an infinite number of sizes of infinity. The smallest infinite cardinal number he called $\aleph_0$, which as you have seen is the cardinal number of the natural numbers. He proved that the set of all subsets of the natural numbers has another cardinal number $\aleph_1$ that is greater than $\aleph_0$. Cantor believed that perhaps $\aleph_1$ was equal to c. Although he labored for years to prove this conjecture, called the *continuum hypothesis,* Cantor was unable to prove it.

In 1900, the noted German mathematician David Hilbert said that the continuum hypothesis was one of the great unanswered questions in mathematics, and for the first half of the twentieth century, mathematicians could not answer this question. However, in 1963, a mathematician at Princeton University named Paul Cohen, made the remarkable discovery that *it is impossible to either prove or disprove the continuum hypothesis.* That is, you can accept it or reject it. If you accept it, then you are doing Cantorian set theory. If you reject it, then you are doing non-Cantorian set theory. Both are equally consistent but different areas of mathematics.

Continuing this pattern, we make sure that x is different from the third number in the third decimal place, say 3, and different from the fourth number in the fourth decimal place, make it 5, and so on. At this point $x = 0.5435$ . . . . By constructing x in this fashion, it cannot be the first number on the list, nor the second, nor the third, and so on. In fact, x will differ from every number on the list in at least one decimal place, so it cannot be any of the numbers on the list.

This means that our assumption that we were able to match the numbers between 0 and 1 with the natural numbers is wrong. So the cardinal number of this set is not $\aleph_0$. Cantor used the number c, for the word *continuum,* for this cardinal number. Because there is a number between 0 and 1 that cannot be matched with a natural number, we can argue as we did in our discussion of the seats and the students that the cardinal number c is greater than the cardinal number $\aleph_0$.

Exercises

In Exercises 1–8, show that each set has cardinal number $\aleph_0$ by showing a one-to-one correspondence between the natural numbers and the given set. Be sure to indicate the general correspondence.

1. {4, 8, 12, 16, 20, . . .}

2. {5, 10, 15, 20, 25, . . .}

3. {8, 11, 14, 17, 20, . . .}

4. {7, 11, 15, 19, 23, . . .}

5. {2, 4, 8, 16, 32, . . .}

6. {3, 9, 27, 81, 243, . . .}

---

* For a good example of this divergence in thinking, see the extensive highlights on non-Euclidean geometry in Chapter 8.

7. {1, 1/2, 1/3, 1/4, 1/5, . . .}

8. {1/2, 2/3, 3/4, 4/5, 5/6, . . .}

In Exercises 9–12, we give an expression describing the number that corresponds to the natural number n. Use this expression to describe a one-to-one correspondence between the natural numbers and one of its subsets. For example, if we gave you the expression 2n, you would write the following correspondence that we gave in Example 1:

$$
\begin{array}{cccc}
1 & 2 & 3 \cdots & n \cdots \\
\updownarrow & \updownarrow & \updownarrow & \updownarrow \\
2 & 4 & 6 \cdots & 2n \cdots
\end{array}
$$

9. $3n$

10. $2n + 3$

11. $3n - 2$

12. $4n + 5$

In Exercises 13–22, describe a one-to-one correspondence between the given set and one of its proper subsets. For example, if we gave you the set {3, 5, 7, 9, 11, . . .}, the nth term is 2n + 1. You could then write the correspondence by matching the elements of {3, 5, 7, 9, 11, . . .} with the elements of the subset {5, 7, 9, 11, 13, . . .}. The general correspondence would match 2n + 1 with 2n + 3.

13. {2, 4, 6, 8, 10, . . .}

14. {5, 10, 15, 20, 25, . . .}

15. {7, 10, 13, 16, 19, . . .}

16. {6, 9, 12, 15, 18, . . .}

17. {2, 4, 8, 16, 32, . . .}

18. {3, 9, 27, 81, 243, . . .}

19. {1, 1/2, 1/3, 1/4, 1/5, . . .}

20. {1/2, 2/3, 3/4, 4/5, 5/6, . . .}

21. {1/2, 1/4, 1/6, 1/8, 1/10, . . .}

22. {1/2, 1/4, 1/8, 1/16, 1/32, . . .}

In Example 3, we showed you how to match the natural numbers with the positive rational numbers in a one-to-one correspondence. Use that matching to answer the following questions.

23. What rational number corresponds to the natural number 12?

24. What rational number corresponds to the natural number 15?

25. What natural number is matched with the rational number 4/5?

26. What natural number is matched with the rational number 1/6?

Further Exercises

27. Communicating Mathematics Explain in your own words the argument we used in Example 4 to show that the cardinal number of the set of numbers between 0 and 1 is not $\aleph_0$.

28. Communicating Mathematics In constructing the number x in Example 4, how would you decide what to put in the 99th decimal place? The one millionth place?

29. How many one-to-one correspondences are there between the set {1, 2, 3} and the set {4, 5, 6}?

30. How many one-to-one correspondences are there between two four-element sets?

31. The arithmetic of infinite cardinal numbers has some peculiar properties. For example, it is a fact that $1 + \aleph_0 = \aleph_0$. Show this by forming the union of a set whose cardinal number is 1 with a disjoint set whose cardinal number is $\aleph_0$ and then showing that the union of the sets is in one-to-one correspondence with the natural numbers.

32. Another strange infinite arithmetic fact is that $\aleph_0 + \aleph_0 = \aleph_0$. Show why this is true by considering the union of two disjoint sets having cardinal number $\aleph_0$.

CHAPTER

2

Logic:*
The Study of What's True or False or Somewhere in Between

A doctor in a remote Peruvian village is using a computer program called an expert system to help her diagnose a patient with severe abdominal pain. At the same time, a graduate student uses an Internet search engine to locate data for his thesis on unemployment caused by corporate mergers. Also at the same time, a consumer applying for an "instant" car loan by phone is answering a series of questions. One thing that binds these three people together is that each is using a product based on an area of mathematics as old as Aristotle and yet as modern as the latest personal computer—logic. This ancient area of mathematics has special relevance today because it is the basis for the design of the hardware and software of modern computer systems.

Logic also guides the actions of the endearing *Star Wars* robot, R2-D2. However, his logic is not the ice-cold logic of a computer; rather, if he is to be humanlike, he must be able to interpret shades of meaning. In order to show human intelligence, a robot uses a modern form of logic called *fuzzy logic* that allows it to compute with statements that are "usually true," "sometimes true," or "not often true."

In this chapter you will learn to represent statements symbolically and to compute with logical statements just as you do algebraic computations. By these computations, you will be able to determine when a statement is true or false and when an argument is valid. We will also introduce you to a form of argument called a *syllogism*, which is the way Aristotle approached logic over two

---

* You can find further resources on logic at www.aw.com/pirnot.

thousand years ago. Finally, we will give you a glimpse of fuzzy logic and show you how to make decisions based on fuzzy reasoning.

Logic is an active area of mathematical research as today's scientists push the frontiers of our understanding of human reasoning and intelligence.

2.1 INDUCTIVE AND DEDUCTIVE REASONING

We will begin this section with three apparently unrelated ramblings.

1. As I was writing this section of this book, my house was being invaded by ants. I noticed that on sunny days, I had an abundance of ants in my bathroom and kitchen, while on rainy days there was hardly an ant to be seen. Today I woke up, saw that it was raining and concluded that I probably wouldn't see many ants today.

2. About a week prior to spring break, our telephone rings and my wife says, "I'll bet that's Karen calling. She probably wants to get together over the break."

3. You notice that a new technology stock, TechniCom, has doubled in price for each of the last three weeks. You decide to invest in TechniCom because you want to double your money.

What do these situations about the ants, our friend Karen, and your TechniCom stock all have in common? In each case, we were using inductive reasoning. We were looking at recurring patterns and drawing general conclusions. Inductive reasoning is one type of reasoning that we use in mathematics, science, and everyday life.

In inductive reasoning, we begin with specific examples.

> **DEFINITION**
>
> **Inductive reasoning** is the process of drawing a general conclusion by observing a pattern in specific instances. This conclusion is called a **hypothesis** or **conjecture.**

Mathematicians and scientists often make conjectures based upon their observations. Mathematicians try to prove that conjectures are true by using the laws of mathematics. Example 1 explains a famous conjecture that was made hundreds of years ago; but mathematicians have not yet been able to prove it.

EXAMPLE 1 Goldbach's Conjecture

In 1742, Christian Goldbach made the conjecture that we could write every even integer greater than 2 as the sum of two prime numbers.* A prime number is a

---

* The primes do not have to be different.

number such as 3, 5, and 7 that can be divided evenly only by itself and one. Illustrate Goldbach's conjecture, by expressing the even integers 20, 48, and 100 as the sum of two prime numbers.

SOLUTION: We can write $20 = 13 + 7$, $48 = 11 + 37$, and $100 = 41 + 59$.

Although no one has yet been able to prove Goldbach's conjecture, many believe it to be true based on inductive reasoning.

In Example 2, we ask you to make a conjecture that mathematicians have been able to prove is true.

EXAMPLE 2   A Divisibility Test for Nine

Consider the numbers a) 72, b) 963, c) 10,854, and d) 7,236,261 which are each evenly divisible by nine. (Verify this.) Add the digits in each number. Do you see any pattern? Make a conjecture.

SOLUTION:

a) $7 + 2 = 9$

b) $9 + 6 + 3 = 18$

c) $1 + 0 + 8 + 5 + 4 = 18$

d) $7 + 2 + 3 + 6 + 2 + 6 + 1 = 27$

Notice that in each case we get a sum that is evenly divisible by 9. Our conjecture is that in order for nine to divide evenly into a number, nine must divide the sum of the digits of the number.

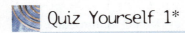

Verify Goldbach's conjecture for a) 38 and b) 46.

> **PROBLEM SOLVING**
>
> Keep in mind that in doing inductive reasoning, you are only making an educated guess. You cannot be sure that your conclusion is true. Recall our discussion of the Always Principle and the Counterexample Principle in Section 1.1.

Example 3 illustrates inductive reasoning in a nonnumerical situation.

EXAMPLE 3   Counting Line Segments

If we join three points A, B, and C in all possible ways with line segments, we get three line segments as we see in Figure 2.1(a) and joining four points, we get six line segments as shown in Figure 2.1(b).

If you were to join ten points, how many line segments would you have?

SOLUTION: We can continue the pattern that we began in Figure 2.1. Figure 2.2(a) shows that in joining five points we get ten line segments and in Figure 2.2(b), with six points, we get 15 line segments.

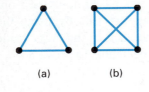

(a)   (b)

FIGURE 2.1   Joining three and four points in all possible ways by line segments.

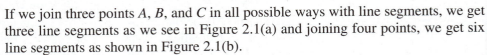

---

* Quiz Yourself answers begin on page 849.

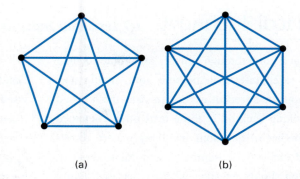

(a)   (b)

FIGURE 2.2   Joining five and six points in all possible ways by line segments.

We summarize what we have seen so far in Table 2.1.

Number of Points	Number of Line Segments	Increase in Number of Line Segments
3	3	
4	6	3
5	10	4
6	15	5
7	21	6
8	28	7
9	36	8
10	45	9

TABLE 2.1   Highlighted boxes summarize what we have observed in counting line segments.

If you examine the increase in the number of line segments from each row to the next, you find increases of 3, 4, and then 5. From this pattern, you might conjecture that for seven points there will be 21 line segments, and for eight points there will be 28 line segments, and so on. So based on this inductive reasoning, we will say that for ten points, there will be 45 line segments. Although we will not prove this now,* this is in fact a correct statement.

Sometimes inductive reasoning can mislead us into thinking that something is true which is not, as we see in Example 4.

Quiz Yourself 2

In Example 3, conjecture how many line segments there will be if we have 12 points.

---

* In the Further Exercises, we will consider why this pattern holds.

Historical Highlight: *IQ Tests and Inductive Reasoning**

As you have gone through school, from time to time you have taken tests that measure your intelligence. Perhaps your most recent experience was the dreaded SATs. Looking back on it, those tests often contained a large number of questions involving pattern recognition and inductive reasoning. If you think about it, perhaps you should question whether that one ability should play such a major role in measuring your intelligence.

The notion of intelligence tests has an interesting history. Prior to the 1800s, people believed that except for a few geniuses and a few people with below-average mental capability, the majority of the population had basically the same intelligence. However, in the nineteenth century, the British scientist Sir Francis Galton proposed that intelligence, like the color of your hair and the shape of your nose, is an inherited trait. Galton's ideas had been influenced by the Belgian statistician Lambert Quetelet, who was the first to apply statistics to the study of human characteristics.† Galton believed that it was possible to measure intelligence.

In France, the psychologist Alfred Binet, who like Galton was interested in measuring intelligence, devised tests to measure a child's mental age. Using Binet's test, the German psychologist Wilhelm Stern suggested that the ratio of a child's mental age divided by his physical age would be a good measure of his intelligence. For example, if a 10-year-old child had a mental age of 11, then his "IQ" would be $11/10 = 1.10 = 110\%$ or an IQ of 110. For a 5-year-old child whose mental age was 6, the IQ would be $6/5 = 1.20 = 120\%$ or an IQ of 120. In the 1960s and 1970s, the notion of IQ fell out of favor because many felt that the questions on IQ tests had both cultural and racial biases.

In 1983, Howard Gardner proposed the notion of multiple intelligences consisting of logical-mathematical, linguistic, spatial, musical, bodily kinesthetic, and interpersonal and intrapersonal intelligences. In recent years, many have accepted Gardner's notion of measuring intelligence. However, others agree with the psychologist Arthur Jensen, who says, "Achievement is the school's main concern. I see no need to measure anything other than achievement itself."

EXAMPLE 4   A False Conjecture

We wish to divide a circle into regions by selecting points on its circumference and drawing line segments from each point to each other point. Figure 2.3 shows the greatest number of regions that we get if we have one point (no line segment is possible for this case), two, three, and four points.

Use inductive reasoning to find the greatest number of regions we would get if we had six points on the edge of the circle.

SOLUTION: This problem seems somewhat similar to what we did in Example 3. Notice that it appears that each time we add another point, we double the number of regions. It is natural to conjecture that if we have five points, there would

---

* This highlight is based on information contained in the essay *The History of IQ Testing* that can be found at the Web site www.ivillagehealth.com.
† Quetelet was the first to discover the normal distribution, or bell-shaped curve, that we discuss in detail in Section 14.5.

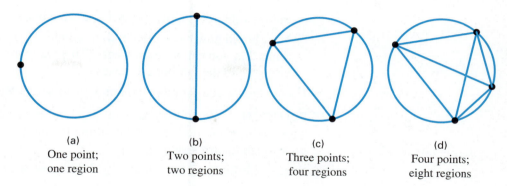

(a)	(b)	(c)	(d)
One point; one region	Two points; two regions	Three points; four regions	Four points; eight regions

FIGURE 2.3   Dividing a circle into regions.

be sixteen regions and with six points we would get 32 regions. However, this is not true. You can try it for yourself by drawing a large circle and pick six points in different ways on the circle. The largest number of regions that you will find is 31, not 32.

Deductive reasoning begins with accepted facts and principles.

In inductive reasoning, we draw a general conclusion by considering a number of specific conclusions. In a sense, deductive reasoning is a reverse way of reasoning from inductive reasoning.

> **DEFINITION**
>
> In **deductive reasoning,** we use accepted facts and general principles to arrive at a specific conclusion.

In mathematics, we often use general mathematical principles to prove some specific fact. In Example 5, we will show you how to explain what at first seems to be a mystifying number trick.

EXAMPLE 5   Explaining a Number Trick by Using Deductive Reasoning

Consider the following number trick:*

1. How many days a week do you eat out?
2. Multiply this number by 2.
3. Add 5 to the number you got in step 2.
4. Multiply the number you obtained in step 3 by 50.
5. If you have already had your birthday this year add 1755; if you haven't, add 1754.

---

* Thanks to Dr. Robert Voytas of the Kutztown University Psychology Department for asking me to explain why this trick works, back in the spring of 1999.

6. Subtract the four-digit year that you were born.

7. You should now have a three-digit number. The first digit is the number of times that you eat out each week. (There is more to this trick that we will investigate in the Further Exercises.)

Use deductive reasoning to explain why this trick works.

SOLUTION: We will use general algebra principles and deductive reasoning to explain this trick. Let's begin by calling the number of days that you eat out n and go through the trick step by step.

Step 2.   Multiplying the number by 2 will give us $2n$.

Step 3.   Adding 5, we get $2n + 5$.

Step 4.   Multiplying by 50 gives us the expression $(2n + 5)50$.

Step 5.   Let's assume that you haven't had your birthday yet, so we will add 1754 to get $(2n + 5)50 + 1754$.

Step 6.   Let's assume that you were born in 1986, so we will subtract 1986 to get $(2n + 5)50 + 1754 - 1986$.

When we simplify the expression in step 6, we get

$$(2n + 5)50 + 1754 - 1986 = 100n + 250 + 1754 - 1986$$
$$= 100n + 2004 - 1986$$
$$= 100n + 18.$$

Notice that when we look at this expression in this form, the hundred's digit is n, which is the number of days that you eat out each week. Notice that the difference $2004 - 1986$ has no effect on the hundred's digit. We will have more to say about this difference in the Further Exercises. ◎

Exercises 2.1

Is each of the following situations an example of inductive or deductive reasoning?

1. It has rained the past three weekends, canceling your softball game. You expect that next Saturday it will rain again.

2. Carla is calculating her income taxes to determine if she will get a refund this year.

3. As you read a murder mystery, you are keeping track of the clues that the author has given in order to predict who the killer will be.

4. You tell your friend Jay to be ready 15 minutes before you actually intend to pick him up because Jay is always late for his appointments.

5. Luis has noticed that the stock market has gone up on the Friday before each of the last three holidays. He plans to buy stock on the Friday before Labor Day in order to cash in on this trend.

6. Marianne is using the rules of algebra to solve a word problem on a quiz.

7. Brett is calculating his expenses for next year in order to determine how large his student loan should be.

8. Latisha noticed that on every true–false quiz so far this semester, her instructor has given twice as many false questions as true questions. On the next quiz, if she is not sure of an answer, she will guess "false."

9. The American Conference team has won the Super Bowl for three straight years. You expect that the American Conference team will win again this year.

10. Emily estimates that if she can average 50 miles per hour, she will reach San Diego in five and one-half hours.

11. Communicating Mathematics   Explain the difference between inductive and deductive reasoning.

12. Communicating Mathematics   Give an example of inductive reasoning.

13. Communicating Mathematics   Give an example of deductive reasoning.

14. Communicating Mathematics   Give an example of inductive reasoning that leads you to draw a false conclusion.

In Exercises 15–22, use inductive reasoning to predict the next term in the sequence of numbers.

15. 1, 4, 7, 10, 13, ?

16. 2, 8, 14, 20, 26, ?

17. 3, 6, 12, 24, 48, ?

18. 5, 15, 45, 135, 405, ?

19. $\dfrac{1}{2}, \dfrac{1}{4}, \dfrac{1}{8}, \dfrac{1}{16}, \dfrac{1}{32}, ?$

20. $\dfrac{1}{2}, \dfrac{2}{3}, \dfrac{3}{4}, \dfrac{4}{5}, \dfrac{5}{6}, ?$

21. 1, 1, 2, 3, 5, 8, 13, ?

22. 0.1, 0.10, 0.101, 0.1010, 0.10101, ?

In Exercises 23–26, use inductive reasoning to draw the next figure in the pattern. There may be several correct answers.

23.

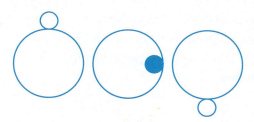

24.

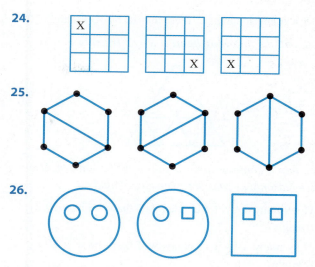

25.

26.

In Exercises 27 and 28, place the next X in a box to continue the pattern.

27.

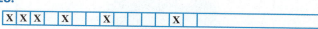

28.

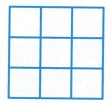

In Exercises 29–32, use inductive reasoning to determine the units digit of the given number.

29. 2^{50} **30.** 3^{37}

31. 4^{95} **32.** 7^{100}

Illustrate Goldbach's conjecture for each of the following numbers.

33. 16   **34.** 18

35. 20   **36.** 26

We will call the following figure a 3 by 3 square.

Notice that it contains nine 1 by 1 squares, four 2 by 2 squares, and one 3 by 3 square, for a total of 14 squares of different sizes. Use inductive reasoning to generalize this pattern to answer Exercises 37 and 38.

37. How many different squares are there in a 5 by 5 square?

38. How many squares are there in a 10 by 10 square?

We will call the following figure a 3 by 3 triangle.

Notice that it contains nine smaller 1 by 1 triangles. Use inductive reasoning to answer Exercises 39 and 40.

39. How many 1 by 1 triangles will there be in a 10 by 10 triangle?

40. How many 1 by 1 triangles will there be in a 20 by 20 triangle?

41. A stack of cannonballs at a Civil War museum has the shape of a pyramid with a square base, as shown in the figure.

How many cannonballs are in the stack?

42. If a stack of cannon balls similar to those shown in Exercise 41 was six layers high, how many cannon balls would be in the stack?

A magic square is a square arrangement of numbers such that if you add the numbers in any row, column, or diagonal, you always get the same sum. In the magic square shown below, called a 3 by 3 square, the sum of any row, column, or diagonal is always 15.

8	1	6
3	5	7
4	9	2

In Exercises 43 and 44, the magic squares use each of the integers from 1 to 16 exactly once. Use deductive reasoning to

a) Determine the total of all of the numbers in the square.

b) Determine the total for each row, column, and diagonal.

c) Complete the magic squares.

Explain your reasoning.

*43. Communicating Mathematics

7			9
			8
13		2	
4	1		14

44. Communicating Mathematics

16	3		13
5			8
	6		12
4		14	

Communicating Mathematics   *In Exercises 45–48, follow the instructions for each "trick" starting with several different numbers of your choice. Make a conjecture as to what result you get in each case. Use algebra and deductive reasoning, as we did in Example 5, to explain why your conjecture is correct.*

45. a) Choose any natural number.
   b) Multiply the number by three.
   c) Add nine to the product you just found.
   d) Divide the results of part c) by three.
   e) Subtract the number that you started with.

46. a) Choose any natural number.
   b) Multiply the number by five.
   c) Add 20 to the product you just found.
   d) Divide the results of part c) by five.
   e) Subtract four.

47. a) Choose any natural number.
   b) Multiply the number by eight.
   c) Add 12 to the product you just found.
   d) Divide the results of part c) by four.
   e) Subtract three.

---

* Exercise numbers circled in red can be used as group exercises.

48. a) Choose any natural number.

b) Multiply the number by 15.

c) Add 20 to the product you just found.

d) Divide the results of part c) by five.

e) Subtract four.

In Exercises 49 and 50, draw the next figure in the sequence.

49.

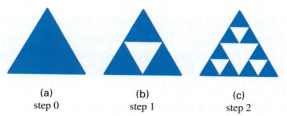

(a) (b) (c)
step 0 step 1 step 2

50.

Further Exercises

51. We will call the rectangle below, which has three rows and four columns, a 3 by 4 rectangle.

How many squares of various sizes can you find in this rectangle?

52. Repeat Exercise 51 for a 6 by 4 rectangle.

53. a) Repeat Exercise 51, but now count rectangles of all types, including squares. b) Repeat Exercise 52, now counting rectangles of all types, including squares. It is important when doing this that you are systematic. Count rectangles of sizes 1 by 1, 1 by 2, 1 by 3, 1 by 4, 2 by 1, 2 by 2, and so on.

54. Communicating Mathematics Can you find some general pattern that would enable you to count rectangles of all various sizes in a 10 by 6 rectangle without actually drawing the rectangle and counting? Explain your thinking.

55. Stacking cannon balls. If a stack of cannon balls (similar to the one shown in Exercise 41) has a rectangular base which is six cannon balls long and four cannon balls wide, and is stacked with as many can-

non balls as possible, how many balls will be in the stack? Explain your reasoning. We can only place a cannon ball on the stack if it is resting on four other cannon balls.

56. Stacking cannon balls. Redo Exercise 55, but now assume the base is seven cannon balls long and five cannon balls wide.

57. Communicating Mathematics Make up a 3 by 3 magic square of your own using the natural numbers from 1 to 9. Explain how you constructed the magic square.

58. Communicating Mathematics Make up a 4 by 4 magic square of your own using the natural numbers from 1 to 16. Explain how you constructed the magic square.

59. Communicating Mathematics In Example 3, explain why the number of line segments increased as it did when we added extra points. For example, if you have four points connected in all possible ways, what number of segments do you need to connect one additional point.

60. Communicating Mathematics Show that the conjecture that we made in Example 4 is true for five points. Draw a figure that has 31 regions if we are using six points.

61. Explaining a trick. In Example 5, we mentioned that there is more to the trick than what we explained. The rest of the trick states that the last two digits of the three-digit number you obtain is your age. However, this part of the trick will only work in the year 2004! Explain why this is so.

 62. Explaining a trick. How would you adjust the trick so that it will work in the year 2006? Explain how you came up with your adjustment.

2.2   STATEMENTS, CONNECTIVES, AND QUANTIFIERS

In the nineteenth century, the British mathematician George Boole* invented a branch of mathematics called *symbolic logic*. Today symbolic logic is important because it is the foundation for the design of computer hardware and software. Boole's logic differed from classical Aristotelian logic, which originated in Greek philosophy, in that Boole believed that he could calculate with logical ideas symbolically—similar to the way that algebraists calculated with numerical quantities. In algebra we can prove that $(x + y)^2 = x^2 + 2xy + y^2$ is a general statement that is true for any numbers x and y. Similarly, Boole wished to develop general methods to allow him to perform routine calculations to determine whether statements were true or false and whether logical arguments were valid.

In studying algebra, you learned that to solve problems, the first step is to represent verbal expressions with symbols. For example, instead of using the sentence "The area of a rectangle equals the product of its length times its width," we write the more concise statement "$A = l \times w$." Similarly, in this section, you will learn to represent English statements symbolically; in Section 2.3 we will investigate when these statements are true or false.

In symbolic logic, statements are either true or false.

As you begin your study of symbolic logic, keep in mind that *we are only concerned with the truth or falsity of the sentences we analyze and not their content*. As you will see, a statement can seem ridiculous and yet have a proper logical form. On the other hand, a statement that seems believable may have an unacceptable logical form.

> **DEFINITION**
>
> A **statement** in logic is a declarative sentence that is either true or false. We represent statements by lowercase letters such as p, q, or r.

The following are examples of statements. Remember that to be a statement, the sentence must be either true or false; however, *we do not need to know which it is.*

---

* See the Historical Highlight on George Boole in Section 1.5.

a) AIDS is a leading killer of women.

b) If you eat less and exercise more, you will lose weight.

c) We have reduced the depletion of the ozone layer.

d) Oprah Winfrey or Steven Spielberg had the highest earnings in the entertainment industry in 1999.

The following are not statements.

e) Beam me up, Scotty! (This is not a declarative sentence.)

f) When did dinosaurs become extinct? (This is not a declarative sentence.)

g) This statement is false. (This is a paradox. It cannot be either true or false. If we assume that this sentence is true, then we must conclude that it is false. On the other hand, if we assume it is false, then we conclude that it must be true.)

Quiz Yourself 3*

Identify each sentence as simple or compound.

a) The 1998 Yankees were the best team in the history of baseball.

b) If you break your lease, then you forfeit your deposit.

c) The world's tallest skyscrapers are in Kuala Lumpur and Chicago.

Notice that statements (a)–(d) fall into two categories. Statements (a) and (c) each express a single idea; if we were to remove any part of these sentences, they would no longer make sense. Such sentences are called *simple statements*. On the other hand, statements (b) and (d) clearly contain several ideas connected to make a more complex sentence. These sentences are called *compound statements*. Notice that words such as *if . . . then* and *or* connect the ideas to form the compound sentences.

> **DEFINITIONS**
>
> A **simple statement** contains a single idea. A **compound statement** contains several ideas combined together. The words used to join the ideas of a compound sentence are called **connectives**.

In logic, we connect ideas using *not, and, or, if . . . then*, and *if and only if*.

Because the English language is so rich, we can use many words to connect ideas. However, the connectives we use in logic generally fall into five categories: *negation, conjunction, disjunction, conditional,* and *biconditional*. We will now discuss each of these connectives individually.

> **DEFINITION** **Negation**
>
> **Negation** is a statement expressing the idea that something is not true. We represent negation by the symbol $\sim$.[†]

---

† Even though a negation does not combine two separate ideas, we will consider a negation to be a compound statement.

EXAMPLE 1   Negating Sentences

Negate each statement.

a) *p*: The optical disk drive is included in the price of the computer.

b) *b*: The blue whale is the largest living creature.

SOLUTION: a) The optical disk drive is *not* included in the price of the computer. We write this negation symbolically as ~*p*.

b) The blue whale is *not* the largest living creature. The symbols ~*b* represent this statement. ◎

DEFINITION   Conjunction

A **conjunction** expresses the idea of *and*. We use the symbol ∧ to represent a conjunction.

EXAMPLE 2   Joining Statements Using *And*

Consider the following statements:

    p: The tenant pays utilities. *d*: A $150 deposit is required.

a) Express the statement "It is not true that: the tenant pays utilities and a $150 deposit is required" symbolically.

b) Write the statement ~*p* ∧ ~*d* in English.

SOLUTION: a) This sentence has the form ~(*p* ∧ *d*).

b) The tenant does not pay utilities and a $150 deposit is not required. ◎

PROBLEM SOLVING

In Example 2(a), we first joined *p* and *d* using *and* before forming the negation, whereas in Example 2(b), we negated *p* and *d* first and then joined these negations using *and*. Even though these statements sound similar, they do not say the same thing. The Order Principle from Section 1.1 warns that changing the order of operations in mathematics *may* change the meaning. In Section 2.3 we will explain the difference between these statements.

DEFINITION   Disjunction

A **disjunction** conveys the notion of *or*. We use the symbol ∨ to represent a disjunction.

EXAMPLE 3   Joining Ideas Using *Or*

Consider the following statements:

u: Urban sprawl will increase. *c*: Commuting will be more difficult.

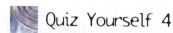

Quiz Yourself 4

Consider the following statements:

c: I will buy a CD player. v: I will buy a VCR. Write each statement in symbolic form.

a) I will not buy a CD player or I will not buy a VCR.

b) I will not buy a CD player and I will buy a VCR.

a) Write the following statement symbolically:

>   Urban sprawl will not increase or commuting will be more difficult.

b) Write the symbolic form $\sim(u \vee c)$ in English.

SOLUTION: a) The symbolic form of this sentence is $(\sim u) \vee c$.

b) When we translate this into English we get:

>   It is not true that: urban sprawl will increase or commuting will be more difficult.

DEFINITION   Conditional

A **conditional** expresses the notion of *if . . . then.* We use an arrow, →, to represent a conditional.

EXAMPLE 4   Connecting Ideas Using *If . . . Then*

Suppose that p represents "The Phillies win the World Series" and s represents "Susan will win an Emmy."

a) We would read the statement $p \rightarrow s$ as "If the Phillies win the World Series, then Susan will win an Emmy."

b) We would write the statement "If the Phillies do not win the World Series, then Susan will not win an Emmy" symbolically as $\sim p \rightarrow \sim s$.

Again, you must be particular about the order in which you use connectives. It would be incorrect to change the order of the negations and the conditional to write the answer for (b) as $\sim (p \rightarrow s)$.*

Although mathematicians use the next connective frequently, it is not as common in ordinary language.

DEFINITION   Biconditional

A **biconditional** represents the idea of *if and only if.* Its symbol is a double arrow, ↔.

EXAMPLE 5   Connecting Ideas with *If and Only If*

Write each biconditional statement in symbolic form.

a) A polygon has three sides if and only if it is a triangle.

b) The software can be returned if and only if the seal is not broken.

SOLUTION: a) Let p represent the statement "A polygon has three sides" and let t represent the statement "The polygon is a triangle." Then this statement has the form $p \leftrightarrow t$.

---

* You will understand the difference in meaning between $\sim(p \rightarrow s)$ and $\sim p \rightarrow \sim s$ when we discuss logically equivalent statements in Section 2.3.

Historical Highlight: *The Evolution of Logic*

"He was undoubtedly the most fertile logician there has ever been. . . ."* This statement describes how remarkable Aristotle was. In the more than two thousand years that have passed since his work, vast numbers of philosophers, logicians, and mathematicians from Greece, the Middle East, India, Western and Eastern Europe, and America have made many important contributions to logic. Yet, in spite of all this, some still argue that Aristotle was the greatest logician of all time. Some of the rules and principles we study in this chapter were discussed by students in Greece thousands of years ago, by Buddhist monks in the tenth century, and by Thomas Aquinas in the thirteenth century, before explorers discovered the new world.

Although we now consider logic to be a branch of mathematics, much of the early work in logic was not associated with mathematics at all. Philosophers and theologians often studied logic because they were interested in understanding correct reasoning in order to discover religious truth that would lead them to live more perfect lives.

Over the centuries, the reasons for studying logic began to change; by the seventeenth century, some mathematicians were viewing it as an area of mathematics. The German mathematician Gottfried Leibniz believed that it was possible to develop a symbolic form of logic that would make philosophy unnecessary:

"I happened unexpectedly upon this remarkable idea. That an alphabet of human thought could be devised, and that everything could be discovered and distinguished by the combination of the letters of this alphabet . . . discussion between two philosophers will not be any longer necessary. It will rather be enough to take pen in hand and . . . say to one another: We Calculate!"

Although logic has not reached Leibniz's expectations, it has proved to be a powerful tool in modern mathematics.

Quiz Yourself 5

Consider the following statements:

f: I fly to Houston.
q: I will qualify for frequent-flyer miles.

Write each statement in symbolic form.

a) If I do not fly to Houston, then I will not qualify for frequent-flyer miles.

b) I fly to Houston if and only if I will qualify for frequent-flyer miles.

b) Let *r* represent the statement "The software can be returned" and let *b* represent the statement "The seal is broken." Then this statement has the form $r \leftrightarrow\ \sim b$.

In working with connectives we will often use the informal terms *not, and, or, if . . . then,* and *if and only if* instead of the more formal terms *negation, conjunction, disjunction,* and so on. By combining these five connectives, we can form extremely complex sentences. Consider these simple statements:

t : Today is Tuesday.
r : It is raining.
s : It is springtime.
h : You are hungry.

Suppose we make a compound statement:

If today is not Tuesday and it is raining, then either you are hungry or it is not springtime.

---

* Innocentius M. Bochenski, *A History of Formal Logic* (Notre Dame Press, 1961).

If you are hungry for lunch on a rainy Wednesday in April, is the previous statement true or false? What if you are not hungry on a sunny Tuesday in March? As you can see, it is often difficult to have a precise understanding of what an English sentence is saying. Notice that the form of this statement is $(\sim t \land r) \to (h \lor \sim s)$. In Section 2.3, we will develop rules for computing with connectives so that we can then easily answer such questions.

SOME GOOD ADVICE

You can use commas to group the components of a compound statement. For example, you would translate "Stock prices will rise and inflation will fall, or bond prices will decrease" as $(s \land i) \lor b$. You would write the statement $s \land (i \lor b)$ into English as "Stock prices will rise, and inflation will fall or bond prices will decrease."

Quantifiers state how many objects satisfy a given property.

In addition to connectives, there are other special words called *quantifiers* that you need to understand in analyzing sentences. **Quantifiers** tell us "how many," and fall into two categories: universal quantifiers and existential quantifiers.

DEFINITION

Universal quantifiers are words such as *all* and *every* that state that all objects of a certain type satisfy a given property.

The following statements contain universal quantifiers:

> *All* citizens over age eighteen have the right to vote.
> *Every* triangle has an interior angle sum of 180°.
> *Each* rider in the horse show must register by August 1.

DEFINITION

Existential quantifiers are words such as *some, there exists,* and *there is at least one* that state that there are one or more objects that satisfy a given property.

The following statements contain existential quantifiers:

> *Some* drivers qualify for reduced insurance rates.
> *There is* a number whose square is 25.
> *There exists* a bird that cannot fly.

Suppose we want to negate the statement "All professional athletes are wealthy." In order to do this, we must understand how to negate a statement that contains quantifiers. To see how to do this, consider the diagrams* in Figure 2.4. We illustrate the original statement to be true in Figure 2.4(a), by drawing the set of all athletes as contained within the set of all wealthy people. In Figure 2.4(b), we move the circle representing all athletes slightly so that part of the circle lies outside of the circle labeled *wealthy*. We have shown that the statement "Not all athletes are wealthy" means the same thing as "Some athletes are not wealthy."

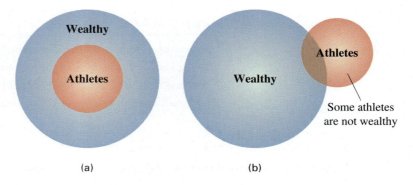

FIGURE 2.4   (a) All athletes are wealthy. (b) Not all athletes are wealthy (some athletes are not wealthy).

Notice that when we negate a statement with a universal quantifier, we get a statement with an existential quantifier.

Let us now investigate how to negate a statement with an existential quantifier. Consider the statement "Some property owners will get a tax rebate," which is illustrated in Figure 2.5(a). To say that "It is false that some property owners will get a tax rebate" means that the circle representing the property owners can have nothing in common with those getting the tax rebate. This is shown in Figure 2.5(b). Thus we can state the negation of the original statement as "All property owners will not get a tax rebate."

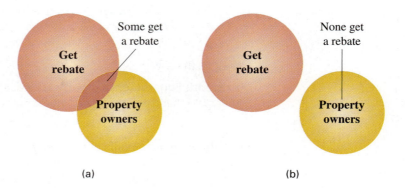

FIGURE 2.5   (a) Some property owners qualify for a tax rebate. (b) It is false that some property owners qualify for a rebate (all do not qualify for a rebate).

---

* We will consider such diagrams, called Euler diagrams, in greater depth in Section 2.6.

From Figure 2.5, we see that when we negate a statement with an existential quantifier, we get a statement with a universal quantifier.

> **Negating Statements with Quantifiers**
>
> The phrase *Not all are* has the same meaning as *Some are not*.
> The phrase *Not some are* has the same meaning as *All are not*.

EXAMPLE 6 Negating Quantified Statements

Negate each quantified statement and rewrite it in English.

a) All customers will get a free dessert.

b) Some computers have a two-year warranty.

SOLUTION: a) We wish to rewrite the statement "Not all customers will get a free dessert." Because this statement has the form *Not all are*, we can rewrite it as *Some are not*. In English we would say "Some customers will not get a free dessert."

b) The negation of the statement is "It is not true that some computers have a two-year warranty." Because this statement has the form *Not some are*, we can rewrite it as *All are not*. In English we would say "All computers do not have a two-year warranty."

In Example 6(b), it is important to understand that *every object* following the word *all* has the stated property. That is, each and every computer fails to have a two-year warranty. Another way to remember this is that *All do not have a property* means *None have the property*.

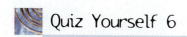

Quiz Yourself 6

Negate each quantified statement and rewrite it in English.
a) All professors are friendly.
b) Some dogs bite.

> **PROBLEM SOLVING**
>
> The Three-Way Principle emphasizes that a picture helps you understand a situation better. Drawing diagrams of quantified statements will help you better understand how to write their negations.

Exercises 2.2

In Exercises 1–10, determine which sentences are statements.

1. Over 40 percent of Americans own computers.

2. The supersonic *Concorde* can fly at over 3,000 miles per hour.

3. Come skiing with me this weekend.

4. Are you going to the concert?

5. Where did you buy your sunglasses?

6. Water quality has steadily improved over the past ten years.

7. You will be a millionaire by the time you are 30.

8. *The Simpsons* is a popular TV show about lawyers.

9. Will you still love me when I'm 64?

10. Romeo, Romeo, wherefore art thou Romeo?

In Exercises 11–20, identify each statement as simple or compound. If a statement is compound, identify the connectives used.

11. If you wish to ride, you must be at least 48 inches tall and have no food, drinks, or Kryptonite.*

12. Laura is satisfied with her performance in the musical.

13. Lee or Nehemiah will lead the conference in scoring.

14. If Hillary supports environmental issues, she will succeed in politics.

15. Alvaro likes opera.

16. Mendy does not know how to inline skate.

17. If *Titanic* had the largest box-office gross of all time, then *Jurassic Park* or *The Phantom Menace* had the second-largest.

18. Jena is going to Europe this summer.

19. Red sky in morning, sailor take warning.

20. I have a dream.

Consider the following statements: r: The Republicans will control Congress. t: Taxes will be cut. s: Social programs will be increased.

In Exercises 21–26, write each statement in symbolic form.

21. The Republicans will control Congress or social programs will not be increased.

22. Taxes will not be cut or Republicans will not control Congress.

23. If the Republicans do not control Congress and taxes are cut, then social programs will not be increased.

24. If social programs are not increased, then the Republicans will not control Congress or taxes will not be cut.

25. Social programs will not be increased if and only if taxes are cut.

26. Republicans will not control Congress or taxes will not be cut, if and only if social programs are increased.

Consider the following statements: t: The radial tires are included. s: The sunroof is extra. w: Power windows are optional.

In Exercises 27–32, translate each statement into words.

27. $t \vee (\sim s)$

28. $\sim s \wedge \sim w$

29. $\sim (s \wedge t)$

30. $\sim (w \wedge t)$

31. $t \rightarrow (s \vee \sim w)$

32. $(s \wedge t) \rightarrow \sim w$

Because the English language is so complex, it is often difficult to see clearly the structure of compound statements. In Exercises 33–40, examine each statement to identify which connectives are being used. You may need to rephrase the sentence to see its structure more clearly.

33. "There was one of two things I had a right to, liberty or death; if I could not have one, I would have the other. . . ." (Harriet Tubman, conductor-in-chief of the Underground Railroad)

34. "Never do today what you can do as well tomorrow." (Aaron Burr)

35. "I have found the best way to give advice to your children is to find out what they want and then advise them to do it." (Harry S. Truman)

36. "I can see, and that is why I can be so happy in what you call the dark, but which to me is golden." (Helen Keller)

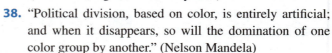

37. "The growing and dying of the moon reminds us of our ignorance, which comes and goes. . . ." (Black Elk, Oglala Sioux holy man)

38. "Political division, based on color, is entirely artificial; and when it disappears, so will the domination of one color group by another." (Nelson Mandela)

39. "Nobody sees a flower—really—it is so small—we haven't time—and to see takes time like to have a friend takes time." (Georgia O'Keeffe)

40. Woman must not depend upon the protection of man, but must be taught to protect herself." (Susan B. Anthony)

In Exercises 41–44, examine each statement based on the 1040 federal income tax form to identify which connectives are being used.

41. If you were a household employee who did not receive a Form W-2 because your employer paid you less than $1,000, be sure to include the amount you were paid on Form 1040-A.

---

* Regulations for riding the Superman the Escape roller coaster at Six Flags Magic Mountain in Los Angeles, California.

42. If you received a Form 1099 showing federal income tax withheld on dividends to interest income, include the amount withheld in the total on line 29a.

43. You were divorced and you made joint estimated tax payments with your former spouse or you changed your name and you made estimated tax payments using your former name.

44. If you were covered by a plan at work, your IRA deduction may be reduced or eliminated, but you can still make contributions to an IRA even if you cannot deduct them.

The following excerpts are taken from a VCR manual. Read them carefully and identify the connectives that they contain.

45. If you select "Movie Advance (Preview)" and press OK, the VCR will fast-motion to the beginning of the next preview and normal playback will resume when it is detected.

46. Input the channel number using the number keys or by pressing CH or SHUTTLE PLUS, then press OK.

In Exercises 47–50, translate each statement taken from the manual for the popular computer game Sim City into symbolic form.

47. Hold down the Option key if you have one or the Control key if you don't.

48. To set the slider bars, you can either click and drag them or just click at your desired setting.

49. If there is water nearby, or even running through your spot, all the better, but not necessary.

50. You, as mayor, must approve and establish these programs, but if your city is doing very well, the City Council may take it upon itself to enact some programs that benefit the city.

Communicating Mathematics *In Exercises 51–58, negate each quantified statement and then rewrite it in English in an alternative way. Draw diagrams to illustrate your thinking. Explain what your diagrams tell you.*

51. All snakes are poisonous.

52. Some auto mechanics are incompetent.

53. Some personal items are not covered by this insurance policy.

54. All married couples must file a joint tax return.

55. Some scientists believe that an asteroid collision led to the extinction of the dinosaurs.

56. All polygons have four sides.

57. All modern art is difficult to understand.

58. Some factories emit toxic wastes.

In Exercises 59–60, draw a diagram to illustrate that each pair of statements in Exercises 59 and 60 do not say the same thing.

59. Some luxury cars do not get good gas mileage.
It is not true that some luxury cars get good gas mileage.

60. All students are not wealthy.
It is not true that all students are wealthy.

In 1937, Claude Shannon showed that computer scientists could use symbolic logic to design computer circuits by using the following approach. Electricity passes through a switch when it is closed and does not flow when the switch is open.

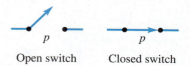

Open switch   Closed switch

As shown in the following diagram, electricity flows through a series circuit only when switches p and q are both closed. A **series** *circuit corresponds to a conjunction, p ∧ q, in logic. A* **parallel** *circuit corresponds to a disjunction, p ∨ q. Electricity will flow through a parallel circuit if either p or q is closed.*

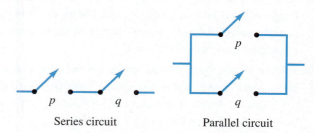

Series circuit   Parallel circuit

We can build more complicated circuits by combining series and parallel circuits. Such circuits can be represented by more complex logical forms. For example, we can represent the circuit

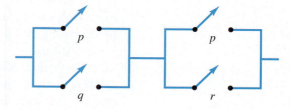

by the logical form $(p \lor q) \land (p \lor r)$.

In Exercises 61–64, represent each circuit with a logical form.

61.

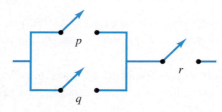

62.

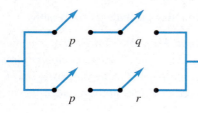

63.

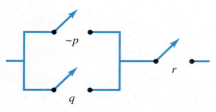

64.

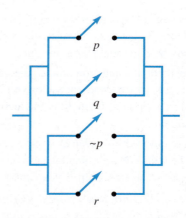

In Exercises 65–68, represent each logical form by a circuit.

65. $(q \wedge \sim p) \vee (q \wedge p)$

66. $(q \vee r) \vee (q \wedge p)$

67. $q \vee (\sim p \wedge r)$

68. $(q \wedge r) \wedge (\sim q \wedge p)$

Further Exercises

69. Many college and university libraries can search their book catalogs automatically. If your library has this capability, search for the keyword "logic" to see how many references there are on this topic.

*(70.) Use the World Wide Web to search for sites that discuss logic. Find several interesting sites—perhaps regarding the history of logic, applications of logic, and so on. Report to your class on the results of your search.

(71.) Think of a real-life situation that you might wish to analyze using symbolic logic. What difficulties do you encounter in doing this?

(72.) Provide arguments for or against the view that "Symbolic logic has nothing to do with real life."

2.3 TRUTH TABLES

In this section, we take a further step in explaining George Boole's approach to logic that we began in Section 2.2. Now that you know how to use connectives to represent compound statements symbolically, you are ready to do logical computations.

In algebra, after you understood how to represent numerical quantities with symbols such as x and y, you learned rules for computing with these quantities. Frequently you were interested in whether you could consider two different-looking

---

* Exercise numbers circled in red can be used as group exercises.

expressions to be the same. For example, you know that even though $(x + y)^2$ and $x^2 + y^2$ look similar, they do not always represent the same quantity. If you substitute numbers for x and y, you may find that the two quantities are different.

Similarly, in Section 2.2 when you saw a pair of logical expressions such as $\sim (p \wedge d)$ and $\sim p \wedge \sim d$, you may have wondered whether this pair of statements have the same logical meaning. In order to find this out, we introduce truth tables. They will give you a precise understanding of when compound statements are true or false.

We will use truth tables to define what we mean by logically equivalent statements and then explain exactly when two logical forms have the same meaning. First, we first present truth tables for negation (*not*), conjunction (*and*), and disjunction (*or*), and then use those tables to construct truth tables for more complex statements. In Section 2.4, we discuss truth tables for the conditional and biconditional.

SOME GOOD ADVICE

In logic, the truth values for statements using connectives such as *not, and, or,* and *if . . . then* are *sometimes* slightly different from everyday usage. It is important that you understand these differences.

Negating a statement reverses its truth value.

Negation works in logic exactly as it does in everyday language. If p is a true statement such as, "$2 + 2 = 4$," then its negation $\sim p$ is false. Also, if q is a false statement such as "George Washington was the king of England," then its negation $\sim q$ is true. We summarize the behavior of negation in Figure 2.6 which is an example of a truth table.

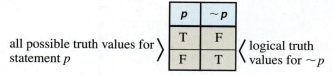

all possible truth values for statement p ⟩ | p | $\sim p$ | ⟨ logical truth values for $\sim p$

p	$\sim p$
T	F
F	T

FIGURE 2.6   The truth table for negation.

The left column indicates *all the possibilities* for statement p, namely that it can be either true or false. The right column indicates what logical values $\sim p$ takes on for the distinct possibilities for p. If p is true, the truth table tells us that $\sim p$ is false. If p is false, then $\sim p$ is true.

EXAMPLE 1   Finding the Truth Value of Negations

Determine whether each statement is true or false and then state its negation. Notice how the negation of each statement has the opposite truth value of the original statement.

a) The Dallas Cowboys are a well-known football team.

b) Mexico is a prosperous country in Asia.

SOLUTION: a) This is true. Therefore its negation, "The Dallas Cowboys are *not* a well-known football team," is false.

b) This is false. So "Mexico is *not* a prosperous country in Asia" is true.

A conjunction is true only when both of its component parts are true.

p	q	$p \wedge q$
T	T	T
T	F	F
F	T	F
F	F	F

FIGURE 2.7   The truth table for conjunction.

The *and* connective also works in logic as it does in everyday life. Suppose you make the following statement to a friend:

"I am a basketball fan and I like rock music."

This statement is of the form $p \wedge q$, where p represents "I am a basketball fan" and q represents "I like rock music." The two left columns in Figure 2.7 list all the possible combinations of truth values for p and q. The right column shows the truth values for $p \wedge q$.

If you think carefully about this, you realize that the only way you are telling the truth is if you are a basketball fan and you also like rock music, that is, if *both* p and q are true. Take a moment to interpret lines 2, 3, and 4 of Figure 2.7 to convince yourself that in each of these cases, the original conjunction is false.

EXAMPLE 2   Finding the Truth Values of Conjunctions

Determine whether each statement is true or false.

a) Stephen King is a well-known writer and the Beatles were a famous rock band.

b) Michael Jordan played first base for the Atlanta Braves and he refuses to do TV commercials.

c) When you are 50 years old you will be a billionaire and Maine is larger than California.

SOLUTION: These statements are of the form $p \wedge q$, so we can use Figure 2.7 to determine their logical values.

a) In this case, both components of the conjunction are true, so by line 1 of Figure 2.7, the entire statement is true.

b) Since both component statements are false, using the fourth line of Figure 2.7, we see that the statement is false.

c) Of course we don't know if you will be a billionaire by age 50, so p is either true or false, but we don't know which. However, we do know that Maine is not larger than California. Therefore we are dealing with either line 2 or 4 of Figure 2.7. In either case the statement is false.

A disjunction is false only when both of its component parts are false.

p	q	$p \lor q$
T	T	T
T	F	T
F	T	T
F	F	F

FIGURE 2.8 The truth table for disjunction.

Mathematicians use the word *or* slightly differently than we use *or* in everyday life. Suppose that your mathematics instructor announces that there will be a quiz next Tuesday or Thursday. You come to class on Tuesday, have a quiz and, not expecting another quiz, you cut Thursday's class. When you return the following week, you find out that you missed a quiz on Thursday. You would probably feel that your instructor had misled you. The problem is that you misinterpreted your instructor's use of *or*. In everyday conversation we use what is called the **exclusive or**. Even though we do not explicitly say it, we understand this version of *or* to mean "one or the other, *but not both*." In mathematics and logic, we use the **inclusive or**, which is defined in Figure 2.8. Notice that on the first line of the table, when *p* is true and *q* is true, the statement $p \lor q$ is true, which differs from the way *or* is usually used in everyday life.

Notice that in Figures 2.7 and 2.8, each truth table has one line that differs from the others. The *and* connective is true only when *both* component statements are true. The *or* connective is false only when *both* component statements are false.

 SOME GOOD ADVICE

If you learn the truth tables for *and* and *or* by remembering the one different line in each table, it will help you do logic calculations more quickly.

As we mentioned in Section 1.1, similar notations often represent similar mathematical ideas. Notice that the *and* symbol, $\land$, in logic, looks somewhat like the intersection symbol, $\cap$, in set theory, and the *or* symbol, $\lor$, is similar to the union symbol, $\cup$. This is not a coincidence, because mathematicians classify set theory and logic as examples of Boolean algebras. Because of their similar structures, there are many parallel ideas that occur in logic and set theory. The following are some of these parallels:

Logic	Set Theory
and	intersection
or	union
not	complement
if . . . then	subset

Often if a result is true for set theory, then a similar result also holds for logic (and vice versa). As you read this chapter, think about how the ideas discussed here remind you of corresponding ideas in set theory.

EXAMPLE 3   Analyzing the Logic of an Advertisement

Consider the following help-wanted ad:

> "Management trainee wanted. Applicant must have four-year degree in accounting or three years of experience working in a financial institution."

Which applicants should be considered for the position if we interpret *or* according to Figure 2.8?

a) Aya has a four-year degree in accounting and has worked two years for a loan company.

b) Heidi has studied accounting in college (but did not graduate) and has worked for five years selling electronics.

c) Monte earned a four-year degree in accounting and has five years of experience working for a credit card company.

SOLUTION: a) Aya's background corresponds to line 2 of Figure 2.8, so she should be considered for the job.

b) Heidi's credentials are reflected in line 4 of Figure 2.8, so she is not qualified.

c) Monte's background corresponds to line 1 of Figure 2.8, so he should also be considered.

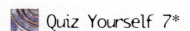

Quiz Yourself 7*

Determine whether each statement is true or false.

a) Exercise strengthens your bones or milk is a good source of calcium.

b) $10^2 < 0$ and $(5 - 3)^2 = 4$.

c) Chocolate is a vitamin or a mineral.

d) You will live to be 70 years old or the earth has a single moon.

We use truth tables to find the logical values of complex statements.

We can use truth tables to analyze the logical behavior of compound statements. However, just as we have to be careful about the order in which we do numerical calculations, we also have to pay attention to the order in which we perform logical operations.

EXAMPLE 4   Finding the Truth Table for a Compound Statement

Compute a truth table for a statement of the form $(\sim p \wedge q) \vee (p \wedge q)$.

SOLUTION: Because there are two variables, the truth table will have four lines. We first list truth values for p and q in the two left columns. Next, we write truth values for p and q underneath these variables wherever they occur in our statement. The parentheses tell us that we need to find the truth values for both $(\sim p \wedge q)$ and $(p \wedge q)$ before we can assign truth values to $\vee$. However, we must first compute truth values for $\sim p$ before we can find truth values for $(\sim p \wedge q)$.

		1		2		4	3		
p	*q*	(~	*p*	∧	*q*)	∨	(*p*	∧	*q*)
T	T	F	T	F	T	T	T	T	T
T	F	F	T	F	F	T	F	F	F
F	T	T	F	T	T	T	F	F	T
F	F	T	F	F	F	F	F	F	F

The numbers across the top of the table indicate the order in which we will compute the truth values. In step 1, we negate p's truth values and then use these values in step 2 to compute truth values for $(\sim p \wedge q)$. Notice that we get a T only on the third line of step 2, since that is the only line in which we are joining two T's with the $\wedge$.

We use the truth values for $(\sim p \wedge q)$ in step 2 and the truth values for $(p \wedge q)$ in step 3 (highlighted) to compute the truth values for the disjunction in step 4. We can do this quickly by looking for the situation $F \vee F$, which is the only case in which the disjunction is false. We will denote the final truth values for this truth table with boldface type.*

		1		2		4	3		
p	*q*	(~	*p*	∧	*q*)	∨	(*p*	∧	*q*)
T	T	F	T	F	T	**T**	T	T	T
T	F	F	T	F	F	**F**	T	F	F
F	T	T	F	T	T	**T**	F	F	T
F	F	T	F	F	F	**F**	F	F	F

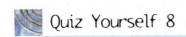

Quiz Yourself 8

Construct a truth table for the statement $(p \wedge \sim q) \vee (\sim p \vee q)$.

You can now use this truth table to find truth values for statements of the form $(\sim p \wedge q) \vee (p \wedge q)$. For example, if p is false and q is true, then the third line of the table tells you that $(\sim p \wedge q) \vee (p \wedge q)$ is true.   ☺

We could have saved time in step 1 of the table in Example 4 if we directly wrote the truth values for $\sim p$ instead of first writing the values for p and then negating them.

In the final column for the truth table in Quiz Yourself 8 you got all T's. A statement that is always true is called a **tautology**.

Because we compute truth tables so frequently, it is important to set them up efficiently. As you have seen, if a statement has a single variable such as p, then the truth table will have two lines because p can be either T or F. If a statement has two variables, there will be four lines corresponding to the cases TT, TF, FT, and FF. In general, the following pattern is true.

> If the statement has k variables, then its truth table will have 2^k lines.

A statement with the variables p, q, and r will have $2^3 = 8$ lines. To fill in the columns quickly under p, q, and r, first place 4 T's and then 4 F's under p. Next place 2 T's, 2 F's, 2 T's, and 2 F's under q. Finally place TFTFTFTF under r. You will see this pattern in Example 5.

EXAMPLE 5   **Constructing a Truth Table for a Statement with Three Variables**

Construct a truth table for $(\sim p \wedge q) \vee (\sim r)$.

---

* We will show you an alternative method for computing truth tables in Example 8 in this section.

Highlight: *The Death of Leibniz's Dream*

For hundreds of years, mathematicians pursued Leibniz's dream of finding a universal method of symbolic reasoning. Indeed, David Hilbert, the greatest mathematician of the early twentieth century, thought that it was possible to develop logical machinery for building all of mathematics. In the 1920s, two British mathematicians, Bertrand Russell and Alfred North Whitehead, tried to develop such a plan; however, these hopes were dashed in 1931 when Kurt Gödel, a mathematician at Princeton University, discovered one of the most remarkable results in all of the history of mathematics. His result, called Gödel's Incompleteness Theorem, showed that by starting with a finite set of assumptions and then proving theorems in a mathematical theory such as elementary arithmetic, there will always exist statements that are true* but that cannot be proved using the theory! This result was a disappointment to those mathematicians who had believed that all mathematics could be developed from logic alone. Leibniz's dream was finally dead.

SOLUTION: Again the parentheses tell us that we must compute $(\sim p \wedge q)$ and also $(\sim r)$ before we can compute the truth values for the disjunction $\vee$. In order to speed up the process, in step 1 we write the truth values directly for $(\sim p)$. Next, in step 2, we use the truth values for $(\sim p)$ and q to calculate the values for $\wedge$. In step 3, we write the truth values for $(\sim r)$. We finish the table by using the results from steps 2 and 3 (highlighted) to find the values for the disjunction in step 4.

			1	2		4	3
p	q	r	$(\sim p$	$\wedge$	$q)$	$\vee$	$(\sim r)$
T	T	T	F	F	T	**F**	F
T	T	F	F	F	T	**T**	T
T	F	T	F	F	F	**F**	F
T	F	F	F	F	F	**T**	T
F	T	T	T	T	T	**T**	F
F	T	F	T	T	T	**T**	T
F	F	T	T	F	F	**F**	F
F	F	F	T	F	F	**T**	T

Logically equivalent statements express the same meaning.

It is easy to confuse statements that seem similar but may, in fact, mean very different things. Therefore, it is important for us to know when statements that look different in logic express exactly the same information and when they do not. We use the following criterion to decide whether two statements mean the same thing.

---

* It would be interesting if we could give you an example of a result that cannot be proved. However, such an example would be very abstract and beyond the scope of this text.

> **DEFINITION**
>
> Two statements are **logically equivalent** if they have the same variables and, when their truth tables are computed, the final columns in the tables are identical.

EXAMPLE 6   Determining When Statements Mean the Same Thing

Coupon SAVE!

Assume that you have an entertainment book containing coupons for discounts for movies, restaurants, and other leisure activities. You are considering eating at either the Pasta Bar or the Deli. Do the following two statements say the same thing?

a) It is not true that: the Pasta Bar accepts coupons and the Deli accepts coupons.

b) The Pasta Bar does not accept coupons or the Deli does not accept coupons.

SOLUTION: If we let p represent "The Pasta Bar accepts coupons" and let d represent "The Deli accepts coupons," then we can write (a) symbolically as

$$\sim (p \wedge d)$$

and (b) as

$$(\sim p) \vee (\sim d).$$

To decide whether these statements say the same thing, all we have to do is make two truth tables.

		2	1			1	3	2
p	d	$\sim$	$(p$	$\wedge$	$d)$	$(\sim p)$	$\vee$	$\sim d$
T	T	**F**	T	T	T	F	**F**	F
T	F	**T**	T	F	F	F	**T**	T
F	T	**T**	F	F	T	T	**T**	F
F	F	**T**	F	F	F	T	**T**	T

Since the final columns (highlighted) in the two truth tables are identical, the two statements are logically equivalent and therefore express exactly the same information.

It is easier to remember mathematical facts if we make connections between them and see analogies. Notice that in Example 6, the logical equivalence of $\sim (p \wedge d)$ and $(\sim p) \vee (\sim d)$ looks very similar to one of DeMorgan's Laws in set theory, namely, $(A \cap B)' = A' \cup B'$. You should expect to see such a parallel because logic and set theory are both Boolean algebras.

Quiz Yourself 9

Prove that $\sim(p \lor q)$ is logically equivalent to $(\sim p) \land (\sim q)$.

DeMorgan's Laws for Logic

If p and q are statements, then:
a) $\sim(p \land q)$ is logically equivalent to $(\sim p) \lor (\sim q)$
b) $\sim(p \lor q)$ is logically equivalent to $(\sim p) \land (\sim q)$

By defining the idea of logical equivalence, we have taken a large step toward treating logic algebraically. In algebra, when we say $2(x + y) = 2x + 2y$, we are making a general statement that holds for an infinite number of specific cases. For example, we can substitute $x = 3$ and $y = 5$ to get the true statement $2(3 + 5) = 2 \cdot 3 + 2 \cdot 5$. Similarly, when we say that $\sim(p \land q)$ is logically equivalent to $(\sim p) \lor (\sim q)$, we are saying that an infinite number of pairs of English language statements are equivalent. For example, if p is the statement "Today is Tuesday" and q is the statement "It is raining," then the following two statements have the same meaning:

"It is not true that: today is Tuesday and it is raining."
"Today is not Tuesday or it is not raining."

Now that we have the notion of logical equivalence, we can reason about the meaning of complex statements based on their form rather than their content. Documents such as apartment leases, warranties, car loan applications, college admission forms, and income tax instructions often are written in a legalistic style using the connectives you have been studying.

EXAMPLE 7 Applying Logic to Legal Documents

Use DeMorgan's Laws to rewrite the following statement, which is based on instructions for filing form 1040-A with the U.S. Internal Revenue Service.

> It is false that: you received interest from a seller-financed mortgage and the buyer used the property as a personal residence.

SOLUTION: It is easy to rewrite this statement if first we represent it in symbolic form. Let r represent "You received interest from a seller-financed mortgage" and let b represent "The buyer used the property as a personal residence." This statement has the form $\sim(r \land b)$.

By DeMorgan's Laws this statement is equivalent to $(\sim r) \lor (\sim b)$. We can now rewrite this in English as "You did not receive interest from a seller-financed mortgage or the buyer did not use the property as a personal residence."

 Quiz Yourself 10

Use one of DeMorgan's Laws to rewrite (in English) the following statement taken from a car warranty:

The car is not more than five years old and has not been driven over 50,000 miles.

SOME GOOD ADVICE

When you use logic to rewrite a statement, the result can sound awkward. You may wish to smooth out the grammar so that the sentence sounds better. *This is usually a mistake!* Unless you are quite careful, you can easily change the meaning of a sentence by rewriting it. In logic, the form of a statement is more important than its literary style.

■

There is an alternative way to construct truth tables.

Some people prefer to use another method to construct truth tables which we will explain in Example 8. We will use this method to recalculate the truth tables in Examples 4 and 5.

EXAMPLE 8   Using an Alternative Method to Construct Truth Tables

a) We will construct a truth table for $(\sim p \wedge q) \vee (p \wedge q)$. Recall from Example 4 that first we calculated truth values for $\sim p$. In the second step we found truth values for $\sim p \wedge q$. Then we computed values for $p \wedge q$, and finally, we calculated the truth values for the disjunction $(\sim p \wedge q) \vee (p \wedge q)$.

In the following truth table, we will make separate columns for each of these four steps. Notice that the truth values in the fourth step are exactly the same as those you obtained in the final step of Example 4.

		1	2	3	4
p	q	$\sim p$	$\sim p \wedge q$	$p \wedge q$	$(\sim p \wedge q) \vee (p \wedge q)$
T	T	F	F	T	T
T	F	F	F	F	F
F	T	T	T	F	T
F	F	T	F	F	F

b) We will now construct a truth table for $(\sim p \wedge q) \vee (\sim r)$. Recall from Example 5 that first we calculated truth values for $\sim p$. Second, we found values for $\sim p \wedge q$. In the third step, we computed truth values for $\sim r$. We then used the values from steps 2 and 3 to find the truth values for the disjunction, $\vee$.

As in part a), we will make separate columns in the truth table for each of these steps.

			1	2	3	4
p	q	r	$\sim p$	$\sim p \wedge q$	$\sim r$	$(\sim p \wedge q) \vee (\sim r)$
T	T	T	F	F	F	F
T	T	F	F	F	T	T
T	F	T	F	F	F	F
T	F	F	F	F	T	T
F	T	T	T	T	F	T
F	T	F	T	T	T	T
F	F	T	T	F	F	F
F	F	F	T	F	T	T

Exercises 2.3

In Exercises 1–8, if p represents a true statement and q represents a false statement, what is the truth value of each statement?

1. $p \wedge q$

2. $p \vee q$

3. $p \wedge (\sim q)$

4. $(\sim p) \wedge q$

5. $\sim (p \vee q)$

6. $\sim (p \wedge q)$

7. $\sim (\sim p \wedge q)$

8. $\sim (p \vee \sim q)$

In Exercises 9–12, use numbers to specify the order in which you would perform the logical operations for each statement.

9. $\sim (p \vee \sim q)$

10. $\sim (\sim p \wedge q)$

11. $p \wedge \sim (p \vee \sim q)$

12. $(p \wedge \sim q) \vee \sim p$

For Exercises 13–16, fill in the missing values in the following truth table.

		2	1	5	3	4		
p	q	(p	∧	~q)	∨	(~p	∨	q)
T	T	T	F	F	T	F	T	T
T	F	T	T	**13.__**	T	F	**15.__**	F
F	T	F	F	F	T	T	T	T
F	F	F	**14.__**	T	**16.__**	T	T	F

For Exercises 17–20, fill in the missing values in the following truth table.

			2	1	4		3		
p	q	r	(p	∧	~r)	∨	(q	∨	r)
T	T	T	T	F	F	**19.__**	T	T	T
T	T	F	T	T	T	T	T	T	F
T	F	T	T	F	**18.__**	T	F	T	T
T	F	F	T	T	T	T	F	F	F
F	T	T	F	F	F	T	T	T	T
F	T	F	F	**17.__**	T	T	T	T	F
F	F	T	F	F	F	T	F	T	T
F	F	F	F	F	T	**20.__**	F	F	F

In Exercises 21–22, determine how many lines will be in the truth table for each statement.

21. $(p \vee q) \wedge (r \vee s) \wedge t$

22. $p \wedge q \wedge r \wedge s \wedge t \wedge u$

In Exercises 23–26, state whether each number is a possibility for the number of the lines in a truth table.

23. 16

24. 36

25. 9

26. 32

27. Communicating Mathematics   Explain the purpose of having eight lines in the truth table for a statement having three variables *p*, *q*, and *r*.

28. Communicating Mathematics   Why does the number of lines in a truth table double if we increase the number of variables in the statement by one?

In Exercises 29–40, construct a truth table for each statement.

29. $p \wedge \sim q$

30. $\sim p \wedge q$

31. $\sim (\sim p \wedge q)$

32. $\sim (p \vee \sim q)$

33. $\sim (p \vee q) \wedge \sim (p \wedge q)$

34. $\sim (p \wedge q) \vee \sim (p \vee q)$

35. $(p \vee r) \wedge (p \wedge \sim q)$

36. $(p \wedge r) \vee (p \wedge \sim q)$

37. $(p \vee q) \wedge (p \vee r)$

38. $(p \wedge q) \vee (p \wedge r)$

39. $\sim (p \vee \sim q) \wedge r$

40. $(p \wedge \sim q) \vee \sim r$

In Exercises 41–44, determine whether we are using the inclusive or or the exclusive or in each sentence.

41. Pay me now or pay me later.

42. You will earn tax rebate if your income is under $23,500 or if you are over 65.

43. The warranty is void if the appliance has been abused or improperly maintained.

44. The penalty is a $500 fine or 40 hours of community service.

In Exercises 45–52, use DeMorgan's Laws to rewrite the negation of each statement.

45. Bill is tall and thin.

46. Yorrel will go to law school or pursue an MBA.

47. Christian will apply for either a loan or work study.

48. Joanna will quit her job and join the Peace Corps.

49. Ken qualifies for a rebate or a reduced interest rate.

50. Pei Li got a larger disk drive and the extended warranty with her computer.

51. The number x is not equal to 5 and s is not odd.

52. The number y divides 10 but it is not even.

The statements in Exercises 53–56 are excerpts from the instructions for filing the Form 1040-A with the Internal Revenue Service. Rewrite them, as we did in Example 7.

53. The earned income tax did not reduce the tax you owe or did not give you a refund.

54. You cannot claim yourself or your spouse as a dependent.

55. You are not single and not the head of a household.

56. It is not true that: you are filing a joint return and are covered by a retirement plan at work.

57. Communicating Mathematics   Why did we introduce the idea of logically equivalent statements?

58. Communicating Mathematics   What advantage do you see in using truth tables to determine logical equivalence rather than using our "common sense" and knowledge of the English language?

In Exercises 59–66, determine which pairs of statements are logically equivalent.

59. $\sim(p \wedge \sim q)$, $\quad\quad\quad\quad (\sim p) \vee q$

60. $\sim(p \vee \sim q)$, $\quad\quad\quad\quad (\sim p) \wedge q$

61. $\sim(p \vee \sim q) \wedge \sim(p \vee q)$, $\quad p \vee (p \wedge q)$

62. $\sim(\sim p \vee \sim q)$, $\quad\quad\quad p \vee q$

63. $\sim(p \vee \sim q) \wedge \sim(p \vee q)$, $\quad (\sim p \wedge q) \wedge (\sim p \wedge \sim q)$

64. $(p \wedge \sim q) \vee \sim(p \wedge q)$, $\quad \sim(p \vee \sim q) \vee (\sim p \vee \sim q)$

65. $p \vee (\sim q \wedge r)$, $\quad\quad\quad (p \vee (\sim q)) \wedge (p \vee r)$

66. $p \wedge (\sim q \vee \sim r)$, $\quad\quad (p \wedge (\sim q)) \vee (p \wedge \sim r)$

Recall from Exercises 61–68 in Section 2.2 that we can represent electrical circuits by logical forms and vice versa. In Exercises 67–70, represent each circuit by a logical form and then rewrite that logical form in an equivalent form. Use truth tables to prove that the second form is equivalent to the first. Draw a circuit that corresponds to the second logical form. If possible, try to choose the second form so that the corresponding circuit has fewer switches than the original circuit.

67.

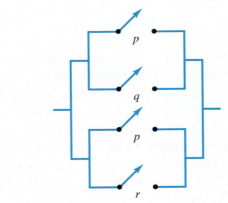

68.

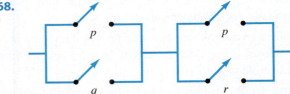

69.

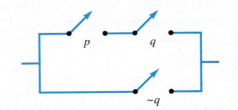

70.

Further Exercises

Because set theory and logic are Boolean algebras, for each result in set theory, we can expect to find a similar result in logic. In Exercises 71–74, give a statement in logic that is similar to each statement in set theory.

71. $A \cap (B \cap C) = (A \cap B) \cap C$

72. $A \cap (B \cup C) = (A \cap B) \cup (A \cap C)$

73. $A \cup (B \cap C) = (A \cup B) \cap (A \cup C)$

74. $A - (B \cup C) = (A - B) \cap (A - C)$

We stated without proof that a truth table for a logical statement with k variables has 2^k lines. The following tree diagram* shows the four lines that are possible for a statement that has variables p and q.

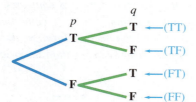

† **75.** Draw a tree diagram to illustrate the eight possibilities for a statement that has variables p, q, and r.

76. Communicating Mathematics   Without drawing a tree diagram, explain why the truth table for a statement with four logical variables has sixteen lines. Explain why the truth table for a statement with k variables has 2^k lines.

77. Communicating Mathematics   In set theory we showed that a set with k elements has 2^k subsets. If a logical statement has variables p, q, and r, explain how the eight subsets of $\{p, q, r\}$ correspond to the eight lines of the truth table for this statement.

78. Make a truth table for $(p \vee q) \wedge (p \vee r)$. Show a correspondence between the subsets of $\{p, q, r\}$ and the lines of the truth table.

79. Show that the *and* connective is unnecessary in the sense that any statement of the form $(p \wedge q)$ can be written in an equivalent form that does not use $\wedge$.

80. Show that the *or* connective is unnecessary in the sense that any statement of the form $(p \vee q)$ can be written in an equivalent form that does not use $\vee$.

The stroke connective‡ has the following truth table:

p	q	$p \mid q$
T	T	F
T	F	T
F	T	T
F	F	T

81. The stroke connective is sometimes called NAND (*not and*). Explain why this is so.

82. Prove that $\sim p$ is logically equivalent to $p \mid p$.

83. Prove that $p \vee q$ is logically equivalent to $(p \mid p) \mid (q \mid q)$.

84. Give a form using only the stroke connective that is equivalent to $p \wedge q$.

85. Research the life of George Boole and write a brief report on your findings.

86. Research the life of Augustus DeMorgan and write a brief report on your findings.

2.4   THE CONDITIONAL AND BICONDITIONAL

Conditional statements are as common in everyday language as they are in mathematics and logic. A software manufacturer tells you, "If the seal is broken, then the product cannot be returned." Or, an amusement park may have a rule that states, "If you are not 48 inches tall, then you cannot ride this roller coaster."

In this section, we will construct truth tables for conditionals and biconditionals that you can use to find truth values of compound statements that use these connectives. As we analyze these statements, notice that sometimes we use conditionals slightly differently in mathematics than in everyday language.

---

* For a review of tree diagrams, see Section 1.1.
† Exercise numbers circled in red can be used as group exercises.
‡ The stroke connective does not correspond to any single English word.

There is only one way a conditional can be false.

In order to understand when a conditional statement is true or false, consider this example. Mr. Rich, the owner of a small factory, has a rush order that must be filled by next Monday and he approaches you with this generous offer:

If you work for me on Saturday, then I'll give you a $100 bonus.

If we let w represent "You work for me on Saturday" and b represent "I'll give you a $100 bonus," then this statement has the form $w \rightarrow b$. We must examine four cases to determine exactly when Mr. Rich is telling the truth and when he is not.

Case 1 (w is true and b is true):

You come to work and you receive the bonus.

In this case, Mr. Rich certainly made a truthful statement.

Case 2 (w is true and b is false):

You come to work and you don't receive the bonus.

Mr. Rich has gone back on his promise, so he has made a false statement.

Case 3 (w is false and b is true):

You don't come to work, but Mr. Rich gives you the bonus anyway.

Think carefully about this case, because it differs from the way we tend to use *if . . . then* in everyday language. When you listen to Mr. Rich, do not read more into his statement than he actually said. You do not expect to get the bonus if you did not come to work because that is your experience in everyday life. However, Mr. Rich never said that. *You are assuming this condition.* Remember that in logic a statement is either true or false. Therefore because Mr. Rich did not say something false, he has told the truth.

Case 4 (w is false and b is false):

You don't come to work and you don't receive the bonus.

In this case, Mr. Rich is telling the truth for exactly the same reason as in Case 3. Since you did not come to work, Mr. Rich can give you the bonus or not give you the bonus. In either case he has not told a falsehood and therefore is telling the truth.

This discussion explains the truth table for the *if . . . then* connective in Figure 2.9.

p	q	$p \rightarrow q$
T	T	T
T	F	F
F	T	T
F	F	T

FIGURE 2.9   Truth table for the conditional.

DEFINITIONS

For a conditional $p \rightarrow q$, statement p is called the *hypothesis* and q is called the *conclusion.**

---

* Some call the hypothesis the *antecedent* and the conclusion the *consequent*.

SOME GOOD ADVICE

A good way to remember the truth table for the conditional is that the *only way a conditional can be false* is if it has a true hypothesis and a false conclusion.

We can build truth tables for statements that combine the conditional with the connectives you have studied earlier.

EXAMPLE 1   Constructing a Truth Table for a Complex Conditional Statement

Construct a truth table for the statement $(\sim p \vee q) \rightarrow (\sim p \wedge \sim q)$.

SOLUTION: Because there are two variables, p and q, the truth table has four lines. After carrying out steps 1 through 5, the values for the hypothesis of the conditional are under the *or* symbol of step 2 (highlighted) and the values for the conclusion are under the *and* symbol of step 5 (highlighted).

In step 6, lines 1 and 3 are the only cases with a true hypothesis and a false conclusion, so we put an F on those lines. The final truth table then looks like this:

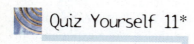

Quiz Yourself 11*

Construct a truth table for the statement $(p \wedge \sim q) \rightarrow (\sim p \vee q)$.

		1	2		6	3	5	4
p	q	$(\sim p$	$\vee$	$q)$	$\rightarrow$	$(\sim p$	$\wedge$	$\sim q)$
T	T	F	T	T	**F**	F	F	F
T	F	F	F	F	**T**	F	F	T
F	T	T	T	T	**F**	T	F	F
F	F	T	T	F	**T**	T	T	T

Many formal and informal contracts use the *if . . . then* connective. We can understand a condition in an agreement by writing it in a symbolic form.

EXAMPLE 2   Analyzing an Agreement

A boutique has the following policy:

> *If* you return clothing within 30 days with the original labels on it,
> *Then* you will be given a full refund on your purchase.

Shandra returned her suit with its original labels, but was not given a refund. What can you deduce from this information?

Historical Highlight: *Polish Logicians*

Logic, like so many areas of mathematics, has been enriched by the contributions of mathematicians from many parts of the world. For example, in Eastern Europe, Kazimierz Twardowski and his student Jan Lukasiewicz developed a school of logic that flourished in Poland in the early part of the twentieth century.

In a paper criticizing some work of Aristotle, Lukasiewicz made basic ideas of logic available in the Polish language, and by doing so prompted many mathematics students at the University of Warsaw to specialize in logic. Among these was Alfred Tarski, who is generally recognized as one of the principal logicians of the twentieth century. In Krakow, Father Jan Salamucha formed a group that attempted to modernize Catholic dogma using mathematical logic.

Unfortunately, World War II ended the golden age of Polish logic. During this period, Tarski went to America, and Salamucha and a number of Jewish logicians were murdered. Lukasiewicz and others who survived the war did not return to Poland after the war ended.

SOLUTION: Let r, l, and f represent the following statements:

r : You return the clothing within 30 days.
l : The clothing has its original labels.
f : You will be given a full refund on your purchase.

We can see that the store's policy has the form $r \wedge l \rightarrow f$.

Since Shandra was not given the refund, that means that the conclusion f was false. For the boutique's policy to be true, this means that the hypothesis $r \wedge l$ must also be false. The hypothesis is an *and* statement, so in order for it to be false, one of the two component parts, r or l, must be false. We know that the clothing has the original labels on it, so l is true; therefore, r is false. The clothing was not returned within 30 days.

The converse, inverse, and contrapositive are three derived forms of a conditional.

There are three typical ways that you might wish to rephrase a conditional; however, always remember that when you rewrite a statement in logic, you also may be changing its meaning.

DEFINITIONS

We can derive the following statements from the conditional $p \rightarrow q$.

The *converse* has the form $q \rightarrow p$.
The *inverse* has the form $\sim p \rightarrow \sim q$.
The *contrapositive* has the form $\sim q \rightarrow \sim p$.

According to this definition, you form the converse of a conditional by interchanging its hypothesis and conclusion. You form the inverse by keeping the hypothesis and conclusion in the same positions as in the original conditional, but you must negate them both. To form the contrapositive, you negate both the hypothesis and conclusion and also switch their positions.

SOME GOOD ADVICE

Before you try to rephrase a statement in words, you should first try to write the statement symbolically. This makes it easier to understand the precise form of the statement before rewriting it in English.

EXAMPLE 3   **Rewriting the Converse, Inverse, and Contrapositive of Statements in Words**

Write, in words, the converse, inverse, and contrapositive of the statement

"If marijuana is legalized, then drug abuse will increase."

SOLUTION: We can write this statement symbolically as $m \rightarrow d$, where

m stands for "Marijuana is legalized"

and

d stands for "Drug abuse will increase."

We first write each of the derived forms symbolically.
The converse has the form $d \rightarrow m$. Translating this into words, we get

"If drug abuse will increase, then marijuana is legalized."

The form of the inverse is $\sim m \rightarrow \sim d$, which we can write as

"If marijuana is not legalized, then drug abuse will not increase."

The form $\sim d \rightarrow \sim m$ represents the contrapositive. We write this as

"If drug abuse will not increase, then marijuana is not legalized."

You must be careful when you rephrase a conditional because the derived statement may not be logically equivalent to the original. We now investigate which derived forms of a conditional are logically equivalent.

Quiz Yourself 12

Write, in words, the inverse and contrapositive of the statement

"If the Federal Reserve raises the prime rate, then interest rates will increase."

EXAMPLE 4   **Equivalence of the Derived Forms of a Conditional**

Which of the statements $p \rightarrow q$, $q \rightarrow p$, $\sim p \rightarrow \sim q$, and $\sim q \rightarrow \sim p$ are logically equivalent?

SOLUTION: We will answer this question by comparing truth tables for these statements.

		conditional	converse			inverse			contrapositive		
p	*q*	$p \rightarrow q$	*q*	$\rightarrow$	*p*	~*p*	$\rightarrow$	~*q*	~*q*	$\rightarrow$	~*p*
T	T	T	T	T	T	F	T	F	F	T	F
T	F	F	F	T	T	F	T	T	T	F	F
F	T	T	T	F	F	T	F	F	F	T	T
F	F	T	F	T	F	T	T	T	T	T	T

Because the final columns under the conditional and contrapositive are identical, these statements are logically equivalent. We also see that neither the converse nor the inverse are equivalent to the original conditional; however, they are equivalent to each other.

Example 5 shows how to rephrase the types of statements found in leases, contracts, warranties, and tax forms without changing their meaning.

EXAMPLE 5   Rewriting Legal Statements

Rewrite the contrapositive of each of the following statements in words.

a) If you wish to contribute $3 to the presidential campaign fund, then check box (b) below.

b) If you do not itemize deductions, then you will be allowed to take the standard deduction.

SOLUTION: In each case we form the contrapositive by negating both the hypothesis and conclusion and then interchanging them.

a) If you do not check box (b) below, then you do not wish to contribute $3 to the presidential campaign fund.

b) If you are not allowed to take the standard deduction, then you do itemize deductions.

There are many ways to express a conditional without using the words *if . . . then*. Each of the following forms expresses the conditional "if *p* then *q*."

q if *p*	Here the *if* still is associated with *p* even though it occurs later in the sentence.
p only if *q*	Recognize that *only if* does not say the same thing as *if*. The *if* condition is the hypothesis; the *only if* condition is the conclusion.
p is sufficient for *q*	The sufficient condition is the hypothesis.
q is necessary for *p*	The necessary condition is the conclusion.

EXAMPLE 6 **Rewriting Statements in** *If . . . Then* **Form**

Write each statement in *if . . . then* form.

a) Your driver's license will be suspended if you are convicted of driving under the influence of alcohol.

b) You will graduate only if you have a 2.5 grade point average.

c) In order to hold your room reservation, it is sufficient to give us your credit card number.

d) In order to qualify for a discount on your airline tickets, it is necessary to pay for them two weeks in advance.

SOLUTION: a) Because the *if* goes with the clause "you are convicted of driving . . . ," that clause is the hypothesis. We can rewrite this sentence as

"If you are convicted of driving under the influence of alcohol, then your driver's license will be suspended."

b) The *only if* goes with the clause "you have a 2.5 grade point average"; therefore, this is the conclusion. We write this sentence as

"If you graduate, then you have a 2.5 grade point average."

c) The phrase *it is sufficient* identifies the hypothesis. We write the sentence as

"If you give us your credit card number, then [we will] hold your room reservation."

d) The necessary condition is the conclusion. Rewriting this sentence, we get

"If you qualify for a discount on your airline tickets then you pay for them two weeks in advance."

Quiz Yourself 13

Write each statement using the *if . . . then* connective.

a) In order to travel to Europe, it is necessary to update your immunization.

b) You will increase your cardiovascular fitness only if you exercise three times a week.

The biconditional means that two statements say the same thing.

p	q	$p \leftrightarrow q$
T	T	T
T	F	F
F	T	F
F	F	T

FIGURE 2.10 Truth table for the biconditional.

The biconditional (*if and only if*) indicates that two statements mean the same thing. For example, we could say that "Today is Tuesday if and only if tomorrow is Wednesday." In algebra we say "$x + 3 = 7$ if and only if $x = 4$." The notation for the biconditional, which is $p \leftrightarrow q$, suggests that we are saying $p \rightarrow q$ and $q \rightarrow p$ at the same time. We show the truth table for the biconditional in Figure 2.10.

You should use truth tables to verify that the biconditional $p \leftrightarrow q$ is logically equivalent to the statement $(p \rightarrow q) \wedge (q \rightarrow p)$.

EXAMPLE 7 **Computing a Truth Table for a Complex Biconditional**

Construct a truth table for the statement $\sim(p \vee q) \leftrightarrow (\sim q \wedge p)$.

Highlight: *Artificial Intelligence and Logic*

"As if driven by some invisible hand, humans have always yearned to understand what makes them think, feel and be, and have tried to re-create that interior life artificially."* This statement describes an area of computer science that relies heavily on logic, known as artificial intelligence (AI). In the past several decades, as computers have become more powerful, scientists have intensified research into how to design computers and computer programs that can imitate human intelligence.

Although computer scientists have made much progress in AI, they are not yet close to duplicating the functions of our remarkable brains. According to one estimation, the calculating power of a desktop computer equals the intelligence of a snail, while the world's largest, most powerful computer might be capable of matching the intelligence of a small rodent.†

The human brain has about 100 billion neurons connected by about a million billion synapses that collectively fire about 10 million billion times per second. To build a computer that is comparable to the human brain, with current technology, would require enough power to light an entire town. Yet our brains use less electricity than a standard light bulb.

Although researchers have not yet reached the goal of building machines with human intelligence, AI has become a multibillion-dollar industry with many commercial applications. Most large industries now use AI in forecasting, planning, scheduling, and manufacturing.

SOLUTION:

		2	1	5	3	4
p	**q**	(~	(p ∨ q))	↔	(~q	∧	p)
T	T	F	T	**T**	F	F	T
T	F	F	T	**F**	T	T	T
F	T	F	T	**T**	F	F	F
F	F	T	F	**F**	T	F	F

Another way to verify that two statements are logically equivalent is to join them with a biconditional and see whether this new statement is a tautology. We will investigate this further in the exercises.

Exercises 2.4

In Exercises 1–8, assume that p *represents a true statement,* q *a false statement, and* r *a true statement. Determine the truth value of each statement.*‡

1. $\sim(p \vee q) \rightarrow \sim p$

2. $(p \wedge \sim q) \rightarrow q$

3. $(p \wedge q) \rightarrow (q \vee r)$

4. $(p \vee \sim q) \rightarrow r$

5. $(\sim p \vee \sim q) \rightarrow r$

6. $r \rightarrow (\sim p \wedge q)$

7. $\sim(\sim p \wedge q) \rightarrow \sim r$

8. $\sim(\sim p \vee r) \rightarrow \sim q$

---

* Daniel Crevier, *AI* (Basic Books, 1993), p. 1.
† David Freeman, *BrainMakers* (Simon & Schuster, 1994), p. 98.
‡ In interpreting these expressions, assume that the negation symbol affects as little of the expression as possible. For example, you should interpret $\sim p \wedge q$ as meaning $(\sim p) \wedge q$ rather than $\sim(p \wedge q)$.

In Exercises 9–20, construct a truth table for each statement.

9. $p \rightarrow \sim q$

10. $\sim p \rightarrow q$

11. $\sim (p \rightarrow q)$

12. $\sim (q \rightarrow p)$

13. $(p \vee r) \rightarrow (p \wedge \sim q)$

14. $(p \wedge q) \rightarrow (p \wedge \sim r)$

15. $\sim (p \vee r) \rightarrow \sim (p \wedge q)$

16. $\sim (p \wedge r) \rightarrow \sim (p \vee q)$

17. $(p \vee q) \leftrightarrow (p \vee r)$

18. $(p \wedge q) \leftrightarrow (p \wedge r)$

19. $(\sim p \rightarrow q) \leftrightarrow (\sim q \rightarrow p)$

20. $(p \rightarrow \sim q) \leftrightarrow (q \rightarrow \sim p)$

In Exercises 21–28, write in words the converse, inverse, or contrapositive as indicated for each statement.

21. If it rains, it pours. (converse)

22. If this appliance fails within 30 days, then it will be fixed free of charge. (converse)

23. If you buy the all-weather radial tires, then they will last for 80,000 miles. (inverse)

24. If you are over 18, then you must register with Selective Service. (inverse)

25. If a geometric figure is an equilateral triangle, then its sides are all equal in length. (contrapositive)

26. If a geometric figure is a quadrilateral, then the sum of its interior angles is 180 degrees. (contrapositive)

27. If x evenly divides 6, then x evenly divides 9. (inverse)

28. If x is an even prime number, then x is divisible by 2. (inverse)

In Exercises 29–32, write the converse, inverse, or contrapositive of each conditional statement based on instructions for filing Form 1040-A with the Internal Revenue Service.

29. Interpreting tax forms. If your gross income is over \$2,250, then you cannot be claimed by someone else as a dependent. (contrapositive)

30. Interpreting tax forms. If you are filing a joint return, then include your spouse's income. (inverse)

31. Interpreting tax forms. If the amount you overpaid is large, then decrease the amount being withheld from your pay. (converse)

32. Interpreting tax forms. If you are a nonresident alien, then you cannot claim an earned income tax credit. (contrapositive)

In Exercises 33–36, write the converse, inverse, and contrapositive of each statement in symbolic form.

33. $(\sim p) \rightarrow q$

34. $p \rightarrow \sim q$

35. $(\sim p) \rightarrow \sim (q \wedge r)$

36. $\sim (p \vee r) \rightarrow q$

In Exercises 37–40, determine which pairs of statements are equivalent. (It is helpful to first write the statements in symbolic form.)

37. If you activate your cell phone before October 1, then you receive 100 free minutes. If you do not receive 100 free minutes, then you do not activate your cell phone before October 1.

38. If you don't register for the conference before August 1, then you must pay a \$30 late fee. If you register for the conference before August 1, then you do not pay a \$30 late fee.

39. If it is raining, then use your headlights. If you use your headlights, then it is raining.

40. If you do not score high in the figure skating compulsories, then you will not qualify for a medal. If you qualify for a medal, then you score high in the figure skating compulsories.

In Exercises 41–48, rewrite each statement using the words if . . . then.

41. I'll take a break if I finish my workout.

42. You can return the video game only if you have not opened the package.

43. To qualify for this deduction, it is necessary for you to complete Form 3093.

44. To reserve a campsite, it is sufficient that you pay a small deposit.

45. You will receive a free cell phone only if you sign up before March 1.

46. I'll go to Florida if I can save \$850.

47. To get a reduction on your auto insurance, it is sufficient that you remain accident-free for three years.

48. To graduate this semester, it is necessary that you complete eighteen credits.

For Exercises 49–54, it is useful to study Example 4 carefully.

49. Communicating Mathematics Give an example of a true conditional statement in words whose converse is also true.

50. Communicating Mathematics Give an example of a false conditional statement in words whose converse is true.

51. Communicating Mathematics Is it possible to give an example of a false conditional statement in words whose converse is also false? Explain.

52. Communicating Mathematics Prove that a biconditional is logically equivalent to the conjunction of two conditionals.

53. Form a symbolic statement involving a disjunction that is equivalent to $p \rightarrow q$.

54. Form a symbolic statement involving a conjunction that is equivalent to $\sim(p \rightarrow q)$.

In Exercises 55–56, assume that a credit card company has the following policy:

If you have a platinum credit card with an outstanding balance of over \$1,000, or you have been a member for at least ten years, then you qualify for the discount loan rate.

Represent the conditions stated in this policy as follows:

$p =$ You have a platinum credit card.
$b =$ You have a balance over \$1,000.
$m =$ You have been a member for at least ten years.
$d =$ You qualify for the discount loan rate.

Based on the partial information given, what can you deduce about each situation? Explain your thinking. (Hint: Recognize that the policy can be written in symbolic form as $((p \wedge b) \vee m) \rightarrow d$.)

*55. Communicating Mathematics Jamie is a platinum credit card member and has an outstanding balance of \$750. He does not qualify for the discount loan rate.

56. Communicating Mathematics Carla has a balance of \$1,245 on her platinum credit card and has been a member for twelve years.

In Exercises 57–58, use your knowledge of logic to rewrite each statement in a simpler, equivalent form using fewer connectives.

57. $(\sim p \vee \sim q) \rightarrow \sim r$ **58.** $\sim q \rightarrow (\sim r \wedge p)$

Further Exercises

Another way to verify that two statements are logically equivalent is to join them with a biconditional and see whether this new statement has a truth table whose final column is all T's. Use this technique in Exercises 59–60.

59. Show that $p \rightarrow q$ is logically equivalent to $(\sim p) \vee q$.

60. Show that $\sim(p \rightarrow q)$ is logically equivalent to $p \wedge \sim q$.

61. Explain why your intuition would suggest the result in Exercise 59.

62. If you assume that the result in Exercise 59 is true, how would DeMorgan's Laws lead you to expect the result in Exercise 60?

63. Use Exercise 59 to rewrite the following statement:

If you buy the all-weather radial tires, then they will last for 80,000 miles.

64. Use Exercise 59 to rewrite the following statement:

If you are over 18, then you must register with Selective Service.

In Exercises 65–66, construct truth tables for each statement. Study these tables carefully and see if you can recognize how to write these statements in simpler, equivalent forms.

65. $p \vee (r \wedge (p \wedge q))$ **66.** $p \wedge (r \vee (p \vee q))$

Exercises 67–70 are based on the exercise sets in earlier sections in which we discussed the correspondence between electrical circuits and logical forms. Draw a circuit that corresponds to each form.

67. $p \rightarrow q$ (Hint: Consider a form that is equivalent to this conditional that uses other connectives.

68. $p \leftrightarrow q$ **69.** $(\sim p) \rightarrow \sim(q \wedge r)$

70. $p \leftrightarrow (q \wedge \sim r)$

---

* Exercise numbers circled in red can be used as group exercises.

2.5   VERIFYING ARGUMENTS

So far we have discussed how to represent statements symbolically using the basic connectives. We have also used truth tables to determine a statement's truth value and to decide whether two statements are logically equivalent. You are now ready to take the next step in symbolic logic, which is to determine when arguments are valid.

■▬▬▬▬▬▬▬▬▬▬▬▬▬▬▬▬▬▬

We use truth tables to determine whether arguments are valid.

In this section we will consider when a collection of statements produces a logical conclusion. Instead of expressing an argument in sentence form, we will represent it symbolically and use truth tables to determine whether it is valid. The following is a simple example of an argument.

1. If Madelyn passed the bar exam, then she is qualified to practice law.
2. Madelyn passed the bar exam.

   ----

3. Therefore, Madelyn is qualified to practice law.

This argument begins with statements 1 and 2, called *premises* and ends with the final statement 3, called a *conclusion*. If the two premises are true, then the conclusion must also be true. Another way of saying this is that *the conclusion follows from the premises*.

DEFINITIONS

An **argument** is a series of statements called **premises** followed by a single statement called the **conclusion**. An argument is **valid** if whenever all the premises are true, then the conclusion must also be true.

We will use a truth table to determine whether the preceding argument is valid. Let us represent the statement "Madelyn passed the bar exam" by p and represent the statement "She is qualified to practice law" by q. Then the argument has the following symbolic form. (The symbol $\therefore$ means *therefore* in the conclusion of an argument.)

$$p \rightarrow q$$
$$p$$
$$--------$$
$$\therefore q$$

We can view this argument as a conditional statement that has the following form:

> ***If***   the first premise is true
> ***and***
> the second premise is true
> ***then***   the conclusion is true.

This argument then has the symbolic form $[(p \rightarrow q) \wedge p] \rightarrow q$. We can verify that this argument is valid by showing that this conditional statement is *always* true, so we will consider its truth table.

			1			**5**	**2**	**4**	**3**
p	*q*	[(*p*	$\rightarrow$	*q*)	$\wedge$	*p*]	$\rightarrow$	*q*
T	T	T	T	T	T	T	**T**	T
T	F	T	F	F	F	T	**T**	F
F	T	F	T	T	F	F	**T**	T
F	F	F	T	F	F	F	**T**	F

Because this statement is a tautology, the argument is valid. The form $[(p \rightarrow q) \wedge p] \rightarrow q$ is called the **law of detachment**. Note that not only have we proved that this particular argument is valid, but also that *any argument having this form is valid*.

> **Verifying an Argument**
>
> We verify an argument by performing the following steps:
>
> 1. Write the argument symbolically.
> 2. Join the premises together using the *and* connective.
> 3. Form a conditional statement using the conjunction from step 2 for the hypothesis and the conclusion of the argument as the conclusion of the conditional.
> 4. If the statement you form in step 3 is a tautology, then the argument is valid. If there are any F's in the final column, then the argument is not valid.

EXAMPLE 1   Determining the Validity of an Argument

Determine whether the following argument is valid.

If you have a preferred customer card, then you qualify for a discount at the video store.
You do not qualify for a discount at the video store.

∴ You do not have a preferred customer card.

SOLUTION: Let *p* stand for "You have a preferred customer card" and *q* represent "You qualify for a discount at the video store." Then this argument has the following form:

$$p \rightarrow q$$
$$\sim q$$
$$\text{--------}$$
$$\therefore \sim p$$

We now consider the truth table for the conditional $[(p \rightarrow q) \wedge (\sim q)] \rightarrow (\sim p)$.

			1			3	2	5	4
p	*q*	[(*p*	→	*q*)	∧	~*q*]	→	~*p*
T	T	T	T	T	F	F	**T**	F
T	F	T	F	F	F	T	**T**	F
F	T	F	T	T	F	F	**T**	T
F	F	F	T	F	T	T	**T**	T

Because the final column of the truth table contains all true values, we conclude that this argument is valid. In essence, we have shown that whenever both of the premises are true, the conclusion is also true.

An argument that has the form shown in Example 1 is an example of the **law of contraposition**.

■

If the truth table for an argument has even a single F, then the argument is invalid.

EXAMPLE 2   Showing That an Argument Is Invalid

Determine whether the following argument is valid.

If Canada is an ally of the United States, then we are friendly with Canada.
We are friendly with Canada.

Therefore, Canada is an ally of the United States.

SOLUTION: If we let *c* represent "Canada is an ally of the United States" and let *f* represent "We are friendly with Canada," then we can write the argument symbolically as follows:

$$c \rightarrow f$$
$$f$$
$$\text{--------}$$
$$\therefore c$$

We next consider a truth table for the statement $[(c \rightarrow f) \wedge f] \rightarrow c$.

			1			2		3	
c	*f*	[(*c*	→	*f*)	∧	*f*]	→	*c*
T	T	T	T	T	T	T	**T**	T
T	F	T	F	F	F	F	**T**	T
F	T	F	T	T	T	T	**F**	F
F	F	F	T	F	F	F	**T**	F

Quiz Yourself 14*

Determine whether the following argument is valid.

If we spend more money for police, then there will be less crime.
We will not spend more money for police.

Therefore, there will not be less crime.

Since an F appears in the third row of the final column of the table, the argument is not valid.

Example 2 illustrates a common invalid argument form called the **fallacy of the converse**.

SOME GOOD ADVICE

You may feel intuitively that the argument in Example 2 is valid because Canada is an ally of the United States. It is very important when analyzing arguments that you do not bring outside information or opinions into the discussion. *The form of the argument is more important than the content of the statements we are making.*

Arguments are often one of several standard forms.

We list some common forms of valid and invalid arguments.

Valid Arguments

Law of Detachment	Law of Contraposition	Law of Syllogism	Disjunctive Syllogism
$p \rightarrow q$	$p \rightarrow q$	$p \rightarrow q$	$p \vee q$
p	$\sim q$	$q \rightarrow r$	$\sim p$
-------	-------	-------	-------
$\therefore q$	$\therefore \sim p$	$\therefore p \rightarrow r$	$\therefore q$

Invalid Arguments

Fallacy of the Converse	Fallacy of the Inverse
$p \rightarrow q$	$p \rightarrow q$
q	$\sim p$
-------	-------
$\therefore p$	$\therefore \sim q$

EXAMPLE 3   **Identifying the Form of an Argument**

Identify the form of each argument and state whether it is valid or invalid.

a) If you want to improve your cardiovascular fitness, then take up cross-country skiing.

You take up cross-country skiing.

Therefore, you want to improve your cardiovascular fitness.

b) If you sleep through your morning math class, then you will be well rested.

If you are well rested, then you will do well on your math test.

Therefore, if you sleep through your morning math class, then you will do well on your math test.

SOLUTION: a) Let *i* represent "You want to improve your cardiovascular fitness" and let *c* represent "You take up cross-country skiing." This argument has the following form:

$$i \rightarrow c$$
$$c$$
$$\overline{}$$
$$\therefore i$$

This is the fallacy of the converse, which is an invalid argument form.

b) Let *s* represent "You sleep through your morning math classes," let *r* represent "You will be well rested," and let *w* represent "You will do well on your math test." The argument then has the following form:

$$s \rightarrow r$$
$$r \rightarrow w$$
$$\overline{}$$
$$\therefore s \rightarrow w$$

This is a syllogism and is therefore valid, even though the reasoning seems to make no sense.

 Quiz Yourself 15

Identify the form of the argument and state whether it is valid.

If more is spent on research, then medical science will advance.
More is not spent on research.

Therefore, medical science will not advance.

You may have been surprised in Example 3(b) that the argument was valid. Keep in mind that *the symbolic form, not the content*, determines the validity of an argument.

We can use logic to analyze more complex arguments than those we have studied so far.

EXAMPLE 4 Analyzing a Complex Argument with Three Variables

Determine whether this argument is valid or invalid.

If we balance the budget or reduce taxes, then there will be more money available to fight pollution.
If we do not balance the budget, then we will not reduce taxes.
We will not reduce taxes.

Therefore, there will be more money to fight pollution.

SOLUTION: Consider the statements b, "We balance the budget"; r, "We reduce taxes"; and m, "There will be more money available to fight pollution." This argument then has the form

$$[((b \lor r) \to m) \land (\sim b \to \sim r) \land (\sim r)] \to m,$$

and, as usual, we construct a truth table to investigate the argument's validity.

			1	2		6	3	5	4	8	7	9	
b	*r*	*m*	[((b ∨ r)	→	m)	∧	(~b	→	~r)	∧	(~r)]	→	m
T	T	T	T	T	T	T	F	T	F	F	F	T	T
T	T	F	T	F	F	F	F	T	F	F	F	T	F
T	F	T	T	T	T	T	F	T	T	T	T	T	T
T	F	F	T	F	F	F	F	T	T	F	T	T	F
F	T	T	T	T	T	F	T	F	F	F	F	T	T
F	T	F	T	F	F	F	T	F	F	F	F	T	F
F	F	T	F	T	T	T	T	T	T	T	T	T	T
F	F	F	F	T	F	T	T	T	T	T	T	F	F

The F in the eighth row of the final column shows that this argument is invalid.

Even though the truth table in Example 4 has T's in seven of eight rows, by the Always Principle, we must say that the argument is invalid.

Highlight: *Logic and National Defense*

Prior to the 1980s the United States based its military defense on the idea of "mutual assured destruction." This meant that we would stockpile so many nuclear missiles that our enemies would not attack us for fear of a devastating counterattack. By the 1980s, some felt that the United States would be protected better by a strong defensive capability, rather than an overwhelming offense. They believed that massive computer programs controlling radar and missiles would be able to protect the United States from a nuclear attack. However, many scientists were concerned that because these systems would rely on millions of lines of computer code, mistakes in the code could render our defenses vulnerable.

The scientists faced with this problem are dealing with the same type of problems that you have studied in this chapter—but on a much larger scale. Computer programs contain statements that use the same connectives you have learned, and verifying that a computer program is correct is somewhat similar to verifying an argument. Although scientists do use methods other than truth tables to investigate the reliability of computer programs, at present they have not completely solved the problem of determining when a large complex program is error-free.

Exercises 2.5

In Exercises 1–18, identify the form of each argument and state whether the argument is valid.

1. If a car has air bags, then it is safe.
This car has air bags.

Therefore, this car is safe.

2. If news on inflation is good, then stock prices will increase.
News on inflation is good.

Therefore, stock prices will increase.

3. If a movie is exciting, then it will gross a lot of money.
This movie grossed a lot of money.

Therefore, it is exciting.

4. If we spend much on national defense, then the country will avoid war.
The country is avoiding war.

Therefore, we are spending much on national defense.

5. This computer has a Zip drive or a large monitor.
This computer does not have a Zip drive.

Therefore, this computer has a large monitor.

6. Either my player is defective or this CD is damaged.
My player is not defective.

Therefore, this CD is damaged.

7. If you pay your taxes late, then you will pay a late penalty.
You do not pay your taxes late.

Therefore, you will not pay a late penalty.

8. If you perform maintenance on your PC, then you violate your warranty.
You do not perform maintenance on your PC.

Therefore, you do not violate your warranty.

9. If you score high on the LSATs, then you will be accepted by the law school of your choice.
If you are accepted by the law school of your choice, then you will impress your friends.

Therefore, if you score high on the LSATs, then you will impress your friends.

10. If June 1 is Monday, then June 2 is Friday.
If June 2 is Friday, then June 5 is Wednesday.

Therefore, if June 1 is Monday, then June 5 is Wednesday.

11. If you do not pay your taxes, then you will not qualify for further government assistance.
You qualify for further government assistance.

Therefore, you did pay your taxes.

12. If you love me, then you will do everything I ask.
You do not do everything I ask.

Therefore, you do not love me.

13. If Phillipe joins the basketball team, then he will not be able to work part-time.
Phillipe did not join the basketball team.

Therefore, he will be able to work part-time.

14. If Sarah gets a raise, then she will be able to afford a bigger apartment.
Sarah gets a raise.

Therefore, Sarah will be able to afford a bigger apartment.

15. If Malik buys a satellite dish, then he will get 128 TV channels.
Malik does not get 128 TV channels.

Therefore, he did not buy a satellite dish.

16. If you cut the amount of fat in your diet, then you will have more energy.

You don't have more energy.

Therefore, you did not cut the amount of fat in your diet.

17. Minxia will take the standard deduction or itemize deductions.

Minxia will not take the standard deduction.

Therefore, she will itemize deductions.

18. Rob will take the $1,000 rebate or the interest-free loan.

Rob will not take the $1,000 rebate.

Therefore, Rob will take the interest-free loan.

19. Communicating Mathematics Explain in your own words what we mean by an argument in logic.

20. Communicating Mathematics Discuss why the technique we explained for showing an argument to be valid is a reasonable approach. Include a discussion of the conditional in your explanation.

In Exercises 21–40, determine whether each form represents a valid argument.

21.

p
$q \rightarrow \sim p$

$\therefore \sim q$

22.

p
$\sim q \rightarrow p$

$\therefore \sim p \vee q$

23.

$\sim r$
$r \rightarrow q$

$\therefore \sim q \wedge r$

24.

p
$\sim q \rightarrow \sim p$

$\therefore q$

25.

$\sim q \rightarrow p$
$r \rightarrow \sim q$

$\therefore \sim p \rightarrow r$

26.

p
$\sim q \rightarrow \sim p$
$(p \wedge q) \rightarrow r$

$\therefore q \rightarrow r$

27.

p
$q \rightarrow \sim p$
$\sim q \rightarrow (r \vee p)$

$\therefore r$

28.

p
$\sim p \rightarrow \sim q$
$q \rightarrow r$

$\therefore r$

29.

r
$r \rightarrow \sim q$
$p \vee q$

$\therefore p$

30.

r
$r \rightarrow q$
$\sim p \vee \sim q$

$\therefore \sim p$

31.

$q \rightarrow \sim p$
$r \rightarrow \sim q$

$\therefore \sim p \rightarrow r$

32.

p
$\sim p \rightarrow \sim q$
$(p \wedge q) \rightarrow r$

$\therefore q \rightarrow r$

33. If the product has a lower price, then the product does not have quality.

If the product does not have a lower price or does not have quality, then the product is not reliable.

The product has a lower price.

Therefore, the product is reliable.

34. If the team wins this game, then it will qualify for the playoffs.

The team will play in a tournament or will not qualify for the playoffs.

The team wins this game.

Therefore, the team will play in a tournament and will qualify for the playoffs.

35. If the test is negative, then you will not require treatment.

If the test is positive, then you will require medication.

You do not require medication.

Therefore, you do not require treatment.

36. If health care is not improved, then the quality of life will not be high.

If health care is improved and the quality of life is high, then the incumbents will be re-elected.

The quality of life is high.

Therefore, the incumbents will be reelected.

37. If Dave is alone tonight, then he will not come to the party.

If Dave is not alone tonight or does not come to the party, then he will work on his term paper.

Dave will not work on his term paper tonight.

Therefore, Dave is not alone tonight.

38. Raul has an IRA.

If Raul has an IRA, then he will not withdraw from his CD.

Raul will withdraw from his CD or invest in bonds.

Therefore, Raul will invest in bonds.

39. Jamie is fluent in Spanish.

If Jamie is fluent in Spanish, then she will work in Madrid.

She will not visit Mexico or she will not work in Madrid.

Therefore, she will visit Mexico.

40. Matthew is studying law.

If Matthew does not study law, then he will not become an attorney.

If Matthew studies law and becomes an attorney, then he will have a rewarding job.

Therefore, if Matthew becomes an attorney, then he will have a rewarding job.

Further Exercises

We have emphasized that the form of a logical argument is more important than its content. Recall in Example 2 that you saw an argument that sounds reasonable, yet has an invalid form. In Exercises 41–44, refer to the tables on page 113 that show forms of some standard valid and invalid arguments. In each exercise explain your thinking.

***41.** Communicating Mathematics   Write an argument that sounds reasonable and also has a valid form.

42. Communicating Mathematics   Write an argument that sounds reasonable but has an invalid form.

43. Communicating Mathematics   Write an argument that sounds unreasonable but has a valid form.

44. Communicating Mathematics   Write an argument that sounds unreasonable and also has an invalid form.

In a complicated argument with many variables, it is not practical to use truth tables because of their size. We can, however, use valid argument forms to reason without using truth tables. For example, consider the following argument:

$$p$$
$$p \rightarrow q$$
$$p \wedge q \rightarrow r$$
$$\sim s \rightarrow \sim r$$

$$\therefore s$$

We assume that all the premises are true, and we reason like this to prove that the argument is valid:

1. We assumed that p and p $\rightarrow$ q are both true, therefore by the law of detachment, we have that q is true.
2. Now both p and q are true, so p $\wedge$ q is also true.
3. By the law of detachment again, p $\wedge$ q true and p $\wedge$ q $\rightarrow$ r true force r to be true.
4. Because the statement $\sim$ s $\rightarrow \sim$ r is equivalent to its contrapositive, we then know that r $\rightarrow$ s is true.
5. Knowing that r and r $\rightarrow$ s are both true, we conclude that s is also true.

Therefore, by assuming that all the premises are true, we were able to reason that the conclusion s also must be true. This means that the argument is valid. In Exercises 45–46, reason similarly to prove that each argument is valid.

45.
$$a \wedge b$$
$$b \rightarrow c$$
$$d \rightarrow \sim c$$

$$\therefore \sim d$$

46.
$$p$$
$$p \rightarrow \sim q$$
$$p \rightarrow r$$
$$\sim s \rightarrow q$$

$$\therefore r \wedge s$$

47. Communicating Mathematics   Write an essay connecting the ideas of this section to the ideas of previous sections. For example, what does the law of contraposition

---

* Exercise numbers circled in red can be used as group exercises.

have to do with the contrapositive? Identify several such connections and explain the connections thoroughly.

48. Communicating Mathematics   We have presented several forms of valid and invalid argument forms in this section. Find a book on logic in your library that lists other forms of logical arguments and explain an argument that we have not discussed.

2.6   USING EULER DIAGRAMS TO VERIFY SYLLOGISMS

In this section we will analyze arguments called syllogisms. Like the arguments you studied in Section 2.5, a **syllogism** consists of a set of statements called *premises* followed by a statement called a *conclusion*. However, syllogisms differ from earlier arguments in that the premises and conclusion of a syllogism may contain quantifiers such as *all*, *some*, and *none*. The arguments in Section 2.5 did not. A syllogism is **valid** if whenever its premises are all true, then the conclusion is also true. If the conclusion of a syllogism can be false even though all the premises are true, then the syllogism is **invalid**.

Perhaps the most famous syllogism is

> All people are mortal.
> Socrates is a person.
> --------------------------------------
> Therefore, Socrates is mortal.

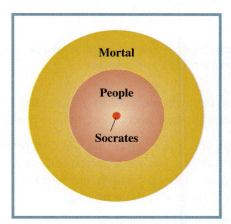

FIGURE 2.11   Euler diagram for the Socrates syllogism.

To determine whether a syllogism is valid, we use a drawing called an **Euler diagram**, which is named after the great eighteenth-century Swiss mathematician Leonhard Euler. The drawing, which looks very much like a Venn diagram in set theory, is drawn so that all the premises are satisfied. We then can determine from the diagram whether the conclusion must also be true. Figure 2.11 shows an Euler diagram for the preceding syllogism.

In Figure 2.11, we represent the first premise, "All people are mortal," by drawing a small circle labeled "People" contained in a larger circle labeled "Mortal." We represent the second premise, "Socrates is a person," by placing a dot labeled "Socrates" inside the circle labeled "People." By drawing the two premises this way, we are forced to put the dot labeled "Socrates" in the circle labeled "Mortal." Which shows that the conclusion "Socrates is mortal" must occur. Because the premises being true force the conclusion also to be true, we see that the syllogism is valid.

We will often use the symbol ∴ for the word *therefore* in the conclusion of a syllogism.

EXAMPLE 1   Determining the Validity of a Syllogism

Use an Euler diagram to determine whether the syllogism is valid.

> All poets are good spellers.
> Dante is not a good speller.
> ----------------------------
> ∴ Dante is not a poet.

SOLUTION: We represent the first premise, "All poets are good spellers," by the diagram in Figure 2.12(a) and add the second premise, "Dante is not a good speller," in Figure 2.12(b).

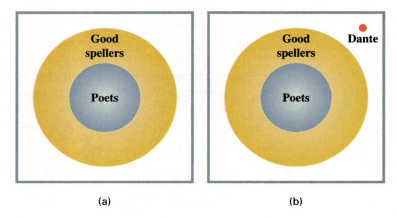

(a)   (b)

FIGURE 2.12   (a) The first premise of the Dante syllogism. (b) The second premise of the Dante syllogism.

We see from Figure 2.12(b) that if Dante is not a good speller, then that forces Dante not to be a poet. Therefore the conclusion is valid.

Perhaps you don't agree with the first premise in Example 1 because you believe that it is possible for a person to be a poet without being a good speller. Your feeling about this premise is not relevant to the validity of the argument. It is important to remember the following:

> It is possible for individual premises or the conclusion to be false and yet the syllogism can be valid.

It is the form of the argument rather than its content that is important; you must rely on Euler diagrams rather than your intuition when deciding whether syllogisms are valid.

EXAMPLE 2   Using an Euler Diagram to Show That a Syllogism is Invalid

Use an Euler diagram to show that the syllogism is invalid.

All tigers are meat eaters.
Simba is a meat eater.

∴ Simba is a tiger.

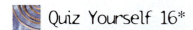 Quiz Yourself 16*

Draw an Euler diagram to determine whether the syllogism is valid.

All credit cards are made of plastic.
This card is not a credit card.

∴ This card is not made of plastic.

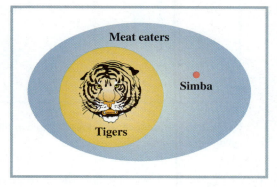

FIGURE 2.13   Euler diagram for an invalid argument.

SOLUTION: Figure 2.13 illustrates the two premises.

We see in Figure 2.13 that it is possible for Simba to be a meat eater without being a tiger. The argument is therefore invalid because the premises do not force the conclusion to hold.   ◎

Example 3 shows how to handle a premise containing the quantifier *none are* or *no*.

EXAMPLE 3   **Analyzing a Syllogism That Contains the Quantifier *No***

Assume that a syllogism begins with the following two premises:

No iMacs have floppy disk drives.
My computer does not have a floppy disk drive.

a) Can we conclude that my computer is an iMac?

b) Can we conclude that my computer is not an iMac?

FIGURE 2.14   Because the conclusion holds in this diagram, you may mistakenly think that the syllogism is valid.

SOLUTION: a) We need to determine the validity of the syllogism

No iMacs have floppy disk drives.
My computer does not have a floppy disk drive.

∴ My computer is an iMac.

Figure 2.14 shows one possible way to draw the first two premises in an Euler diagram. We illustrate the first premise by drawing two disjoint circles to represent iMacs and computers that have floppy disk drives. The rectangle represents the set of all computers.

From Figure 2.14, you might conclude that my computer is an iMac. *That would be a mistake.* As we show in Figure 2.15, there is another way to draw the premise "My computer does not have a floppy disk drive."

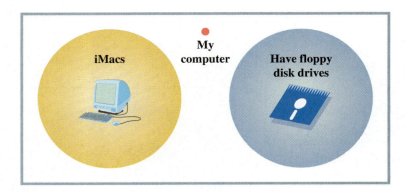

FIGURE 2.15 The conclusion does not hold in this diagram so the syllogism *cannot* be valid.

From Figure 2.15 we see that the syllogism is invalid, because the given premises do not force the conclusion to hold. Therefore, you cannot conclude that my computer is an iMac.

b) Figure 2.14 shows that the premises can hold and yet we cannot conclude that my computer is not an iMac. Therefore, based on the given premises, we cannot conclude that my computer is not an iMac.

The quantifier *some* can be misleading in syllogisms because there are several different ways to represent *some* in Euler diagrams. In Figure 2.16, each diagram represents the premise "Some A's are B's."

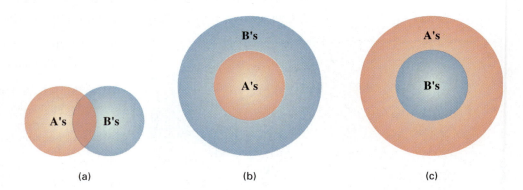

(a)   (b)   (c)

FIGURE 2.16 Euler diagrams illustrating that "Some A's are B's."

Notice in Figure 2.16(b) that the fact that "All A's are B's" does not rule out the possibility that "Some A's are B's."

EXAMPLE 4 Analyzing a Syllogism Having the Quantifier *Some*

Is the syllogism valid?

All cars manufactured after 1998 have driver-side air bags.

Some cars with driver-side air bags have passenger-side air bags.

∴ Some cars with passenger-side air bags were manufactured after 1998.

SOLUTION: Because a single diagram can show a syllogism to be invalid, our strategy will be to find such a diagram. After looking at several diagrams, if we feel that it is impossible to show the argument invalid, then we will say that it is valid. Figure 2.17 shows three ways to illustrate the two premises.

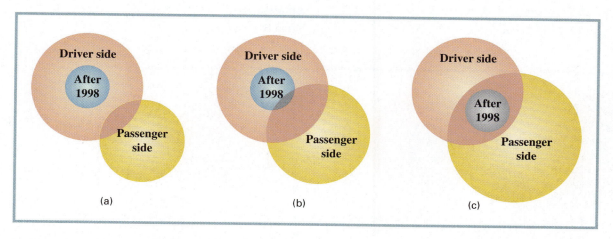

(a)   (b)   (c)

FIGURE 2.17 Three ways to illustrate the premises of the air bag syllogism.

By luck, the first diagram we drew, Figure 2.17(a), shows that *it is possible* that no cars manufactured after 1998 have passenger-side air bags. Therefore the conclusion does not follow from the premises, so the syllogism is invalid.

In Example 4, we cannot conclude that no cars manufactured after 1998 have passenger-side air bags. Clearly the diagrams in Figures 2.17(b) and 2.17(c) show that *it is possible* for some cars manufactured after 1998 to have passenger-side air bags. All we are able to say in Example 4 is that even though the conclusion sounds reasonable, it does not follow from the premises.

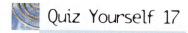 Quiz Yourself 17

Determine whether the syllogism is valid.

All firefighters are brave.
Some women are firefighters.

∴ Some women are brave.

PROBLEM SOLVING

From the Counterexample Principle, you know that one Euler diagram can show a syllogism to be invalid. In testing the validity of a syllogism, it is usually a good strategy to try to draw several diagrams, with the idea that one of them may show the syllogism to be invalid. If you are convinced that no diagram can show the syllogism to be invalid, then you can conclude that the syllogism is valid.

The problem with drawing Euler diagrams, particularly for syllogisms using the quantifier *some*, is that whatever diagram you draw will show extra conditions that were not stated as a premise. These extra conditions can mislead you in determining the validity of a syllogism.

EXAMPLE 5 **Extra Conditions Are Always Present in an Euler Diagram**

Each Euler diagram in Figure 2.18 illustrates the following premises:

All A's are B's.
Some B's are C's.

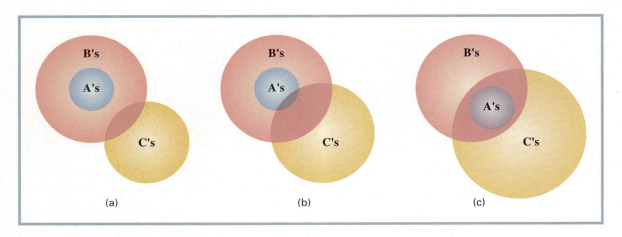

FIGURE 2.18 Euler diagrams illustrating "All A's are B's" and "Some B's are C's."

State a condition that is present in each diagram that has not been stated as a premise.

SOLUTION: There are many possible conditions. We will give one for each diagram.

a) Because the circles containing A's and C's do not intersect, we have drawn the extra condition that "No A's are C's."

b) Because the circles containing A's and C's do intersect, we have drawn the condition that "Some A's are C's."

c) Because we drew the circle containing A's inside the circle containing C's, we have the additional condition that "All A's are C's."

In Figure 2.18, several other extra conditions appear in the diagrams that were not stated as premises. It may be hard to show that a syllogism is valid because by the Always Principle, the conclusion must hold in every Euler diagram that illustrates the premises. You must be very careful to consider a variety of diagrams before you accept a syllogism as valid.

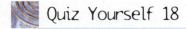

Quiz Yourself 18

Find one additional condition in Figure 2.18, other than those given in Example 5, that was not stated as a premise.

Exercises 2.6

In Exercises 1–16, determine whether each syllogism is valid or invalid.

1. All original parts are under warranty.
This part is an original part.

∴ This part is under warranty.

2. All original parts are under warranty.
This part is under warranty.

∴ This part is an original part.

3. All imports are subject to a surcharge.
This jacket is subject to a surcharge.

∴ This jacket is an import.

4. All imports are subject to a surcharge.
This necklace is an import.

∴ This necklace is subject to a surcharge.

5. All vitamins are healthful.
Milk is healthful.

∴ Milk is a vitamin.

6. All vitamins are healthful.
Caffeine is a vitamin.

∴ Caffeine is healthful.

7. All athletes are fit.
Julio is not an athlete.

∴ Julio is not fit.

8. All athletes are fit.
Sheena is not fit.

∴ Sheena is not an athlete.

9. All politicians are honest.
Walt is not honest.

∴ Walt is not a politician.

10. All salespeople are sincere.
Emily is not sincere.

∴ Emily is not a salesperson.

11. Some large corporations are concerned about the environment.
General Motors is not a large corporation.

∴ General Motors is not concerned about the environment.

12. Some children love cookies.
Dani does not love cookies.

∴ Dani is not a child.

13. Some biologists believe the Loch Ness monster exists.
All who believe the Loch Ness monster exists are irrational.
Antawn is not irrational.

∴ Antawn is not a biologist.

14. Some investors are wealthy.
All wealthy people are happy.

∴ Some investors are happy.

15. Some mammals are large.
All large animals are dangerous.

∴ Some mammals are dangerous.

16. Some mammals are large.
Some dangerous animals are large.

∴ Some mammals are dangerous.

In Exercises 17–20, complete each syllogism so that it is valid and the conclusion is true. There may be several correct answers.

17. Some taxes are unfair.
All unfair taxes should be abolished.

∴.

18. All honest politicians should be supported.
Marika should not be supported.

∴.

19. No team that plays in a domed stadium has won the Super Bowl.
Some teams that wear red uniforms have won the Super Bowl.

∴.

20. Some mathematicians are fine musicians.
All fine musicians are intelligent.

∴.

Further Exercises

*⟨**21.**⟩ Communicating Mathematics   Give an example of a valid syllogism that has a false statement for its conclusion.

⟨**22.**⟩ Communicating Mathematics   Give an example of an invalid syllogism that has a true statement for its conclusion.

In Exercises 23–26, write a valid syllogism that would be illustrated by each Euler diagram. For example, the Socrates syllogism would be illustrated by a diagram similar to the one in Exercise 23.

⟨**23.**⟩
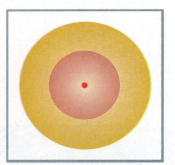

⟨**24.**⟩
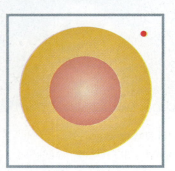

⟨**25.**⟩
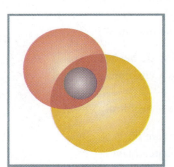

⟨**26.**⟩
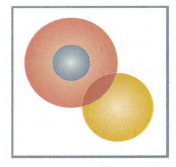

CHAPTER SUMMARY

SECTION 2.1

inductive reasoning

Inductive reasoning is the process of drawing a general conclusion by observing a pattern in specific instances. This conclusion is called a hypothesis or conjecture.

deductive reasoning

In deductive reasoning, we use accepted facts and general principles to arrive at a specific conclusion.

SECTION 2.2

statement

A statement in logic is a declarative sentence that is either true or false.

simple statement, compound statement, connectives

A simple statement is one such that if we were to remove any part of it, the statement would no longer make sense. A compound statement contains several ideas joined together by words called connectives.

negation

Negation, represented symbolically by ~, expresses the idea that something is not true.

conjunction, disjunction

Conjunction, represented by ∧, expresses the idea *and*. Disjunction, represented by ∨, expresses the idea *or*.

conditional, biconditional

The conditional, represented by →, expresses the idea *if . . . then*. The biconditional, represented by ↔, expresses the idea *if and only if*.

quantifier

Quantifiers are words in statements that involve the notion of "how many."

universal quantifier, existential quantifier

Universal quantifiers are words such as *all* and *every* that state that all objects of a certain type satisfy a given property. Existential quantifiers are words such as *some, there exists,* and *there is at least one* that state that there are one or more objects that satisfy a given property.

SECTION 2.3

truth tables for negation, conjunction, and disjunction

Negation (*not*)	
p	*~p*
T	F
F	T

Conjunction (*and*)		
p	*q*	*p ∧ q*
T	T	T
T	F	F
F	T	F
F	F	F

Disjunction (*or*)		
p	*q*	*p ∨ q*
T	T	T
T	F	T
F	T	T
F	F	F

tautology

A statement is a tautology if it is always true.

setting up truth tables

A truth table for a statement with k variables will have 2^k lines. Fill in half of the values under the first variable as true and the remaining half as false. In the second column, alternate placing half as many T's and then half as many F's as were placed in the first column. Continue this process until you reach the last column, which will contain alternating T's and F's.

logically equivalent

Statements are logically equivalent if they contain the same variables and have identical truth tables.

DeMorgan's Laws for logic

$\sim(p \wedge q)$ is logically equivalent to $(\sim p) \vee (\sim q)$.
$\sim(p \vee q)$ is logically equivalent to $(\sim p) \wedge (\sim q)$.

SECTION 2.4   **truth tables for conditional and biconditional**

Conditional (*if . . . then*)		
p	q	$p \rightarrow q$
T	T	T
T	F	F
F	T	T
F	F	T

Biconditional (*if and only if*)		
p	q	$p \leftrightarrow q$
T	T	T
T	F	F
F	T	F
F	F	T

converse, inverse, contrapositive

For the conditional $p \rightarrow q$, the converse has the form $q \rightarrow p$, the inverse has the form $\sim p \rightarrow \sim q$, and the contrapositive has the form $\sim q \rightarrow \sim p$. A conditional and its contrapositive are logically equivalent.

alternate ways of expressing conditionals

A statement in the form "If p then q" can be expressed alternately as "q if p," "p only if q," "p is sufficient for q," and "q is necessary for p."

SECTION 2.5   **argument, premises, conclusion, valid**

An argument is a series of statements called premises followed by a statement called the conclusion. The argument is valid if whenever all the premises are true, then the conclusion must also be true.

verifying an argument

Write the argument symbolically and join the premises together using the *and* connective. Form a conditional statement using this conjunction for the hypothesis and the conclusion of the argument as the conclusion of the conditional. If the final column of the truth table for this conditional contains all T's, then the argument is valid; otherwise it is invalid.

some valid argument forms

Law of Detachment	Law of Contraposition	Law of Syllogism	Disjunctive Syllogism
$p \rightarrow q$	$p \rightarrow q$	$p \rightarrow q$	$p \vee q$
p	$\sim q$	$q \rightarrow r$	$\sim p$
-------	-------	-------	-------
$\therefore q$	$\therefore \sim p$	$\therefore p \rightarrow r$	$\therefore q$

some invalid argument forms

Fallacy of the Converse	Fallacy of the Inverse
$p \rightarrow q$	$p \rightarrow q$
q	$\sim p$
-------	-------
$\therefore p$	$\therefore \sim q$

SECTION 2.6   **syllogism**

A syllogism consists of a set of statements called premises followed by a statement called a conclusion. The premises and conclusion of a syllogism may contain quantifiers such as *all, some,* and *none.*

valid, invalid

A syllogism is valid if whenever its premises are all true, then the conclusion is also true. If the conclusion of a syllogism can be false even though all the premises are true, then the syllogism is invalid.

Euler diagram

An Euler diagram is a visual way of deciding whether a syllogism is valid or invalid.

CHAPTER TEST

SECTION 2.1

1. Are the following situations illustrating inductive or deductive reasoning?

   a) For the last three Mondays, your history professor has shown a film. You expect that next Monday you will also have a film in class.

   b) You are following a series of clues on a scavenger hunt to obtain a list of items.

2. Explain the difference between inductive and deductive reasoning.

3. Use inductive reasoning to predict the next term in the following sequences   a) 2, 7, 12, 17, 22, ...   b) 3, 4, 7, 11, 18, 29, ...

4. Use inductive reasoning to draw the next figure in the pattern.

5. Illustrate Goldbach's conjecture for the number 48.

6. (a) Follow the instructions for this "trick" starting with several different numbers of your choice. (b) Make a conjecture as to what result you get in each case. (c) Use algebra and deductive reasoning to explain why your conjecture is correct.

   a) Choose any natural number.

b) Multiply the number by 8.

c) Add 12 to the product you just found.

d) Divide the results of part c) by 4.

e) Subtract three.

SECTION 2.2

7. Which of the following are statements? Explain your answers.

a) Why do these things always happen to me?

b) The distance from Los Angeles to New York City is two thousand miles.

c) Bring me back a pizza.

8. Let b represent the statement "I will buy a laptop computer" and let s represent the statement "I will sell my old computer." Write each statement in symbolic form.

a) I will not buy a laptop computer or I will sell my old computer.

b) It is not true that: I will buy a laptop computer and not sell my old computer.

9. Let f represent "Antonio is fluent in Spanish" and let l represent "Antonio has lived in Spain for a semester." Write each statement in English.

a) $\sim(f \wedge \sim l)$

b) $\sim f \vee \sim l$

10. Negate each quantified statement and then rewrite it in English in an alternate way.

a) All writers are passionate.

b) Some graduates received several job offers.

SECTION 2.3

11. Let p represent some true statement and let q represent some false statement. What is the truth value of each statement?

a) $p \wedge (\sim q)$

b) $\sim(p \vee q)$

c) $\sim(p \wedge q)$

12. How many rows will be in the truth table for each statement?

a) $\sim(p \vee q) \wedge \sim(r \vee p)$

b) $(p \vee q) \wedge (r \vee s) \wedge t$

13. Construct a truth table for each statement.

a) $\sim(p \vee \sim q)$

b) $\sim(p \vee \sim q) \wedge \sim r$

14. Negate each statement and then rewrite the negation using DeMorgan's Laws.

a) You qualify for the extended warranty or the free maintenance contract.

b) I will not sign the lease or I will not accept the housing agreement.

15. Which pairs of statements are logically equivalent?

a) $\sim(p \wedge \sim q), (\sim p) \vee q$

b) $\sim(p \vee \sim q) \wedge \sim(p \vee q), p \vee (p \wedge q)$

SECTION 2.4

16. Assume that p represents a true statement, q a false statement, and r a true statement. What is the truth value of each statement?

a) $\sim(p \vee q) \rightarrow \sim p$

b) $p \wedge q \leftrightarrow q \vee r$

c) $((\sim p) \vee (\sim q)) \rightarrow r$

17. Construct a truth table for each statement.

a) $\sim p \rightarrow q$

b) $\sim(p \wedge r) \leftrightarrow \sim(p \vee q)$

18. Write in words the converse, inverse, and contrapositive for the statement "If the disk drive has been damaged, then we cannot recover the data."

19. Rewrite each statement using the words *if . . . then*.

a) You qualify for the tax rebate only if you earned less than $6,500 last year.

b) To be an astronaut, it is necessary to have a pilot's license.

SECTION 2.5

20. Identify the form of each argument.

a) If you invest in this stock, then you will risk your savings.

You do not invest in this stock.

Therefore, you do not risk your savings.

b) If Felicia enjoys spicy food, then she will enjoy this Cajun chicken.

Felicia does not enjoy this Cajun chicken.

Therefore, Felicia does not enjoy spicy food.

21. Determine whether the form represents a valid argument.

$$\sim p$$
$$q \rightarrow p$$
$$(p \vee q) \rightarrow r$$
$$--------------$$
$$\therefore r$$

22. Use a truth table to determine whether the argument is valid.

If the product has a higher price, then the product has quality.

If the product has a higher price or has quality, then the product is reliable.

The product has a higher price.

$\therefore$ The product is reliable.

SECTION 2.6

23. Use an Euler diagram to determine whether the syllogism is valid or invalid.

Some used cars are expensive.

All expensive cars are safe.

This car is not safe.

$\therefore$ This is not a used car.

Of Further Interest: FUZZY LOGIC

To say "It is warm today" may not mean the same thing on the first day of March as it does in the middle of August. Statements we make in everyday life have shades of meaning and, in that sense, differ from the statements that we have been studying in symbolic logic. Although it may be warm today, perhaps it is not as warm as it was two weeks ago. To simply say "It is warm today" ignores exactly how warm it is. The strict condition that statements must be either true or false makes symbolic logic unsuitable for representing many real-life situations.

Statements in fuzzy logic have truth values between 0 and 1.

In order to apply the techniques of logic more widely, mathematicians developed *fuzzy logic*. This term might seem contradictory, but, as you will see shortly, it is a perfectly acceptable, well-developed area of mathematics. Once we define the meaning of fuzzy statements, and explain how fuzzy connectives behave, we can work with them as we did for other (nonfuzzy*) statements.

> **DEFINITIONS**
>
> In **fuzzy logic**, a **statement** is a declarative sentence that has an associated **truth value** between 0 and 1 inclusive.

EXAMPLE 1   The Truth Values of Statements in Fuzzy Logic

Here are some examples of statements in fuzzy logic.

a) "I like peach ice cream," with a truth value of 0.9.

b) "Tiger Woods is a great golfer," with a truth value of 0.92.

c) "Maine is a large state," with a truth value of 0.45.

d) "*Gone with the Wind* is a great movie," with a truth value of 0.73.

e) "You find mathematics interesting," with a truth value of 0.7.

The way we assigned truth values in Example 1 is somewhat arbitrary. You may want to assign a truth value of 0.85 to the statement "*Gone with the Wind* is a great movie" because you feel that it is a greater movie than we do. *This is perfectly OK and is really the point of fuzzy*

---

* Some use the term crisp statements to describe the statements we studied earlier.

logic. If we could not differ in our assignment of truth values, we would then both be forced to say that *Gone with the Wind* was a great movie, but we could not express the difference of our opinions of exactly how great the movie was.

Negation, conjunction, and disjunction in fuzzy logic are defined differently than the corresponding connectives in nonfuzzy logic.

We will now consider how connectives work in fuzzy logic. In Example 1, we assigned a truth value of 0.9 to "I like peach ice cream." Since this expresses the degree to which I like peach ice cream, that also says that $1 - 0.9 = 0.1$ indicates the degree to which I *do not like* peach ice cream. The next definition explains how to calculate the truth value of negations in fuzzy logic.

> ### DEFINITION
>
> If p is a statement in fuzzy logic, then the truth value of the **negation** of p, written $\sim p$, is
>
> $$1 - \text{the truth value of } \boldsymbol{p}.$$

Quiz Yourself 19*

Calculate the truth values for the negations of statements (d) and (e) in Example 1.

EXAMPLE 2   Calculating the Truth Values of Negations in Fuzzy Logic

Calculate the truth values for the negations of statements (b) and (c) in Example 1.

SOLUTION: b) "Tiger Woods is not a great golfer" has a truth value of $1 - 0.92 = 0.08$.

c) "Maine is not a large state" has a truth value of $1 - 0.45 = 0.55$.

It is easy to compute conjunctions and disjunctions in fuzzy logic.

> ### DEFINITIONS
>
> Suppose that p and q are two statements in fuzzy logic.
>
> a) The truth value of the **disjunction** of p and q, written $p \vee q$, is the *maximum* of the truth values for p and q.
> b) The truth value of the **conjunction** of p and q, written $p \wedge q$, is the *minimum* of the truth values for p and q.

EXAMPLE 3   Calculating Disjunctions and Conjunctions in Fuzzy Logic

Let h represent the statement "Foreign cars have high resale value" and let s represent the statement "Foreign cars are safer than domestic cars." Assume that h has a truth value of 0.85 and that s has a truth value of 0.73. Find the truth values of the following statements.

Quiz Yourself 20

Let *l* represent the statement "I like living in a large city" with a truth value of 0.82 and let *e* represent the statement "I like living on the East Coast" with a truth value of 0.45. Find the truth value of each statement.

a) $l \wedge e$ b) $l \vee e$

a) "Foreign cars have high resale value *and* are safer than domestic cars."

b) "Foreign cars have high resale value *or* are safer than domestic cars."

SOLUTION: a) The statement "Foreign cars have high resale value and are safer than domestic cars" has the form $h \wedge s$, and so has a truth value equal to the minimum of 0.85 and 0.73, which is 0.73.

b) The statement "Foreign cars have high resale value or are safer than domestic cars" has the form $h \vee s$, and therefore has a truth value equal to the maximum of 0.85 and 0.73, which is 0.85.

We can use our knowledge of fuzzy connectives to analyze more complex statements.

EXAMPLE 4   Computing the Truth Value of a Statement with Three Variables

Let *h, s,* and *c* be defined as follows:

> *h* : "Foreign cars have high resale value," with truth value 0.85.
> *s* : "Foreign cars are safer than domestic cars," with truth value 0.73.
> *c* : "Foreign cars cost more than domestic cars," with truth value 0.6.

Find the truth value of the statement

> "Foreign-made cars have high resale value, but it is not true that they are safer or cost more than domestic cars."

SOLUTION: This statement has the following form:

$$\underset{0.85}{h} \quad \wedge \quad \sim \quad (\underset{0.73}{s} \quad \vee \quad \underset{0.6}{c})$$

We have written the truth values of *h, s,* and *c* to make the calculations easier to follow.

The truth value of $s \vee c$ is the maximum of 0.73 and 0.6, which is 0.73. The truth value of $\sim(s \vee c)$ is then $1 - 0.73 = 0.27$. The truth value of the conjunction of $\sim(s \vee c)$ and *h* is the minimum of 0.27 and 0.85, namely 0.27.

It is awkward to always talk about the minimum of the truth values of *h* and *s*, the maximum of the truth values of *h* and *s*, and so on, so for simplicity's sake, in calculating the truth value of $h \vee s$, we will write $0.85 \vee 0.73 = 0.85$, and the truth value of $h \wedge s$ as $0.85 \wedge 0.73 = 0.73$. We will handle negation similarly.

With this agreement in mind, we can calculate the truth value of $\sim(h \vee \sim(s \wedge \sim h))$:

$$\sim(0.85 \vee \sim(0.73 \wedge \sim 0.85)) =$$
$$\sim(0.85 \vee \sim(0.73 \wedge 0.15)) =$$
$$\sim(0.85 \vee \sim 0.15) =$$
$$\sim(0.85 \vee 0.85) = \sim 0.85 = 0.15$$

Quiz Yourself 21

Let *p* have a truth value of 0.65, let *q* have a truth value of 0.80, and let *r* have a truth value of 0.55. Find the truth value of each statement.

a) $\sim(p \wedge q) \vee (\sim r)$

b) $((\sim p) \wedge q) \vee r$

Highlight: *A Fuzzy Logic Air Conditioner**

Mitsubishi Heavy Industries of Japan manufactures a fuzzy logic–controlled air conditioner that analyzes room characteristics to maintain a stable air temperature. A sensor determines whether anyone is in the room, and fuzzy logic is used to control temperature cycling and direction of air flow to achieve temperature stabilization. This technology avoids the annoying characteristic that traditional air conditioners have of turning on and off frequently. By smoothing out temperature variation, the air conditioner consumes less power and reduces the cost of cooling by more than 20 percent.

We use an equivalent form of the conditional in ordinary logic to define the conditional in fuzzy logic.

It is easy to use truth tables to prove that the conditional $p \rightarrow q$ is logically equivalent to $(\sim p) \vee q$. We use this result in defining how to compute the truth value of conditionals in fuzzy logic.

> **DEFINITION**
>
> If p and q are two statements in fuzzy logic, then the truth value of the **conditional**, $p \rightarrow q$, is defined to be the truth value of $(\sim p) \vee q$.

EXAMPLE 5   Computing the Truth Values of Conditionals

Let p have a truth value of 0.75 and let q have a truth value of 0.46. Find the truth value of each statement.

a) $p \rightarrow q$ b) $q \rightarrow p$ c) $\sim q \rightarrow \sim p$

SOLUTION: a) The truth value of $p \rightarrow q$ is the same as the truth value of $(\sim p) \vee q$, which is $(\sim 0.75) \vee 0.46 = 0.25 \vee 0.46 = 0.46$.

b) To calculate the truth value of $q \rightarrow p$, instead we calculate the truth value of $(\sim q) \vee p$, which is $(\sim 0.46) \vee 0.75 = 0.54 \vee 0.75 = 0.75$.

c) The truth value of $\sim q \rightarrow \sim p$ is found by calculating the truth value of $\sim(\sim q) \vee (\sim p)$. We simplify this by replacing $\sim(\sim q)$ with q. The truth value of $q \vee (\sim p)$ is $0.46 \vee (\sim 0.75) = 0.46 \vee 0.25 = 0.46$. It should not surprise you that this is the same truth value that $p \rightarrow q$ had. Why?   ◉

We will now show how to use fuzzy logic to make decisions. Deciding on what to do when your car needs an expensive repair is certainly a situation in which the factors you consider are not clear-cut. It is a perfect example of a problem that we can analyze using fuzzy logic.

Quiz Yourself 22

Let p have a truth value of 0.65, let q have a truth value of 0.80, and let r have a truth value of 0.55. Find the truth value of each statement.

a) $p \rightarrow q$

b) $(p \wedge q) \rightarrow r$

c) $\sim(p \vee q) \rightarrow \sim r$

---

* Maureen Caudill, "Using Neural Nets: Fuzzy Decisions," *AI Expert* (April 1990) p. 61.

EXAMPLE 6*   Making Decisions Using Fuzzy Logic

Assume that your car has broken down and you are trying to decide whether to repair it, buy a newer used car, or buy a brand-new car. In Table 2.1,† we list the characteristics of a car that are important to you and assign a number between 0 and 1 to indicate how important each characteristic is to you.

Characteristic	Importance
Cost	0.70
Reliability	0.60
Size (large enough)	0.45
Good gas mileage	0.65
Safety	0.75

TABLE 2.1   Fuzzy truth values describe how important each characteristic is to you.

Table 2.2 lists the degree to which each option satisfies these characteristics.

	Repair	Newer Used Car	New Car
Cost	0.80	0.65	0.30
Reliability	0.40	0.60	0.95
Size (large enough)	0.90	0.75	0.50
Good gas mileage	0.55	0.80	0.85
Safety	0.45	0.60	0.75

TABLE 2.2   Fuzzy truth values describe how well each option satisfies the characteristics that are important to you.

The ideal option for you is the one that scores the highest in the characteristics that are important to you. In other words, we want to consider fuzzy statements of the following type:

"If cost is important to you, then repairing your car satisfies this condition."
"If reliability is important to you, then buying a new car satisfies this condition."

We thus want truth values for fuzzy statements of the form $i \rightarrow c$, where i is a statement that a characteristic is important to you and c is a statement that an option possesses that characteristic. Recall that $i \rightarrow c$ has the same truth value as $(\sim i) \vee c$. Let us look at the truth value for the statement

"If cost is important to you, then repairing your car satisfies this condition."

---

* The method used in this example is based on Ronald R. Yager, "Concepts, Theory and Techniques: A New Methodology for Ordinal Multiobjective Decisions Based on Fuzzy Sets," *Decision Sciences* 12(4) (October 1981), pp. 589–600.
† In Tables 2.1 and 2.2, the assignment of values between 0 and 1 is somewhat arbitrary. Another person might make a different assignment of values.

Quiz Yourself 23

Calculate the truth value of each statement.

a) "If reliability is important to you, then buying a new car satisfies this condition."

b) "If being large enough is important to you, then buying a newer used car satisfies this condition."

Here i is the statement "Cost is important to you," which from Table 2.1 has a truth value of 0.70. The statement c is "Repairing your car satisfies this condition (cost)," which, from Table 2.2, has a truth value of 0.80. The truth value of $(\sim i) \vee c$ is thus $(\sim 0.70) \vee 0.80 = 0.30 \vee 0.80 = 0.80$.

We will now construct a truth table for all the statements $(\sim i) \vee c$ for each option: repair, buy a used car, buy a new car.

	i	$\sim i$	Repair c	Repair $(\sim i) \vee c$	Used c	Used $(\sim i) \vee c$	New c	New $(\sim i) \vee c$
Cost	0.70	0.30	0.80	0.80	0.65	0.65	0.30	0.30
Reliability	0.60	0.40	0.40	0.40	0.60	0.60	0.95	0.95
Size	0.45	0.55	0.90	0.90	0.75	0.75	0.50	0.55
Gas mileage	0.65	0.35	0.55	0.55	0.80	0.80	0.85	0.85
Safety	0.75	0.25	0.45	0.45	0.60	0.60	0.75	0.75
conjunction of all $(\sim i) \vee c$				**0.40**		**0.60**		**0.30**

In order for an option to be satisfactory, you wish to have all of the conditions satisfied as well as possible. Therefore, you are interested in cost AND reliability AND size AND gas mileage AND safety. Thus, to finish making your decision, you calculate the *conjunction* of all the expressions $(\sim i) \vee c$ for each option. The conjunction of all the expressions $(\sim i) \vee c$ for repairing your car is the minimum of the truth values in this column, which is 0.40 (highlighted). Similarly we obtain a value of 0.60 (highlighted) for buying a newer used car and 0.30 (highlighted) for buying a new car. Since buying a newer used car has the highest final rating, the best choice for you *according to this rating method* is to buy a newer used car.

The method we used in Example 6 is a real way to use fuzzy logic to make decisions. If you do not agree totally with this process for making the final decision, we invite you to criticize this method and to suggest alternatives in the exercises.

Exercises

In Exercises 1–8, assign a truth value between 0 and 1 to each fuzzy statement. Of course, the assignments will vary from student to student.

1. *The Simpsons* is a television show intended for adults.

2. Physicians earn high salaries.

3. Madonna is a good singer.

4. Abraham Lincoln was our best president.

5. The United States should increase its foreign aid.

6. We should spend more money to shelter the homeless.

7. Money brings happiness.

8. Baseball is a difficult sport.

In Exercises 9–12, calculate the truth value of the negation of each fuzzy statement. The truth value of each fuzzy statement is given in parentheses.

9. Picasso was a famous artist. (0.95)

10. Mark McGwire is known mainly for his prowess at pitching. (0.15)

11. Learning a musical instrument can teach self-discipline. (0.75)

12. Space research is vital to national defense. (0.40)

In Exercises 13–16, consider the following fuzzy statements:

p : Home buyers choose a home because of its location. (0.75)
q : Home buyers choose a home because of its cost. (0.85)

Determine the truth value of each statement.

13. Home buyers choose a home because of its location or cost.

14. Home buyers choose a home because of its location and cost.

15. Home buyers choose a home because of its location and not its cost.

16. It is not true that home buyers choose a home because of its location or its cost.

In Exercises 17–24, assume that p *has a truth value of 0.27,* q *has a truth value of 0.64, and* r *has a truth value of 0.71. Find the truth value of each statement.*

17. $p \wedge \sim q$

18. $\sim (p \vee q)$

19. $\sim (p \vee \sim q)$

20. $(p \vee r) \wedge (p \wedge \sim q)$

21. $\sim (p \vee \sim q) \wedge \sim r$

22. $p \rightarrow r$

23. $q \rightarrow (p \vee r)$

24. $\sim (p \vee q) \rightarrow \sim (q \wedge r)$

In Exercises 25–28, use the method described in Example 6 to evaluate each situation.

25. You are trying to decide which of two job offers to accept. The first job is a retail sales trainee for a large consumer products company and the second is a marketing

data analyst for a communications company. In the first table, we list the characteristics of a job that are important to you; in the second table, we list the degree to which each job satisfies these characteristics. Which job should you take?

Characteristic	Importance
Salary	0.70
Interesting	0.80
Work with people	0.60
Flexible hours	0.75

	Marketing Trainee	Data Analyst
Salary	0.60	0.65
Interesting	0.50	0.80
Work with people	0.90	0.30
Flexible hours	0.80	0.60

26. You are trying to decide whether to buy a house or continue to rent an apartment. In the first table, we list the characteristics that are important to you; in the second table, we list the degree to which each situation satisfies these characteristics. What should you do?

Characteristic	Importance
Cost	0.70
Close to job	0.60
Adequate space	0.55
Near city attractions	0.65

	House	Apartment
Cost	0.80	0.65
Close to job	0.40	0.80
Adequate space	0.90	0.75
Near city attractions	0.55	0.90

27. Your younger sister is having a difficult time deciding which of three colleges to attend: Little Small College

(LSC), Good Old State (GOS), and Mom's Favorite University* (MFU). In the first table, we list features of each school that are important to your sister and rank these qualities; in the second table, we list the degree to which each school satisfies these characteristics. What should she do?

Characteristic	Importance
Size (not too large)	0.60
Cost	0.80
Academics	0.75
Social life	0.90
Close to home	0.30

	LSC	GOS	MFU
Size (not too large)	0.95	0.60	0.40
Cost	0.40	0.80	0.30
Academics	0.80	0.70	0.65
Social life	0.55	0.75	0.70
Close to home	0.70	0.65	0.95

28. You are trying to decide which of three investments to make: A, B, or C. In the first table, we list the characteristics of investing that are important to you; in the second table, we list the degree to which each choice satisfies these characteristics. Which investment is your best choice?

Characteristic	Importance
Past performance	0.85
Safety	0.60
Liquidity	0.75
Minimum amount to invest	0.35
Management fee	0.55

	A	B	C
Past performance	0.80	0.65	0.90
Safety	0.45	0.60	0.80
Liquidity	0.60	0.75	0.75
Minimum amount to invest	0.85	0.60	0.65
Management fee	0.45	0.80	0.75

Further Exercises

†**29.** Communicating Mathematics Suppose that in fuzzy logic we decided only to allow truth values of 0 or 1. Explain how the definitions in fuzzy logic would compare with the definitions in symbolic logic that we discussed in earlier sections of this chapter.

30. Communicating Mathematics Why would it not make sense to discuss truth tables in fuzzy logic?

31. Communicating Mathematics Choose a situation you will face in which you must make a decision, and make the decision using the method in Example 6. Begin by identifying characteristics that are important to you and assign them levels of importance between 0 and 1, inclusive. Next, specify to what degree each choice satisfies

each characteristic. Finally, mimic the calculations in Example 6 to make your decision.

32. Communicating Mathematics Search the Internet to find some applications of fuzzy logic at various universities and commercial sites and report on what you find.

33. Communicating Mathematics Do you have any criticisms of the decision-making method that we used in Example 6? Are you comfortable with the way in which decisions are made by this method? Can you suggest any changes? If so, try to implement these changes and then reconsider the examples and exercises of this section to see whether your method results in different decisions being made.

---

* The reason that this is Mom's favorite university is that Mom would like to see your sister go to school close to home.
† Exercise numbers circled in red can be used as group exercises.

CHAPTER

3

Graph Theory:*
The
Mathematics of
Relationships

- A sales manager traveling to branch offices wishes to minimize the cost of the trip.
- As fog rolls into Chicago's O'Hare airport, air traffic controllers struggle to reschedule flights to get passengers to their destinations on time.
- An anthropologist seeks to trace the tribal alliances among tribes in New Guinea.
- As winter approaches, a public works administrator in Buffalo, New York, studies a street map to locate the most efficient routes for its snowplows.
- A political scientist studying a congressional subcommittee wants to identify the member who exerts the greatest influence.

Although these problems seem unrelated, each of these people is looking for the relationship of objects to each other. The sales manager is interested in relationships between cities in terms of the cost of traveling between them. The Buffalo administrator has to know how the sections of the city are connected by streets. The political scientist needs to measure how the committee members are related by their ability to influence each other.

In such diverse areas as sociology, biology, psychology, political science, and urban studies, researchers must understand the relationships among the objects that they study. In this chapter, you will learn the basic terminology and techniques of an area of

_____
* You can find more resources on graph theory at www.aw.com/pirnot.

mathematics called graph theory, and you will see how to model relationships in order to solve problems.

Later in the chapter, you learn the best way to schedule a large project, such as building a space colony, that consists of numerous smaller subtasks. There are many dependencies among the subtasks in this project and once you understand these relationships thoroughly, you can schedule the project efficiently. Using this schedule, you then can identify which parts of the project are critical and require further resources in order to speed the project along.

3.1   GRAPHS, PUZZLES, AND MAP COLORING

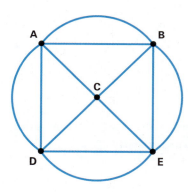

FIGURE 3.1   Puzzle tracing.

The topics we investigate in this chapter are quite different from the mathematics you have studied before. Here we are interested in constructing mathematical models that describe the relationships that occur among a collection of objects. Although our first few examples may lead you to believe that graph theory is concerned with children's games, this field has many theorems that have important practical applications as you will see later in this chapter.

Consider the puzzle in Figure 3.1. Attempt to solve this problem before reading on. Place your pencil at any dot on this figure and try to trace it completely without lifting your pencil and without tracing any part of any line twice. Do you think you can do it?

Instead of tackling such problems using trial and error, you will soon learn to apply a simple test to determine whether there is a solution to tracing problems such as this. But first, we need to introduce some terminology so we can describe this problem more precisely.

Graphs consist of vertices and edges.

> **DEFINITIONS**
>
> A **graph** consists of a finite set of points, called **vertices**, and lines, called **edges**, that join pairs of vertices.

Figure 3.1 is an example of a graph. Its vertices are the points A, B, C, D, and E and the twelve connecting lines are the edges. Generally, we use capital letters to designate vertices. If there is only one edge joining a pair of vertices, then we label that edge by the vertices it connects. For example, in Figure 3.1 we can refer to the edge joining vertices A and C as AC or CA; the order is not important. To refer to edge AB would be confusing because there are two edges joining vertices A and B. In this case we might label one edge e_1 and the other e_2.

The graph in Figure 3.2(a) has five vertices: A, B, C, D, and E. Although edges BC and AD intersect, the point of intersection is not a vertex. We will not consider a point of intersection of a pair of edges to be a vertex unless we place a solid dot there and label it as a vertex.

In drawing a graph, the important information is what vertices we connect by edges. The placement of the vertices and the shape of the edges are unimportant. We could have also drawn Figure 3.2(a) as in Figure 3.2(b), because this graph contains exactly the same vertices and edges.

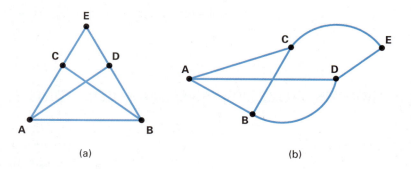

(a)   (b)

FIGURE 3.2   Two ways to draw the same graph.

Graphs represent relationships among objects.

We will now introduce several situations that we can model by graphs.

One of the most famous and yet elementary applications of graph theory originated in Koenigsberg, Prussia, during the eighteenth century. The Pregel River divided Koenigsberg into four distinct sections, as shown on the map in Figure 3.3 on the following page.

Seven bridges connected the four portions of Koenigsberg. It was a popular pastime for the citizens of Koenigsberg to start in one section of the city and take a walk visiting all sections of the city, trying to cross each bridge exactly once and to return to the original starting point. This problem is called the **Koenigsberg bridge problem**.

Historical Highlight: *The Development of Graph Theory*

In 1735, the Swiss mathematician Leonhard Euler solved the Koenigsberg bridge problem, and by doing so, he was the first person to work in graph theory. He wrote almost nine hundred papers, producing mathematics at such an incredible rate that his biographer Arago said he calculated without effort: "as men breathe, or as eagles sustain themselves in the wind." Incredibly, he wrote about four hundred of these papers during the last seventeen years of his life, when he was totally blind. One afternoon, while working on a problem, Euler suffered a stroke and with the words "I die" he ceased his calculations.

After Euler's work, the progress of graph theory was erratic. The English mathematician Arthur Cayley (1821–1895) became interested in the four-color problem (p. 153) and subsequently wrote several papers applying graph theory to chemistry.

In modern times, the area of mathematics known as operations research has greatly influenced the development of graph theory. Operations research applies mathematics to study the operation of systems—here *system* could mean a cancer cell, an industrial plant, or even the U.S. government. Graph theory is particularly useful in describing the interrelationships of the components of these complex systems.

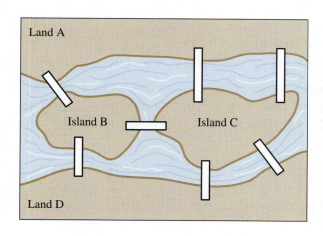

FIGURE 3.3   Map of Koenigsberg.

It might not be apparent that this is a graph theory problem; however, realize that any two bodies of land that are joined by a bridge can be represented in a graph as two vertices connected by an edge. In Figure 3.4, we model Koenigsberg using vertices A, B, C, and D to represent the bodies of land, and seven edges to represent the bridges.

In 1735, Leonhard Euler discovered a brilliant, yet simple way to determine when a graph can be traced.* To **trace** a graph means to begin at some vertex and draw the entire graph without lifting your pencil and without going over any edge more than once. Before giving the solution to this graph-tracing problem, we introduce a few more definitions.

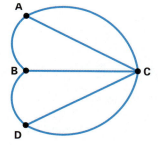

FIGURE 3.4   Graph model of Koenigsberg.

DEFINITIONS

A graph is **connected** if it is possible to travel from any vertex to any other vertex of the graph by moving along successive edges. A **bridge** in a connected graph is an edge such that if it were removed the graph is no longer connected.

---

* A very readable translation of Euler's original paper is J. R. Newman, editor, "Leonhard Euler and Koenigsberg Bridges," *Scientific American* (July 1953), pp. 66–70.

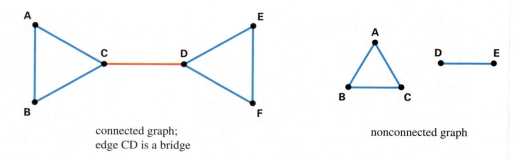

connected graph;
edge CD is a bridge

nonconnected graph

FIGURE 3.5   Examples of a connected graph and a nonconnected graph.

In Figure 3.5 we see examples of a connected graph, a graph that is not connected, and a bridge.

> **DEFINITIONS**
>
> A vertex of a graph is **odd** if it is an endpoint of an odd number of edges of the graph. Similarly, a vertex is **even** if it is an endpoint of an even number of edges.

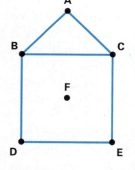

FIGURE 3.6   Odd and even vertices.

EXAMPLE 1   Odd and Even Vertices in a Graph

In Figure 3.6, determine which vertices are odd and which are even.

SOLUTION: Since B and C are endpoints of three edges, they are both odd. Vertices A, D, and E are endpoints of two edges, so they are even. Vertex F is also even because it is the endpoint of zero edges.   ◎

Quiz Yourself 1*

Is this graph connected? List its odd and even vertices. Are any edges bridges?

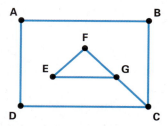

We make two observations regarding a graph G that can be traced.

Observation 1: If A is a vertex in G such that we neither begin nor end the tracing at A, then A must be even.

---

* Quiz Yourself answers begin on page 849.

This is easy to see. Assume that we are tracing the entire graph and are neither beginning nor ending at vertex A. Eventually we must come into vertex A by means of one edge, call it e_1, and then leave by a different edge, call it e_2. See Figure 3.7(a).

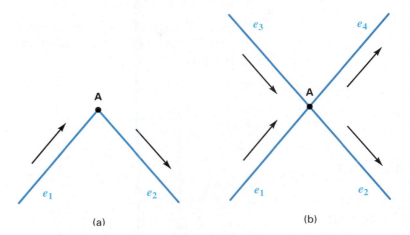

FIGURE 3.7 Every time we come into A by one edge, we must leave by another.

If no other edges are joined to A, then A is even because it is the endpoint of exactly two edges. If more edges are joined to A, then we will come into A by means of a third edge, call it e_3, and leave by a fourth edge, e_4. See Figure 3.7(b). Should A be the endpoint of more than four edges, remember that each time we come into A by one edge, we also must leave by a different edge. From this discussion, you can see that A must be the endpoint of an even number of edges and therefore it is even.

This means that an odd vertex can only be used as a starting or ending point in tracing a graph. This conclusion leads us to our second observation.

Observation 2: If a graph can be traced, then it can have at most two odd vertices.

This follows directly from Observation 1. One odd vertex could be the starting point and another one could be the ending point. Since no other vertices can be starting or ending points, all other vertices must be even.

Euler's theorem tells when a graph can be traced.

We now paraphrase Euler's theorem, which tells us when a graph can be traced.

> **Euler's Theorem**
>
> A graph can be traced if (1) it is connected and (2) it has either no odd vertices or two odd vertices.

If a graph has two odd vertices, the tracing must begin at one of these and end at the other. If all the vertices are even, then the graph tracing must begin and end at the same vertex. It does not matter at which vertex this occurs.

EXAMPLE 2   Tracing Graphs

Which of the graphs in Figure 3.8 can be traced? Recognize that Figure 3.8(a) is the puzzle graph in Figure 3.1 and Figure 3.8(b) is the graph of Koenigsberg and its bridges.

Quiz Yourself 2

Using Euler's theorem, state which of the following graphs can be traced. For a graph that can be traced, list a sequence of vertices that describes how to trace the graph. For a graph that cannot be traced, state which part of Euler's theorem fails.

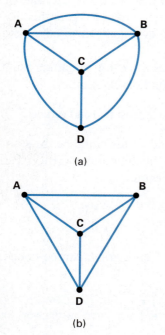

(a)

(b)

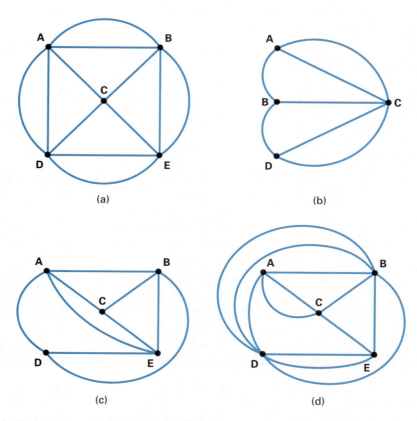

(a)

(b)

(c)

(d)

FIGURE 3.8   (a) Puzzle graph. (b) Koenigsberg graph. (c) and (d) can be traced by Euler's theorem.

SOLUTION: a) Notice that vertices A, B, D, and E are all odd; therefore, by Euler's theorem, this graph cannot be traced.

b) All vertices are odd; this means that the Koenigsberg bridge graph cannot be traced.

c) This graph has two odd vertices, C and D. One way to trace this graph is to begin at D and follow this sequence of vertices:

$$D, A, B, D, E, B, C, A, E, C$$

d) All vertices in this graph are even. Can you find a sequence of vertices describing how to trace this graph?

The Koenigsberg bridge problem is typical of situations that we can model by a graph. In such situations there is a collection of objects, like the four bodies of land, and there is a relationship among these objects, like the connection formed by the bridges. As we will see shortly, there are many diverse situations in which modeling the relationship among a set of objects gives us insight as to how to solve a problem.

Building a Graph Model

If we have a collection of objects with a relationship among them, then we develop a graph model as follows:

1. Represent each object by a vertex. Choose names that remind you what the vertices represent.
2. For each pair of related objects, join the two corresponding vertices with an edge.

EXAMPLE 3   Some Relationships among Objects in a Set

For each set, state a relationship among the objects that you could model with a graph.

a) a set of people

b) a set of countries

c) a set of animals

d) a set of towns

SOLUTION: We will give one relationship for each set, but of course you may think of others that are equally valid.

a) Two people are related if they have the same blood type.

b) Two countries are related if they share a common boundary.

c) Two animals are related if they compete for the same food.

d) Two towns are related if they are connected by a fiber optic cable.

Before discussing further applications of Euler's theorem, we introduce more terminology.

DEFINITIONS

A **path** in a graph is a series of consecutive edges in which no edge is repeated. The number of edges in a path is called its **length**. A path containing all the edges of a graph is called an **Euler path**. An Euler path that begins and ends at the same vertex is called an **Euler circuit**. A graph with all even vertices contains an Euler circuit and is called an **Eulerian graph**.

Highlight: *Graph Models in the Social Sciences*

Graph theory is used in many areas that people do not think of as normally involving mathematics. In the social sciences, graph theory models situations dealing with relationships among individuals and groups of people. Graph theory is an important tool of anthropologists investigating family and tribal structures, the transmission of myths, social organization, or the evolution of

thought and communication in primitive societies.* For example, Figure 3.10 depicts an alliance structure among tribes in the central highlands of New Guinea.

Although we will not delve into the intricacies of this model, we present it to show how a graph can help us organize nonnumerical data in order to understand their complexity.

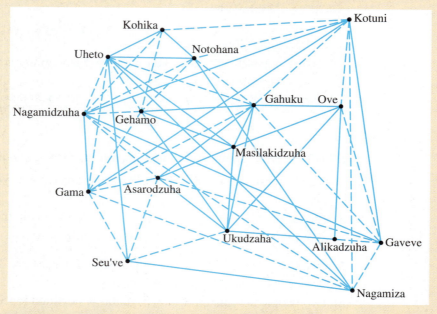

FIGURE 3.10   Graph depicting New Guinea tribal alliances.

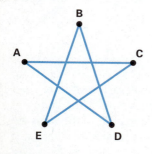

FIGURE 3.9   A graph containing an Euler circuit.

EXAMPLE 4   An Euler Circuit

Find some paths in the graph shown in Figure 3.9.

SOLUTION: There are many paths in this graph. For example, the path ACEB from A to B has length 3. Also, the path ACEBDA is an Euler path of length 5 that is also an Euler circuit because it begins and ends at the same vertex.

_____
* P. Hage and F. Harary, *Structural Models in Anthropology* (Cambridge University Press, 1983).

Quiz Yourself 3

a) Find an Euler path in this graph.

b) Is there an Euler circuit? Explain.

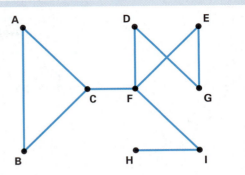

We use Fleury's algorithm to find Euler circuits.

Although we can use Euler's theorem to find out whether Euler circuits exist for a given graph, it does not tell us how to find them. Until now, we have used trial and error; however, in a large graph, this is not efficient. A method called **Fleury's algorithm** provides a systematic technique for finding Euler circuits. An **algorithm** is a series of steps that we follow to accomplish something. You might think of an algorithm as a recipe to get some mathematical task done.

> **Fleury's Algorithm**
>
> If a graph is connected and has all even vertices, we can find an Euler circuit for it as follows: Begin at any vertex and travel over consecutive edges according to the following rules:
>
> 1. After you have traveled over an edge, erase it. If all the edges for a particular vertex have been erased, then erase that vertex also.
>
> 2. Travel over an edge that is a bridge only if there is no alternative.

EXAMPLE 5   Using Fleury's Algorithm to Find an Efficient Route

Assume you are doing maintenance work along pathways joining locations A, B, and so on in a theme park, as shown in Figure 3.11. Find an Euler circuit in this graph in order to make your job efficient by not retracing pathways. Assume that you leave from and return to building C.

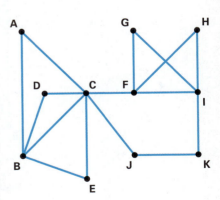

FIGURE 3.11 Graph showing paths in a theme park.

SOLUTION: We will solve this problem by using Fleury's algorithm to find an Euler circuit for this graph.

- *Step 1:* We begin at vertex C, traverse edge CJ, and erase it giving the graph in Figure 3.12(a).
- *Step 2:* We leave vertex J by means of edge JK, erase it, and, since all of J's edges have been removed, we also erase J. The path so far is CJK. The resulting graph is shown in Figure 3.12(b).

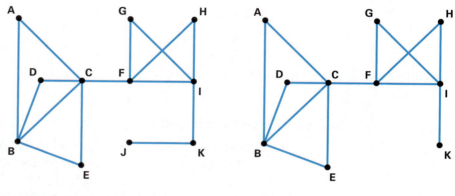

FIGURE 3.12   (a) Step 1.   (b) Step 2.

- *Step 3:* We next traverse edges KI and IF, erasing both and also vertex K (Figure 3.13(a)). The path constructed to this point is CJKIF.
- *Step 4:* We now are at vertex F. If we were to traverse the bridge FC and erase it, we would disconnect the graph and not be able to finish our circuit. So instead we traverse FG, GI, IH, and HF. Our path is now CJKIFGIHF and, after removing appropriate edges and vertices, the graph looks like Figure 3.13(b).
- *Step 5:* We have no choice now but to traverse FC. We follow this with CA and AB. The path is now CJKIFGIHFCAB and the graph looks like Figure 3.13(c).

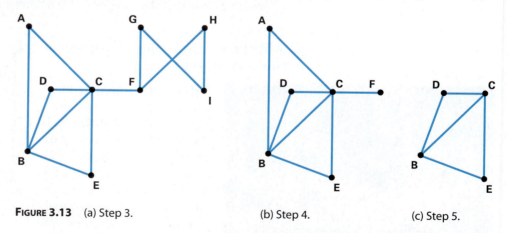

FIGURE 3.13   (a) Step 3.   (b) Step 4.   (c) Step 5.

We can finish the circuit by traveling over BE, EC, CD, DB, and BC. The final circuit is CJKIFGIHFCABECDBC. Notice that we have traversed every edge exactly once and ended at our starting point, vertex C. If you were to follow this circuit, you would cover each path exactly once in performing maintenance on the paths in the park.

In Example 5, some of the decisions we made were arbitrary, and there are certainly many different ways to construct an Euler circuit for this graph.

We can add edges to convert a non-Eulerian graph to an Eulerian graph.

Being Eulerian is a desirable property for a graph, but of course if a graph has any odd vertices, it cannot be Eulerian. In this case, we duplicate some edges in the graph to change the odd vertices to even so that the graph becomes Eulerian. This is called **Eulerizing the graph**. Example 6 shows a practical application of this technique.

EXAMPLE 6   Designing a Shuttle Bus Route

The Metrodelphia Transportation Authority is planning a new shuttle bus system to transport shoppers around the business district of the city, which is shown in the map in Figure 3.14. If possible, we wish to design a route that travels over all the streets of the zone exactly once and returns to the starting location. If that is not possible, we would like to minimize traveling over any streets more than once. Find such a bus route.

SOLUTION: We can model this map with the graph shown in Figure 3.15. We represent each intersection by a vertex and each section of street joining two intersections by an edge.

We labeled the *odd* vertices in this graph A, B, and so on. To Eulerize this graph, we duplicate some edges so that the new graph has only even vertices. We show this in Figure 3.16.

If we begin our route at the upper right corner of the graph and follow the edges as they are numbered, we will traverse all the edges of the graph and return to our starting point. Notice that the five pairs of duplicate edges—AB, CG, and so on—represent streets that must be traveled twice. Although we will not prove it, there is not a better way to Eulerize the original graph to reduce the number of streets that are traveled twice.

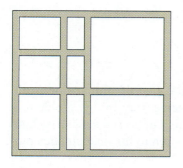

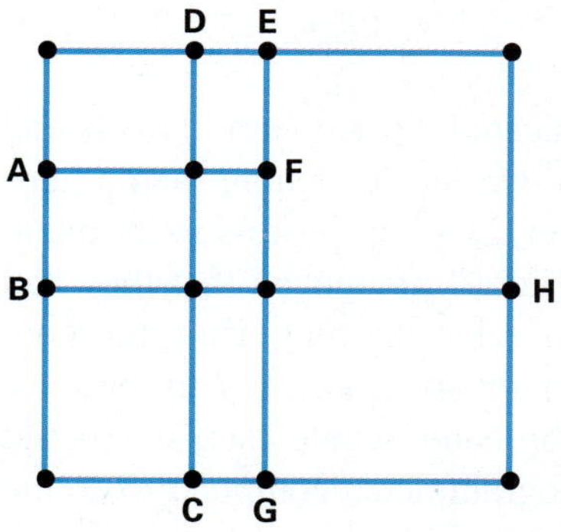

FIGURE 3.14 Map of Metrodelphia commercial zone.

FIGURE 3.15 Graph representing Metrodelphia's business district.

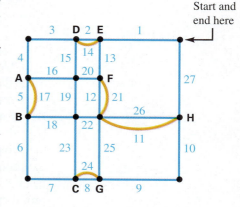

FIGURE 3.16 Shuttle bus route for Metrodelphia.

SOME GOOD ADVICE

When Eulerizing a graph, you can only duplicate edges that are *already present* in the graph. In Example 6, you are not allowed to insert a *single* edge from F to H to Eulerize the graph.

 Quiz Yourself 4

Add edges to this graph to make it Eulerian.

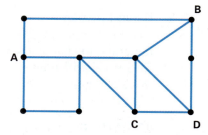

The four-color problem is another application of graph theory.

We now turn our attention to an interesting problem called the **four-color problem**. Although we state this as a puzzle, like the Koenigsberg bridge problem, it has interesting real-life applications. In 1852, Francis Guthrie, a student at University College, London, first posed this famous question to his mathematics professor, Augustus DeMorgan.

> **The Four-Color Problem**
>
> Using at most four colors, is it always possible to color a map so that any two regions sharing a common border receive different colors?

Unable to find an answer, DeMorgan communicated the problem to his friend Sir William R. Hamilton at Trinity College, Dublin, Ireland. For more than a hundred years this problem remained unsolved. However, in 1976, professors Kenneth Appel and Wolfgang Haken of the University of Illinois announced that they had solved this problem by proving that it is possible to color any map using no more than four colors. However, their proof was treated with some skepticism, because the proof was not done in the traditional method—by hand, where each step could be checked for validity. Instead Appel and Haken programmed a computer to do the proof; the completion of the proof required 1,200 hours of computer time!

For an illustration of the statement of this problem, consider the map of South America in Figure 3.17.

FIGURE 3.17   Map of South America.

Using at most four colors, we wish to color this map so that we use a different color for any two countries having a common border. For example, we cannot use the same color for Chile and Peru; however, we can use the same color for Paraguay and Colombia.

EXAMPLE 7   Solving the Four-Color Problem for South America

Model the map of South America by a graph and use this graph to color the map using at most four colors.

SOLUTION: In this problem, we have a set of countries, some of which are related in that they share a common border. Therefore we can model this situation by a graph. We will represent each country by a vertex; if two countries share a common border, we draw an edge between the corresponding vertices. This graph appears in Figure 3.18.

Note that we connect the vertices representing Peru and Colombia with an edge, because they share a common boundary. We do not connect the vertices representing Argentina and Peru, because they have no boundary in common.

We can rephrase the map-coloring question now as follows: Using four or fewer colors, can we color the vertices of a graph so that no two vertices of the same edge receive the same color? It is easier to think about coloring a graph than it is to think about coloring the original map.

There are several ways to color the graph in Figure 3.18. However, unlike tracing graphs, there is no particular procedure for accomplishing this coloring except trial and error. One possible coloring for the graph appears in Figure 3.19. You may wish to color this graph in a different way; however, notice that you cannot do it using fewer than four colors.

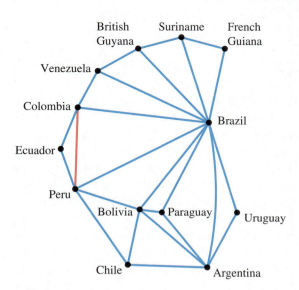

FIGURE 3.18   Graph model of map of South America.

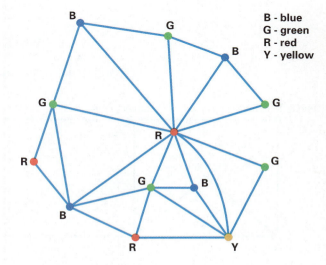

FIGURE 3.19   Coloring of graph of South America.

Quiz Yourself 5

Color the following "map" by first modeling it with a graph and then coloring the graph, as we did in Example 7.

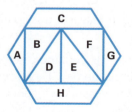

Example 8 shows how we can use map coloring for a practical purpose.

EXAMPLE 8   Using Graph Theory to Schedule Committees

Each member of a city council usually serves on several committees to oversee the operation of various aspects of city government. Assume that council members serve on the following committees: police, parks, sanitation, finance, development, streets, fire department, and public relations.

Use Table 3.1, which lists committees having common members, to determine a conflict-free schedule for the meetings. We do not duplicate information in Table 3.1. That is, because police conflicts with fire department, we do not also list that fire department conflicts with police.

Committee	Has Members in Common With:
Police	Public relations, fire department
Parks	Streets, development
Sanitation	Fire department, parks
Finance	Police, public relations
Development	Streets
Streets	Fire department, public relations
Fire department	Finance

TABLE 3.1   Committees that have common members.

SOLUTION: We first will model the information in this table with the graph in Figure 3.20. We join two committees by an edge provided they have a conflict. This problem is similar to the map-coloring problem. If we color this graph, then all vertices having the same color represent committees that can meet at the same time. We show one possible coloring of the graph in Figure 3.20.

From Figure 3.20, we see that the police, streets, and sanitation committees have no common members and therefore can meet at the same time. Public relations, development, and fire can meet at a second time. Finance and parks can meet at a third time.

FIGURE 3.20   Graph model of committees with common members.

Exercises 3.1

In Exercises 1–8, determine whether the graph is connected. Which vertices are odd? Which vertices are even?

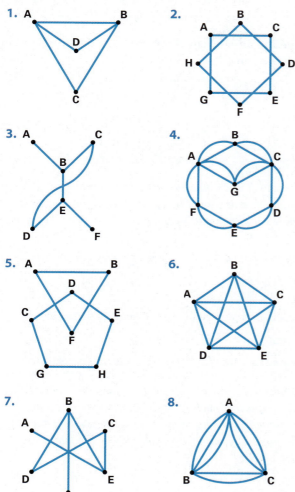

1.

2.

3.

4.

5.

6.

7.

8.

In Exercises 9–16, try to give an example of each graph that we describe. If, after several tries, you cannot find the graph that we have requested, explain why you think that it may be impossible to find that example. (In the answer key, we will give examples which are as simple as possible in the sense that they have the fewest number of vertices and edges.) The degree of a vertex is the number of edges that are joined to that vertex.

9. A graph with four even vertices

10. A graph with four odd vertices

11. A graph with three odd vertices

12. A graph with four vertices of degree two and two vertices of degree three

13. A connected graph with one even vertex and four odd vertices

14. A graph with one odd vertex

15. A graph with six vertices of degree three

16. A graph with five vertices and the largest number of edges so that no edges are repeated. That is, there will be only one edge joining A and B, only one edge joining A and C, and so on.

(17.) Communicating Mathematics   Make up several examples of graphs. By examining these graphs, explain why the sum of the degrees of all the vertices in a graph must be an even number.

(**18.**) Communicating Mathematics   Use Exercise 17 to explain why a graph cannot have an odd number of odd vertices.

For Exercises 19–26, use Euler's theorem to decide whether the specified graph can be traced. If the graph cannot be traced, tell which conditions of the theorem fail.

19. The graph in Exercise 1   **20.** The graph in Exercise 2

21. The graph in Exercise 3   **22.** The graph in Exercise 4

23. The graph in Exercise 5   **24.** The graph in Exercise 6

25. The graph in Exercise 7   **26.** The graph in Exercise 8

27. Communicating Mathematics   Assume that a graph is traceable. Explain why we cannot encounter an odd vertex in the middle of tracing that graph.

28. Communicating Mathematics   If we are to trace a graph with two odd vertices, explain why we must start at one odd vertex and end at the other.

In Exercises 29–34, if the given graph is Eulerian, find an Euler circuit in it. If the graph is not Eulerian, first Eulerize it and then find an Euler circuit. Write your answer as a sequence of vertices, as we did in Example 5. There are many

---

* Exercise numbers circled in red can be used as group exercises.

possible correct answers to these exercises. We will provide only one in the answer key.

29.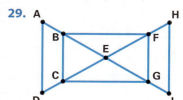

30. 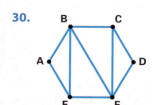   **31.**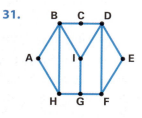

32.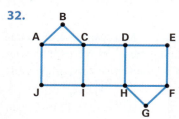

33.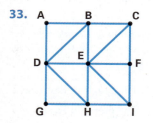

34.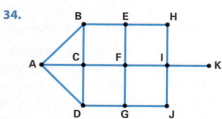

35. Finding an efficient route. An ice cream vendor wishes to travel over each of the streets indicated in the following map, but does not want to travel over any part of the route more than once. Can this be done? Explain your answer.

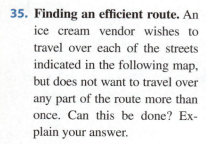

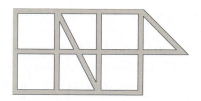

36. Finding an efficient route. Repeat Exercise 35 for the following map.

In Exercises 37–38, model each street map by a graph and design an efficient way of traveling through all the streets so that there are a minimal number of streets that are traveled more than once.

37. **38.**

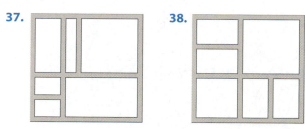

In Exercises 39–42, use the map of the continental United States to represent each group of states with a graph as we did in Example 7. Recall that vertices representing two states will be joined by an edge if and only if the states share a common border.

Figure for Exercises 39–42

39. Map coloring. Washington, Oregon, Idaho, Montana, Wyoming

40. Map coloring. Ohio, Kentucky, Tennessee, Arkansas, Missouri, Illinois, Indiana, Iowa, Minnesota, Wisconsin

41. Map coloring. Texas, Oklahoma, Arkansas, Louisiana, Mississippi, Alabama, Tennessee, North Carolina, South Carolina, Georgia, Florida

42. Map coloring. California, Nevada, Utah, Arizona, New Mexico, Colorado, Kansas, Oklahoma, Arkansas, Louisiana

In Exercises 43–46, we give you a group of states. Is it possible to begin in one of the states in the group and travel through all of the states without ever crossing the same boundary between two states twice? Hint: Think of the graphs that you drew in Exercises 39–42.

43. Use the states that we listed in Exercise 39.

44. Use the states that we listed in Exercise 40.

45. Use the states that we listed in Exercise 41.

46. Use the states that we listed in Exercise 42.

47. Finding an efficient route. The following is a floor plan of a suite of offices. Assume all the doors indicated are open. Is it possible for a security guard to enter the suite from the hallway, pass through each door locking it behind him, and then exit without ever having to open a door that has been previously locked? (*Hint:* Model this with a graph where you consider the set of objects to be the rooms and the hallway and that any two rooms are related if they are connected by an open door.)

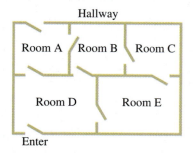

48. Finding an efficient route. The following is another floor plan for a suite of offices. The situation is the same as in Exercise 47, except there is now only one door by which the security guard can enter the suite. Place an exit door in one of the rooms A, B, or C so that the guard can enter by the door marked "Enter," pass

through each door and lock it behind him, and then exit by the door you've placed in the plan. Explain why this is the only possible place to locate the door.

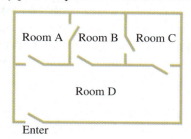

Use the technique presented in Example 8 to answer Exercises 49–52. We do not list duplicates in the tables of information.

49. Avoiding conflicts. The chief of protocol is arranging a state department luncheon for the ambassadors of 12 countries. It is important that ambassadors not on friendly terms be seated at different tables.

Ambassador From:	Is Not Friendly with Ambassador From:
Russia	United States, Poland, Canada, Latvia, Britain
Germany	Latvia, Canada
Mexico	United States, Britain
Italy	Spain, Portugal
Poland	Canada, France
Britain	United States
Latvia	Spain, Canada
Portugal	Spain, United States

Determine a satisfactory seating arrangement for the luncheon using as few tables as possible.

50. Avoiding conflicts. The designer of the new Jungle World entertainment complex wants to include several large enclosures in which wild animals can roam freely. If one animal can harm another, then those two animals cannot share the same enclosure. (See table to the right.) Use this information to determine the smallest number of enclosures necessary to contain the animals. Also, state how you could assign the animals.

Animal	Cannot Be Placed With:
Tiger	Zebra, leopard, rhinoceros, giraffe, antelope, ostrich
Leopard	Zebra, boar, antelope, giraffe, ostrich
Crocodile	Ostrich, heron
Boar	Tiger, crocodile, zebra

51. Scheduling meetings. A college's faculty has a number of committees that meet Tuesdays between 11:00 and 12:00. To avoid conflicts, it is important not to schedule two committee meetings at the same time if the two committees have faculty members in common. Use the following table, which lists possible conflicts, to determine an acceptable schedule for the meetings.

Committee	Has Members in Common With:
Academic standards	Academic exceptions, scholarship, faculty union
Computer use	University advancement, event scheduling
Campus beautification	Curriculum, faculty union, event scheduling
Affirmative action	Academic exceptions, scholarship
University advancement	Parking, curriculum, academic standards
Parking	Academic standards, affirmative action
Faculty union	Computer use, event scheduling
Scholarship	Campus beautification

52. Scheduling meetings. Repeat Exercise 51 using the following table.

Committee	Has Members in Common With:
Academic standards	Scholarship, university advancement
Computer use	University advancement, affirmative action
Campus beautification	Curriculum, academic festival, faculty union, event scheduling, scholarship
Affirmative action	Scholarship
University advancement	Curriculum, academic festival, faculty union
Faculty union	Computer use, event scheduling, academic festival
Long-range planning	Campus beautification, computer use

Further Exercises

53. Draw a graph that can be colored with only two colors.

54. Communicating Mathematics   Draw a graph that cannot be colored with two colors but can be colored with three. Can you state what configuration of vertices will force us to use at least three colors in coloring a graph?

55. Draw a graph that cannot be colored with three colors, but can be colored with four.

56. Communicating Mathematics   Can you state what configuration of vertices will force us to use at least four colors in coloring a graph?

57. Different notes on a trumpet are obtained by moving its three valves up and down. The given table indicates the eight possible positions for these three valves and a note

that would sound if the valves were in the indicated position. Is it possible to play one of these notes and then, by changing only one valve at a time, to play all the other notes without repeating a note twice? (*Hint:* Consider the notes as the objects of a set and that two notes are related if one can be obtained from the other by moving only one valve.)

58. If an instrument has four valves, there are sixteen possible positions for the valves. Repeat Exercise 57 for this situation.

Valve 1	Up	Up	Up	Up	Down	Down	Down	Down
Valve 2	Up	Up	Down	Down	Up	Up	Down	Down
Valve 3	Up	Down	Up	Down	Up	Down	Up	Down
Note	C	A	B	Eb	F	D	E	C#

3.2   THE TRAVELING SALESPERSON PROBLEM

Some problems in mathematics are so simple to state that a child can understand them, yet the greatest mathematicians in the world cannot solve them. This is the case with a famous and difficult problem in graph theory called **the traveling salesperson problem (TSP)**. The TSP gets its name from the problem of determining the most efficient way for a salesperson to schedule a trip to a series of cities and then return home.

For an example of a TSP, suppose that Danielle, who is regional sales manager for a publishing company, lives in Philadelphia and must make visits next week to branch offices in New York City, Cleveland, Atlanta, and Memphis (see Figure 3.21). In order to determine which would be her cheapest trip, she has obtained prices of flights between each pair of

FIGURE 3.21   Cities Danielle must visit.

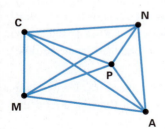

Figure 3.22 Graph representing cities Danielle will visit.

cities.* Danielle could start at Philadelphia and fly first to New York, then to Cleveland, Atlanta, and finally to Memphis before returning home. Or, she could first fly to Cleveland, then Atlanta, Memphis, New York, and then home. If Danielle wants to find the cheapest trip, she will have to spend some effort in considering all possible trips through the four cities and then return home.

In keeping with the spirit of this chapter, we model this situation with a graph. We represent each city by a vertex, and join two vertices by an edge if Danielle might fly from one to the other. In this graph, every pair of vertices will be joined by an edge. For clarity, we will represent each city by a letter—P for Philadelphia, N for New York City, and so on. This gives us the graph in Figure 3.22.

We now model Danielle's possible trips by paths in this graph. For instance, we can represent the trip (Philadelphia, New York, Cleveland, Atlanta, Memphis, home) by the path PNCAMP, and the trip (Philadelphia, Cleveland, Atlanta, Memphis, New York, home) by the path PCAMNP. If we knew the prices of all the possible flights, we could then solve Danielle's problem. But first we need to develop some more graph theory.

DEFINITIONS

A path that passes through all the vertices of a graph exactly once is called a **Hamilton path**. If a Hamilton path begins and ends at the same vertex, then it is called a **Hamilton circuit**.

PROBLEM SOLVING

Remember the "splitting hairs" principle in Section 1.1. Although the definitions of *Hamilton path* and *Euler path* sound similar, they are not the same. In producing a Hamilton path you do not have to trace every edge, as with an Euler path.

EXAMPLE 1   Hamilton Paths and Circuits

Find a Hamilton path in each graph shown in Figure 3.23.

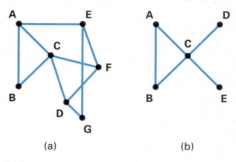

(a)   (b)

Figure 3.23 Hamilton paths and circuits.

SOLUTION: a) In Figure 3.23(a), the path ABCFDGE is a Hamilton path. If we return to vertex A, then ABCFDGEA is a Hamilton circuit.

b) The graph in Figure 3.23(b) has no Hamilton paths or circuits. If you start at any vertex, you will see that you cannot follow edges to visit every vertex exactly once. For example, if you start at vertex A and then go to vertices B and C, you

---

* We will assume that direction is not important here and that a one-way flight between two cities costs the same in either direction.

are in trouble. If you go to D, then you cannot get to E without passing through C a second time. Similarly, if you go to E, you cannot get to D. You may start your path at other vertices but you will find that no matter how you try you cannot find a Hamilton path.

You might expect, now that you have seen graphs with and without Hamilton paths, that we will next state a theorem telling exactly when a graph has Hamilton paths. Unlike the situation with Euler's theorem for graph tracing, there is no hard-and-fast rule for determining when a graph has a Hamilton path. If you think about it, you see that the reason we could not find a Hamilton path in the graph in Example 1(b) was that there were not enough edges. When we went to vertex D, we were stuck. If we had more edges to use, we might have avoided going back through C a second time to get to vertex E. Often we will be dealing with graphs which have every possible edge.

> **DEFINITION**
>
> A **complete graph** is one in which every pair of vertices is joined by an edge. A complete graph with n vertices is denoted by K_n.*

EXAMPLE 2   Complete Graphs

Each graph shown in Figure 3.24 is complete.

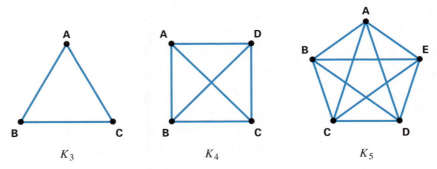

K_3 $\qquad$ K_4 $\qquad$ K_5

FIGURE 3.24   Complete graphs.

Tree diagrams help us find Hamilton circuits systematically.

Often we need to find all the Hamilton circuits in a graph. It is easy to do this in complete graphs.

EXAMPLE 3   Finding Hamilton Circuits in K_4

Find all Hamilton circuits in K_4.

---

* The notation K for complete graphs is chosen to honor the twentieth-century Polish mathematician Kasimir Kuratowski, who discovered several important theorems in graph theory.

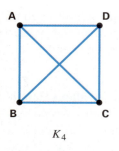

K_4

FIGURE 3.25 K_4.

SOLUTION: The path ABCDA is a Hamilton circuit in the graph of Figure 3.25. For the sake of counting circuits, we will not consider the path BCDAB to be a different Hamilton circuit. Any two circuits that pass through the same vertices in the same order, regardless of the starting vertex, will be considered to be the same. Therefore, in forming a Hamilton circuit, we can always assume it begins at vertex A.

If we are systematic, we can easily list all the Hamilton circuits. We will begin at vertex A. From A we can go to either B, C, or D. If we go to C, then we can go next to either B or D. It is easy to visualize all the possibilities using a device called a *tree diagram*. We show how to do this in Figure 3.26.

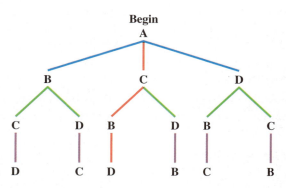

FIGURE 3.26 Finding Hamilton circuits in K_4.

The place marked "Begin" means that we are beginning at vertex A. The three branches below leading to B, C, and D indicate that after starting at vertex A, we can move to either vertex B, vertex C, or vertex D. The path we have highlighted in red shows that we can form a Hamilton path beginning at A, and then going to vertices C, B, and D, in that order. Returning to vertex A, we get the Hamilton circuit

ACBDA.

Tracing through the six branches of the tree, we see that K_4 has six Hamilton circuits:

ABCDA, ABDCA, ACBDA, ACDBA, ADBCA, ADCBA

Notice in Example 3 that the Hamilton circuits occur in pairs. Consider the circuit ABCDA. If we list these vertices in reverse order, we get the circuit ADCBA. In any graph, once we have found one Hamilton circuit, we automatically have another one by listing the vertices in reverse order.

Let us see if we can generalize what happened in Example 3. In constructing the tree, we assumed that the Hamilton paths always began at vertex A. This left us with three choices for the second vertex of the path which meant the tree now had three branches. Since there were now only two vertices left to choose, this gave us two new branches for each of the three existing branches. The tree at this stage had six branches. We now had only one vertex left to finish the Hamilton path, so the final tree had $3 \times 2 \times 1 =$ six branches.

If we were solving this problem for K_5, we would begin each path with vertex A. This would leave 4 vertices for the second vertex of the path. The tree at this point would have 4 branches. Because there are 3 vertices remaining, we would add 3 branches to each of these 4 branches, to give us a tree with 12 branches. Now only 2 vertices remain. We would indicate this in the tree by

Historical Highlight: *William Rowan Hamilton**

Hamiltonian circuits, that you study in this section, are named after whom many consider to be the greatest Irish mathematician of all time, William Rowan Hamilton. Hamilton was born in Dublin, Ireland, in 1805, and, except for short trips, he spent his entire life in that city. His primary education was done at home, and at an early age he showed a great proficiency at learning different languages. It is said that by the time he was thirteen, he knew as many languages as his age.

In his early teens, he took part in a contest in Dublin with an American youth named Zerah Colburn, who was known as a "lightning calculator." When Hamilton came in second best in the contest, it made him dedicate himself fiercely to studying mathematics. He read the works of Euclid in the original Greek, Isaac Newton in Latin, and the eminent French mathematician Pierre Simon Laplace in French. At the age of

17, he discovered an error in Laplace's work, and he sent his discovery to the president of the Royal Irish Academy, who declared that at such a young age, Hamilton was already a first-rate mathematician.

He entered Trinity College in Dublin, and distinguished himself not only in mathematics, but also in poetry. He often said that although he was a mathematician by profession, he was a poet by heart.

In 1827, he presented a long technical paper on the geometry of optics, which so impressed the faculty of Trinity College, that he was appointed a professor of astronomy although he was still an undergraduate student!

In addition to graph theory, he made significant contributions to many other areas of mathematics. His work, which seemed very theoretical at the time, has become useful in many applied areas of mathematics, science, and engineering.

adding 2 branches to each of the 12 branches, giving a tree with 24 branches. For each of these 24 paths, we have only one remaining vertex to choose, so we would draw a single branch from each of the 24 branches to get a final tree with 24 branches. Notice that the number 24 is the product

$$4 \times 3 \times 2 \times 1.$$

You can see now that if we were solving this problem for K_n, we would start each Hamilton path at vertex A. We would then draw $n - 1$ branches corresponding to the $n - 1$ remaining vertices. We would then draw $n - 2$ branches from each of these $n - 1$ branches, giving us $(n - 1)(n - 2)$ branches. Continuing in this fashion, we would get a final tree having $(n - 1)$ $(n - 2)\,(n - 3)\,(n - 4)\,\ldots\,3 \times 2 \times 1$ branches.

The Number of Hamilton Circuits in K_n

K_n has $(n - 1)\,(n - 2)\,(n - 3)\,(n - 4)\,\ldots\,3 \times 2 \times 1$ Hamilton circuits. This number is written $(n - 1)!$ and is called $(n - 1)$ *factorial.*

---

* This highlight on Hamilton is based on material from David M. Burton, *The History of Mathematics: An Introduction* (McGraw-Hill, 1999).

Table 3.2 shows the number of Hamilton circuits in K_n for various values of n.

n	Number of Hamilton Circuits in K_n
3	$2! = 2 \cdot 1 = 2$
4	$3! = 3 \cdot 2 \cdot 1 = 6$
5	$4! = 4 \cdot 3 \cdot 2 \cdot 1 = 24$
10	$9! = 362,880$
15	$14! = 87,178,291,200$
20	$19! = 121,645,100,408,832,000$

TABLE 3.2   Number of Hamilton circuits in K_n.

Quiz Yourself 6*

How many Hamilton circuits are there in K_7?

As you can see, as n increases, the number of Hamilton circuits in K_n grows at an incredible rate.

In solving a TSP problem by brute force, we consider all possible Hamilton circuits.

Now that we have introduced some basic ideas regarding Hamilton circuits, we are almost ready to solve the TSP posed earlier. Recall that Danielle was deciding which order for visiting her branch offices would be the cheapest. In Figure 3.27, we redraw the graph that models her possible trips, except now on each edge we have indicated the cost of traveling between the cities represented by the vertices of the edge.

> **DEFINITIONS**
>
> When we assign numbers to the edges of a graph, the graph is called a **weighted graph** and the numbers on the edges are called **weights**. The **weight of a path** in a weighted graph is the sum of the weights of the edges of the path.

In Danielle's problem, the weights represent money; in other problems, the weights might represent distance or time.

EXAMPLE 4   Solving a TSP Using Brute Force

Use Figure 3.27 to find the sequence of cities for Danielle to visit that will minimize her total travel cost.

SOLUTION: Another way to state this problem is that we wish to find the Hamilton circuit that has the smallest weight. For example, if Danielle chooses the circuit PMACNP, the cost of her trip will be $280 + 170 + 290 + 350 + 210 =$ $1,300. Our plan for a solution is simple, but tedious. We will calculate the

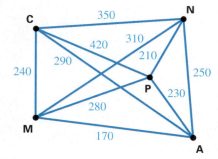

FIGURE 3.27   Graph model of Danielle's possible trips. Weights on edges denote cost of travel between cities.

_____
* Quiz Yourself answers begin on page 849.

Hamilton Circuit	Weight ($)
PACMNP	1,280
PACNMP	1,460
PAMCNP	1,200
PAMNCP	1,480
PANCMP	1,350
PANMCP	1,450
PCAMNP	1,400
PCANMP	1,550
PCMANP	1,290
PCNAMP	1,470
PMACNP	1,300
PMCANP	1,270

TABLE 3.3 Hamilton circuits in G and their weights.

weight of each Hamilton circuit in the graph. The circuit with the smallest weight will be our solution. Since this graph has five vertices there will be 4! = 24 Hamilton circuits. However, in Table 3.3 we have listed only twelve of these circuits, because clearly each circuit has the same weight as its reverse.

From Table 3.3, we see that the smallest weight is $1,200, which corresponds to the circuit PAMCNP. Thus Danielle should start at Philadelphia and then travel to Atlanta, Memphis, Cleveland, and New York City in that order and then return home. ◎

In Example 4, our solution was not clever or sophisticated. We simply considered every possible Hamilton circuit and chose the best one. This approach to the TSP is called **brute force**. Realize how impractical this method would have been if Danielle visited fifteen cities. In that case, from Table 3.2 we see that we would have had to consider 14! = 87,178,291,200 Hamilton circuits. Suppose that we were doing this by hand and could examine one circuit per minute. Now, multiplying 60 minutes per hour times 24 hours per day times 365 days per year gives us 525,600 minutes per year. If we divide 14! by this number, we get

$$\frac{14!}{525,600} = \frac{87,178,291,200}{525,600} \approx 165,864 \text{ years}$$

to complete the solution by hand!

Of course, you would probably argue that we should use a computer to solve such a large problem. Suppose that we had a fast computer that could examine 100 circuits per second. There are 31,536,000 seconds in a year, so the computer could examine 31,536,000 × 100 = 3,153,600,000 Hamilton circuits per year. Thus it would still take the computer

$$\frac{14!}{3,153,600,000} = \frac{87,178,291,200}{3,153,600,000} \approx 28 \text{ years}$$

to complete the solution!

Recall that an algorithm is a series of steps that finds a solution for a problem. We will formalize what we did in Example 4.

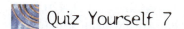

Quiz Yourself 7

Use the brute-force algorithm to find a Hamilton circuit with minimum weight in the graph.

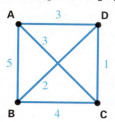

The Brute-Force Algorithm for Solving the TSP

Step 1: List all Hamilton circuits in the graph.
Step 2: Find the weight of each circuit found in Step 1.
Step 3: The circuits with the smallest weights tell us the solution to the TSP.

There are algorithms that give good approximations to solutions to the TSP.

It would be nice if there were an algorithm that would solve a TSP much faster than the brute-force algorithm. Unfortunately, at present there is no such algorithm and, in fact, mathematicians do not know whether it is even possible to find such an algorithm.

In a situation such as this, mathematicians ask a different question: "Is there an algorithm that, although it may not find the best solution, does find a pretty good approximation to the best solution?" In fact, a number of algorithms will do this. One rather intuitively simple approach is, when constructing a Hamilton circuit, always to choose the next edge that has the smallest weight. This method is called the **nearest-neighbor algorithm**.

The Nearest-Neighbor Algorithm for Solving the TSP

Step 1: Start at any vertex *X*.

Step 2: Of all the edges connected to *X*, choose any one that has the smallest weight. (There may be several with smallest weight.) Select the vertex at the other end of this edge. This vertex is called the *nearest neighbor* of *X*.

Step 3: Choose subsequent *new* vertices as you did in Step 2. When choosing the next vertex in the circuit, choose one whose edge with the current vertex has the smallest weight.

Step 4: After all vertices have been chosen, close the circuit by returning to the starting vertex.

EXAMPLE 5   Solving a TSP Using the Nearest-Neighbor Algorithm

Use the nearest-neighbor algorithm to schedule Danielle's trip.

SOLUTION: We redraw the graph in Figure 3.28, showing the possibilities for her trip.

Danielle will begin her trip in Philadelphia, so we start the Hamilton circuit with vertex P. There are four edges connected to P with weights of 210, 420, 280, and 230. According to the nearest-neighbor algorithm, we select the edge with weight 210, which means that N is the next vertex in our circuit. We do not want to choose P again, so we consider the edges connected to N of weights 350, 310, and 250. Since 250 is the smallest weight, we select A as the next vertex in the circuit. So far the circuit has vertices P, N, A.

Because we do not wish to travel back to Philadelphia or New York, from A we consider edges to Cleveland and Memphis with weights 290 and 170. This tells us to select vertex M next. At this point, we have no choice but to select vertices C and then P to finish the circuit. Thus the circuit is PNAMCP. The weight of this circuit is 210 + 250 + 170 + 240 + 420 = \$1,290.

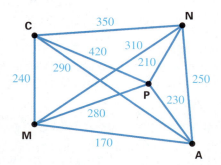

FIGURE 3.28   Graph model of Danielle's possible trips.

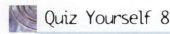

Quiz Yourself 8

Redo Example 5, but this time assume that Danielle is based in Memphis and must visit each of the other four cities. Again, use the nearest-neighbor algorithm to find a Hamilton circuit and state its weight.

Notice from Examples 4 and 5 that although the nearest-neighbor algorithm did not find the best solution to Danielle's problem, it did find a reasonably inexpensive way to schedule her trip. In Quiz Yourself 8, the nearest-neighbor algorithm did find the best solution. Of course, we cannot predict when the nearest-neighbor algorithm will find the best solution to the TSP.

We will investigate another way to find an approximate solution to the TSP. In looking at Figure 3.28, it seems a mistake not to try to use edges like AM (weight 170) and PN (weight 210). If we focus our attention on always choosing the "best edge," rather than trying to construct the circuit vertex by vertex, we might be able to find a pretty good approximation to the best solution to the TSP.

> **The Best-Edge Algorithm for Solving the TSP**
>
> **Step 1:** Begin by choosing any edge with the smallest weight.
> **Step 2:** Choose any remaining edge in the graph with the smallest weight.
> **Step 3:** Keep repeating Step 2; however, do not allow a circuit to form until all vertices have been used. Also, since the final Hamilton circuit cannot have three edges joined to the same vertex, never allow this to happen during the construction of the circuit.

EXAMPLE 6   Solving a TSP Using the Best-Edge Algorithm

Use the best-edge algorithm to schedule Danielle's trip.

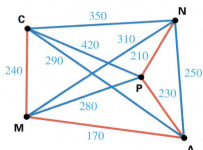

FIGURE 3.29 First four edges selected for Danielle's trip using the best-edge algorithm.

SOLUTION: We begin by selecting edge AM, which has the smallest weight, 170. Next choose edges NP (weight 210), AP (weight 230), and CM (weight 240). We have highlighted the edges we have already selected in Figure 3.29.

The edge with the next smallest weight is AN; however, selecting it forms a circuit, so we do not choose it. We do not choose MP, AC, or MN for the same reason. Selecting the last remaining edge, CN, forms a Hamilton circuit. Notice that it has a weight of 1,200, which also makes it the best solution to Danielle's problem. ☺

The fact that the best-edge algorithm found the solution for Danielle's TSP in Example 6 is fortunate, but there is no guarantee that this will happen for other TSPs.

Exercises 3.2

For many of these exercises, there may be several possible correct answers. We will provide only one in the answer key.

In Exercises 1–4, find a Hamilton circuit in each graph that begins with the specified edge.

1. a) AD   b) EB

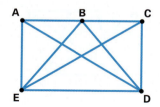

2. a) AC   b) BD

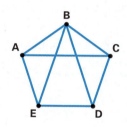

3. a) AD   b) EA

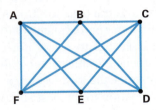

4. a) AE   b) FD

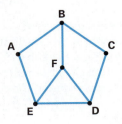

In Exercises 5–8, it helps organize your solution if you construct a tree similar to the one shown in Figure 3.26.

5. Find all the Hamilton circuits that begin with edge AB in the graph shown in Exercise 1.

6. Find all the Hamilton circuits that begin with edge AF in the given graph.

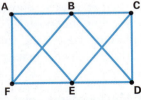

Figure for Exercise 6

7. How many Hamilton circuits are in K_7?

8. How many Hamilton circuits are in K_8?

9. Draw K_6.

10. Draw K_7.

11. Find the weight of the specified paths in the following graph.

a) AGEDCB   b) ABCDGEF

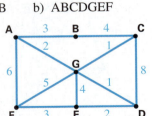

12. Find the weight of the specified paths in the following graph.

a) AGBHCI   b) FGHIDJ

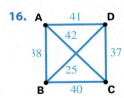

In Exercises 13–16, use the brute-force algorithm to find a Hamilton circuit that has minimal weight. Recall that we found all Hamilton circuits in K_4 in Example 3.

13.

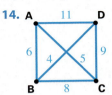

14.

15.

16.

In Exercises 17–20, use the nearest-neighbor algorithm to find a Hamilton circuit that begins at vertex A in each graph.

17.

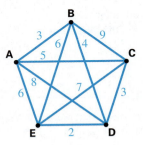

18.

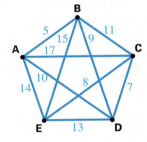

19.

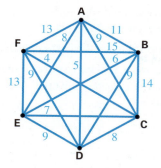

20.

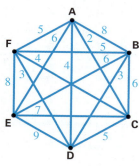

In Exercises 21–24, use the best-edge algorithm to find a Hamilton circuit in each graph. List the circuit beginning at vertex A.

21. the graph of Exercise 17

22. the graph of Exercise 18

23. the graph of Exercise 19

24. the graph of Exercise 20

In Exercises 25–28, explain why each graph cannot have any Hamilton circuits.

25.

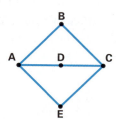

26.

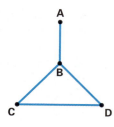

27.

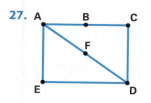

28.

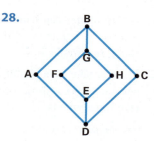

In Exercises 29–30, we have added some cities to Danielle's territory. Instead of drawing a graph, we have listed the cost between pairs of cities in a table. Again Danielle will start at Philadelphia, visit all the cities, and then return home. Recall that the original cities in this problem were (P)hiladelphia, (N)ew York, (A)tlanta, (M)emphis, and (C)leveland.

29. Finding the cheapest route. Assume that (R)aleigh and (B)oston are to be added to the trip.

	P	N	A	M	C	R	B
P	0	210	230	280	420	240	430
N	210	0	250	310	350	320	180
A	230	250	0	170	290	90	510
M	280	310	170	0	240	120	500
C	420	350	290	240	0	270	380
R	240	320	90	120	270	0	290
B	430	180	510	500	380	290	0

a) If you were to draw a graph as we did in Example 4 and then use the brute-force algorithm, how many Hamilton circuits would you have to consider?

b) If you could examine one Hamilton circuit per minute, how many hours would it take you to solve this TSP using the brute-force algorithm?

c) Use the nearest-neighbor algorithm to find a Hamilton circuit beginning at Philadelphia.

d) Use the best-edge algorithm to find a Hamilton circuit beginning at Philadelphia.

30. Finding the cheapest route. Redo Exercise 29; however, in addition to Raleigh and Boston, assume that we have also added (D)allas to the trip. The cost of travel between Dallas and each of the other cities is given in the following table.

	P	N	A	M	C	R	B
D	510	530	410	370	450	260	560

31. If a computer can examine 1,000 Hamilton circuits per second, how long would it take to examine all of the Hamilton circuits in K_{10}?

32. If a computer can examine 1,000,000 Hamilton circuits per second, how long would it take to examine all of the Hamilton circuits in K_{20}?

Use the map to answer Exercises 33–34. Assume that all blocks are the same length.

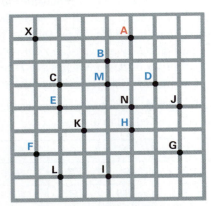

Figure for Exercises 33–34

*(33.) Finding the most efficient route. There is a florist shop located at the position marked A on the map. Assume that deliveries must be made to locations B, D, E, H, F, and M, and then the delivery truck must return to the store.

a) Use the nearest-neighbor algorithm to devise a deliv-ery route that will minimize the number of blocks traveled. You may assume that all streets are two-way streets. (*Hint:* First make a table of the dis-

tances between the various locations, similar to what we did in Exercise 29.)

b) Repeat part (a) using the best-edge algorithm.

(34.) Finding the most efficient route. Repeat Exercise 33, but now assume that the deliveries must be made to B, E, H, I, J, and L.

Use the map for Exercises 35–36.

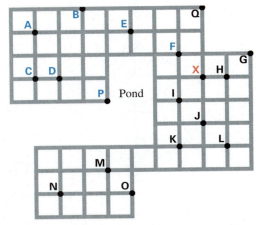

Figure for Exercises 35–36

(35.) Finding the most efficient route. There is a package delivery company located at the position marked X on the map.

a) Use the nearest-neighbor algorithm to devise an effi-cient route to make deliveries to locations A, B, C, D, E, F, and P. The route starts and ends at X.

b) Repeat part (a) using the best-edge algorithm.

(36.) Finding the most efficient route. Repeat Exercise 35, but now deliveries must be made to locations G, H, I, J, K, L, and N.

Further Exercises

If a weighted graph has vertices A, B, C, and so on, a variation of the nearest-neighbor algorithm is to use the algorithm to find a Hamilton circuit starting at A, then find another one starting at B, then find another one starting at C, and so on. Then pick the Hamilton circuit from among these that has the smallest weight.

We will call this algorithm the extended nearest-neighbor algorithm.

(37.) Redo Exercise 17 using the extended nearest-neighbor algorithm.

(38.) Redo Exercise 19 using the extended nearest-neighbor algorithm.

---

* Exercise numbers circled in red can be used as group exercises.

39. Communicating Mathematics Explain why the extended nearest-neighbor algorithm may give you a better solution to a TSP than the nearest-neighbor algorithm.

40. Draw a weighted graph in which the brute-force, nearest-neighbor, and best-edge algorithms all give different Hamilton circuits.

41. Communicating Mathematics Explain why there must always exist Hamilton circuits in K_n.

42. Communicating Mathematics When asked to find all Hamilton circuits in a complete graph, Michael found fifteen. How do we know he has made a mistake?

43. Communicating Mathematics Explain why there are $(n - 1)!$ Hamilton circuits in K_n.

44. Communicating Mathematics Compare the nearest-neighbor algorithm and the best-edge algorithm with the brute-force algorithm. What are the advantages and disadvantages of each?

Communicating Mathematics *In Exercises 45–48, make up a TSP problem using the scenario that we give you. Then solve your problem using one of the methods that you studied in this section.*

45. Consider a situation involving minimizing the cost of laying pipes—perhaps for water fountains at a zoo, irrigation pipes for a homeowner's yard, sewer lines for a community, or some similar situation of your choosing.

46. Imagine that you are laying out a network of computers, or some other electronic devices.

47. Map out a trip in which you visit five or six cities. Perhaps you wish to visit certain baseball stadiums, or several of your favorite college football teams. Or, maybe you wish to visit certain of the great cities in the United States such as New York, Chicago, Philadelphia, Los Angeles, and so on. Use some source to obtain real airline prices for your trip.

48. Consider some situation that we have not suggested in which you think that the TSP is relevant.

49. Use the Internet to investigate the current status of research on the TSP problem.

50. Find a real application of the TSP and write an essay on that application. You might use the Internet, or books and articles on operations research to find your example.

3.3 DIRECTED GRAPHS

Consider the relationship *is the mother of* for a set of people. If Annette is the mother of Kathy, then Kathy is not the mother of Annette, so we see that *this relationship does not go in both directions*. In order to model such a relationship by a graph, we assign directions to the edges of the graph. When an edge has a direction it is called a **directed edge**. A graph in which all edges are directed is called a **directed graph**. One application of directed graphs is to model the flow of information within a group of people or within an organization.

Directed graphs model the flow of information.

EXAMPLE 1   Modeling the Spread of Rumors

Suppose we wish to model how rumors are spread among four friends: Buffy, Cordelia, Willow, and Xander. Assume that we have gathered the following information:

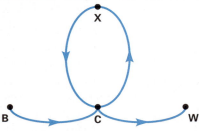

1. If Buffy hears a rumor, she will communicate it to Cordelia, but Cordelia will not tell rumors she hears to Buffy.

2. Cordelia and Xander will tell any rumors they hear to each other.

FIGURE 3.30   A directed graph modeling the spread of rumors.

3. Cordelia tells Willow rumors she hears, but Willow does not relay rumors to Cordelia.

Model this situation with a directed graph.

SOLUTION: Since rumors do not always flow in both directions, we can model this situation with the directed graph shown in Figure 3.30.

   The arrows on the edges in Figure 3.30 show the direction in which rumors travel. For instance, the directed edge from B to C shows that Buffy communicates rumors to Cordelia. Likewise, the absence of a directed edge from C to B indicates that Cordelia never tells rumors to Buffy.

   To obtain information from a directed graph, it is often useful to identify certain paths in the graph.

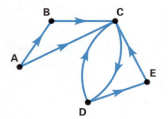

FIGURE 3.31   A directed graph.

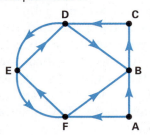
DEFINITIONS

Suppose that X and Y are vertices in a directed graph. If it is possible to begin at X, follow a sequence of edges in the directions indicated, and end at Y, then we call the sequence of edges encountered a **directed path from X to Y**. We will denote a directed path by the sequence of vertices encountered along the path. The **length** of a directed path is the number of edges along that path. (Recall that edges cannot be repeated in a path.)

EXAMPLE 2   Finding Paths in a Directed Graph

Consider the directed graph shown in Figure 3.31.

a) What is the length of the directed path ACDE?

b) Is ABCE a directed path?

SOLUTION: a) ACDE is a directed path of length 3 from A to E.

b) ABCE is *not* a directed path from A to E because CE has the wrong direction.

■

Directed graphs model how members of a set exert influence.

An interesting application of directed graphs is to model how members of a set influence each other. Since influence usually is not exerted equally in both directions, we use directed graphs as models in these situations.

EXAMPLE 3   Modeling Influence

A group of parents wants the town council to install a traffic light at a dangerous intersection. They feel that their best strategy is to obtain the support of the most influential member of the council and, from past experience, they know which council members influence others.

Council Member	Council Members Influenced by This Person
Alvarez	Cohen, Ellis, Ferraro
Baker	Ferraro
Cohen	Baker, Ellis
Davis	Baker, Cohen, Ellis
Ellis	Baker
Ferraro	Cohen, Ellis

TABLE 3.4   Influence on town council members.

Use Table 3.4 to determine which council member has the most influence.

SOLUTION: Since influence is not exerted in both directions (Alvarez influences Cohen but Cohen does not influence Alvarez), we use the directed graph shown in Figure 3.32 to model this relationship among council members.

The directed edges in the graph indicate who influences whom. For instance, the directed edge from A to E reflects that Alvarez has influence over Ellis. Since Alvarez and Davis each exert direct influence over three people, we might say that both are equally influential.

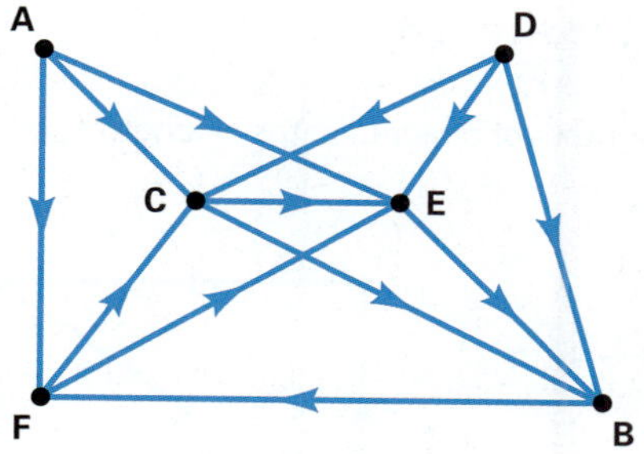

However, let us consider what we will call two-stage influence. We observe that Alvarez influences Cohen, who in turn influences Ellis. Therefore, we will say that Alvarez had two-stage influence over Ellis. In terms of our model, if there is a directed path of length 2 from vertex X to vertex Y, then council member X exerts two-stage influence over Y.

FIGURE 3.32   Directed graph modeling influence on town council.

Let us now determine the number of ways that Alvarez can influence Ellis, either

		A	B	C	D	E	F
	A	0	2	2	0	3	1
	B	0	0	1	0	1	1
From	C	0	2	0	0	1	1
	D	0	3	1	0	2	1
	E	0	1	0	0	0	1
	F	0	2	1	0	2	0

(column group header: **To**)

TABLE 3.5 One- and two-stage influence among council members.

Row	Sum of Entries
A	8
B	3
C	4
D	7
E	2
F	5

TABLE 3.6 Sum of one- and two-stage influence exerted by council members.

Most Influential

Alvarez
Davis
Ferraro
Cohen
Baker
Ellis

Least Influential

FIGURE 3.33
Ranking of council members according to one- and two-stage influence.

directly or in two stages. In Figure 3.32, in addition to the directed edge AE, there are also the directed paths ACE and AFE. Therefore there are three ways in which Alvarez can influence Ellis in one or two stages.

We would like to determine this number for every other pair of council members. For convenience, we have recorded the number of directed paths of length 1 or 2 between each pair of vertices in Table 3.5.

We saw that there are three paths of length 1 or 2 from A to E; therefore, we put a 3 in row A, column E of our table. The entry of row C, column B is 2, because there are two paths of length 1 or 2 from C to B—namely, CB and CEB. In general, the entry made in row X, column Y tells us the number of paths of length 1 or 2 from vertex X to vertex Y in the graph.

Now that we have determined the amount of one- or two-stage influence, we can rank the council members. Although Alvarez and Davis influence Ellis directly and also by way of Cohen, we see that Alvarez also influences Ferraro who influences Ellis. There are three ways in which Alvarez exerts influence over Ellis but only two ways in which Davis can. Therefore it is reasonable to say that Alvarez exerts more influence over Ellis than Davis can.

Thus we will rank the council members according to the number of ways they can exert their influence in one or two stages. If we add the entries in row A of the table, we obtain

$$0 + 2 + 2 + 0 + 3 + 1 = 8,$$

which means there are eight ways Alvarez exerts influence over other council members in either one or two stages. Adding the entries for each of the other rows, we obtain Table 3.6.

According to the way we are measuring influence, the council members would be ranked as in Figure 3.33.

Quiz Yourself 10

List in a table the number of directed paths of length 1 or 2 between each pair of vertices in the given graph.

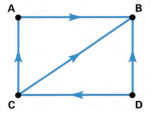

Highlight: *Operations Research and Graph Theory*

Building networks of communication lines, designing complex industrial processes, developing complicated airline schedules, scheduling thousands of tasks to refurbish ships, and delivering thousands of packages efficiently are examples of situations that require extensive organization. Mathematicians working in an area known as *operations research* are employed to find solutions to such problems that use critical resources as efficiently as possible. Often graph theory helps to represent these problems abstractly and organize large amounts of information. Then the mathematicians use computers to analyze the data and find solutions.

As in Example 3, we often wish to rank objects with respect to some property. For example, we could rank football teams by using their number of

victories. Assume Ohio State and Nebraska have played the same number of games, but Ohio State has four victories and Nebraska has three. We could then rank Ohio State above Nebraska. Suppose instead that each team has four victories. In order to break a tie in the ranking order, we could consider the notion of two-stage dominance; that is, if Ohio State has beaten Michigan and Michigan has defeated Indiana, then Ohio has demonstrated two-stage dominance over Indiana. If Ohio State has more two-stage dominances over other teams than Nebraska has, then we could rank Ohio State above Nebraska. If the two teams are still tied with regard to the number of two-stage dominances, we might then consider the number of three-stage dominances, and so forth.

Keep this example in mind when solving the exercises at the end of this section in which you are asked to rank objects of a set.

SOME GOOD ADVICE

Realize that when building mathematical models, *you* decide what features you feel are important in the model. You may not agree with the way a model is constructed. For example, should two-stage dominance count the same as one-stage dominance in the last discussion? If not, then you may want to change the method of building the model.

Directed graphs model disease transmission.

We now use directed graphs to answer a different type of question.

EXAMPLE 4   **Modeling the Spread of a Disease**

A small town in southwestern New Mexico is facing a crisis. Eight people have reported to the town's medical center suffering from hanta virus. Officials have isolated these people and hope that no others have contracted this virus. After interviewing the patients, health officials determined how the virus spread within this group. They believe the virus was introduced by one person in town and then communicated to others in the town. Use the information in Table 3.7 to determine whether or not all the infected people in the town have been isolated.

Patient	Others within the Group Who Could Have Contracted the Virus from This Patient
Amanda	Dustin, Jackson
Brian	Caterina, Frank, Ina
Caterina	Frank
Dustin	Caterina
Frank	Louisa
Ina	Brian, Frank
Jackson	Amanda, Caterina, Frank
Louisa	Caterina

TABLE 3.7   How a virus might have spread in a town.

SOLUTION: We can easily model this situation using the directed graph in Figure 3.34. We represented each of the eight patients with a vertex and drew a directed edge from vertex X to vertex Y, if X could have transmitted the virus to Y.

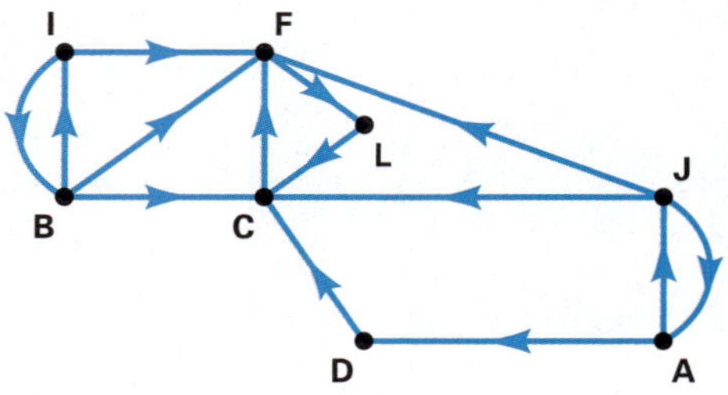

FIGURE 3.34   Directed graph representing spread of a virus.

Since Amanda could have given the virus to Jackson, we draw a directed edge from A to J. Similarly, since Brian could have transmitted the virus to Frank, there is a directed edge drawn from B to F. The directed paths in the graph show how the infection might have spread within the group. For example, the directed path ADCF shows that it was possible for the virus to travel from Amanda to Dustin to Caterina, and finally to Frank. It is important to realize that if the virus spread from X within the group to Y, then there must be a directed path in the graph from vertex X to vertex Y.

We see that it was impossible for the virus to have started with Amanda and spread within the group to Brian since there is no directed path from A to B. In fact, checking each of the eight people, we see that it was impossible for the virus to start with any one of them and then spread to all the others. Therefore if our information and assumption is correct, there is at least one other person in town who has the virus but who has not yet been identified.   ☺

We have presented a few examples of modeling with graphs. A classic book devoted entirely to applications of directed graphs to sociology is F. Harary et al., *Structural Models: An Introduction to the Theory of Directed Graphs* (Wiley, 1965).

Exercises 3.3

For many of these exercises there may be several correct answers. We will provide only one in the answer key.

In Exercises 1–4, use each graph to find the requested items, if it is possible. If it is not possible to find a requested item, explain why not.

1. a) two different directed paths from A to E

b) a directed path from A to C

c) a directed path of length 3 from A to E

d) a directed path of length 2 from A to E

e) a directed path of length 5 from A to A

2. a) two different directed paths from A to E

b) a directed path from A to C

c) a directed path of length 3 from A to E

d) a directed path of length 2 from A to E

e) a directed path of length 5 from A to A

3.

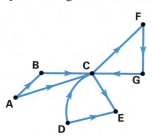

a) a directed path from B to E

b) a directed path from A to G

c) a directed path of length 5 from A to E

d) a directed path from F to D

e) a directed path of length 5 from A to C

4.

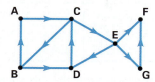

a) two different directed paths from D to E

b) a directed path from B to F

c) a directed path from F to C

d) a directed path of length 5 from A to E

e) a directed path of length 5 from A to A

5. Construct a table that displays the number of directed paths of length 1 or 2 between each pair of vertices in the graph of Exercise 1.

6. Construct a table that displays the number of directed paths of length 1 or 2 between each pair of vertices in the graph of Exercise 2.

7. Construct a table that displays the number of directed paths of length 1 or 2 between each pair of vertices in the graph of Exercise 3.

8. Construct a table that displays the number of directed paths of length 1 or 2 between each pair of vertices in the graph of Exercise 4.

9. Modeling the spread of rumors. The following directed graph shows how rumors spread among five neighbors: Alicia, Bob, Kara, Ted, and Mike. Which neighbors could start a rumor that would eventually spread to all the others?

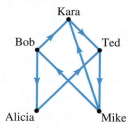

10. Modeling the flow of information. Four news organizations made public a top-secret report. The following graph indicates how this information could have passed among the four news organizations. Determine which organization first obtained this report.

11. Modeling the flow of paperwork through a bureaucracy. In a large corporation, five signatures are needed on a form. It is known that certain managers will not give their approval before others. This situation is depicted in the following directed graph. A directed edge from vertex A to vertex B indicates that A must sign the form before B. If a secretary must hand-carry the form from office to office, in which sequence can all the signatures be obtained?

12. Modeling the spread of disease. The following directed graph models the spread of a flu virus among a group of passengers on a cruise ship. Are there any people who could have introduced the flu to the group? Explain.

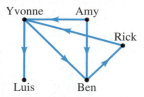

For Exercises 13–16, use Figure 3.34 in Example 4 which models the spread of the Hanta virus.

13. Communicating Mathematics   What do you see in the graph that tells you that the virus could not have been introduced by either I or B?

14. Communicating Mathematics   What do you see in the graph that tells you that the virus could not have been introduced by J, D, or A?

15. Communicating Mathematics   What would have to be present in the graph for it to be possible that the virus was introduced by F? Is there one edge whose direction you could change to make this happen?

16. Communicating Mathematics   Is there one edge whose direction you could change so that the graph would indicate that the virus could have been introduced by D? Explain.

17. Modeling a food chain. The African hare and gazelle are vegetarians that feed primarily on grass. Gazelles

and hares are eaten by lions, cheetahs, and humans. Draw a directed graph to model this food chain.

18. Modeling a communications network. An emergency broadcast system has been set to transmit emergency messages among several cities. The following table indicates how this system has been set up. Draw a directed graph modeling this emergency network. Can a message originate in one city and spread throughout the network? Explain.

City	Can Broadcast To:
Philadelphia	New York, Detroit
New York	Philadelphia, Boston
Boston	Philadelphia, New York
Dallas	Los Angeles, Phoenix
Detroit	Philadelphia, Dallas
Los Angeles	Dallas, Phoenix
Phoenix	Dallas, Los Angeles

19. Ranking teams. Let us assume that in the National Football League over a period of several weeks the following happens:

Team	Steelers	Bengals	Steelers	Dolphins	Dolphins
Beats	Redskins	Broncos	Broncos	Steelers	Raiders

Draw a directed graph to model these results.

20. Ranking teams. Let us assume that in a series of debating contests the following results are obtained:

School	Yale	Rutgers	Berkeley	Berkeley	Texas
Defeats	Texas	UCLA	St. John's	UCLA	Michigan

Draw a directed graph to model these results.

21. Modeling influence. The following influence has been observed among five committee members:

Member	Pratt	Quigley	Ross	Stickle	Thums
Influences	No one	Stickle, Thums	Pratt, Stickle	Pratt	Pratt, Ross

Use a directed graph to model this situation and determine a ranking order for the committee members using one- and two-stage influence, as we did in Example 3.

22. **Ranking teams.** In a round-robin singles pool tournament, each of six competitors plays each other once. The results of the tournament are shown in the following table.

Player	Defeated
Carla	Rob, Sara, Orlando, Matt
Rob	Tanya, Orlando, Matt
Tanya	Carla, Sara, Matt
Sara	Rob, Matt
Orlando	Tanya, Sara
Matt	Orlando

Use a directed graph to model this situation and determine a ranking order for the players using one- and two-stage dominance.

23. **Modeling consumer preferences.** In a "paired-comparison" test, an individual is asked to compare five brands, two at a time, and for each pair, the individual is to indicate a preference. The following results are obtained:

Brand	A	B	C	D	E
Preferred Over:	B, C	E	B, D, E	A, B, E	A

Use a directed graph to determine a ranking order for the five brands using one- and two-stage preference.

24. **Modeling consumer preferences.** Repeat Exercise 23 using the following table:

Brand	A	B	C	D	E
Preferred Over:	B, C, D	E	B, D, E	B, E	A

Further Exercises

*25. **Modeling consumer preferences.** When purchasing food products, a consumer would consider ease of preparation, nutritional value, price, and taste. Conduct a consumer survey by asking five people you know to complete a copy of the following ballot.

For each of the following pairs, circle the quality that is more important to you:

Ease of preparation	Nutritional value
Ease of preparation	Price
Ease of preparation	Taste
Nutritional value	Price
Nutritional value	Taste
Price	Taste

On the third line of the ballot, if more people chose taste over ease of preparation, then we can say that taste is preferred over ease of preparation. Use the results of your survey to determine which quality is preferred for each pair and then summarize your results by means of a directed graph.

26. When shopping for an automobile, a buyer may consider each of the following features: economy, safety, comfort, style, manufacturer, and availability of service. Assume you are a car buyer and compare each possible pair of features, indicating which of the two is more important to you. After completing this paired-comparison test, determine a ranking order by importance of the six features.

27. Communicating Mathematics   In modeling influence, we counted two-stage influence as equal with one-stage influence. Criticize this approach. Can you suggest another method? How would this affect our tables that summarize one- and two-stage influence? Explain.

28. Communicating Mathematics   Think of a situation that we have not discussed in this section in which you feel directed graphs could be used as models. Describe the application and explain your reasons why you think it can be represented by a directed graph.

---

* Exercise numbers circled in red can be used as group exercises.

CHAPTER SUMMARY

graph, vertices, edges

A graph consists of a finite set of points, called vertices, and lines, called edges, that join pairs of vertices.

connected graph, bridge

A graph is connected if it is possible to travel from any vertex of the graph to any other vertex of the graph by moving along successive edges. If there is an edge in a connected graph such that if we were to remove it the graph would become disconnected, that edge is called a bridge.

odd, even vertices

A vertex of a graph is odd if it is an endpoint of an odd number of edges of the graph. Similarly, a vertex is even if it is an endpoint of an even number of edges.

Euler's theorem

A graph can be traced provided:

1. It is connected.

2. It has no odd vertices or two odd vertices.

3. If it has two odd vertices, the tracing must begin at one of these and end at the other.

4. If all the vertices are even, then the tracing must begin and end at the same vertex; it does not matter at which vertex this occurs.

building a graph model

We can model the relationship among a collection of objects as follows:

1. Represent each object by a vertex. Choose names that remind you of what the vertices represent.

2. For each pair of related objects, join the two corresponding vertices with an edge.

path, length of path

A path in a graph is a series of consecutive edges in which no edge is repeated. The number of edges in a path is called its length.

Euler path, Euler circuit, Eulerian graph

A path containing all the edges of a graph is called an Euler path. An Euler path that begins and ends at the same vertex is called an Euler circuit. A graph containing an Euler circuit is called an Eulerian graph.

Fleury's algorithm

If a connected graph has all even vertices, we can find an Euler circuit for it as follows: Begin at any vertex and travel over consecutive edges subject only to the following rules:

1. After you have traveled over an edge, erase it. If all the edges for a particular vertex have been erased, then erase that vertex also.

2. Travel over an edge that is a bridge only if there is no alternative.

Eulerizing a graph

If a graph does not have all even vertices, we duplicate some existing edges in order to make all vertices even and thus make the graph Eulerian.

map coloring (the four-color problem)
It is always possible to color a map with four or fewer colors so that any two regions sharing a common border receive different colors.

SECTION 3.2   **the traveling salesperson problem (TSP)**
The traveling salesperson problem gets its name from the problem of determining the most efficient way for a salesperson to schedule a trip to a series of cities and then return home.

Hamilton path, Hamilton circuit
A path that passes through all the vertices of a graph exactly once is called a Hamilton path. A Hamilton path that begins and ends at the same vertex is called a Hamilton circuit.

complete graph
A complete graph is one in which every pair of vertices is joined by an edge. A complete graph with n vertices, denoted by K_n, has $(n - 1)!$ Hamilton circuits.

weighted graph, weight of a path
When we assign numbers to the edges of a graph, the graph is called a weighted graph. The weight of a path is the sum of the weights of the edges of the path.

the brute-force algorithm for solving the TSP
Step 1: List all Hamilton circuits in the graph.
Step 2: Find the weight of each circuit found in Step 1.
Step 3: The circuits with the smallest weights are solutions to the TSP.

the nearest-neighbor algorithm for solving the TSP
Step 1: Start at any vertex X.
Step 2: Of all the edges connected to X, choose any one that has the smallest weight. (There may be several with the smallest weight.) Select the vertex at the other end of this edge. This vertex is called the *nearest neighbor* of X.
Step 3: Choose subsequent *new* vertices as you did in Step 2. When choosing the next vertex in the circuit, choose one whose edge with the current vertex has the smallest weight.
Step 4: After all vertices have been chosen, close the circuit by returning to the starting vertex.

the best-edge algorithm for solving the TSP
Step 1: Begin by choosing any edge with the smallest weight.
Step 2: Choose any remaining edge in the graph with the smallest weight.
Step 3: Keep repeating Step 2; however, do not allow a circuit to form until all vertices have been used and never allow three edges to be joined to the same vertex.

SECTION 3.3   **directed edge, directed graph**
An edge in a graph that is given a direction is called a directed edge. A graph in which all edges are directed is called a directed graph.

directed path from X to Y, length of directed path
If we can begin at vertex X in a directed graph and follow a sequence of edges in the directions indicated ending at vertex Y, that sequence of edges is called a directed path from X to Y. The length of a directed path is the number of edges along that path.

directed graphs as models
Directed graphs model relationships among objects when that relationship is not necessarily exerted in both directions. Influence, dominance, and disease transmission are several examples of situations that can be modeled by directed graphs.

CHAPTER TEST

SECTION 3.1

1.

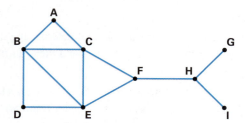

Use the preceding graph to answer the following questions.

a) How many edges does the graph have?

b) Which vertices are odd? Which are even?

c) Is the graph connected?

d) Does the graph have any bridges?

2. Explain how graphs are used to model a collection of objects in which some of the objects are related to each other. Give an example.

3. Which of the following graphs can be traced? Explain your answer by referring to Euler's theorem.

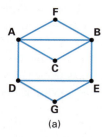   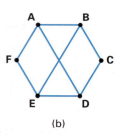

    (a)   (b)

4. Use Fleury's algorithm to find an Euler circuit in the following graph. Describe the circuit by listing the vertices on the path.

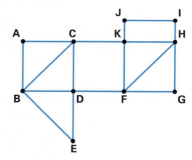

5. Model the following street map with a graph and design an efficient way of traveling through all the streets

so a minimal number of streets are traveled more than once.

6. Model the following "map" of countries A, B, C, . . . , M by a graph and then devise a way of coloring the map using a minimal number of colors.

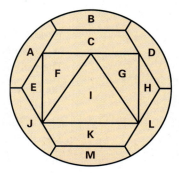

SECTION 3.2

7. Find all Hamilton circuits that begin at vertex A in K_5 and pass next through vertex B.

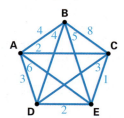

K_5

Use the following weighted graph to answer Questions 8, 9, and 10. Your answers may differ from ours.

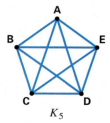

8. Use the brute-force algorithm to find a Hamilton circuit that has minimal weight.

9. Use the nearest-neighbor algorithm to find a Hamilton circuit that begins at vertex A.

10. Use the best-edge algorithm to find a Hamilton circuit that begins at vertex A.

SECTION 3.3

11. Use the following directed graph to find the requested items, if possible. If it is not possible to find an item, explain why not.

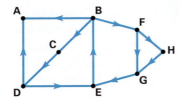

a) a directed path from C to B

b) a directed path of length greater than 5 from C to B

c) a directed path from H to F

d) two different directed paths from F to B

12. When are directed graphs rather than nondirected graphs used as models?

13. A committee has been formed to study ways to bring about social justice. The influence exerted among committee members is listed in the following table. Use a graph to determine one- and two-stage influence among these people and determine who is the most influential committee member.

Committee Member	Committee Members Influenced by This Person
Pham	Stein, Vaccaro, Jackson
Vaccaro	Jackson
Stein	Vaccaro, Bartkowski
Robinson	Stein, Jackson
Bartkowski	Vaccaro
Jackson	Stein, Bartkowski

Of Further Interest: SCHEDULING PROJECTS USING PERT

Congress has commissioned a committee of scientists to develop a timetable for constructing and populating an orbiting space colony. Table 3.8 lists the ten major tasks needed to complete the project, along with the time needed for each task.

As members of this committee, our goal is to schedule these tasks efficiently and to determine whether we can shorten the total time required for the project by reducing the amount of time needed to build the life-support systems. The model we will use in determining the schedule is a special type of directed graph called a **PERT* diagram**. In determining the total time required for the project, it is tempting to simply add the time required for each task and conclude that we need 75 months to construct and populate the colony. But the tasks need not be done one at a time—some can be worked on simultaneously—so the total time can be less than 75 months. Before devising a schedule, we need to know which tasks must wait for others to be completed. Table 3.9 on the facing page lists which tasks must precede others.

We can display the data in Tables 3.8 and 3.9 in the PERT diagram shown in Figure 3.35. We represent each task by a vertex containing the number of months necessary for completion of that task. The beginning and end of the project are shown in our diagram, and we assume that these require no time. We have drawn a directed

Task	Time Required (Months)
1. Train construction workers	6
2. Build shell	8
3. Build life-support systems	14
4. Recruit colonists	12
5. Assemble shell	10
6. Train colonists	10
7. Install life-support systems	4
8. Install solar-energy systems	3
9. Test life-support and energy systems	4
10. Bring colonists to the colony	4

TABLE 3.8   Ten tasks of space colony project.

---

* PERT (Program Evaluation and Review Technique) is an organizational technique invented to aid in scheduling the construction of the Polaris submarine.

Task	Preceding Tasks
1. Train construction workers	None
2. Build shell	None
3. Build life-support systems	None
4. Recruit colonists	None
5. Assemble shell	1, 2
6. Train colonists	2, 3, 4
7. Install life-support systems	1, 2, 3, 5
8. Install solar-energy systems	1, 2, 5
9. Test life-support and energy systems	1, 2, 3, 5, 7, 8
10. Bring colonists to the colony	1, 2, 3, 4, 5, 6, 7, 8, 9

TABLE 3.9 Precedence of the ten space colony tasks.

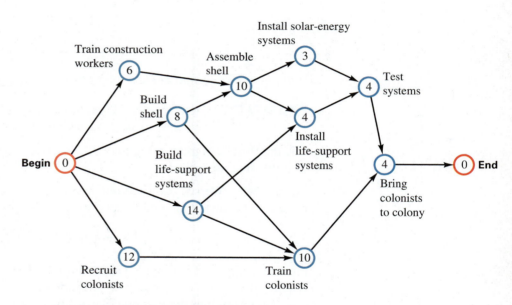

FIGURE 3.35 A PERT diagram for the space colony project.

edge from vertex X to vertex Y if the task represented by vertex X *immediately* pre-cedes the task represented by Y. For example, Table 3.9 shows that training construction workers precedes assembling the shell in space, so we have drawn a directed edge from the vertex "Train construction workers" to the vertex "Assemble shell."

Although building the shell on Earth does come before installing the life-support systems, we do not draw a directed edge to show this. Edges in the diagram already indicate that "Build shell" precedes "Assemble shell," which precedes "Install life-support systems." This sequence of directed edges shows that "Build shell" comes before "Install life-support systems." We will return to the solution of this problem after Quiz Yourself 11.

> **SOME GOOD ADVICE**
>
> If your first attempt at drawing a PERT diagram does not show the relationship between the tasks clearly, take the time to redraw it so that the information stands out.

Quiz Yourself 11*

The following table lists the times required to complete the tasks of a particular project and the dependency among these tasks. Draw a PERT diagram for this project.

Task	Days Required	Preceding Tasks
A	3	None
B	4	A
C	6	A
D	2	A, B, C
E	5	A, B, C, D
F	7	A, B, C, D, E

We now return to our space colony scheduling problem.

EXAMPLE 1   **Determining When to Install the Life-Support Systems in the Space Colony**

Use the PERT diagram in Figure 3.35 to determine the earliest time we could install life-support systems.

SOLUTION: We see from the diagram that several different sequences of tasks must precede installing the life-support systems. For example, the sequence

Begin, Train construction workers, Assemble shell

must be completed before we can install the life-support systems. This sequence requires $0 + 6 + 10 = 16$ months to complete.

By examining the diagram, we see that the most time-consuming sequence of tasks preceding the installation of life-support systems is the sequence

Begin, Build shell, Assemble shell,

which requires $0 + 8 + 10 = 18$ months. This means that we can begin installing the life-support systems in the nineteenth month of the project.

Note that an efficient schedule would have construction of the shell and the life-support systems taking place simultaneously, because neither of these tasks depends on the other. We can schedule other tasks using reasoning similar to that in Example 1. In order to simplify our discussion, we introduce the following definition.

DEFINITION

Suppose T is a task in a PERT diagram. Let us consider all directed paths from "Begin" to T. If we add the time along each of these paths, any path requiring the most time to complete is called a **critical path** for task T.

Using this new terminology, we can say that the path

Begin, Build shell, Assemble shell, Install life-support systems is a critical path for the task "Install life-support systems."

Scheduling a Task in a PERT Diagram

To determine when to schedule a task T in a PERT diagram, do the following:

1. Find a critical path for task T.
2. Add all times along this critical path, with the exception of the time required for task T. This sum gives us the time to be allowed before scheduling T.

EXAMPLE 2   Scheduling the Testing of the Colony's Systems

Use the preceding scheduling procedure to determine when to start the testing of the colony's systems.

SOLUTION: From our PERT diagram in Figure 3.35, we see that a critical path for this task is

Begin, Build shell, Assemble shell, Install life-support systems, Test systems

Examining this path, we see that $0 + 8 + 10 + 4 = 22$ months must be allowed *before* the testing of the colony's systems can begin. Therefore, this task should be scheduled to begin in the 23rd month.

Table 3.10 lists a schedule for all the tasks of the space colony project.

Task	Month Task Begins
1. Train construction workers	1st
2. Build shell	1st
3. Build life-support systems	1st
4. Recruit colonists	1st
5. Assemble shell	9th
6. Train colonists	15th
7. Install life-support systems	19th
8. Install solar-energy systems	19th
9. Test life-support and energy systems	23rd
10. Bring colonists to the colony	27th

TABLE 3.10   Schedule for the space colony tasks.

EXAMPLE 3 Scheduling the Space Colony Project

Determine the time needed for the entire space colony project.

SOLUTION: To do this, we must find a critical path for the vertex "End." Such a path is

Begin, Build shell, Assemble shell, Install life-support systems, Test systems, Bring colonists to the colony, End,

which is highlighted by the red edges in Figure 3.36.

From this diagram, we find that we need $0 + 8 + 10 + 4 + 4 + 4 + 0 = 30$ months to complete the project.

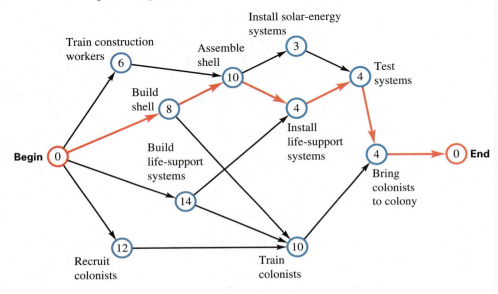

FIGURE 3.36   Critical path for the space colony project.

 Quiz Yourself 12

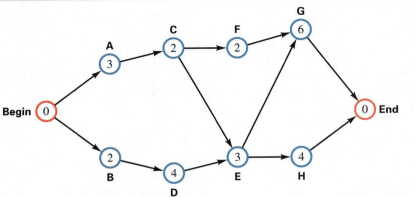

a) Find a critical path for G.   b) What is a critical path for "End"?

c) Assuming that the numbers in the vertices represent days, when should task H be scheduled?

d) How long will the whole project take?

Let us now consider whether it is possible to shorten the total length of the project by reducing the time spent in building the life-support systems. Referring to Figure 3.36, we see that the task of building the life-support systems *does not lie on a critical path* for the vertex "End." Therefore, reducing this time would in no way decrease the total length of the project.

EXAMPLE 4   Using PERT to Organize a Concert

Assume that you are in charge of organizing a concert to raise money to aid victims of a recent earthquake. Your job is to develop a schedule to complete the project in the shortest possible time. The project tasks and their dependencies are given in Table 3.11, and are displayed in the PERT diagram in Figure 3.37.

Task	Time Required (weeks)	Tasks That Must Precede This One
1. Obtain permits from the city	2	None
2. Raise funds from local agencies	1	None
3. Canvass local merchants for advertising support	4	None
4. Hire performers	3	Obtain permits, raise funds, canvass merchants
5. Rent an auditorium	2	Obtain permits
6. Print the program	1	All of the above
7. Advertise the concert	2	All of the above except print programs

TABLE 3.11   Tasks for concert project.

SOLUTION: By finding critical paths for each of the vertices in the PERT diagram in Figure 3.37, we easily obtain the following schedule of the tasks:

Task	Week Task Begins
1. Obtain permits	1
2. Raise funds	1
3. Canvass merchants	1
4. Rent auditorium	3
5. Hire performers	5
6. Advertise concert	8
7. Print program	8

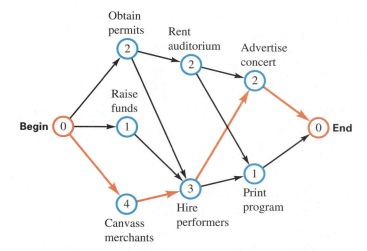

FIGURE 3.37   PERT diagram for concert project.

We determine the shortest time period to complete this project by finding a critical path for the vertex "End." Such a path (marked in red) is

Begin, Canvass merchants, Hire performers, Advertise concert, End.

Thus the entire concert project can be completed in nine weeks, which is the sum of the times along this path. ⊚

Exercises

Some of these exercises may have several correct answers. We will give only one in the answer key.

In Exercises 1–4, assume that the time is measured in days.

1. Use the following PERT diagram to answer the following questions.

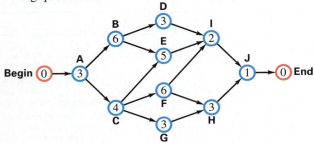

a) Find a critical path for task I.

b) Find a critical path for task E.

c) On what day will task H begin?

d) When will task I be completed?

e) What is the least number of days needed to complete this project?

f) Find a critical path for the vertex "End."

2. Use the following PERT diagram to answer the following questions.

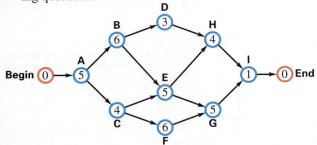

a) Find a critical path for task H.

b) Find a critical path for task G.

c) On what day will task G begin?

d) When will task H be completed?

e) What is the least number of days needed to complete this project?

f) Find a critical path for the vertex "End."

3. Use the following PERT diagram to answer the following questions.

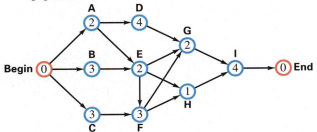

a) Find a critical path for task G in the PERT diagram.

b) Find a critical path for task H.

c) What is a critical path for "End"?

d) When should task F be scheduled?

e) When should we schedule task G?

f) How many days are required for the whole project?

4. Use the following PERT diagram to answer the following questions.

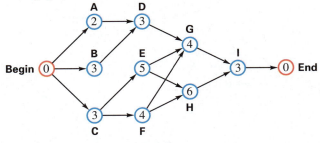

a) Find a critical path for task G.

b) Find a critical path for task H.

c) What is a critical path for "End"?

d) When should task H be scheduled?

e) When should we schedule task G?

f) How many days are required for the whole project?

In Exercises 5–8, use the given PERT diagrams to schedule the tasks so each task is completed in the least possible amount of time. Assume the numbers in the vertices refer to days.

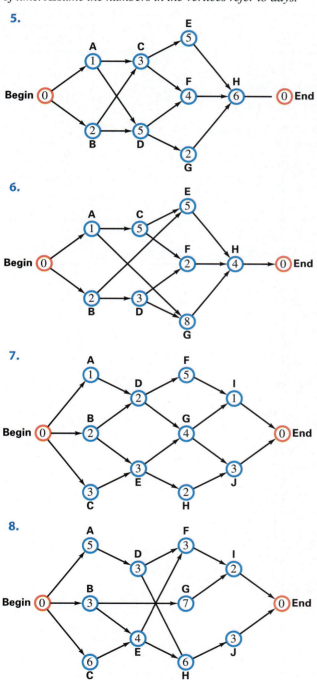

5.

6.

7.

8.

9. Planning a festival. The Earth Day Committee has begun planning for an Earth Day festival. We list the subtasks for this project and the dependencies among them in the following table. Draw a PERT diagram for this project and construct a schedule for the tasks.

Task	Tasks That Precede This Task	Time Needed for This Task (Weeks)
1. Get funding	none	2
2. Get permits	none	1
3. Decide on program	1	2
4. Rent tents	1, 2, 3	2
5. Arrange for speakers, entertainers, etc.	1, 2, 3, 4	4
6. Advertise	1, 3, 5	2
7. Set up tents, booths, etc.	1, 2, 3, 4	1
8. Set up festival	1, 2, 3, 4, 5, 6, 7	1

10. Organizing a project. Assume that you need to complete a senior group project in order to graduate. Since you have a heavy workload, it is important that you plan carefully so you will graduate on time. You need to choose an advisor for this project and a group of students to work with. The subtasks for this project and the dependencies among them are listed in the following table. Draw a PERT diagram for this project and give a schedule for the tasks.

Task	Tasks That Precede This Task	Time Needed for This Task (Weeks)
1. Decide on group	none	2
2. Choose advisor	1	1
3. Choose project	1, 2	4
4. Divide responsibilities for project among group	1, 2, 3	2
5. Do research	1, 2, 3	4
6. Develop rough outline of project	1, 2, 3	2
7. Refine outline	1, 2, 3, 4, 5, 6	2
8. Complete project	1, 2, 3, 4, 5, 6, 7	4
9. Arrange for presentation	1, 2, 3, 4, 5, 6, 7, 8	1

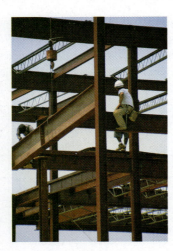

12. Organizing a health program. A developing nation plans to improve available health care for its citizens. Use the following table to draw a PERT diagram for this project and then schedule the tasks so the project can be completed as efficiently as possible.

Task	Preceding Tasks	Time Required (Months)
1. Appropriate money	None	3
2. Build health clinics	1	8
3. Construct hospitals	1	18
4. Educate citizens in health practices	1	12
5. Recruit students	1	8
6. Train doctors	1, 3, 5	30
7. Train aides	1, 2, 5	15
8. Inoculate	1, 2, 6, 7	4
9. Conduct follow-up study	All of the above	2

11. Building a library. Waldenville is planning to build a new library. The subtasks for this project and their dependencies are listed in the following table. Draw a PERT diagram for this project and give a schedule for the tasks.

Task	Tasks That Precede This Task	Time Needed for This Task (Months)
1. Get funding	none	3
2. Choose contractor	1	2
3. Draw plans	1	4
4. Grade land	1, 2, 3	1
5. Lay underground utilities, sewers, etc.	1, 2, 3, 4	1
6. Build building	1, 2, 3, 4, 5	8
7. Install utilities in building	1, 2, 3, 4, 5, 6	2
8. Install computer network in building	1, 2, 3, 4, 5, 6	1
9. Purchase furniture	1, 2, 3, 4, 5, 6	2
10. Inspect building	1, 2, 3, 4, 5, 6, 7, 8, 9	1

13. Organizing an advertising campaign. The state health department is developing a new advertising campaign to reduce smoking in young adults. A series of newspaper ads, billboards, and TV and radio commercials are to be produced. Use the following table to draw a PERT diagram for this project and then schedule the tasks so the project can be completed efficiently.

Task	Preceding Tasks	Time Required (Months)
1. Conduct survey	None	3
2. Develop budget	1	1
3. Hire advertising	1	6
4. Set up production schedule	1, 2, 3	1
5. Produce ads	1, 2, 3	8
6. Disseminate ads	1, 2, 3, 5	2
7. Evaluate results	1, 2, 3, 4, 5, 6	6

Further Exercises

*（14.) **Planning an innovative house.** A local electric company wants to build an experimental house that will conserve energy. Company officials feel the project breaks down naturally into the following tasks (the time required for each task is given in parentheses).

1. Draw plans for house (2 months)
2. Design energy systems for house (6 months)
3. Develop new insulation techniques (3 months)
4. Purchase land (4 months)
5. Build conventional shell of house (6 months)
6. Install new energy systems (2 months)
7. Install new insulation (1 month)
8. Finish the remaining conventional parts of the house (3 months)
9. Landscape (1 month)
10. Perform tests to determine energy usage in house (8 months)

Determine a reasonable list of dependencies for these tasks, then draw a PERT diagram and develop a schedule for these tasks.

（15.) Communicating Mathematics   Write a brief report on a real-life project that can be organized using PERT.†

（16.) Communicating Mathematics   Select a project that interests you. Identify the subtasks and draw a PERT diagram to develop a schedule for the project.

---
* Exercise numbers circled in red can be used as group exercises.
† Books on operations research often contain sections that discuss PERT. PERT is sometimes called the critical path method, or CPM.

CHAPTER 4

Numeration Systems:*
Does It Matter How We Name Numbers?

One of the most important inventions in the history of science and mathematics is something that is all around us and works so smoothly that we hardly even notice it — our common number system, called the Hindu – Arabic numeration system. Without such an efficient, easy-to-use numeration system, we could not write numbers concisely or do computations smoothly. Mathematics, as we know it, would not exist and as a consequence, modern science and technology would be impossible.

In this chapter, you will take a journey from ancient times to the present, following the evolution of numeration systems. You will visit the ancient Babylonians, the Chinese, and the Romans who invented systems that, although primitive and cumbersome, contained the germ of elegant ideas that would blossom forth in more sophisticated systems thousands of years later.

In Section 4.2, we discuss our common Hindu – Arabic system in great detail. The design of this system is the reason that, as a child, you were able to learn to add, subtract, multiply, and divide so simply. Toward the end of this section, we will peek in on an intellectual struggle in which the abacists battle the algorists — a battle that was eventually decided by the way the opposing sides wrote their numbers.

In Section 4.3, we will speculate what mathematics might have been like had we been born with only five fingers instead of ten. Or what if we learned to count with 12 fingers? Or even 16? However,

_____
* You can find more resources on numeration systems at www.aw.com/pirnot.

you will see that such systems are not merely flights of mathematical fancy, but in fact have concrete, real-world applications. In this section, you will also get a glimpse of the cryptic, binary language of digital devices such as DVD players, personal digital assistants, and cell phones.

Later in the chapter, you will study an arithmetic which is done on the faces of strangely numbered clocks. As you will see, this "clock arithmetic" is critical in serious situations which depend on our ability to send secure, unbreakable coded messages such as in banking and military applications.

4.1   THE EVOLUTION OF NUMERATION SYSTEMS

We can easily take a brilliant idea for granted because of its inherent simplicity. Such is the case with our familiar system for representing numbers that is the result of thousands of years of development culminating in the Hindu–Arabic system that we use today. In this section, we will examine some of the early attempts made by various cultures to develop practical numeration systems.

Primitive societies needed only simple, counting systems. An early sheep-herder might have scratched tally marks in the dirt, tied knots in a vine, cut notches in a tally stick, or matched a pile of small pebbles with his sheep in order to keep track of them. Such early counting methods as these eventually led to the invention of the abstract concept of number. The number ten could represent ten sheep, ten pebbles, or ten people.

DEFINITIONS

A **number** tells us how many objects we are counting; a **numeral** is a symbol which represents a number.

Although the *number* ten would mean the same amount in each culture, the *numeral* for ten (the symbols we use to write the number) varied greatly. For example, Babylonians used the symbol ⟨, Egyptians used ∩, Greeks used the letter iota ι, Romans used X, and we, of course, represent ten by the symbols 10.

In addition to inventing numerals for counting, early cultures also developed methods for performing mathematical calculations such as addition, subtraction, and multiplication.

Because it was cumbersome to calculate in these early systems, as trade, finance, and science advanced, there was a corresponding need to develop more sophisticated numeration systems. Sophisticated does not mean more complicated, and as you will see, as numeration systems evolved, they became easier to use.

In this section, we will show you several of these early numeration systems. We begin with the simple grouping systems used by the Egyptians and the Romans, and then explain the multiplicative grouping system of the Chinese, which has some similarities to the Hindu–Arabic system that we commonly use today. We will discuss the Hindu–Arabic system in detail in Section 4.2.

In a simple grouping system, we represent a number as the sum of the values of its numerals.

The Egyptian hieroglyphic system, which is over 5,500 years old,* is an example of a **simple grouping** numeration system. In the Egyptian system, we begin by representing powers of ten by various symbols, as in Table 4.1.

Number	Power of 10	Symbol	Name
1	10^0	I	Stroke
10	10^1	∩	Heel bone
100	10^2	9	Scroll
1,000	10^3	ℒ	Lotus flower
10,000	10^4	⌇	Pointing finger
100,000	10^5	⌒	Fish or Tadpole
1,000,000	10^6	⚼	Astonished Person

TABLE 4.1 Egyptian numeration symbols.

We then form numerals by combining copies of these symbols. The value of the numeral is then the sum of the values of the symbols. For example, we could write the number 325 in Egyptian notation as 999∩∩IIIII. Because the order of symbols is not important in Egyptian notation, we could also write 325 as II∩99∩9III.

EXAMPLE 1   Converting between Egyptian and Hindu–Arabic Notation

a) Convert ℒℒℒ9999∩∩IIIII to Hindu–Arabic notation.

b) Write 1,230,041 in Egyptian notation.

SOLUTION: a) We have 3 thousands, 4 hundreds, 2 tens, and 5 ones, so this Egyptian numeral represents

$$3,000 + 400 + 20 + 5 = 3,425.$$

b) We will represent 1,000,000 by ⚼, 200,000 by ⌒⌒, 30,000 by ⌇⌇⌇, 40 by ∩∩∩∩, and 1 by I to get

⚼⌒⌒⌇⌇⌇∩∩∩∩I.

As you can see, the Egyptian system is not an efficient way to represent numbers. The Egyptian numeral IIIIIIIII requires nine times as many symbols as

---

* Jan Gullberg, *Mathematics: From the Birth of Numbers* (W. W. Norton & Company, 1997), p. 34.

Historical Highlight: *Solving the Mystery of Egyptian Hieroglyphics**

In 1798, Napoleon sailed with a large army to conquer Egypt and disrupt the lucrative British trade routes to India. Although his plan failed and he was severely defeated, this military disaster turned out to be a scientific triumph for France and the rest of Europe. Napoleon had taken with him a large number of scholars, including mathematicians, to study Egypt's culture. Returning to France, they brought back a wealth of information about the achievements of the ancient Egyptians. European scholars were fascinated with the new-found knowledge about this ancient civilization. However, much of the material was written in a cursive form of hieroglyphics called demotic script which no one had been able to translate previously.

As fortune would have it, Napoleon brought back not only this puzzle, but also the key to its solution. While in Egypt, his soldiers found a polished stone, now called the Rosetta stone, which contained three sections of writing—Greek on the bottom third of the stone, demotic script in the middle, and ancient hieroglyphics on the top. Scholars, believing that the stone contained three versions of the exact same text, set about using their knowledge of Greek to translate the other two cryptic sections.

At this point, another key person enters our story. Shortly after Napoleon returned to France, the mathematician Jean-Baptiste Fourier, who had accompanied Napoleon to Egypt, was showing some hieroglyphics to an 11-year-old boy named Jean Francois Champollion. When Fourier stated that no one could read the hieroglyphics, the young boy replied boldly, "I will do it when I am older." From that point on, Champollion dedicated his life to translating hieroglyphics, and by 1822 he had completely translated the upper portion of the Rosetta stone. Legend has it that when Champollion realized that he had solved the mystery of hieroglyphics, he said, "I've got it," and fainted.

Our knowledge of Egyptian mathematics comes mainly from two papyri—The Rhind Papyrus and the Golenischev, or Moscow papyrus. The Rhind papyrus, named after a Scotsman named A. Henry Rhind who purchased it in 1858, is a rich source of knowledge about early Egyptian mathematics. The scroll was written by a scribe named Ahmes in 1650 B.C. and was about 18 feet long and one foot high. Ahmes stated that the document contained "a thorough study of all things, insight into all that exists, [and] knowledge of all obscure secrets." This was an exaggerated promise, because, when the papyrus was translated, it was found to contain a collection of mathematical exercises and rules for doing multiplication and division.

we use to represent nine using the Hindu–Arabic system. The number 68 requires 14 symbols, ∩∩∩∩∩∩IIIIIIII. Imagine how awkward it would be to balance your checkbook or total your restaurant bill using this system!

Quiz Yourself 1†

a) Write the Egyptian numeral ℓℓ⌢999∩∩∩∩IIIIII in Hindu–Arabic notation.

b) Write the number 241,536 using Egyptian notation.

---

* This highlight is based on David M. Burton, *The History of Mathematics: An Introduction* (WCB-McGraw Hill, 1999), pp. 31–35.
† Quiz Yourself answers begin on page 849.

It is straightforward to add and subtract in the Egyptian hieroglyphic system.

EXAMPLE 2   Adding in the Egyptian System

Add 𝟿𝟿∩∩∩∩∩∩∩∩‖‖‖‖‖‖ and 𝟿𝟿∩∩∩∩‖‖‖‖‖‖ using Egyptian notation.

SOLUTION: In order to add these two numbers, we simply group all of the symbols together (the addition sign is not part of Egyptian notation).

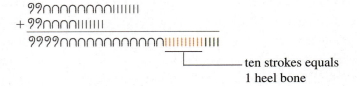

ten strokes equals
1 heel bone

However, we can rewrite ten strokes as one heel bone to give us

ten heel bones
equals 1 scroll

Finally, we replace ten heel bones by one scroll, as follows, to get the final answer:

𝟿𝟿𝟿𝟿𝟿∩∩∩‖‖‖‖.

EXAMPLE 3   Subtracting Using Egyptian Notation

Subtract ∩∩∩∩∩∩‖‖ from 𝟿𝟿∩∩∩‖‖‖‖ using Egyptian notation.

SOLUTION: We compute:

$$
\begin{array}{r}
𝟿𝟿∩∩∩‖‖‖‖ \\
- \quad ∩∩∩∩∩∩‖‖ \\
\hline
\end{array}
$$

We can subtract two strokes from four strokes, however, we cannot subtract six heel bones from three heel bones unless we "borrow." That is, we convert one scroll to ten heel bones as follows:

Now we can perform the subtraction as indicated above. Four strokes minus two strokes is two strokes, 13 heel bones minus six heel bones is seven heel bones, and we have one additional scroll.

Notice in Examples 2 and 3 how the Egyptian calculations remind us of the way we carry and borrow in performing addition and subtraction using Hindu–Arabic notation.

SOME GOOD ADVICE

You can check your work in performing Egyptian arithmetic by converting all the numbers to Hindu–Arabic notation and then redoing the computations.

Power of 2	Value
2^0	1
2^1	2
2^2	4
2^3	8
2^4	16
2^5	32
2^6	64

TABLE 4.2   Powers of 2.

Quiz Yourself 2

Perform the following operations using Egyptian notation.

a) 999∩∩∩∩||||||| + 99999999∩∩|||||

b) 𝄃𝄃𝄃𝄃𝄃∩∩∩∩|| − 𝄃𝄃9999∩∩∩||||||||||

The Egyptians also had a method for doing multiplication, as we show in Example 4.* This method is based on the fact that any positive integer can be written as the sum of powers of two (see Table 4.2). For example, $19 = 1 + 2 + 16$ and $81 = 1 + 16 + 64$.

EXAMPLE 4   **Computing Area Using the Egyptian Method of Doubling**

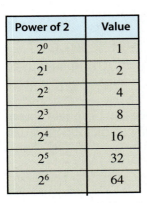

A craftsman is tiling the rectangular wall of a temple with tiles. If the wall measures 13 feet by 21 feet, how many square feet must be covered? Use the Egyptian method of doubling to compute the area.

SOLUTION: First we write 13 as $1 + 4 + 8$. To find the area, we will calculate 13×21, which we can now write as $(1 + 4 + 8) \times 21$. Consider the values in Table 4.3 which we obtained by repeatedly doubling 21.

Power of 2	Times 21
1	21
2	$21 + 21 = 42$
4	$42 + 42 = 84$
8	$84 + 84 = 168$
16	$168 + 168 = 336$

TABLE 4.3   Powers of 2 times 21.

Quiz Yourself 3

a) Represent 25 as the sum of powers of 2.

b) Use the Egyptian method of doubling to calculate 25×43.

---

* In explaining this doubling method, we will use Hindu–Arabic numerals rather than the cumbersome Egyptian numerals. You can imagine how tedious it would be if we were to represent all of the numbers in Tables 4.2 and 4.3 and all of our calculations in Example 4 using the Egyptian numerals.

Therefore, we see that 13×21 is equal to

$$(1 + 4 + 8) \times 21* = 1 \times 21 + 4 \times 21 + 8 \times 21.$$

Using Table 4.3, we can write this as $21 + 84 + 168 = 273$. So, the craftsman must use 273 square feet of tile for the temple wall.

The Roman numeration system is a more sophisticated simple grouping numeration system.

Number	Roman Numeral
1	I
5	V
10	X
50	L
100	C
500	D
1,000	M

TABLE 4.4 Roman numerals.

The Roman numeration system, which was developed between 500 B.C. and 100 A.D., has several improvements over the Egyptian system. The Romans used letters of the alphabet as numerals to represent certain numbers, as we show in Table 4.4.

The first advantage that the Roman system had over the Egyptian system was that the Romans used a subtraction principle that allowed them to represent numbers more concisely than the Egyptians could.

Evaluating Roman Numerals

In Roman notation, we add the values of the numerals from left to right, provided we never have a numeral with a smaller value than the numeral to its right.

For example, DCLXXVIII has the value

$$500 + 100 + 50 + 10 + 10 + 5 + 1 + 1 + 1 = 678.$$

Notice that the values of the numerals either stay the same or decrease, but never increase.

However, *if the value of a numeral is ever less than the value of the numeral to its right,* then the value of the left numeral is subtracted from the value of the numeral to its right. For example,

$$\text{IV represents } 5 - 1 = 4,$$

$$\text{IX represents } 10 - 1 = 9,$$

$$\text{XL represents } 50 - 10 = 40,$$

$$\text{and CM represents } 1,000 - 100 = 900.$$

There are two restrictions on this subtraction principle:

1. We can only subtract the numerals I, X, C, and M. For example, we can not use VL to represent 45.

2. We can only subtract numerals from the next two higher numerals. For instance, we can only subtract I from V and X, therefore, we cannot use IC to represent 99.

---

* We are using the property that multiplication distributes over addition here.

Historical Highlight: *The Algorists versus the Abacists**

In 1299, the rulers of Florence, Italy, passed a law forbidding bankers from using Hindu–Arabic numerals. They were forced to continue using Roman numerals instead. Although Hindu–Arabic numerals had been introduced to Europe over 500 years earlier and made computations more efficient, we find them being outlawed as late as the fourteenth century.

In order to understand why such a law might be necessary, we first must understand how computations were done using Roman numerals. Archaeologists have found stone and marble counting boards called abaci† (plural of abacus) which were tablets with four grooves cut into them, as shown in Figure 4.1.

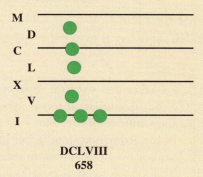

DCLVIII
658

FIGURE 4.1   A Roman abacus showing 658.

The grooves corresponded to powers of 10, namely, 1, 10, 100, and 1,000. The spaces between the grooves represented the intermediate values, 5, 50, and 500. The **abacist** (person computing with an abacus) would place small stones or metal counters in the grooves and spaces to represent Roman numerals, as we show in Figure 4.1. Then he would compute, perhaps adding several numbers, by rapidly sliding stones back and forth replacing two intermediate values with one higher value. For example, he would replace two counters in the five space with one counter in the ten groove. An uneducated person of the time could understand the concrete nature of such calculations in which counters were moved about and regrouped on the board.

On the other hand, Hindu–Arabic calculations, which were done with pen and paper, were called pen reckoning, and those who practiced this art were called **algorists.** The public was suspicious of this new method in which one ink mark could represent one counter but another mark could represent two, three, or even four counters. To make matters worse, algorists had a symbol that stood for no counters at all! In medieval times, the notion of zero was confusing to many people and was unnecessary to those who used Roman numerals. Furthermore, unscrupulous merchants and bankers could cheat uneducated customers by changing a Hindu–Arabic numeral such as a 0 to a 6 or a 1 to a 4.

Thus in Europe, the use of Roman numerals continued well into the sixteenth century. In Germany, the counters were called *Rechen-pfennig* which means "reckoning penny." These Rechen-pfennig were minted in different forms and often contained sayings such as "Here today, gone tomorrow" or "The word of God endures forever." Some of these counters showed an abacus on one side and an example of a Hindu–Arabic computation on the other, indicating that the struggle between the abacists and the algorists had not yet ended. Because of the efficiency of the Hindu–Arabic notation, the algorists eventually won the battle, which is probably why you do not calculate with counting boards and pebbles today.

---

* This highlight is based on the article by Barbara E. Reynolds, "The Algorists vs. the Abacists: An Ancient Controversy on the Use of Calculators," *The College Mathematics Journal,* Vol. 24, No. 3, May 1993, pp. 218–223.
† Chinese and Japanese abaci were invented much later and are considerably different from the Roman ones. We will discuss these other abaci in Section 4.2.

Notice that the subtraction principle allows us to write multiples of 4 and 9 more efficiently than we could using Egyptian notation. For example, instead of writing 99 as LXXXXVIIII, we can write it as XCIX.

EXAMPLE 5   **Translating between Roman Numerals and Hindu–Arabic Notation**

a) Convert MCMXLIII to Hindu–Arabic notation.

b) Write 492 in Roman numerals.

SOLUTION: a) The diagram below shows how to interpret this numeral. Remember, every time we see a smaller numeral to the left of a larger numeral, we must subtract.

$$
\begin{array}{l}
\text{M\ \ CM\ XL\ \ III} \\
1,000 \longrightarrow \\
1,000 - 100 = 900 \longrightarrow \\
50 - 10 = 40 \longrightarrow \\
3 \longrightarrow
\end{array}
$$

Thus, MCMXLIII represents $1,000 + 900 + 40 + 3 = 1,943$.

b) You can write 400 as CD, 90 as XC, and 2 as II to express 492 as CDXCII.

A second advantage the Roman system had over the Egyptian system was that it used a multiplication principle that is similar to those found in more advanced numeration systems, such as the Chinese system that you will study next and also our Hindu–Arabic system. In the Roman system, a bar above a symbol means to multiply the value of the symbol by 1,000. Also, bracketing a symbol by two vertical lines multiplies the value of the symbol by 100.* Thus $\overline{\text{X}}$ represents $10 \times 1,000 = 10,000$, $|\text{V}|$ represents $5 \times 100 = 500$, and $|\overline{\text{L}}|$ represents $50 \times 1,000 \times 100 = 5,000,000$.

Quiz Yourself 4

a) Identify two errors made in writing the Roman numeral LDIL.

b) Convert DXLVIII to Hindu–Arabic notation.

c) Convert |IX| and $\overline{\text{V}}$ to Hindu–Arabic notation.

The Chinese numeration system is a multiplicative numeration system.

The traditional Chinese numeration system uses multiples of powers of 10 such as 10, 100, and 1,000 to represent numbers and is based upon the notation shown in Table 4.5. These symbols originated during the Han dynasty which extended from 206 B.C. to 220 A.D. Thus, these symbols have not changed much during the past 2,000 years.†

This system is an example of a **multiplicative system** because we form numerals by writing products of integers between 1 and 9, inclusive, and powers of 10. Traditionally, the ancient Chinese wrote numerals vertically. For example, to

---

* See Jan Gullberg, *Mathematics: From the Birth of Numbers* (W. W. Norton & Company, 1997), p. 39.
† See Jan Gullberg, *Mathematics: From the Birth of Numbers* (W. W. Norton & Company, 1997), p. 45.

Chinese Numeral	Value
⁓	1
⼆	2
三	3
⊖	4
五	5
六	6
七	7
八	8
九	9
十	10
百	100
千	1,000

TABLE 4.5   Chinese numerals.

write 300, they would write

$$三 \leftarrow 3$$

$$百 \leftarrow \text{times } 100,$$

and they would write 5,000 as

$$五 \leftarrow 5$$

$$千 \leftarrow \text{times } 1,000.$$

However, the Chinese now write these symbols horizontally, and we will also follow this modern practice. (There are also other features of modern Chinese notation that we will not go into here but will discuss in the exercises.) Therefore, we will write 300 as 三百 and 5,000 as 五千.

EXAMPLE 6   **Translating from Chinese to Hindu–Arabic Notation**

Express each of the following in Hindu–Arabic notation.

a) 九百⊖十⼆   b) 五千⼆十八

SOLUTION: a) As you can see in Figure 4.2(a), the first two symbols represent the product of 9 and 100, or 900. The second two symbols represent 4 times 10, or 40, and the last figure represents 2. Thus the number we are representing is 942.

Notice how we write the units numeral without multiplying by a power of 10.

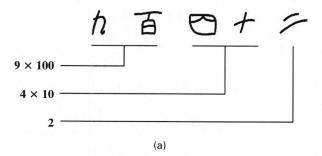

(a)

FIGURE 4.2(A)   Translating Chinese numerals.

b) As we see from Figure 4.2(b), this numeral represents 5 times 1,000, plus 2 times 10 plus 8, or 5,028.

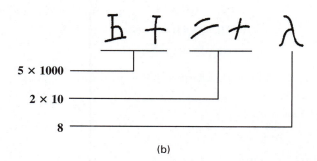

(b)

FIGURE 4.2(B)   Translating another Chinese numeral.

EXAMPLE 7 Translating from Hindu–Arabic to Chinese Notation

Write the following numbers using Chinese notation.

a) 694

b) 9,056

SOLUTION: a) To express this number, we need 6 hundreds, 9 tens, and 4 units. We write this as

六百 九十 四

$$6 \times 100 + 9 \times 10 + 4$$

Again, in using this notation, we do not multiply the units symbol, 4, by a power of 10. Also, we have separated the groups of symbols for emphasis.

b) Here we need 9 thousands, 5 tens, and 6 units. We write this as

九千五十六

The Chinese did not have a symbol for zero in the traditional notation, nor did they use the concept of place value, which we will discuss in Section 4.2. Therefore, when the Chinese wrote 六 (6), they had to specify whether they meant 6 tens, 6 hundreds, or 6 thousands, etc., which required them to use an extra symbol. As you will see in Section 4.2, the concept of place value allows us to avoid these extra symbols and express numbers more efficiently.

Exercises 4.1

In Exercises 1–6, write the Egyptian numerals using Hindu–Arabic numerals.

1. 𓍢𓍢𓍢𓎆𓎆𓏺𓏺𓏺𓏺
2. 𓎆𓍢𓍢𓍢𓍢𓎆𓎆𓏺𓏺
3. 𓂭𓂭𓆼𓍢𓍢𓍢𓎆𓎆𓏺𓏺𓏺𓏺
4. 𓆐𓆐𓂭𓂭𓂭𓆼𓆼𓍢𓎆𓎆𓏺𓏺𓏺
5. 𓁨𓁨𓆐𓂭𓂭𓆼𓍢𓍢𓎆
6. 𓁨𓆐𓆐𓂭𓂭𓂭𓍢𓍢𓎆𓏺𓏺𓏺

In Exercises 7–12, write each Hindu–Arabic numeral using Egyptian numerals.

7. 245
8. 362
9. 3,245
10. 23,416
11. 245,310
12. 2,036,042

Perform each of the following addition problems using Egyptian notation.

13. 𓍢𓍢𓍢𓎆𓎆𓏺𓏺𓏺𓏺𓏺 + 𓍢𓍢𓍢𓍢𓍢𓍢𓍢𓍢𓎆𓎆𓏺𓏺𓏺𓏺𓏺𓏺𓏺𓏺
14. 𓍢𓎆𓎆𓎆𓎆𓎆𓎆𓎆𓎆𓏺𓏺𓏺𓏺𓏺𓏺 + 𓍢𓍢𓍢𓍢𓍢𓍢𓎆𓎆𓎆𓎆𓏺𓏺𓏺𓏺𓏺𓏺𓏺
15. 𓁨𓂭𓂭𓂭𓂭𓆼𓍢𓍢𓎆 + 𓁨𓆐𓂭𓂭𓂭𓂭𓂭𓂭𓂭𓍢
16. 𓁨𓂭𓂭𓂭𓍢𓍢𓍢𓍢𓍢𓍢 + 𓁨𓂭𓂭𓂭𓂭𓆼𓍢𓍢𓍢𓍢𓍢𓎆
17. 𓍢𓍢𓍢𓎆𓎆𓎆𓎆𓎆𓎆𓎆𓎆𓏺𓏺 + 𓍢𓍢𓍢𓍢𓍢𓍢𓎆𓎆𓎆𓎆𓏺𓏺𓏺𓏺𓏺
18. 𓆼𓆼𓆼𓆼𓆼𓍢𓍢𓍢𓍢𓍢𓍢 + 𓂭𓆼𓆼𓆼𓆼𓆼𓆼𓍢𓍢𓍢𓍢𓍢𓎆

Perform each of the following subtraction problems using Egyptian notation.

19. 𓍢𓍢𓍢𓍢𓍢𓍢𓎆𓎆𓏺𓏺𓏺𓏺𓏺𓏺𓏺𓏺 − 𓍢𓍢𓍢𓎆𓎆𓎆𓎆𓎆𓏺𓏺𓏺𓏺
20. 𓆼𓍢𓍢𓎆𓎆𓎆𓏺𓏺𓏺 − 𓍢𓍢𓍢𓎆𓎆𓏺𓏺𓏺𓏺
21. 𓂭𓂭𓂭𓍢𓍢𓍢𓍢𓍢 − 𓂭𓂭𓆼𓆼𓆼𓆼𓍢𓍢𓍢𓎆𓎆𓎆𓎆𓎆𓏺𓏺𓏺𓏺
22. 𓁨𓆐𓂭𓂭𓂭 − 𓁨𓂭𓂭𓆼
23. 𓍢𓍢𓎆𓎆𓎆𓏺𓏺 − 𓍢𓎆𓎆𓎆𓎆𓏺𓏺𓏺𓏺𓏺
24. 𓍢 − 𓎆𓎆𓏺𓏺𓏺𓏺𓏺
25. An Egyptian merchant has a warehouse that contains 𓆐𓆐𓂭𓂭𓂭𓆼𓆼𓆼𓆼𓍢𓍢𓍢𓍢𓍢𓎆𓎆𓎆𓎆 square feet of storage. If he purchases another warehouse containing 𓆐𓆼𓆼𓎆𓎆𓎆𓎆𓎆𓎆𓎆𓎆 square feet of storage, what is the

total square feet of storage that he now has? Do the calculation using Egyptian notation and double check your answer using Hindu–Arabic notation.

26. An ancient Egyptian merchant had on hand ⌒𝄢𝄢𝄢𝄢999∩∩∩∩∩ bushels of wheat and sold ⌒𝄢𝄢𝄢99999∩∩∩ to another merchant. How many bushels of wheat does he have remaining? Do the calculation using Egyptian notation and double check your answer using Hindu–Arabic notation.

Use the Egyptian method of doubling to calculate the following products.

27. 14 × 43

28. 11 × 57

29. 12 × 107

30. 21 × 126

31. 35 × 121

32. 64 × 103

In Exercises 33–44, write each Roman numeral using Hindu–Arabic numerals.

33. LIX

34. LXI

35. DLXIV

36. CLXIX

37. MCMLXIII

38. MDCXXXVI

39. $\overline{\text{V}}$MMDXLIV

40. $\overline{\text{X}}$MMMCDLIV

41. |D|CCLXII

42. |M|DLVII

43. |$\overline{\text{V}}$|MCDXX

44. |$\overline{\text{L}}$|MMMDX

In Exercises 45–54, write each numeral in Roman notation. (There may be several correct answers.)

45. 44

46. 96

47. 278

48. 947

49. 444

50. 999

51. 4,795

52. 3,247

53. 89,423

54. 98,546

Frequently, Roman numerals are used today in movie credits to specify the date that a movie was released. Translate each of the following movie dates into Roman notation.

55. *Gone with the Wind* (1939)

56. *On the Waterfront* (1954)

57. *Forrest Gump* (1994)

58. *Titanic* (1997)

In Exercises 59–64, write each Chinese numeral as a Hindu–Arabic numeral. (Recall that we are writing Chinese numerals horizontally rather than in the traditional vertical form.)

59. 四百三十六

60. 入千五百二十五

61. 五十六七

62. 四十三百六十四

63. 九千九百九十九

64. 四千六百九十七

In Exercises 65–72, write each numeral using Chinese numerals.

65. 68

66. 83

67. 495

68. 726

69. 2,835

70. 3,926

71. 9,846

72. 8,754

73. The oldest discovery of Chinese written numerals is from the Yin dynasty (1523–1027 B.C.). Express these dates using Chinese numerals.

74. When Marco Polo visited China in 1274, he was impressed by the grandeur of the court of Kublai Khan. Express this date in Chinese numerals.

Further Exercises

75. Communicating Mathematics Do the words "number" and "numeral" mean the same thing? Explain.

76. In the Egyptian numeration system, whenever we have ten of one symbol we regroup using a single symbol representing the next highest power of ten. For example, we replace ten ∩s by one 𝒯. If we follow this rule, what is the largest number that you can express using the Egyptian notation that we discussed in this section? Write your answer using Hindu–Arabic notation.

77. Suppose that Egyptian numeration was based on five rather than ten. That is, ∩ would represent five strokes and 𝒯 would represent five ∩s, etc. In this case, what would be the largest number that you could express in this system? Write your answer in Hindu–Arabic notation.

*(78.) Invent an Egyptian type of numeration system using the following symbols: $\vee$, $\otimes$, ∇, $\propto$, $\approx$, $\uparrow$, $\diamond$, where these symbols represent units, tens, hundreds, and so on, respectively.

   a) Write 3,142 and 7,203 using this notation.

   b) Find 3,142 + 7,203 similar to the way we added in Example 2.

   c) Find 7,203 − 3,142 similar to the way we subtracted in Example 3.

79. Communicating Mathematics Explain two advantages of the Roman numeration system over the Egyptian system.

(80.) Write the number 1,999 in Roman numerals as many ways as you can. (Hint: Realize that it is not mandatory to use the subtraction principle in writing Roman numerals.)

81. Communicating Mathematics Why is there no need for zero in the traditional Chinese numeration system?

82. Communicating Mathematics The modern Chinese numeration system has a symbol for zero. How would that allow the Chinese to write some numerals more compactly? Give several examples.

(83.) Communicating Mathematics Research the Ionic Greek numeration system, which is an example of a ciphered numeration system. Write a paragraph explaining how this system works. What is one advantage and one disadvantage of this system?

(84.) Experiment with the Microsoft Excel spreadsheet ROMAN function which converts Hindu–Arabic numerals to Roman numerals. For example, ROMAN(499,0) equals CDXCIX. If you change the 0 to 1, 2, 3, or 4, the form of the Roman numeral becomes increasingly simplified. For example, ROMAN(499,2) equals XDIX and ROMAN(499,4) equals ID. Practice using this function to convert several numerals using different forms.

(85.) Communicating Mathematics What are some of the ways that options 1, 2, 3, and 4 for the ROMAN function violate the classical rules for writing Roman numerals?

4.2 PLACE VALUE SYSTEMS

With a few modifications, the ancient Chinese might have invented a place value system hundreds of years before the development of the Hindu–Arabic system.†

> **DEFINITION**
>
> In a **place value system**, also called a **positional system**, the placement of the symbols in a numeral determines the value of the symbols.

---

* Exercise numbers circled in red can be used as group exercises.
† As we mentioned in Section 4.1, modern Chinese write numerals horizontally and also have the concept of place value.

For example, in the numeral 35, the 3 represents three tens, or 30, and the 5 represents five ones, or 5. However, in the numeral 53, the 3 now represents 3 and the 5 represents 50.

Had the ancient Chinese realized that the placement of the symbols could tell the reader which power of ten was being multiplied, then instead of writing 543 as in Figure 4.3(a), they could have written 543 more efficiently as in Figure 4.3(b).

$$5 \times 100 \quad + \quad 4 \times 10 \quad + \quad 3 \times 1 \qquad\qquad 100\text{'s} \quad 10\text{'s} \quad 1\text{'s}$$

(a)   (b)

FIGURE 4.3   (a) Chinese notation (b) with the concept of place value.

With this modification, the rightmost symbol represents units, the symbol to its left represents 10s, the next symbol to the left represents 100s, etc. As you will see later, in order to have a true place value system, it also would have been necessary to invent a symbol for zero.

We will now discuss several place value systems, beginning with the ancient Babylonian system.

The Babylonians developed an early example of a place value system.

The Babylonians* had a primitive place value system that was based on powers of 60, and hence is called a *sexagesimal* system. There were two symbols in the Babylonian system: **▐**, which represented one, and **◀**, which represented ten. These symbols were written in wet clay with wedge-shaped sticks and when the clay hardened, there remained a permanent record of the calculations. To write small numbers, this system worked much like the Egyptian system. For example, the number 23 could be written as **◀◀▐▐▐**. However, to represent larger numbers, the Babylonians used several groups of these symbols, separated by spaces, and multiplied the value of these groups by increasing powers of 60, as we illustrate in Example 1.

EXAMPLE 1   Converting Babylonian Numerals to Hindu–Arabic Numerals

Convert **▐▐   ◀◀◀▐   ◀◀▐▐▐** to Hindu–Arabic notation.

---

* Ruins of ancient Babylon are located almost 60 miles south of modern-day Baghdad in Iraq. The Babylonian civilization lasted from 2000 B.C. to about 600 B.C.

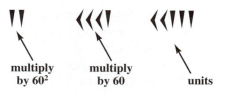

FIGURE 4.4 Babylonian notation.

SOLUTION: We interpret the three groups of numerals as follows: the right group represents units, the middle group represents 60s, and the left group represents 60^2s as shown in Figure 4.4.

Thus, this notation represents the number

$$2 \times 60^2 + 31 \times 60 + 23 = 2 \times 3,600 + 31 \times 60 + 23 = 9,083$$

Because the early Babylonians did not have a symbol for zero, it could be hard to tell exactly how many spaces were between groups of symbols. For example, it could be difficult to determine whether ❚ ⟨❚❚❚❚ represents $1 \times 60 + 14$ or $1 \times 60^2 + 14$.*

The Babylonians used the symbol ⍦ to indicate subtraction. Thus the numeral ⟨⟨⍦❚❚❚❚, represents $20 - 3 = 17$.

In order to convert Hindu–Arabic numerals to Babylonian numerals, we must divide by powers of 60, similar to the way that we convert seconds to hours and minutes. For example, in order to convert 7,717 seconds to hours and minutes, we first divide by 3,600 (there are 3,600 seconds in one hour) to get the number of whole hours.

$$
\begin{array}{r}
2 \text{ hours} \\
3600\overline{)7717} \\
\underline{7200} \\
517 \text{ seconds}
\end{array}
$$

Thus, we have 2 whole hours with 517 seconds left over. Now, we divide 517 by 60 to find the whole number of minutes.

$$
\begin{array}{r}
8 \text{ minutes} \\
60\overline{)517} \\
\underline{480} \\
37 \text{ seconds}
\end{array}
$$

So, we see that 7,717 seconds corresponds to 2 hours, 8 minutes, and 37 seconds.

---

* A much later Babylonian numeration system did have a symbol for zero which avoids this problem.

EXAMPLE 2   Converting from Hindu–Arabic to Babylonian Notation*

Write 12,221 as a Babylonian numeral.

SOLUTION: First divide by $3{,}600 = 60^2$, which is the largest power of 60 that will divide into 12,221.

$$
\begin{array}{r}
\overline{3 \text{ times } 60^2} \\
3600\overline{)12221} \\
\underline{10800} \\
1421 \text{ units left over}
\end{array}
$$

The quotient 3 tells us how many 60^2s are present in the number. Next, divide 1,421 by 60.

$$
\begin{array}{r}
\overline{23 \text{ times } 60} \\
60\overline{)1421} \\
\underline{1380} \\
41 \text{ units left over}
\end{array}
$$

The quotient 23 tells us that there are 23 60s in the number, and the remainder 41 tells us that there are 41 units left over. The number 12,221 can now be written as $3 \times 60^2 + 23 \times 60 + 41$. We write this in Babylonian notation as

 Quiz Yourself 5†

a) Convert  to a Hindu–Arabic numeral.
b) Convert 7,573 to Babylonian notation.

You might wonder why the Babylonians chose such a strange base as 60 for their mathematical system. Some speculate that it was due to the fact that they did fraction calculations by combining unit fractions such as $\frac{1}{2}, \frac{1}{3}, \frac{1}{4}, \frac{1}{5}$, and so on. For example, they would write $\frac{7}{12}$ as $\frac{1}{3} + \frac{1}{4}$. The number 60 was a convenient number to use in these situations because it has many different divisors. In computing a sum such as $\frac{1}{12} + \frac{1}{5} + \frac{1}{10} + \frac{1}{15} + \frac{1}{15} = \frac{31}{60}$, the denominator 60 would frequently arise. Thus, using the base 60 made it easier for them to do fractional computations. You can see the influence of a base 60 system today in that we measure an hour as 60 minutes or $3{,}600 = 60^2$ seconds. Also, in geometry, we have 360 degrees in a circle.

---

* We discuss an alternate method for making this type of conversion in Example 2 of Section 4.3.
† Quiz Yourself answers begin on page 849.

Historical Highlight: *Mayan Mathematics and Astronomy*

The Maya Indians lived on the Yucatan Peninsula in Central America from about 200 B.C. to 1540 A.D. During the height of their culture, from about 290 to 925 A.D., the Mayans made remarkable contributions to mathematics, astronomy, and art. They invented a numeration system based on the number 20. The system used bars and dots, with each bar representing 5 dots. The Mayans would count as shown in Figure 4.5.

After 19, the Mayans used positioning of symbols to represent larger numbers. They wrote numerals vertically, with the lowest position representing units, the next higher representing 20s, the next position representing 20×18, the next representing $20 \times 18 \times 20$, as shown in Figure 4.6. They used the symbol ⬯, which looked somewhat like a seashell, to represent zero.

You may think it strange that in Figure 4.6 we multiplied 14 by 20×18 rather than 20 squared. However, the Mayans had a good reason for doing this because their calendar consisted of 18 *uinals* ("months") of 20 days each, plus five extra days added on at the end of the year.

•	$1 \times 20 \times 18 \times 20^2$	$= 144{,}000$
⬯	$0 \times 20 \times 18 \times 20$	$= \quad 0$
•••• (over bars)	$14 \times 20 \times 18$	$= \quad 5{,}040$
••• (over bar)	8×20	$= \quad 160$
(three bars)	15	$= \quad 15$
		$\overline{\qquad\qquad}$
		$149{,}215$

FIGURE 4.6   Representing a large number with Mayan notation.

They had a remarkably good estimate of the length of a year—365.242000 days versus our current estimate of 365.242198 days. Mayan astronomers also observed the movements of the sun, moon, and planets and accurately predicted solar eclipses. Their accomplishments were truly remarkable considering that the Mayans did not know how to make glass, so they did not have the advantage of telescopes as did later astronomers.

•	••	•••	••••	—	•	••	•••	••••	—
1	2	3	4	5	6	7	8	9	10

•	••	•••	••••	—	•	••	•••	••••
11	12	13	14	15	16	17	18	19

FIGURE 4.5   Counting with Mayan numerals.

The Hindu–Arabic numeration system is a place value system based on ten.

The oldest known examples of Hindu–Arabic numerals were found on a stone column in India, and are believed to have been written about 250 B.C. That Hindu system had no symbol for zero, and scholars are not certain as to when our

current Hindu–Arabic system became fully developed. The Persian mathematician al-Khowarizmi became aware of Hindu notation and in 825 he wrote a book whose title translated into English means "The Book of al-Khowarizmi on Hindu Number." After this, Hindu notation was adopted by the Arabs, and in 1202, the Italian mathematician Leonardo Fibonacci, after studying in the Middle East, wrote a book on arithmetic and algebra, titled *Liber Abaci* (a Book on the Abacus), which helped spread Hindu–Arabic numerals throughout Europe.

The Hindu–Arabic system is a place value system based on ten, unlike the Babylonian system which is based on 60, and the Mayan system which is based on 20 (see Historical Highlight). One feature of the Hindu–Arabic system is that all numbers can be written using only the digits* 0, 1, 2, . . . , 9. Thus, we do not need special symbols for 10, 100, 1,000, etc., as did some of the earlier numeration systems. Also, the invention of zero, used as a place holder, makes it easy to distinguish between numbers such as 5,001, 501, and 51. Contrast this approach with that of the Babylonians who used spaces between groups of symbols.

We can write Hindu–Arabic numerals in *expanded form* to show explicitly how each digit is multiplied by a power of 10. For example,

$$6{,}582 = 6 \times 10^3 + 5 \times 10^2 + 8 \times 10^1 + 2 \times 10^0$$

(Recall that $10^0 = 1$, $10^1 = 10$, $10^2 = 10 \times 10 = 100$, $10^3 = 10 \times 10 \times 10 = 1{,}000$, etc.)

EXAMPLE 3   Writing Hindu–Arabic Numbers in Expanded Form

a) Write 53,024 in expanded form.

b) Write 530,024 in expanded form.

c) Write $4 \times 10^3 + 0 \times 10^2 + 2 \times 10^1 + 5 \times 10^0$ using Hindu–Arabic notation.

SOLUTION: a) $53{,}024 = 5 \times 10^4 + 3 \times 10^3 + 0 \times 10^2 + 2 \times 10^1 + 4 \times 10^0$.

b) $530{,}024 = 5 \times 10^5 + 3 \times 10^4 + 0 \times 10^3 + 0 \times 10^2 + 2 \times 10^1 + 4 \times 10^0$.

c) $4 \times 10^3 + 0 \times 10^2 + 2 \times 10^1 + 5 \times 10^0 = 4$ thousands + 0 hundreds + 2 tens + 5 ones = 4,025.

We can use expanded notation to explain the algorithms to perform the arithmetic operations that you learned in elementary school.

EXAMPLE 4   Using Expanded Notation to Explain the Addition Algorithm

Calculate $4{,}625 + 814$ using expanded notation.

---

* The word *digit* is the Latin word for finger. Because we have ten fingers, it is not surprising that many numeration systems are based on the number ten.

SOLUTION: Using expanded notation, we can write this problem as:

$$
\begin{array}{rl}
4{,}625 = & 4 \times 10^3 + 6 \times 10^2 + 2 \times 10^1 + 5 \times 10^0 \\
+\quad 814 = +\ & 8 \times 10^2 + 1 \times 10^1 + 4 \times 10^0 \\
\hline
& 4 \times 10^3 + 14 \times 10^2 + 3 \times 10^1 + 9 \times 10^0
\end{array}
$$

The computations in the 10^0 and 10^1 places are straightforward; however, when we add 6×10^2 and 8×10^2, we get 14×10^2. We cannot express 14 using a single place, so instead, we think of 14 as $10 + 4$. Then we can write $14 \times 10^2 = (10 + 4) \times 10^2 = 10 \times 10^2 + 4 \times 10^2$. (We are using the fact that multiplication distributes over addition here.) This gives us 4 ten squareds and an additional ten cubed, making 5 ten cubeds. In expanded notation, we can write this as

$$5 \times 10^3 + 4 \times 10^2 + 3 \times 10^1 + 9 \times 10^0,$$

which equals 5,439.

The way we handled the 14 in the answer in Example 4 explains why you were taught to "put down the 4 and carry the 1 to the next column to the left" in performing this addition.

EXAMPLE 5   Using Expanded Notation to Explain the Subtraction Algorithm

Calculate $728 - 243$ using expanded notation.

SOLUTION: Using expanded notation, we can write this problem as:

$$
\begin{array}{rl}
728 = & 7 \times 10^2 + 2 \times 10^1 + 8 \times 10^0 \\
-\ 243 = -\ & 2 \times 10^2 + 4 \times 10^1 + 3 \times 10^0
\end{array}
$$

The computations in the 10^2 and 10^0 places would be straightforward, except for the fact that we cannot directly subtract 4×10^1 from 2×10^1.

To remedy this, we express one of the ten squareds as 10×10^1. This gives us 12 tens and 6 ten squareds. We can now rewrite the subtraction as

$$
\begin{array}{rl}
728 = & 6 \times 10^2 + 12 \times 10^1 + 8 \times 10^0 \\
-\ 243 = -\ & 2 \times 10^2 + 4 \times 10^1 + 3 \times 10^0 \\
\hline
& 4 \times 10^2 + 8 \times 10^1 + 5 \times 10^0
\end{array}
$$

We now can write $4 \times 10^2 + 8 \times 10^1 + 5 \times 10^0 = 485$.

When you learned to do subtraction, you probably were taught to do this same "borrowing," but in a less drawn out fashion without using expanded notation.

Hindu–Arabic notation allows us to simplify computations.

An important benefit of Hindu–Arabic notation is that we can do basic numerical calculations very simply using pencil and paper rather than an abacus or counting board. In Example 6, we explain the galley method for doing multiplication,

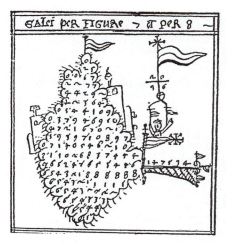

An example of using the galley method to do division.*

which you will see is a forerunner of the multiplication method that we use today. This method was popular in Italy in the fifteenth century; however, because printers found it cumbersome to typeset galleys (particularly when doing division), this method eventually gave way to our more modern way of doing multiplication.

EXAMPLE 6   Multiplying Using the Galley Method

Multiply 685 and 49 using the galley method.

SOLUTION: We begin by constructing the rectangle, divided into triangles, which is called a galley,† in Figure 4.7(a). Then we compute the partial products in each box of the galley as shown in Figure 4.7(b). For example, we put the 2 and 4 in the triangles of the upper left-hand box because the product of 6 and 4 is 24.

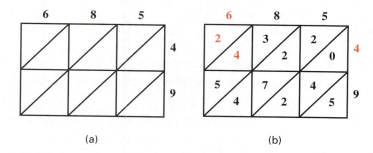

(a)   (b)

FIGURE 4.7   A galley to multiply 685 and 49.

Next we add the numbers along the diagonals, starting at the bottom right, as shown in Figure 4.8. If the sum along a diagonal is 10 or greater, then place the units digit at the end of the diagonal and carry the tens digit to the channel to the left. For example, in the third diagonal from the right, the sum is $4 + 7 + 2 + 2 = 15$. We put the 5 at the end of the channel and carry the 1 to the next channel to the left. We read the product by following the numbers as indicated by the red arrow.

The final product is therefore 33,565.

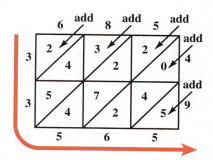

FIGURE 4.8   Computing the product by adding partial products.

---

* From Frank J. Swetz (Ed.), *Five Fingers to Infinity* (Open Court Pub. Co., 1994).
† It is also possible to do division using the galley method, and when doing so, the pattern of numbers resembles a ship, or galley. This method is also sometimes called the *gelosia* method. Gelosia in Italian means window.

The galley method we used in Example 6 is much like the multiplication algorithm that we use today. In Figure 4.8, you see that the channels, from right to left, correspond to powers of 10—units, tens, hundreds, thousands, and so on.

Without drawing the galley, we might think of this calculation as follows:

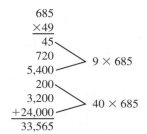

$$
\begin{array}{r}
685 \\
\times 49 \\
\hline
45 \\
720 \\
5,400 \\
200 \\
3,200 \\
+24,000 \\
\hline
33,565
\end{array}
$$

9 × 685

40 × 685

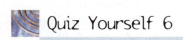

Quiz Yourself 6

Multiply 328 × 39 using the galley method. Show the galley as well as showing your final answer.

Notice that the columns in this multiplication, from right to left, correspond to the channels in the galley in Figure 4.8 with the numbers slightly rearranged.

Napier's rods are a variation on the galley method.

In the seventeenth century, the English mathematician John Napier developed a device called **Napier's rods** or **Napier's bones** for doing multiplication. This device consists of a series of strips, each labeled at the top with a digit, 0, 1, 2, . . . , 9. The remainder of each strip lists all multiples of the label at the top, as shown in Figure 4.9. There is an additional strip called the *index* which is also shown in Figure 4.9.

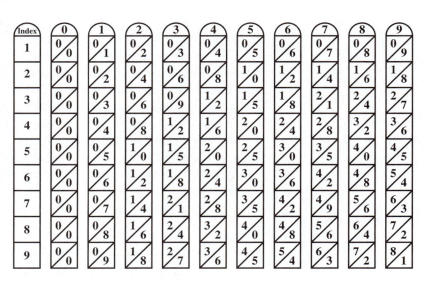

FIGURE 4.9   Napier's rods.

Highlight: *John Napier**

In the sixteenth century, a time long before the advent of graphing calculators and before computers were invented, Napier's rods were an important calculating device. A person with a set of these rods did not need to learn multiplication tables in order to do a lengthy multiplication. All that one had to do was line up the rods, read off the partial products, and then do simple addition to compute the product.

Napier, the eighth Baron of Merchiston, Scotland, was interested in many things. As an amateur theologian, he reasoned that the Roman Pope was the anti-Christ, and from studying the Bible's Book of Revelation, he deduced that God would destroy the world by the year 1700. Napier mistakenly considered these religious "discoveries" to be his greatest contributions to mankind.

As an inventor, Napier invented a pump for removing mine water and, although he never constructed them, he anticipated the design of the modern machine gun, the tank, and the submarine.

Although he probably would have disagreed, Napier's greatest contributions were in the field of mathematics. He had a great interest in simplifying computations, and in addition to his rods, he spent 20 years developing tables of logarithms† which greatly simplified scientific computations. His discovery of logarithms led to the invention of the slide rule, one of the principal computational tools of scientists and engineers during the twentieth century.

The noted French mathematician Pierre de Laplace said that by shortening the time it took to do lengthy calculations, Napier had doubled the life of astronomers. In 1620, the great mathematician and astronomer Johannes Kepler, who discovered the laws of planetary motion, gave a public tribute to Napier and his work, not realizing that Napier had died three years earlier.

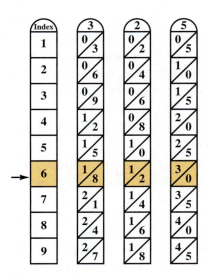

FIGURE 4.10   Using Napier's rods to compute 6 × 325.

In order to calculate a product such as 6 × 325, you select strips for 3, 2, and 5 and place them side by side next to the index, as shown in Figure 4.10.

We then use the small galley, formed by the three boxes next to the 6 in the index, to compute the product 1,950 as we did in Example 6. See Figure 4.11.

We will investigate how to use Napier's rods to do more complex multiplications in Exercises 67 and 68.

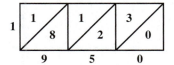

FIGURE 4.11   6 × 325 = 1,950.

---

* This highlight is based on David M. Burton, *The History of Mathematics: An Introduction* (WCB-McGraw Hill, 1999), pp. 325–330.
† Your calculator probably has keys labeled "log" or "ln" which are today's electronic equivalents of Napier's logarithms tables.

Exercises 4.2

Write these Babylonian numerals as Hindu–Arabic numerals.

1. 𒌋𒌋𒌋𒑲

2. 𒌋𒌋𒌋𒌋𒌋𒑲𒑲𒑲𒑲

3. 𒌋𒌋𒌋𒑲𒑲𒑲

4. 𒌋𒌋𒌋𒌋𒑲𒑲𒑲𒑲𒑲

5. 𒌋𒑲𒑲 𒌋𒑲

6. 𒌋𒑲 𒌋𒌋𒑲𒑲𒑲

7. 𒑲𒑲 𒌋𒌋𒑲𒑲𒑲𒑲 𒌋𒌋𒑲𒑲𒑲𒑲𒑲𒑲

8. 𒑲𒑲 𒌋𒑲𒑲𒑲 𒌋𒌋𒌋𒑲𒑲𒑲

Write each number using Babylonian notation.

9. 8,235

10. 7,331

11. 18,397

12. 26,411

13. 123,485

14. 227,597

15. 188,289

16. 173,596

What does 5 represent in each of the following numerals?

17. 37,521

18. 53,184

19. 105,000

20. 5,023,671

Write each number using expanded notation.

21. 25,389

22. 37,248

23. 278,063

24. 820,634

25. 1,200,045

26. 3,002,608

Write each number using standard Hindu–Arabic notation.

27. $5 \times 10^3 + 3 \times 10^2 + 6 \times 10^1 + 8 \times 10^0$

28. $8 \times 10^3 + 2 \times 10^2 + 1 \times 10^1 + 4 \times 10^0$

29. $3 \times 10^5 + 7 \times 10^4 + 0 \times 10^3 + 0 \times 10^2 + 8 \times 10^1 + 2 \times 10^0$

30. $6 \times 10^5 + 0 \times 10^4 + 8 \times 10^3 + 2 \times 10^2 + 0 \times 10^1 + 4 \times 10^0$

31. $3 \times 10^6 + 7 \times 10^4 + 5 \times 10^2 + 2 \times 10^0$

32. $6 \times 10^6 + 3 \times 10^2 + 8 \times 10^1 + 2 \times 10^0$

Perform the following additions and subtractions using expanded notation as we did in Example 4 and Example 5.

33. 2,863 + 425

34. 5,264 + 583

35. 3,482 + 2,756

36. 7,843 + 1,692

37. 926 − 784

38. 835 − 362

39. 5,238 − 1,583

40. 3,417 − 2,651

In Exercises 41–46, a) Use the galley method to perform the multiplication, and b) rewrite the calculations in a more conventional fashion, as we did in the discussion following Example 6.

41. 4 × 235

42. 6 × 382

43. 23 × 876

44. 56 × 371

45. 293 × 465

46. 473 × 628

Use Napier's rods to perform each multiplication.

47. 3 × 628

48. 7 × 346

49. 8 × 492

50. 5 × 728

51. 6 × 924

52. 4 × 834

53. Communicating Mathematics   Why might we confuse 65 and 3,605 when using Babylonian notation?

54. Communicating Mathematics   Give an example of two numbers, other than 65 and 3,605, that might be confused when using Babylonian notation. Why might they be confused?

55. Communicating Mathematics   Why is zero important in a place value system?

56. Communicating Mathematics   Is the Babylonian numeration system a true place value system? Explain.

Further Exercises

57. Communicating Mathematics   Explain two differences of the Babylonian numeration system from the Roman system.

58. Communicating Mathematics   State a feature that is present in the Hindu–Arabic numeration system that is lacking in both the early Babylonian and ancient Chinese systems.

The modern Chinese numeration system is a base 10 place value system that uses the symbol ○ for zero. Rewrite each of the following numerals using this modern notation.

Realize that now you do not need special symbols for 10, 100, 1,000, etc.

59. 囗百三十六   **60.** 入千五百二十五

61. 五千六十七   **62.** 囗十三百囗

63. 九千九   **64.** 囗十九十

65. Communicating Mathematics What difficulty does the ancient Chinese numeration system have in representing very large numbers? How does the Hindu–Arabic system resolve this problem?

66. Communicating Mathematics Is it possible to represent very large numbers in the Babylonian system? Explain.

****67.** Communicating Mathematics Devise a method for using Napier's rods to multiply numbers having several digits, such as 324 × 615. Hint: Think of 324 as 300 + 20 + 4. Use this method to multiply 324 × 615.

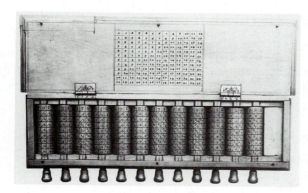

68. Use the method that you invented in Exercise 67 to multiply 426 × 853.

69. Communicating Mathematics Discuss how the galley method for doing multiplication is similar to the method of using Napier's rods.

70. Communicating Mathematics Place the following types of numeration systems in order beginning with the earliest.

a) place value system

b) multiplicative grouping

c) tally marks

d) simple grouping.

Give an example of each type of system. Explain why you decided to order the systems as you did.

Use the partially completed galleys to determine the numbers that are being multiplied.

71.    **72.**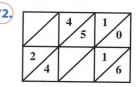

Translate each of the following Mayan numerals to Hindu–Arabic notation.

73. **74.** **75.** **76.**

4.3   CALCULATING IN OTHER BASES

The fact that we have ten fingers is no doubt the reason that throughout history numeration systems based on ten are so common. However, as you saw in Section 4.2, it is possible to have place value systems using other bases such as the Babylonian system that uses base 60 and the Mayan system that essentially uses a base of 20. Indeed, as you will see soon, we can construct a place value system using any integer greater than one. Such numeration systems are not just

* Exercise numbers circled in red can be used as group exercises.

mathematical curiosities. Various societies around the world use 2, 3, 4, 5, or 8 as bases for their numeration systems. We come in daily contact with applications such as computers, DVD players, and cell phones that are based on numeration systems using bases of 2, 8, and 16.

Although we will emphasize the base 5 numeration system in this section, you will find that the principles we explain are easy to carry over to systems having other bases.

The principles of a base 5 system are similar to those of a base 10 system.

We begin by explaining how to represent numbers in a base 5 system. It is useful to compare what we are going to do with what you have already learned about our base 10 system. Recall that we only use the digits 0, 1, 2, 3, . . . , 9 and can write a number such as 3,542 in expanded notation as $3 \times 10^3 + 5 \times 10^2 + 4 \times 10^1 + 2 \times 10^0$. Similarly, in the base 5 system, we only use the digits 0, 1, 2, 3, 4 and write a number such as $2,431_5$ in expanded notation as

$$2 \times 5^3 + 4 \times 5^2 + 3 \times 5^1 + 1 \times 5^0.$$

Note that the subscript 5 indicates that the numeral is a base 5 numeral, not a base 10 numeral. The numeral 32_5 represents 3 *fives* and 2 units, whereas 32 represents 3 *tens* and 2 units. Also, you should read 32_5 as "three two base five," not "thirty-two base five." The word "thirty" implies a base of ten.

EXAMPLE 1   Converting to Decimal Notation

a) Convert $40,312_5$ to decimal (base 10) notation.

b) Convert $7,036_8$ to decimal notation.

SOLUTION: a) To convert to decimal notation, we write $40,312_5$ in expanded notation as

$$
\begin{aligned}
40,312_5 &= 4 \times 5^4 + 0 \times 5^3 + 3 \times 5^2 + 1 \times 5^1 + 2 \times 5^0 \\
&= 4 \times 625 + 0 \times 125 + 3 \times 25 + 1 \times 5 + 2 \times 1 \\
&= 2,500 + 0 + 75 + 5 + 2 \\
&= 2,582
\end{aligned}
$$

b) In base 8, we use the digits, 0, 1, 2, . . . , 7 and express numbers in expanded notation using powers of 8. Thus,

$$
\begin{aligned}
7,036_8 &= 7 \times 8^3 + 0 \times 8^2 + 3 \times 8^1 + 6 \times 8^0 \\
&= 7 \times 512 + 0 \times 64 + 3 \times 8 + 6 \times 1 \\
&= 3,584 + 0 + 24 + 6 \\
&= 3,614
\end{aligned}
$$

Quiz Yourself 7*

Convert 354_6 to decimal notation.

When counting in base 10, we count until we reach nine and then, because we do not have a single symbol to represent ten, we write 10, indicating one times the base plus zero units. Similarly, in a base 5 system, we count 1, 2, 3, 4, and now, since the next number is the base, five, we write 10_5. Continuing this pattern, when we reach 14_5, the next number would require us to write five in the units place, which we cannot do. Thus we write 20_5. Therefore, we would count in base 5 as follows:

$$1, 2, 3, 4, 10_5, 11_5, 12_5, 13_5, 14_5, 20_5, 21_5, 22_5, \ldots, 44_5.$$

Now if we increase the units place by one, we must write a zero and carry a one to the fives place. However, we cannot write a single symbol for five in the fives place, so we must write a zero and carry a one to the five squared place. Thus, the number following 44_5 is 100_5. This represents the decimal number 25.

In a base 8 system, called an **octal** system, we would count:

$$1, 2, 3, 4, 5, 6, 7, 10_8, 11_8, 12_8, \ldots, 17_8, 20_8, 21_8, \ldots, 77_8, 100_8.$$

However, in counting in a base 16 system, called a **hexadecimal** system, we run into a problem. If we start counting in the usual way, 1, 2, 3, 4, 5, 6, 7, 8, 9, we cannot write 10_{16} next because 10_{16} represents $1 \times 16 + 0 \times 1$, which is decimal sixteen, not ten. To solve this problem, it is customary to use the symbols A, B, C, D, E, and F to represent the decimal numbers 10, 11, 12, 13, 14, and 15. Thus in base 16, we would count:

$$1, 2, 3, \ldots, 8, 9, A, B, C, D, E, F, 10_{16}, 11_{16}, \ldots, 1F_{16}, 20_{16}, \ldots, FF_{16}, 100_{16}.$$

A numeration system that is used heavily in designing digital devices such as computers and DVD players is the base 2, or **binary** system. In this system, we use only the digits 0 and 1 and count as follows:

$$1, 10_2, 11_2, 100_2, 101_2, 110_2, 111_2, 1,000_2, 1,001_2, \ldots$$

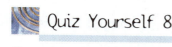

Quiz Yourself 8

Count to the decimal number ten using a base 3 system.

SOME GOOD ADVICE

It will help you to count in other bases if you practice thinking of the numbers as being written in expanded form. For example, think of 17_8 as $1 \times 8^1 + 7 \times 8^0$ and 101_2 as $1 \times 2^2 + 0 \times 2^1 + 1 \times 2^0$.

We will now explain how to convert a decimal numeral to a base 5 numeral. In this process, which we explain in Example 2, we first determine the number of units, next the number of fives, then the number of twenty-fives, and so on.

EXAMPLE 2   **Converting from Decimal to Base 5 Notation**

Write 384 as a base 5 numeral.

SOLUTION: The following division shows that when dividing 5 into 384, we get a quotient of 76 with 4 units left over.

$$\begin{array}{r} 76 \\ 5\overline{)384} \\ 380 \\ \hline \end{array}$$
4 units left over

We next divide 76 by 5. This is equivalent to dividing the original number by 25, which is five squared. From this division, we see that after dividing by 25, there is 1 five left over. This means that in the base 5 numeral, there will be a 1 in the fives place.

$$\begin{array}{r} 15 \\ 5\overline{)76} \\ 75 \\ \hline \end{array}$$
1 five left over

We continue this process and express the divisions more compactly, as in Figure 4.12.

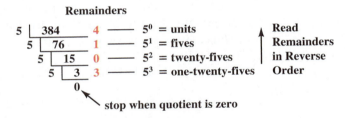

Remainders

$$
\begin{array}{lll}
5 \mid 384 & 4 \rule[0.5ex]{1em}{0.4pt} & 5^0 = \text{units} \\
\quad 5 \mid 76 & 1 \rule[0.5ex]{1em}{0.4pt} & 5^1 = \text{fives} \\
\qquad 5 \mid 15 & 0 \rule[0.5ex]{1em}{0.4pt} & 5^2 = \text{twenty-fives} \\
\qquad\quad 5 \mid 3 & 3 \rule[0.5ex]{1em}{0.4pt} & 5^3 = \text{one-twenty-fives} \\
\qquad\qquad 0 &
\end{array}
$$

Read Remainders in Reverse Order

stop when quotient is zero

FIGURE 4.12 Repeatedly dividing by 5 to find a base 5 numeral.

In Figure 4.12, we continued dividing each quotient by 5 until we obtained a quotient of 0. Reading the remainders in reverse order will give us the base 5 numeral for 384, which is $3{,}014_5$.

It is important to remember, in Example 2, that we stopped the process when we reached *a quotient of zero*—not a remainder of zero.

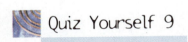

Quiz Yourself 9

Convert the decimal number 113 to base 5.

PROBLEM SOLVING

In Chapter 1, we recommended that to understand mathematical ideas, it is useful to make analogies with situations that you have seen before. If you thoroughly understand the concept of place value in our Hindu–Arabic numeration system, you can easily modify the techniques you have used before to understand how to compute in other place value systems.

■

Arithmetic operations in other bases are similar to base 10 operations.

Imagine that you had grown up in a base 5 world. As a small child, perhaps you learned to count by watching Sesame Street. Instead of hearing repeatedly the ditty

"one, two, three, four, five, six, seven, eight . . . nine . . . ten,"

you may have learned to count

"one, two, three, four, fen, fenone, fentwo, thirfeen, fourfeen, twenfy, twenfy one, etc."

It is doubtful that you would have used the cumbersome name "one zero base five" for five or "one one base five" for six, etc. We have invented the shorter, fanciful names "fen, fenone, fentwo" for numbers such as 10_5, 11_5, and 12_5. Do not bother to memorize these make-believe names.

Your early days of elementary school would have been much easier because you would have had fewer number facts to memorize. You would not have had to learn that $7 + 8 = 15$, because problems using the digits 7 and 8 would never arise. Also some number facts would be written differently in a base 5 world. For example, you would write the decimal fact that $4 + 4 = 8$ as $4 + 4 = 13_5$. Using our made-up terminology above, you would say

"four plus four equals thirfeen."

You would have spent a good deal of time memorizing the addition facts that we list in Table 4.6. For example, in Table 4.6, we see that $3 + 4 = 12_5$

 Quiz Yourself 10

Find $2 + 4$ and $2 + 3$ in Table 4.6.

+	0	1	2	3	4
0	0	1	2	3	4
1	1	2	3	4	10_5
2	2	3	4	10_5	11_5
3	3	4	10_5	11_5	12_5
4	4	10_5	11_5	12_5	13_5

TABLE 4.6   Base 5 addition facts.

We will use the addition facts in Table 4.6 to do the addition in Example 3.

EXAMPLE 3   **Adding in Base 5**

Add 342_5
$+ 223_5$.

$$
\begin{array}{r}
\textbf{1} \\
342_5 \\
+\ 223_5 \\
\hline
0_5
\end{array}
\quad \longleftarrow\ \ 2 + 3 = 10_5
$$

write 0, carry 1

(a)

$$
\begin{array}{r}
\textbf{11} \\
342_5 \\
+\ 223_5 \\
\hline
20_5
\end{array}
\quad\ 1 + 4 + 2 = 12_5
$$

write 2, carry 1

(b)

FIGURE 4.13 Adding in base 5.

SOLUTION: We add much like we do in performing decimal addition. First we add the units, $2 + 3 = 5$ decimal which we express as 10_5. We place the 0 in the units place and carry the 1 to the fives place, as shown in Figure 4.13(a).

We next add the numerals in the fives place: $1 + 4 + 2 = 12_5$ as shown in Figure 4.13(b). Note how we write down the 2 and carry the 1 to the five squared place. We finish by adding the numerals in the five squared place, carrying as necessary.

$$
\begin{array}{r}
111 \\
342_5 \\
+\ \ 223_5 \\
\hline
1120_5
\end{array}
$$

EXAMPLE 4 Subtracting in Base 5

Subtract
$$
\begin{array}{r}
424_5 \\
-\ 143_5 \\
\end{array}
$$

SOLUTION: Subtraction in base 5 is also much like base 10 subtraction. We begin by subtracting three units from four units to get one unit.

In the fives place, we cannot subtract 4 fives from 2 fives, so we borrow 1 twenty-five and write it as 5 fives. Thus we now have 7 fives in the fives place which we have written as 12_5, as shown in Figure 4.14. We complete the subtraction by subtracting 1 twenty-five from the 3 remaining twenty-fives.

$$
\begin{array}{r}
3 \\
\not{4}^{\,1}2\ 4_5 \\
-\ 1\ 4\ 3_5 \\
\hline
2\ 3\ 1_5
\end{array}
$$

Borrowing 1 twenty-five we now have 12_5, or decimal 7 fives. Subtracting, we get 3 fives

FIGURE 4.14 Subtracting in base 5 may require borrowing.

Quiz Yourself 11

a) Add $315_6 + 524_6$

b) Subtract $325_6 - 131_6$

SOME GOOD ADVICE

Although it is useful to check base 5 computations by converting all numbers to base 10, you will increase your skill in working in other bases if you try to do all computations by thinking in base 5 even though it may seem a little difficult at first.

In order to understand how to multiply in other bases, we first need to know our multiplication tables. The multiplication facts will look different from what you are used to seeing in base 10. For example, in Table 4.7, we see that $3 \times 4 = 22_5$, which is the base 5 representation of decimal 12. You should verify the other base 5 multiplication facts given in Table 4.7.

$\times$	0	1	2	3	4
0	0	0	0	0	0
1	0	1	2	3	4
2	0	2	4	11_5	13_5
3	0	3	11_5	14_5	22_5
4	0	4	13_5	22_5	31_5

TABLE 4.7 Base 5 multiplication facts.

Quiz Yourself 12

Find 2×4 and 4×4 in Table 4.7.

Before giving an example of multiplication, it is useful to review the discussion following Example 6 of Section 4.2, explaining the meaning of galley multiplication in base 10. Recall that we wrote all of the partial products in multiplying 685 and 49 as follows:

$$
\begin{array}{r}
685 \\
\times 49 \\
\hline
45 \\
720 \\
5{,}400 \\
200 \\
3{,}200 \\
+24{,}000 \\
\hline
33{,}565
\end{array}
$$

— units times units ⎤
— units times tens ⎦ Combine into first partial product
— units times hundreds

— tens times units ⎤
— tens times tens ⎦ Combine into second partial product
— tens times hundreds

Notice how

$$\text{units times units} = \text{units,}$$

$$\text{units times tens} = \text{tens,}$$

$$\text{tens times tens} = \text{hundreds, and so on.}$$

Also, when we do this multiplication in the usual way, we combine the first three products into a single partial product and the second three products into a second partial product. We will follow this pattern in doing base 5 multiplication.

EXAMPLE 5 Multiplying in Base 5

Multiply

$$
\begin{array}{r}
134_5 \\
\times \ 32_5 \\
\hline
\end{array}
$$

SOLUTION: We first multiply 2 units time 4 units to get 8 units in base 10, or 13_5 units. We write down 3 units and carry 1 to the fives place.

$$
\begin{array}{r}
1 \quad\text{— 5 units expressed} \\
\text{as 1 five} \\
134_5 \\
\times\ 32_5 \\
\hline
3_5 \\
\text{— 3 units}
\end{array}
$$

We next multiply 2 units times 3 fives to get 6 fives, plus the 1 five we carried, gives 7 fives, which we write as 12_5. We write the 2 and carry the 1 to the five squared column.

$$
\begin{array}{r}
\text{— 5 fives expressed} \\
11 \quad\text{as 1 five squared} \\
134_5 \\
\times\ 32_5 \\
\hline
23_5 \\
\text{— 2 fives}
\end{array}
$$

We complete the first partial product by multiplying the 2 units times 1 five squared and add the 1 five squared, which was carried, to get 3 five squareds.

$$
\begin{array}{r}
134_5 \\
\times\ 32_5 \\
\hline
323_5
\end{array}
$$

We compute the second partial product in a similar way. We begin by multiplying 3 fives times 4 units to get 22_5 fives. We write 2 fives and carry a 2 to the five squared column.

$$
\begin{array}{r}
2 \\
134_5 \\
\times\ 32_5 \\
\hline
323_5 \\
2
\end{array}
$$

We complete the second partial product as follows:

$$
\begin{array}{r}
122 \\
134_5 \\
\times\ 32_5 \\
\hline
323_5 \\
1012
\end{array}
$$

We then add the partial products to get the final product.

$$
\begin{array}{r}
134_5 \\
\times\ 32_5 \\
\hline
323_5 \\
1012 \\
\hline
10443_5
\end{array}
$$

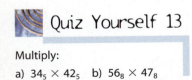

Quiz Yourself 13

Multiply:

a) $34_5 \times 42_5$ b) $56_8 \times 47_8$

Before doing division, you must have a very good understanding of multiplication. For example, consider how we do the first step in the following decimal division.

$$\begin{array}{r} 2 \\ 18\overline{)4721} \\ \underline{36} \\ 11 \end{array}$$

Based upon years of experience in working with the decimal system and our ability to make good estimations, we see that 18 divides into 47 twice, but not 3 times. In dividing

$$23_5\overline{)4132_5}$$

we do not have the same deep background that enables us to find the trial quotient, so we have to use another approach, as we show in Example 6.

EXAMPLE 6   Dividing in Base 5

Perform the division

$$23_5\overline{)4132_5}$$

SOLUTION: In order to be able to estimate trial quotients, we have to know the multiples of 23_5, which we list below. (You should verify these facts.)

$$23_5 \times 0 = 0$$
$$23_5 \times 1 = 23_5$$
$$23_5 \times 2 = 101_5$$
$$23_5 \times 3 = 124_5$$
$$23_5 \times 4 = 202_5$$

From this list we see that 23_5 divides into 41_5 once but not twice. So we start the division by placing a 1 in the quotient.

$$\begin{array}{r} 1 \\ 23_5\overline{)4132_5} \\ \underline{23} \\ 13 \end{array}$$

Remember, that when subtracting $41_5 - 23_5$, we are doing a base 5 subtraction. We next bring down the digit 3 and add another digit to the quotient. From the above list, we see that 23_5 divides into 133_5 three times with a remainder of 4.

$$\begin{array}{r} 13 \\ 23_5\overline{)4132_5} \\ \underline{23} \\ 133 \\ \underline{124} \\ 4 \end{array}$$

Next, we bring down the 2 and complete the division.

$$
\begin{array}{r}
131_5 \\
23_5\overline{)4132_5} \\
\underline{23} \\
133 \\
\underline{124} \\
42 \\
\underline{23} \\
14_5
\end{array}
$$

Thus, 23_5 divides into 4132_5, with a quotient of 131_5 and a remainder of 14_5.

SOME GOOD ADVICE

In order to check to see if we have properly done a division, such as the one we did in Example 6, you can multiply the divisor 23_5 by the quotient 131_5, and then add the remainder 14_5. The result should be the dividend 4132_5.

Binary, octal, and hexadecimal notation are closely related.

As we have mentioned, digital devices such as computers, DVD players, and cell phones use the binary system. You might think of the 1s and 0s as representing "on" and "off" commands controlling microscopic switches on electronic chips. The values 0 and 1 are called **bits,** which is a contraction of the words **binary digit.** A video card in your PC, for example, may use the 16-bit binary pattern 1011001101011110 as the command to change the pen color to red in a drawing package. Because it is tedious to remember such long strings of bits, often we rewrite these commands in a form that is easier for humans to remember.

Realize that it requires three binary spaces to count from 0 to 7 (000_2, 001_2, 010_2, . . . , 111_2), whereas we can count from 0 to 7 using only one octal space. So we can express the above 16-bit binary command using fewer octal numerals, as we show in Example 7.

EXAMPLE 7

a) Write the binary command 1011001101011110 using octal notation.

b) Write the binary command 1011001101011110 using hexadecimal notation.

SOLUTION: a) We begin by grouping the bits in 1011001101011110 in groups of three *beginning from the right.* We then interpret each group of three bits as an octal number as shown below. (For the sake of efficiency, we often omit the subscripts that indicate the base.)

$$\text{binary} \longrightarrow 1 \quad 011 \quad 001 \quad 101 \quad 011 \quad 110$$

$$\text{octal} \longrightarrow 1 \quad 3 \quad 1 \quad 5 \quad 3 \quad 6$$

We see that we can represent the binary command 1011001101011110 more succinctly as the octal command 131536. Notice how much easier it is to remember the octal form of the command rather than the binary form.

b) In converting binary to hexadecimal, we recognize that we can count from 0 to 15 (decimal) using 4 binary places (0000 to 1111); however, we can count from 0 to 15 (decimal) using only one hexadecimal place (0 to F).

Thus, we first group the bits in 1011001101011110 in groups of four, again beginning from the right. We then interpret each group of four bits as a hexadecimal number, as we show below.

$$\text{binary} \longrightarrow \quad 1011 \quad 0011 \quad 0101 \quad 1110$$

$$\text{hexadecimal} \longrightarrow \quad B(11) \quad\quad 3 \quad\quad 5 \quad\quad E(14)$$

We can now represent the binary command 1011001101011110 more concisely as the hexadecimal command B35E.

Historical Highlight: *From the Abacus to the Computer*

The earliest mechanical calculating device is the abacus, which dates back to 300 B.C. Early abaci, similar to the Roman abacus we described in the Highlight in Section 4.2, were widely used in Europe prior to the adoption of written Hindu–Arabic numerals. Some believe that Christians introduced the abacus to China beginning about 1200 A.D. and later to Japan and Korea.

We show a typical Chinese abacus, or suan-pan, in Figure 4.15.

The suan-pan consists of beads that slide on wires. The wires represent (from right to left) powers of $10 - 10^0$, 10^1, 10^2, 10^3, and so on. Beads below the horizontal bar represent 1 and beads above the bar represent 5. Beads next to the bar are active and are used in representing the number, whereas beads away from the bar are inactive. The suan-pan in Figure 4.15 shows how to represent the number 9,073.

As mathematics and science progressed, so did the interest in doing calculations mechanically. In addition to Napier's rods and the slide rule, others invented machines for doing mathematics. In 1642, the noted French mathematician Blaise Pascal invented an adding machine that introduced principles that were used in later calculators (see Figure 4.16(a)). Later in the seventeenth century, Gottfried Leibniz, a German, improved on Pascal's design (see Figure 4.16(b)).

The British mathematician Charles Babbage designed several complicated calculating machines, cul-

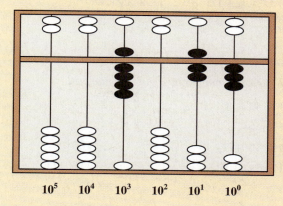

FIGURE 4.15   A Chinese suan-pan representing 9,073.

Continued

Historical Highlight: *From the Abacus to the Computer* *(Continued)*

minating with his design of an analytical engine in 1826 (see Figure 4.17).

Although Babbage was unable to actually construct his machines,* his invention laid the groundwork for the construction of today's computers.

The first electronic computer was constructed by J. Presper Eckert and John Mauchly at the University of Pennsylvania in the early 1940s. The computer was so large (over 30 tons) and so expensive (about one half million dollars), that some of the leading businessmen of the time felt that there would be a world-wide demand for only about a half dozen of these costly machines. Little did they imagine that computers would become so small and cheap, and such a commonplace part of modern-day life.

FIGURE 4.16 Pascal's and Leibniz's calculators.

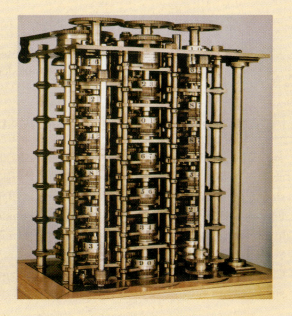

FIGURE 4.17 Babbage's analytic engine.

Exercises 4.3

List the numbers that precede and follow each of the given numbers in the given base.

1. 24_5
2. 200_5
3. 500_6
4. 55_6
5. 1011_2
6. 1100_2
7. 67_8
8. 77_8
9. FE_{16}
10. 100_{16}

Write each number as a base 10 numeral.

11. 432_5
12. 243_5
13. 504_6
14. 555_6
15. 100111_2
16. 100101_2
17. 1110101_2
18. 1100111_2
19. 267_8
20. 137_8
21. 704_8
22. 561_8

---

* Many years later, IBM constructed perfectly working models of Babbage's machines using his designs.

23. $2F4_{16}$

24. $18E_{16}$

25. $D08_{16}$

26. $C3B_{16}$

Convert each decimal number to a numeral in the given base.

27. 334 base 5

28. 1,298 base 5

29. 1,838 base 6

30. 3,968 base 6

31. 103 base 2

32. 51 base 2

33. 94 base 2

34. 107 base 2

35. 3,403 base 8

36. 2,297 base 8

37. 3,149 base 8

38. 2,431 base 8

39. 2,792 base 16

40. 2,219 base 16

41. 3,562 base 16

42. 3,827 base 16

Perform each addition or subtraction.

43. $3412_5 + 231_5$

44. $3215_6 + 423_6$

45. $4013_5 + 1242_5$

46. $2067_8 + 2443_8$

47. $2A18_{16} + 43B_{16}$

48. $BF2E_{16} + A35_{16}$

49. $11011_2 + 10101_2$

50. $100111_2 + 10111_2$

51. $2412_5 - 321_5$

52. $1325_6 - 453_6$

53. $A83_{16} - 43B_{16}$

54. $6C2E_{16} - A35_{16}$

55. $111011_2 - 10101_2$

56. $100101_2 - 10011_2$

Perform each multiplication or division.

57. $41_5 \times 23_5$

58. $24_5 \times 31_5$

59. $302_5 \times 43_5$

60. $413_5 \times 34_5$

61. $3412_5 \div 24_5$

62. $2143_5 \div 32_5$

63. $4132_5 \div 42_5$

64. $4402_5 \div 14_5$

Write each binary number first as an octal number and then as a hexadecimal number.

65. 1011101101_2

66. 1011110111_2

67. 1000111101_2

68. 1100101101_2

69. 1111101001_2

70. 1010100101_2

Write each number as a binary number.

71. 246_8

72. 573_8

73. $A3E_{16}$

74. $B8C_{16}$

Many electronic devices store and communicate information using the ASCII coding system, which is an acronym for American Standard Code for Information Interchange. The ASCII coding system uses seven-bit binary codes to represent characters. For example, the binary equivalents of the numbers 65 to 90 represent the capital letters A to Z. Using ASCII coding, we represent A by 1000001, B by 1000010, C by 1000011, and so forth. In Exercises 75–78, translate each binary string into English. (First group the bits into seven-bit groups.)

75. 1000011100000110011101000100101 1001

76. 1001000100010110011001001100100 1111

77. 10011001001111101011010 00101

78. 10101001010010101010110101001001 000

Further Exercises

*(79.) Some are proponents of a base 12, or duodecimal, system.† Invent a base 12 numeration system. Give careful thought to what additional symbols you will need in this system. Count from 1 to 25 using your system.

(80.) Communicating Mathematics   Locate a book or article that discusses the history of numeration systems and give a brief report on societies that use bases other than 10.

Consider a base 6 numeration system that uses the symbols $\mp$, $\oplus$, $\neq$, $\$$, @, *and* %, *corresponding to 0, 1, 2, 3, 4, and 5, respectively.*

(81.) Count to decimal 20 using this system.

(82.) Translate $\oplus \mp \neq \$$ to a decimal number.

(83.) Translate the decimal number 30 to this system.

---

* Exercise numbers circled in red can be used as group exercises.

† You may find it interesting to research the Duodecimal Society.

84. Add $\oplus \neq \mp \%$ and $\%\$@$ in this system.

85. Communicating Mathematics   Which base is larger if the equation $1202_b = 10122_c$ holds? Explain your answer.

86. Communicating Mathematics   What can you say about the number of places needed to represent a number as the size of the base increases?

87. Communicating Mathematics   How many basic addition and multiplication facts would you need to memorize in a base 8 system? In a base 16 system? Why is that the case?

88. Communicating Mathematics   How many basic addition and multiplication facts would you need to memorize in a base b system? Why is that the case?

89. What base would you be using to change seconds to hours and minutes? What about changing pennies to nickels and quarters?

90. Communicating Mathematics   Think of other situations similar to those we mentioned in Exercise 89 in which you change one unit of measurement into another unit of measurement. State what base you are using when doing the conversion.

Either obtain or simulate a suan-pan as described in the Highlight. You can simulate a suan-pan by drawing the frame and wires on cardboard and then use pennies to represent the beads.

91. Communicating Mathematics   Devise a method for doing addition on the suan-pan. Explain your method by finding $3,462 + 8,275$ on the suan-pan.

92. Communicating Mathematics   Devise a method for doing subtraction on the suan-pan. Explain your method by finding $4,381 - 2,514$ on the suan-pan.

CHAPTER SUMMARY

SECTION 4.1

number, numeral
A number tells us how many objects we are counting; a numeral is a symbol that represents a number.

simple grouping system
In a simple grouping system, the value of a number is the sum of the values of its numerals.

the Egyptian numeration system
The Egyptian numeration system is a simple grouping system.

addition and subtraction in the Egyptian system
Addition and subtraction in the Egyptian system is based on grouping and regrouping its basic symbols.

method of doubling
The Egyptians performed multiplication by using a method of doubling that is based on the fact that we can write any positive integer as the sum of powers of 2.

the Roman numeration system
The Roman numeration system is a more sophisticated simple grouping numeration system than the Egyptian system. It has a *subtraction principle* that allows us to represent

numbers more concisely and a *multiplication principle* that makes it easier to represent large numbers.

the Chinese numeration system

The Chinese numeration system is an example of a *multiplicative system.* We form Chinese numerals by writing products of integers between 1 and 9, inclusive, and powers of 10.

SECTION 4.2

place value system, positional system

In a place value system, also called a positional system, the placement of the symbols in a numeral determines the value of the symbols.

the Babylonian numeration system

The Babylonians had a primitive place value system that was based on powers of 60, and hence is called a *sexagesimal* system.

the Hindu–Arabic numeration system

The Hindu–Arabic numeration system is a place value system based on 10.

expanded form

We write Hindu–Arabic numerals in expanded form to show how each digit in the numeral is multiplied by a power of 10.

arithmetic operations in the Hindu–Arabic system

Expanded notation helps us to understand the algorithms to perform arithmetic operations. It helps us to understand processes such as carrying and borrowing when doing addition and subtraction, as well as the way we compute partial products when multiplying.

the galley method

In the galley method, we construct a rectangle, divided into triangles, and then compute partial products in each box of the galley. We add along the diagonals of the galley to compute the final product.

Napier's rods

Napier's rods are a variation on the galley method. Placing these rods side by side, we can read off the partial products quickly when doing multiplication.

SECTION 4.3

base 5

A base 5 system uses the digits 0, 1, 2, 3, and 4. In expanded notation, we use powers of 5, rather than powers of 10.

binary, octal, hexadecimal systems

The binary system has a base of two; the octal system has a base of eight; the hexadecimal system has a base of sixteen.

arithmetic operations

We perform arithmetic operations in other bases similar to the way we do these operations in base 10.

CHAPTER TEST

SECTION 4.1

1. Write 𒀭𒍝𒈾𒈾𒇻𒇻𒇻𒇻𒀮𒀮𒅅𒅅 using Hindu–Arabic numerals.

2. Find 𒅅999∩∩|| − 9∩∩∩∩∩||||| using Egyptian notation.

3. Use the Egyptian method of doubling to calculate 37×53.

4. Write 4,795 in Roman notation.

5. Write 七千五百九十三 as a Hindu–Arabic numeral.

6. Do the words "number" and "numeral" mean the same thing? Explain.

7. Explain two advantages of the Roman numeration system over the Egyptian system.

SECTION 4.2

8. Write 11,292 using Babylonian notation.

9. Subtract $4,237 - 2,673$ using expanded notation.

10. Use the galley method to multiply 46×103.

11. Why is zero important in a place value system?

SECTION 4.3

12. Write 342_5 and $B3D_{16}$ as base 10 numerals.

13. Write decimal 3,403 in base 8 notation.

14. Add $10111_2 + 11001_2$.

15. Divide $4312_5 \div 23_5$.

16. Write 1011100010_2 first as an octal number and then as a hexadecimal number.

Of Further Interest: MODULAR SYSTEMS

In the numeration systems that you studied in Section 4.3, you can count forever without repeating numbers. For example, in base 5, after you count to 4444_5, you count 10000_5, 10001_5, and so on, indefinitely. However, there are systems that you use every day where this is not the case. The clock in your bedroom counts the hours from 1 to 12 and then returns to 1 to start counting all over again. Similarly, if we number the days of the week, Monday, Tuesday, . . . , Sunday as 1, 2, . . . , 7, then we can count the days of our lives as 1, 2, 3, 4, 5, 6, 7, 1, 2, 3, 4, 5, 6, 7, and so on. In the same way, your car odometer counts the miles from 0 to 999,999 and then rolls over and starts counting again, 0, 1, 2, ...

Each of these systems is an example of a modulo m system which we will study next. These systems have many properties in common with the familiar system of integers. As you will see, in modulo m systems, we can count, perform arithmetic operations, and solve equations. Additionally, modulo m systems have interesting, real-life applications.

We use a clock to visualize the operations in a modulo m system.

> **DEFINITIONS**
>
> If m is an integer greater than 1, then a **modulo m system*** consists of the numbers 0, 1, 2, . . . , $m - 1$. Counting and arithmetic operations are performed in a manner corresponding to movements on an m-hour clock. The number m is called the **modulus** of the system.

FIGURE 4.18 A 12-hour clock.

In order to understand a modulo 12 system, we draw a 12-hour clock in Figure 4.18. Notice how we have replaced the 12 on the clock with the number 0—the reason for doing this will become clear shortly. The numbers 0, 1, 2, . . . , 11 now form a modulo 12 system. If we were to begin at 0 and count to 53 in this system, at what number would we stop? Of course, we could begin counting on the clock, 1, 2, 3, 4, 5, 6, 7, 8, 9, 10, 11, 0, 1, 2, and so on until we counted 53 numbers. However, there is an easier way. Notice that if we count on the clock to multiples of 12, such as 12, 24, 36, and then 48, we return to the 0 position (see Figure 4.19). If we now count an additional 5, we reach 53,

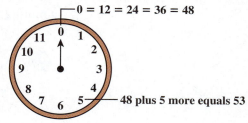

FIGURE 4.19 Counting to 53 on a 12-hour clock.

---

* Modulo m systems are also called **modular arithmetic systems,** or simply **modular systems.**

which is position 5 on this 12-hour clock. Thus, counting to 53 in a modulo 12 system will give us 5.

> **SOME GOOD ADVICE**
>
> Because multiples of 12 are the same as 0 on a 12-hour clock, a quicker way to determine the 5 would have been to divide 53 by 12, and keep the remainder.

We can visualize the days of the week, Monday (1), Tuesday (2), Wednesday (3), Thursday (4), Friday (5), Saturday (6), and Sunday (7) on the 7-hour clock shown in Figure 4.20. Notice that we replace the 7 by 0 on this clock. The numbers 0, 1, . . . , 6 form a modulo 7 system.

If we count to 45 on this clock, we see that we reach 0 at each multiple of 7, namely 7, 14, 21, 28, 35, and 42. If we now count 3 more, we reach 45. This means that 45 in a modulo 7 system is the same as 3. Notice again, that dividing 45 by 7 and keeping the remainder 3 is a quicker way to solve this problem.

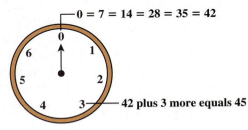

FIGURE 4.20   Counting to 45 on a 7-hour clock.

We can express the fact that 53 occupies the same position as 5 on a 12-hour clock more precisely.

Quiz Yourself 14*

a) Count to 75 on a 12-hour clock.

b) Count to 41 on a 7-hour clock.

> **DEFINITION**
>
> We say that a **is congruent to** b **modulo** m, written $a \equiv b \pmod{m}$, provided that m evenly divides $a - b$. Note, if you find it more convenient to divide m into $b - a$, this is also acceptable.

Because 12 divides $53 - 5 = 48$ evenly, this means that $53 \equiv 5 \pmod{12}$. We read this, 53 is congruent to 5 modulo 12. Similarly, because $45 - 3 = 42$ is a multiple of 7, we write $45 \equiv 3 \pmod{7}$.

> **PROBLEM SOLVING**
>
> Remember the Splitting Hairs Principle from Section 1.1. The three lines $\equiv$ look somewhat like the equal sign but do not mean exactly the same thing. When we write $a \equiv b \pmod{m}$, we are not saying that a and b are equal in the usual sense, but in terms of an m-hour clock, we can consider them to be the same.

EXAMPLE 1   Determining When Numbers Are Congruent

Determine which statements are true.

a) $39 \equiv 15 (\text{mod } 12)$

b) $33 \equiv 19 (\text{mod } 8)$

c) $25 \equiv 48 (\text{mod } 7)$

d) $11 \equiv 35 (\text{mod } 6)$

SOLUTION: a) $39 - 15 = 24$, which is evenly divisible by 12. Therefore, $39 \equiv 15 (\text{mod } 12)$ is a true statement.

b) $33 - 19 = 14$, which is not evenly divisible by 8. Thus, $33 \equiv 19 (\text{mod } 8)$ is false.

c) Instead of dividing 7 into $25 - 48$, we consider $48 - 25 = 23$. Because 7 does not divide evenly into 23, this statement is false.

d) Since 6 divides evenly into $35 - 11 = 24$, this statement is true.

Quiz Yourself 15

Determine which statements
are true.

a) $11 \equiv 33 (\text{mod } 12)$

b) $27 \equiv 59 (\text{mod } 8)$

When we work in a modulo system, we group all numbers that are congruent to each other into a set called a **congruence class**. For example, consider the 5-hour clock shown in Figure 4.21(a).

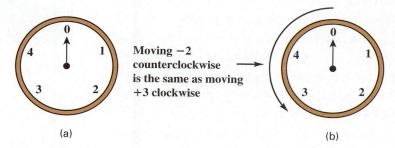

FIGURE 4.21   (a) A 5-hour clock; (b) −2 is congruent to 3 modulo 5.

The congruence classes in this modulo-5 system would be:

Congruence class containing 0: 0, 5, 10, 15, 20, ...

Congruence class containing 1: 1, 6, 11, 16, 21, ...

Congruence class containing 2: 2, 7, 12, 17, 22, ...

Congruence class containing 3: 3, 8, 13, 18, 23, ...

Congruence class containing 4: 4, 9, 14, 19, 24, ...

There are only these five congruence classes. The congruence class containing 5 would be the same as the congruence class containing 0, and 6's class would be the same as 1's class, and so on. Any two numbers in the same congruence class differ by a multiple of 5 and thus are congruent modulo 5. Numbers in different classes are not congruent modulo 5.

You might wonder why we are not discussing negative numbers. Realize that in a modulo m system, we do not need negative numbers. For example, on a 5-hour clock, −2 corresponds to moving 2 in a counterclockwise direction. If we do this, we stop at 3. Thus −2 is congruent to 3 modulo 5. See Figure 4.21(b).

In doing calculations in a modulo m system, we can use any two numbers that are in the same congruence class interchangeably. Thus, if in doing a modulo 5 computation you get 19, you can replace it with 4 because $19 \equiv 4 \pmod{5}$. We will use this principle frequently when we perform arithmetic operations in modulo m systems.

Historical Highlight: *Cryptography**

In 1994, an international group of criminals was the first ever to breach the security of Citicorp's computerized cash management system. Over several months, a Russian transferred $12 million to accomplices in Finland, Israel, Holland, Switzerland, and the United States. Although $12 million may seem like a lot, to put this amount in perspective, at the same time, Citicorp was moving about $500 *billion daily* through various computer networks.

The magnitude of electronic banking, coupled with easy access to computer networks by the ordinary citizen, points to the necessity of having secure, impenetrable communications between banks and customers. Additionally, in military situations, life and death literally depend on our ability to send messages secretly and securely.

The need for secret communications is nothing new. We can trace coded messages, or **ciphers,** back to the Egyptians around 2000 B.C., then to the early Greeks and Romans, and eventually to Western civilization.

Code breaking played an important part in the allied victory in World War II. In the 1920s, the Germans invented a coding device called the Enigma machine, consisting of a keyboard and an elaborate system of rotors (see Figure 4.22). An operator would set the rotors and then type in a message in order to create an elaborate cipher. A person with another Enigma machine, and knowledge of the rotor settings, could then easily decode the message.

FIGURE 4.22 The Enigma machine.

During the 1930s, Marian Rejewski led a group of Polish mathematicians who eventually solved Enigma's mystery. After the fall of Poland in 1939, they passed their information on to the allies who used it to decode numerous German military messages.

Today, we use powerful computers and sophisticated algorithms to code and decode messages. One popular method for encrypting messages is the RSA algorithm named after its inventors Ronald Rivest, Adi Shamir, and Leonard Adelman. The RSA algorithm is based on modular arithmetic.

Let's consider a very simplified example of how the RSA algorithm encodes a message. The algorithm uses the equation

$$c = m^e \pmod{n}.$$

* This highlight is based on material taken from Lance J. Hoffman, *Building in Big Brother* (Springer-Verlag, 1995), pp. 11–14.

Historical Highlight: *Cryptography* *(Continued)*

Here m is the message, and c is the coded form of the message. We choose the numbers n and e according to certain rules that we will not explain here. To illustrate this, suppose that we want to code the letter C which we will represent by the number 3. Let $n = 35$ and $e = 5$. (Do not be concerned how we picked 35 and 5.) Then to encode 3, we compute $c = 3^5$ (mod 35) $= 243$ (mod 35) $= 33$. So 33 is the coded form of the letter C.

This process is an example of a one-way function in mathematics. Although it is easy for us to explain to you how to encode the number 3, it is much more difficult for you to decode the 33. In practice, when doing the encoding, the number n is a huge number having five or six hundred digits. Unless you know how to factor n (see Chapter 5), it is impossible to break the code. The RSA algorithm has been used in many popular technologies such as the Netscape Navigator and Microsoft Explorer network browsers and Mastercard and Visa credit cards.

Operations in a modulo *m* system are closely related to operations on the integers.

It is easy to add, subtract, and multiply in a modulo m system. For example, in a modulo 7 system, $6 + 4 = 10$. However, because $10 \equiv 3 \pmod 7$, we can write $6 + 4 \equiv 3 \pmod 7$. It helps to see this computation if you draw the clock in Figure 4.23. If we begin at 0 and count to 6 and then count 4 more, we see that we stop at 3.

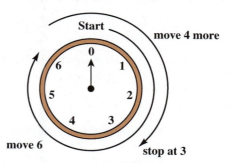

FIGURE 4.23 A 7-hour clock showing $6 + 4 \equiv 3 \pmod 7$.

All modulo m operations are done in a similar way.

Performing arithmetic operations in a modulo *m* system

In order to add, subtract, and multiply* in a modulo m system:

1) Perform the operation as usual.
2) Replace the result in 1) by one of the numbers $0, 1, 2, \ldots, m - 1$, which is congruent to the result in part 1).

---

* We do division differently. See Exercise 47.

EXAMPLE 2 Adding in a Modulo *m* System

a) Find $10 + 9$ (mod 12). b) Find $7 + 3$ (mod 8).

SOLUTION: a) To find $10 + 9$ (mod 12) is asking us in a succinct way to compute $10 + 9$ in a modulo 12 system. Now, $10 + 9 = 19$; however, $19 \equiv 7$(mod 12). Therefore, we can write $10 + 9 \equiv 7$(mod 12).

b) Here $7 + 3 = 10$. But $10 \equiv 2$(mod 8), so $7 + 3 \equiv 2$(mod 8).

Doing subtraction requires a slight variation of our approach as we see in Example 3.

EXAMPLE 3 Subtracting in a Modulo *m* System

a) Find $2 - 5$ (mod 12). b) Find $3 - 7$ (mod 8).

SOLUTION: a) We begin by calculating $2 - 5 = -3$. However, as we pointed out earlier, we do not need negative numbers in a modulo *m* system. On a 12-hour clock, to represent -3, we think of beginning at 0 and moving counterclockwise three numbers as shown in Figure 4.24. Because we stop at 9, we can say $2 - 5 \equiv 9$(mod 12).

b) First calculate $3 - 7 = -4$. If you visualize an 8-hour clock, you see that -4 is the same as 4. Therefore, $3 - 7 \equiv 4$(mod 8).

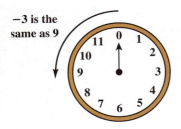

−3 is the same as 9

FIGURE 4.24 A 12-hour clock showing $2 - 5 = -3 = 9$.

> **PROBLEM SOLVING**
>
> To check a subtraction problem in the integers, such as $8 - 5 = 3$, you can perform the addition $3 + 5 = 8$. Similarly, to check subtraction in a modulo *m* system, we do a corresponding addition. In Example 3, we can check that $2 - 5 \equiv 9$(mod 12) by showing $9 + 5 \equiv 2$(mod 12).

Multiplication is straightforward in a modulo *m* system.

EXAMPLE 4 Multiplying in a Modulo *m* System

a) Find 9×7 (mod 12). b) Find 5×8 (mod 9).

SOLUTION: a) We first calculate $9 \times 7 = 63$. Next, we divide 63 by 12 to get a remainder of 3. Therefore, $9 \times 7 \equiv 3$(mod 12).

b) We begin by calculating $5 \times 8 = 40$. Dividing 40 by 9, we get a remainder of 4. Thus, $40 \equiv 4$ (mod 9), so $5 \times 8 \equiv 4$(mod 9).

We can use trial and error to solve congruences.

In a modulo *m* system, we solve congruences instead of solving equations. For example, we might want to solve $5 + x \equiv 3$(mod 7). Because we only have seven numbers to try, namely 0, 1, 2, . . . , 6, we can try each of these in turn to see if it solves the congruence.

$$5 + 0 \equiv 3(\text{mod } 7) \quad (\text{false; no solution})$$
$$5 + 1 \equiv 3(\text{mod } 7) \quad (\text{false; no solution})$$
$$5 + 2 \equiv 3(\text{mod } 7) \quad (\text{false; no solution})$$
$$5 + 3 \equiv 3(\text{mod } 7) \quad (\text{false; no solution})$$
$$5 + 4 \equiv 3(\text{mod } 7) \quad (\text{false; no solution})$$
$$5 + 5 \equiv 3(\text{mod } 7) \quad (\text{true; 5 is a solution})$$
$$5 + 6 \equiv 3(\text{mod } 7) \quad (\text{false; no solution})$$

Thus we see that 5 is the only solution for this congruence.

EXAMPLE 5   Solving Congruences Using Trial and Error

Solve a) $5 - x \equiv 7(\text{mod } 8)$ b) $2x \equiv 4(\text{mod } 6)$ c) $2x \equiv 3(\text{mod } 6)$

SOLUTION: a) When we substitute the numbers 0 to 7 for x, we find that 6 is the only number that makes the congruence true.

$$5 - 0 \equiv 7(\text{mod } 8)$$
$$5 - 1 \equiv 7(\text{mod } 8)$$
$$5 - 2 \equiv 7(\text{mod } 8)$$
$$5 - 3 \equiv 7(\text{mod } 8)$$
$$5 - 4 \equiv 7(\text{mod } 8)$$
$$5 - 5 \equiv 7(\text{mod } 8)$$
$$5 - 6 \equiv 7(\text{mod } 8) \quad \text{true}$$
$$5 - 7 \equiv 7(\text{mod } 8)$$

Thus 6 is the only solution of the congruence $5 - x \equiv 7(\text{mod } 8)$.

b) To solve this congruence, we substitute $0, 1, \ldots, 5$ for x in the congruence $2x \equiv 4(\text{mod } 6)$.

$$2 \times 0 \equiv 4(\text{mod } 6)$$
$$2 \times 1 \equiv 4(\text{mod } 6)$$
$$2 \times 2 \equiv 4(\text{mod } 6) \quad \text{true}$$
$$2 \times 3 \equiv 4(\text{mod } 6)$$
$$2 \times 4 \equiv 4(\text{mod } 6)$$
$$2 \times 5 \equiv 4(\text{mod } 6) \quad \text{true}$$

Here we see that both 2 and 5 are solutions for this congruence.

c) If you substitute $0, 1, 2, \ldots, 5$ for x in the congruence, you will find that none of these numbers makes the congruence true. (Verify this.) Therefore, this congruence has no solutions.

In Example 6, you must solve a pair of congruences simultaneously.

 Quiz Yourself 16

Solve the following congruences.

a) $4 - x \equiv 7(\text{mod } 8)$.

b) $3x \equiv 0(\text{mod } 9)$.

EXAMPLE 6   Solving a Pair of Congruences

Maria is arranging a collection of antique plates for display in her antique store. If she arranges the plates in stacks of five, she has three left over. If she arranges the plates in stacks of four, she has two left over. What is the smallest number of plates that she can have?

SOLUTION: Let us say that x is the number of plates. When Maria stacks the plates five at a time, she has three left over, which tells us that if we divide x by five, the remainder is three. In terms of congruences, this means that $x \equiv 3 \pmod 5$. We conclude that x is in the congruence class containing 3 modulo 5 which is

$$3, 8, 13, \mathbf{18}, 23, 28, 33, 38. \ldots$$

Similarly, when Maria stacks the plates four at a time, she has two left over, which tells us that $x \equiv 2 \pmod 4$. This means that x must also be in the congruence class containing 2 modulo 4, which is

$$2, 6, 10, 14, \mathbf{18}, 22, 26, 30, \ldots .$$

The smallest number that is found in both of these classes is 18, so the smallest number of plates that Maria has to display is 18.

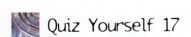

Quiz Yourself 17

Find the smallest positive integer that satisfies $x \equiv 5 \pmod 7$ and $x \equiv 6 \pmod 9$.

An interesting application of modular arithmetic appears on many of the products that we use each day. Consider the 12-digit Universal Product Code (UPC) found underneath the bar code on a box of Honey Nut Cheerios shown in Figure 4.25(a). The first digit, 0, identifies the product. The next five digits, 16000, identify the manufacturer. The next group of digits, 66590, provides information about the product. The last digit, in this case 3, is called a check digit. The check digit provides a way of determining if a UPC is a valid code.

FIGURE 4.25   Honey Nut Cheerios product code.

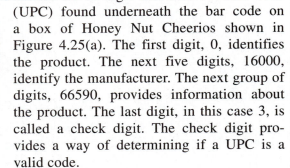

Example 7 explains how to compute the check digit for a UPC.

EXAMPLE 7   Using Modular Arithmetic to Find Check Digits

If the 12-digit UPC is of the form $a_1a_2a_3a_4a_5a_6a_7a_8a_9a_{10}a_{11}a_{12}$,* then we compute the check digit in two steps.

1. First we compute the expression

$$3a_1 + a_2 + 3a_3 + a_4 + 3a_5 + a_6 + 3a_7 + a_8 + 3a_9 + a_{10} + 3a_{11}$$

---

* Don't let this notation scare you. All we are saying is that the UPC consists of 12 numbers. The first number is called a_1, the second is called a_2, and so forth.

That is, we multiply the first digit in the UPC by three; add the second digit; then add three times the third digit; add the fourth digit, and so on.

2. Choose the check digit, a_{12}, so that the sum in part 1 plus a_{12} is congruent to 0 modulo 10.

   a) Verify that the UPC for Honey Nut Cheerios in Figure 4.25 is a valid code.

   b) Suppose that in the middle of typing this UPC you accidentally typed the pattern 655 instead of 665. Show why this would not be a valid UPC.

SOLUTION: a) In the Honey Nut Cheerios UPC, which is 016000665903, the first digit, a_1, is 0, the second digit, a_2, is 1, the third digit, a_3, 6, and so on. Therefore,

$$3a_1 + a_2 + 3a_3 + a_4 + 3a_5 + a_6 + 3a_7 + a_8 + 3a_9 + a_{10} + 3a_{11} =$$

$$3 \times 0 + 1 + 3 \times 6 + 0 + 3 \times 0 + 0 + 3 \times 6 + 6 + 3 \times 5 + 9 + 3 \times 0 = 67$$

To find the check digit, a_{12}, we must find a number such that $67 + a_{12} \equiv 0 \pmod{10}$. It is easy to see that if $a_{12} = 3$, then $67 + 3 \equiv 0 \pmod{10}$. Therefore, the check digit for this UPC is 3, so the UPC is a valid code.

b) If you had typed 016000655903 for the UPC, then in computing $3a_1 + a_2 + 3a_3 + a_4 + 3a_5 + a_6 + 3a_7 + a_8 + 3a_9 + a_{10} + 3a_{11}$ you would calculate $3 \times 0 + 1 + 3 \times 6 + 0 + 3 \times 0 + 0 + 3 \times 6 + 5 + 3 \times 5 + 9 + 3 \times 0 = 66$. This would mean that to have $66 + a_{12} \equiv 0 \pmod{10}$, the check digit would have to be 4, not 3. This means that you typed an invalid UPC.

Identification numbers occur in many other places, such as on bank checks, U.S. postal money orders, drivers licenses, express shipping labels, airline tickets, and ISBN numbers on books. Although there are many different ways to calculate check digits other than the method that we explained in Example 6, most methods are based on modular arithmetic. We will discuss several of these other methods in the exercises.

Exercises

In Exercises 1–4, count to the number that is specified on the clock which is indicated.

1. 43 on a 12-hour clock

2. 21 on a 6-hour clock

3. 39 on a 7-hour clock

4. 69 on an 8-hour clock

5. In a Jules Verne novel, Phineas Fogg took an imaginary trip around the world in 80 days. If he were counting the days of his trip on a 7-hour clock, what would the clock show when he landed?

6. Columbus' first voyage to America took 71 days. If he were counting the days of his voyage on a 7-hour clock, what would the clock show when he landed in America?

Determine which of the following statements are true.

7. $35 \equiv 59 \pmod{12}$

8. $31 \equiv 14 \pmod{7}$

9. $11 \equiv 43 \pmod{8}$

10. $29 \equiv 18 \pmod{6}$

11. $46 \equiv 78 \pmod{10}$

12. $53 \equiv 75 \pmod{11}$

13. How many congruence classes would there be in a modulo 8 system?

14. How many congruence classes would there be in a modulo 10 system?

In Exercises 15–18, write each of the congruence classes.

15. The class containing 5 in a modulo 7 system.

16. The class containing 8 in a modulo 12 system.

17. The class containing 6 in a modulo 10 system.

18. The class containing 3 in a modulo 9 system.

Perform the following operations.

19. $8 + 9 \pmod{12}$

20. $4 + 7 \pmod 8$

21. $3 + 4 \pmod 5$

22. $6 + 8 \pmod{11}$

23. $4 - 9 \pmod{12}$

24. $5 - 6 \pmod 8$

25. $3 - 4 \pmod 5$

26. $3 - 8 \pmod{11}$

27. $4 \times 9 \pmod{12}$

28. $5 \times 6 \pmod 8$

29. $2 \times 5 \pmod 7$

30. $9 \times 3 \pmod{14}$

Solve the following congruences.

31. $3 + x \equiv 1 \pmod{12}$

32. $9 + x \equiv 6 \pmod{12}$

33. $6 + x \equiv 4 \pmod 8$

34. $5 + x \equiv 2 \pmod 6$

35. $3 - x \equiv 4 \pmod 7$

36. $5 - x \equiv 7 \pmod{10}$

37. $2 - x \equiv 8 \pmod 9$

38. $3 - x \equiv 4 \pmod 5$

39. $3x \equiv 5 \pmod 7$

40. $5x \equiv 3 \pmod 8$

41. $4x \equiv 2 \pmod{10}$

42. $6x \equiv 4 \pmod 8$

In Exercises 43–46, find the smallest positive integer that solves all of the given congruences.

43. $x \equiv 3 \pmod 7$, $x \equiv 4 \pmod 5$

44. $x \equiv 1 \pmod 6$, $x \equiv 7 \pmod 8$

45. $x \equiv 2 \pmod{10}$, $x \equiv 2 \pmod 8$, $x \equiv 0 \pmod 6$

46. $x \equiv 5 \pmod 9$, $x \equiv 1 \pmod{11}$, $x \equiv 7 \pmod 8$

47. Packaging candy. A package designer for a chocolate manufacturer is designing boxes to hold a certain number of candies. When arranging the pieces 6 in a row, there will be 4 left over. When arranging the pieces 5 in a row, there will be three left over. What is the smallest number of pieces that the manufacturer might be considering to place in the box?

48. Arranging a marching band. A marching band is considering different configurations for its upcoming halftime show. When the members are arranged 8 or 10 in a row, there are two members left over. If they are arranged 12 in a row, there are 6 left over. What is the smallest number of members that the band can have?

Further Exercises

49. When we do usual division of integers, to say that $12/4 = 3$ is the same as saying that $12 = 4 \cdot 3$. More generally, if $a/b = x$, then $a = b \cdot x$. Use this idea to devise a way to perform division in a modulo m system. That is, how would you explain how to solve a congruence such as $4/6 \equiv x \pmod 8$? Use the method you devise to find

   a) $5/3 \pmod 8$

   b) $2/6 \pmod 8$

50. Solve the congruence $2/x \equiv 5 \pmod 6$.

While taking a class on the culture of China, you have learned about the Chinese zodiac in which people fall into one of 12 categories depending on the year of their birth. The categories, numbered 0 to 11, correspond to the following animals: (0) monkey, (1) rooster, (2) dog, (3) pig, (4) rat, (5) ox, (6) tiger, (7) rabbit, (8) dragon, (9) snake, (10) horse, *(11) goat. Those who believe in this zodiac think that the year of a person's birth influences both their personality and fortune in life.*

To find your zodiac sign, divide the year of your birth by 12. The remainder then determines your sign. The remainder, when we divide 1985 by 12, is 5; therefore, a person born in 1985 is an ox according to the Chinese zodiac. Use this information to do Exercises 51 and 52.

51. Estimating a person's age. You would like to date Chris who is in your Chinese culture class. However, before asking for a date you would like to find out Chris' age without asking directly. Chris is at least 18 and seems to be less than 30 years old. If according to the Chinese zodiac Chris is a dog, then in what year was Chris born?

52. Estimating a person's age. A person who is between 40 and 50 years old in the year 2003, and who according to the Chinese zodiac is a pig, was born in what year?

$10 \times 0 =$	0
$9 \times 2 =$	18
$8 \times 0 =$	0
$7 \times 1 =$	7
$6 \times 7 =$	42
$5 \times 9 =$	45
$4 \times 5 =$	20
$3 \times 1 =$	3
$2 \times 1 =$	2
Total	137

TABLE 4.8 Table for calculating ISBN numbers.

If you look at the back cover of this textbook you will see a 10-digit number called an ISBN (International Standard Book Number). This book's ISBN is 0-201-79511-6. The first digit, 0, indicates that the book is from an English speaking country. The digits 201 identify the publisher (in this case, Addison Wesley), and the third group of digits 79511 identifies the book. The last digit, 6, is a check digit. Finding check digits for ISBNs is more complicated than finding check digits for UPCs. We calculate an ISBN check digit in three steps.

1. *Multiply the first digit by 10, the second digit by 9, the third digit by 8, continuing until you multiply the ninth digit by 2 as in Table 4.8. Then find the total of these products. In this example the total is 137.*

2. *Divide the total by 11, keep the remainder, which is 5, and call it r.*

3. *Find the check digit c such that $r + c = 11$. In this case $c = 6$.*

Find the missing digit d in each of the following ISBNs.

53. 0-201-73d02-4

54. 0-618-05d60-X   (note: X means 10)

55. 0-534-368d0-5

56. 0-201-d0930-9

****57.** Read the paper "Modular Arithmetic in the Marketplace" by Joseph Gallian and Steven Winters, *The American Mathematical Monthly*, **95** (1988), pp. 548–551. Report on a check digit scheme that we have not discussed in this section.

58. Read the paper "The Mathematics of Identification Numbers" by Joseph Gallian, *The College Mathematics Journal*, **22** (1991), pp. 194–202. Report on a check digit scheme that we have not discussed in this section.

* Exercise numbers circled in red can be used as group exercises.

CHAPTER

5

Number Theory and the Real Number System:* Understanding the Numbers All Around Us

Atoms are very tiny things. In fact, they are so small that it would take about ten million of them to make a length that is the width of the capital "T" in the word "Theory" in the title of this chapter.

Imagine that we were able to line up all the atoms in your body in a straight line. Think of them as tiny beads next to each other on a very fine thread. If we were to take that thread and stretch it out in a straight line, how far do you think that it would reach? Would it stretch across the room? Would it stretch from New York City to Chicago? Could it stretch around the world? Once? Twice? One hundred times? We will answer this question later in this chapter and the answer will astound you.

Have you ever thought about what it would be like to take a trip around our solar system? If a rocket ship were to travel at 25,000 miles per hour, how long would it take to go to Mars? Or how long would it take to travel to a more distant planet such as Neptune?

On a less fanciful note, do you have any understanding of how the national debt has changed in the last 50 years? Since 1950, the United States population has not quite doubled, but the national debt has grown from 256 billion to 5.8 trillion dollars. How can we make these numbers understandable?

In your everyday life, there are numbers that are much less dramatic. These numbers measure your gas mileage, your credit card

---

* For further resources on number theory and the real number system, see www.aw.com/pirnot.

debt, the number of square feet in your house, and count the ounces in the box of cereal that you eat each morning. They also measure the speed of your modem, calculate the time it takes to download a video from the Internet, and tell you the space remaining on your computer's disk drive.

In this chapter, you will learn about different number systems and how to work in them. In the Of Further Interest section, you will see interesting lists of numbers called sequences and study a particular sequence that occurs in many beautiful patterns in nature.

Because numbers are all around us, it is important that we understand them. The author John Allen Paulos wrote a book about a condition that he calls *innumeracy,* which is the mathematical equivalent of illiteracy. After studying this chapter, you can certainly expect to be more "numerate."

5.1   NUMBER THEORY

From the time you were a small child, you have known about a set of numbers that has fascinated the greatest minds in the history of mathematics for thousands of years. The numbers that we use for counting, {1, 2, 3, . . .}, are called the **natural numbers,** or **counting numbers.** This simple set of numbers has elegant properties and patterns that have intrigued mathematicians from ancient times to the present. The study of the natural numbers and their properties is called **number theory.**

Prime numbers are the building blocks of the integers.

One question that people have studied since the time of the ancient Greeks is exactly which natural numbers can be written as a product of *other* natural numbers and which cannot. For example, $30 = 2 \cdot 3 \cdot 5$, whereas $17 = 17 \cdot 1$. The fact that we cannot write 17 as a product of natural numbers without using 17 greatly interests mathematicians. In order to investigate this question, we have to explain what it means for one natural number to divide another.

> **DEFINITIONS**
>
> If a and b are natural numbers then we say a **divides** b, written $a|b$, provided there is a natural number q such that $b = qa$. Other ways of stating this is that a is a **divisor** of b, a is a **factor** of b, or that b is a **multiple** of a. A way of testing to see if a divides b is to actually divide b by a and see if you get a zero remainder.

For example, we can say that 5 divides 30, because there is a natural number, namely 6, such that $30 = 5 \cdot 6$. We can write this fact more concisely as $5 | 30$. Also, $17 | 17$ because $17 = 17 \cdot 1$. When we write a natural number as a product of natural numbers, we say that we have **factored** the number. Several factorizations of 72 are $8 \cdot 9$ and $2 \cdot 2 \cdot 2 \cdot 3 \cdot 3$. Numbers like 17, that have only trivial factorizations, are particularly important in number theory.

DEFINITIONS

A natural number, greater than one,* which has only itself and one as factors, is called a **prime number.** A natural number, greater than one, which is not prime is called **composite.**

Numbers like 2, 5, 17, and 29 are prime numbers because their only divisors are themselves and 1. On the other hand, numbers like 4, 33, 87, and 102 are composite, because you can find divisors for them which are other than 1 and the number itself.

One very well-known way to generate a list of primes is called the **Sieve of Eratosthenes** which is named after the Greek mathematician Eratosthenes of Alexandria, whom we discuss in the highlight. We explain his process in Example 1.

EXAMPLE 1   Finding a List of Prime Numbers

Use the Sieve of Eratosthenes to find all prime numbers less than 50.

SOLUTION: First we list all natural numbers from 1 to 50 in Figure 5.1, and then we will systematically cross off all composite numbers from this list, leaving only prime numbers.

First we cross off 1, because it is not prime. We next circle 2, which is prime and then cross off all other multiples of 2, namely 4, 6, 8, 10, . . . , 50. The next number in the list that we did not cross off is 3, which is prime. We circle it, and cross off all other multiples of 3 that remain—9, 15, 21, . . . , 45.

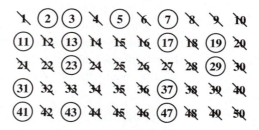

FIGURE 5.1   Using the Sieve of Eratosthenes.

---

* There are technical reasons, that we will not get into, as to why mathematicians consider one to be neither prime nor composite.

Next we circle 5, a prime, and cross off its remaining multiples, 25 and 35. Finally, we circle 7 and cross off its one remaining multiple, 49.

Because 7 is less than the square root of 50, but the next prime, 11, is greater than the square root of 50, we can stop looking for composites at this point. If you think about it for a minute, you will recognize that it would be impossible to find two factors that are 11 or greater whose product is less than 50. This means that all remaining numbers in the list are prime, so we circle them (see Figure 5.1). It would be a good idea for you to reproduce the steps that we described to get Figure 5.1 on your own from scratch.

 SOME GOOD ADVICE

When doing problems, you will frequently need to know some of the smaller prime numbers. It is a good idea to memorize the following primes: 2, 3, 5, 7, 11, 13, 17, 19, 23, 29, 31, 37, 41, 43, 47.

EXAMPLE 2 Identifying Prime Numbers

Determine whether the following numbers are prime or composite.

a) 83   b) 87   c) 127

SOLUTION: a) The number 83 is prime because if you try to find its divisors, you will not find any other than 1 and 83 itself. A good way to see this is to start dividing 83 by primes* such as 2, then 3, then 5, and so on until you get to the square root of 83. The square root of 83 is greater than 9 but less than 10 (because $9^2 = 81$ and $10^2 = 100$). As in Example 1, if you do not find a divisor of 83 by the time you get to square root of 83, you will not find another divisor until you reach 83.

b) The number 87 factors as $3 \cdot 29$. Again, in testing 87, you only need to check primes up to the square root of 87, which is between 9 and 10.

c) In testing 127, we divide by the primes 2, 3, 5, 7, and 11, none of which divide 127. The next prime is 13, which is greater than the square root of 127, so therefore 127 is prime.

In checking larger numbers for primality, it is useful to have some quick tests for divisibility so that we don't actually have to do a lengthy division. For example, it would be nice to be able to look at the number 21,021, and have a quick test to see whether or not 3 divides into it without actually having to do the division. Table 5.1 lists some divisibility tests. We will examine why these tests work in the exercises.

 Quiz Yourself 1†

Identify the following numbers as either prime or composite.

a) 89   b) 187

c) 143   d) 211

---

* You do not have to consider possible composite divisors such as 6, because if 6 were to divide 83, then 2 and 3 would also divide 83.

† Quiz Yourself answers begin on page 849.

Number Is Divisible by	Test	Example
2	The last digit of the number is divisible by 2.	2 divides 13,57**8** because 2 divides 8.
3	The sum of the digits is divisible by 3.	3 divides 21,021 because 3 divides $2 + 1 + 0 + 2 + 1$.
4	The number formed by the last two digits is divisible by 4.	102,7**36** is divisible by 4 because 4 divides 36.
5	The last digit is 0 or 5.	607,89**5** is divisible by 5.
6	The number is divisible by both 2 and 3.	802,674 is divisible by both 2 and 3, so it is divisible by 6.
8	The number formed by the last three digits is divisible by 8.	8 divides 230,**264** because 8 divides 264.
9	The sum of the digits is divisible by 9.	2,081,763 is divisible by 9 because 9 divides $2 + 0 + 8 + 1 + 7 + 6 + 3$.
10	The number ends in 0.	12,865,89**0** is divisible by 10.

TABLE 5.1 Divisibility tests for some small numbers.

Although there are divisibility tests for 7, and also 11, they are hard to remember, and you are really better off simply dividing by 7 or 11 rather than using the test.

EXAMPLE 3 **Applying Divisibility Tests**

Test the number 11,352 for divisibility by

a) 2 b) 3 c) 4 d) 5 e) 6 f) 8 g) 9 h) 10

SOLUTION: a) The last digit of 11,352 is divisible by 2, so 2 divides the number.

b) The sum of the digits is $1 + 1 + 3 + 5 + 2 = 12$ which is divisible by 3, so 11,352 is also divisible by 3.

c) The last two digits of 11,352 form the number 52, and 52 is divisible by 4, so therefore, 11,352 is also divisible by 4.

d) The last digit is neither 0 nor 5, so 5 does not divide this number.

e) Both 2 and 3 divide 11,352, therefore 6 also divides this number.

f) The number formed by the last three digits is 352, which is divisible by 8, so 8 divides 11,352.

g) In part b), we saw that the sum of the digits of 11,352 is 12. Because 9 does not divide 12, therefore 9 does not divide 11,352.

h) The number does not end in 0, so 10 does not divide the number.

In chemistry, we form compounds by combining simpler objects called atoms. For example, a molecule of table salt is formed by combining one

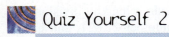
Quiz Yourself 2

Test the number 30,690 for divisibility by

a) 2   b) 3
c) 4   d) 5
e) 6   f) 8
g) 9   h) 10

sodium and one chlorine atom. Similarly, in mathematics, objects are built from simpler objects. For example, $120 = 2 \cdot 2 \cdot 2 \cdot 3 \cdot 5$ and $84 = 2 \cdot 2 \cdot 3 \cdot 7$. The Fundamental Theorem of Arithmetic states that every natural number greater than 1 is "built" by multiplying a unique combination of prime numbers. This theorem appeared in another way by Euclid in his *Elements* (see Exercise 78) and was stated in its present form by the great German mathematician Karl Friedrich Gauss.

> **The Fundamental Theorem of Arithmetic**
>
> Every natural number greater than 1 is a unique product of prime numbers, except for the order of the factors. (Product could mean a single prime number.)

Example 4 shows us one way to find the prime factorization of a natural number by using factor trees.

EXAMPLE 4   Factoring a Natural Number

Factor 4,620.

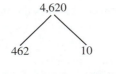

FIGURE 5.2   A first factorization of 4,620.

SOLUTION: In order to factor 4,620, we first try to think of a way to write it as a product of two smaller numbers. For example, $4,620 = 462 \cdot 10$. It really doesn't matter how you write it as a product, *what is important is that you somehow express 4,620 as a product of smaller, simpler numbers*. We can represent this preliminary factorization graphically by the diagram in Figure 5.2.

We call this a **tree diagram.** As we add more branches, you will see that this diagram looks like an upside down tree. Next, we factor 462 and 10 and add new branches to the tree as in Figure 5.3(a). Using the divisibility test, we see that 3 divides 231.

If we now multiply all the primes at the ends of the branches (shown in red) in Figure 5.3(b), we see that $4,620 = 2 \cdot 3 \cdot 7 \cdot 11 \cdot 2 \cdot 5 = 2^2 \cdot 3 \cdot 5 \cdot 7 \cdot 11$.   🌀

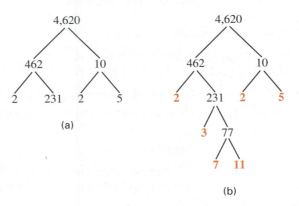

FIGURE 5.3   Completing the factorization of 4,620, using a factor tree.

Historical Highlight: *Eratosthenes*

One of the Greeks who made great contributions to the field of number theory was Eratosthenes (276–194 B.C.). He studied at Plato's academy in Athens and was considered to be one of the most learned men of antiquity. Eratosthenes was a scholar in many areas including geography, philosophy, history, astronomy, mathematics, and literary criticism. In fact, he had so many varied interests that his friends called him Pentathis, a name given to a champion in five athletic events. His enemies, however, called him Beta, the second letter of the Greek alphabet, charging that although he worked in many fields, he was outstanding in none.

Many would argue that Eratosthenes was certainly first rate in both mathematics and geography. He was the first to try to put geography on a sound scientific basis. In his work, *Geographia,* he argued that the world was round—1,700 years before Columbus made his famous voyage in 1492. He also produced the most accurate map of the known world of his time, and was the first to use a grid of lines of longitude and latitude that is still used in maps today. Although writers of mathematics textbooks remember him for his prime sieve, he is probably more famous for devising a method for measuring the circumference of the earth.*

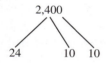

FIGURE 5.4 A possible first step in factoring 2,400.

If you were factoring a number like 2,400 and recognized that it is the product of 24 and 10 and 10, then you could begin drawing your tree with three branches, as in Figure 5.4. As we said, it really doesn't matter how you start your tree, the important thing is to express 2,400 as a product of smaller natural numbers.

It is important to understand that the Fundamental Theorem of Arithmetic tells us that although we may factor a natural number several different ways, we will always end up with the same factorization when we are finished.

 ## Quiz Yourself 3

a) Draw a factor tree to factor 1,560.

b) Write the factorization.

We use prime factorization to find the GCD and LCM of numbers.

In many applications of number theory, we need to find the greatest common divisor and least common multiple of natural numbers.

> **DEFINITION**
>
> The **greatest common divisor, GCD,** of two natural numbers is the largest natural number that divides both numbers.

---

* The method we explain in Example 3 of Section 8.1 is essentially the one that Eratosthenes invented.

For small numbers, you can find the GCD by looking at the numbers and identifying the largest natural number that divides both numbers. For example, to find the GCD of 24 and 40, we list the divisors of both numbers. The common divisors are highlighted in red.

Divisors of 24: **1**, **2**, 3, **4**, 6, **8**, 12, 24

Divisors of 40: **1**, **2**, **4**, 5, **8**, 10, 20, 40

From these lists, we see that the largest number dividing both 24 and 40 is 8.

For larger numbers, it is tedious to list all divisors of both numbers, so we use the approach shown in Example 5, which involves factoring both numbers.

EXAMPLE 5   Using Prime Factorization to Find the GCD

Find the greatest common divisor of 600 and 540.

SOLUTION: We first factor both numbers using factor trees to get

$$600 = 2 \cdot 2 \cdot 2 \cdot 3 \cdot 5 \cdot 5 \text{ and } 540 = 2 \cdot 2 \cdot 3 \cdot 3 \cdot 3 \cdot 5.$$

If we think about what the words "greatest common divisor" mean, we recognize that we need to find as many primes as possible that divide *both* numbers at the same time. We see that two 2s will divide both 600 and 540, however, three 2s will not. Similarly, only one 3 divides both numbers and only one 5 divides both numbers. Therefore, the GCD of 600 and 540 is $2 \cdot 2 \cdot 3 \cdot 5 = 60$.

Another way of thinking about what we did in Example 5 is to write $600 = 2^3 \cdot 3^1 \cdot 5^2$ and $540 = 2^2 \cdot 3^3 \cdot 5^1$. Then in forming the GCD, we multiplied the 2^2, the 3^1, and 5^1, which were the *smallest powers* of the primes that divide both numbers. So the GCD is $2^2 \cdot 3^1 \cdot 5^1$.

PROBLEM SOLVING

The meaning of the words "greatest common divisor" tells you how to compute the GCD of two numbers. The word "greatest" tells you to find the *largest* natural number that divides evenly into *both* numbers. That is why, for instance, we used 3^1, rather than 3^3, in constructing the GCD of 600 and 540, because 3^1 divides both numbers; however, 3^3 does not divide 600.

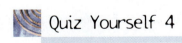

Quiz Yourself 4

Find the GCD of 126 and 588.

Next we will discuss the least common multiple of two numbers.

DEFINITION

The **least common multiple, LCM,** of two natural numbers is the smallest natural number that is a multiple of both numbers.

The LCM of 8 and 6 is 24 because 24 is the smallest natural number that is divisible by both 6 and 8. The LCM of 15 and 10 is 30.

We can use a prime factorization approach, similar to what we did in Example 5, to find the LCM of two numbers.

EXAMPLE 6   Using Prime Factorization to Find the LCM

Find the least common multiple of 600 and 540.

SOLUTION: As in Example 5, we start by factoring both numbers. Recall that

$$600 = 2 \cdot 2 \cdot 2 \cdot 3 \cdot 5 \cdot 5 \text{ and } 540 = 2 \cdot 2 \cdot 3 \cdot 3 \cdot 3 \cdot 5.$$

Now think about the words "least common multiple." Here we recognize that we need to find the *smallest* natural number that is *divisible by all the prime factors* of 600 and 540. However, we don't want to use any more primes in our number than we absolutely have to. The LCM will have to have three 2s, three 3s, and two 5s. So the LCM of 600 and 540 is

$$2 \cdot 2 \cdot 2 \cdot 3 \cdot 3 \cdot 3 \cdot 5 \cdot 5 = 5,400.$$

Although this number seems large, realize that we cannot omit any of the prime factors. For example, if we omit one of the 3s, then 540 will not divide into the number. If we omit one of the 5s, then 600 will not divide into the number.   ◎

Another way of thinking about what we did in Example 6 is to write $600 = 2^3 \cdot 3^1 \cdot 5^2$ and $540 = 2^2 \cdot 3^3 \cdot 5^1$. Then in forming the LCM, we multiplied the 2^3, the 3^3, and 5^2, which were the *highest powers* of the primes that divide both numbers. So the LCM is $2^3 \cdot 3^3 \cdot 5^2$.

> **Finding the GCD and LCM by Using Factorization**
>
> To find the GCD and LCM of two numbers, do this:
>
> 1. Factor both numbers and write each as a product of powers of primes.
> 2. To calculate the GCD, multiply the *smallest* powers of any primes that are common to both numbers.
> 3. To calculate the LCM, multiply the *largest* powers of all primes that occur in either number.

We find applications of the GCD in situations where we want to take several larger objects and represent them as a collection of smaller objects all of the same size. For example, a clerk in a store may want to pile different types of dishes on piles of same size, or a cabinet maker might want to cut several boards of different lengths to make smaller boards all of the same size.

The LCM occurs in situations where we have two types of objects and we want to take some of each to get a larger object of the same size. For example,

Highlight: *Using Computer Algebra Systems to Find Prime Numbers*

It is possible to use a computer algebra system to factor very large numbers, and in the process find some extremely large prime numbers. Here is a Maple worksheet using the ifactor function which factors natural numbers into a product of integers.

>**ifactor**(34253423523432523633533363227338328271);
  $(3)^2(41)$ (17541514568497)   (10181733812491)
              (51974317)

If you were to try to factor this number, no doubt you would have found the factors 3, 3, and then 41. However, the next prime that you would have to find is 51,974,317! Can you imagine how long it would take you to find this factor by hand? Also notice that Maple claims that both 17,541,514,568,497 and 10,181,733,812,491 are prime numbers. (As you might guess, I have not bothered to check these claims.) This raises the interesting philosophical question as to whether or not we can trust that the computer is correct. Mathematicians debate as to whether or not we should accept a computer result that may be impossible to verify with hand calculations.

You will see that something very interesting happens when we try to factor the next number.

>**ifactor**(4543545362645463452645645264264234523
            6424624624245264235613);
        Warning, computation interrupted.

In this case, Maple did not finish this computation after a considerable length of time. It is possible that we are trying to factor a very large prime number. However, I was doing this factorization on a very old 25-MHz computer. It may be possible that if you have access to a much faster computer with a computer algebra system on it that you may be able to determine the factorization of this number.

one person has every fifth day off and another may have every sixth day off, and we ask when will both be off together.

EXAMPLE 7   Scheduling Airport Security

If several flights leave at the same time from an airport, it is important that there is enough security available to handle the increased number of passengers. Assume that flights for Philadelphia leave every 45 minutes and flights for New York City leave every 25 minutes. If flights to Philadelphia and New York have just departed, how long will it be before this will happen again?

SOLUTION: Flights to Philadelphia will leave again in 25, 50, 75, . . . minutes, and flights to New York will leave in 45, 90, 135, . . . minutes. We want to know the smallest number of minutes until there are flights departing to both cities. That is, we want to know the LCM of 25 and 45.

Now, $25 = 5^2$ and $45 = 3^2 \cdot 5$, so the LCM of these numbers is $3^2 \cdot 5^2 = 225$. Therefore, the airport will need increased security in 225 minutes, which is 3 hours and 45 minutes. So if the person in charge of security knows this, she may want to ask someone to come in early or stay late and certainly will not schedule any personnel for a lunch break at that time.

Exercises 5.1

Determine whether or not each of the following statements is true.

1. 8 divides 56.

2. 21 is a multiple of 2.

3. 27 is a multiple of 6.

4. 5 is a divisor of 35.

5. 9 is a factor of 96.

6. 14 divides 42.

7. 63 is a multiple of 7.

8. 6 is a factor of 76.

Restate each of the following statements in two other ways.

9. 4 divides 48.

10. 21 is a multiple of 3.

11. 28 is a multiple of 7.

12. 3 is a factor of 15.

13. 9 is a factor of 72.

14. 13 divides 39.

15. Use the Sieve of Eratosthenes to find all prime numbers between 51 and 100, inclusive.

16. Use the Sieve of Eratosthenes to find all prime numbers between 101 and 120, inclusive.

Estimate the square root of each of the following numbers, n, by finding a natural number a such that $a < \sqrt{n} < a + 1$.

17. 95

18. 138

19. 153

20. 229

Identify the following numbers as either prime or composite. If a number is composite, factor it as a product of two smaller natural numbers (not necessarily prime numbers).

21. 231

22. 89

23. 113

24. 153

25. 197

26. 143

27. 119

28. 137

Check each of the following numbers for divisibility by each of these numbers: 2, 3, 4, 5, 6, 8, 9, 10. State the numbers that divide the given number.

29. 141,270

30. 18,036

31. 47,385

32. 476,376

Find the smallest natural number that is divisible by:

33. 2, 3, 5, and 6

34. 2, 4, 8, and 5

35. 4, 5, and 10

36. 2, 3, 4, 5, 6, and 10

Provide a counterexample to show that each of the following statements is false.

37. If 2 and 4 divide *a*, then 8 divides *a*.

38. If 3 and 6 divide *a*, then 18 divides *a*.

39. If 10 and 4 divide *a*, then 40 divides *a*.

40. If 4 and 6 divide *a*, then 24 divides *a*.

Factor each of the following natural numbers.

41. 980

42. 396

43. 9,900

44. 2,106

45. 621

46. 805

47. 319

48. 403

Find the GCD and LCM of each of the following pairs of natural numbers.

49. 20, 24

50. 60, 72

51. 56, 70

52. 66, 110

53. 216, 288

54. 675, 1,125

55. 147, 567

56. 275, 363

57. Communicating Mathematics   Explain how you use the prime factorization method to get the GCD of two natural numbers *a* and *b*.

58. Communicating Mathematics   Explain how you use the prime factorization method to get the LCM of two natural numbers *a* and *b*.

*There is another way to find the GCD of two natural numbers called the **Euclidean algorithm**. The Euclidean algorithm goes as follows:*

Suppose that we wish to find the GCD of two numbers such as 24 and 88.

First we divide the smaller number, 24, into the larger number, 88, as shown in Figure 5.5.

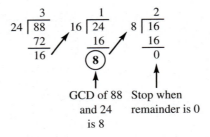

FIGURE 5.5   The Euclidean algorithm.

Next we divide the remainder, 16, into the previous divisor, 24, to get a new remainder, 8. We continue this process, always dividing the remainder into the previous divisor. So, next we divide 8 into 16. At this point, the remainder is 0, so we stop. The previous remainder, 8, is the GCD of 24 and 88.

Use the Euclidean algorithm to find the GCD of each of the following pairs of natural numbers.

59. 280, 588

60. 84, 1,200

61. 495, 1,575

62. 99, 1,155

It can be proved (see Exercise 84) that if you multiply the GCD and LCM of two natural numbers a and b, the product is the same as the product a · b. In Exercises 63–66, first find the GCD of the two numbers and then divide the GCD into the product of the two numbers to find the LCM. To find the GCD, you can use the Euclidean algorithm.

63. 12, 27

64. 16, 56

65. 90, 120

66. 17, 178

67. Scheduling flights. Assume that flights for Miami leave every 35 minutes and flights for Dallas leave every 20 minutes. If flights to Miami and Dallas have just departed, how many minutes will it be before this will happen again?

68. Scheduling nurses shifts. Dani and Malik are nurses who occasionally work in the emergency room. Dani works in the emergency room every 6 weeks and Malik works in the emergency room every 8 weeks. If Dani and Malik have just worked together in the emergency room, in how many weeks will they work in the emergency room together again?

69. Storing medical supplies. In a medical supply room, there are 36 packages of type O positive and 30 packages of type AB negative blood. It is important not to confuse these two types of blood. If we wish to stack piles of each type of blood on a shelf, with only one type of blood in a pile, what is the largest number of packets of blood that we can have in each pile if each pile is of the same size?

70. Displaying store merchandise. A sporting goods store has 20 instructional videos on skiing and 12 videos on snow boarding. The owner of the store wishes to display the videos on a shelf with stacks of the same size and each stack consisting of only one type of video. What is the most number of videos on each stack that will accomplish this?

71. Tiling a museum floor. In the Central America room of the Ancient Civilization Museum, we wish to tile the floor with replicas of ancient Aztec tiles. If the floor measures 33 feet by 21 feet, what is the largest size square tile that we can use to tile the floor without having to cut any tiles?

72. Scheduling lawn service. At the Berkshire Country Club, the lawn service cuts the grass every 8 days and the pest service sprays for insects every 30 days. If both services have just been to the country club, in how many days will both be there again?

73. Monitoring pollution. The Environmental Protection Agency is monitoring the Lackawanna River to determine the cause of a fish kill. A steel mill releases hot water into the river every 48 hours and a plastics plant discharges pollutants into the river every 54 hours. If both hot water and pollutants from the plastics plant have just been discharged, in how many hours will both be released into the river again?

Further Exercises

*(74.) Communicating Mathematics   Search the Internet for information about GIMPS (the Great Internet Mersenne Prime Search). A Mersenne prime is a prime of the form $2^n - 1$. For example, $7 = 2^3 - 1$ is a Mersenne prime. As of December 2001, the largest known Mersenne prime was $2^{13466917} - 1$ which was discovered by a 20-year-old student. If you were to expand this number, it would take over 4 million digits and would fill a whole book! Write a brief report on your findings.

75. Communicating Mathematics   Explain why the divisibility test for 4 works by considering the number 36,824. It helps to think of 36,824 as $36,800 + 24$.

76. Communicating Mathematics   Explain why the divisibility test for 3 works by considering the number 5,712. It helps to think of the following equations:

$$5,712 = 5(1,000) + 7(100) + 1(10) + 2$$
$$= 5(999 + 1) + 7(99 + 1) + 1(9 + 1) + 2$$
$$= 5 \cdot 999 + 7 \cdot 99 + 1 \cdot 9 + (5 + 7 + 1 + 2)$$

There are many famous conjectures† regarding prime numbers. In Exercises 77 and 78, we investigate two of these most famous conjectures.

77. A pair of **twin primes** is a pair of numbers that differ by two that are both prime. For example, 17 and 19 are a pair of twin primes. Also 41 and 43 are another pair of twin primes. The **twin prime conjecture** states that there are an infinite number of twin primes. Find the next three twin primes that are greater than 43.

78. **Goldbach's conjecture** states that any even number greater than two can be written as the sum of two primes. For example, $16 = 11 + 5$, $26 = 13 + 13$, and $40 = 3 + 37$. Write each of the following as a sum of two primes‡

a) 80   b) 100   c) 200

(79.) Communicating Mathematics   If both p and q divide the natural number n, then does $p \cdot q$ divide the number n? Either explain why this must be true, or else provide a counterexample to this statement.

(80.) Communicating Mathematics   Explain why the rules that we stated for divisibility in Exercises 37 to 40 failed.

(81.) Communicating Mathematics   Explain how you would apply the Euclidean algorithm to find the GCD of three numbers.

82. Communicating Mathematics   The divisibility test for 6 is that the number must be divisible by both 2 and 3. Explain why the divisibility test for 8 is *not* that the number must be divisible by both 2 and 4.

83. Devise a divisibility test for 15.

(84.) Communicating Mathematics   If a and b are natural numbers, explain why the GCD of a and b times the LCM of a and b is equal to the product of a and b. Use the notion of prime factorization of a and b to give your explanation.

85. Communicating Mathematics   Euclid stated that "If a number be the least that is measured by prime numbers, it will not be measured by any other prime number except those originally measuring it."§ Give an interpretation of this statement in terms of the Fundamental Theorem of Arithmetic.

86. Communicating Mathematics   Research the technique for checking addition called "casting out nines." How is it similar to our test for divisibility by nine?

---

* Exercise numbers circled in red can be used as group exercises.
† You can think of a conjecture as an educated guess that has not yet been proved.
‡ There may be several ways to do this.
§ David M. Burton, *The History of Mathematics*, 4th edition (WCB McGraw-Hill, 1999), p. 171.

5.2 THE INTEGERS

Although the natural numbers were known in various forms to many ancient peoples, zero did not appear until many years later. The Hindus are credited with inventing zero, which later became part of our Hindu–Arabic system of notation.

> **DEFINITION**
>
> We adjoin zero to the set of natural numbers to form the set {0, 1, 2, 3, . . .}, called the set of **whole numbers.**

Many mathematicians were slow to accept zero, arguing that we could not have a number that represented nothing. The Greek mathematician Diophantus said that to solve the equation $3x + 20 = 8$ and get a solution of -4 was absurd. As late as the 16th century, many European mathematicians would use negative numbers in doing calculations, but would not accept zero or a negative number as an answer to a problem.

Now, of course, once we learn the rules for calculating with negative numbers, we can calculate with them as easily as we do with positive numbers.

> **DEFINITION**
>
> The set of **integers** is the set {. . ., $-3, -2, -1, 0, 1, 2, 3,$ . . .}.

In this section, we will investigate the rules for adding, subtracting, multiplying, and dividing integers.

SOME GOOD ADVICE

Keep in mind that when working with the integers, it is important to know *why* a rule works as well as *what* the rule tells you to do.

We can explain addition rules by considering movement on a number line.

A common model that we can use to explain the addition of integers is movement on the number line. In this model, we think of integers as points on the number line, as shown in Figure 5.6.

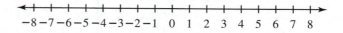

FIGURE 5.6 Representing integers on the number line.

Then, beginning at 0, we interpret positive numbers as movement to the right and negative numbers as representing movement to the left. We illustrate this in Example 1.

EXAMPLE 1   Representing Integer Addition as Movement on the Number Line

Perform the following additions.

a) $(+3) + (+5)$ b) $(+9) + (-12)$ c) $(-149) + (137)$

SOLUTION: a) Starting at 0, think of $+3$ as a movement three places to the right. Then think of $+5$ as an additional movement of 5 places to the right, as in Figure 5.7. Thus, we see that $(+3) + (+5) = +8$.

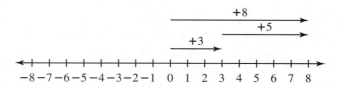

FIGURE 5.7 $(+3) + (+5) = +8$.

b) To calculate $(+9) + (-12)$, we first move 9 places to the right and then move 12 places to the left, as we show in Figure 5.8.

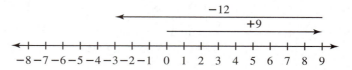

FIGURE 5.8 $(+9) + (-12) = -3$.

Quiz Yourself 5*

Find

a) $(+4) + (6)$

b) $(+8) + (-13)$

c) $(-3) + (-7)$

d) $(+145) + (-123)$

c) It is not practical to draw a number line when calculating with such large numbers; however, you still can imagine movements along the number line. If you visualize moving 149 units to the left and then 137 units to the right, you see that the net effect is that you have moved $149 - 137 = 12$ units to the left, so $(-149) + (+137) = -12$. ◎

We define subtraction of integers in terms of addition.

Mathematicians consider the set of integers as an abstract system having only two operations—addition and multiplication. At first, they do not mention subtraction and division, considering these operations to be somewhat secondary operations.

Now we will define subtraction in terms of addition, and later we will define division in terms of multiplication. But first, we need a definition.

---

* Quiz Yourself answers begin on page 849.

Historical Highlight: *Negative Numbers**

It took mathematicians a long time to accept the notion of negative numbers. Historians tell us that there is no evidence of negative numbers in the mathematics of ancient civilizations such as the Babylonians and the Greeks. In fact, Greek mathematics was so heavily based on geometry, which involved the measuring of lengths and distances, that they had no need for negative numbers.

By the 7th century, Indian bookkeepers used negative numbers to represent debt and Indian mathematicians such as Brahmagupta showed that they had a clear idea of how to use negative numbers in doing scientific calculations.

During the Renaissance, the Italian mathematician Girolamo Cardano used negative numbers in 1545 in his work, *Ars Magna*, which was a text on algebraic equations. However, a century later, other European mathematicians were still slow to accept negative numbers. The Frenchman Rene Descartes called negative numbers "false roots" and his contemporary Blaise Pascal was convinced that numbers "less than zero" could not exist. The German Gottfried Leibniz said that although negative numbers were useful in doing calculations, they could lead to absurd conclusions and misconceptions. It was not until the 18th century that negative numbers were generally accepted, although even then there were still mathematicians who tried to avoid using them.

DEFINITION

Two integers x and y are **opposites†** if $x + y = 0$.

For example, -8 is the opposite of $+8$, $+13$ is the opposite of -13, and 0 is the opposite of 0.

DEFINITION

If a and b are integers, then $a - b = a + (-b)$.

This definition means that in order to compute $3 - 8$, you should think instead of $3 + (-8)$, which is -5. To compute $(-6) - (-19)$, think $(-6) + ($ the opposite of $-19)$. That is, $(-6) + (+19) = +13$.

Quiz Yourself 6

Convert the following subtraction problems to addition problems and find the answers.

a) $(+5) - (+12)$

b) $(-3) - (-9)$

SOME GOOD ADVICE

You may have heard in the past that "two negatives make a positive." This is not a rule in mathematics, but rather a memory device that *sometimes works* and *sometimes does not*, depending on the situation in which you try to apply it. For example, $(-6) + (-2)$ does not make a positive. Rather than thinking of "two negatives make a positive," it is better to think of $-(-6)$ as the opposite of -6, or $+6$, and think of $3 - (-5)$ as $3 + (+5)$.

---

* This highlight is based on material in Jan Gullberg, *Mathematics: From the birth of Numbers* (W. W. Norton, 1997), pp. 72–73.
† Numbers which are opposites are also called additive inverses of each other.

We can use a money model to explain integer multiplication.

You can understand the rules for multiplying integers, if you consider the following model. Call today day zero and assume that you have $100 in your wallet. Also assume that each day one of two things happens, either you receive $5 in the mail from your rich uncle, or you spend $5 for a pizza. We assume that you are making no other changes in your finances.

We will now consider the product of integers a and b. The integer a will represent changes in days—a positive a represents days in the future, a negative a represents days in the past. The number b will represent changes in your finances—a positive b means you receive $5, a negative b means you spend $5.

For example, we interpret $(+3)(-5)$ to mean that for the next three days, you spend $5 each day on pizza. The product $(-4)(+5)$ means for the past four days, you have been receiving $5 from your uncle. With this in mind, consider Table 5.2.

	Interpretation of $a \cdot b$	Multiplication Rule
$(+3)(+5)$	For the next three days, you receive $5 from your uncle. So, in 3 days you will gain $15. Thus $(+3)(+5) = +15$.	A positive times a positive is a positive.
$(+3)(-5)$	For the next three days, you spend $5 on pizza. So, in 3 days you will lose $15. Thus $(+3)(-5) = -15$.	A positive times a negative is a negative.
$(-3)(+5)$	For the past three days, you have received $5, so three days ago, you had $15 less. Thus, $(-3)(+5) = -15$.	A negative times a positive is a negative.
$(-3)(-5)$	For the past three days, you have spent $5, so three days ago, you had $15 more. Thus $(-3)(-5) = +15$.	A negative times a negative is a positive.

TABLE 5.2 Rules for multiplying integers.

We can state these rules for multiplying integers more succinctly.

Rules for Multiplying Integers

If a and b are integers, then:
a) If a and b have the same sign, then $a \cdot b$ is positive.
b) If a and b have opposite signs, then $a \cdot b$ is negative.

EXAMPLE 2   Applying the Rules for Multiplying Integers to the Stock Market

a) Assume that stock for Itech, an international technology company, has been losing value at $3 per day for each of the last 8 days. How much more was the stock worth 8 days ago?

b) If the stock continues to lose its value at $3 per day, how much less will the stock be worth in 4 days?

SOLUTION: a) To find the worth 8 days ago, we represent "8 days ago" by -8 and "losing value at $3 per day" by -3. We then calculate the product $(-8)(-3) = +24$. So the stock was worth $24 more 8 days ago.

b) The change in value in 4 days will be $(+4)(-3) = -12$, so the stock will be worth $12 less.

Quiz Yourself 7

Find the following products.

a) $(-4)(-6)$ b) $(+8)(-7)$

We define division in terms of multiplication.

Like subtraction, we consider division to be somewhat of a secondary operation, and we define division in terms of multiplication.

> **DEFINITION**
>
> If $a, b,$ and c are integers, then $a/b = c$ means that $a = b \cdot c$.

Try to imagine that you did not know any division facts, but were comfortable with doing multiplication. If you knew the definition of division, in order to compute 8/2, you would consider the equation $\frac{8}{2} = c$, which according to the definition of division means that you should consider $8 = 2 \cdot c$ instead. From your knowledge of multiplication, you would know that 4 is a solution for this equation; therefore, $\frac{8}{2} = 4$. In Example 3, we will use the definition of division to explain the rules for dividing signed numbers.

EXAMPLE 3   Deriving the Rules for Dividing Signed Numbers

Use the definition of division to find each of the following quotients.

a) $\dfrac{+6}{+2}$ b) $\dfrac{+12}{-6}$ c) $\dfrac{-15}{+3}$ d) $\dfrac{-20}{-5}$

SOLUTION: In each case, we will rewrite the division problem as a corresponding multiplication problem.

a) To solve $\frac{+6}{+2} = c$, we solve $+6 = (+2) \cdot c$ instead. The solution is clearly $+3$, so we can say $\frac{+6}{+2} = +3$. This illustrates that a *positive divided by a positive is a positive.*

b) Instead of solving $\frac{+12}{-6} = c$, we solve $+12 = (-6) \cdot c$. We see that $c = -2$. Thus, $\frac{+12}{-6} = -2$. This shows that a *positive divided by a negative is a negative.*

c) To compute $\frac{-15}{+3}$, we must solve the equation $-15 = (+3) \cdot c$. In this case, $c = -5$, so $\frac{-15}{+3} = -5$. Thus, a *negative divided by a positive is a negative*.

d) Lastly, to find $\frac{-20}{-5}$, we solve $-20 = (-5) \cdot c$. Because $c = +4$, we can say $\frac{-20}{-5} = +4$. This illustrates that a *negative divided by a negative is a positive*.

We can state the rules for dividing integers that we derived in Example 3 more concisely.

> **Rules for Dividing Integers**
>
> If a and b are integers and $b \neq 0$, then:
>
> a) If a and b have the *same sign*, then $\dfrac{a}{b}$ is positive.
>
> b) If a and b have *opposite signs*, then $\dfrac{a}{b}$ is negative.

Notice that it is permissible to divide *into* zero. For example, $\frac{0}{5} = 0$. However, we do not allow you to divide *by* zero. For an explanation of why we cannot divide by zero, see Exercises 91 and 92.

EXAMPLE 4   Calculating Quotients of Integers

Find each of the following:

a) $\dfrac{+12}{-6}$ b) $\dfrac{-20}{-5}$ c) $\dfrac{+16}{+8}$ d) $\dfrac{-15}{+3}$

SOLUTION: According the rules for division that we just stated, the signs of b) and c) will be positive and the signs of a) and d) will be negative.

a) $\dfrac{+12}{-6} = -2$ b) $\dfrac{-20}{-5} = 4$ c) $\dfrac{+16}{+8} = 2$ d) $\dfrac{-15}{+3} = -5$

 Quiz Yourself 8

Find

a) $\dfrac{+24}{-6}$ b) $\dfrac{-30}{-6}$ c) $\dfrac{+26}{+2}$ d) $\dfrac{-25}{+5}$

 PROBLEM SOLVING

Recall the Analogies Principle in Section 1.1 that says it is important to make connections between mathematical ideas. You should keep in mind that once you know your multiplication rules, the division rules are exactly the same. If you think about this, you will realize that this is not a coincidence, because we derived the division rules from the rules for multiplication.

Exercises 5.2

Calculate each of these sums using movements on the number line as we did in Example 1.

1. $(+8) + (-5)$

2. $(-9) + (-4)$

3. $(+4) + (+6)$

4. $(-3) + (+7)$

5. $(-128) + (+137)$

6. $(+47) + (+32)$

7. $(+57) + (-38)$

8. $(-38) + (-53)$

Rewrite the following subtraction problems as addition problems and then find the answers.

9. $(+18) - (-5)$

10. $(-8) - (-13)$

11. $(+6) - (+19)$

12. $(-3) - (+14)$

13. $(-28) - (+37)$

14. $(+41) - (+13)$

15. $(+32) - (-18)$

16. $(-28) - (-23)$

Communicating Mathematics   *Explain each of the following calculations using a money model similar to the ones given in Table 5.2. For example, $(-2)(+7)$ would mean for the past 2 days, you have been receiving \$7. Explain what the change in your finances would be in each instance.*

17. $(+4)(-6)$

18. $(+5)(+8)$

19. $(-3)(+7)$

20. $(-6)(-3)$

Find the following products.

21. $(-5)(-7)$

22. $(+5)(+3)$

23. $(-7)(+8)$

24. $(+3)(-8)$

25. $(+8)(+6)$

26. $(-3)(-9)$

27. $(+19)(-2)$

28. $(-6)(+7)$

29. $(-9)(+7)$

30. $(+5)(+9)$

31. $(-8)(-9)$

32. $(+7)(-3)$

Write each of the following division problems as a corresponding multiplication problem and then solve it as we did in Example 3.

33. $\dfrac{+16}{-2}$

34. $\dfrac{+24}{+6}$

35. $\dfrac{-14}{-2}$

36. $\dfrac{-26}{+2}$

37. $\dfrac{-40}{+8}$

38. $\dfrac{+15}{-3}$

39. $\dfrac{+12}{+3}$

40. $\dfrac{-27}{-3}$

Perform the following divisions.

41. $\dfrac{-24}{+8}$

42. $\dfrac{+20}{+5}$

43. $\dfrac{-30}{-2}$

44. $\dfrac{+28}{-4}$

45. $\dfrac{+15}{+3}$

46. $\dfrac{-14}{+7}$

47. $\dfrac{+30}{-5}$

48. $\dfrac{-22}{-11}$

49. $\dfrac{-18}{-3}$

50. $\dfrac{+20}{+5}$

51. $\dfrac{-40}{+10}$

52. $\dfrac{+25}{-5}$

Perform each of the following calculations.

53. $(-2)(-3) + (+4)(-8)$

54. $(-2)(+5) + (-4)(+7)$

55. $(-4)(-2) + (+4)(-5)$

56. $(+8)(-2) + (-3)(-5)$

57. $(-5)(-6) + (+2)(-7)$

58. $(+6)(+5) + (-3)(-9)$

59. $(-3)(+6) + (-4)(+8)$

60. $(-4)(-6) + (+7)(-8)$

Communicating Mathematics   *If it is possible, give an example of each of the following. If it is not possible, explain why it is not possible. There may be many possible answers.*

61. An integer which is also a whole number

62. An integer which is not a whole number

63. A negative integer which is also a whole number

64. A whole number which is not a natural number

65. An integer which is not a natural number

66. A natural number which is not a whole number

67. A natural number which is not an integer

68. An integer which is a natural number

Determine whether the following statements are true or false. Remember the Always Principle from Section 1.1 when deciding on your answer. If a statement is false, then provide a counterexample.

69. The product of two negative integers is a positive integer.

70. The product of two whole numbers is a positive integer.

71. The sum of a negative and a positive integer is a negative integer.

72. The product of a negative integer and a positive integer is negative.

73. The quotient of two negative integers is negative.

74. The quotient of a negative integer and a positive integer is negative.

75. The product of a natural number and a positive integer is positive.

76. The quotient of two natural numbers is a natural number.

The following chart gives some heights of mountains and depths of ocean trenches. Use this information to do Exercises 77–80.

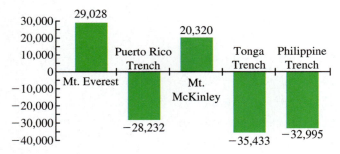

77. Comparing extreme distances in geography. How much higher is the top of Mount Everest than the bottom of the Puerto Rico Trench?

78. Comparing extreme distances in geography. How much higher is the top of Mount McKinley than the bottom of the Tonga Trench?

79. Comparing extreme distances in geography. How much higher is the bottom of the Puerto Rico Trench than the bottom of the Tonga Trench?

80. Comparing extreme distances in geography. How much higher is the bottom of the Philippine Trench than the bottom of the Tonga Trench?

81. Comparing extreme temperatures. According to the World Almanac, the highest temperature ever recorded in the world was 136 degrees Fahrenheit in Libya on September 13, 1922, and the lowest temperature recorded was − 129 degrees in Antarctica on July 21, 1983. What is the difference between these two records?

82. Comparing extreme temperatures. According to the World Almanac, the highest temperature recorded in California was 134 degrees Fahrenheit on July 10, 1913, and the lowest temperature recorded in California was − 45 degrees on January 20, 1937. What is the difference between these two records?

83. The distance an elevator travels. The ground floor of a skyscraper is labeled floor zero. However, there are floors below labeled − 1, − 2, and so on. If an elevator goes from floor − 7 to floor 59, how many floors has the elevator traveled?

84. Temperature change in a laboratory. The temperature in a cryogenics laboratory chamber has dropped from 14 degrees Fahrenheit to − 63 degrees. How much has the temperature dropped?

85. Change in historical dates. Confucius was born in 551 B.C., and the Ming Dynasty began in China in 1368 A.D. How far apart did these two events occur? (Warning: In doing this calculation, there is no year zero.)

86. Change in historical dates. Euclid was born in 323 B.C. and Karl Friedrich Gauss was born in 1777 A.D. How far apart were their birthdates? (Warning: In doing this calculation, there is no year zero.)

87. Temperature change. The present temperature is 65 degrees and the weather bureau is predicting that as a cold front comes in, the temperature will drop 7 degrees per hour over the next four hours. What will the temperature be at that time?

88. Tracking the stock market. Assume that the Dow Jones Average for the stock market began the week at 9,782 points. On Monday the market went down 102 points, then on Tuesday rose 78 points, rose 11 points on Wednesday, declined 14 points on Thursday, and rose 12 points on Friday. What was the Dow Jones Average at the close on Friday?

Further Exercises

A magic square* is a square arrangement of numbers as shown below. Because this square has three rows and three columns, it is called a 3 by 3 square.

8	1	6
3	5	7
4	9	2

In a magic square, if you add the numbers in any row, or column, or diagonal, you always get the same sum. For example, in this magic square, the sum of any row, column, or diagonal is always 15.

† **89.** Is the following a magic square? Explain.

− 8	− 1	4	3
5	2	− 7	− 2
− 5	− 4	7	1
6	1	− 6	− 3

90. Complete the entries in the following magic square.

6	a	− 8	3
− 5	b	1	− 2
− 1	c	− 3	2
− 6	5	d	e

91. Communicating Mathematics   Use the definition of division in terms of multiplication to explain why we cannot divide 8 by 0. What does it mean to say $\frac{8}{0} = x$?

92. Communicating Mathematics   Use the definition of division in terms of multiplication to explain why we cannot divide 0 by 0. What does it mean to say $\frac{0}{0} = x$?

---

* Notice that this magic square uses the digits 1 to 9. A 4 by 4 magic square would customarily use the natural numbers from 1 to 16. In our exercises, we will relax this rule so that we only have to use consecutive integers.
† Exercise numbers circled in red can be used as group exercises.

Use the following block of numbers to answer Exercises 93–96.

1	2	3	4	5	6	7	8
9	10	11	12	13	14	15	16
17	18	19	20	21	22	23	24
25	26	27	28	29	30	31	32
33	34	35	36	37	38	39	40
41	42	43	44	45	46	47	48
49	50	51	52	53	54	55	56
57	58	59	60	61	62	63	64

93. Communicating Mathematics   What is the relationship of the sum of the numbers in the solid box, which we will call a 3 by 3 box, to the number in the center of the box?

94. Communicating Mathematics   Does the relationship you saw in Exercise 93 hold for other boxes of the same size? Can you explain why this does or does not happen?

95. Communicating Mathematics   Choose a box of numbers of size 4 by 4 and determine a pattern for finding the sum of the numbers in the box.

96. Communicating Mathematics   Without doing any calculations, state how to find the sum of numbers in a 5 by 5 box. Do the same for a 6 by 6 box.

97. Give an example of an equation that we cannot solve with a whole number but that we can solve with an integer.

98. Give an example of an equation that we cannot solve with an integer.

5.3   THE RATIONAL NUMBERS

There was a time when people only knew about natural numbers—they did not comprehend zero, and negative numbers were unheard of. It is interesting to think about why new numbers are needed and how they come into existence. One way of thinking about how number systems evolve, is that while working in one system, we pose a problem that we cannot solve using the numbers in that system. So we have to invent new numbers.

For example, suppose that you knew only about the natural numbers and were asked to solve the equations $x + 2 = 6$ and $x + 6 = 2$. On the surface, they seem to be very much the same problem. However, the first equation has a solution within the natural numbers, namely $x = 4$, but the second equation does not. If you want to solve the second equation, you need to invent a new number, -4. The need for these new numbers leads us to develop the system of integers that you studied in Section 5.2.

Similarly, if, while working in the system of integers, you wish to solve the equation $6x = -2$, you find that there are no solutions. So you have to invent another number, $\frac{-2}{6} = -\frac{1}{3}$. To solve such equations, we need the system of rational numbers.

> **DEFINITIONS**
>
> The set of **rational numbers,** which we will denote by Q, is the set of all numbers that can be written in the form $\frac{a}{b}$, where the a and b are integers, and $b \neq 0$. The top number a is called the **numerator** and the bottom number b is called the **denominator.**

Numbers such as $\frac{1}{2}$, $\frac{7}{13}$, $\frac{-4}{5}$, and $\frac{-9}{-20}$ are examples of rational numbers. Also, integers such as 5, -3, and 0 are rational numbers because they *can be* written in the form $\frac{5}{1}$, $\frac{-3}{1}$, and $\frac{0}{1}$. Also, as you will see later, many decimal numbers are also rational numbers.

We cross multiply to test for equality of rational numbers.

One of the most important things to know in any mathematical system is when two different-looking objects are the same. There is a straightforward way to determine if two rational numbers are equal.

> **DEFINITION**
>
> We define **equality of rational numbers** as follows:
>
> $$\frac{a}{b} = \frac{c}{d}^* \text{ if and only if } ad = bc$$

FIGURE 5.9 $ad = bc$.

Computing the products in the equation $ad = bc$ is called **cross multiplying** or calculating the **cross product.** Figure 5.9 helps us to remember this equation.

> **SOME GOOD ADVICE**
>
> Although there are other methods for determining when rational numbers are equal, it is best to learn this one method well and use it all the time. Many of the rules that we use in working with rational numbers are based on this definition of equality.

EXAMPLE 1   Testing for Equality of Rational Numbers

Determine which of the following pairs of rational numbers are equal.

a) $\dfrac{3}{8} = \dfrac{12}{32}$ b) $\dfrac{39}{116} = \dfrac{35}{108}$ c) $\dfrac{1{,}419}{1{,}892} = \dfrac{165}{220}$

SOLUTION: a) Calculating the cross product we get $3 \cdot 32 = 8 \cdot 12$, or $96 = 96$, so the two rational numbers are equal.

b) When we cross multiply, we get $39 \cdot 108 = 4{,}212$ and $116 \cdot 35 = 4{,}060$, so the two rational numbers are not equal.

c) It is unlikely that you would ever have to deal with such awkward numbers; however, we gave this example to make the point that the cross multiplication method works even for "bad" numbers.

Cross multiplying, we get $1{,}419 \cdot 220 = 312{,}180$ and $1{,}892 \cdot 165 = 312{,}180$. Thus the two rational numbers are equal. ◎

Quiz Yourself 9†

Determine whether or not the following pairs of rational numbers are equal.

a) $\dfrac{51}{187}$, $\dfrac{45}{165}$

b) $\dfrac{31}{91}$, $\dfrac{63}{147}$

---

* For the remainder of this section, assume that any expression such as $\frac{a}{b}$ is a rational number.
† Quiz Yourself answers begin on page 849.

Reducing fractions is based on the notion of equality.

We use the following rule to simplify rational numbers.

Canceling Common Factors

$$\frac{a \cdot c}{b \cdot c} = \frac{a}{b}$$

This rule says that we can cancel the same *factor* from both the numerator and denominator of a rational number. This process of canceling common factors is called **reducing the rational number.** When all common factors have been canceled, we say that the number is in **lowest terms.**

EXAMPLE 2   Reducing Rational Numbers

Reduce the following rational numbers.

a) $\dfrac{48}{72}$ b) $\dfrac{1848}{2112}$

SOLUTION: a) There are several ways to do this. For example, using the divisibility rules, we recognize that 4 divides both the numerator and the denominator and then cancel 4 to get the number $\frac{12}{18}$, as we show in the first step of Figure 5.10. Next we see that 6 is a common factor of both 12 and 18 and cancel again to get $\frac{2}{3}$. Notice that you stop this reducing process when you see that the numerator and denominator have no factors in common.

$$\frac{48}{72} = \frac{\cancel{4} \cdot 12}{\cancel{4} \cdot 18} = \frac{12}{18} = \frac{\cancel{6} \cdot 2}{\cancel{6} \cdot 3} = \frac{2}{3}$$

FIGURE 5.10   Reducing $\frac{48}{72}$ by canceling common factors.

Quiz Yourself 10

Reduce $\frac{252}{336}$.

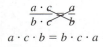

$$\frac{a \cdot c}{b \cdot c} \diagup \frac{a}{b}$$

$a \cdot c \cdot b = b \cdot c \cdot a$

FIGURE 5.11   The cancellation rule is valid because of the definition of equality of rational numbers.

b) Here it is difficult to see all the common factors of the numerator and denominator, so we do the cancellation in stages. First we see that 4 is a factor that we can cancel. So we write $\frac{1{,}848}{2{,}112} = \frac{\cancel{4} \cdot 462}{\cancel{4} \cdot 528} = \frac{462}{528}$. We have made progress, but we are not done. We can reduce further.

Both 2 and 3 are factors of the numerator and denominator, so 6 is also a common factor that we can cancel. So, $\frac{462}{528} = \frac{\cancel{6} \cdot 77}{\cancel{6} \cdot 88} = \frac{77}{88}$. However, now we see that we can cancel 11 to reduce the number further. $\frac{77}{88} = \frac{\cancel{11} \cdot 7}{\cancel{11} \cdot 8} = \frac{7}{8}$. We have written $\frac{1{,}848}{2{,}112}$ in lowest terms as $\frac{7}{8}$.

It is one thing to *know* how to use a rule, it is another to understand *why* we can use the rule. The cancellation rule states that $\frac{a \cdot c}{b \cdot c} = \frac{a}{b}$. Why is this equation true? If we use the definition of equality and cross multiply, we see that we get an equality, as we show in Figure 5.11.

PROBLEM SOLVING

When extending this cancellation rule to algebra, students sometimes mistakenly cancel terms rather than factors. For example, it is tempting to cancel 5 from the numerator and denominator of a quotient such as $\frac{x+5}{y+5}$ to get $\frac{x+\cancel{5}}{y+\cancel{5}} = \frac{x}{y}$. You can convince yourself that this is not valid by considering an example as the Three Way Principle in Section 1.1 suggests. If you substitute 2 for x and 3 for y, you see that the equation $\frac{x+5}{y+5} = \frac{x}{y}$ becomes the equation $\frac{2+5}{3+5} = \frac{2}{3}$, which is false. Therefore this cancellation of terms is not valid.

We will next consider operations on rational numbers.

We can add or subtract rational numbers that have the same denominators.

When we add rational numbers that have the same denominator, such as $\frac{3}{5} + \frac{8}{5}$, we can think that "three of one thing, plus eight of that same thing is eleven of that thing." So, $\frac{3}{5} + \frac{8}{5} = \frac{3+8}{5} = \frac{11}{5}$.

When adding rational numbers with different denominators, the addition is more complicated because, we can't, so to speak, add apples and oranges. However, we can add rational numbers with different denominators if first we rewrite both fractions in a form where they have common denominators.

To do this, suppose we wish to add $\frac{a}{b} + \frac{c}{d}$. We can rewrite this addition so that both numbers have the common denominator $b \cdot d$. That is,

$$\frac{a}{b} + \frac{c}{d} = \frac{a \cdot d}{b \cdot d} + \frac{c \cdot b}{d \cdot b} = \frac{a \cdot d + c \cdot b}{b \cdot d}.$$

This gives us the rule for adding rational numbers. The rule for subtracting is essentially the same.

Adding and Subtracting Rational Numbers

$$\frac{a}{b} + \frac{c}{d} = \frac{a \cdot d + b \cdot c}{b \cdot d} \qquad \frac{a}{b} - \frac{c}{d} = \frac{a \cdot d - b \cdot c}{b \cdot d}$$

We illustrate these rules in Example 3.

EXAMPLE 3 **Adding and Subtracting Rational Numbers**

Find a) $\frac{5}{6} + \frac{3}{4}$ b) Find $\frac{11}{18} - \frac{3}{8}$

SOLUTION: a) We apply the rule for adding rational numbers.

$$\frac{5}{6} + \frac{3}{4} = \frac{5 \cdot 4 + 6 \cdot 3}{6 \cdot 4} = \frac{20 + 18}{24} = \frac{38}{24} = \frac{19}{12}.$$

b) We apply the rule for subtracting rational numbers.

$$\frac{11}{18} - \frac{3}{8} = \frac{11 \cdot 8 - 18 \cdot 3}{18 \cdot 8} = \frac{34}{144} = \frac{17}{72}.$$

If you can recognize the LCM of the two denominators, you may find it quicker to add rational numbers by rewriting the numbers as quotients having a common denominator. In Example 3(a), if you recognize that the LCM of 6 and 4 is 12, then you could write

$$\frac{5}{6} + \frac{3}{4} = \frac{5 \cdot 2}{6 \cdot 2} + \frac{3 \cdot 3}{4 \cdot 3} = \frac{10}{12} + \frac{9}{12} = \frac{19}{12}.$$

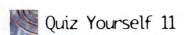 Quiz Yourself 11

Find $\frac{3}{8} + \frac{5}{6}$ and $\frac{13}{16} - \frac{11}{24}$.

It is up to you to decide in each situation, which method is the fastest and most reliable for you.

If you wish to add three numbers, you can add the first two numbers, get an answer, and then add the third. To compute an expression like $x - y - z$, first subtract $x - y$ and then subtract z from that result.

SOME GOOD ADVICE

Why is $\frac{a}{b} + \frac{c}{d} = \frac{a + c}{b + d}$ an incorrect way to do rational number addition?

Recall that the Always Principle from Section 1.1 states that in order for a mathematical statement to be true, it must always be true 100% of the time, *without exception*. If we perform the addition $\frac{1}{2} + \frac{1}{2}$ incorrectly, as stated above, we get

$$\frac{1}{2} + \frac{1}{2} = \frac{1 + 1}{2 + 2} = \frac{2}{4} = \frac{1}{2},$$

which is clearly false.

To multiply rational numbers, we multiply numerators and denominators.

We will now explain the multiplication rule for rational numbers. In Figure 5.12(a) we visualize the fraction $\frac{3}{4}$ and in Figure 5.12(b) we represent $\frac{5}{6}$.

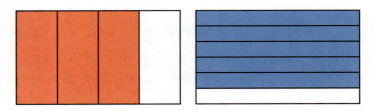

FIGURE 5.12   (a) Representation of $\frac{3}{4}$, (b) representation of $\frac{5}{6}$.

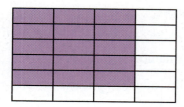

FIGURE 5.13   Representation of $\frac{3}{4} \times \frac{5}{6} = \frac{15}{24}$.

If we overlap these two figures, we get Figure 5.13, which shows that if we take $\frac{5}{6}$ of the $\frac{3}{4}$, we get 15 of the 24 sections, or $\frac{15}{24}$, which equals $\frac{5}{8}$.

Another way of looking at Figure 5.13 is $\frac{5}{6} \cdot \frac{3}{4} = \frac{5 \cdot 3}{6 \cdot 4} = \frac{15}{24} = \frac{5}{8}$. This helps us to remember the multiplication rule for rational numbers.

Multiplying Rational Numbers

$$\frac{a}{b} \cdot \frac{c}{d} = \frac{a \cdot c}{b \cdot d}$$

EXAMPLE 4   Multiplying Rational Numbers

Find a) $\frac{4}{3} \cdot \frac{3}{15}$ b) $\frac{18}{25} \cdot \frac{10}{81}$ c) $\left(\frac{9}{4}\right) \cdot \left(-\frac{12}{5}\right)$

SOLUTION: a) Applying the multiplication rule, we get $\frac{4}{3} \cdot \frac{3}{15} = \frac{4 \cdot 3}{3 \cdot 15} = \frac{12}{45} = \frac{4}{15}$.

b) You can use the multiplication rule to get $\frac{18}{25} \cdot \frac{10}{81} = \frac{180}{2,025}$ and then reduce this number; however, this is a lot of unnecessary work.

   If you recognize that eventually you are going to cancel common factors, it is easier to do that first before multiplying the numerators and denominators. You can cancel 9 from the first numerator and second denominator, and then cancel 5 from the first denominator and the second numerator. (You can also do the cancellation in one step if you prefer.) Finally, multiply the simpler numbers $\frac{2}{5}$ and $\frac{2}{9}$ to get $\frac{4}{45}$, as we show in Figure 5.14.

$$\frac{\overset{2}{\cancel{18}}}{25} \cdot \frac{10}{\underset{9}{\cancel{81}}} = \frac{2}{\underset{5}{\cancel{25}}} \cdot \frac{\overset{2}{\cancel{10}}}{9} = \frac{2}{5} \cdot \frac{2}{9} = \frac{4}{45}$$

↑ cancel 9   ↑ cancel 5   ↑ now multiply

FIGURE 5.14   Canceling before multiplying saves work.

c) *The rules for multiplying signed rational numbers are the same as those for multiplying integers,* so the product of a positive number and a negative number is a negative number. We will first multiply without regard to signs, remembering that the final product is negative.

$$\frac{9}{\underset{1}{\cancel{4}}} \cdot \frac{\overset{3}{\cancel{12}}}{5} = \frac{9}{1} \cdot \frac{3}{5} = \frac{27}{5}$$

So the answer is $-\frac{27}{5}$.*

Quiz Yourself 12

Find

a) $\dfrac{24}{9} \cdot \dfrac{21}{20}$

b) $\left(-\dfrac{8}{15}\right) \cdot \left(-\dfrac{25}{12}\right)$

---

* Although you may recall that $-\frac{27}{5}, \frac{-27}{5}$, and $\frac{27}{-5}$ are all equal, we will tend to write the answer as $-\frac{27}{5}$.

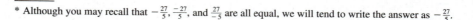

Division of rational numbers is based upon multiplication.

In order to understand the rule for dividing rational numbers, consider how we might think about dividing $\dfrac{\frac{3}{4}}{\frac{5}{8}}$. What makes this division complicated is that we have $\frac{5}{8}$ in the denominator. One way to get rid of the $\frac{5}{8}$ is to multiply it by its reciprocal, $\frac{8}{5}$. However, if we multiply the denominator by $\frac{8}{5}$, we must also multiply the numerator by $\frac{8}{5}$. Multiplying by $\frac{8}{5}$ gives us a one in the denominator and the product $\frac{3}{4} \cdot \frac{8}{5}$ in the numerator as we see below.

$$\frac{\frac{3}{4} \cdot \frac{8}{5}}{\frac{5}{8} \cdot \frac{8}{5}} = \frac{\frac{3}{4} \cdot \frac{8}{5}}{1} = \frac{3}{\cancel{4}_{1}} \cdot \frac{\cancel{8}^{2}}{5} = \frac{6}{5}$$

Instead of going through this lengthy process each time we divide, we use the following rule, which tells us in effect to "invert the denominator and multiply."

> **Dividing Rational Numbers**
>
> $$\frac{\frac{a}{b}}{\frac{c}{d}} = \frac{a}{b} \cdot \frac{d}{c}$$

EXAMPLE 5   Dividing Rational Numbers

Perform the following divisions: a) $\dfrac{\frac{25}{12}}{\frac{10}{3}}$ b) $\dfrac{-\frac{11}{6}}{\frac{7}{9}}$

SOLUTION: a) Applying the division rule, we invert the denominator and multiply.

$$\frac{\frac{25}{12}}{\frac{10}{3}} = \frac{\cancel{25}^{5}}{\cancel{12}_{4}} \cdot \frac{\cancel{3}^{1}}{\cancel{10}_{2}} = \frac{5}{4} \cdot \frac{1}{2} = \frac{5}{8}$$

b) *The rules for division of rational numbers are the same as the rules for division of integers,* so a negative number divided by a positive will result in a negative number. We will remember that fact and divide without regard to signs. Again, we invert and multiply to get $\dfrac{\frac{11}{6}}{\frac{7}{9}} = \frac{11}{\cancel{6}_{2}} \cdot \frac{\cancel{9}^{3}}{7} = \frac{11}{2} \cdot \frac{3}{7} = \frac{33}{14}$, so the answer is $-\frac{33}{14}$.

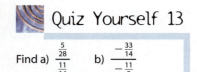

Quiz Yourself 13

Find a) $\dfrac{\frac{5}{28}}{\frac{11}{14}}$ b) $\dfrac{-\frac{33}{14}}{-\frac{11}{8}}$

Note that you have to be careful how you interpret a quotient such as $\dfrac{\frac{4}{8}}{16}$. In this case, the longer horizontal bar tells us that we are dividing $\frac{4}{8}$ by 16, so you can think of $\dfrac{\frac{4}{8}}{16}$ as $\dfrac{\frac{4}{8}}{\frac{16}{1}}$. You might find it enlightening to compute $\dfrac{\frac{4}{8}}{16}$ and $\dfrac{4}{\frac{8}{16}}$. You will see that you do not get the same result.

Example 6 requires us to use several operations on rational numbers at the same time.

EXAMPLE 6   Using the Scale Function in a Graphics Program

Suppose you are drawing a landscaping plan using a graphics package, such as Adobe Illustrator, and have drawn a rectangle measuring $\frac{8}{3}$ inches long by $\frac{5}{6}$ inches wide representing a rectangular Japanese rock garden.

a) If you scale down the length by a factor of $\frac{3}{4}$, and increase the width by a factor of $\frac{9}{8}$, what are the new dimensions of the sides of the garden in your drawing?

b) How does the area of the new rectangle compare with the area of the original rectangle in your drawing?

SOLUTION: a) We will multiply the length of the original rectangle by the scaling factor of $\frac{3}{4}$ to get a length of $\frac{8}{3} \cdot \frac{3}{4} = \frac{24}{12} = 2$ inches. We multiply the width of the original rectangle by the scaling factor $\frac{9}{8}$ to get a new width of $\frac{5}{6} \cdot \frac{9}{8} = \frac{45}{48} = \frac{15}{16}$ inches.

b) To answer this question, we will divide the area of the new rectangle by the area of the old rectangle. The area of the new rectangle is $2 \cdot \frac{15}{16} = \frac{30}{16} = \frac{15}{8}$ square inches. The area of the old rectangle is $\frac{8}{3} \cdot \frac{5}{6} = \frac{40}{18} = \frac{20}{9}$ square inches. Dividing the new area by the old area, we get

$$\frac{\dfrac{15}{8}}{\dfrac{20}{9}} = \frac{\overset{3}{\cancel{15}}}{8} \cdot \frac{9}{\underset{4}{\cancel{20}}} = \frac{27}{32}$$

So the area of the new garden in the drawing is $\frac{27}{32}$ the area of the old garden in the drawing.

Mixed numbers help us understand the size of rational numbers.

In working with rational numbers, you could get an answer such as $\frac{123}{8}$. A rational number such as this, in which the numerator is greater than the denominator, is called an **improper fraction.***

$$
\begin{array}{r}
15 \\
8\overline{)123} \\
120 \\
\hline
3
\end{array}
$$

15 ← Quotient tells us that 8 divides into 123 fifteen whole times

3 ← Remainder tells us that 3 eighths are left over.

FIGURE 5.15 Converting $\frac{123}{8}$ to $15\frac{3}{8}$.

Although the answer may be correct in its present form, we have to think a little to understand the size of this number. However, if we divide 123 by 8 to get the quotient 15 and remainder 3 (see Figure 5.15), then we see that $\frac{123}{8} = 15\frac{3}{8}$, which is not quite 15 and one half.

We can use this example to give a general rule for writing an improper fraction as a mixed number.

Converting an Improper Fraction to a Mixed Number

To convert the improper fraction $\frac{a}{b}$ to a mixed number, perform the division

$$
\begin{array}{r}
q \\
b\overline{)a} \\
\cdot \\
\cdot \\
\hline
r
\end{array}
$$

Then write $\frac{a}{b}$ as $q + \frac{r}{b} = q\frac{r}{b}$.

EXAMPLE 7   Converting an Improper Fraction to a Mixed Number

Convert the following improper fractions to mixed numbers.

a) $\frac{45}{6}$ b) $\frac{133}{8}$

SOLUTION: a) In a), if we divide 45 by 6, we get the quotient 7 and remainder 3. Thus $\frac{45}{6} = 7\frac{3}{6} = 7\frac{1}{2}$.

b) Similarly, in b), $\frac{133}{8} = 16\frac{5}{8}$.

We can also convert mixed numbers into improper fractions. For example, we think of $5\frac{3}{4}$ as $5 + \frac{3}{4} = \frac{5 \cdot 4}{4} + \frac{3}{4} = \frac{23}{4}$. This calculation is the basis for the following conversion rule.

Converting a Mixed Number to an Improper Fraction

The mixed number $q\frac{r}{b}$ equals the improper fraction $\frac{q \cdot b + r}{b}$.

---

* Rational numbers with a positive numerator and a positive denominator are often called **fractions.** In this discussion, we will only talk about fractions.

Historical Highlight: *Sophie Germain*

Sophie Germain was born in France in 1776, at a time when women were discouraged from studying mathematics. Because women were barred from enrolling at the Ecole Polytechnique in Paris, she secretly obtained lecture notes of various professors at this prestigious university. Students were allowed to make written comments on the lectures of their professors and, even though she was not a student, she submitted her commentaries using the pseudonym M. Leblanc.

Unaware that his "student" was a woman, Joseph Lagrange, one of the eminent mathematicians of the time, had high praise for the work of M. Leblanc.

When Lagrange discovered Leblanc's true identity, he extolled Germain as one of the promising young mathematicians of the time. In 1816, she was awarded a prize from the French Academy for her paper on the mathematics of elastic surfaces.

Don't let this formula intimidate you. All we are saying is you multiply q times b, and add r to get the numerator. Then use b for the denominator.

EXAMPLE 8   Converting a Mixed Number to an Improper Fraction

Convert the following mixed numbers to improper fractions.

a) $5\frac{2}{7}$ b) $8\frac{3}{11}$

SOLUTION: a) $5\frac{2}{7} = \frac{5 \cdot 7 + 2}{7} = \frac{37}{7}$ b) $8\frac{3}{11} = \frac{8 \cdot 11 + 3}{11} = \frac{91}{11}$.

In calculating with mixed numbers, you will often find it useful to first convert the numbers to improper fractions before doing the computation. For example, to compute $3\frac{1}{2} \times 2\frac{2}{5}$, rewrite it as $\frac{7}{2} \times \frac{12}{5} = \frac{42}{5} = 8\frac{2}{5}$.

> **Quiz Yourself 14**
>
> a) Convert $\frac{23}{4}$ to a mixed number.
>
> b) Convert $2\frac{3}{5}$ to an improper fraction.

Rational numbers have repeating decimal representations.

If you were to divide 3/16 on your calculator,* the answer would look something like 0.1875000. We will now discuss how to write rational numbers in decimal form and then how to write a decimal number as the quotient of two integers.

EXAMPLE 9   Writing a Rational Number in Decimal Form

Write each of the following rational numbers as a decimal number.

a) $\frac{5}{8}$ b) $\frac{7}{16}$ c) $\frac{82}{111}$

SOLUTION: a) If you divide the numerator by the denominator, you will get $\frac{5}{8} = 0.625$.

b) Dividing, we find that $\frac{7}{16} = 0.4375$.

———————

* Some calculators will also do computations in fraction form.

c) Although this seems like a strange choice for an example, as you will see, it is very interesting. If you divide $\frac{82}{111}$ on your calculator, you might get an answer that looks something like 0.7387387387. This is not quite correct!

If you do the division by hand, as in Figure 5.16, you will see that the digits in the quotient will repeat indefinitely. Once you have calculated a quotient of 0.738 and get a remainder of 82, the calculations that you just did will repeat all over again to give you the quotient 0.738738.

$$
\begin{array}{r}
.738 \\
111\overline{\smash{)}82.000000} \\
777 \\
\hline
430 \\
333 \\
\hline
970 \\
888 \\
\hline
820 \\
\end{array}
$$

We are again dividing 111 into 820, so the digits in the quotient repeat all over again.

FIGURE 5.16 $\frac{82}{111}$ has a repeating decimal expansion.

The true decimal expansion of $\frac{82}{111}$ is actually the infinite repeating decimal 0.738738738738 . . . . We usually write the part of a decimal that repeats with a bar over it, so we would write $\frac{82}{111}$ as $0.\overline{738}$.

When dividing a rational number to get a decimal expansion, we can have only a finite number of possible remainders, so at some point, the type of repetition that we saw in Example 9(c) must always occur.

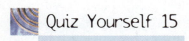

Quiz Yourself 15

Write $\frac{21}{33}$ as a decimal number.

> **Decimal Expansions of Rational Numbers**
>
> When writing a rational number in decimal form, we always get either a terminating expansion, as in a) and b) in Example 9, or an infinite repeating expansion, as in c). (We could think of a terminating expansion as a repeating expansion in which 0 repeats from some point on.)

We will now explain how to write decimal numbers as quotients of integers. If a number has a terminating expansion such as 0.124, we remember that we read this number as 124 thousandths (see Figure 5.17). So we can write 0.124 as $\frac{124}{1,000}$, and then reduce this number to $\frac{31}{250}$ if we wish. Similarly, we could write 0.2375 as $\frac{2,375}{10,000}$.

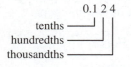

FIGURE 5.17 Reading 0.124.

To write a repeating decimal such as $0.\overline{36}$ as a quotient of integers is a little more complicated, as we explain in Example 10.

EXAMPLE 10   Writing a Repeating Decimal as a Quotient of Integers

Write $x = 0.\overline{36}$ as a quotient of integers.

SOLUTION: The problem in writing 0.36363636363636 . . . as a quotient of integers is that we have to deal with the infinite "tail" of 363636 . . . . The technique that we use is to create another number with the same tail that x has. We then subtract the two numbers to get a number without a repeating infinite tail.

Consider the number $100 \cdot x = 36.36363636 \ldots$. If we subtract $100 \cdot x - x$, we get

$$
\begin{aligned}
100 \cdot x &= 36.36363636\ldots \\
- \qquad x &= -0.36363636\ldots, \\
\hline
99 \cdot x &= 36
\end{aligned}
$$

so the infinite tails have disappeared. Solving $99 \cdot x = 36$, we find that $x = \frac{36}{99} = \frac{4}{11}$. Thus, $0.36363636363636\ldots = \frac{4}{11}$. ◎

If in Example 10 we were dealing with a number such as $x = 0.\overline{634}$, then we would subtract $1{,}000 \cdot x - x$ to get rid of the infinite tail.

A number with a nonrepeating decimal expansion is not a rational number, and we will discuss such numbers in the next section.

Quiz Yourself 16

a) Write 0.2548 as a quotient of integers.

b) Write $x = 0.\overline{54}$ as a decimal number.

Exercises 5.3

Write all answers in this exercise set in lowest terms.

Which of the following pairs of rational numbers are equal?

1. $\dfrac{2}{3}, \dfrac{8}{12}$

2. $\dfrac{5}{6}, \dfrac{10}{18}$

3. $\dfrac{12}{14}, \dfrac{14}{16}$

4. $\dfrac{11}{6}, \dfrac{18}{20}$

5. $\dfrac{22}{14}, \dfrac{30}{21}$

6. $\dfrac{5}{16}, \dfrac{10}{32}$

7. $\dfrac{5}{14}, \dfrac{15}{42}$

8. $\dfrac{9}{7}, \dfrac{54}{42}$

9. $\dfrac{32}{14}, \dfrac{18}{12}$

10. $\dfrac{9}{42}, \dfrac{3}{14}$

Reduce each fraction.

11. $\dfrac{15}{35}$

12. $-\dfrac{30}{48}$

13. $\dfrac{-24}{72}$

14. $\dfrac{60}{135}$

15. $\dfrac{-77}{126}$

16. $\dfrac{72}{108}$

17. $\dfrac{225}{350}$

18. $-\dfrac{132}{96}$

19. $\dfrac{143}{154}$

20. $\dfrac{-120}{216}$

Perform the following operations. Express your answer as a positive or negative quotient of two integers in reduced form.

21. $\dfrac{2}{3} + \dfrac{1}{2}$

22. $\dfrac{3}{4} + \dfrac{1}{6}$

23. $\dfrac{1}{6} - \dfrac{1}{2}$

24. $\dfrac{13}{16} - \dfrac{5}{8}$

25. $\dfrac{7}{16} - \dfrac{1}{3}$

26. $\dfrac{2}{9} - \dfrac{3}{8}$

27. $\dfrac{5}{24} + \dfrac{7}{18}$

28. $\dfrac{5}{12} + \dfrac{3}{14}$

29. $\dfrac{3}{4} + \dfrac{5}{6} + \dfrac{7}{8}$

30. $\dfrac{1}{3} + \dfrac{2}{5} + \dfrac{5}{6}$

31. $\dfrac{1}{8} - \dfrac{2}{3} + \dfrac{1}{2}$

32. $\dfrac{2}{9} - \dfrac{2}{27} + \dfrac{1}{4}$

Perform the following operations. Express your answer as a positive or negative quotient of two integers in reduced form.

33. $\dfrac{2}{3} \cdot \dfrac{1}{2}$

34. $\dfrac{5}{16} \cdot \dfrac{4}{15}$

35. $\dfrac{1}{6} \div \dfrac{1}{2}$

36. $\dfrac{5}{16} \div \dfrac{3}{8}$

37. $\dfrac{7}{8} \div \left(-\dfrac{5}{24}\right)$

38. $\dfrac{2}{33} \div \dfrac{6}{55}$

39. $\dfrac{7}{18} \cdot \left(-\dfrac{3}{4}\right)$

40. $-\dfrac{7}{32} \cdot \dfrac{8}{35}$

41. $\left(\dfrac{7}{18} \cdot \left(-\dfrac{3}{4}\right)\right) \div \left(\dfrac{7}{9}\right)$

42. $\left(\dfrac{14}{25} \div \dfrac{4}{5}\right) \cdot \left(\dfrac{10}{3}\right)$

43. $\left(\dfrac{11}{30} \div \left(-\dfrac{1}{6}\right)\right) \cdot \left(\dfrac{15}{4}\right)$

44. $\left(\left(\dfrac{7}{4} \cdot \dfrac{8}{21}\right) \div \left(\dfrac{2}{5}\right)\right)$

Perform the following operations. Express your answer as a positive or negative quotient of two integers in reduced form.

45. $\dfrac{5}{3} \cdot \left(\dfrac{8}{15} + \dfrac{2}{3} \right)$

46. $\dfrac{8}{9} \cdot \left(\dfrac{5}{6} + \dfrac{1}{4} \right)$

47. $\dfrac{22}{27} \cdot \left(\dfrac{3}{11} + \dfrac{2}{3} \right)$

48. $\dfrac{10}{9} \cdot \left(\dfrac{2}{5} - \dfrac{2}{3} \right)$

49. $\dfrac{7}{30} \div \left(\dfrac{1}{6} - \dfrac{3}{14} \right)$

50. $\dfrac{11}{40} \div \left(\dfrac{3}{5} - \dfrac{1}{4} \right)$

Each of the following divisions represents a conversion of an improper fraction to a mixed number. Write that conversion as an equation with the improper fraction on the left and the mixed number on the right.

51.
$$\begin{array}{r} 15 \\ 5\overline{)79} \\ 75 \\ \hline 4 \end{array}$$

52.
$$\begin{array}{r} 4 \\ 8\overline{)35} \\ 32 \\ \hline 3 \end{array}$$

53.
$$\begin{array}{r} 9 \\ 3\overline{)29} \\ 27 \\ \hline 2 \end{array}$$

54.
$$\begin{array}{r} 17 \\ 8\overline{)143} \\ 136 \\ \hline 7 \end{array}$$

55.
$$\begin{array}{r} 7 \\ 12\overline{)95} \\ 84 \\ \hline 11 \end{array}$$

56.
$$\begin{array}{r} 8 \\ 15\overline{)126} \\ 120 \\ \hline 6 \end{array}$$

Convert each improper fraction to a mixed number.

57. $\dfrac{27}{4}$

58. $\dfrac{139}{8}$

59. $\dfrac{121}{15}$

60. $\dfrac{214}{12}$

61. $\dfrac{1,036}{17}$

62. $\dfrac{2,087}{40}$

Convert each mixed number to an improper fraction.

63. $2\dfrac{3}{4}$

64. $5\dfrac{3}{8}$

65. $9\dfrac{1}{6}$

66. $4\dfrac{5}{12}$

67. $11\dfrac{2}{3}$

68. $15\dfrac{3}{5}$

Write each rational number as a terminating or repeating decimal

69. $\dfrac{3}{4}$

70. $\dfrac{5}{8}$

71. $\dfrac{3}{16}$

72. $\dfrac{27}{32}$

73. $\dfrac{16}{3}$

74. $\dfrac{1}{6}$

75. $\dfrac{9}{11}$

76. $\dfrac{16}{33}$

77. $\dfrac{4}{13}$

78. $\dfrac{4}{7}$

Write each decimal as a quotient of two integers in reduced form.

79. 0.64

80. 0.075

81. 0.836

82. 3.45

83. 2.345

84. 0.0083

85. 12.2

86. 4.068

87. $0.\overline{4}$

88. $0.3\overline{8}$

89. $0.\overline{189}$

90. $0.\overline{21}$

91. $0.3\overline{18}$

92. $0.4\overline{18}$

93. $0.\overline{384615}$

94. $0.2\overline{142857}$

95. Living expenses. Andre spends $\frac{1}{3}$ of his paycheck on rent, $\frac{1}{4}$ on food, and $\frac{1}{6}$ on utilities. What fractional part of his paycheck does he have left for other expenses?

96. Scholarships allocation. At Central State College, $\frac{1}{8}$ of the students have athletic scholarships and $\frac{1}{3}$ of the students have academic scholarships. (Assume that no students have both.) What fractional part of the student body have one or the other of these scholarships?

97. Dividing the purse for an auto race. At the Grand Speedway, the winner of a six-car race receives $\frac{1}{3}$ of the purse, the person who comes in second receives $\frac{1}{4}$ of the purse, and the remainder is split equally among the other drivers. What fractional part of the purse does each of the last four drivers receive?

98. Advertising budget. The Great Downhill Ski company spends $\frac{1}{3}$ of its advertising budget on the print media, $\frac{2}{5}$ of the budget on TV ads, and $\frac{1}{6}$ of the budget on the radio. What part of the budget remains for other types of advertising?

99. Tiling walls of a museum. The director of the New Central American Art Museum is redecorating the walls of a room with replicas of ancient Inca tiles. The tiles are 8 and $\frac{1}{2}$ inches square. If the room is 17 feet high, how many rows of tiles will be needed? Will any of the tiles have to be cut?

100. Dividing a bonus. At the law firm of Goode, Best, and Associates, the annual bonus is determined by the number of hours billed by each associate. If Matt has billed $\frac{1}{3}$ of the hours, Marcia has billed $\frac{1}{4}$ of the hours, and Joe has billed $\frac{1}{5}$ of the hours, what fractional part of the bonus is available for the remaining associates?

101. Modifying a recipe. In the Fanny Farmer Cookbook, the recipe for Hungarian Goulash, which serves four, calls for (among other ingredients) $1\frac{1}{2}$ tablespoons of lemon juice. If you wish to increase this recipe to serve ten, how much lemon juice should you use?

102. **Modifying a recipe.** In the Fanny Farmer Cookbook, the recipe for one cup of horseradish cream sauce (to accompany roast beef) calls for $\frac{3}{4}$ cup of heavy cream. If you wish to make $1\frac{1}{2}$ cups of this sauce, how many cups of heavy cream should you use?

103. **Measuring stock changes.** Until recently, stocks used to be quoted using fractions rather than decimals. Suppose that an airline stock opened at $37\frac{1}{8}$ and closed at $35\frac{3}{4}$. How much did the stock drop?

104. **Buying art supplies.** Janelle is an artist who purchases paint solvent in $2\frac{3}{8}$-liter cans. If she buys the larger can, which is $1\frac{1}{3}$ times as large as the smaller can, how many liters will it contain?

105. **Unit pricing.*** Milan can buy a $5\frac{1}{4}$-ounce tube of toothpaste for $2.60 or a $7\frac{1}{8}$-ounce tube of the same toothpaste for $3.60. Which size is the better buy? Explain. (Hint: Convert the mixed numbers to decimals and then calculate the price per ounce for each size.)

106. **Unit pricing.** Janita can buy a $37\frac{1}{2}$-ounce jar of orange juice for $1.75 or a $44\frac{3}{4}$-ounce jar of the same juice for $2.70. Which size is the better buy? Explain. (Hint: Convert the mixed numbers to decimals and then calculate the price per ounce for each size.)

107. **Minimizing waste on a decorating project.** Denny is placing $3\frac{3}{8}$-foot strips of gold foil on the inside of a dome of a cathedral and wishes to minimize the waste caused by having left-over pieces that are too short. How many $3\frac{3}{8}$ strips can he cut from a 12-foot strip? How much waste will be left over from each 12-foot strip? If Denny buys the strips in 15-foot lengths, will there be more or less waste per 15-foot strip than for a 12-foot strip?

108. **Estimating a painting job.** Kirk is estimating the cost of a painting job. A room measures $10\frac{1}{4}$ feet on two walls and $13\frac{1}{2}$ feet on the other two walls. Also the room is $8\frac{3}{4}$ feet high. What is the total area of the four walls of the room?

Further Exercises

109. **Making a bookshelf.** Antonio is making a student bookshelf from boards and cinder blocks. He has a piece of weathered pine that is $10\frac{7}{8}$ feet long and wishes to cut it into 4 pieces of equal length. If we ignore the width of the saw's cut, how long will each shelf be? Express your answer in terms of feet.

110. **Making a bookshelf.** Redo Exercise 109 if we take into account that each saw cut will be $\frac{1}{8}$ of an inch wide.

111. Suppose that you wish to hang a mirror that is $40\frac{1}{2}$ inches wide on a wall that is 6 feet (72 inches) wide. You wish to center the mirror and put two hooks into the wall to support the mirror at $\frac{1}{3}$ and $\frac{2}{3}$ distance from one side of the mirror to the other side (see Figure 5.18). How far should the hooks be placed from the edges of the wall?

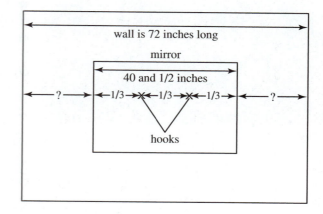

FIGURE 5.18 Hanging a mirror.

---

* You may find it interesting to go to a market and find examples of products where the smaller size costs less per unit than does the larger "economy" size.

112. Publishing a textbook. A publisher wishes to increase the content in a new edition of a textbook without increasing the number of pages. The current printed area on each page is $5\frac{1}{4}$ inches wide and $8\frac{1}{4}$ inches long. If we increase the width of the printed area to $6\frac{1}{8}$ inches, and increase the length to $8\frac{1}{2}$ inches, by how many square inches will the printed area increase per page?

113. Communicating Mathematics We know in general that we cannot cancel terms from the numerator and denominator of a quotient. However, even though it is dangerous to do so, in certain situations, by coincidence, the calculation is correct. If you cancel 5 from the numerator and denominator of $\frac{x+5}{y+5}$ and get the true equation $\frac{x+5}{y+5} = \frac{x}{y}$, what can you conclude about x and y? Think carefully about how we decide when two quotients are equal.

114. Communicating Mathematics Explain why the decimal expansion of a rational number has to repeat.

Communicating Mathematics *Draw diagrams as we did to illustrate the product $\frac{5}{6} \cdot \frac{3}{4} = \frac{15}{24}$ in the discussion prior to Example 4 for each of the following products. Explain how these diagrams illustrate each product.*

115. $\dfrac{1}{2} \cdot \dfrac{4}{5}$ **116.** $\dfrac{1}{4} \cdot \dfrac{2}{3}$

117. $\dfrac{2}{3} \cdot \dfrac{2}{5}$ **118.** $\dfrac{3}{8} \cdot \dfrac{2}{3}$

Perform each of the following divisions using the technique that we used to calculate $\dfrac{\frac{3}{4}}{\frac{5}{8}}$ as we did in the discussion prior to Example 6.

119. $\dfrac{\frac{3}{2}}{\frac{1}{4}}$ **120.** $\dfrac{\frac{5}{6}}{\frac{2}{3}}$ **121.** $\dfrac{\frac{5}{12}}{\frac{3}{8}}$ **122.** $\dfrac{\frac{3}{10}}{\frac{2}{5}}$

5.4 THE REAL NUMBER SYSTEM

About 500 B.C., a school of ancient Greek mathematicians, led by Pythagoras of Samos (see Historical Highlight), discovered an amazing number, the square root of two, which we write as $\sqrt{2}$. Their discovery was based on the Pythagorean theorem that we illustrate in Figure 5.19(a). This theorem states that for a right triangle, the square of the length of the hypotenuse (the longest side) is equal to the sum of the squares of the lengths of the other two sides. Or, $c^2 = a^2 + b^2$. If the right triangle has two sides of length one, as in Figure 5.19(b), then $c^2 = 1^2 + 1^2 = 2$, or $c = \sqrt{2}$.

What is amazing about this discovery is that we cannot write $\sqrt{2}$ as the quotient of two integers, and therefore

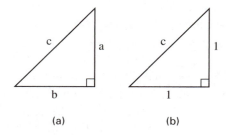

(a)   (b)

FIGURE 5.19 (a) $c^2 = a^2 + b^2$, (b) $c^2 = 1^2 + 1^2 = 2$.

it is not a rational number. Prior to that time, the Greeks believed that all numbers were rational. We will give Pythagoras' proof that $\sqrt{2}$ is not rational in Exercise 87.

■

Irrational numbers have nonrepeating decimal expansions.

You saw in Section 5.3 that the rational numbers are precisely those numbers that have repeating* decimal expansions. This means that any number that is not rational, such as $\sqrt{2}$, must have a nonrepeating decimal expansion.

> **DEFINITION**
>
> An **irrational number** is a number that is not a rational number and therefore has a nonrepeating decimal expansion.

A number such as 5.12112111211112 . . . is an example of an irrational number. Although there is a pattern to the digits in this expansion, there is no single block of numbers that repeats from some point on. Other examples of irrational numbers would be 15.1234567891011121314 . . . and 0.1020030004000050000 . . . .

If you use your calculator to find $\sqrt{2}$, you may get an answer like $\sqrt{2} = 1.414213562$. Because this is a terminating (and hence repeating) decimal, what your calculator is telling you is not quite true. In reality, your calculator gave you a very close rational number approximation which is a little less than $\sqrt{2}$. *In fact, any number of the form $\sqrt{n}$, where n is not a perfect square, will be an irrational number.*

> **DEFINITIONS**
>
> A number, such as $\sqrt{x}$, is called a **radical**. The symbol $\sqrt{}$ is called a **radical sign** and x is called the **radicand**.

Some might argue that in this day of powerful calculators and computers that there is really no reason to work with a number like $\sqrt{2}$ when the approximation 1.414213562 would do just as well. Instead of doing calculations with radicals, we could convert all radicals to decimals and use a calculator to do decimal computations. However, to others there is something disturbing about working with numbers that are "almost the same" as $\sqrt{2}$, $\sqrt{3}$, $\sqrt{6}$ and so forth. Because scientific and mathematical formulas often contain radicals, you should understand the following rules that are essential to working with radicals.

---

* Recall that we consider terminating decimals to be repeating decimals with the repeated digit being zero.

Multiplying and Dividing Radicals

If $a \geq 0$ and $b \geq 0$, then

$$\sqrt{ab} = \sqrt{a}\sqrt{b}, \text{ and if } b \neq 0, \text{ then } \sqrt{\frac{a}{b}} = \frac{\sqrt{a}}{\sqrt{b}}.$$

EXAMPLE 1   Multiplying and Dividing Radicals

Use the rules for multiplying and dividing radicals to rewrite each of the following. Use a calculator to verify your answers. (Remember that your calculator is only giving you approximations instead of exact results.)

a) $\sqrt{6 \cdot 5}$ b) $\sqrt{8}\sqrt{3}$ c) $\sqrt{\dfrac{36}{12}}$ d) $\dfrac{\sqrt{7}}{\sqrt{15}}$

SOLUTION: a) $\sqrt{6 \cdot 5} = \sqrt{6}\sqrt{5}$. Using a calculator, $\sqrt{6 \cdot 5} = \sqrt{30} = 5.477225575$, and $\sqrt{6}\sqrt{5} = 5.477225575$. Note when calculating $\sqrt{6}\sqrt{5}$, calculate this product* by first calculating $\sqrt{6}$; do not clear the calculator. Then press the multiplication key. Next calculate $\sqrt{5}$ and then press "Enter" ("=" key).

b) $\sqrt{8}\sqrt{3} = \sqrt{8 \cdot 3} = \sqrt{24}$. Both sides of the equation have a decimal approximation of 4.898979486.

c) $\sqrt{\dfrac{36}{12}} = \dfrac{\sqrt{36}}{\sqrt{12}} = \dfrac{6}{\sqrt{12}}$. Both sides of the equation have an approximate decimal value of 1.732050808.

d) $\dfrac{\sqrt{7}}{\sqrt{15}} = \sqrt{\dfrac{7}{15}}$. Calculating both sides, we get 0.6831300511.

We can use the product and quotient rules for radicals to simplify radicals. In simplifying a radical, we wish to eliminate all perfect squares (numbers such as $2 \cdot 2 = 4$, $3 \cdot 3 = 9$, $5 \cdot 5 = 25$, and so on) under the radical. There are usually several different ways to do calculations that will give you the same simplification.

EXAMPLE 2   Simplifying Radicals

Simplify each of the following.

a) $\sqrt{18}$ b) $\sqrt{15}\sqrt{5}$ c) $\sqrt{\dfrac{54}{12}}$ d) $\dfrac{\sqrt{20}}{\sqrt{15}}$

SOLUTION: a) We recognize that 9 is a factor of 18 and 9 is a perfect square. So, $\sqrt{18} = \sqrt{2 \cdot 9} = \sqrt{2}\sqrt{9} = \sqrt{2} \cdot 3 = 3\sqrt{2}$.†

b) We notice that 5 is a factor of 15 and 5 also appears in the second radical sign. So if we combine the two radicals into one radical, then we can simplify the 5 squared in the single radical. Thus,

$$\sqrt{15}\sqrt{5} = \sqrt{15 \cdot 5} = \sqrt{3 \cdot 5 \cdot 5} = \sqrt{3 \cdot 25} = \sqrt{3}\sqrt{25} = \sqrt{3} \cdot 5 = 5 \cdot \sqrt{3}.$$

---

* Practice duplicating our calculations until you are confident in working with your calculator.
† Although we will not do so, you can use a calculator to verify that $\sqrt{18} = 3\sqrt{2}$.

c) There are several ways to do this. We will begin by using the quotient rule for radicals. Thus

$$\sqrt{\frac{54}{12}} = \frac{\sqrt{54}}{\sqrt{12}} = \frac{\sqrt{6\cdot9}}{\sqrt{3\cdot4}} = \frac{\sqrt{6}\cdot3}{\sqrt{3}\cdot2} = \sqrt{\frac{6}{3}}\cdot\frac{3}{2} = \sqrt{2}\cdot\frac{3}{2} = \frac{3}{2}\cdot\sqrt{2}.$$

d) Here we see that if we combine the two radicals into a single radical, then we will can cancel the 5 from the numerator and denominator. So

$$\frac{\sqrt{20}}{\sqrt{15}} = \sqrt{\frac{20}{15}} = \sqrt{\frac{4}{3}} = \frac{\sqrt{4}}{\sqrt{3}} = \frac{2}{\sqrt{3}}.$$

Prior to the invention of calculators, people had to use pencil and paper computations to make rational number estimates of numbers such as $\frac{2}{\sqrt{3}}$. It was easier to do such calculations if quotients did not have radicals in the denominator. Shortly we will discuss how to rewrite such quotients such as $\frac{2}{\sqrt{3}}$ in this preferred form.

 Quiz Yourself 17*

Simplify the following:

a) $\sqrt{45}$ b) $\sqrt{\dfrac{15}{48}}$

 SOME GOOD ADVICE

In doing a calculation, you will find it to your advantage to try to work with smaller numbers whenever possible. For example, if you write $\sqrt{72}\sqrt{48}$ as $\sqrt{72\cdot48} = \sqrt{3,456}$, then you have the problem of finding the factors in 3,456.

On the other hand, if you recognize that $\sqrt{72} = \sqrt{2\cdot36} = 6\sqrt{2}$ and $\sqrt{48} = \sqrt{3\cdot16} = 4\sqrt{3}$, then you can do this calculation quickly as $\sqrt{72}\sqrt{48} = 6\sqrt{2}\cdot4\sqrt{3} = 24\sqrt{6}$.

Radicals often occur in engineering applications involving stresses on structures.

EXAMPLE 3   Calculating the Strength of a Cable

Engineers often use the Pythagorean theorem to compute forces on objects. For example, suppose we wish to hang a sign on the wall of a building as in Figure 5.20. If we know that the downward force on the cable is 60 pounds and the horizontal force is 40 pounds, what is the minimal strength of a cable from point A to point B that is needed to support the sign?

SOLUTION: We can visualize the forces as in Figure 5.21 and then apply the Pythagorean theorem which gives us the equation $x^2 = 60^2 + 40^2$. We rewrite this equation as $x^2 = 3,600 + 1,600 = 5,200$. Therefore, $x = \sqrt{5,200} = \sqrt{100\cdot52} = \sqrt{100}\sqrt{52} = 10\sqrt{52} \approx 72.11$. So the cable must be able to support a minimum of 72.11 pounds.†

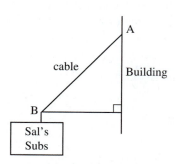

FIGURE 5.20 Hanging a sign.

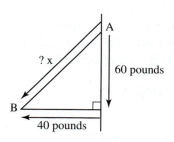

FIGURE 5.21 $x^2 = 60^2 + 40^2$.

---

* Quiz Yourself answers begin on page 849.
† If you were doing this calculation in a real-life situation, you need to round up to 72.2. We will not concern ourselves with this in the exercises.

Rationalizing the denominator makes a quotient more understandable.

Sometimes when we calculate with radicals, we get an answer such as $\frac{2}{\sqrt{3}}$. We have no difficulty evaluating this expression with modern calculators; however, without using a calculator, it is a little difficult to estimate the size of this number. If we write the number in another way, without a radical in the denominator, we can get a better intuitive understanding of the number. The process of writing a quotient so that it has no radicals in the denominator is called **rationalizing the denominator.** We illustrate this technique in Example 4.

EXAMPLE 4   Rationalizing Denominators

Rationalize the denominator in the following expressions.

a) $\dfrac{2}{\sqrt{3}}$ b) $\dfrac{7}{\sqrt{12}}$

SOLUTION: a) We want to write this quotient so that it has a perfect square in the radical in the denominator. So we multiply both the denominator and numerator by $\sqrt{3}$ to get

$$\frac{2}{\sqrt{3}} = \frac{2 \cdot \sqrt{3}}{\sqrt{3} \cdot \sqrt{3}} = \frac{2 \cdot \sqrt{3}}{\sqrt{3 \cdot 3}} = \frac{2 \cdot \sqrt{3}}{\sqrt{9}} = \frac{2 \cdot \sqrt{3}}{3} = \frac{2}{3}\sqrt{3}.$$

b) If we factor 12, we find that $12 = 2 \cdot 2 \cdot 3$. The product $2 \cdot 2$ is already a perfect square, so we need another factor of 3 to have the radical contain a perfect square. Thus we will multiply the numerator and denominator by $\sqrt{3}$. So we have

$$\frac{7}{\sqrt{12}} = \frac{7 \cdot \sqrt{3}}{\sqrt{12} \cdot \sqrt{3}} = \frac{7 \cdot \sqrt{3}}{\sqrt{36}} = \frac{7 \cdot \sqrt{3}}{6} = \frac{7}{6}\sqrt{3}.$$

Quiz Yourself 18

Rationalize the denominator in $\frac{3}{\sqrt{15}}$.

We add and subtract radicals with the same radicand.

It is a little more complicated to add and subtract radicals. If we add $5\sqrt{3} + 4\sqrt{3}$, we can think "five of something plus four of something equals nine of something." Thus, $5\sqrt{3} + 4\sqrt{3} = 9\sqrt{3}$. However, if we want to add $5\sqrt{2} + 4\sqrt{7}$, we are, so to speak, adding apples and oranges. We cannot add five of one thing to four of another thing. In order to add (or subtract) expressions with radicals, all the radicals must have the same radicand. We illustrate this in Example 5.

EXAMPLE 5   Adding Expressions Containing Radicals

Perform the operations.

a) $3\sqrt{2} + 6\sqrt{2}$ b) $7\sqrt{5} - 9\sqrt{5}$ c) $5\sqrt{3} + 8\sqrt{12}$
d) $17\sqrt{5} - 2\sqrt{45}$ e) $\sqrt{45} + \sqrt{12}$

Historical Highlight: *The Pythagoreans**

Shortly before 500 B.C., Pythagoras founded a school in the Greek settlement of Crotona, Italy, that was to have a profound influence on mathematics for the next 2,500 years. Prior to that time, Pythagoras had spent his adult life wandering around the ancient world, studying in places such as Phoenicia, Egypt, and Babylonia.

The school, which had about 300 students (including about 30 women), was somewhat like a secret society or fraternity. Members shared all goods in common and the school strictly regulated their diet and way of life. All students studied the same subjects, or mathemata—number theory, music, geometry, and astronomy. In addition, they studied logic, grammar, and rhetoric.

Beginners, called acoustici, listened silently as Pythagoras lectured to them from behind a curtain. After three years of obedient study, they could become one of the full members of the society, who were called the mathematici. At this time, the secrets of the society were revealed to them. Many of these "secrets" are the mathematical theorems that we still study today.

Like other secret societies, the Pythagoreans had strange rituals and beliefs. They would not eat beans, nor drink wine. They believed that one's soul could leave the body and live in another person or animal. Accordingly, they would not eat meat or fish fearing that by doing so, they might consume the residence of another's soul.

Around 500 B.C., there was a political revolt in Crotona. The school was burned and Pythagoras was killed. There are conflicting reports of his death. One romantic account tells that as he fled for his life, Pythagoras stopped at the edge of a field of beans, and allowed his enemies to kill him rather than trample the sacred plants.

SOLUTION: a) This addition is straightforward because both radicands are the same. So, $3\sqrt{2} + 6\sqrt{2} = 9\sqrt{2}$.

b) Similarly, because the radicands are the same, the subtraction is quite simple. Thus we get, $7\sqrt{5} - 9\sqrt{5} = -2\sqrt{5}$.

c) Here the radicands are different. However, we see that 12 has a factor of 4, so we can simplify $\sqrt{12}$. Therefore,

$$5\sqrt{3} + 8\sqrt{12} = 5\sqrt{3} + 8\sqrt{4 \cdot 3} = 5\sqrt{3} + 8\sqrt{4}\sqrt{3}$$
$$= 5\sqrt{3} + 8 \cdot 2\sqrt{3} = 5\sqrt{3} + 16\sqrt{3} = 21\sqrt{3}.$$

d) Again the radicals are different, but we see that 45 is 9 times 5, so we can simplify $\sqrt{45}$. We will use the product rule for radicals. So,

$$17\sqrt{5} - 2\sqrt{9 \cdot 5} = 17\sqrt{5} - 2\sqrt{9}\sqrt{5}$$
$$= 17\sqrt{5} - 2 \cdot 3\sqrt{5} = 17\sqrt{5} - 6\sqrt{5} = 11\sqrt{5}.$$

e) We rewrite $\sqrt{45} + \sqrt{12}$ as $3\sqrt{5} + 2\sqrt{3}$ and see that because the radicands are different, we cannot simplify this expression any further.   ◎

Quiz Yourself 19

Find:
a) $3\sqrt{5} + 8\sqrt{5}$
b) $3\sqrt{20} - 2\sqrt{45}$

---

* This highlight is based on material from David Burton, *The History of Mathematics* (WCB McGraw-Hill, 1999), pp. 86–90.

Number systems share many common properties.

Figure 5.22 summarizes the relationships between the number systems that you have studied in this chapter.

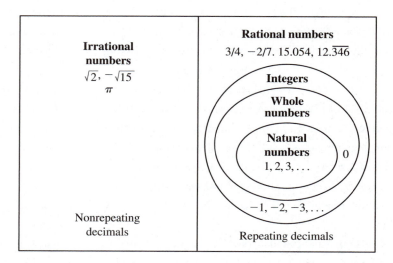

Figure 5.22   The set of real numbers.

We can think of the set of real numbers, and its subsets, as coherent *mathematical systems* which satisfy certain properties. The first property that we will consider is closure.

DEFINITION

A set is **closed** under an operation, if whenever we use the operation to combine any two elements of the set, we get an element in the set.

EXAMPLE 6   The Closure Property

Which of the following sets are closed with respect to a given operation?

a) The rational numbers under addition.

b) The natural numbers under subtraction.

c) The irrational numbers under multiplication.

SOLUTION: a) The sum of two rational numbers is a rational number, so the rational numbers are closed under addition.

b) If we compute the difference, $5 - 8 = -3$, we do not get a natural number. Thus, the natural numbers are not closed under subtraction.

c) Be careful! The product $\sqrt{2} \cdot \sqrt{2} = 2$, so the product of two irrational numbers need not be irrational. Therefore the set of irrational numbers is not closed under multiplication.   ◎

PROBLEM SOLVING

In checking whether a property such as closure holds for an operation, remember the Always Principle from Section 1.1. If we say that the property holds, then it must hold for every possibility *without even one exception.* For example, the set of rational numbers is not closed under division because we cannot divide by the rational number 0.

The next property we will discuss is commutativity. An operation is commutative if the order in which we combine elements does not matter.

DEFINITIONS

Addition and multiplication are **commutative.** That is, for any real numbers a and b,

$$a + b = b + a \qquad \text{and} \qquad a \cdot b = b \cdot a$$

EXAMPLE 7   Investigating Commutativity

Which of the operations are commutative on the given set? If the operation is commutative, give an example illustrating commutativity. If the operation is not commutative, provide a counterexample.

a) Addition on the set of integers.

b) Subtraction on the set of integers.

c) Multiplication on the set of rational numbers.

d) Division on the set of rational numbers.

SOLUTION: a) The order in which we add two integers does not matter, so addition is commutative on the set of integers. For example, $3 + 5 = 5 + 3$.

b) Subtraction is not commutative on the set of integers. For example $8 - 2 = 6$, but $2 - 8 = -6$.

c) Multiplication is commutative on the set of rational numbers. As an example, we see that $\frac{2}{3} \cdot \frac{3}{5} = \frac{3}{5} \cdot \frac{2}{3}$.

d) Division is not commutative on the set of rational numbers. Consider the following counterexample, $\frac{3}{4} \div \frac{9}{8} = \frac{3}{4} \cdot \frac{8}{9} = \frac{24}{36} = \frac{2}{3}$. However, $\frac{9}{8} \div \frac{3}{4} = \frac{9}{8} \cdot \frac{4}{3} = \frac{36}{24} = \frac{3}{2}$.   ◎

Associativity is another important property that many operations have. If an operation is associative, then it doesn't matter how we regroup the numbers when performing the operation.

DEFINITION

Addition and multiplication are **associative** on the set of real numbers. That is, for any real numbers $a, b,$ and $c,$

$$(a + b) + c = a + (b + c) \qquad \text{and} \qquad (a \cdot b) \cdot c = a \cdot (b \cdot c)$$

EXAMPLE 8 Investigating Associativity

Which of the operations are associative on the given set? If the operation is associative, give an example illustrating associativity. If the operation is not associative, provide a counterexample.

a) Addition on the set of integers.

b) Subtraction on the set of integers.

c) Multiplication on the set of rational numbers.

d) Division on the set of integers.

SOLUTION: a) Addition is associative on the set of integers. For example, $(2 + 3) + 5 = 2 + (3 + 5)$.

b) Subtraction is not associative on the set of integers, because $(8 - 3) - 2 = 5 - 2 = 3$, whereas $8 - (3 - 2) = 8 - 1 = 7$.

c) Multiplication is associative on the set of rational numbers. For example, if you simplify both sides of this equation, you will see that $(\frac{2}{3} \cdot \frac{5}{7}) \cdot \frac{9}{8} = \frac{2}{3} \cdot (\frac{5}{7} \cdot \frac{9}{8})$.

d) Division is not associative on the set of integers, because $\dfrac{\frac{16}{8}}{2} = \frac{2}{2} = 1$.

However, $\dfrac{16}{\frac{8}{2}} = \frac{16}{4} = 4$.

 SOME GOOD ADVICE

It is easy to remember the difference between the commutative and associative properties if you make analogies* between mathematical terminology and ordinary English words. "Commutative" reminds us of the word "commuter" who is a person who goes back and forth between two places. With the commutative property, we have numbers change places in an expression. Similarly, the word "associate" in English means to associate or group things together.

The next two properties are related to each other, and are found in many common number systems. An identity element for an operation preserves the identity of other numbers when combined with those numbers using the operation. When we use an operation to combine a number and its inverse, we get the identity element for that operation.

---

* Recall the Analogies Principle in Section 1.1.

DEFINITIONS

Zero is an **identity element for addition,** and 1 is an **identity element for multiplication.** That is, for any real number x, we have

$$x + 0 = 0 + x = x \quad \text{and} \quad x \cdot 1 = 1 \cdot x = x.$$

Also, if x is a real number, then $-x$ is the **additive inverse** for x and, if $x \neq 0$, then $\frac{1}{x}$ is the **multiplicative inverse** for x. That is,

$$x + (-x) = (-x) + x = 0 \quad \text{and} \quad x \cdot \frac{1}{x} = \frac{1}{x} \cdot x = 1.$$

EXAMPLE 9   **Finding Inverses in a Mathematical System**

Find the additive and multiplicative inverses for each of the following numbers.

a) 5   b) -8 c) 0   d) $\frac{3}{4}$ e) $\sqrt{2}$ f) π

SOLUTION: a) The additive inverse for 5 is a number that when added to 5 gives us 0. Clearly, -5 is the additive inverse for 5. The multiplicative inverse for 5 is a number that when multiplied by 5 gives us 1. Because $\frac{1}{5} \cdot 5 = 1$, we see that $\frac{1}{5}$ is the multiplicative inverse for 5.

b) Since $8 + (-8) = 0$, we have that 8 is the additive inverse for -8. The reciprocal of -8, namely $\frac{1}{-8} = -\frac{1}{8}$ is the multiplicative inverse for -8 because $-8 \cdot (-\frac{1}{8}) = 1$.

c) Zero is the additive inverse for zero because $0 + 0 = 0$. Zero has no multiplicative inverse. We cannot find a number, call it a, such that $a \cdot 0 = 1$, because any number times 0 will be 0.

d) Because $\frac{3}{4} + (-\frac{3}{4}) = 0$, $-\frac{3}{4}$ is the additive inverse for $\frac{3}{4}$. Also, since $\frac{3}{4} \cdot \frac{4}{3} = 1$, we have that $\frac{4}{3}$ is the multiplicative inverse for $\frac{3}{4}$.

e) The number $-\sqrt{2}$ is the additive inverse for $\sqrt{2}$ and $\frac{1}{\sqrt{2}}$ is the multiplicative inverse for $\sqrt{2}$.

f) Clearly $-\pi$ is the additive inverse for π and $\frac{1}{\pi}$ is the multiplicative inverse for π.

   The last property that we study is distributivity, which relates addition and multiplication.

DEFINITION

Multiplication is **distributive** over addition. This means, for any real numbers a, b, and c,

$$a \cdot (b + c) = a \cdot b + a \cdot c.$$

EXAMPLE 10 Using Distributivity

Use distributivity to rewrite each of the following expressions. Then evaluate the expression.

a) $2 \cdot (3 + 4)$ b) $5 \cdot 3 + 5 \cdot 4$ c) $2 \cdot (3 - 5)$

SOLUTION: a) $2 \cdot (3 + 4) = 2 \cdot 3 + 2 \cdot 4 = 6 + 8 = 14$.

b) $5 \cdot 3 + 5 \cdot 4 = 5 \cdot (3 + 4) = 5 \cdot 7 = 35$.

c) $2 \cdot (3 - 5) = 2 \cdot (3 + (-5)) = 2 \cdot 3 + 2 \cdot (-5) = 6 - 10 = -4$.

Exercises 5.4

Which of the following numbers are rational and which are irrational?

1. $\dfrac{3}{8}$ **2.** 5.0136

3. 1.23456789101112 . . .   **4.** $\sqrt{10}$

5. 3.1416   **6.** 0.10110111

7. 0.101101110 . . .   **8.** $\sqrt{81}$

9. Give an example to show that $\sqrt{n}$ can be a rational number.

10. Give an example to show that $\sqrt{n}$ can be an irrational number.

In Exercises 11–20, simplify the radical.

11. $\sqrt{18}$ **12.** $\sqrt{27}$ **13.** $\sqrt{12}$ **14.** $\sqrt{20}$

15. $\sqrt{75}$ **16.** $\sqrt{50}$ **17.** $\sqrt{48}$ **18.** $\sqrt{24}$

19. $\sqrt{189}$ **20.** $\sqrt{240}$

In Exercises 21–30, if possible, combine the radicals into a single radical.

21. $3\sqrt{5} + 8\sqrt{5}$ **22.** $2\sqrt{3} - 4\sqrt{2}$

23. $5\sqrt{7} - 3\sqrt{5}$ **24.** $3\sqrt{7} + 2\sqrt{7}$

25. $\sqrt{20} + 6\sqrt{5}$ **26.** $5\sqrt{12} - 4\sqrt{3}$

27. $5\sqrt{12} + 13\sqrt{18}$ **28.** $4\sqrt{27} - 2\sqrt{3}$

29. $\sqrt{50} + 2\sqrt{75}$ **30.** $\sqrt{28} + 2\sqrt{63}$

Perform the indicated operation and simplify if possible.

31. $\sqrt{18}\sqrt{2}$ **32.** $\sqrt{15}\sqrt{5}$ **33.** $\sqrt{12}\sqrt{15}$

34. $\sqrt{72}\sqrt{10}$ **35.** $\sqrt{28}\sqrt{21}$ **36.** $\sqrt{27}\sqrt{33}$

37. $\dfrac{\sqrt{24}}{\sqrt{6}}$ **38.** $\dfrac{\sqrt{54}}{\sqrt{6}}$ **39.** $\dfrac{\sqrt{32}}{\sqrt{18}}$

40. $\dfrac{\sqrt{45}}{\sqrt{20}}$ **41.** $\dfrac{\sqrt{96}}{\sqrt{72}}$ **42.** $\dfrac{\sqrt{63}}{\sqrt{112}}$

In Exercises 43–52, rationalize the denominator and simplify.

43. $\dfrac{3}{\sqrt{5}}$ **44.** $\dfrac{4}{\sqrt{3}}$ **45.** $\dfrac{12}{\sqrt{6}}$

46. $\dfrac{21}{\sqrt{14}}$ **47.** $\dfrac{10}{\sqrt{22}}$ **48.** $\dfrac{8}{\sqrt{10}}$

49. $\dfrac{4\sqrt{6}}{\sqrt{18}}$ **50.** $\dfrac{3\sqrt{11}}{\sqrt{33}}$ **51.** $\dfrac{3\sqrt{10}}{\sqrt{72}}$

52. $\dfrac{3\sqrt{5}}{\sqrt{135}}$

Communicating Mathematics *Decide whether the following statements are true or false. Explain your answers.*

53. $\sqrt{2} = 1.414213562$.

54. Every irrational number has a nonterminating decimal expansion.

55. If n is a natural number, then $\sqrt{n}$ is irrational.

56. A real number cannot be both rational and irrational.

57. $\sqrt{3}\sqrt{5}$ is irrational.

58. $\dfrac{\sqrt{72}}{\sqrt{8}}$ is irrational.

59. The sum of a rational number and an irrational number is an irrational number.

60. The product of two irrational numbers is always irrational.

Exercises 61–64 are based on Example 3. You are given the downward and horizontal force and must find the minimum strength of the cable to support the sign.

61. downward force: 12 pounds; horizontal force: 5 pounds

62. downward force: 30 pounds; horizontal force: 40 pounds

63. downward force: 120 pounds; horizontal force: 80 pounds

64. downward force: 8 pounds; horizontal force: 8 pounds

In Exercises 65–72, (a) find a rational number between the two given numbers; (b) find an irrational number between the two given numbers. Give your answers in decimal form. There will be many correct answers for these exercises. Hint: Write all numbers in decimal form before answering the questions.

65. 0.43 and 0.44

66. 1.245 and 1.246

67. $0.\overline{4578}$ and $0.45\overline{78}$

68. $0.\overline{123}$ and $0.123\overline{1}$

69. $\frac{4}{7}$ and $\frac{5}{7}$

70. $\frac{5}{8}$ and $\frac{3}{4}$

71. 0.12112111211112 . . . and $0.\overline{12}$

72. $0.\overline{123}$ and 0.1234567891011121314 . . .

In Exercises 73–76, put the numbers in order from smallest to largest.

73. a) 0.345345   b) $0.\overline{345}$
c) $0.3\overline{45}$ d) 0.34534534

74. a) $0.\overline{261}$ b) $0.2\overline{61}$ c) $0.2\overline{6}$ d) 0.2626

75. a) $\frac{4}{9}$ b) $\frac{5}{9}$ c) $0.4\overline{54}$ d) $0.5\overline{54}$

76. a) $\frac{5}{13}$ b) $0.3\overline{84}$ c) $\frac{4}{13}$ d) $0.\overline{38}$

State which property of the real numbers we are illustrating in Exercises 77–86.

77. $3(4 + 5) = 3 \cdot 4 + 3 \cdot 5$ **78.** $6(8 + 2) = (8 + 2)6$

79. $3 + (6 + 8) = (6 + 8) + 3$

80. $5(3 \cdot 2) = (5 \cdot 3)2$

81. $3 + (6 + 8) = (3 + 6) + 8$

82. $5(3 \cdot 2) = (3 \cdot 2)5$

83. $7 + 0 = 7$ **84.** $8 + (-8) = 0$

85. $4 + (2 + (-1)) = 4 + ((-1) + 2)$

86. $7(8 - 2) = 7 \cdot 8 - 7 \cdot 2$

Further Exercises

*(87.) Communicating Mathematics** We will now give a proof that $\sqrt{2}$ is irrational. Explain why each of the steps in the proof is valid.

a) Assume that $\sqrt{2}$ can be written as a quotient of integers $\frac{a}{b}$, where the quotient is in reduced form. So, $\sqrt{2} = \frac{a}{b}$.

b) $2 = \dfrac{a^2}{b^2}$ c) $2 \cdot b^2 = a^2$

d) a^2 is even, that is, divisible by 2.

e) a cannot be odd, so it must be even.

f) $a = 2 \cdot k$ for some integer k.

g) $2 \cdot b^2 = (2k)^2$.   h) $2 \cdot b^2 = 4k^2$.

i) $b^2 = 2k^2$.   j) b^2 is even.

k) b is even.

l) We have a contradiction, so our assumption, that $\sqrt{2}$ can be written as a quotient of integers $\frac{a}{b}$, is false. Therefore $\sqrt{2}$ is not rational.

(88.) **Communicating Mathematics** Do a similar proof to show that $\sqrt{3}$ is irrational.

We know that $\sqrt{28}$ is greater than 5 and less than 6 because $5^2 = 25$ and $6^2 = 36$. If you next square 5.0, 5.1, 5.2, and so on, you will find that $\sqrt{28}$ is between 5.2 and 5.3. Without using the square root key on your calculator, use this method to estimate the following square roots to two decimal places. Do not round your answer.

89. $\sqrt{71}$ **90.** $\sqrt{95}$

91. $\sqrt{106}$ **92.** $\sqrt{150}$

---
* Exercise numbers circled in red can be used as group exercises.

93. Find the length of line segment AB in the given figure.

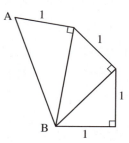

94. Using the pattern that is shown in Exercise 93, construct a length of $\sqrt{5}$.

95. The triple of natural numbers 3, 4, 5 is called a Pythagorean triple. Why do you think that this is so? Hint: Consider right triangles.

96. Find another triple of natural numbers that is a Pythagorean triple.

5.5 EXPONENTS AND SCIENTIFIC NOTATION

In this section, we will be able to answer the question that we posed at the very beginning of this chapter. If we were able to unravel all the atoms in your body, as though you were a great big ball of thread and then stretch out all those atoms in a straight line, how far would it reach? To be able to represent both the tiny size of an atom as well as the vastness of the size of the universe, we have to represent real numbers in a new way.

■━━━━━━━━━━━━━━━━━

All exponent rules are based on the definition of exponents.

In order to write such very large and very small numbers efficiently, we will use exponential notation. Some examples of exponential notation are, $10^3 = 10 \cdot 10 \cdot 10 = 1,000$ and $2^5 = 2 \cdot 2 \cdot 2 \cdot 2 \cdot 2 = 32$. In general, we have the following definition.

> **DEFINITION**
>
> If a is any real number and n is a counting number, then
>
> $$a^n = \underbrace{a \cdot a \cdot a \cdot \ldots \cdot a}_{\substack{\text{product} \\ \text{of } n \text{ as}}}$$
>
> The number a is called the **base,** and the number n is called the **exponent.**

This simple definition is the basis for all the rules for exponents that we will discuss in this section.

EXAMPLE 1 Evaluating Expressions with Exponents

Write each of the following expressions in another way using the definition of exponents then evaluate the expression.

a) $3 \cdot 3 \cdot 3 \cdot 3 \cdot 3$ b) 2^4 c) 0^6

d) $(-2)^4$ e) -2^4 f) 5^1

SOLUTION: a) $3 \cdot 3 \cdot 3 \cdot 3 \cdot 3 = 3^5 = 243$.

b) $2^4 = 2 \cdot 2 \cdot 2 \cdot 2 = 16$.

c) $0^6 = 0 \cdot 0 \cdot 0 \cdot 0 \cdot 0 \cdot 0 = 0$.

d) $(-2)^4 = (-2) \cdot (-2) \cdot (-2) \cdot (-2) = +16$.

e) This is not the same as d). Notice that here the exponent, 4, takes precedence over the negative sign. So first we multiply the four 2s and then insert the negative sign. Thus,

$$-2^4 = -2 \cdot 2 \cdot 2 \cdot 2 = -16$$

f) $5^1 = 5$.

Suppose in a physics class you encountered the expression $x^3 \cdot x^4$. If you did not know the rules for working with exponents that we are going to introduce, you could simplify this expression by using only the definition of exponents. You could reason as follows:

$$x^3 \cdot x^4 = \underbrace{(x \cdot x \cdot x)}_{3 \ xs} \cdot \underbrace{(x \cdot x \cdot x \cdot x)}_{4 \ xs} = \underbrace{x \cdot x \cdot x \cdot x \cdot x \cdot x \cdot x}_{7 \ xs} = x^7$$

Thus, $x^3 \cdot x^4 = x^7$.

By thinking of a simple example such as this, you can remember the following rule for rewriting certain products.

> **Product Rule for Exponents**
>
> If x is a real number and m and n are natural numbers, then
>
> $$x^m x^n = x^{m+n}$$

Notice in this rule that all the bases are the same. This rule would not apply to an expression such as $x^3 y^5$ because the bases are different.

EXAMPLE 2   Applying the Product Rule for Exponents

Use the product rule for exponents to rewrite each of the following expressions, if possible.

a) $2^5 \cdot 2^9$ b) $3^2 \cdot 5^4$

SOLUTION: a) $2^5 \cdot 2^9 = 2^{5+9} = 2^{14}$

b) We cannot apply the product rule here. The product $3^2 \cdot 5^4$ contains two 3s and four 5s as factors and we cannot combine these six factors into a single expression. We could, however, rewrite this as $3 \cdot 3 \cdot 5 \cdot 5 \cdot 5 \cdot 5 = 5{,}625$.

Quiz Yourself 20*

Use the product rule to rewrite $3^4 \cdot 3^2$.

---

* Quiz Yourself answers begin on page 849.

Suppose that you wished to simplify the expression $(y^3)^4$ but did not recall the rule for doing so. Again you could use the definition of exponents and your common sense to do the simplification. Think of $(y^3)^4$ as (y^3) multiplied by itself 4 times. That is,

$$(y^3)^4 = (y^3)(y^3)(y^3)(y^3).$$

But now you can write each (y^3) as $y \cdot y \cdot y$. So we have

$$(y^3)^4 = (y \cdot y \cdot y)(y \cdot y \cdot y)(y \cdot y \cdot y)(y \cdot y \cdot y) = y^{12}.$$

We see that in this example, we multiplied the exponents to write the simplification. This leads us to the following rule* for exponent expressions which are raised to powers.

> **Power Rule for Exponents**
>
> If x is a real number and m and n are natural numbers, then
>
> $$(x^m)^n = x^{m \cdot n}$$

Quiz Yourself 21

Use the power rule for exponents to simplify each of the following.

a) $(5^4)^3$ b) $((-2)^2)^4$

EXAMPLE 3 **Applying the Power Rule for Exponents**

Use the power rule for exponents to simplify each of the following.

a) $(2^3)^2$ b) $((-2)^3)^4$

SOLUTION: a) $(2^3)^2 = 2^{3 \cdot 2} = 2^6 = 64$.

b) $((-2)^3)^4 = (-2)^{3 \cdot 4} = (-2)^{12} = 4,096$.

 PROBLEM SOLVING

The Three Way Principle in Section 1.1 suggests that a good way to remember what exponent rule to use in a particular situation is to think of simple examples. Then use the definition of exponents to recall how we derived the rule. For example, to remember whether you add or multiply exponents in simplifying the expression $(a^8)^7$, think what you would do with $(x^3)^2$. If you write this expression as $x^3 \cdot x^3 = x \cdot x \cdot x \cdot x \cdot x \cdot x = x^6$, then you recall that you are supposed to multiply exponents in this situation.

We often have to divide expressions containing exponents. For example, suppose that we wish to simplify the expression $\frac{x^7}{x^3}$. Using the definition of exponents, we can rewrite the numerator and denominator as follows:

$$\frac{x^7}{x^3} = \frac{x \cdot x \cdot x \cdot x \cdot \cancel{x} \cdot \cancel{x} \cdot \cancel{x}}{\cancel{x} \cdot \cancel{x} \cdot \cancel{x}} = \frac{x \cdot x \cdot x \cdot x}{1} = x \cdot x \cdot x \cdot x = x^4$$

This example leads us to the following quotient rule for exponents.

---

* We will explain some other power rules for exponents in the exercises.

Quotient Rule for Exponents

If x is a nonzero real number and both m and n are natural numbers, then

$$\frac{x^m}{x^n} = x^{m-n}$$

This rule is fine provided m is greater than n. However, if $m = n$, then $m - n = 0$, and we have not defined what x^0 means. Also, if $m < n$, then we have $m - n < 0$, and we have not discussed what it means to raise x to a negative power. We will take care of these definitions before we give examples of the quotient rule.

DEFINITIONS

If $a \neq 0$, then $a^0 = 1$, and if n is a natural number, then $a^{-n} = \frac{1}{a^n}$.

These definitions tell us that $3^0 = 1$ and $4^{-2} = \frac{1}{4^2} = \frac{1}{16}$.

EXAMPLE 4   Applying the Quotient Rule for Exponents

Use the quotient rule to simplify the following expressions and write your answer as a single number.

a) $\dfrac{3^7}{3^5}$ b) $\dfrac{7^5}{7^8}$ c) $\dfrac{17^9}{17^9}$

SOLUTION: a) $\dfrac{3^7}{3^5} = 3^{7-5} = 3^2 = 9$

b) $\dfrac{7^5}{7^8} = 7^{5-8} = 7^{-3} = \dfrac{1}{7^3} = \dfrac{1}{343}$ c) $\dfrac{17^9}{17^9} = 17^0 = 1$

Quiz Yourself 22

Simplify and evaluate these expressions.

a) $\dfrac{6^8}{6^{10}}$ b) $\dfrac{(-8)^{11}}{(-8)^9}$

c) $\dfrac{11^0}{11^0}$

We can extend the exponent rules to all integer exponents.

We explained the rules for working with exponents by using very simple examples in which the exponents m and n were both natural numbers. Without proving it, we will simply state that the *rules don't change* when you are working with other types of exponents (see Figure 5.23).

For example, if you are dealing with the expression $3^{-4} \cdot 3^6$, you can rewrite it as $3^{-4} \cdot 3^6 = 3^{-4+6} = 3^2 = 9$. In other mathematics courses, you may see rational exponents, or even irrational exponents! However, if you remember that the rules are the same when dealing with these more complicated examples, you will be able to do your computations confidently.

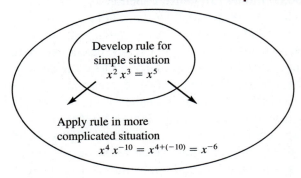

FIGURE 5.23   The rules don't change as we work in more complicated situations.

EXAMPLE 5   **Using Exponent Rules When the Exponents Are not Positive**

Assume that the exponent rules extend to these expressions and then simplify as in Example 4.

a) $5^{-4} \cdot 5^7$ b) $(2^{-3})^2$ c) $\dfrac{(-3)^{-2}}{(-3)^{-7}}$ d) $\dfrac{12^{-3}}{12^{-3}}$

SOLUTION: a) We apply the product rule to get $5^{-4} \cdot 5^7 = 5^{-4+7} = 5^3 = 125$.

b) Here we apply the power rule:

$$(2^{-3})^2 = 2^{-3 \cdot 2} = 2^{-6} = \frac{1}{2^6} = \frac{1}{64}.$$

c) Now we use the quotient rule to get

$$\frac{(-3)^{-2}}{(-3)^{-7}} = (-3)^{-2-(-7)} = (-3)^{-2+7} = (-3)^5 = -243.$$

Quiz Yourself 23

Simplify

a) $3^8 \cdot 3^{-6}$ b) $(2^{-2})^{-1}$

d) We'll use the definition of a zero exponent to simplify

$$\frac{12^{-3}}{12^{-3}} = 12^{-3-(-3)} = 12^{-3+3} = 12^0 = 1.$$

We use scientific notation to represent very large and very small numbers.

As we mentioned at the beginning of this section, scientists often have to deal with very large and very small numbers. In order to be able to understand and calculate with such extreme quantities, mathematicians have developed scientific notation. Before giving the formal definition, we will illustrate scientific notation with several examples.

5,865,696,000,000.

Moving the decimal point 12 places to the left makes the number smaller

(a)

5.865696 × 10¹²

Multiplying by 10¹² restores the size of the number

(b)

FIGURE 5.24   (a) Moving the decimal point to the left, (b) Multiplying by 10¹².

Astronomers often measure distances in light years, which is the distance that light travels in one year. A light year is approximately 5,865,696,000,000 miles. To convert such a very large number to scientific notation, we keep dividing the number by 10 until the number is between 1 and 10. We do this by moving the decimal point to the left and keeping track of how many places we have moved the decimal point. In this case, we need to move the decimal point 12 places to the left (see Figure 5.24(a)). Because moving the decimal point to the left makes the number smaller, we must multiply by 10^{12} to restore the size of the number (see Figure 5.24(b)).

In physics, the electric charge on an electron is 0.00000000000000000016 coulombs. To write this number in scientific notation, we move the decimal point 19 places to the right to get 1.6, which, in effect, multiplies the number by 10^{19}. To restore the number to its small size, we must divide by 10^{19}, or what is equivalent, multiply by 10^{-19}. Thus $0.00000000000000000016 = 1.6 \times 10^{-19}$.

> **DEFINITION**
>
> A number is written in **scientific notation** if it is of the form $a \times 10^n$, where $1 \le a < 10$ and n is any integer.

Remember that in order for a number of form $a \times 10^n$ to be in scientific notation, the a must be greater than or equal to one and less than ten. Therefore, numbers such as 14.36×10^3 and 0.634×10^8 would not be in scientific notation form. We can rewrite these numbers in scientific notation as 1.436×10^4 and 6.34×10^7.

If your calculator has the capability to express numbers in scientific notation, it probably has one of the following keys:

$$\boxed{\text{EE}} \qquad \boxed{\text{EEx}} \qquad \boxed{\text{EXP}}$$

To enter 6.34×10^7 on your calculator you might enter:

$$\boxed{6} \quad \boxed{.} \quad \boxed{3} \quad \boxed{4} \quad \boxed{\text{EE}} \quad \boxed{7}$$

Calculators also display scientific notation in different ways. One popular way to indicate 6.34×10^7 is to display 6.34 E 7. The E 7 tells us that the exponent of 10 is 7 in scientific notation. Another popular method uses a space instead of the E to display 6.34×10^7 as 6.34   7.

> **Rules for Converting a Decimal Number to Scientific Notation**
>
> If the number x is very large,
> a) Make the number smaller by moving the decimal point to the left until the number is between one and ten.
> b) Restore the size of x by multiplying the number you obtained in part a) by ten raised to the power which is the number of places that you moved the decimal point to the left.
>
> If x is very small,
> c) Make the number larger by moving the decimal point to the right until the number is between one and ten.
> d) Restore the size of x by multiplying the number you obtained in part c) by ten raised to the power which is the *negative* of the number of places that you moved the decimal point to the right.

EXAMPLE 6 Converting from Standard Notation to Scientific Notation

Rewrite the numbers in the following statements in scientific notation.

a) The length of Manhattan Island is approximately 708,660 inches.

b) A particular computer takes 0.0000000034 of a second to perform an operation.

SOLUTION: a) Move the decimal point in 708,660. to the left five places to get 7.08660. Now multiply this number by 10^5 to get 7.08660×10^5.

b) Move the decimal point in 0.0000000034 to the right 9 places to get 3.4. Now multiply this number by 10^{-9} to get 3.4×10^{-9}.

EXAMPLE 7   Converting from Scientific Notation to Standard Notation

Rewrite the numbers in the following statements in standard notation.

a) The precise length of a year is 3.1556925511×10^7 seconds.

b) The diameter of an atom is about 2.0×10^{-10} meters.

SOLUTION: a) We move the decimal point in 3.1556925511 to the right seven places to get 31,556,925.511.

b) Move the decimal point in 2.0 ten places to the left to get 0.0000000002 meters.

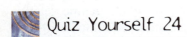

Quiz Yourself 24

a) Convert 53,728.41 to scientific notation.

b) Convert 2.45×10^{-3} to standard notation.

We use rules of exponents to multiply and divide numbers in scientific notation.

It is easy to multiply and divide numbers written in scientific notation. For example, $(3 \times 10^2) \cdot (4 \times 10^5) = (3 \times 4) \cdot (10^2 \times 10^5) = 12 \times 10^7$. Also, $\frac{8 \times 10^6}{2 \times 10^4} = \frac{8}{2} \times \frac{10^6}{10^4} = 4 \times 10^{6-4} = 4 \times 10^2$.

PROBLEM SOLVING

Notice that in multiplying and dividing numbers written in scientific notation, you are not doing anything new. You are simply applying properties such as associativity and commutativity and the rules for working with rational numbers that you have learned earlier.

We will now use scientific notation to solve some applied problems.

EXAMPLE 8   Comparing Computing Capacity

Carlos' new hard disk has 1.8 gigabytes of storage, which is 1.8×10^9 bytes. His old hard disk had 300 megabytes of storage, which is 300×10^6 bytes. How many times larger is the capacity of his new hard disk than his old hard disk?

SOLUTION: We must divide 1.8×10^9 by 300×10^6. Now 300×10^6 is not in scientific notation. However, if we divide 300 twice by 10 and then increase the exponent by 2, we can rewrite 300×10^6 as 3×10^8. Now we will divide 1.8×10^9 by 3×10^8. So,

$$\frac{1.8 \times 10^9}{3 \times 10^8} = \frac{1.8}{3} \times \frac{10^9}{10^8} = 0.6 \times 10^{9-8} = 0.6 \times 10^1 = 0.6 \times 10 = 6.$$

Therefore, Carlos' new disk has 6 times the storage capacity of his old disk.

EXAMPLE 9   **Finding the Total Length of all the Atoms in a Person's Body**

If all the atoms in a person weighing 175 pounds were lined up in a straight line, how long would that line be? Compare this length with the distance to the nearest star, Proxima Centauri. Use the following information* to answer the question:

The person's body contains 3.4×10^{27} atoms.

The diameter of an atom is 2×10^{-10} meters (a meter is slightly more than 39 inches).

The distance to the nearest star, Proxima Centauri is 4.6 light years.

A light year is 9.46×10^{15} meters.

SOLUTION: a) We first multiply the number of atoms in the person's body by the diameter of a single atom.

$$(3.4 \times 10^{27})(2 \times 10^{-10}) = (3.4 \times 2)(10^{27} \times 10^{-10}) = 6.8 \times 10^{27+(-10)} = 6.8 \times 10^{17}$$

Thus we find that total length of the atoms in the body is 6.8×10^{17} meters.

b) Next we find the distance, in meters, to Proxima Centauri. To do this we multiply $4.6 \times 9.46 \times 10^{15}$. This gives us

$$4.6 \times 9.46 \times 10^{15} = 43.516 \times 10^{15} = 4.3516 \times 10^{16}.$$

c) Finally, we divide the length we found in part a) by the distance we found in part b), to get

$$\frac{6.8 \times 10^{17}}{4.3516 \times 10^{16}} = \frac{6.8}{4.3516} \times \frac{10^{17}}{10^{16}} \approx 1.56 \times 10 = 15.6.$$

Thus we have found that the total length of the atoms in the person's body is roughly 16 times the distance to the nearest star!

---

* This information is taken from Cesare Emiliani, *The Scientific Companion* (Wiley, 1988), pp. 1–10.

Generally, we will not be giving you exercises as complicated as Example 9; however, we just wanted to show you the power of using the rules of exponents and scientific notation.

Exercises 5.5

Evaluate each expression.

1. $2 \cdot 2 \cdot 2 \cdot 2 \cdot 2$ **2.** 5^3

3. -2^4 **4.** 0^3

5. -3^2 **6.** 5^{-2}

7. 9^1 **8.** 3^{-4}

9. 3^0 **10.** 5^1

11. 0^6 **12.** $(-5)^2$

Use the rules for exponents to first rewrite each of the following expressions and then evaluate the new expression.

13. $3^2 \cdot 3^4$ **14.** $(-2)^3 \cdot (-2)^5$

15. $(7^2)^3$ **16.** $8^0 \cdot 8^2$

17. $5^4 \cdot 5^{-6}$ **18.** $4^2 \cdot 4^3$

19. $(3^2)^{-3}$ **20.** $2^{-4} \cdot 2^{-3}$

21. $(-3)^{-2}(-3)^3$ **22.** $(3^2)^4$

23. $\dfrac{5^9}{5^7}$ **24.** $(7^{-1})^{-3}$

25. $\dfrac{6^{-2}}{6^{-4}}$ **26.** $\dfrac{(-3)^6}{(-3)^9}$

27. $\dfrac{-3^6}{-3^9}$ **28.** $(2^{-3})^{-2}$

29. $(3^{-4})^0$ **30.** $(7)^{-2} \cdot (7)^6$

31. $\dfrac{11^{-2}}{11^{-3}}$ **32.** $(4^2)^{-1}$

33. $(-2^2)^{-3}$ **34.** $\dfrac{5^{-2}}{5^{-2}}$

35. $\dfrac{2^9}{4^3}$ Hint: rewrite 4.

36. $9^3 \cdot 27^{-2}$ Hint: Rewrite 9 and 27.

37. Communicating Mathematics   Explain the difference in evaluating -2^4 and $(-2)^4$. How does that affect your answer when calculating these expressions?

38. Communicating Mathematics   Provide a counterexample to show that $(x + y)^2 = x^2 + y^2$ is not true.

How do you evaluate the left side of the equation versus the right side of the equation?

Give a simple example, as we did in the text, to explain how you might remember to rewrite each of the following.

39. $a^m a^n$ **40.** $(a^m)^n$

41. $\dfrac{a^m}{a^n}$

42. Communicating Mathematics   What do we mean when we say "the rules don't change"?

Rewrite each of the following numbers in scientific notation.

43. 4,356,000   **44.** 3,200,000,000

45. 783   **46.** 0.000258

47. 0.0024   **48.** 28

49. 382,400   **50.** 0.02

51. 0.4   **52.** 0.000045

53. 0.008   **54.** 8,056

Rewrite each of the following numbers in standard notation.

55. 3.25×10^4 **56.** 4.7×10^8

57. 1.78×10^{-3} **58.** 7.41×10^{-8}

59. 6.3×10^1 **60.** 9.7×10^1

61. 6.27×10^{-2} **62.** 4.36×10^2

63. 4.5×10^{-7} **64.** 8×10^{-7}

65. 1×10^6 **66.** 1×10^{-5}

Each of the following numbers is not written in scientific notation. Rewrite them so that they are in scientific notation.

67. 23.81×10^6 **68.** 426.5×10^5

69. 0.84×10^3 **70.** 0.03×10^8

Use scientific notation to perform the following operations. Leave your answer in scientific notation form.

71. $(3 \times 10^6)(2 \times 10^5)$ **72.** $(4 \times 10^3)(2 \times 10^4)$

73. $(1.2 \times 10^{-3})(3 \times 10^5)$

74. $(5.2 \times 10^6)(1.45 \times 10^5)$

75. $(7.24 \times 10^{-5})(3.6 \times 10^8)$

76. $(4.2 \times 10^{-2})(1.83 \times 10^{-4})$

77. $(8 \times 10^{-2}) \div (2 \times 10^3)$

78. $(4.8 \times 10^4) \div (1.6 \times 10^{-3})$

79. $\dfrac{(5.44 \times 10^8)(2.1 \times 10^{-3})}{(3.4 \times 10^6)}$

80. $\dfrac{(1.4752 \times 10^{-2})(5.7 \times 10^4)}{(4.61 \times 10^{-3})}$

81. $\dfrac{(9.6368 \times 10^3)(4.15 \times 10^{-6})}{(1.52 \times 10^4)}$

82. $\dfrac{(3.5445 \times 10^{-3})(2.8 \times 10^{-5})}{(8.34 \times 10^6)}$

Rewrite each number using scientific notation before performing the operation. Leave your answer in scientific notation form.

83. $67,300,000 \times 1,200$

84. $83,600 \times 4,200,000$

85. $6,800,000 \times 2,300,000$

86. $1,750,000 \times 3,400,000$

87. 0.00016×0.0025

88. 0.000325×0.000008

89. 0.00165×0.0004

90. 0.00085×0.000016

In Exercises 91–108, use scientific notation to do all calculations. Give your answer in scientific notation. Use the fact that one million is $1,000,000 = 10^6$, one billion is $1,000,000,000 = 10^9$, and one trillion is $1,000,000,000,000 = 10^{12}$.

91. **The national debt.** In 1950 the population of the United States was 151 million and the national debt was 256 billion dollars. Divide the national debt by the size of the population to determine the amount owed in 1950 per person to the national government.

92. **The national debt.** Repeat Exercise 91 for the year 2000 when the population was 281 million and the national debt was 5.807 trillion.

93. **Comparing populations.** In 2001, the world's population was 6.157 billion and the U.S. population was 278 million. How many times larger was the world's population than the U.S. population?

94. **Comparing astronomical distances.** The distance from the Earth to the moon is 239,000 miles and the distance from the Earth to the sun is 93 million miles.

How many times greater is the distance from the earth to the sun compared with the distance from the Earth to the moon?

95. **The defense budget.** In 2001, the U.S. population was 278 million and the defense budget was 281 billion dollars. How much was spent on defense for each person in the United States?

96. Convert one million seconds to days. (There are $60 \times 60 \times 24$ seconds in a day.)

97. Convert one billion seconds to years.

98. On July 4, 2006, how many seconds old will the United States be? (Ignore leap years.)

99. **Comparing weights.** How many times heavier is a 130-pound person than a mosquito? Assume that there are 454 grams to a pound and that a mosquito weighs 1×10^{-3} grams.

    a) First find the number of grams that the person weighs.

    b) Divide the number that you found in part a) by 1×10^{-3}.

    c) Interpret your answer in part b).

100. **Land per person in the United States.** According to the 1950 census, there were approximately 42.6 people per square mile of land in the United States. The total land area of the United States at that time was about 3.55×10^6 square miles. Use this information to estimate the population of the United States in 1950 to the nearest million.

101. **Land per person in the United States.** By 2000, the U.S. population had grown to approximately 278 million and the land area of the United States was about 3.537×10^6 square miles. How many people per square mile were there in the United States in 2000?

102. Health care expenditures. According to the National Center for Health Statistics, in 1990, the United States spent 695.6 billion dollars for health care. In 1999, this expenditure had risen to 1.21 trillion dollars. How many times larger was the expenditure for health care in 1999 than in 1990?

103. Space flight.* The sun is about 93,000,000 miles away. If a rocket ship travels from earth to the sun at approximately 25,000 miles per hour, how many hours will the flight take?

104. Space flight. In 1977, scientists sent the spaceship Voyager II to Neptune, which is about 2.8 billion miles away. If the spacecraft averaged about 25,000 miles per hour, how long did the journey take?

105. Comparing astronomical distances. The Earth is 93 million miles from the sun and Pluto is 3.6 billion miles from the sun. How many more times is Pluto's distance from the sun than the Earth's?

106. Comparing disk drives. The author is writing this text on a Centris 650 computer that has a 230-megabyte hard disk. His son, Mike, plays games on the Internet using a computer that has a 20-gigabyte hard disk. How many times more disk space does Mike have on his computer than his dad has?

107. Estimating time for computer computations. Linda has a computer that can perform 2.6 million operations per second. If she is running a calculation that requires 56.42 billion calculations, how many seconds will this calculation take? How many hours?

108. Distance light travels. Light travels at 186,000 miles per second. How far will light travel in one day?

Further Exercises

109. Communicating Mathematics What is the point of using scientific notation?

† **110. Communicating Mathematics** Does $(x^m)^n = x^{(m^n)}$? Consider simple examples.

Communicating Mathematics *Look at simple examples to complete the following rules of exponents.*

111. $\left(\dfrac{a}{b}\right)^n =$

112. $(ab)^n =$

113. Communicating Mathematics Use almanacs, encyclopedias, science textbooks, or other sources to locate examples, like the ones we gave you in this section, that involve very large or very small numbers. Make up a question similar to Exercises 103–120 and answer that question. For example, if a dollar bill is 0.003 inches thick, how high would a stack of one-dollar bills be that would represent the U.S. federal budget in 2002?

114. You are allowed to enter only three digits in your calculator, but can use any other combination of keystrokes to form numbers. For example, you might enter $4*(5^6)$ or $(5^6)^8$. On my calculator, $4^{(4^4)}$ causes an overflow. Using the restrictions above, find the largest number that your calculator will display without causing an overflow.

---

* In dealing with astronomical distances, we have simplified these problems in the sense that we are assuming the distances are straight-line distances. In reality, a spaceship would not fly in a straight line through space. However, to take these factors into account would complicate the problem.
† Exercise numbers circled in red can be used as group exercises.

CHAPTER SUMMARY

SECTION 5.1

natural numbers, number theory
The numbers that we count with, $\{1, 2, 3, \ldots\}$, are called the natural numbers, or counting numbers. The study of the natural numbers and their properties is called number theory.

divisor, factor, multiple
If a and b are natural numbers, then we say a divides b, written $a|b$, if there is a natural number q such that $b = qa$. We also say that a is a divisor of b, a is a factor of b, or that b is a multiple of a.

factoring numbers
When we write a natural number as a product of natural numbers, we have factored the number.

prime number, composite number
A natural number, greater than one, which has only itself and one as factors, is called a prime number. A natural number which is not prime is called composite.

Sieve of Eratosthenes
The Sieve of Eratosthenes is a way of "sieving out" composites to find prime numbers.

divisibility tests
Table 5.1 gives simple ways to test a number for divisibility by 2, 3, 4, 5, 8, 9, and 10.

fundamental theorem of Arithmetic
Every natural number greater than 1 is a unique product of prime numbers, except for the order of the factors.

factor tree
We use a factor tree to find the prime factorization of a number.

greatest common divisor, least common multiple
The greatest common divisor, GCD, of two natural numbers is the largest natural number that divides both numbers. The least common multiple, LCM, of two natural numbers is the smallest natural number that is a multiple of both numbers. We can find the GCD and LCM of two numbers by using their prime factorizations.

SECTION 5.2

whole numbers, integers
The whole numbers is the set $\{0, 1, 2, 3, \ldots\}$. The set of integers is $\{\ldots, -3, -2, -1, 0, 1, 2, 3, \ldots\}$.

adding whole numbers
We can explain addition rules by considering movement on a number line.

subtracting whole numbers
If a and b are integers, then $a - b = a + (-b)$.

multiplying whole numbers
We use a money model to explain integer multiplication. If a and b are integers, then: If a and b have the same sign, then $a \cdot b$ is positive. If a and b have opposite signs, then $a \cdot b$ is negative.

dividing whole numbers

If a, b, and c are integers, then $\frac{a}{b} = c$ means that $a = b \cdot c$. If a and b have the same sign, then $\frac{a}{b}$ is positive. If a and b have opposite signs, then $\frac{a}{b}$ is negative.

SECTION 5.3   **rational numbers, numerator, denominator**

The set of rational numbers, **Q**, is the set of all numbers that can be written in the form $\frac{a}{b}$, where the a and b are integers, and $b \neq 0$. The top number a is called the numerator and the bottom number b is called the denominator.

equality of rational numbers, cross product

$$\frac{a}{b} = \frac{c}{d} \text{ if and only if } ad = bc.$$

$ad = bc$ is called cross multiplying or calculating the cross product.

reducing a rational number, canceling common factors

We reduce a rational number by canceling common factors as follows:

$$\frac{a \cdot c}{b \cdot c} = \frac{a}{b}$$

When all common factors have been canceled, the number is in lowest terms.

adding and subtracting rational numbers

$$\frac{a}{b} + \frac{c}{d} = \frac{a \cdot d + b \cdot c}{b \cdot d} \qquad \frac{a}{b} - \frac{c}{d} = \frac{a \cdot d - b \cdot c}{b \cdot d}$$

multiplying and dividing rational numbers

$$\frac{a}{b} \cdot \frac{c}{d} = \frac{a \cdot c}{b \cdot d} \qquad \frac{\dfrac{a}{b}}{\dfrac{c}{d}} = \frac{a}{b} \cdot \frac{d}{c}$$

converting an improper fraction to a mixed number

To convert the improper fraction $\frac{a}{b}$ to a mixed number, divide a by b to get the quotient q and remainder r. Then $\frac{a}{b} = q\frac{r}{b}$.

converting a mixed number to an improper fraction

The mixed number $q\frac{r}{b}$ equals the improper fraction $\frac{q \cdot b + r}{b}$.

decimal expansions of rational numbers

Rational numbers have repeating decimal expansions. A terminating expansion is a repeating expansion in which zero repeats from some point on.

SECTION 5.4   **irrational numbers**

An irrational number is a number that is not a rational number and has a nonrepeating decimal expansion.

multiplying and dividing radicals

If $a \geq 0$ and $b \geq 0$, then

$$\sqrt{ab} = \sqrt{a}\sqrt{b} \text{ and if } b \neq 0, \text{ then } \sqrt{\frac{a}{b}} = \frac{\sqrt{a}}{\sqrt{b}}.$$

properties of the real number system

The real numbers satisfy the following properties for any real numbers a, b, and c.

Closure: $a + b$ and $a \cdot b$ are real numbers.

Commutativity: $a + b = b + a$ and $a \cdot b = b \cdot a$.

Associativity: $(a + b) + c = a + (b + c)$ and $(a \cdot b) \cdot c = a \cdot (b \cdot c)$.

Identity element: $a + 0 = 0 + a = a$ and $a \cdot 1 = 1 \cdot a = a$.

Inverse property: $a + (-a) = (-a) + a = 0$ and if $a \neq 0$, $a \cdot \frac{1}{a} = \frac{1}{a} \cdot a = 1$.

Distributivity: $a \cdot (b + c) = a \cdot b + a \cdot c$.

SECTION 5.5

exponents

If a is any real number and n is a counting number, then

$$a^n = \underbrace{a \cdot a \cdot a \cdot \ldots \cdot a}_{\text{product of } n \text{ as}}$$

rules for exponents

If x is a real number and m and n are natural numbers, then

$$x^m x^n = x^{m+n} \qquad \text{Product Rule}$$

$$(x^m)^n = x^{m \cdot n} \qquad \text{Power Rule}$$

$$\frac{x^m}{x^n} = x^{m-n}, x \neq 0 \qquad \text{Quotient Rule}$$

$$a^0 = 1 \text{ and If } a \neq 0, \text{ then } a^{-n} = \frac{1}{a^n}$$

scientific notation

A number is written in scientific notation if it is of the form $a \times 10^n$, where $1 \leq a < 10$ and n is any integer.

CHAPTER TEST

SECTION 5.1

1. Use the Sieve of Eratosthenes to find all prime numbers between 70 and 90.

2. Estimate the square root of 180 by finding a natural number a such that $a < \sqrt{180} < a + 1$.

3. Are 191 and 441 prime or composite? If the number is composite, factor it into a product of primes.

4. Check 1,080,036 for divisibility by: 3, 4, 5, 6, 8, 9, and 10.

5. Find the GCD and LCM of 396 and 330.

6. Explain how you use prime factorization to get the GCD and the LCM of two natural numbers.

SECTION 5.2

7. Find the following:

   a) $-5 + 14$ b) $-13 - (-24)$

   c) $(-12)(-6)$ d) $\dfrac{48}{-3}$

8. Use a money example to illustrate how to compute the product $(-4)(+3)$.

9. Use the definition of division in terms of multiplication to compute $\dfrac{-24}{-8}$.

10. If the high temperature in northern Alaska is 17° Fahrenheit at the beginning of October and at the beginning of November it is $-3°$, what is the difference in these two temperatures?

11. Use the definition of division in terms of multiplication to explain why we cannot divide 5 by 0.

SECTION 5.3

12. Are the numbers $\frac{28}{65}$ and $\frac{14}{35}$ equal? Explain your answer.

13. Perform the following computations:

    a) $\dfrac{4}{9} \cdot \left(\dfrac{3}{4} - \dfrac{1}{3} \right)$ b) $\dfrac{3}{7} \div \left(\dfrac{2}{3} + \dfrac{3}{14} \right)$

14. Convert $\frac{53}{8}$ to a mixed number.

15. Find $(3\frac{1}{2})(4\frac{1}{7})$.

16. Write the following numbers as quotients of integers.

    a) 0.375   b) $0.\overline{63}$

17. If a recipe for a hot chili sauce that serves 16 requires 2 and $\frac{1}{2}$ cups of hot peppers and 3 and $\frac{1}{4}$ cups of tomatoes, how much peppers and tomatoes would be required to make an amount of sauce to serve 24 people?

18. Explain by giving a specific example to show why we invert and multiply when dividing rational numbers.

SECTION 5.4

19. In terms of their decimal expansions, what is the difference between a rational and an irrational number?

20. Simplify the following: a) $\sqrt{108}$ b) $\frac{\sqrt{15}}{\sqrt{5}}$. Rationalize the denominator if necessary.

21. Give an example to show that division of rational numbers is not associative.

SECTION 5.5

22. Evaluate each expression:

    a) $3^6 \cdot 3^{-2}$ b) $(2^4)^{-2}$ c) $\dfrac{6^8}{6^5}$ d) $\dfrac{8^{-6}}{8^{-4}}$

23. Explain the difference between evaluating -2^4 and $(-2)^4$.

24. Write 0.000456 and 1,230,000 in scientific notation.

25. Rewrite 1.325×10^6 and 8.63×10^{-5} in standard notation.

26. Simplify $\frac{(3.6 \times 10^3)(2.8 \times 10^{-5})}{(4.2 \times 10^4)}$ and write the answer in scientific notation.

27. In 2001, the world's population was 6.157 billion and Mexico's population was 102 million. How many times larger was the world's population than Mexico's population? Express your answer in scientific notation.

Of Further Interest: SEQUENCES

Congratulations. When you signed your contract for your new job, you agreed to a monthly salary of $2,560 per month; however, today your supervisor gave you a pleasant surprise. During your first year, you will receive a salary increase of $15 each month. Since your salary will be different every month, you wonder what your total income will be for the next year. You could get out your calculator and add the amounts

$$2560, 2575, 2590, 2605, \ldots,$$

but, as you will soon see, there is a faster way to do this.

In spite of your good news, you don't seem to be feeling so well. In fact, several days ago, you picked up a single virus which has been doubling ever since. The following list of numbers describes the growth of the virus in your body.

$$1, 2, 4, 8, 16, 32, 64, \ldots.$$

You have just had a good experience and a bad experience with lists of numbers called sequences. We will study the patterns found in several different types of sequences in this section.

DEFINITION

A **sequence** is a list of numbers that follows some rule or pattern. The numbers in the list are called the **terms** of the sequence. We often name the terms $a_1, a_2, a_3, a_4, a_5, \ldots$. We read this list "$a$ sub one, a sub two, a sub three, and so on."

In the sequence of monthly salaries, the rule to get each successive term in the sequence is to add 15 to the previous number. In this sequence, $a_1 = 2,560$, $a_2 = 2,575$, $a_3 = 2,590$, and so forth. In the sequence that describes the virus growth, the rule is that you get successive terms by doubling the previous term. Here $a_1 = 1$, $a_2 = 2$, $a_3 = 4$, $a_4 = 8$, $a_5 = 16$, and so on.

Terms in an arithmetic sequence differ by a fixed constant.

The salary sequence is an example of an arithmetic sequence.

DEFINITION

An **arithmetic sequence** is a sequence in which each term after the first term differs from the preceding term by a fixed constant amount called the **common difference** of the sequence.

EXAMPLE 1   **Writing Terms of Arithmetic Sequences**

Find the common difference for the following arithmetic sequences. Then write the next three terms of the sequence.

a) 3, 7, 11, 15, . . .   b) 5, 3, 1, − 1, . . .

SOLUTION: a) We can find the common difference of an arithmetic sequence by subtracting the first term from the second term, so the common difference is $7 - 3 = 4$. The next three terms of the sequence are 19, 23, and 27.

b) The common difference is $3 - 5 = -2$. Therefore, the next three terms of the sequence are $-3, -5$, and -7.

We can rewrite sequence a) in Example 1 as 3, 3 + 4, 3 + 8, 3 + 12, . . . and then again as 3, 3 + 4, 3 + 2·4, 3 + 3·4, . . . From this we see that in the first term, we have added 4 *zero* times, in the second term, we have added 4 *one* time, in the third term, we have added 4 *two* times, and so on. In general, in each term we have added one less 4 than the number of the term. From this pattern, we see that the 16th term would have fifteen 4s added, or would equal $3 + 15 \cdot 4 = 63$. This leads to the following formula for the nth term of an arithmetic sequence.

> **FORMULA 1:**
>
> **The nth Term of an Arithmetic Sequence**
>
> The nth term of an arithmetic sequence with first term a_1 and common difference d is
>
> $$a_n = a_1 + (n - 1)d$$

EXAMPLE 2   **Finding the nth Term of Arithmetic Sequences**

a) Find the 12th term of the arithmetic sequence whose first term is 2,560 and whose common difference is 15.

b) Find the 20th term of the arithmetic sequence whose first term is 5 and whose common difference is − 2.

SOLUTION: a) Here $n = 12$, $a_1 = 2{,}560$, and $d = 15$. Using Formula 1 we have,

$$a_{12} = a_1 + (n - 1)d = 2{,}560 + (12 - 1)15 = 2{,}560 + 11 \cdot 15 = 2{,}725.$$

b) Now, $n = 20$, $a_1 = 5$, and $d = -2$. Using Formula 1 we have,

$$a_{20} = a_1 + (20 - 1)d = 5 + (20 - 1)(-2)$$
$$= 5 + 19(-2) = 5 - 38 = -33.$$

 Quiz Yourself 25*

Find the 16th term in the arithmetic sequence 5, 8, 11, 14, . . .

As in the salary example, we sometimes want to add the first n terms in an arithmetic sequence. If we look carefully at a concrete example, we will see a pattern that leads us to a general formula for finding such sums.

Suppose that we wish to find the sum of the first six terms of the arithmetic sequence 2, 5, 8, 11, 14, 17, ... Rather than finding this sum directly, we will look at it in a slightly different way. Consider the sum 2 + 5 + 8 + 11 + 14 +

17 and the same sum written in reverse order as $17 + 14 + 11 + 8 + 5 + 2$. If we place these sums under one another and add, we get

$$
\begin{array}{r}
2 + 5 + 8 + 11 + 14 + 17 \\
+ 17 + 14 + 11 + 8 + 5 + 2. \\
\hline
19 + 19 + 19 + 19 + 19 + 19
\end{array}
$$

What you are seeing is that we got the sum, 19, which is $a_1 + a_6$, six times. Therefore, because we have added every term twice, $6(a_1 + a_6)$ is exactly twice the sum that we originally wanted. Therefore, $\frac{6(a_1 + a_6)}{2}$ is the sum we want. This example leads us to the following formula.

> **FORMULA 2:**
>
> **The Sum of the First n Terms of an Arithmetic Sequence**
>
> The sum of the first n terms of an arithmetic sequence is given by
>
> $$\frac{n(a_1 + a_n)}{2}$$

EXAMPLE 3   Summing Terms in an Arithmetic Sequence

Find the sum of the first 20 terms in the arithmetic sequence 3, 7, 11, 15, ...

SOLUTION: We will use Formula 2, $\frac{n(a_1 + a_n)}{2}$, but in order to do so, we first need to find a_{20}. We know that $a_1 = 3$ and $d = 4$, so using Formula 1 for finding the nth term of a sequence we get

$$a_{20} = a_1 + (20 - 1)d = 3 + 19 \cdot 4 = 3 + 76 = 79.$$

Now we can now apply Formula 2. We know that $a_1 = 3$, $a_{20} = 79$, and $n = 20$. Therefore, the sum is

$$\frac{20(3 + 79)}{2} = \frac{20(82)}{2} = 820.$$

We can now answer the question we posed at the beginning of this section.

EXAMPLE 4   Finding the Yearly Total of an Arithmetic Sequence of Monthly Salaries

Assume that you receive \$2,560 as a salary this month, and for the next 11 months, you receive a monthly raise of \$15. What is the total salary that you receive over the 12 months?

SOLUTION: The monthly salaries form an arithmetic sequence with the first term $a_1 = 2,560$ and the monthly difference $d = 15$. In Example 2, part a) we found that for this sequence, $a_{12} = 2,725$.

So, using Formula 2, the sum of your salaries over 12 months is

$$\frac{n(a_1 + a_n)}{2} = \frac{12(2,560 + 2,725)}{2} = \frac{\overset{6}{\cancel{12}} \cdot 5,285}{2} = 6 \cdot 5,285 = 31,710$$

Quiz Yourself 26

Find the sum of the first 16 terms in the sequence 5, 8, 11, 14, ...

■

We multiply by a fixed constant to get new terms in a geometric sequence.

You have seen that we generate each new term in an arithmetic sequence by adding the same fixed constant to the previous term. We generate each new term in a geometric sequence by multiplying the previous term by the same fixed constant. The virus sequence, 1, 2, 4, 8, 16, 32, . . . that we discussed earlier is an example of a geometric sequence because to get each new term, we multiply the previous term by 2.

> **DEFINITION**
>
> A **geometric sequence** is a sequence in which each term, after the first term, is a nonzero constant multiple of the preceding term. The constant by which we are multiplying is called the **common ratio.**

EXAMPLE 5   Writing Terms in a Geometric Sequence

Find the common ratio of each of the following geometric sequences. Then write the next three terms of the sequence.

a) 2, 6, 18, 54, . . .   b) 3, − 6, 12, − 24, . . .

SOLUTION: a) We can find the common ratio of a geometric sequence by dividing the second term of the sequence by the first term. So the common ratio of this sequence is $\frac{6}{2} = 3$. The next three terms of the sequence are 162, 486, and 1,458.

b) The common ratio is $-\frac{6}{3} = -2$. The next three terms of the sequence are 48, − 96, and 192.

The following example leads to a formula for computing the nth term of a geometric sequence. Let's rewrite the terms of the sequence 2, 6, 18, 54, . . . from Example 5a) in a slightly different form as 2, $2 \cdot 3^1$, $2 \cdot 3^2$, $2 \cdot 3^3$, ...

Recall that 3 is the common ratio. We see that in the 1st term, there are *zero* 3s, in the second term there is *one* 3, in the third term there are *two* 3s, and so on. In general, in the nth term, there are $n - 1$ factors which are 3. This gives us the following formula.

> **FORMULA 3:**
>
> **The nth Term of a Geometric Sequence**
>
> The nth term of a geometric sequence with common ratio r is
>
> $$a_n = a_1 \cdot r^{n-1}.$$

We will use this formula in Example 6.

EXAMPLE 6   **Finding the *n*th Term of Geometric Sequences**

a) Find the 8th term of the geometric sequence whose first term is 5 and whose common ratio is 3.

b) Find the 6th term of the geometric sequence whose first term is 3 and whose common ratio is -2.

c) Find the number of viruses in your body after 18 hours if initially you have 1 and the virus doubles every hour.

SOLUTION: a) Here $n = 8$, $a_1 = 5$, and $r = 3$. Using Formula 3 we have,

$$a_8 = a_1 \cdot r^{8-1} = 5 \cdot 3^7 = 10{,}935$$

b) Now, $n = 6$, $a_1 = 3$, and $r = -2$. Again using Formula 3 we have,

$$a_6 = a_1 \cdot r^{6-1} = 3 \cdot (-2)^5 = 3 \cdot (-32) = -96.$$

c) In this case, $n = 18$, $a_1 = 1$, and $r = 2$, so

$$a_{18} = a_1 \cdot r^{18-1} = 1 \cdot (2)^{17} = 1 \cdot 131{,}072 = 131{,}072.$$

So there are 131,072 copies of the virus in your body after 18 hours.

> **Quiz Yourself 27**
>
> Find the 12th term in the sequence $2, 6, 18, 54, \ldots$.

We will use the following formula for finding the sum of the first *n* terms of a geometric sequence. We will use algebra to explain why this formula is valid in Exercise 38.

> **FORMULA 4:**
>
> **The Sum of the First *n* Terms of a Geometric Sequence**
>
> The sum of the first *n* terms of a geometric sequence is
>
> $$\frac{a_1 \cdot (r^n - 1)}{r - 1}.$$

Example 7 illustrates Formula 4.

EXAMPLE 7   **Finding the Sum of the First *n* Terms of a Geometric Sequence**

Find the sum of the first 6 terms of the geometric sequence whose first term is 4 and whose common ratio is 3.

SOLUTION: For this sequence, we have $n = 6$, $a_1 = 4$, and $r = 3$. Using Formula 4, we get,

$$\frac{a_1 \cdot (r^n - 1)}{r - 1} = \frac{4 \cdot (3^6 - 1)}{3 - 1} = \frac{4 \cdot (729 - 1)}{3 - 1} = \frac{4 \cdot 728}{2} = \frac{2{,}912}{2} = 1{,}456.$$

Example 8 shows the dramatic difference between arithmetic and geometric growth.

Highlight: *The Mathematics of Chain Letters*

You have perhaps received a letter assuring you of riches that goes something like this:

This letter is not a joke. It really works! Follow these instructions carefully and do not break the chain. Send one dollar to the person who is first on the list, cross off that person's name and add your own name to the bottom of the list. Then send the new list to five people asking them to follow these same instructions. If you do not break the chain, after several weeks, you will receive in the mail over one million dollars.

At first, this plan seems like it might work, provided that no one breaks the chain. In step 1, you send five letters, then each of these five people send five more letters for 25 or 5^2 letters. At the next stage, these 25 peo-ple send five letters each to get $125 = 5^3$ letters. If the list has 15 names, it would seem that by the time your name gets to the top of the list, you would be on easy street.

The problem, however, is that 5, 25, 125, 625, . . . is a geometric sequence with $a_1 = 5$ and $r = 5$. If the original list has 15 names, by the time your name reaches the top of the list and you start cashing in, $a_{15} = 5(5^{14}) = 5^{15}$. However, 5^{15} is roughly 30 billion which is about 5 times the size of the whole world's population! Even if no one breaks the chain, you will probably never get to see any money because we run out of people before your name ever gets to the top of the list.

Schemes, such as chain letters, that rely on geometric growth, or "pyramid," cannot work and are usually illegal.

EXAMPLE 8   Comparing Arithmetic and Geometric Growth

Suppose that you have won a lottery and can take your choice of either of the following two options.

a) You receive $50,000 this month, $60,000 next month, $70,000 the following month, $80,000 the month after that, and so on, for 30 months.

b) You receive one cent this month, two cents next month, four cents the following month, eight cents the month after that, and so on for 30 months.

Which option is the better deal? (Before you do the calculations, decide which option you would choose.)

SOLUTION: With option a) we are summing the first 30 terms of an arithmetic sequence in which $a_1 = 50,000$, $d = 10,000$, and $n = 30$. In order to use Formula 2, we need to know the 30th term, a_{30}. Using Formula 1, we get

$$a_{30} = a_1 + (n - 1)d = 50,000 + (30 - 1)10,000$$
$$= 50,000 + 290,000 = 340,000$$

Now we use Formula 2 to get the sum of the 30 months of payments as follows:

$$\frac{n(a_1 + a_n)}{2} = \frac{30(50,000 + 340,000)}{2} = \frac{30 \cdot 390,000}{2} = 5,850,000.$$

So, with option (a) you will receive $5,850,000.

With option b), we are summing the first 30 terms of a geometric sequence in which $a_1 = 1$, $r = 2$, and $n = 30$. We will use Formula 4 to get

$$\frac{a_1 \cdot (r^n - 1)}{r - 1} = \frac{1 \cdot (2^{30} - 1)}{2 - 1} = \frac{1,073,741,823}{1} = 1,073,741,823.$$

The number 1,073,741,823 represents cents, so we convert this to dollars to get 10,737,418.23 dollars. Therefore we see that the second option will give you almost $11 million dollars versus the $5,850,000 of the first option.   ◎

Example 8 shows how remarkably fast quantities can grow if we are using a geometric sequence versus an arithmetic sequence.

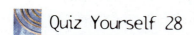

Quiz Yourself 28

Find the sum of the first 8 terms in the sequence 2, 6, 18, 54, ...

We add consecutive pairs of terms to generate terms in a Fibonacci sequence.

In 1202, Leonardo of Pisa, also known a Fibonacci, wrote a book *Liber Abaci* in which he introduced the Hindu–Arabic numeration system to Europe. His intention was to explain this new system in order to replace the more tedious Roman system of numeration for doing calculations. He included in his text the following problem that has given rise to perhaps the most famous sequence in the history of mathematics.

EXAMPLE 9   The Fibonacci Sequence

Assume that every pair of baby rabbits will mature during their second month and produce a pair of baby rabbits during their third month. If we place a single pair of adult rabbits in a pen, how many rabbits (both babies and adults) will there be in the pen in the eighth month? We are assuming that once adults start producing babies that they do so every month from then on.

SOLUTION: Figure 5.25 will help you to understand the problem.

If we count the pairs of rabbits each month, we get the following sequence: 1, 2, 3, 5, 8, ...

The pattern that is beginning to emerge is that from the third term on, each term in the sequence is the sum of the previous two terms. Study Figure 5.25 carefully to see why this must be so. If we continue this pattern, there will be $5 + 8 = 13$ pairs in the sixth month, $8 + 13 = 21$ pairs in the seventh month, and $13 + 21 = 34$ pairs in the eighth month.   ◎

The pattern that we saw in Example 9 is essentially the Fibonacci sequence.

FIGURE 5.25   The growth of rabbits leads to the Fibonacci sequence.

DEFINITION

The **Fibonacci sequence** is the sequence that begins

$$1, 1, 2, 3, 5, 8, 13, 21, 34, 55, 89, \ldots$$

where each term after the second is the sum of the previous two terms. We often label the terms in the Fibonacci sequence by $F_1, F_2, F_3, F_4, \ldots$ rather than $a_1, a_2, a_3, a_4, \ldots$

$\frac{1}{1} = 1$
$\frac{2}{1} = 2$
$\frac{3}{2} = 1.5$
$\frac{5}{3} = 1.66$
$\frac{8}{5} = 1.6$
$\frac{13}{8} = 1.625$
$\frac{21}{13} = 1.615$
$\frac{34}{21} = 1.619$
$\frac{55}{34} = 1.617$
$\frac{89}{55} = 1.618$
$\frac{144}{89} = 1.618$
$\frac{233}{144} = 1.618$

TABLE 5.3 Quotients obtained by dividing consecutive Fibonacci numbers.

The Fibonacci sequence occurs in nature in many different ways. The seeds of some flowers, for example, daisies and sunflowers, are arranged in spirals going in two different directions. If you count the number of spirals going in one direction and then count the number of spirals going in the other direction, you will often find that the numbers you get are a pair of consecutive numbers in the Fibonacci sequence, such as 21 and 34, or 34 and 55 (see Figure 5.26(a)). The hexagonal "bumps" on the skin of a pineapple and the woody leaf-like structures on a pine cone also occur in two types of spirals (see Figure 5.26(b)). Again, if you count the spirals, you will get a pair of Fibonacci numbers. Later in this section, we will show you how the Fibonacci sequence describes the pattern in the shell of the chambered nautilus (see Figure 5.29(b)).

So much research has been done on the Fibonacci sequence that there is a Fibonacci Association and even a research journal called The Fibonacci Quarterly devoted to publishing research on its properties.

One of the remarkable patterns that was discovered about the Fibonacci sequence is that if you go through the sequence, dividing each term by the term preceding it, you get the numbers which we see in Table 5.3.* It seems as though these numbers are settling down and going toward some fixed number. And,

FIGURE 5.26   The Fibonacci sequence appears in (a) the spirals in a sunflower and (b) the patterns on a pineapple.

---

* We used a spreadsheet to generate these numbers and in many cases we are approximating the true quotient, which has an infinite, repeating expansion.

FIGURE 5.27 The architectures of the Parthenon and the United Nations building are based on the golden ratio.

in fact they are. It was proved about 200 years ago that the number these quotients are approaching is $\frac{\sqrt{5}+1}{2}$, which we will call ϕ (Phi), and approximate it by 1.618.

The number ϕ is often called the **golden ratio** and was known to the ancient Greeks, although they obtained it in a different way. They believed that any rectangle whose one side was $\frac{\sqrt{5}+1}{2}$ times as long as its other side was perfectly proportioned, and they used this principle in art and architecture. Such a rectangle is called a **golden rectangle.** For example, they used golden rectangles in designing the Parthenon (see Figure 5.27(a)). In modern times, architects used golden rectangles to determine the rectangular shapes in the sides of the United Nations building (see Figure 5.27(b)). The pleasing shape of the golden rectangle is also found in art and in sculpture from ancient to modern times.

An interesting property of a golden rectangle is that if you cut off square (a) on one end of the rectangle, as we show in Figure 5.28, the smaller rectangle that remains is again a golden rectangle. If you then cut off square (b), again, the part that remains is a golden rectangle. If you continue doing this, and connect

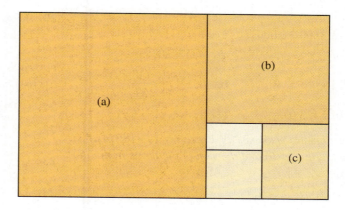

FIGURE 5.28 Cutting off squares from a golden rectangle gives other golden rectangles.

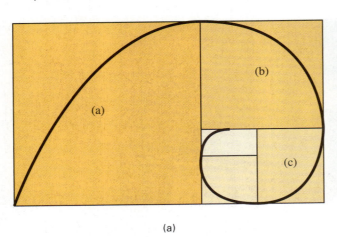

(a)

FIGURE 5.29   (a) Spiral generated from golden rectangles, (b) Chambered nautilus shell.

the opposite diagonals of the squares that you are cutting off with a smooth curve, you generate a spiral that is much like the shape of a chambered nautilus shell (see Figure 5.29).

Exercises

In Exercises 1–12, identify the sequence as either arithmetic or geometric. List the next two terms of each sequence.

1. 5, 8, 11, 14, . . .

2. 11, 7, 3, − 1, . . .

3. 8, 24, 72, 216, . . .

4. 4, 12, 20, 28, . . .

5. $1, \dfrac{1}{2}, \dfrac{1}{4}, \dfrac{1}{8}, \ldots$

6. 0.1, 0.01, 0.001, 0.0001, . . .

7. 10, 5, 0, − 5, . . .

8. 4, 17, 30, 43, . . .

9. 2, 4, 8, 16, . . .

10. $1, \dfrac{1}{3}, \dfrac{1}{9}, \dfrac{1}{27}, \ldots$

11. 1.5, 2.0, 2.5, 3.0, . . .

12. 1, − 1, 1, − 1, . . .

For each arithmetic sequence in Exercises 13–18: (a) find the specified term a_n; (b) find the sum of the terms from a_1 to a_n, inclusive.

13. 5, 8, 11, 14, . . . ; Find a_{11}

14. 11, 17, 23, 29, . . . ; Find a_9

15. 2, 8, 14, 20, . . . ; Find a_{15}

16. − 6, − 2, 2, 6, . . . ; Find a_{22}

17. 1, 1.5, 2.0, 2.5, . . . ; Find a_{20}

18. 3, 3.25, 3.5, 3.75, . . . ; Find a_{11}

19. In Exercise 13, find the sum of the terms from a_{14} to a_{21}, inclusive.

20. In Exercise 14, find the sum of the terms from a_{21} to a_{37}, inclusive.

For each geometric sequence in Exercises 21–26: (a) find a_n; (b) find the sum of the terms from a_1 to a_n.

21. 1, 3, 9, 27, . . . ; Find a_{11}

22. 3, 6, 12, 24, . . . ; Find a_9

23. $1, \dfrac{1}{2}, \dfrac{1}{4}, \dfrac{1}{8}, \ldots$; Find a_7

24. 2, − 4, 8, − 16, . . . ; Find a_{10}

25. 2, 0.2, 0.02, 0.002, . . . ; Find a_6

26. 5, 50, 500, 5000, . . . ; Find a_9

In Exercises 27–30, we give you two terms in the Fibonacci sequence and you must find the specified term.

27. $F_{11} = 89$ and $F_{13} = 233$. Find F_{12}.

28. $F_{22} = 17{,}711$ and $F_{23} = 28{,}657$. Find F_{21}.

29. $F_{13} = 233$ and $F_{15} = 610$. Find F_{14}.

30. $F_{23} = 28{,}657$ and $F_{24} = 46{,}368$. Find F_{25}.

31. Finding total salary. Redo Example 4, but now assume that your starting salary is $2,875 and you are getting an increase of $35 per month. Find the amount that you will earn in a year.

32. Displaying products. There is a pyramid of cans against the wall of a supermarket. There are 11 cans on the first row, 10 on the second row, 9 on the third row, and so on. What is the total number of cans on the stack?

33. Growth in a bank account. You have $1,200 in your bank account that is paying an interest rate of 3.5% (0.035) per year. How much money will you have in that account in six years?

34. Chain letters. Reconsider the discussion of chain letters in the highlight. Assume that you receive a chain letter and at each stage, the person who receives the letter must send the letter to eight people. If you send your letters and the chain is unbroken, at what stage will the number of letters sent at that stage exceed the population of the world, which in 2001 was about 6.1 billion people?

35. Height of a bouncing ball. A ball is dropped from a height of 8 feet. The ball always bounces $\frac{7}{8}$ of the distance from which it was dropped. What will be the height of the ball after the fifth bounce?

36. Gambling strategy. A sure-fire way to never lose when betting at a casino game is to use the following strategy: if you lose, then always double your bet and play again until you win. Assume that you have $2,000. If you are gambling at a roulette wheel and you begin with a $3 bet and continue to lose, at what stage will you be unable to continue doubling your bet?

Further Exercises

*37. Find a formula to add the terms from a_k to a_n in an arithmetic sequence. This pattern is exactly like the pattern for adding the first n terms. Look at examples and add the terms from a_k to a_n in two different ways, as we did in the discussion prior to Formula 2.

38. Derive Formula 4 by explaining the following steps.

a) The first n terms of a geometric sequence are

$$a + ar + ar^2 + ar^3 + ar^4 + \ldots + ar^{n-1}.$$

b) Rewrite this as $a(1 + r + r^2 + r^3 + r^4 + \ldots + r^{n-1})$.

c) But $1 + r + r^2 + r^3 + r^4 + \ldots + r^{n-1} = \frac{1 - r^n}{1 - r}$.

d) Therefore,

$$\begin{aligned}
a_1 + a_1 r + a_1 r^2 &+ a_1 r^3 + a_1 r^4 + \ldots + a_1 r^{n-1} \\
&= a_1(1 + r + r^2 + r^3 + r^4 + \ldots + r^{n-1}) \\
&= a_1 \frac{1 - r^n}{1 - r} = \frac{a_1(r^n - 1)}{r - 1}
\end{aligned}$$

You are given the population and the growth rate as of 2001 for the following countries. Assume that the growth rate remains the same from year to year. Determine the size of the population in 2010.

39. Brazil: population = 174 million   growth rate = 1.3%

40. China: population = 1,273 million   growth rate = 1%

41. Communicating Mathematics   By looking at concrete examples, derive a formula for the sum of the first n Fibonacci numbers, $F_1 + F_2 + F_3 + \ldots + F_n$.

42. Communicating Mathematics   By looking at concrete examples, derive a formula for the sum of the squares for two consecutive Fibonacci numbers, F_n and F_{n+1}.

The Binet form of Fibonacci numbers uses the formula

$$F_n = \frac{\left(\dfrac{(1 + \sqrt{5})}{2}\right)^n + \left(\dfrac{(1 - \sqrt{5})}{2}\right)^n}{\sqrt{5}}.$$

---

* Exercise numbers circled in red can be used as group exercises.

For example, to calculate F_8, substitute 8 for n in the above formula. Using a calculator, due to roundoff errors, the author got 21.00736 for F_8 whose value is actually 21. Use this formula and a calculator to approximate the Fibonacci numbers in Exercises 43–46.

43. F_5

44. F_7

45. F_9

46. F_{11}

47. Find examples of the golden ratio in art and architecture.

48. Locate a book or other source that deals with the Fibonacci sequence and find a new property of the Fibonacci sequence that we have not discussed in this section.

CHAPTER 6

Algebraic Models:* How Do We Approximate Reality?

What is a model? To an automotive engineer in Detroit, it might be a series of computer graphics images describing the effect of wind resistance on the gas mileage of a sleek, futuristic car. To an office designer, it might be a computer rendering of plans for a new office building. A meteorologist at a midwestern research university might consider a model to be a series of complex equations describing the steamy conditions on a summer evening that can lead to a devastating tornado. In each case, the model builder seeks to represent reality in a simplified way. By stripping away unimportant details, they can understand relationships between the variable quantities in a situation and use their models to predict future events.

In this chapter we will use equations to represent reality just as the engineer, office designer, and meteorologist do. You will see that these equations are more than just arrangements of numbers and *x*'s and *y*'s. These algebraic models will enable you to solve real-world problems, although less complex than the ones mentioned here. In Sections 6.1 and 6.2, we build simple models by using linear equations. Although data rarely conforms exactly to a linear equation, you will see that linear equations are often "good enough" to give a reasonably accurate description of actual data.

In Section 6.3, we will use quadratic equations to model situations where the graph of the data is curved in a way that makes linear models inappropriate. The discussion of exponential and logistic models in Section 6.4 will show you how to represent the

---

* Further resources on algebraic models can be found at www.aw.com/pirnot.

explosive growth typical in bank accounts, populations, and the spread of disease.

By the end of this chapter, you will have studied a variety of algebraic equations and will have learned when they are appropriate models and when they are not.

6.1   LINEAR EQUATIONS

Algebra is an effective tool in modeling real-world situations because we can use abstract reasoning to describe relationships between different quantities. You will see in this section that relatively simple equations can give us good approximations of real situations. It is usually simpler to manipulate equations than to manipulate concrete objects, such as heavy office furniture or automobile parts. Therefore, we can spot design errors and make changes more quickly and cost effectively than if we were working with actual objects.

Linear equations have a constant rate of change between the variables.

In spite of its simplicity, a linear equation often provides us with a reasonably good model for a complex set of data.

> **DEFINITIONS**
>
> A **linear equation*** in two variables is an equation that can be written in the form
>
> $$Ax + By = C,$$
>
> where A, B, and C are real numbers and A and B are not both zero. When a linear equation is written in this form, it is in **standard form**.

The equations $-3x + y = 6$ and $5p = 6q - 4$ are examples of linear equations in two variables. In a linear equation in two variables, there is a constant rate of change between the two variables. To understand what we mean by this, let's rewrite the first equation in the form $y = 6 + 3x$. Notice that if we increase x by a certain amount, then y increases by three times as much. For example, if we increase x from 10 to 15, which is an increase of 5, then y increases from 36 to 51, which is an increase of 15. If we increase x from 1,000 to 1,007, then y increases from 3,006 to 3,027, which is an increase of 21.

A **solution** for an equation is a number or numbers such that if we substitute them for the variables in the equation, the resulting statement is true. Solutions of linear equations are ordered pairs of numbers. For example, the **ordered**

---

* Linear equations contain variables to the first power only.

pair (3, 15)* is a solution for the equation $-3x + y = 6$ because if we substitute 3 for x and 15 for y, we get the true statement $-3(3) + 15 = 6$. Two equations are **equivalent** if they have the same solutions. In solving an equation, we often use the following rules to rewrite it in a simpler, equivalent form. These rules apply to all types of equations.

> **Rewriting Equations in an Equivalent Form**
>
> 1. Adding or subtracting the same expression from both sides of an equation gives an equivalent equation.
> 2. Multiplying or dividing both sides of an equation by the same nonzero expression gives an equivalent equation.

For example, to solve an equation† such as

$$0.5x + 0.7 = 0.35x + 1.3,$$

we might first multiply both sides by 100 to get the equivalent equation

$$50x + 70 = 35x + 130.$$

Subtracting 70 and $35x$ from both sides gives us the equivalent equation

$$15x = 60,$$

whose solution is $x = 4$.

SOME GOOD ADVICE

When solving equations, you may have been told that "when a quantity is taken from one side of an equation to the other, plus signs change to minus signs and vice versa." It is unwise to rely on memory devices such as this. Our advice for working with equations can be traced back to the early Greeks who had rules such as "equals added to equals give equals." Remember when rewriting equations, you are *adding, subtracting, multiplying, or dividing* both sides of the equation by the same quantity.

We can use linear equations‡ to model many consumer purchases. The pattern shown in renting Jet-Skis in Example 1 also occurs in car rental agreements, long-distance calling plans, health club memberships, and many service contracts.

EXAMPLE 1 Comparing Jet-Ski Rental Plans

A store that rents Jet-Skis has two rental plans. In plan A, the customer pays a base fee of $35.00 plus $12.50 per hour. In plan B, the customer pays a base fee

---

* In the pair (3, 15), the number 3 is called the *first coordinate* and the number 15 is called the *second coordinate*. These coordinates are also called the *x*-coordinate and *y*-coordinate, respectively.
† This is a linear equation in a single variable.
‡ Unless we state otherwise in this section, linear equations will always have two variables.

of $22.50 plus $15 per hour. For what number of hours do plans A and B charge the same amount?

SOLUTION: The cost of renting a Jet-Ski for n hours using plan A is the base fee of $35 plus n times $12.50. If we let c represent the total rental cost of a Jet-Ski, we can model plan A by the equation

$$c = 35 + n(12.50).$$

Similarly, we model plan B by the equation

$$c = 22.50 + n(15).$$

We want to know when these two costs are equal, so we will solve the equation

$$35 + n(12.50) = 22.50 + n(15).$$

Multiplying both sides of this equation by 100 and simplifying, we get

$$3{,}500 + 1{,}250n = 2{,}250 + 1{,}500n.$$

Next subtract 2,250 and 1,250n from each side, which gives us

$$1{,}250 = 250n.$$

Dividing by 250, we see that $n = 5$ is the solution.

We can now use this simple model to decide which rental plan is better. We see that if we rent a Jet-Ski for five hours, it does not matter which plan we use. If we rent for less than five hours, plan B is better. If we rent for more than five hours, plan A is cheaper.

Notice the constant rate of change between the variables in Example 1. For instance, in plan A, each additional hour that we rent causes our cost to increase by $12.50.

Many tax laws are also based on a constant rate of change, as illustrated in Example 2.

EXAMPLE 2   Modeling Income Tax with a Linear Equation

According to a particular tax law, a wage earner must pay $1,380 plus 22 percent of income over $8,000. If a person pays $13,260 in taxes, model this situation with a linear equation.

SOLUTION: Using the strategy of choosing good names from Section 1.1, we will represent income by I. We do not have to model the situation all in one step. In fact, mathematicians often break a problem into smaller parts before developing a complete solution.

We can model the phrase "the amount of income over $8,000" algebraically by the expression $I - 8{,}000$. You can verify that this expression is correct by testing it with several incomes over $8,000. The phrase "22 percent of the amount

Quiz Yourself 1*

It costs $40 a day to rent a car, plus $0.15 per mile. Let *c* represent the cost of renting a car for one day if it is driven *m* miles. Write an equation to describe this rental agreement.

of income over $8,000" can then be represented by $0.22(I - 8,000)$. We are told that the income tax paid is $1,380 more than this, so the tax paid is

$$0.22(I - 8,000) + 1,380.$$

We set this equal to $13,260 to get the equation

$$0.22(I - 8,000) + 1,380 = 13,260.$$

In Example 2, we described the amount of income over $8,000 by the expression $I - 8,000$. You may have been tempted to write $8,000 - I$ instead. You can decide easily which is correct by making a numerical example. If the income were $40,000, then $I - 8,000 = 32,000$; however, $8,000 - I = -32,000$, which does not make sense. Therefore $I - 8,000$ is the correct representation.

In many applications, we need to express one variable in an equation in terms of the others.

EXAMPLE 3   Converting Fahrenheit Temperatures to Celsius

The following equation converts the temperature in degrees Celsius, C, to the temperature in degrees Fahrenheit, F:

$$F = \frac{9}{5} C + 32 \tag{1}$$

Solve this equation for C; that is, write an equation of the form $C = \ldots$.

SOLUTION: We can solve for C by rewriting equation (1) in the following series of steps.

Quiz Yourself 2

Solve the equation $P = 2L + 2W$, which expresses the perimeter of a rectangle in terms of its length and width, for *W*.

$$F - 32 = \frac{9}{5} C \quad \text{(subtract 32 from both sides)}$$

$$\frac{5}{9} (F - 32) = C \quad \text{(multiply both sides by } \tfrac{5}{9}\text{)}$$

We can now use this equation to convert degrees Fahrenheit to degrees Celsius. For example, a nice warm 95° Fahrenheit day at the beach converts to a chilly sounding $C = \frac{5}{9}(95 - 32) = 35°$ Celsius.

We use intercepts to graph a linear equation.

The graph of a linear equation in two variables is a straight line. Therefore, we can draw such graphs by plotting two solutions to the equation and drawing a straight line through these points. Drawing the graph of a linear equation gives us a way of visualizing the relationship between the two variables in the model.

In some applications, you know two data points that satisfy a linear equation, and you can use those points to graph the line. However, in other situations,

---

Historical Highlight: *René Descartes and Analytical Geometry*

The mathematician and philosopher René Descartes was born in La Haye, France, in 1596. Because of his poor health during childhood, his early teachers allowed him the luxury of remaining in bed as late as he pleased. This became a lifelong habit to which he attributed his later success, because it gave him time to think and reflect on philosophy and mathematics. While a student, he began to doubt whether his studies were leading him to true knowledge, which prompted him to begin a search for a means to learn the truth. In 1637 he published one of his most famous works, *Discourse on Method*, in which he proposed these basic principles for finding the truth:

1) Accept nothing as true which you do not fully understand.

2) Divide a complicated problem into several parts.

3) Arrange thoughts in order, beginning with the simplest and proceeding to the complex.

4) Review all calculations and arguments so thoroughly that nothing is omitted.

Descartes had such faith in his method that he wrote,

Those long chains of reasoning, each step simple and easy, . . . , have led me to surmise that all things we human beings are competent to know are interconnected in the same manner, and that none are so remote as to be beyond our reach or so hidden that we cannot discover them. . . .

In an appendix to *Discourse on Method*, Descartes developed what today we call analytical geometry, in which we represent objects such as points, lines, circles, and planes as numbers and equations.* The advantage that analytical geometry has over the traditional geometry developed by the ancient Greeks is that we can use algebraic methods such as solving equations to solve geometric problems.

Descartes also made important contributions to chemistry, anatomy, embryology, medicine, astronomy, and meteorology. His work in biology was so important that some have referred to him as the "father of modern biology." He also made significant discoveries in several areas of physics, particularly in optics, and he is credited with being the first person to satisfactorily explain the rainbow.

Unfortunately, this genius met with an untimely end while tutoring Queen Christine of Sweden, who insisted on rising at five o'clock to study philosophy with him. The early hour and the severe Swedish winter were too much for Descartes—he caught pneumonia and died on February 11, 1650.

you have the linear equation and must find solutions before graphing. It doesn't matter which points you choose; however, there are two particularly simple points, called the intercepts, that are convenient to use in drawing the graph of a linear equation.

> **DEFINITIONS**
>
> The **x-intercept** of the graph of a linear equation is the point where the graph crosses the *x*-axis. The **y-intercept** is the point where the graph crosses the *y*-axis. (See Figure 6.1.)

*As he lay in bed watching a fly walk on the ceiling, Descartes realized that he could describe the fly's position by the pair of numbers telling its distance from two perpendicular walls.

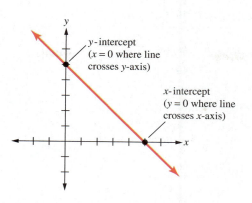

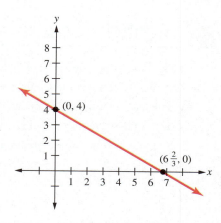

FIGURE 6.1 Plotting *x*- and *y*-intercepts.

FIGURE 6.2 Using intercepts to graph $3x + 5y = 20$.

SOME GOOD ADVICE

You may mistakenly try to calculate the *x*-intercept by setting *x* = 0. Keep in mind that the term "*x*-intercept" means the point where the line "intercepts" or crosses the *x*-axis, so the *y*-coordinate of the intersection point is 0.

In Example 4, we use simple algebra to find intercepts.

EXAMPLE 4   Using Intercepts to Graph a Linear Equation

Find the intercepts and graph the equation $3x + 5y = 20$.

SOLUTION: To find the *x*-intercept, we set $y = 0$. This gives us $3x + 5(0) = 20$, or $3x = 20$. Thus, $x = 6\frac{2}{3}$, so the *x*-intercept of the line is the point $(6\frac{2}{3}, 0)$.

Setting $x = 0$, we get $3(0) + 5y = 20$. This simplifies to $5y = 20$, which means that the *y*-intercept is the point $(0, 4)$. We graph this line in Figure 6.2.

EXAMPLE 5   Interpreting Intercepts in a Financial Situation

Suppose you have saved $2,205 to use for living expenses that you expect to be $315 per month. Express this information in an equation using the variables *r* (for remaining cash) and *m* (for the number of months you are spending the money). Interpret the meaning of the intercepts for the graph of this equation.

SOLUTION: You are starting with $2,205, and each month you will reduce this amount by $315. Therefore the equation

$$r = 2{,}205 - 315m$$

describes your financial state at the end of *m* months. The *r*-intercept corresponds to the time when $m = 0$; that is,

$$r = 2{,}205 - 315(0) = 2{,}205.$$

At this point, you still have all your money.

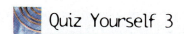

Quiz Yourself 3

Find the intercepts and draw the graph of the equation $4x + 3y = 18$.

Highlight: *Using Technology to Draw Graphs**

Many equations are too complicated to graph accurately by merely plotting a few points. Fortunately, technology such as graphing calculators and computer algebra systems makes our job easier. In Figure 6.3, we used a computer algebra system to graph the equation

$$y = \frac{2x}{\sqrt{x^4 + 10}} - 2\frac{(x^2 + 6)(x^3)}{(x^4 + 10)^{3/2}}$$

over the interval $x = -10$ to $x = 10$.

   We have also plotted three points on the graph of this equation. If we happened to choose only these three points to graph the equation, we would be completely fooled as to what the graph looked like and might draw a straight line. You might ask then, "How many points should we plot? Ten? Twenty?" It is unlikely that you would get an accurate graph by simply plotting 10 or 20 points. In the area of mathematics called calculus, mathematicians use sophisticated techniques to graph equations properly, but this approach can require complicated algebra and tedious hand calculations.

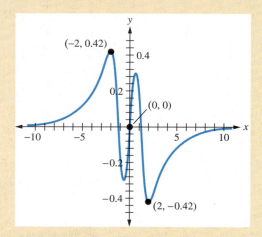

FIGURE 6.3   A graph of a complex equation drawn by a computer algebra system.

   To draw this graph, the computer algebra system plotted 107 points† that had fifteen decimal places of accuracy. The system plotted many more points where the graph is turning than where the graph is relatively straight. For example, between $x = -10$ and $x = -8$, the system chose only five points in drawing the graph, but between $x = -3$ and $x = -1$, it used 23 points.

   To find the m intercept, we set $r = 0$. This is the time at which you run out of savings. To find this time, we will solve the equation

$$0 = 2{,}205 - 315m.$$

Adding $315m$ to each side of the equation, we get $315m = 2{,}205$, and then, dividing both sides by 315, we obtain $m = 7$. Therefore you can expect to deplete your savings at the end of seven months.

   When using linear equations as models, we often want to know the steepness of an equation's graph. An economist may want to know how rapidly inflation is growing or, perhaps during a drought, government officials would want to know how rapidly a town's water supply is being depleted. We measure the steepness of a line by its slope.

---

* For tutorials on how to use technology to draw graphs go to www.aw.com/pirnot.
† It is not possible to see the 107 points that the system used. However, when using a computer algebra system, it is possible to ask the system to print out the points it used for the graph.

DEFINITION

If (x_1, y_1) and (x_2, y_2) are two points on a line and $x_1 \neq x_2$, then the **slope** of the line is defined as

$$m = \frac{\text{rise}}{\text{run}} = \frac{\text{change in } y}{\text{change in } x} = \frac{y_2 - y_1}{x_2 - x_1}.$$

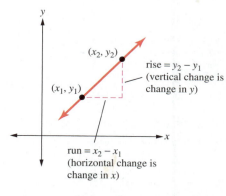

(x_2, y_2)

(x_1, y_1)

rise $= y_2 - y_1$
(vertical change is change in y)

run $= x_2 - x_1$
(horizontal change is change in x)

FIGURE 6.4   The rise and run determine a line's slope.

You can remember the meaning of the words *rise* and *run* by thinking about their use in everyday language. We watch the sun rise and see a cake or a hot air balloon rising. In each case something is going up, so rise corresponds to a vertical change. For the word *run*, think of a person running across a field or a train running down a track. In these cases, *run* means a horizontal change. Figure 6.4 illustrates the relationship between slope, rise, and run.

Quiz Yourself 4

Find the slope of the line through the points (7, 5) and (11, 10).

EXAMPLE 6   Finding the Slope of a Line

Calculate the slope of the line containing the points (1, 5) and (6, 45).

SOLUTION: The slope of the line is

$$\text{slope} = \frac{\text{rise}}{\text{run}} = \frac{\text{change in } y}{\text{change in } x} = \frac{45 - 5}{6 - 1} = \frac{40}{5} = 8.$$

SOME GOOD ADVICE

In the definition of slope, if the run is equal to 1, then the slope is simply $y_2 - y_1$. This means that we can interpret the slope of a line as the amount of change in y corresponding to a change of 1 in x.

By estimating the rise and the run (including whether it is positive or negative), we can get a good idea of the slope of the line without doing any calculations. In Figure 6.5, line (a) has a positive slope, since both rise and run are positive. Line (b) has negative slope, since the run is positive but the rise is negative.

Notice that if a line is horizontal, then its rise is zero, so slope $= \frac{\text{rise}}{\text{run}} = \frac{0}{\text{run}}$, which is equal to zero. Similarly, if a line is vertical, then slope $= \frac{\text{rise}}{\text{run}} = \frac{\text{rise}}{0}$, which is undefined because we cannot divide by zero.

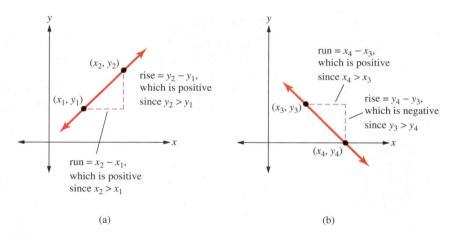

FIGURE 6.5   (a) A line with a positive slope rises from left to right. (b) A line with a negative slope falls from left to right.

The slope-intercept form of a linear equation tells us geometric information about its graph.

> **DEFINITION**
>
> A linear equation is in **slope-intercept form** if it is written in the form $y = mx + b$. The number m is the slope of the line that is the graph of the equation and $(0, b)$ is the y-intercept.

The equation $y = 2x + 3$ is in slope-intercept form. The slope is 2 and the y-intercept is $(0, 3)$. Example 7 shows that we can recognize useful geometric information about a line if we write its equation in slope-intercept form.

EXAMPLE 7   **Using the Slope-Intercept Form of Linear Equations to Evaluate Alternatives**

You are considering renting a car from two rental companies. Save-U charges an $89 base fee plus $35 per week; Cheap Car charges a $119 base fee plus $27 per week. Model each of these rental plans by a linear equation in slope-intercept form. Graph the equations and estimate at which point renting from Cheap Car begins to save you money.

SOLUTION: We will model Save-U's and Cheap Car's pricing structures by the following equations

$$y = 35x + 89 \qquad \text{(Save-U's pricing)}$$

and

$$y = 27x + 119 \qquad \text{(Cheap Car's pricing)}$$

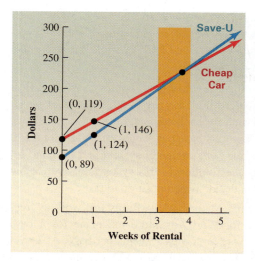

FIGURE 6.6   The cost of Cheap Car drops below the cost of Save-U during the fourth month.

From these equations, we can see that the slope of Save-U's graph is 35 and the slope of Cheap Car's graph is 27. This means that although Save-U's price is initially cheaper, because of the steeper slope of its graph, eventually Save-U's price becomes more expensive.

We graph each equation by plotting two points. From Save-U's equation, we see that the y-intercept is $(0, 89)$, which gives us one point on the graph. For a second point, we set $x = 1$ to get the equation

$$y = 35(1) = 124.$$

Thus, $(1, 124)$ is a second point on the graph of Save-U's equation.

To graph Cheap Car's equation, we can use the y-intercept, $(0, 119)$. For a second point, we again set $x = 1$ to get the equation

$$y = 27(1) + 119 = 146.$$

Therefore, $(1, 146)$ is a second point on Cheap Car's graph.

In Figure 6.6 we use the points $(0, 89)$ and $(1, 124)$ to graph Save-U's equation and $(0, 119)$ and $(1, 146)$ to graph Cheap Car's equation.

Until $x = 3$, Save-U's graph lies below the graph of Cheap Car, so up to three weeks, renting from Save-U is the better deal. By week four, the graph of Cheap Car is below the graph of Save-U, and from that point on, it is cheaper to rent from Cheap Car.

We also can use algebra to solve Example 7. The point where the graphs cross tells us when Save-U's price is equal to Cheap Car's price. This gives us the equation $35x + 89 = 27x + 119$. Solving this equation, we find that $x = 3\frac{3}{4}$.

The fact that we got the same solution both algebraically and geometrically demonstrates Descartes's reasoning in developing analytical geometry—what we can represent geometrically we can represent algebraically, and vice versa.

It is important when selecting a model that the model is appropriate for the situation you are modeling. When you write a linear equation in slope-intercept form as $y = mx + b$, realize that an increase of 1 for x causes an increase of m for y; an increase of 2 for x causes an increase of $2m$ for y; an increase of 5 for x causes an increase of $5m$ for y; and so on. The point is that a change in x causes an increase of m times that change for y.

If the situation you are modeling does not satisfy this property, then a linear equation is not a good choice for a model. In subsequent sections of this chapter, we discuss several real-life situations that do not satisfy this property; in those cases, we use nonlinear equations to model them.

Exercises 6.1

In Exercises 1–6, solve each equation by applying the rules for rewriting equations.

1. $3x + 4 = 5x - 6$

2. $4 - 2x = 9x + 13$

3. $4 - 2y = 8 + 3y$

4. $5y - 6 = 14y + 12$

5. $\frac{1}{2}x + 4 = \frac{3}{4}x - 6$

6. $\frac{1}{2}x - 6 = \frac{1}{5}x + 3$

7. $\frac{1}{3}y + 4 = \frac{1}{4}y + 3$

8. $\frac{5}{6}y + 1 = \frac{1}{3}y - 2$

9. $0.2x + 6 = 3x - 0.4$ **10.** $0.25x + 0.35 = 0.2x - 4$

11. $0.3y + 2 = 0.5y - 3$ **12.** $0.4y - 0.2 = 0.6y + 3$

In Exercises 13–24, solve each equation for the stated variable.

13. $P = 2l + 2w;$ solve for w

14. $m = \dfrac{a + b}{2}$; solve for a

15. $z = \dfrac{x - \mu \,*}{\sigma}$; solve for μ

16. $z = \dfrac{x - \mu}{\sigma}$; solve for σ

17. $A = P(1 + rt)$; solve for r

18. $A = \dfrac{1}{2}h(b + B)$; solve for b

19. $2x + 3y = 6$, for x **20.** $4x - 5y = 3$, for y

21. $V = lwh$, for l **22.** $A = \dfrac{1}{2}hb$, for b

23. $S = 2\pi rh + 2\pi r^2$, for h

24. $A = 2lw + 2lh + 2hw$, for w

In order to solve Exercises 25–28, first write an equation that describes each situation. Use meaningful names for the variables.

25. Computing health club charges. A health club charges a yearly membership fee of $95 and members must pay $2.50 per hour to use its facilities. How many hours did Paola use the club last year if her bill was $515?

26. Computing video club charges. A video club has a $5 membership fee and members can buy videos at $12 per video. How many videos did Marcus buy if his annual bill was $233?

27. Computing charges for word processing. Thiep does word processing and charges $16 for documents up to ten pages and $1.30 per page extra for pages beyond ten. If he billed a customer $44.60, how long was the customer's document?

28. Computing overtime. Sarah works at a factory job where she is paid $9 per hour up to 40 hours. If she works over 40 hours, she is paid 1.5 times her usual wage. If she earned $468 last week, how many hours did she work?

We can describe each situation in Exercises 29–32 by a pair of linear equations as we did in Example 7. Use the algebraic method that we described after Example 7 to determine when both deals are the same.

29. Communicating Mathematics Shaun can work at Best Deal Electronics for a base pay of $225 per week plus a $45 commission for each computer system that he sells. At Circuit Town his base pay would be $400 with a $20 commission for each computer system that he sells. Interpret what the solution to this system tells you. How should Shaun decide which position to take?

30. Communicating Mathematics Emily can work as freelance editor for Wild Adventure magazine for a base salary of $20 per hour plus 25 cents per page. At Travel World magazine, she can earn $22 per hour plus 20 cents per page. Interpret what the solution to this system tells you. How should Emily decide which position to take?

31. Communicating Mathematics Cassandra is comparing two long distance phone plans. BT&T charges $12.75 per month and 7 cents per minute. Cingleton charges $14.15 per month and 5 cents per minute. Interpret what the solution to this system tells you. How should Cassandra decide which plan to take?

32. Communicating Mathematics Nugyen is considering two satellite TV systems. Global Communications charges $240 for installation and $39 per month. World Satellite charges $350 for installation and $28 per month. Interpret what the solution to this system tells you. How should Nugyen decide which system to buy?

In Exercises 33–36, write a linear equation to describe each situation before answering the question. Use good names for the variables.

33. Computing net pay. Dani earns a gross pay from which the following deductions are subtracted:

$10 per capita tax
6.75% social security tax
14% federal income tax
2.1% state tax
$17 unemployment tax

What net pay corresponds to her gross pay of $400?

34. Redo Exercise 33, but now assume that Dani's net pay is $640 and then find her gross pay.

---

* μ and σ are the lowercase Greek letters mu and sigma.

35. Computing the price of a car. Christian intends to buy a used car. In addition to the base cost of the car, he must pay the following:

$35 title and tag fee
6% sales tax
1.2% uninsured motorist fund
$45 automobile club fee

What is his total cost for a car that has a base price of $2,200?

36. Redo Exercise 35, but now assume that Christian's total cost for the car is $3,750 and then find the base price of the car.

In Exercises 37–44, write an equation that models each situation.

37. Modeling nutritional requirements. Assume that a cup of lettuce contains 10 calories and a tomato contains 25 calories. Write a linear equation that describes the amount of lettuce and number of tomatoes in a salad containing 200 calories.

38. Calculating earnings. Alicia has two part-time jobs. Her job in a diner pays $5.60 per hour and her work in a pet store pays $7.35 per hour. She earns $133 in a given week.

39. Determining rental payments. Jared furnished his new apartment by renting a living-room set and a TV from a local "rent to own" store. He has l more payments on the living-room set of $22 each, and t payments remaining on the TV of $13 each. The total he still owes is $341.

40. Finding interest on investments. Matthew has two investments. His investment in a biomedical stock pays 8 percent interest and his investment in an Internet company pays 7 percent. The total earned on his two investments is $350.

41. Investing in stock. Tamyra has invested $14,500 in the stock market. Some of the stock is Time-Warner that cost $35 a share and the rest is in Dell computers that cost $55 a share.

42. Grading. RJ's psychology professor is grading the course based on a contract system. A project is worth 40 points and a report on a journal article is worth 10 points. RJ earned 250 toward his final grade.

43. Manufacturing furniture. Justin owns a small woodworking company. He knows that it takes 9 board feet to construct an end table and 15 to construct a coffee table. He has 342 board feet available.

44. Advertising a business. Kelly is deciding how to adver-
tise a new gourmet restaurant that she is opening. A radio ad costs $300 and a TV ad costs $700. She has $7,500 to spend on the ads.

45. Communicating Mathematics Explain how to remember how to calculate x- and y-intercepts.

46. Communicating Mathematics What advantage do you see in using intercepts to graph lines rather than other points?

In Exercises 47–58, graph each equation by first finding the x- and y-intercepts.

47. $3x + 2y = 12$

48. $x - 5y = 10$

49. $4x - 3y = 16$

50. $5x + 4y = -20$

51. $\frac{1}{3}x + \frac{1}{2}y = 3$

52. $x - \frac{1}{5}y = 2$

53. $\frac{1}{6}x - 2y = \frac{3}{4}$

54. $x - \frac{1}{4}y = 2$

55. $0.2x = 4y + 1.6$

56. $0.3y = 1.2x + 0.6$

57. $0.4x - 0.3y = 1.2$

58. $0.2x + 0.5y = 2$

In Exercises 59–66, find the slope of the line passing through the two given points.

59. $(2, 5)$ and $(6, 8)$

60. $(4, 1)$ and $(7, 3)$

61. $(3, 6)$ and $(8, 2)$

62. $(9, 1)$ and $(6, 4)$

63. $(3, -4)$ and $(5, 1)$

64. $(8, -5)$ and $(9, 2)$

65. $(6, 5)$ and $(6, 8)$

66. $(2, 3)$ and $(7, 3)$

In Exercises 67–70, list all the lines in the given figure which satisfy the stated condition.

67. slope is positive

68. slope is negative

69. slope does not exist

70. slope is zero

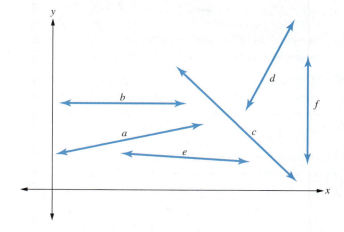

FIGURE FOR EX. 67–70

71. Communicating Mathematics Explain why the slope of a horizontal line is 0.

72. Communicating Mathematics Explain why a vertical line has no slope.

In Exercises 73–76, state the y-intercept and slope of the graph of each equation.

73. $y = 4x - 3$

74. $y = 3x + 5$

75. $y = -5x - 3$

76. $y = -2x + 5$

In Exercises 77–78, write an equation of a line in slope-intercept form. Interpret the meaning of the slope and the y-intercept in terms of the conditions stated.

77. Communicating Mathematics You owe $262.35 on a DVD player and are making payments of $23.85 on this debt.

78. Communicating Mathematics You currently have $300 in a vacation savings account. You are periodically depositing $37.50 in this account. (Ignore interest.)

79. Comparing car rental plans. You are considering renting a car as in Example 7. In this case, Save-U charges $183 base fee plus $32 per week, while Cheap Car charges $119 base fee plus $43 per week. Represent each of these rental plans by a linear equation.

a) Estimate graphically during which week renting from Save-U begins to save you money.

b) Explain why there must be a point at which the Save-U rental agreement becomes the better deal.

80. Comparing car rental plans. Solve Exercise 79 algebraically using the method we discussed following Example 7.

81. Comparing commuting costs. You travel a toll road that requires a 35-cent token. If you purchase a special express sticker for $8, it is possible to use the express lane, for which tokens are only 30 cents. At what point is the express plan cheaper?

82. Comparing food plans. College students can purchase points that can be used in the food service areas instead of cash. If you initially pay the basic food service fee of $35, then points can be purchased for 23 cents each; otherwise, points cost 30 cents each. How many points must you use in order for it to be cheaper to pay the basic food service fee?

Further Exercises

*83. Comparing investments. You wish to invest $5,000 that a wealthy relative has given to you. There is a certificate of deposit that pays 6.1 percent interest. You are in the 14 percent federal income tax bracket, however, and will have to pay federal income tax on interest earned by the CD. There are several bonds that you could also purchase that are exempt from federal income tax; however, they pay less interest. What interest do you need to earn on these bonds to equal the return that you would get from the certificate of deposit?

84. Comparing investments. Redo Exercise 83, except now you are considering investing in bonds that are exempt from both the federal tax of 14 percent and the 2.1 percent state tax.

The table shows the average number of currency units per dollar in 1997 according to the International Monetary Fund.

In Exercises 85–88, use this information to write a linear equation that makes each conversion.

85. Converting currency. a) dollars to yen b) yen to dollars

86. Converting currency. a) dollars to rupees b) rupees to dollars

87. Converting currency. euros to drachmas

88. Converting currency. yen to new pesos

Country	One Dollar Equals (rounded to nearest unit)
Japan	108 yen
France	0.9 euros
India	45 rupees
Greece	366 drachmas
Mexico	9 new pesos

---

* Exercise numbers circled in red can be used as group exercises.

89. Wheelchair accessibility. In order to comply with the Americans with Disabilities Act, the town library must be made wheelchair-accessible. Consider the given diagram of the entrance, which has enough room to construct a ramp as shown. Presently there are three steps, each six inches high and six inches wide, leading up to this entrance. How far should point P be from point B (at the base of the first step) for the ramp to have a slope of 8 percent (that is, 0.08)?

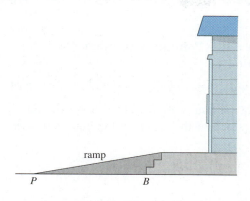

FIGURE FOR EX. 89

90. Wheelchair accessibility. Repeat Exercise 89, but now assume that there are six steps instead of three.

91. Communicating Mathematics   Explain when it is appropriate to use a linear equation as a model. Give an example of a situation we have not covered in this section that you feel fits a linear equation as a model.

92. Communicating Mathematics   Give an example for which you feel it would not be appropriate to use a linear equation as a model. Explain why you feel this way.

6.2   MODELING WITH LINEAR EQUATIONS

Like the simplifications we make when building physical models, we also simplify mathematical models. The engineer we described at the start of the chapter was interested only in improving the aerodynamics of the car; therefore, she could ignore features such as the sound system and seats that did not affect air resistance. Similarly we may decide to omit certain features of a mathematical model in order to make it simpler. Although a simple model allows us to see fundamental relationships more clearly, a more complex model may be more accurate. If a model is too simple, we can always refine it by adding features to make it more accurate.

For the linear equations we use as models in this section, realize that we do not always give you the information in exactly the same form. We have seen in Section 6.1 that we can describe the same linear equation in several different ways.

Method of Specifying Equation	Information Provided
Write the equation in standard form	$3x + 2y = 6$
State the x- and y-intercepts of the graph of the equation	x-intercept is $(2, 0)$ y-intercept is $(0, 3)$
Specify the slope and y-intercept of the graph of the equation	Slope is $-\frac{3}{2}$ and y-intercept is 3. Slope-intercept form of the equation is $y = -\frac{3}{2}x + 3$.

In addition to these methods, there are several other ways we can specify information that determines a linear equation as a model.

A point and a slope determine a line.

We can find a linear equation if we know the slope of its graph and a point on its graph.

EXAMPLE 1 **Using the Slope and a Point to Determine a Linear Equation**

Find a linear equation of the line with slope 3 passing through the point (4, 5).

SOLUTION: We will assume that we can write the equation of the line in slope-intercept form, $y = mx + b$. The slope of the graph of the equation is $m = 3$, so we can rewrite this equation as

$$y = 3x + b.$$

We need to find b. Because (4, 5) lies on the line, we substitute 4 for x and 5 for y, to get

$$5 = 3(4) + b.$$

Solving for b gives us $-7 = b$. The equation we want is therefore

$$y = 3x - 7.$$

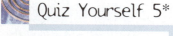

Quiz Yourself 5*

Find an equation of the line passing through (2, −3) having slope −4. Write your answer in slope-intercept form.

EXAMPLE 2 **Building a Model Using a Point and a Slope**

As part of the 1999 settlement of lawsuits against tobacco companies, the U.S. government required cigarette manufacturers to pay for antismoking education programs. Assume that, as a result of such programs, the smoking rate among 18- to 25-year-olds is dropping at a rate of 0.6 percent per year and that after three years the smoking rate is 38.8 percent. If we model this reduction by a linear equation, what would we predict the smoking rate among 18- to 25-year-olds to be in ten more years?

SOLUTION: We will model this decrease in the smoking rate by a linear equation in slope-intercept form written as

$$r = mt + b, \tag{1}$$

where t represents the time in years from the beginning of the campaign, and r represents the rate of smoking in this age group. In order to determine the relationship between the time t and the smoking rate r, we must first find m and b.

The change in the smoking rate per year is -0.6 percent per year. This means that the slope of the graph of (1) is $m = -0.6$. Substituting this value for m, we get

$$r = -0.6t + b. \tag{2}$$

To find b, we will use the fact that three years into the campaign, the smoking rate was 38.8 percent. This means that the pair $(3, 38.8)$ is a solution to equation (2). Substituting 3 for t and 38.8 for r, we get

$$38.8 = (-0.6)3 + b. \tag{3}$$

We rewrite equation (3) as $38.8 = -1.8 + b$. Solving for b, we get $b = 40.6$. A linear equation describing this reduction in smoking is therefore

$$r = -0.6t + 40.6.$$

We now use this equation to predict the smoking rate when $t = 13$. Substituting the value $t = 13$, we get

$$r = (-0.6)13 + 40.6 = -7.8 + 40.6 = 32.8.$$

Therefore, if the decline in smoking continues at the same rate, in ten more years the smoking rate among 18- to 25-year-olds will be 32.8 percent.   ◎

Although the model we developed in Example 2 was simple to construct and to use, you may feel that it has several drawbacks that outweigh its simplicity. First, you could argue that a linear equation is not appropriate to describe accurately the relationship between the time and the smoking rate because a change in time does not produce a corresponding change in the smoking rate. Or, you could argue that we have not considered factors other than the advertising campaign that affect the smoking rate.

It also may seem unreasonable to use the model over such a long period of time. As a result of such concerns, a researcher studying this problem might develop a more complex model than a linear equation, or might modify the model every few years to make it more accurate.

We can use two points to find a linear equation.

In modeling data, we often use two data points to write a linear equation. Example 3 shows how to find a linear equation if we know two points on its graph.

EXAMPLE 3   Using Two Points to Find a Linear Equation

Find an equation of the line passing through the points $(4, 2)$ and $(8, 5)$.

SOLUTION: We will write the equation in slope-intercept form as $y = mx + b$, so we need to find m and b. Using points $(4, 2)$ and $(8, 5)$, we find the slope

$$m = \frac{\text{rise}}{\text{run}} = \frac{5 - 2}{8 - 4} = \frac{3}{4}.$$

Substituting for m, we get

$$y = \frac{3}{4}x + b.$$

Because $(4, 2)$ lies on the graph,* we substitute 4 for x and 2 for y to get

$$2 = \frac{3}{4}(4) + b.$$

Solving this equation, we find that $b = -1$. The equation we want is therefore

$$y = \frac{3}{4}(x) - 1.$$

Quiz Yourself 6

Find an equation of the line passing through $(-3, 5)$ and $(6, 0)$. Write the equation in slope-intercept form.

> **PROBLEM SOLVING**
>
> In Example 3, we used the Convert a New Problem into an Older One Strategy that we discussed in Section 1.1. We did not approach Example 3 as a brand-new problem. Rather, we recognized that once we had the slope, we could solve the problem as we did in Example 1. A good problem solver often modifies a technique that is used in one situation and applies it in another.

EXAMPLE 4 Finding a Linear Equation for a Model Based on Two Data Points

Sheena sells gourmet pastry on the Internet. In the fourth month of operation, she sold 480 dozen pastries and in the seventh month she sold 792 dozen. Assume that we can model the increase in her business by a linear equation. She estimates that with her current equipment, she can bake a maximum of 1,500 dozen pastries per month. If she wishes her business to keep growing, during what month will she exceed her current capacity to produce baked goods?

SOLUTION: We will model Sheena's situation by a linear equation of the form

$$d = mt + b, \tag{4}$$

where t is the time in months that her business has been in operation and d is the number of dozens of pastries she can sell. We will use the points $(4, 480)$ and $(7, 792)$ to find the slope of the graph of equation (4) as follows:

$$m = \frac{\text{rise}}{\text{run}} = \frac{792 - 480}{7 - 4} = \frac{312}{3} = 104.$$

Substituting for m in equation (4), we get

$$d = 104t + b.$$

Now we can use either $(4, 480)$ or $(7, 792)$ to find b. As a rule of thumb, whenever possible, we use smaller numbers rather than larger ones, so we will use $(4, 480)$. Substituting 4 for t and 480 for d gives us the equation

---

* We could have also used $(8, 5)$; however, we generally use the point with the simpler coordinates.

$$480 = 104(4) + b.$$

Solving this equation for b, we find that $b = 64$. Thus a linear equation describing the increase in Sheena's sales is

$$d = 104t + 64.$$

We now want to find when Sheena will be able to sell 1,500 dozen pastries, so we set $d = 1{,}500$ and solve the equation $1{,}500 = 104t + 64$ in the usual way.

$$1{,}500 = 104t + 64$$
$$1{,}436 = 104t \qquad \text{(subtract 64)}$$
$$\frac{1{,}436}{104} = t \qquad \text{(divide by 104)}$$

Therefore, $t = \frac{1{,}436}{104} \approx 13.8$, which means that she will exceed her capacity to produce pastries during the fourteenth month.

The line of best fit gives the best linear approximation of data.

The next example illustrates model building with real data. Real data are often messier to work with than the example data in problems because the data points do not fall on a straight line. Instead they may have a linear pattern and we must find a line that best fits the data.

EXAMPLE 5   **Modeling Compact Disc Sales with a Linear Equation**

The number of compact discs sold in 1996, 1997, and 1998 were 779, 753, and 847 million, respectively.* We have plotted the points $(96, 779)$, $(97, 753)$, and $(98, 847)$ in Figure 6.7.

Although these points do not lie on a straight line, we still wish to use a linear equation to model the relationship between the year and the number of CDs sold.

a) Model the data with an equation of the line passing through points A and C.

b) Use the data to predict the number of compact discs sold in 2000.

c) There were 942.5 million CDs sold in 2000. Evaluate the accuracy of this model.

SOLUTION: In order to keep the numbers in our computations small, we will think of 1996 as year 0, 1997 as year 1, and so on. Using the technique of Example 4, we find an equation of the line through $(0, 779)$ and $(2, 847)$ to be

$$n = \frac{68}{2}t + 779,$$

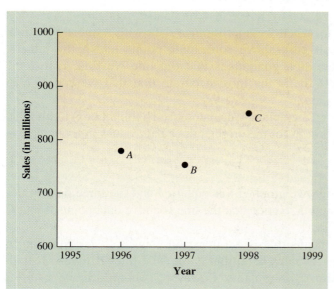

FIGURE 6.7   CD sales from 1996 to 1998.

---

* Source: Recording Industry Association of America, Washington, D.C., as reported in the *World Almanac* (2002).

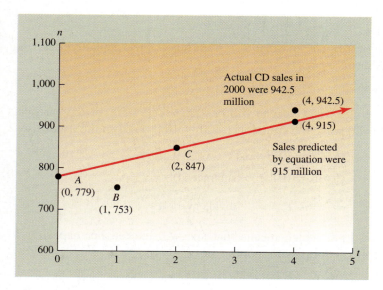

FIGURE 6.8   Graph of a linear equation modeling CD sales.

where t is the time in years from 1996 and n is the number of millions of compact discs sold. Notice in Figure 6.8 that this line passes through points A and C, but lies above point B. Therefore, we expect that this equation would overestimate the number of CD sales in 1997.

c) This equation predicts that in 2000, when $t = 4$,

$$n = \frac{68}{2}(4) + 779 = 136 + 779 = 915.$$

According to the Recording Industry Association of America, 942.5 million CDs were sold in 2000. Thus we see in Figure 6.8 that this equation underestimates the sales in 2000, and we conclude that it is probably not a very reliable model.

Keep in mind that there is nothing wrong with the mathematical computations in Example 5. However, we made a poor decision in choosing the line through points A and C to model the data. It may very well be the case that no linear equation will model the data accurately.

In Example 5, we imposed a strong condition by requiring that the line modeling CD sales pass *through* two of the data points. There is a technique in statistics called **linear regression** that we can use to find an equation of a line to approximate data. Because a set of real data points rarely lies on a straight line, when we approximate data with a linear equation, the line will pass above some data points and below others. Linear regression guarantees that the line we find, called the **line of best fit,** is the most accurate line that we can find to model the data.

We used a graphing calculator (see Highlight box) to find that the line of best fit for the original three data values in Example 5 is

$$y = 34x + 759.$$

In Figure 6.9, we graphed this equation and the three data values from Example 5. Notice that the line is above data points A and C and below B.

This means that if we insist on using a linear equation to model the CD sales data, this is the best line that we can find. The question still remains, however: "Should we be using a linear equation to model the data?" In Section 6.3 we will consider some alternative modeling methods that might produce better results.

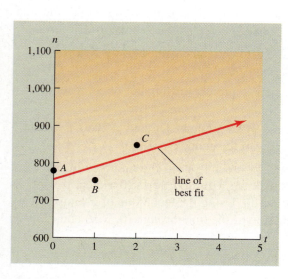

FIGURE 6.9   Line of best fit for CD sales data.

Highlight: *Finding the Line of Best Fit with a Graphing Calculator*

The computations necessary to find the line of best fit for a set of data points are so tedious that researchers rarely do them with pencil-and-paper calculations. Many calculators have a built-in capability to find the coefficients for the line of best fit, which we will now illustrate.

We will find the line of best fit for the data points (0, 779), (1, 753), and (2, 847). Calculator screen 1 shows that we have entered the *x*-coordinates of the data

points in a list called L1 and the *y*-coordinates in a list called L2. Screen 2 shows a menu of choices for statistical computations, and we have chosen LinReg(ax + b) for linear regression.

After we select the command to perform linear regression, screen 3 shows that we told the calculator to use lists L1 and L2 to get the coordinates of the data points. Screen 4 indicates the slope *a* and *y*-intercept *b* for the equation $y = ax + b$ for the line of best fit.

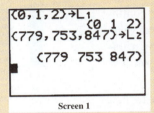

Screen 1

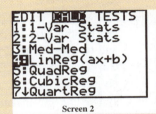

Screen 2

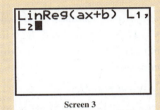

Screen 3

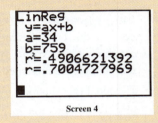

Screen 4

Exercises 6.2

In Exercises 1–8, find a linear equation whose graph has the indicated slope and passes through the given point.

1. slope 3, point (2, 1)

2. slope 4, point (5, 2)

3. slope 4, point (−2, 6)

4. slope 3, point (−5, 3)

5. slope −2, point (4, 3)

6. slope −3, point (6, 4)

7. slope −5, point (−6, −1)

8. slope −8, point (−4, −9)

In Exercises 9–16, find a linear equation whose graph passes through the given points. Write the equation in slope-intercept form.

9. (2, 3) and (5, 9)

10. (3, 5) and (5, 17)

11. (17, 12) and (9, 10)

12. (14, 10) and (5, 7)

13. (11, −4) and (−8, 2)

14. (2, −6) and (−14, 9)

15. (−6, −8) and (−4, −1)

16. (−2, −9) and (−1, −4)

In Exercises 17–26, it is useful to think of slope as representing the average rate of change of one variable corresponding to a change in the other variable. In order to keep the numbers you work with small, represent the first year for which you have data as year 0. For example, in Exercise 17, the year 1990 is year 0. In Exercise 24, the year 1994 is year 0.

17. Life expectancy. In 1990, the life expectancy for a female born in the United States was 78.8 years and was increasing at a rate of 0.3 years per year. Assume that this rate of increase remained constant.

a) Model this situation by a linear equation.

b) Use the equation in part (a) to estimate the life expectancy of a female born in the United States in 2000.

---

* See www.aw.com/pirnot for a tutorial on finding the line of best fit and other types of regression.

18. Life expectancy. In 1990, the life expectancy for a male born in the United States was 71.8 years and was increasing at a rate of 0.1 years per year. Assume that this rate of increase remained constant.

a) Model this situation by a linear equation.

b) Use the equation in part (a) to estimate the life expectancy of a male born in the United States in 2000.

19. College enrollments. In 1999, there were 1,984,000 males enrolled in a four-year public college; over the next several years this number increased at a rate of 25 thousand per year.

a) Model this situation by a linear equation.

b) Use the equation in part (a) to estimate the number of males enrolled in college in 2010.

20. College enrollments. In 1999, there were 2,309,000 females enrolled in a four-year public college; over the next several years this number increased at a rate of 49 thousand per year.

a) Model this situation by a linear equation.

b) Use the equation in part (a) to estimate the number of females enrolled in college in 2010.

21. College tuition. According to U.S. Department of Education statistics, in 2000, tuition at a public four-year college averaged $3,506 and increased at an average rate of $157 over the next several years.

a) Model this situation by a linear equation.

b) Use the equation in part (a) to estimate the tuition at a four-year public college in 2010.

22. College tuition. According to U.S. Department of Education statistics, in 2000, tuition at a private four-year college averaged $15,531 and increased at an average rate of $1,031 over the next several years.

a) Model this situation by a linear equation.

b) Use the equation in part (a) to estimate the tuition at a four-year private college in 2010.

23. Crime statistics. The graph shows the decrease in violent crimes in the United States from 1996 to 1999, according to the FBI *Uniform Crime Reports*, 1999.* Assume that this rate of decrease continues for the next several years and can be modeled by a linear equation of the form $v = mt + b$.

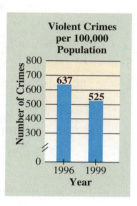

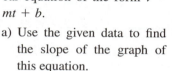

FIGURE FOR EX. 23

a) Use the given data to find the slope of the graph of this equation.

b) Use the slope from part (a) and either of the data points to find a linear equation that models this decrease in violent crime. (1996 = year 0.)

c) Use the equation you found in part (b) to predict the number of violent crimes per 100,000 U.S. inhabitants in 2005.

24. Poverty data. The graph shows the decrease in the percentage of Americans living below the poverty level from 1994 to 1999, according to the U.S. Department of Commerce. Assume that the rate of decrease continues for the next several years and can be modeled by a linear equation of the form $p = mt + b$.

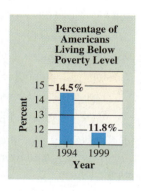

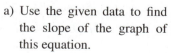

FIGURE FOR EX. 24

a) Use the given data to find the slope of the graph of this equation.

b) Use the slope from part (a) and the data from 1994 to

---

* Data in Exercises 23–26 are as reported in the *World Almanac* (2002).

write an equation that describes this decrease in poverty. (1994 = year 0.)

c) Use the equation you found in part (b) to predict the percentage of Americans living below the poverty level in 2005.

25. Employment data. According to the Bureau of Labor Statistics, the demand for database administrators will increase from 87,000 in 1998 to 155,000 in 2008. Assume that this projection is correct. Model these data with a linear equation and estimate the demand for database administrators in 2003.

26. Employment data. According to the Bureau of Labor Statistics, the demand for desktop publishing specialists will increase from 26,000 in 1998 to 44,000 in 2008. Assume that this projection is correct. Model these data with a linear equation and estimate the demand for desktop publishing specialists in 2003.

27. Sleep data. According to the National Sleep Foundation, in 2002, Americans averaged 6.9 hours of sleep each weeknight. In 1900, they averaged about 8 hours and 30 minutes of sleep per weeknight. Model these data by a linear equation and use it to predict the year when Americans will not be sleeping at all on a weeknight.

28. According to Nielsen Media Research, in 2000, there were 68.5 million households that subscribed to basic cable TV—up from 64.7 million in 1996. Model these data by a linear equation and use it to predict the year in which 100 million American households will be subscribing to basic cable TV.

*(29.) **U.S. labor force data.** The graph shows the increase in the civilian labor force from 1998 to 2001, according to the U.S. Department of Labor. Assume that the rate of increase continues for the next several years and can be modeled by a linear equation of the form $l = mt + b$.

a) What is the average yearly increase from 1998 to 2001?

b) Use the result from part (a) as the slope and the point that represents the 1998 data to write an equation that describes this increase in the labor force.

_____
* Exercise numbers circled in red can be used as group exercises.

FIGURE FOR EX. 29

c) Use the equation you found in part (b) to predict the size of the U.S. civilian labor force in 2005.

(30.) **Medical data.** The graph below shows the number of hospice patients per 10,000 population per year from 1992 to 1998, according to the National Center for Health Statistics. Assume that the rate of increase continues for the next several years and can be modeled by a linear equation of the form $c = mt + b$.

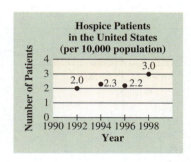

FIGURE FOR EX. 30

a) What is the average yearly increase from 1992 to 1998?

b) Use the result from part (a) as the slope and the point that represents the 1992 data to write an equation describing the change in hospice patients from 1992 to 1998.

c) Use the equation you found in part (b) to predict the expected number of hospice patients per 10,000 population per year in 2005.

Further Exercises

31. **Line of best fit* for labor force data.** We can use the four data points in Exercise 29 to find an equation of the line of best fit to be $y = 1.25t + 131.95$. Complete the following table, which compares the actual data values, your estimate of the data values using the equation you found in Exercise 29(b), and the data values that would be predicted by using the equation of the line of best fit.

33. Communicating Mathematics   Discuss the appropriateness of using a linear equation as a model for the CD sales data in Example 5.

34. Communicating Mathematics   Discuss the appropriateness of using a linear equation as a model for the U.S. civilian labor force data in Exercise 29.

	1998 (Year 0)	1999 (Year 1)	2000 (Year 2)	2001 (Year 3)
Actual data	131.5	133.5	135.2	135.1
Values predicted by your model				135.1
Values predicted by line of best fit				

32. **Line of best fit for medical data.** Repeat Exercise 31 for the data given in Exercise 30. The line of best fit for these data points is given by the equation $y = 0.145t + 1.94$. Complete the following table.

	1992 (Year 0)	1994 (Year 2)	1996 (Year 4)	1998 (Year 6)
Actual data	2.0	2.3	2.2	3.0
Values predicted by your model			2.67	
Values predicted by line of best fit				

6.3 MODELING WITH QUADRATIC EQUATIONS

Most real-life data sets do not follow linear patterns, so we must use other types of equations to model them. In this section we study certain types of quadratic equations in two variables and construct models using them.

A **quadratic equation** is an equation of degree two. Although there are different types of quadratic equations, in this section we will restrict our discussion to quadratic equations of the form $y = ax^2 + bx + c$, where a, b, and c are real numbers. The equations $y = 2x^2 - 3x + 5$ and $y = \frac{3}{2}x^2 + 6x - 17$ are examples of the type of quadratic equations we will be using as models.

---

* For a tutorial on using a TI-83 to find the line of best fit, see www.aw.com/pirnot.

As with linear equations in two variables, solutions to quadratic equations will be ordered pairs of numbers.

EXAMPLE 1   Verifying Solutions for Quadratic Equations in Two Variables

Determine whether each ordered pair is a solution for the quadratic equation $y = 2x^2 - 3x + 5$.

a) $(4, 25)$ b) $(1, 2)$

SOLUTION: a) Substituting 4 for x and 25 for y gives us the true statement $25 = 2(4)^2 - 3(4) + 5$. This tells us that $(4, 25)$ is a solution.

b) When we substitute 1 for x and 2 for y, we get the false statement $2 = 2(1)^2 - 3(1) + 5$. So $(1, 2)$ is not a solution.   ◎

The graph of $y = ax^2 + bx + c$ is a parabola.

If we plot all the solutions to a quadratic equation, we get a graph that is a geometric figure called a **parabola**. We show two parabolas in Figure 6.10. If a is positive, the graph opens up as in Figure 6.10(a). If a is negative, the parabola opens down as in Figure 6.10(b). If a parabola is opening up, the lowest point on the parabola is called the **vertex** of the parabola. Similarly, for a parabola opening down, the highest point is called the vertex.

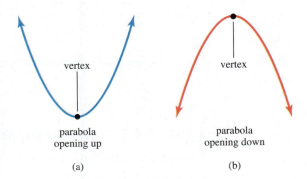

(a)   (b)

FIGURE 6.10   The graph of $y = ax^2 + bx + c$ is a parabola.

We will frequently use the following fact about quadratic equations without proof.

DEFINITION

The vertex of the graph of the quadratic equation $y = ax^2 + bx + c$ occurs when

$$x = \frac{-b}{2a}.$$

EXAMPLE 2 **Finding the Vertex of a Parabola**

Find the vertex of the graph of $y = 2x^2 - 4x + 5$.

SOLUTION: In this equation, $a = 2$, $b = -4$, and $c = 5$. The x-coordinate of the vertex is $x = \frac{-b}{2a} = \frac{-(-4)}{2(2)} = 1$. Substituting this value for x, we get the y-coordinate of the vertex, $y = 2(1)^2 - 4(1) + 5 = 2 - 4 + 5 = 3$. The vertex of the parabola, which is the graph of $y = 2x^2 - 4x + 5$, is therefore the point $(1, 3)$. ◎

We use the quadratic formula to find the intercepts of a parabola.

In order to graph quadratic equations, we need another tool. We find x- and y-intercepts for the graphs of quadratic equations exactly as we did for linear equations. Recall that to find the x-intercept of a linear equation, we set y equal to 0 and then solve the resulting equation. Doing this for a quadratic equation, we get an equation of the form $ax^2 + bx + c = 0$, which is a quadratic equation in a single variable. Such an equation may or may not have solutions. If it has solutions, they can always be found using one of the most famous formulas in algebra, called the **quadratic formula**.

> **The Quadratic Formula**
>
> The solutions of the quadratic equation $ax^2 + bx + c = 0$ will be
>
> $$x = \frac{-b + \sqrt{b^2 - 4ac}}{2a} \quad \text{and} \quad x = \frac{-b - \sqrt{b^2 - 4ac}}{2a}.$$

We often combine these two formulas into the single formula

$$x = \frac{-b \pm \sqrt{b^2 - 4ac}}{2a}.$$

EXAMPLE 3 **Using the Quadratic Formula to Solve an Equation**

Use the quadratic formula to solve the equation $x^2 + 5x - 84 = 0$.

SOLUTION: In this equation, $a = 1$, $b = 5$, and $c = -84$. We substitute these values into the quadratic formula to get the solutions:

$$x = \frac{-5 \pm \sqrt{5^2 - 4(1)(-84)}}{(2)(1)}$$

$$= \frac{-5 \pm \sqrt{25 + 336}}{2}$$

$$= \frac{-5 \pm \sqrt{361}}{2} = \frac{-5 \pm 19}{2}.$$

Quiz Yourself 7*

Use the quadratic formula to solve $x^2 + 7x - 44 = 0$.

The solutions are therefore

$$x = \frac{-5 + 19}{2} = 7 \quad \text{and} \quad x = \frac{-5 - 19}{2} = -12.$$

SOME GOOD ADVICE

In solving a quadratic equation in a single variable, we must calculate the quantity $b^2 - 4ac$, which is called the **discriminant** of the equation. There are three cases that can arise:

- $b^2 - 4ac$ is greater than 0—in this case, there are two distinct solutions to the equation.

- $b^2 - 4ac$ equals 0—in this case, there is only one solution to the equation.

- $b^2 - 4ac$ is less than 0—in this case, there are no real number solutions, since we cannot take the square root of a negative number and get a real number.

EXAMPLE 4   Graphing a Quadratic Equation in Two Variables

Graph the equation $y = x^2 + x - 6$, which is a parabola.

a) Does the parabola open up or down?

b) Find the vertex of the parabola.

c) Find the x- and y-intercepts of the graph.

d) Draw the graph.

SOLUTION: a) Since the coefficient for x^2 is 1, the parabola is opening up.

b) For this equation, $a = 1$, $b = 1$, and $c = -6$. The x-coordinate of the vertex is

$$x = \frac{-b}{2a} = \frac{-1}{2(1)} = -\frac{1}{2}.$$

The y-coordinate of the vertex is

$$y = \left(-\frac{1}{2}\right)^2 + \left(-\frac{1}{2}\right) - 6 = \frac{1}{4} - \frac{1}{2} - 6 = -6\frac{1}{4}.$$

c) To find the x-intercepts of the graph, we set y equal to 0 and get the equation

$$0 = x^2 + x - 6.$$

Using the quadratic formula, we obtain

$$x = \frac{-1 \pm \sqrt{1^2 - 4(1)(-6)}}{(2)(1)} = \frac{-1 \pm \sqrt{25}}{2} = \frac{-1 \pm 5}{2},$$

so $x = 2$ or $x = -3$. The x-intercepts of the graph are therefore $(2, 0)$ and $(-3, 0)$.

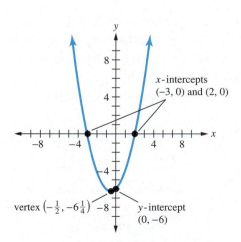

Quiz Yourself 8

Graph the equation $y = x^2 + x - 12$.

To find the y-intercept, set $x = 0$. Thus, $y = 0^2 + 0 - 6 = -6$. The y-intercept is therefore $(0, -6)$.

d) With this information, it is now easy to draw a reasonable graph as shown in the margin on page 345.

Highlight: *Solving Equations*

The methods for solving quadratic equations appeared in Babylon as far back as 2000 B.C. Ancient Greek and Hindu mathematicians also had algebraic methods to solve quadratic equations, but it was the work of a ninth-century Arabic mathematician named Mohammed ibn-Musal al-Khowarizmi who made the term *algebra* a household word. In his book *Al-jabr wa'l muqabalah*, he introduced the algebraic methods of earlier mathematicians to Europe. His elementary presentation of algebra was used to solve equations, particularly quadratic equations. Some scholars have interpreted his title as referring to balancing and completing algebraic expressions, which are processes we commonly use in algebra.

Having mastered the quadratic equations, it was natural for mathematicians to consider *cubic equations*—those that have an x^3 term in them. Was there a mechanical process, similar to the quadratic formula, that could solve cubics? In 1545, the Italian Girolamo Cardano published his famous book *Ars Magna*, an essay on algebra, in which he showed not only how to solve cubics, but also *quartic equations* (equations with an x^4 term). At this point, it seemed that eventually mathematicians would invent methods to solve equations having any power of x. This turned out not to be the case.

In the nineteenth century, Niels Henrik Abel from Norway and Evariste Galois of France proved the remarkable result that there are *quintic equations* (equations having an x^5 term) that cannot be solved using only addition, subtraction, multiplication, division, and the taking of roots. Thus they proved that once we get beyond quartic equations, we cannot find mechanical formulas similar to the quadratic formula to find solutions.

We will now apply our knowledge of quadratic equations to model building. After the release of a new book, movie, or video, there are four periods in its life cycle.

- Stage 1: Sales increase rapidly. With a new video, for example, the number of purchasers grows rapidly soon after the video is introduced.
- Stage 2: The sales are still growing, but the increase from week to week is not as great as in the early phase.

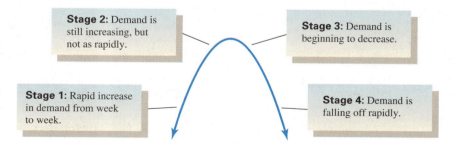

Stage 2: Demand is still increasing, but not as rapidly.

Stage 3: Demand is beginning to decrease.

Stage 1: Rapid increase in demand from week to week.

Stage 4: Demand is falling off rapidly.

FIGURE 6.11 A parabola models the product life cycle for a new product.

- Stage 3: The product is still selling, but now each week's sales are a little lower than the week before.
- Stage 4: The market is saturated and sales are now dropping rapidly.

We can represent this situation by a parabola opening downward, as in Figure 6.11.

We use this pattern to model sales of a particular video in Example 5.

EXAMPLE 5   Modeling Video Sales with a Quadratic Equation

Assume that a producer knows that the demand for a video by a popular artist can be modeled by the equation $S = -13n^2 + 169n$. Here n is the number of weeks since the introduction of the video and S is the dollar value in thousands of the videos sold during week n.

a) When do we expect the sales for the video to peak?

b) After sales have peaked, when does this model predict that the sales will sink below $100,000 per week?

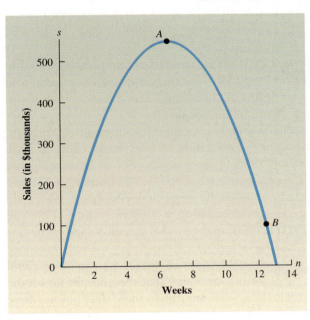

FIGURE 6.12   Parabola modeling video sales.

SOLUTION: a) The equation is quadratic with a negative n^2 coefficient, so its graph is a parabola opening downward, as shown in Figure 6.12.

Point A, the vertex of the parabola, shows where the weekly sales are the greatest. Since $S = -13n^2 + 169n$, $a = -13$, $b = 169$, and $c = 0$. Therefore, the first coordinate of the vertex is

$$n = \frac{-b}{2a} = \frac{-169}{2(-13)} = \frac{-169}{-26} = 6.5.$$

Therefore, sales should peak during the seventh week.

Substituting 6.5 for n, we find the greatest dollar amount for weekly sales to be

$$S = -13(6.5)^2 + 169(6.5) = -13(42.25) + 169(6.5)$$

$$= -549.25 + 1{,}098.5 = \$549.25.$$

Therefore, the highest weekly sales should be about $549,250.

b) Point B shows where the sales, after peaking, drop to $100,000 per week. To find this point we must answer the question: When will $S = 100$? This gives us the quadratic equation

$$100 = -13n^2 + 169n.$$

We must put the equation in the form $an^2 + bn + c = 0$, so subtracting 100 from both sides we get

$$0 = -13n^2 + 169n - 100.$$

Using the quadratic formula, we find

$$n = \frac{-169 \pm \sqrt{169^2 - 4(-13)(-100)}}{2(-13)}$$

$$= \frac{-169 \pm \sqrt{28{,}561 - 5{,}200}}{-26} = \frac{-169 \pm \sqrt{23{,}361}}{-26}.$$

Thus, $n = 0.62$ and $n = 12.38$. Since we are interested in the point after sales have peaked, we ignore $n = 0.62$ and use $n = 12.38$ weeks as our solution. According to this model, by the thirteenth week, sales will have dropped below $100,000.

Recall in Section 6.2 that we stated that the number of compact discs sold in 1996, 1997, and 1998 were 779, 753, and 847 million, respectively. We also told you that the best linear model that fit this data was the linear equation $y = 34x + 759$. This line was the best line to approximate the data.

Similarly, we can use a technique called **quadratic regression*** to model data with a quadratic equation called the **parabola of best fit.** In order to keep the numbers simple, we considered 1996 to be year 0, 1997 to be year 1, and 1998 to be year 2. We entered the data points $(0, 779)$, $(1, 753)$, and $(2, 847)$ into a graphing calculator and selected the quadratic regression command. We found that the parabola of best fit is given by the equation $y = 60x^2 - 86x + 779$. You can check that the pairs $(0, 779)$, $(1, 753)$, and $(2, 847)$ are all solutions to this equation.

Just as we can find a line that passes through two points, we can always find a parabola that passes through three points, provided that they do not lie on the same line. If we use quadratic regression with more than three data points, there is no guarantee that the parabola will pass through any of the data points.

Exercises 6.3

In Exercises 1–8, solve each quadratic equation.

1. $x^2 - 10x + 16 = 0$ **2.** $x^2 - 7x + 12 = 0$

3. $2x^2 - 5x + 3 = 0$ **4.** $6x^2 - 11x + 4 = 0$

5. $3x^2 + 7x - 6 = 0$ **6.** $2x^2 + x - 3 = 0$

7. $5x^2 - 17x - 12 = 0$ **8.** $4x^2 + 12x + 9 = 0$

In Exercises 9–18, answer the following questions for each quadratic equation. Then draw the graph.

Is the equation's graph opening up or down?

What is the vertex of the graph?

What are the x-intercepts?

What is the y-intercept?

9. $y = -x^2 + 6x - 8$ **10.** $y = x^2 - 4x - 5$

11. $y = -4x^2 + 8x + 5$ **12.** $y = x^2 + 5x + 4$

13. $y = 4x^2 - 4x - 2$ **14.** $y = -2x^2 - 10x - 8$

15. $y = 3x^2 + 7x - 6$ **16.** $y = 2x^2 + x - 3$

17. $y = -x^2 + 7x - 12$ **18.** $y = -4x^2 - 12x - 9$

---

* See www.aw.com/pirnot for a tutorial on doing quadratic regression on the TI-83.

19. Movie attendance. Major movie studios try to release a "blockbuster" movie each summer. Assume that these statistics describe ticket sales for such a movie:

> Week 2   5 million tickets sold
> Week 4   7 million tickets sold
> Week 6   8 million tickets sold

We did quadratic regression to find that $A = -0.125x^2 + 1.75x + 2$ is an equation of the parabola of best fit for these data.

a) Show that the graph of this equation passes through the three data points regarding attendance.

b) According to this model, what is the maximum value that A can attain?

20. Movie attendance. Repeat Exercise 19, but now the statistics are as follows:

> Week 3   6 million tickets sold
> Week 5   8 million tickets sold
> Week 7   9 million tickets sold

Use the model $A = -0.125x^2 + 2x + 1.125$.

21. Business profits. New business owners typically experience a loss for a certain amount of time before the losses bottom out. Eventually the businesses show a profit. This pattern of profit and loss suggests a situation that could be modeled by a quadratic equation whose graph is a parabola opening up. Assume that you have begun a business videotaping weddings and that over the first three months of business you sustain losses of $200, $130, and $70. (We can rephrase this by saying your cumulative profit at the end of month one was −$200, at the end of month two was −$330, and at the end of

month three was −$400). Using quadratic regression, we can show that $P = 30x^2 - 220x - 10$ is the best quadratic equation that fits these data.

a) Verify that this equation fits the given data.

b) Using this model, at the end of what month do you expect to sustain your greatest cumulative loss? What is this loss?

c) At the end of what month do you expect to show your first cumulative profit? (By that, we mean that your total profits from the time the business started are greater than your total losses.)

d) What will your cumulative profit be by the end of month 10?

e) How much do you expect to earn during month 10?

22. Video sales. Assume that the sales for the video in Example 5 is modeled by the equation $S = -4n^2 + 16n - 12$, where S is the number of millions of dollars of sales in week n.

a) When do we expect the sales for the video to peak?

b) According to this model, what is the largest value for S?

c) When do we expect the sales to drop to zero?

23. Presidential popularity ratings. After a U.S. president makes a politically unpopular decision, his approval rating usually drops. After some time, the approval rating rises again. We will model this drop in approval, bottoming out, and then rise in approval by a quadratic equation. Assume that before the president signs an unpopular tax bill, the approval rating is at 48 percent. One week after the president signs the bill, the rating is 41 percent, at two weeks it is at 39 percent, and at three weeks it is 42 percent. Using quadratic regression we can show that $A = 2.5x^2 - 9.5x + 48$ is the best quadratic equation that fits the data.

How many weeks after the signing of the tax bill will the approval rating be back to what it was before the signing of the tax bill?

24. Presidential popularity ratings. Repeat Exercise 23, but now assume that the rating before the tax bill signing is 50 percent, one week after the signing it is 44 percent, two weeks after the signing it is 40 percent, and after three weeks it is 38 percent. Using quadratic regression we can show that $A = x^2 - 7x + 50$ is the best quadratic equation that fits this data.

We model many physical relationships using quadratic equations.

25. Falling bodies. A plane is dropping emergency food supplies to relieve famine in a less-developed country. Crates are dropped, and the height of the crate above the ground at time t is given by the equation $H = 160 - 16t^2$.

a) Graph this equation.

b) Are there any values for t that it would not make sense to use for this equation because of the physical characteristics of a falling crate?

c) Find the time at which the crate will strike the ground.

26. Falling bodies. Repeat Exercise 25, but now use the model $H = 320 - 16t^2$.

27. Rocket flight. Assume that if a model rocket is fired directly upward (with a certain velocity), its distance above

the ground at time t is given by the equation $d = 100t - 16t^2$.

a) Graph this equation.

b) Are there any values for t that it would not make sense to use for this equation because of the physical characteristics of a rocket?

c) Find the time at which the rocket reaches its highest point.

d) Find the time at which the rocket returns to the ground.

28. Rocket flight. Repeat Exercise 27, but now use the equation $d = 144t - 16t^2$ to model the rocket's flight.

Further Exercises

29. Communicating Mathematics   Returning to Exercises 25 and 26, explain why it is reasonable for the graphs representing the falling crates to be shaped as they were. How does this conform to your intuition? Would a linear equation have been an acceptable model?

30. Communicating Mathematics   Returning to Exercises 27 and 28, explain why it is reasonable for the graphs representing the rocket flights to be shaped as they were. How does this conform to your intuition? Would a linear equation have been an acceptable model?

31. Running a race. The equation $d = 0.15t^2 + 8t$ describes the distance an athlete who is running a hundred-yard dash has traveled in t seconds.

a) Graph this equation.

b) Are there any values for t that it would not make sense to use in this equation due to the physical characteristics of this problem?

c) From the graph, determine where the runner is running more slowly and where the runner is running faster.

d) How long does it take for the runner to finish the race?

32. Communicating Mathematics   Would it make sense to use the model in Exercise 31 for a person running a mile race? Explain.

33. Communicating Mathematics   The accompanying figure shows a time-lapse photo of a bouncing golf ball. The photographer captured all the images of the golf ball by using a strobe light that flashed once every 0.03 seconds. It is known from basic physics that the model for the path of a bouncing ball is a quadratic equation. Discuss how you would obtain the numerical data that then could be used to do quadratic regression to obtain the equation.

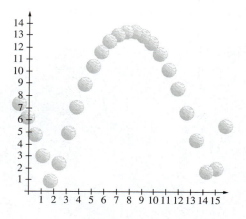

FIGURE FOR Ex. 33

34. Communicating Mathematics Recall that for a quadratic equation $y = ax^2 + bx + c$, the expression $b^2 - 4ac$ is called the discriminant. The value of the discriminant determines how many x-intercepts the graph of the equation will have.

a) If the discriminant is positive, how many x-intercepts will the graph of the equation have?

b) If the discriminant is 0, how many x-intercepts will the graph of the equation have?

c) If the discriminant is negative, how many x-intercepts will the graph of the equation have?

35. Communicating Mathematics Could a quadratic equation have a graph that has three x-intercepts? Explain your answer.

*36. Communicating Mathematics This exercise is based on Exercises 29 and 31 in Section 6.2. In Exercise 29 from Section 6.2, you were given the line of best fit for four data points. The equation was $y = 1.25t + 131.95$. We can use those same four data points to find the parabola of best fit whose equation is $y = -0.525t^2 + 2.825t + 131.425$.

a) Add a line to the table that you began in Exercise 31 from Section 6.2 as follows:

	1998 (Year 0)	1999 (Year 1)	2000 (Year 2)	2001 (Year 3)
Actual data	131.5	133.5	135.2	135.1
Values predicted by your model	131.5	132.7	133.9	135.1
Values predicted by line of best fit	131.95	133.2	134.45	135.7
Values predicted by parabola of best fit				

b) Does the line of best fit or the parabola of best fit seem to model the data better? Explain your answer.

6.4 EXPONENTIAL EQUATIONS AND GROWTH

You may be surprised to know that population growth in less-developed countries, bankruptcy from credit card debt, the spread of AIDS, loan sharking, and the soundness of the social security system can all be modeled using the same type of mathematical equation. You can begin to understand the relationships that we model in this section by recalling the way money grows in a bank account.†

Money compounded in a bank account grows exponentially.

Suppose that you deposit $1,000 (called the *principal*) in an account paying 8 percent interest per year. To keep our calculations simple, we will assume that you deposit the money at the beginning of the year and the interest is paid all at once at the end of the year. At the end of the first year, you have in the account the original $1,000 deposit plus the interest, which is 8% × $1,000 = 0.08 × $1,000 = $80. So, at the beginning of the second year, your account has $1,080 earning 8 percent interest. At the end of the second year, you have the $1,080

---

* Exercise numbers circled in red can be used as group exercises.
† We cover compound interest in much greater depth in Chapter 11.

plus interest, which is $8\% \times \$1,080 = 0.08 \times \$1,080 = \$86.40$. Thus the total at the end of the second year is $\$1,166.40$.

To compute the amount for successive years, we perform similar calculations. That is, we leave the interest in the account and then calculate the new interest for subsequent years based on the total of the principal and interest previously earned. The process of earning interest on interest is called **compounding**.

Table 6.1 shows how compounding works and shows the pattern in the increased value of the account.

Year	Beginning Balance	+	Interest for Current Year	=	Balance at End of Year
1	1,000	+	$0.08 \times 1,000$	=	$1,000(1.08) = 1,080$
2	1,080	+	$0.08 \times 1,080$	=	$1,080(1.08) = 1,000(1.08)(1.08)$
3	1,166.40	+	$0.08 \times 1,166.40$	=	$1,166.40(1.08) = 1,000(1.08)(1.08)(1.08)$

TABLE 6.1 Calculating compound interest.

From Table 6.1, we see the pattern that the balance at the end of each year is 1.08 times the balance at the end of the previous year, which gives us the following rule for calculating compound interest.

The Compound Interest Formula

If we invest the amount P, called the *principal*, in an account earning a yearly interest rate r and we compound the interest for n years, then the amount in the account, A, is

$$A = P(1 + r)^n.*$$

We apply this formula in Example 1.

Quiz Yourself 9†

EXAMPLE 1 Using the Compound Interest Formula

Suppose you deposit $\$1,000$ in an account that is compounded annually at a rate of 8 percent. Find the amount in this account after fifteen years.

SOLUTION: The principal P is $\$1,000$, the number of years n is 15, and the rate r is 0.08. Thus the value of the account at the end of fifteen years is

$$P(1 + r)^n = 1,000(1 + 0.08)^{15} = 1,000(1.08)^{15} = \$3,172.17.$$

Assume that you have invested $\$2,500$ in an account paying 10 percent interest that is compounded yearly. What is the value of this account after three years?

---

* To raise a quantity to a power on your calculator, you must use the y^x key (or the ^ key). For example, if you wish to compute $(1 + 0.08)^{12}$, you first enter 1.08, then press the y^x key (or the ^ key), then press 12, and then the = key or Enter.
† Quiz Yourself answers begin on page 849.

Year	Balance at End of Year
1	$1{,}000(1.08)^1 = 1{,}080$
2	$1{,}000(1.08)^2 = 1{,}166.40$
3	$1{,}000(1.08)^3 = 1{,}259.71$
4	$1{,}000(1.08)^4 = 1{,}360.48$
5	$1{,}000(1.08)^5 = 1{,}469.32$

TABLE 6.2   Growth of an account over five years.

In Example 1, we compounded the interest once a year. Many financial institutions compound interest more frequently. To learn how to do this, see Section 11.2.

Let's plot the amount in the account described in Example 1 over a period of years to see better how the value of the account is growing. In Table 6.2, we first calculate the amount in this account for each of the first five years and plot these values in Figure 6.13, which shows the relationship between time and the amount in the account.

Because the points in Figure 6.13 appear to be close to lying on a straight line, you might want to use the line of best fit to approximate these data. However, if you look very closely, you will see a slight upward curving, so you may decide instead to find the parabola of best fit.

We used a graphing calculator to find the linear equation that best models the data to be

$$A = 97.272n + 975.366,$$

and the best quadratic model is given by

$$A = 3.74n^2 + 74.84n + 1001.54,$$

where A is the amount in the account after n years.

We graphed the line and parabola of best fit in Figure 6.14. In addition, we graphed the original five data points and also the points (30, 10,062.65), (35, 14,785.34), and (40, 21,724.52), which correspond to the value of the account after 30, 35, and 40 years, respectively.

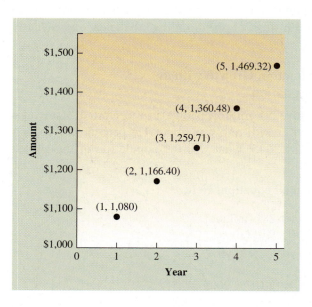

FIGURE 6.13   The balance in a bank account over five years.

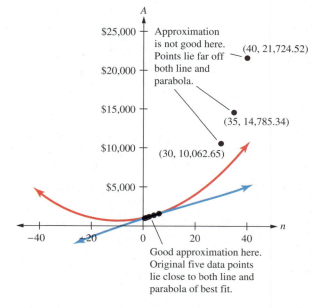

FIGURE 6.14   Graph comparing the balance in a bank account and the line and parabola of best fit.

As time increases, the points representing the amounts in the bank account are farther and farther from both the line and parabola of best fit. The equation describing these data is neither linear nor quadratic.

The type of equation we used to calculate compound interest is also useful for modeling other important phenomena, so we will give it a name.

> **DEFINITION**
>
> An **exponential equation** is an equation of the form
>
> $$y = a \cdot b^x.$$

The compound interest formula, $A = P(1 + r)^n$, is one example of an exponential equation.

> **SOME GOOD ADVICE**
>
> We use an exponential equation to model situations in which *the rate of change of a quantity is proportional to its size*. For example, the more money you have in a bank account, the more interest you earn; the larger the population of a country, the more children will be born to increase the population.

An exponential equation is a simple model of population growth.

We can use an exponential equation to model population growth.* In order to do this, we use an initial population instead of an initial bank balance, and a yearly growth rate instead of an interest rate.

EXAMPLE 2   **Modeling Population Growth with an Exponential Equation**

The 2000 census counted roughly 281 million people in the United States at a time when the yearly population growth rate was about 1.2 percent per year. If this rate of growth continues until 2045, what will the population of the United States be then?

SOLUTION: We use an exponential equation. Here

$$P = 281 \text{ (million)},$$
$$r = 0.012,$$

and

$$n = 2045 - 2000 = 45.$$

___

* We will discuss a more complex model of population growth later in this section.

Therefore,

$$P(1 + r)^n = 281(1 + 0.012)^{45}$$
$$= 281(1.012)^{45} \approx 480.7 \text{ (million)}.$$

Therefore, using an exponential model, we expect that there will be almost one-half billion people living in the United States in 2045.

When using a mathematical model as in Example 2, it is always important to consider the appropriateness of the model. For example, you should ask whether it is reasonable to assume that the growth rate in the United States will stay at 1.2 percent for 45 years.

Often in the media, researchers make dire predictions based on mathematical models regarding such things as population growth, global warming, and the bankruptcy of the social security system. These predictions may or may not be valid, depending on the assumptions made by the researcher. As an educated person, you should have an understanding of what a mathematical model can tell us and what it cannot. If the assumptions underlying a model are not sound, then its predictions are worthless.

We use the log function to find the time it takes a quantity that is growing exponentially to double.

In predicting future bank balances, population, or similar quantities that are growing exponentially, we often want to know, "How long will it take for this quantity to double?"

Before we can answer this question, we need another tool. Many calculators have a key labeled either "log" or "log x," which stands for the *common logarithmic function*.* Pressing this key has the effect of reversing the operation of raising 10 to a power. For example, suppose that you compute $10^5 = 100,000$ on your calculator. If you next press the "log" key, the display will show "5." If you enter 1,000, which is 10 raised to the third power, and press the "log" key, the display will show "3." Practice finding the log of powers of 10, such as 100 and 1,000,000.

The log function has an important property that will help us solve equations of the form $a = b^x$ for x. For example, we may want to find a value of x such that $5 = 3^x$. Although it may seem strange to you, it is possible to find a number, although not an integer, such that if we raise 3 to that power, we will get 5. In order to do that, we need the following property of the log function.

Exponent Property of the Log Function
$$\log y^x = x \log y$$

* We discuss the log function in greater detail in Section 11.2.

In order to understand this property, verify the following:

$$\log 3^5 = 5 \log 3 \quad \text{and} \quad \log 8^2 = 2 \log 8$$

EXAMPLE 3   Using the Log Function to Solve an Equation

Solve the equation $5 = 3^x$.

SOLUTION: We take the log of both sides of the equation to get

$$\log 5 = \log 3^x.$$

We then use the exponent property of the log function to rewrite the equation as

$$\log 5 = x \log 3.$$

If we divide both sides of the equation by log 3, we get

$$\frac{\log 5}{\log 3} = x.$$

We now find $x = \dfrac{\log 5}{\log 3} \approx \dfrac{.69897}{.47712} \approx 1.46.$

If you use your calculator, you will find that $3^{1.46}$ is approximately 5. (Because of the way we rounded numbers, you will not get 5 exactly as a result.)   ☺

We now can find how long it takes for a quantity to double.

EXAMPLE 4   Doubling a Population

In 2001, Cambodia's population was 12 million, with an annual growth rate of roughly 2 percent. In what year will Cambodia's population double, assuming that the growth rate remains the same?

SOLUTION: We want to know when the population will be 24 million. We will use the growth equation $A = P(1 + r)^n$, where P is Cambodia's population in 2001 of 12 (million), r is 0.02, and A is the future population of 24 (million). So, we want to solve the equation

$$24 = 12(1 + 0.02)^n \qquad \text{for } n.$$

First, we divide both sides of the equation by 12 to get

$$2 = (1 + 0.02)^n.$$

Next, we take the log of both sides, giving us

$$\log 2 = \log (1.02)^n.$$

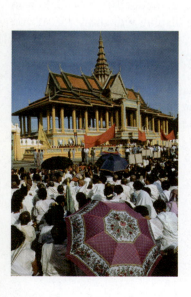

Using the exponent property for the log, we get

$$\log 2 = n \log (1.02).$$

If we divide both sides of the equation by log 1.02, we find that

$$n = \frac{\log 2}{\log 1.02} \approx 35.$$

According to our calculations, the population of Cambodia will double in 35 years.

Limited resources may cause a population to grow according to a logistic model.

Many people were alarmed at the rapid growth of AIDS in the early 1980s. In 1992, a group of medical experts called the Global AIDS Policy Coalition reported the seriousness of the problem and the dire global repercussions of a deadly disease that was exploding throughout the world. Figure 6.15 contains a table of data and a graph from this report showing the increase in the number of AIDS cases in North America from 1985 through 1991.

As with money and population, we might expect that the increase in the number of AIDS victims corresponds to the number of people who have the disease and therefore we can model it by an exponential equation. In order to build this model, we need an initial number of AIDS cases. We arbitrarily choose 70 thousand, which was the number present in 1987. We next need the rate of growth from 1987 to 1988. Since the amount in 1988 is 104 thousand,

$$\frac{\text{number of cases in 1988} - \text{number of cases in 1987}}{\text{number of cases in 1987}} = \frac{104 - 70}{70} = \frac{34}{70} = 0.486.$$

Year	Cumulative AIDS Cases in North America (in thousands)
1985	22
1986	41
1987	70
1988	104
1989	143
1990	180
1991	219

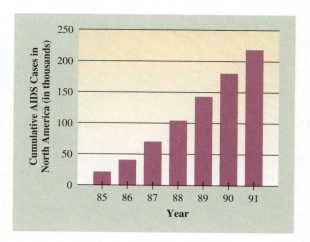

FIGURE 6.15   The growth of AIDS cases in North America, 1985–1991.

The rate of growth from 1987 to 1988 was therefore 48.6 percent. Thus we will use the equation $A = 70(1 + 0.486)^n$ to predict the growth of AIDS cases. Testing this equation for $n = 4$, which corresponds to the year 1991, we get $A = 70(1.486)^4 = 341.33$ thousand cases. Since this is well above the 219 thousand cases that actually occurred in 1991, it suggests that we have overlooked something in constructing our model.

What we have assumed is that the disease will always grow at the same rate. This might be true if we had an unlimited pool of new victims who could contract the disease, but this is not the case. There are only so many among the population who are likely to contract the disease. As more among the population acquire AIDS, there are fewer new people who can contract the disease.

Years	Rate of Growth of AIDS (percent per year, rounded to nearest %)
1985–1986	86
1986–1987	71
1987–1988	49
1988–1989	38
1989–1990	26
1990–1991	22

We see in the table to the left that although the disease was still growing rapidly, as time went by, the rate of increase of new cases slowed.* Because *the rate of growth was not the same* from year to year, an exponential equation is not a good model for the AIDS data.

A logistic model takes into account limits on population growth.

In studying the growth of populations, demographers often use a logistic model. A *logistic model* takes into account the fact that as populations grow, there are limits in terms of space, food, and so on that prevent the population from following a true exponential growth pattern.

Recall that in computing compound interest, we used equations of the type

balance at end of year $= (1 + \text{rate})(\text{balance at end of previous year})$.

Translating this into population growth, we would write

population at end of year $= (1 + \text{growth rate})(\text{population at end of previous year})$.

In both cases, we assumed that the money in the bank account and the population would grow at the same rate from year to year. However, with restrictions on the amount of food, water, space, and so on, any environment can sustain only a limited population. As the population grows, the percentage of this capacity that has been used up influences the rate of growth. If a population has a growth rate of 3 percent, in the beginning it is reasonable to multiply the existing size of the population by $(1 + 0.03)$ to estimate the population the next year. However, as

---

* There are, of course, numerous other factors that could be contributing to this decline, but in order to keep our model simple, we will focus on this one factor only.

the capacity to sustain further population diminishes, we must reduce the growth rate accordingly.

In a logistic model, it is common to represent the maximum capacity that an environment can support by 1, or 100 percent. For our model we will define a quantity P_n, read "P sub n," that represents the percentage of the maximum capacity that has been attained by the population in year n. For example, to say $P_5 = 0.40$ means that at the end of year 5, the population we are studying has attained 40 percent of its maximum capacity. If this same population had an original growth rate of 0.03, to calculate P_6 we would reduce this growth rate by the factor $1 - P_5 = 0.60$ to get a growth rate of

$$0.03 \times (1 - P_5) = 0.03 \times 0.6 = 0.018.$$

The idea is that since 40 percent of the capacity for growth already has been used up by the population, future growth rate can only be 60 percent of what it was originally.

We are now ready to give a precise definition of a logistic growth model.

DEFINITIONS   **Logistic Growth Model**

Assume that a population is growing originally at rate r. We let P_n denote the percentage of the maximum capacity that the population has attained in year n. Moreover, P_n satisfies the following equation

$$P_{n+1} = [1 + r(1 - P_n)]P_n.$$

This collection of equations for $n = 0, 1, 2, \ldots$ is called a **logistic model.**

To calculate P_{n+1}, we reduce the growth rate by multiplying it by $1 - P_n$. We will refer to this quantity as the *rate reduction factor.* So the logistic growth equation can be written as

$$P_{n+1} = [1 + r(\text{rate reduction factor})]P_n.$$

It is useful to recompute the rate reduction factor each time we calculate a new value for P_{n+1}. We illustrate how to use the logistic growth model in Example 5.

EXAMPLE 5   **Using the Logistic Growth Model to Predict Population Growth**

Assume that a population is growing initially at a rate of 3 percent per year. Also assume that at the end of the fifth year, the population is at 40 percent of its maximum size. At what percentage of its maximum size will the population be one year later?

SOLUTION: We are told that $P_5 = 0.40$, $r = 0.03$, and we wish to find P_6. Figure 6.16 shows how to use the logistic growth model.

$$P_{5+1} = [1 + 0.03(1 - P_5)]P_5$$

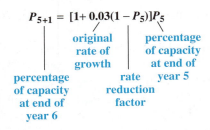

percentage of capacity at end of year 6

original rate of growth

rate reduction factor

percentage of capacity at end of year 5

FIGURE 6.16   Using P_5 to calculate P_6.

At the end of year 5, the rate reduction factor will be $1 - 0.4 = 0.6$; thus

$$
\begin{aligned}
P_6 &= [1 + (0.03)(1 - P_5)]P_5 \\
&= [1 + (0.03)(0.6)](0.4) \\
&= (1 + 0.018)(0.4) \\
&= (1.018)(0.4) = 0.4072.
\end{aligned}
$$

So we see that at the end of the sixth year, the population is at 0.4072, or 40.72 percent of its maximum capacity.

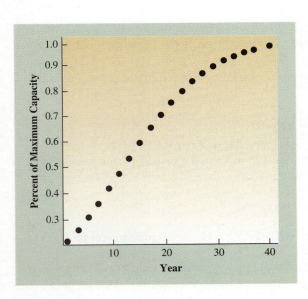

FIGURE 6.17   Graph illustrating logistic growth.

Notice that in Example 5, we did not compute P_6 by substituting 6 into some equation, as we did in previous models. Instead, we described P_6 in terms of P_5. If we had not been given P_5 explicitly, then we would have to calculate it by finding P_4. But to find P_4 would require knowing P_3, and so on. If we keep going, eventually we would need to know P_0. In our population examples, we will always assume that P_0 is the percentage of the capacity we have before the first year begins. An equation that is defined so that to find a given value requires knowing previous values is called a **recursive equation.**

In Figure 6.17, we have plotted the population growth for a hypothetical population over a period of 40 years based on a logistic growth model. It is easy to see that at first the population grows rapidly in a way similar to exponential growth, but as we reach year 15, because the population is at about 60 percent of its maximum capacity, the growth slows considerably.

EXAMPLE 6   Using the Logistic Model to Predict Future Population

Biologists have chosen a nearly extinct type of lemur to populate a small island. They have introduced a number of lemurs that constitute 20 percent of the population that the island is capable of supporting. If the growth rate of the lemur population is 10 percent, find what percentage of the maximum lemur population will be on the island in three years.

SOLUTION: In order to find the percentage of the maximum population at the end of three years, we will find P_1, P_2, and finally P_3.

P_0 is the initial percentage of the maximum population of lemurs, which is 0.20. We will round off calculations to three

decimal places to keep them from becoming too unwieldy. The first rate reduction factor is $1 - P_0 = 1 - 0.20 = 0.80$.

$$P_1 = [1 + (0.10)(\text{rate reduction factor})]P_0$$
$$= [1 + (0.10)(0.8)](0.20)$$
$$= (1 + 0.08)(0.20) = (1.08)(0.20) = .216$$

The rate reduction factor is now $1 - P_1 = 1 - 0.216 = 0.784$.

$$P_2 = [1 + (0.10)(0.784)](0.216)$$
$$= (1 + 0.0784)(0.216) = (1.0784)(0.216) = .233$$

The new rate reduction factor is $1 - P_2 = 1 - 0.233 = 0.767$.

$$P_3 = [1 + (0.10)(0.767)](0.233)$$
$$= (1 + 0.0767)(0.233) = (1.0767)(0.233) = .251$$

At the end of three years, the island will have slightly more than 25 percent of its maximum lemur population. ☉

Quiz Yourself 11

Assume that a population is growing initially at a rate of 4 percent per year. Also assume that at the end of the third year, the population is at 60 percent of its maximum size. At what percentage of its maximum size will the population be one year later?

Exercises 6.4

In Exercises 1–4, you are given the principal in a bank account at the beginning of a year and a rate of interest which is compounded annually. Calculate the amount in the account at the end of the year. *

1. $1,000; 5 percent
2. $3,000; 6 percent
3. $4,000; 2.5 percent
4. $3,000; 6.5 percent

In Exercises 5–12, you are given the principal in a bank account, a yearly interest rate, and the time the money is in the account. Assuming that no withdrawals are made, use the compound interest formula to compute the amount in the account after the specified time period. Assume compounding is done annually.

5. $5,000; 5 percent; five years
6. $7,500; 7 percent; six years
7. $4,000; 8 percent; two years
8. $8,000; 4 percent; three years
9. $10,000; 3 percent; 8 years
10. $12,000; 4.5 percent; 10 years

11. $15,000; 3.5 percent; 12 years
12. $20,000; 6 percent; 20 years

In Exercises 13–18, you are given the population and the growth rate as of 2001 for each country. Assume that the growth rate remains the same from year to year. Use an exponential function to model population growth and determine the size of the population in the specified year.

13. India: population = 1,030 million; growth rate = 1.6 percent; year, 2010
14. Nigeria: population = 127 million; growth rate = 2.6 percent; year, 2010
15. Cambodia: population = 12 million; growth rate = 2.3 percent; year, 2021
16. Brazil: population = 174 million; growth rate = 0.91 percent; year, 2006
17. China: population = 1,273 million; growth rate = 0.88 percent; year, 2010
18. Japan: population = 127 million; growth rate = 0.17 percent; year, 2016

---

* In calculating the amount in the account, we will round the answer *down* to the nearest cent. For example, if the amount in the account is $11,255.47764, we will express the answer as $11,255.47.

In Exercises 19–26, solve each equation for x.

19. $5^x = 20$

20. $2^x = 15$

21. $3^x = 10$

22. $10^x = 3$

23. $10^x = 3.2$

24. $8^x = 4.65$

25. $(3.4)^x = 6.85$

26. $(15.7)^x = 155.5$

27. Compound interest. If you have $10,000 in a bank account that is paying an interest rate of 5 percent, which is being compounded annually, how many years will it take to double if the interest rate stays the same?

28. Compound interest. For the bank account in Exercise 27, how many years will it take to triple?

29. Population growth. In 2000, the population of the United States was 281 million and the growth rate was 1.2 percent. Assuming that the growth rate stays the same, in what year will the population be double what it was in 2000?

30. Population growth. Use the information given in Exercise 16. Assuming that the growth rate remains the same, in what year will Brazil's population double?

If the rate of growth is negative, then we refer to exponential decay instead of exponential growth. Exercises 31–34 involve exponential decay.

31. Population reduction. Large, highly populated countries are often concerned about reducing the size of their population. Assume that in the year 2005 China had a population of 1,320 million. If China could maintain a growth rate of -0.5 percent, what would the population of China be in 2020?

32. Population reduction. Repeat Exercise 31 for India with a population of 1,092 million in 2005 and a growth rate of -0.6 percent.

33. Radioactive decay. Assume that a radioactive material decays with an annual growth rate of -0.35 percent. How many years will it take 100 pounds of the material to decay to 50 pounds?

34. Radioactive decay. Repeat Exercise 33, but now assume that the rate of growth is -0.14 percent.

In Exercises 35–38, assume that a population is growing initially at the specified rate. You are given a value for P_n and are to use the logistic growth model to compute the value of P_{n+1}. Explain what your calculations tell you.

35. rate $= 3$ percent; $P_4 = 0.36$

36. rate $= 5$ percent; $P_7 = 0.48$

37. rate $= 4.5$ percent; $P_8 = 0.72$

38. rate $= 5.5$ percent; $P_3 = 0.51$

In Exercises 39–46, redo the calculations of Example 6 regarding the growth of the lemur population. Use the specified initial growth rate and the given value for P_0 to find P_3. Your answers may vary from ours due to roundoff error.

39. rate $= 8$ percent; $P_0 = 0.30$

40. rate $= 12$ percent; $P_0 = 0.40$

41. rate $= 10$ percent; $P_0 = 0.25$

42. rate $= 15$ percent; $P_0 = 0.35$

43. rate $= 4.5\%$; $P_0 = 0.60$

44. rate $= 4.25\%$; $P_0 = 0.20$

45. rate $= 8.5\%$; $P_0 = 0.50$

46. rate $= 22.5\%$; $P_0 = 0.60$

When a drug such as a pain-killer or an antibiotic is introduced into the body, the kidneys work to eliminate the drug. We will assume that after an hour, 15 percent of any drug in the body is eliminated. After another hour, 15 percent of the remaining drug is eliminated, and so on.

47. Modeling drug concentration. If you receive an injection of 500 mg of Novocain, how much of the drug will remain in your body after three hours?

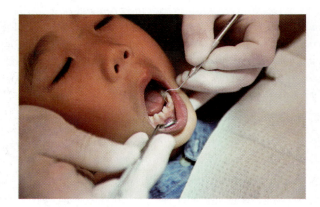

48. Modeling drug concentration. If you take a 500-mg dose of erythromycin, how much of the drug will be in your body after four hours?

49. Modeling drug concentration. If a patient experiences some numbness as long as 150 mg of Novocain remain in the body, after how many complete hours should a patient stop feeling numbness if initially injected with 500 mg of Novocain?

50. Modeling drug concentration. If your physician wishes to keep the level of erythromycin in your body above 200 mg and you are given an initial dose of 500 mg, at what time (to the nearest hour) should you take your next dose?

Further Exercises

Use the logistic growth model to solve Exercises 51–54. You may use a trial-and-error method to find your answer.

*51. Assume that Carl dammed a stream on his land and intends to use the pond that has formed for bass fishing. He stocks the pond with 200 adult bass and believes that at maximum capacity, the pond could support 800 adult bass. Since he wishes the bass population to grow, he will not fish from the pond until it has reached a population of 300 fish. Assuming a yearly growth rate of 18 percent, during which year will he be able to begin fishing?

52. Redo Exercise 51; however, now assume that the pond can support 1,000 adult bass and Carl wishes to begin fishing when the adult population has reached 250.

53. Suppose that the pond in Exercise 51 can support 1,000 fish, the growth rate is 25 percent, and Carl wishes to have 300 fish in the pond by the beginning of the fourth year. Assume that he must purchase fish for stocking in multiples of 50. What is the smallest amount that he can purchase to achieve his goal?

54. Redo Exercise 53, but now use a growth rate of 30 percent and assume that Carl wants 400 fish in the pond by the beginning of the fourth year.

55. Communicating Mathematics   When is an exponential model appropriate? Give examples.

56. Communicating Mathematics   How is a logistic model different from an exponential model?

6.5 PROPORTIONS AND VARIATION

In this section we will discuss some simple, but important ideas, ratio, proportion, and variation that are useful as models in both science and everyday life. We encounter the notion of ratio in many situations. For example, if you drive 86.4 miles and use 2.7 gallons of gasoline, then the quotient $\frac{86.4}{2.7} = 32$ miles per gallon is an example of a ratio.

> **DEFINITIONS**
>
> A **ratio** is a quotient of two numbers. We write the ratio of the numbers a to b as $a:b$ or $\frac{a}{b}$. A **proportion** is a statement that two ratios are equal.

The statement $\frac{3}{4} = \frac{6}{8}$ is an example of a proportion. Notice in the proportion $\frac{3}{4} = \frac{6}{8}$ if you cross multiply $\frac{3}{4} \diagup\hspace{-0.9em}\diagdown \frac{6}{8}$, you get the products $3 \cdot 8 = 4 \cdot 6$. This illustrates a general principle that you will use in working with proportions.

---

* Exercise numbers circled in red can be used as group exercises.

Cross Multiplication Principle

If $\frac{a}{b} = \frac{c}{d}$, then $a \cdot d = b \cdot c$. The quantities $a \cdot d$ and $b \cdot c$ are called **cross products.**

Sometimes one of the quantities in a proportion is unknown and we have to solve for it as we see in Example 1.

EXAMPLE 1   Estimating Gasoline Needed for a Trip

Assume that last week you drove your Honda 86.4 miles and used 2.7 gallons of gasoline. If you plan to take your car on a 168-mile camping trip to the mountains next weekend, how much gasoline do you expect to use?

SOLUTION: We can solve this problem by setting up the following proportion which compares the ratio of gas to distance this week with the ratio of gas to distance on the upcoming trip.

$$\frac{2.7}{86.4} = \frac{x}{168}$$

If we now cross multiply, we get the equation $2.7(168) = 86.4x$, or $453.6 = 86.4x$. Next we divide both sides of this equation by 86.4 to get $5.25 = x$. Therefore, you will use 5.25 gallons on your camping trip.

SOME GOOD ADVICE

In Example 1, we set up the proportion

$$\frac{gas\ used\ last\ week}{miles\ driven\ last\ week} = \frac{gas\ used\ on\ trip}{miles\ driven\ on\ trip}.$$

It would not have mattered if we had started instead with a proportion of the form

$$\frac{miles}{gas} = \frac{miles}{gas},$$

because we would have obtained the same cross products.

However, it is important to be consistent. Whatever, comparison you make on the left hand side of the equation, you must make the comparison in the same way on the right hand side of the equation.

Quiz Yourself 12*

If you can type 30 pages of your term paper in 2.5 hours, how long will it take you to type the entire paper which is 54 pages long?

An interesting application of proportions occurs in estimating the size of wildlife populations. If, for example, we wished to survey the population of leatherback turtles, we could capture a number of turtles, tag them, and release them back to the wild. At a later time, after the tagged turtles have had sufficient time to mix thoroughly with the population, we would capture a new sample of

---

turtles. The proportion of tagged turtles in the new sample allows us to estimate the total turtle population. We illustrate this method in Example 2.

EXAMPLE 2   **The Capture–Recapture Method for Estimating Population Size**

Marine biologists around the world capture 832 female breeding leatherback turtles, tag them, and release them back into the wild. Several months later, the biologists take a second sample of 900 female breeding leatherback turtles of which 7 have been tagged. Use these data to estimate the population of female leatherback breeding turtles.

SOLUTION: We assume that the ratio of all tagged turtles to the total number of all turtles in the population is equal to the ratio of tagged turtles in the second sample to the total number in the second sample. In other words, we can say that

$$\frac{\text{number of tagged turtles in population}}{\text{number of turtles in population}} = \frac{\text{number of tagged turtles in sample}}{\text{number of turtles in sample}}.$$

Let n be the number of turtles in the population. Also, there are 832 tagged turtles in the population and 7 tagged turtles in the sample of 900 turtles. Thus the above equation becomes

$$\frac{832}{n} = \frac{7}{900}.$$

If we cross multiply we get $7n = 832 \cdot 900 = 748{,}800$. Dividing both sides by 7 gives us that $n \approx 106{,}971$. Thus from this survey, the number of female leatherback breeding turtles is almost 107,000.   ◎

Quiz Yourself 13

What would the number of turtles in the population have been in Example 2 if the second sample contained 1,000 turtles of which 8 had been tagged?

Direct variation relates quantities that increase or decrease in the same way.

In many applications, we use the mathematical notion of variation to describe a relationship between several variable quantities. For example, a long distance phone plan may charge you 5 cents per minute. The equation $c = 0.5t$ models the relationship between time, t, and cost, c. This is an example of direct variation. As one of the variables increases or decreases, the other variable changes in exactly the same way.

In another type of variation, known as inverse variation, as one quantity increases, the other quantity decreases. The number of gallons of heating oil used per month and the outside average monthly temperature might be related in this way because the higher the temperature, the fewer gallons of heating oil you would use, and vice versa. We will now make these notions more precise.

DEFINITIONS

We say that y **varies directly** as x, or that y is **directly proportional** to x, if $y = kx$, where k is a nonzero constant. The constant k is called **the constant of variation** or **the constant of proportionality.**

EXAMPLE 3   Solving a Direct Variation Problem

Suppose that y varies directly as x and that $y = 50$ when $x = 15$. Find y when $x = 24$.

SOLUTION: Often the first step in solving a variation problem is to use the information that you are given about x and y to calculate the constant of variation. Because we are told that y varies directly as x, we begin with the equation $y = kx$. If we substitute the values 50 for y and 15 for x, then we get the equation $50 = k \cdot 15$. Dividing by 15, we find that $k = \frac{50}{15} = \frac{10}{3}$.

The second step is to use our knowledge that $k = \frac{10}{3}$ to rewrite the equation $y = kx$ as $y = \frac{10}{3}x$. Then substituting 24 for x gives us $y = \frac{10}{3} \cdot 24 = \frac{240}{3} = 80$. ◎

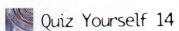

Quiz Yourself 14

Suppose that p varies directly as q and that $p = 154$ when $q = 22$. Find p when $q = 19$.

Sometimes in dealing with direct variation, there may be more than two related quantities. This more complex relationship is called *joint variation*. However, as you will see in Example 4, the two-step solution process that we used in Example 3 is essentially the same.

EXAMPLE 4   Calculating the Strength of a Beam

The strength of a beam is directly proportional to its width and the square of its depth. If a beam that is 3.5 inches wide and 6 inches deep can support a weight of 1,200 pounds, how much could a beam of the same length which is 3 inches wide and 7 inches deep support?

SOLUTION: We begin with the equation $s = kwd^2$ and substitute the values given to us for w, d, and s to find the constant of variation k. That is,

$$1,200 = k(3.5)(6)^2, \text{ or } 1,200 = k \cdot 126.$$

Dividing both sides of this equation by 126, we get

$$k = \frac{1200}{126} = \frac{200}{21}.$$

We can now use this value for k to rewrite the equation $s = kwd^2$ as $s = \frac{200}{21}wd^2$. If we substitute the values 3 for w, and 7 for d, we find that

$$s = \frac{200}{21} \cdot 3 \cdot (7^2) = 1,400 \text{ pounds.}$$ ◎

Inverse variation relates quantities that increase or decrease in opposite ways.

If we say that two quantities vary inversely, this means that as one increases, the other decreases.

> **DEFINITIONS**
>
> We say that y **varies inversely** as x, or that y is **inversely proportional** to x, if $y = \frac{k}{x}$, where k is a nonzero constant.

A good example of inverse variation is the relationship between the distance you are from a sound source and how loud the sound will be that you hear. For example, if you are at a concert and sitting next to a large speaker, the sound level can be painful and perhaps you would want to move to make the sound less intense. That is, the greater the distance you are from the speakers, the less loud the sound is to you. Example 5 illustrates this inverse relationship.

EXAMPLE 5   Measuring the Loudness of a Sound

It is known that the loudness of a sound is inversely proportional to the square of the distance from the sound source. Assume that at a concert when you are sitting 20 feet from the speakers, the sound level is at 120 decibels (a measure of the loudness of a sound). If you were to move 50 feet from the speakers, how loud would the music sound?

SOLUTION: We are assuming that the perceived loudness of the sound is inversely proportional to the square of the distance that you are sitting from the speakers, so we can use the equation $L = \frac{k}{d^2}$ to model this situation.

First we will use the values that we are given for L and d to find the constant of variation, k. Because $L = 120$ when $d = 20$, we get $120 = \frac{k}{20^2} = \frac{k}{400}$. If we multiply both sides of this equation by 400, we find that $120 \cdot 400 = k$, so $k = 48,000$.

Now we can use the fact that $k = 48,000$ to find the value of L when $d = 50$. Substituting for k and d, we get $L = \frac{48,000}{50^2} = \frac{48,000}{2,500} = 19.2$ decibels.   ◎

It is possible to have a combination of both direct variation and inverse variation present in a relationship between several quantities. We will refine the model of the strength of a beam in Example 4 to illustrate this *combined variation.*

EXAMPLE 6   Calculating the Strength of a Beam

As we stated in Example 4, the strength of a beam is directly proportional to its width and the square of its depth. However, as you probably know from practical experience, the longer a beam is, the less weight it will support. That is, the strength of a beam varies inversely as its length. Suppose a wooden beam that is used to support lighting for an outdoor Shakespearean theater is 10 feet long, 3 inches wide, and 4 inches deep and will support a load of 600 pounds.

a) If the length of the beam is increased to 15 feet, how many pounds will the beam support?

b) If we wish the 15-foot beam to support the same weight of 600 pounds, how wide should the beam be to give us the same strength?

SOLUTION: a) We will use the equation $s = k\frac{w \cdot d^2}{l}$ to model this situation. As usual, we first find the constant of variation, k. Substituting the values 3 for w, 4 for d, 10 for l, and 600 for s, we get

$$600 = k\frac{3 \cdot 4^2}{10}, \text{ or } 6,000 = k48.$$

Thus, $k = \frac{6,000}{48} = 125$. We can then rewrite the strength equation as $s = 125\frac{w \cdot d^2}{l}$. If we now keep the values 3 for w, and 4 for d, but substitute 15 for l, we get

$$s = 125\frac{3 \cdot 4^2}{15} = \frac{125 \cdot 3 \cdot 16}{15} = 400.$$

Thus the longer beam will support only 400 pounds.

b) To solve this problem, we will use the values 4 for d, 15 for l, and 600 for s, and we must solve for the value for w. Substituting, we get the equation,

$$600 = 125\frac{w \cdot 4^2}{15} = \frac{125 \cdot w \cdot 16}{15},$$

or,

$$600 \cdot 15 = 125 \cdot 16 \cdot w$$

Therefore, $9,000 = 2,000 \cdot w$, or $w = \frac{9,000}{2,000} = \frac{9}{2} = 4.5$. So, the longer beam should be 4.5 inches wide in order to support 600 pounds.   ◎

Exercises 6.5

Solve for x *in the following proportions.*

1. $24 : x = 18 : 3$

2. $35 : 4 = x : 2$

3. $\frac{50}{4} = \frac{x}{5}$

4. $\frac{x}{8} = \frac{14}{4}$

5. $x : 12 = 3 : 2$

6. $8 : 10 = 32 : x$

7. $\frac{30}{40} = \frac{27}{x}$

8. $\frac{150}{x} = \frac{60}{40}$

In Exercises 9–18, set up a proportion to solve the given problem.

9. Calculating drug dosage. The dosage of a particular drug is proportional to the patient's body weight. If the dosage for a 150-pound adult is six milligrams, what would the dosage be for Maria who weighs 65 pounds?

10. Estimating the amount of deck sealing liquid. If it requires 1.5 gallons of WaterSeal to waterproof a 320-square-foot deck, how much WaterSeal does Josh need to seal his deck that has 480 square feet of area?

11. Calculating a speed limit. While vacationing in Europe, Elton rented a car whose speedometer is calibrated in both miles and kilometers per hour. The marking on the speedometer shows that 30 miles per hour corre-

sponds to 48 kilometers per hour. If the speed limit in France is 100 kilometers per hour, what is the speed limit in miles per hour?

12. Estimating a distance. While driving from Grenoble to Switzerland, Elton sees a road sign that says "Lake Geneva 56 kilometers." How many miles does he still have to drive? (See Exercise 11.)

13. Buying fertilizer. Garth's front lawn is a rectangle measuring 120 by 40 square feet. If a 25-pound bag of "Weed 'N Feed" will treat 2,000 square feet, how many bags must Garth buy to treat his lawn? (Assume that he must buy whole bags.)

14. Mowing grass. Caroline has a part-time job on campus mowing grass. If it takes her 1 and $\frac{1}{2}$ hours to mow a 60,000-square-foot lawn, how long will it take her to mow the rectangular lawn in front of the student center that is 200 feet wide and 650 feet long?

15. Calculating an occupation tax. Last year, Allida paid an occupation tax of $411 on her income of $34,250. What tax will she pay this year on an income of $36,480?

16. Estimating paint needed. Jose needed five gallons of paint to paint a room that measures 20 feet by 20 feet.

How much paint would he need to paint a room that measures 30 feet by 30 feet?

17. Cost of a carpet. Edmund is a hotel manager who paid $864 to have a carpet installed in a conference room that measures 18 feet by 27 feet. How much would he have to pay to have the same carpet installed in a room that measures 24 by 33 feet?

18. Spraying mosquitoes. The wildlife service used 476 gallons of spray to treat 8.5 square miles of forest to kill the mosquito that carries the West Nile virus. How many gallons of spray would be needed to treat 14 square miles?

Use the capture–recapture method in Exercises 19–22. Use the word equation stated in Example 2.

19. Estimating a wildlife population. Biologists capture, tag, and release 400 bald eagles. Several months later, of 240 bald eagles that are captured, 8 are tagged. Estimate the population of bald eagles.

20. Estimating a wildlife population. Marine biologists capture, tag, and release 100 Florida manatees. Several months later, of 90 captured manatees, 5 have tags. Estimate the population of Florida manatees.

21. Estimating a wildlife population. It is estimated that there are 1,000 grizzly bears living in a certain region. Assume that biologists capture, tag, and release 55 bears. Several months later, they capture a sample of 95 bears. How many would you expect to find tagged?

22. Estimating a wildlife population. A lake contains 1,530 largemouth bass. Employees of the state fish commission catch, tag, and release 60 of the bass. If two months later, they recapture 106 largemouth bass, how many would you expect to find tagged?

23. Communicating Mathematics   Assume that last semester at Urbanopolis City College, out of a student body of 5,250 there were 1,470 students who made the dean's list. If this semester the percentage who made the dean's list is the same and 1,554 made the dean's list, what is the current size of the student body at UCC? In solving this problem, Justin set up the equation

$$\frac{5{,}250}{1{,}470} = \frac{1{,}554}{x}.$$

What answer did Justin get and why is it obviously incorrect? Explain what is wrong with Justin's approach.

24. Communicating Mathematics   If the ratio of PCs to Macintoshes on East Central State's campus is $7:2$, what is the ratio of PCs to all computers on campus? (Assume that there are no other computers except PCs and Macs.) Explain your thinking. You may want to give an example to support your conclusion.

25. Communicating Mathematics   Suppose that in solving a problem, you encounter the ratio $m:n$. What would the ratio $n:(m + n)$ represent? Explain your answer.

26. Communicating   Mathematics   What   is $(m:n) \times (n:m)$? Explain. Give an example.

Solve each of the following variation problems by first writing the variation as an equation and finding the constant of variation. Then answer the question that we ask.

27. Assume that y varies directly as x. If $y = 37.5$ when $x = 7.5$, what is the value for y when $x = 13$?

28. Assume that y varies inversely as x. If $y = 10$ when $x = 4$, what is the value for y when $x = 6$?

29. Assume that r varies inversely as s. If $r = 12$ when $s = \frac{2}{3}$, what is the value for r when $s = 8$?

30. Assume that d varies directly as the square of t. If $d = 24$ when $t = 4$, what is the value for d when $t = 10$?

31. Assume that a varies directly as the square of b. If $a = 16$ when $b = 6$, what is the value for a when $b = 15$?

32. Assume that y varies jointly as x and z. If $y = 60$ when $x = 4$ and $z = 5$, what is the value for y when $x = 6$ and $z = 2.5$?

33. Assume that D varies inversely as C. If $D = \frac{3}{4}$ when $C = 2$, what is the value for D when $C = 24$?

34. Assume that A varies directly as the square of r. If $A = 314$ when $r = 10$, what is the value for A when $r = 6$?

35. Assume that r varies jointly as x and y. If $r = 12.5$ when $x = 2$ and $y = 5$, what is the value for r when $x = 8$ and $y = 2.5$?

36. Assume that m varies inversely as n. If $m = 6$ when $n = \frac{2}{3}$, what is the value for m when $n = 15$?

37. Assume that y varies jointly as w and x^2. If $y = 504$ when $w = 4$ and $x = 6$, what is the value of y when $w = 10$ and $x = 14$?

38. Assume that r varies jointly as s and t^2. If $r = 5,600$ when $s = 14$ and $t = 8$, what is the value of r when $s = 6$ and $t = 22$?

39. Assume that y varies directly as w and inversely as x. If $y = 4$ when $x = 10$ and $w = 6$, what is the value of y when $x = 15$ and $w = 3$?

40. Assume that p varies directly as q and inversely as r. If $p = 6$ when $q = 8$ and $r = 5$, what is the value of p when $q = 4$ and $w = 6$?

41. Assume that y varies jointly as x^2 and w, and inversely as z. If $y = 15$ when $x = 2.5$, $w = 8$, and $z = 20$, what is the value of y when $x = 8$, $w = 7$, and $z = 14$?

42. Assume that d varies jointly as a^2 and b, and inversely as c. If $d = 288$ when $a = 6$, $b = 10$, and $c = 4$, what is the value of d when $a = 20$, $b = 11$, and $c = 4$?

43. **Calculating a sales tax.** The amount of sales tax on an item varies directly as the cost of the item. If Drew pays a tax of $2.34 on a toaster that costs $45, how much sales tax would he pay on an electric frying pan that costs $65?

44. **Estimating water usage.** Assume that the amount of water that Mario uses for irrigation at his vineyard is inversely proportional to the amount of rainfall. If he uses 30,000 gallons during a month in which there is 3 inches of rain, how much water would he use in a month that has 5 inches of rain?

45. **Determining the maximum strength of a spring.** Hooke's law states that the length that a spring can be stretched is directly proportional to the force applied to the spring. However, if too much force is applied to the spring, the spring can be stretched to that point at which it can no longer return to its original shape. If a force of 8 pounds stretches a spring 6 inches and stretching the spring beyond 10 inches will ruin the spring, what is the maximum force that can be applied to the spring without damaging it.

46. **Finding the distance a sky diver falls.** The distance that a body falls varies directly as the square of the time that it is falling. If Eileen jumps from a plane and falls 144 feet in the first three seconds, how many feet will she fall in five seconds?

47. **Adjusting a photographer's lighting.** The illumination from a light source is inversely proportional to the square of the distance from the light source. Jonathan is taking a portrait with his light source set 4 feet from his subject and finds that the illumination is twice as bright as it should be. If he wants to reduce the illumination to one-half of what it is now, at what distance should he place his light?

48. **Calculating the amount of heating oil needed.** Assume that the amount of heating oil used during a given month is inversely proportional to the outside temperature. If Andrea's pet store uses 504 gallons of oil in a month that has an average temperature of 42° F, how much oil would she expect to use in a month (of the same length) with an average daily temperature of 36° F?

49. **Determining pool attendance.** In the summer, the monthly attendance at the local pool varies directly as the temperature and inversely as the number of days of rain during the month. If during a given month the average daily temperature is 88° F and it rains eight days, the total attendance for the month is 3,200. What attendance should we expect for a month (of the same length) if it rains 12 days and the average daily temperature is 92° F?

50. **Estimating the length of a trip.** Speed varies inversely with time spent traveling. If a trip to the seashore takes 2 and $\frac{1}{2}$ hours while traveling at an average rate of 54 miles per hour, how long would the same trip take traveling at a rate of 63 miles per hour?

51. **Finding gas pressure.** If the temperature is held constant, the pressure of the gas in a container varies inversely as the volume of the container. Assume that the pressure in a large piston filled with gas is 4 pounds per square inch when the volume of the gas is 120 cubic inches. If the volume of the gas in the piston is reduced to 75 cubic inches, what is the pressure of the gas?

52. **Finding gas pressure.** Redo Exercise 51. Assume now that the pressure in a large piston filled with gas is 12 pounds per square inch when the volume of the gas is 6.5 cubic inches. If the pressure of the gas in the piston is increased to 48 pounds per square inch, what is the volume of the gas?

Further Exercises

*(53.) Communicating Mathematics   If y varies directly as x, does x vary directly or inversely as y? Explain your answer. Give an example.

(54.) Communicating Mathematics   If y varies inversely as x, does x vary directly or inversely as y? Explain your answer. Give an example.

As you saw in Example 6, the strength of a beam is directly proportional to its width and the square of its depth, and inversely proportional to its length. An equation describing this combined variation is

$$s = k\,\frac{w \cdot d^2}{l}$$

Use this information to explain how the strength of a beam would change in each of the following situations. It may help if you make up concrete examples.

(55.) Communicating Mathematics   The width of the beam is doubled.

(56.) Communicating Mathematics   Both the width and the length of the beam are doubled.

(57.) Communicating Mathematics   Both the length and the depth of the beam are tripled.

(58.) Communicating Mathematics   The length, width, and depth of the beam are all doubled.

CHAPTER SUMMARY

SECTION 6.1

linear equation, standard form
A linear equation in two variables is an equation that can be written in the form $Ax + By = C$, where A, B, and C are real numbers and A and B are not both zero. This form is called standard form.

solution, equivalent equations
A solution for an equation is a number or numbers such that if we substitute them for the variables in the equation, the resulting statement is true. Two equations are equivalent if they have the same solutions.

rewriting equations in an equivalent form
Adding or subtracting the same expression from both sides of an equation gives an equivalent equation. Multiplying or dividing both sides of an equation by the same nonzero expression gives an equivalent equation.

points and ordered pairs
Points in the plane correspond to ordered pairs of numbers written in the form (x, y). The number x is called the first coordinate, and the number y is called the second coordinate.

the graph of a linear equation
The graph of a linear equation in two variables is a straight line. We graph a line by first plotting two points.

---

* Exercise numbers circled in red can be used as group exercises.

intercepts

The points where the line crosses the axes are called the intercepts. To find the x-intercept, set $y = 0$ and solve for x. To find the y-intercept, set $x = 0$ and solve for y.

the slope of a line

If (x_1, y_1) and (x_2, y_2) are two points on a line and $x_1 \neq x_2$, then the slope of the line is defined as

$$m = \frac{\text{rise}}{\text{run}} = \frac{\text{change in } y}{\text{change in } x} = \frac{y_2 - y_1}{x_2 - x_1}.$$

slope-intercept form

The linear equation $y = mx + b$ is in slope-intercept form. The number m is the slope of the line and $(0, b)$ is the y-intercept.

SECTION 6.2 **geometric information that determines a line**

We can use either a point and a slope to determine an equation of a line or we can use two points to determine a line.

linear regression, line of best fit

We use linear regression to find the line of best fit to model a set of data points.

SECTION 6.3 **quadratic equation, parabola**

A quadratic equation has degree two. The graph of a quadratic equation of the form $y = ax^2 + bx + c$, where a, b, and c are real numbers, is a parabola.

vertex

The vertex of the graph of the quadratic equation $y = ax^2 + bx + c$ occurs when $x = \dfrac{-b}{2a}$.

quadratic formula

The solutions of the quadratic equation $ax^2 + bx + c = 0$ will be

$$x = \frac{-b + \sqrt{b^2 - 4ac}}{2a} \qquad \text{and} \qquad x = \frac{-b - \sqrt{b^2 - 4ac}}{2a}$$

discriminant

For a quadratic equation of the form $ax^2 + bx + c = 0$, the quantity $b^2 - 4ac$ is called the discriminant of the equation. If the discriminant is positive, there are two distinct solutions to the equation. If the discriminant is zero, there is only one solution. If the discriminant is negative, there are no solutions.

quadratic regression, parabola of best fit

We use quadratic regression to find a parabola of best fit to model a set of data points.

SECTION 6.4 **compounding**

Compounding is a method of computing interest on an account for a given time period based on the total of the principal and interest earned in previous time periods.

calculating compound interest

If the principal P is invested at interest rate r per time period and we compound the interest for n time periods, the amount in the account is $P(1 + r)^n$.

exponential equation, exponential model

An exponential equation is an equation of the form $y = a \cdot b^x$. In an exponential model, the rate of growth is proportional to the amount present.

logistic model

Assume that a population is growing originally at rate r. We let P_n denote the percentage of the maximum capacity that the population has attained in year n. Moreover, P_n satisfies the following equation:

$$P_{n+1} = [1 + r(1 - P_n)]P_n.$$

This collection of equations for $n = 0, 1, 2, \ldots$ is called a logistic model. A logistic model takes into account limits on population growth by reducing the growth rate as the population grows.

recursive equation

A recursive equation is defined so that to find a given value requires knowing other previous values.

SECTION 6.5   **ratio, proportion**

A ratio is a quotient of two numbers. We write the ratio of the numbers a to b as $a:b$ or $\frac{a}{b}$. A proportion is a statement that two ratios are equal.

cross multiplication

If $\frac{a}{b} = \frac{c}{d}$, then $a \cdot d = b \cdot c$. The quantities $a \cdot d$ and $b \cdot c$ are called cross products.

direct variation

We say that y varies directly as x, or that y is directly proportional to x, if $y = kx$, where k is a nonzero constant. The constant k is called the constant of variation or the constant of proportionality.

inverse variation

We say that y varies inversely as x, or that y is inversely proportional to x, if $y = \frac{k}{x}$, where k is a nonzero constant.

CHAPTER TEST

SECTION 6.1

1. Solve the following equations.

a) $\frac{2}{3}x + 2 = \frac{1}{6}x + 4$

b) $0.3x - 2 = 3.5x - 0.4$

2. Solve $A = P(1 + rt)$ for r.

3. Minxia works at a factory job where she is paid $5 per hour up to 40 hours. If she works over 40 hours, she is paid twice her usual wage. Assume that she always works at least 40 hours per week.

a) Model this situation with a linear equation.

b) How much will she earn if she works 46 hours in a given week?

4. Model the following situation with a linear equation. Do not try to solve it. Anton has two part-time jobs. His job in a camera shop pays $5.80 per hour and his work as a data-entry technician pays $11.35 per hour. He earns $160.40 in a given week

SECTION 6.2

5. Graph $3x + 5y = 20$ by plotting the intercepts and drawing a line through them.

6. Find the slope of the line passing through $(2, 5)$ and $(6, 8)$.

7. Match the line in the diagram with the given information regarding its slope. Assume that the scale on both axes is the same.

   line 1: positive slope; less than 1
   line 2: negative slope; between -1 and 0
   line 3: slope does not exist
   line 4: slope greater than 1

8. You are considering buying a cellular phone. CellPrime charges a $29 installation fee plus $11 per month, while CellOne charges a $39 base fee plus $8.50 per month. In which month does renting from CellOne begin to save you money?

9. Explain when it is appropriate to use a linear function as a model.

10. Find an equation of a linear equation whose graph passes through $(3, 4)$ and $(6, 9)$.

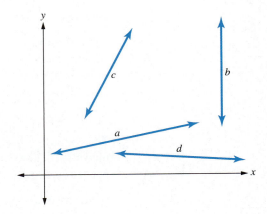

FIGURE FOR QUESTION 7

11. According to the U.S. Bureau of the Census, in 1980, 6.2 percent of the U.S. population were foreign-born and by 2000, this number had risen to 10.4 percent. Use this information to define a linear equation modeling the number of foreign-born people in the U.S. population. Then use this equation to predict the percentage of the U.S. population that will be foreign-born in 2020.

12. What is linear regression?

SECTION 6.3

13. Answer the following questions for the graph of the equation

$$y = -2x^2 - 10x - 8.$$

a) Is its graph opening up or down?

b) What is the vertex of the graph?

c) What are the x-intercepts?

d) What is the y-intercept?

e) Graph the equation.

14. What is quadratic regression?

15. Assume that the equation $A = -0.125x^2 + 2x + 1.125$ models attendance of a recently released movie. During what week will the movie attain its highest attendance?

SECTION 6.4

16. If $10,000 is placed in an investment paying 4.8 percent yearly interest compounded annually, how much will be in the account after five years?

17. How long will it take for the account in question 16 to double?

18. Assume that a population is growing initially at a 3 percent rate and that we are using a logistic growth model to describe the population growth. If $P_3 = 0.50$, what is P_4?

19. What is the difference between an exponential model and a logistic model?

SECTION 6.5

20. Solve for x in the following proportions.

a) $25:8 = x:2$

b) $\dfrac{30}{4} = \dfrac{x}{5}$

21. If it requires 3.5 gallons of sealer to coat a 840-square-foot driveway, how much sealer will be needed to coat a 1,500-square-foot driveway?

22. Marine biologists capture, tag, and release 180 penguins. Several months later, of 55 penguins that are captured, 12 are tagged. Estimate the population of penguins.

23. Assume that y varies inversely as x. If $y = 14$ when $x = 3$, what is the value for y when $x = 18$?

24. Assume that d varies jointly as a^2 and b, and inversely as c. If $d = 144$ when $a = 3$, $b = 7$, and $c = 14$, what is the value of d when $a = 8$, $b = 11$, and $c = 16$?

25. The strength of a beam is directly proportional to its width and the square of its depth, and inversely proportional to its length. How would the strength of the beam change if the width and length are doubled?

Of Further Interest: DYNAMICAL SYSTEMS

Why are weather predictions sometimes so unreliable? With the vast technology we have to gather and analyze data, it would seem possible to build mathematical models that are more reliable than they are at present. The problem in modeling weather is not only that it is much more complex than the situations we discussed in this chapter, but also that many models of systems are unstable. This means that small changes in what is put into the model may result in very large changes in what comes out of the model.

We can use a coin to illustrate an unstable state. A coin can be in one of three resting states: head, tail, and on edge. If the coin is very close to being in the head state (perhaps the coin is held with one edge touching a table and the opposite side lifted $\frac{1}{10}$ of an inch), when the coin is released, it will go to the head state. There are obviously many states close to the head state such that if the coin is in one of these states the coin approaches the head state.

Contrast the head state with the edge state. If we place the coin almost on edge and release it, the coin will not move toward the edge state, but rather it will approach either the head state or the tail state. If a system is close to a *stable state* then the system will move toward that state. In contrast, if a system is close to an *unstable state*, the system's behavior may be very unpredictable.

The systems we study in this section are similar to the bank account and logistic growth models you studied in Section 6.4. In the banking example, knowing the amount in the account at time n allowed us to compute the amount at time $n + 1$. Similarly, in the logistic growth model, knowing the lemur population at time n enabled us to predict the population at time $n + 1$.

In both of these cases, we could think of the model as giving us a sequence of numbers where we can calculate each number in the sequence provided that we know the previous one. For the banking example, before we had the formula for calculating compound interest, we calculated the amounts

$$A_0 = 1{,}000,$$
$$A_1 = 1{,}080,$$
$$A_2 = 1{,}166.40,$$
$$A_3 = 1{,}259.71,$$

and so on.

For the logistic growth model of the lemurs, we had percentages of maximum population attained, which we called

$$P_0 = 0.20,$$
$$P_1 = 0.216,$$
$$P_2 = 0.233,$$
$$P_3 = 0.251,$$

and so on.

■

Each state of a dynamical system is determined by its previous state.

We can describe both the banking example and the logistic growth example by a mathematical model called a dynamical system.

> **DEFINITION**
>
> A **dynamical system*** is a sequence of numbers A_0, A_1, A_2, A_3, and so on such that for each value of n,
>
> $$A_{n+1} = \text{an expression involving } A_n.$$

We now look at Examples 1 and 6 from Section 6.4 as dynamical systems.

EXAMPLE 1   Some Dynamical Systems

Model each situation with a dynamical system.

a) $1,000 is placed in a bank account that pays 8 percent interest compounded yearly.

b) An island is populated with lemurs that have a 10 percent growth rate. The island is initially populated with 20 percent of the island's capacity to sustain these lemurs.

SOLUTION: a) We have seen that each year the account has 1.08 times the money of the previous year. The initial amount is $1,000, thus

$$A_0 = 1,000,$$
$$A_1 = 1.08\, A_0,$$
$$A_2 = 1.08\, A_1,$$
$$A_3 = 1.08\, A_2,$$

and so on, and, in general,

$$A_{n+1} = 1.08\, A_n.$$

This system is described by the equations

$$A_{n+1} = 1.08\, A_n, \qquad n = 0, 1, 2, \ldots, \text{ where } A_0 = 1,000.$$

---

* There are many different categories of dynamical systems, depending upon what properties the system satisfies. For example, systems such that A_{n+1} depends only on A_n are sometimes called *first-order systems*. A *second-order system* would be one in which A_{n+1} depends on both A_n and A_{n-1}. We will not deal with such systems for now, and in order to make the discussion clearer, we have slightly simplified the terminology.

b) For this system,

$$P_0 = 0.20,$$
$$P_1 = [1 + (0.10)(1 - P_0)]P_0,$$
$$P_2 = [1 + (0.10)(1 - P_1)]P_1,$$
$$P_3 = [1 + (0.10)(1 - P_2)]P_2,$$

and so on, and, in general,

$$P_{n+1} = [1 + (0.10)(1 - P_n)]P_n.$$

This system is described by the equations

$$P_{n+1} = [1 + (0.10)(1 - P_n)]P_n, \qquad n = 0, 1, 2, \ldots, \text{ where } P_0 = 0.20.$$

Like the changes in a bank account and changes in the lemur population, if we had enough data and sufficiently complicated equations to describe weather changes, we could view the present weather as an initial state W_0. Then W_1 would be the weather one hour later, W_2 the weather two hours later, and so on. If our model were accurate (and stable), then $W_{8,760}$ would be the state of the weather in 8,760 hours, or one year from now.

Just as equations can stand alone as mathematical objects without attempting to explain them in real-life terms, so can dynamical systems. It is common to identify a dynamical system with a rule that defines it. Here are two examples.

$$(1) \quad A_{n+1} = 5A_n + 3$$
$$(2) \quad A_{n+1} = 2A_n^2 + A_n - 5$$

System (1) is an example of a *linear system* and system (2) is called *nonlinear*. Can you see why these names would be appropriate? Notice that a dynamical system is really an infinite number of equations. For example, system (1) corresponds to the collection of equations

$$A_1 = 5A_0 + 3,$$
$$A_2 = 5A_1 + 3,$$
$$A_3 = 5A_2 + 3,$$

and so on.

Dynamical systems model many phenomena.

If you have ever been prescribed antibiotics by a physician, you have had personal experience with a dynamical system.

EXAMPLE 2   Modeling Antibiotics in the Blood by a Dynamical System

Assume that your doctor has prescribed a 500-mg dose of an antibiotic to be taken three times a day. Assume that the drug enters your bloodstream instantly and that over an eight-hour period, your body eliminates 60 percent of the drug. Model the amount of drug in your blood by a dynamical system. Use this model to compute the amount of the drug in your system after three doses.

SOLUTION: Because you are taking the drug every eight hours, each unit of time in our model corresponds to an eight-hour time interval. Let us represent the amount of drug present at time interval n by D_n. At $n = 5$, for example, we see that your body has eliminated 60 percent of the amount of drug that was present eight hours earlier, so we see that before you take your next dose, your blood contains

$$D_5 = 0.40(D_4).$$

However, you take a dose of the drug that puts 500 mg of the drug into your body. So we get

$$D_5 = 0.40(D_4) + 500.$$

What is true for 5 is true for any time, so we see that

$$D_{n+1} = 0.40(D_n) + 500.$$

The only question that we still must answer is, what should be the value of D_0? If we want our first dose to correspond to time 1, the second to time 2, and so on, then we could say that at time 0, which is eight hours before we took our first dose, there is no drug in the bloodstream. Therefore,

$$D_{n+1} = 0.40(D_n) + 500, \qquad n = 0, 1, 2, \ldots,$$

with the initial value $D_0 = 0$, describes this dynamical system.

We now calculate D_3, but in order to do this, we need to first compute D_0, D_1, and D_2.

$$D_0 = 0,$$
$$D_1 = 0.40(D_0) + 500 = 0.40(0) + 500 = 500,$$
$$D_2 = 0.40(D_1) + 500 = 0.40(500) + 500 = 700,$$
$$D_3 = 0.40(D_2) + 500 = 0.40(700) + 500 = 780.$$

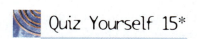

Quiz Yourself 15*

Redo Example 2, but now assume that the body eliminates 80 percent of the drug between doses and that each dose is 400 mg.

Once a dynamical system reaches an equilibrium value, it stays at that value.

In Example 2, the amount of drug in the bloodstream is increasing, and we might wonder what are the effects of taking the drug for a long period of time. Is there the possibility that we could become poisoned by the amount of drug in the bloodstream? In order to look at this question more closely, we generate a table of values for D_n.

n	D_n (rounded to one decimal place)
1	500
2	700
3	780
4	812
5	824.8
6	829.9
7	832.0
8	832.8
9	833.1
10	833.3

We see that as n gets larger, D_n appears to be getting closer to 833.3. In fact, as you will see shortly, if D_n ever equals $\frac{5,000}{6} = 833.3333\ldots$, then from that point on, all subsequent values of D_n will also equal 833.333. . . .

The value $\frac{5,000}{6}$ is an example of an equilibrium value for the dynamical system in Example 2.

> **DEFINITION**
>
> An **equilibrium value** for a dynamical system is a number a such that if $A_n = a$, then A_{n+1} will also equal a.

Equilibrium values for a system are important because they frequently tell us about the long-term behavior of the system. It is easy to find equilibrium values for some systems. For example, consider the system $A_{n+1} = 5A_n + 3$. To say that both $A_n = a$ and $A_{n+1} = a$, by substitution, means that $a = 5a + 3$. When we solve this for a, we get $a = -\frac{3}{4}$ as an equilibrium value for this system.

> **Finding Equilibrium Values for Dynamical Systems**
>
> To find an equilibrium value a for a dynamical system, we do the following:
>
> 1. Write the equation $A_{n+1} =$ an expression involving A_n.
> 2. Substitute a for both A_n and A_{n+1} in the equation in step (1).
> 3. Solve the equation in step (2) for a.

An equilibrium value may be unstable.

An equilibrium value for a system may or may not be stable. You saw in the drug example that the numbers seemed to get closer and closer to 833.333 . . .   and, as you will see in a moment, this is indeed what is happening. In Example 3, we look at another similar dynamical system but you will see very different behavior.

EXAMPLE 3   A Nonstable Equilibrium Value for a Dynamical System

a) Find an equilibrium value for the dynamical system $A_{n+1} = 4A_n - 5$.

b) Is this equilibrium value stable?

SOLUTION: a) Substituting a for A_n and A_{n+1} in the equation $A_{n+1} = 4A_n - 5$, we get $a = 4a - 5$. Solving this equation, we find that $a = \frac{5}{3} \approx 1.6$.

b) We now generate a table of values for A, using two values for A_0 that are extremely close to $\frac{5}{3}$.

n	A_n	A_n	A_n
0	$\frac{5}{3}$	1.66	1.67
1	$\frac{5}{3}$	1.64	1.68
2	$\frac{5}{3}$	1.56	1.72
3	$\frac{5}{3}$	1.24	1.88

continued

continued

n	A_n	A_n	A_n
4	$\frac{5}{3}$	$-.04$	2.52
5	$\frac{5}{3}$	-5.16	5.08
6	$\frac{5}{3}$	-25.64	15.32
7	$\frac{5}{3}$	-107.56	56.28
8	$\frac{5}{3}$	-435.24	220.12
9	$\frac{5}{3}$	$-1,745.96$	875.48
10	$\frac{5}{3}$	$-6,988.84$	3,496.92

Notice that although we take values very close to $\frac{5}{3}$ as initial values A_0 for this system, we see that by the time we compute A_{10}, the values we are getting are very far away from the initial value. Thus the equilibrium value $\frac{5}{3}$ is *not stable*. Much like the coin standing on edge, if we begin even a little bit away from $\frac{5}{3}$, we wind up with values which are quite far away from $\frac{5}{3}$.

There is a theorem, which we will state without proof, that tells us when an equilibrium value is stable and when it is not.

Quiz Yourself 16

Find equilibrium values for the given dynamical systems and determine whether these equilibrium values are stable.

a) $A_{n+1} = 3A_n - 1$

b) $B_{n+1} = 0.5B_n - 3$

Stability of Equilibrium Values for Dynamical Systems

Suppose that a is an equilibrium value for the system

$$A_{n+1} = mA_n + b.$$

1. If $-1 < m < 1$, then the equilibrium value a is stable.
2. If $m < -1$ or $m > 1$, then a is unstable.
3. If $m = -1$, then the values for A_n will oscillate between two values.

Dynamical systems have many applications. An interesting model of one of the causes of war was given by Richardson in his book *Arms and Insecurity: A Mathematical Study of the Causes and Origins of War.** Example 5 is based on Richardson's work.

EXAMPLE 4   Modeling an Arms Race with a Dynamical System

In the early part of the twentieth century, France-Russia and Germany-Austria-Hungary were opposing alliances. According to Richardson, the dynamical system†

$$D_{n+1} = \frac{5}{3}D_n - \frac{380}{3}, \qquad \text{where } D_0 = 199,$$

models total annual defense spending between the two alliances beginning in 1909 (year 0 in our model).

---

* L. F. Richardson, *Arms and Insecurity: A Mathematical Study of the Causes and Origins of War* (Boxwood Press, 1960).
† For a derivation of this system, see James T. Sandefur, *Discrete Dynamical Systems* (Clarendon Press, 1990), p. 76.

a) Find an equilibrium value for this system.

b) Is this equilibrium value stable?

c) What does this model predict will happen as n increases?

Year = n	Amount Spent on Defense = D_n
0	199
1	205
2	215
3	231.7
4	259.5
5	305.7
6	382.9
7	511.5
8	725.8

TABLE 6.3 Growth in defense spending, according to Richardson's model of the European arms race.

SOLUTION: a) We solve the equation $a = \frac{5}{3}a - \frac{380}{3}$ to find an equilibrium value. Multiplying this equation by 3, we get $3a = 5a - 380$. This simplifies to $380 = 2a$, so the equilibrium value is 190.

b) Because the coefficient of D_n is $\frac{5}{3}$, which is greater than 1, the equilibrium value 190 is unstable.

c) The equilibrium value of 190 is not stable. Therefore, if we begin with the value $D_0 = 199$, as n increases, the values for D_n might be quite far away from 190. We calculate some values for D_n in Table 6.3.

You can see in Table 6.3 that as n increases, D_n (the amount spent on defense) also increases rapidly. Since it is impossible for both sides to increase defense expenditures indefinitely, eventually one or the other will be in a position where they feel threatened and will declare war.

Exercises

In Exercises 1–6, for each dynamical system, calculate the value of A_n.

1. $A_{n+1} = 2A_n - 1$; $A_0 = 3$; find A_2

2. $A_{n+1} = 3A_n + 2$; $A_0 = 1$; find A_2

3. $A_{n+1} = -3A_n + 4$; $A_0 = -2$; find A_3

4. $A_{n+1} = 2.5A_n - 3$; $A_0 = 2$; find A_3

5. $A_{n+1} = 1.8A_n - 2$; $A_0 = 4$; find A_4

6. $A_{n+1} = -0.8A_n - 2$; $A_0 = 1.5$; find A_4

In Exercises 7–12, model each situation with a dynamical system. Use your model to answer the question.

7. **Compound interest.** You deposit $1,000 in a bank account paying 5 percent yearly interest that is compounded annually. How much will be in your account at the end of two years?

8. **Compound interest.** You deposit $1,500 in a bank account paying 6 percent yearly interest that is compounded annually. How much will be in your account at the end of three years?

9. **Wildlife growth.** An island is populated with lemurs that have a growth rate of 8 percent. The island is initially populated with 30 percent of the island's capacity to sustain these lemurs. What percentage of the lemur population's maximum capacity will be attained at the end of two years?

10. **Wildlife growth.** Repeat Exercise 9, but now assume the growth rate is 12 percent and the island is initially populated with 20 percent of its maximum capacity.

11. **Antibiotic level.** You take a 250-mg dose of an antibiotic every four hours. Your body eliminates 40 percent of the drug in a four-hour period. How much antibiotic will be in your bloodstream after three doses?

12. **Antibiotic level.** You take a 1,000-mg dose of an antibiotic every twelve hours. Your body eliminates 75 percent of the drug in a twelve-hour period. How much antibiotic will be in your bloodstream after three doses?

In Exercises 13–16, find the equilibrium value for each dynamical system. Comment on whether the value you find is stable or unstable.

13. $A_{n+1} = 2A_n + 3$

14. $A_{n+1} = 4A_n - 5$

15. $B_{n+1} = 0.25B_n + 4$

16. $B_{n+1} = 0.10B_n - 2$

Further Exercises

Living plants and animals all contain the chemical element carbon. A certain percentage of that carbon is radioactive, and scientists believe that the percentage has remained constant for thousands of years. Radioactive carbon decays, so that when an animal dies, a tiny bit of the radioactive carbon is lost each year. It is known that the amount of radioactive carbon that remains in a fossil at the end of a year is approximately 0.99988 of the amount that was present at the beginning. Thus the following dynamical system describes radioactive carbon decay in a fossil:

$$C_{n+1} = 0.99988 \cdot C_n \qquad \text{for } n = 0, 1, 2, 3, \ldots.$$

This system behaves exactly like the compound interest situation (except the amount of radioactive carbon is decreasing), so it is easy to see that after k years, the amount of radioactive carbon in the fossil will be $C_k = 0.99988^k \cdot C_0$. We will always assume that the amount of radioactive carbon at time 0 will be 1; that is, $C_0 = 1$. Use this model to answer Exercises 17–18. You can solve these equations using the log function, as you did in Section 6.4.

17. Carbon dating. A fossilized bone is found that contains 90 percent of the original radioactive carbon that was present. To the nearest 100 years, how old is the bone?

18. Carbon dating. A leaf of a fossilized plant is found that contains 60 percent of the original radioactive carbon that was present. To the nearest 100 years, how old is the plant?

Assume that the logistic equation $P_{n+1} = [1 + r(1 - P_n)]P_n$ models the growth of wild turkeys on a large parcel of state game land. If we wish to allow hunting on this land, we adjust this equation by subtracting some number from P_{n+1} to account for the turkeys killed by hunting. In the following questions, suppose that the game land is capable of supporting a maximum of 1,000 turkeys.

19. Managing wildlife. Assume that currently there are 500 turkeys and we wish to allow 80 to be harvested per year. The growth rate of the turkeys is 10 percent per year. Modify the growth equation accordingly and use it to predict the turkey population at the end of three years.

20. Managing wildlife. Assume that currently there are 750 turkeys and we wish to allow 100 to be harvested per year. The growth rate of the turkeys is 10 percent per year. Modify the growth equation accordingly and use it to predict the turkey population at the end of three years.

CHAPTER 7

Modeling with Systems of Linear Equations and Inequalities:*
What's the Best Way to Do It?

It may surprise you that not all mathematics was invented a long, long time ago. Linear programming, which we discuss later in this chapter, was invented in the 1940s in order to organize the United States' military effort during World War II. The question that we want to answer in a real-life linear programming problem is, "What is the best way to do something?"

The San Francisco Police Department used linear programming to schedule police officers better, thus saving the department $11 million per year.† Bethlehem Steel asked, "What was the best way to design a new steel production facility?" Mathematical researchers used integer programming to design a facility that saved the company $8 million per year.‡

Researchers use linear programming to make a choice from a large set of alternatives. In order to determine how to make the best choice, you will have to learn to solve systems of equations and systems of inequalities. In this chapter, you will solve a variety of real-life problems including how to choose the best diet, make the best investment, and schedule a factory most effectively.

The topics that you will study in this chapter are certainly among the most widely used areas of applied mathematics.

---

* For more resources on systems of equations and inequalities, see www.aw.com/pirnot.
† Taylor, P., and S. Huxley, "A Break from Tradition for the San Francisco Police: Patrol Officer Scheduling Using an Optimization-Based Decision Support System," *Interfaces* **19**(1989, No. 1), p. 24.
‡ Vasko, F., et al. "Selecting Optimal Ingot Size for Bethlehem Steel," *Interfaces* **19**(1989, No. 1), pp. 68–84.

7.1   SYSTEMS OF LINEAR EQUATIONS

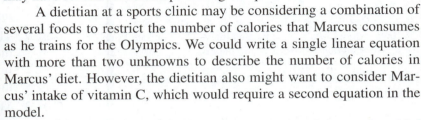

You saw many examples in Section 6.2 where we used a linear equation in two unknowns to model a *single* relationship between two quantities. In more complex situations, however, we often have more than two quantities and may have several relationships among the quantities.

A dietitian at a sports clinic may be considering a combination of several foods to restrict the number of calories that Marcus consumes as he trains for the Olympics. We could write a single linear equation with more than two unknowns to describe the number of calories in Marcus' diet. However, the dietitian also might want to consider Marcus' intake of vitamin C, which would require a second equation in the model.

To describe the diet more completely, we might need a third equation to model saturated fat and a fourth to describe the amount of complex carbohydrates. To model Marcus' diet completely could easily require several dozen linear equations.

There are models that are much more complex than this. For example, models that describe the U.S. economy or predict the weather can have thousands of equations and variables.

A collection of linear equations that are related to each other, such as those we have mentioned for Marcus' diet, the economy, and the weather, is called a **system of linear equations.** For the time being, we will consider very simple systems of equations that have only two equations and two variables (or unknowns).

A system of linear equations represents a pair of lines.

A **solution** of a pair of linear equations is an ordered pair of numbers that satisfies both equations. For example, if you substitute 5 for x and 3 for y in the system

$$2x + 4y = 22$$
$$x - 6y = -13,$$

you will get

$$2 \cdot 5 + 4 \cdot 3 = 22$$
$$5 - 6 \cdot 3 = -13.$$

Since the ordered pair (5, 3) makes both equations true, it is a solution for the system. Although (7, 2) makes the first equation true in the system, it does not make the second equation true (verify). Therefore, it is not a solution for the system.

 Quiz Yourself 1*

Which ordered pair is a solution for the system

$$3x - 5y = 4$$
$$2x + 4y = 10$$?

a) (2, 3)

b) (3, 1)

c) (−3, −1)

---

* Quiz Yourself answers begin on page 849.

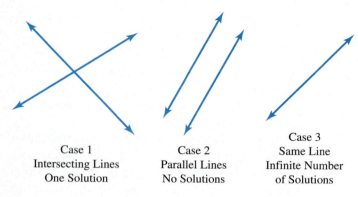

Case 1
Intersecting Lines
One Solution

Case 2
Parallel Lines
No Solutions

Case 3
Same Line
Infinite Number
of Solutions

FIGURE 7.1 A geometric representation of the three cases that may arise in solving a system of linear equations.

Recall from Chapter 6 that the graph of a linear equation in two unknowns is a straight line. Therefore, if you graph a system of two linear equations in two unknowns you will get a *pair* of lines. There are three possible situations, as we show in Figure 7.1.

Case 1: The lines intersect in a single point. The ordered pair that represents this point is the *unique solution* for the system.

Case 2: The lines are distinct parallel lines and therefore don't intersect at all. Since the lines have no common points, this means that the system has *no solutions.*

Case 3: The two lines are the same line. Since the lines have an infinite number of points in common, this means that the system will have *an infinite number of solutions.*

We use the elimination method to solve systems of linear equations.

We will illustrate these three cases by using a technique called the **elimination method** to solve systems of linear equations.* The basic strategy of the elimination method is to replace the system of equations by a single equation which is easier to solve, as we see in Example 1.

EXAMPLE 1 **Solving a System Which Has One Solution**

Solve the system

$$3x + 4y = 10$$
$$5x - 6y = 4 \quad . \tag{1}$$

SOLUTION: To use the elimination method, we will multiply both equations in the system by some constants so that one of the variables appears with *opposite coefficients* in the two equations. If we then add these two equations, the variable with the opposite coefficients will drop out leaving us with a single equation having one unknown.

In system (1) if we multiply the top equation by 3 and the bottom equation by 2, we get opposite coefficients for *y*. This gives us the system

$$9x + 12y = 30$$
$$10x - 12y = 8 \quad . \tag{2}$$

---

* We will explain another method for solving systems of equations, called the substitution method, in the exercises.

Now, adding corresponding sides of both equations in (2) causes the y to drop out, giving us

$$9x + 12y = 30$$
$$+\ 10x - 12y =\ \ 8$$
$$\overline{19x + 0\ \ \ = 38.}$$

Solving for x, we find that $x = 2$.

Because we found a single value for x, we know that we are dealing with Case 1.

There will also be a unique value for y, which we will find next.

When you are using the elimination method, once you find one unknown, you have two choices. If the number you find is not easy to work with, you can go back to the original system, eliminate the other variable, and solve for the second unknown.

If the value you find is easy to use, as in this case, then substitute it in either equation of the original system to find the second unknown. Substituting 2 for x in the first equation, we get

$$9(2) + 12y = 30.$$

This simplifies to $12y = 12$, so $y = 1$. Thus the solution for this system is $(2, 1)$. You should verify that $(2, 1)$ solves both equations.

 SOME GOOD ADVICE

René Descartes' invention of analytic geometry (see Section 6.1) allows us to use geometric intuition to understand algebra and also to use algebra to solve geometric problems. If you look at a situation from both points of view, you often gain insight into how to solve a problem or to check your solution.

If the geometric and algebraic representations of a solution do not agree, then you must go back and find the error.

In Figure 7.2, we graph system (1) from Example 1 on a graphing calculator* where you can see that the lines appear to intersect at the point $(2, 1)$.

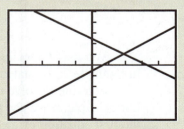

FIGURE 7.2   Graph of system (1).

Quiz Yourself 2

Use the elimination method to solve the following system:

$$5x + 2y = -10$$
$$2x + 3y = 7$$

---

* For more on using a graphing calculator to solve systems of equations, see the technology section at the web site for this text at www.aw.com/pirnot.

EXAMPLE 2   Trying to Solve a System That Has No Solutions

Solve the system

$$-6x + 4y = 5$$
$$3x - 2y = 1 \cdot$$

SOLUTION: If we multiply the bottom equation by 2 and add the equations to eliminate x from the system, we get a surprising result.

$$
\begin{array}{r}
-6x + 4y = 5 \\
+ \quad 6x - 4y = 2 \\
\hline
0 \; + 0 \; = 7.
\end{array}
$$

We have assumed that there is a solution to this system. By doing this, we are now forced to accept that $0 + 0 = 7$! Our only way out of this predicament is to conclude that our assumption is wrong, so a solution for this system does not exist.*

In trying to eliminate one of the variables, we obtained a false statement. We conclude that there are no points common to both lines and therefore we are dealing with Case 2.

◎

A system which has no solutions, like the one in Example 2, is said to be *inconsistent*.

EXAMPLE 3   Solving a System Having Infinitely Many Solutions

Solve the system

$$2x - y = 3$$
$$4x - 2y = 6$$

SOLUTION: If we multiply the top equation by -2 and add the equations, we eliminate x from the system. Again we get a surprising result.

$$
\begin{array}{r}
-4x + 2y = -6 \\
+ \quad 4x - 2y = \quad 6 \\
\hline
0 \; + 0 \; = \quad 0
\end{array}
$$

You may be tempted to confuse this with Case 2. However, unlike the result in Example 2, the statement $0 + 0 = 0$ is a true statement.

In trying to solve for x, we obtained a statement which in no way restricts x. For example, if we choose any value for x, such as $x = 5$, and substitute 5 for x in the first equation of the system, we get $-4(5) + 2y = -6$. Solving for y, gives us $y = 7$. Thus $(5, 7)$ is one of the many pairs that are solutions for the original system.

---

* If you have studied the logic chapter, you can look at what we have done as: If there is a solution, then $0 + 0 = 7$. The contrapositive states: If $0 + 0 \neq 7$, then there is no solution.

If in solving the system we obtain a true statement that does not involve either x or y, this tells us that there are an infinite number of ways to solve the system, so the two lines must be the same.

A system that has an infinite number of solutions, such as the system in Example 3, is said to be *dependent*.

We summarize what you saw happening in Examples 1 to 3.

Quiz Yourself 3

a) Solve the system

$$5x - 3y = 3$$
$$-10x + 6y = 4 \, .$$

b) What is your geometric interpretation of the result from part a)?

Using the Elimination Method to Solve Systems of Linear Equations

Case 1 — We find a single value for both x and y. The pair (x, y) corresponds to the point of intersection of the lines represented by the equations in the system.

Case 2 — We obtain an obviously false statement. We conclude that there are no solutions to this system, which represents a pair of distinct parallel lines.

Case 3 — We obtain a statement which is always true and that does not contain either x or y. Any value can be used for x as the first coordinate of a solution of the system. There are an infinite number of solutions to the system, and the two equations of the system represent the same line.

Example 4 shows that it can be worthwhile to simplify the coefficients of a system before you solve it.

EXAMPLE 4   Simplifying Coefficients Before Solving a System

Solve the system
$$0.25x + 0.4y = 1.8$$
$$0.5x + 0.2y = 2.4 \, .$$

SOLUTION: Rather than deal with decimal computations, we multiply the first equation by 100 and the second equation by 10 to get the system

$$25x + 40y = 180$$
$$5x + 2y = 24 \, .$$

If we now multiply the second equation by -5, we get

$$25x + 40y = 180$$
$$-25x - 10y = -120 \, .$$

Adding these equations results in the equation $30y = 60$, which means $y = 2$. Now substituting 2 for y in the equation $5x + 2y = 24$, we get $5x + 2(2) = 24$. When we solve for x, we find that $x = 4$. So, $(4, 2)$ is a solution for this system.

In Example 4, when we obtained the equation $25x + 40y = 180$, we could have divided both sides of the equation by 5 to get the equivalent equation $5x + 8y = 36$ which would have simplified our computations.

SOME GOOD ADVICE

Whenever you encounter decimals or fractions as coefficients in an equation, it is a good idea to multiply both sides of the equation by a suitable constant to simplify the coefficients. Doing this decreases your chance of making computational errors.

A system of equations models a set of relationships between two quantities.

Now that you know how to solve systems of equations, you can use them to model situations involving several relationships among a set of variables. The next several examples illustrate this.

EXAMPLE 5   Modeling Nutritional Requirements

Yuki is taking two types of food with her on her hike in the Grand Canyon. Each gram of food A provides 4 calories and 0.08 grams of protein, and each gram of food B provides 6 calories and 0.2 grams of protein. If she requires 2,200 calories and 50 grams of protein daily, how much of each should she consume to satisfy these requirements?

SOLUTION: We will assume that Yuki will eat a grams of food A and b grams of food B each day.

We will first write a linear equation to describe the calories that Yuki consumes each day. Since each gram of A provides 4 calories and each gram of B provides 6 calories, the number of calories provided by a grams of A and b grams of B is $4a + 6b$. The number of calories must equal 2,200, so we get the equation shown in Figure 7.3.

Similarly, the number of grams of protein in this diet is $0.08a + 0.2b$, and since we wish to have 50 grams of protein daily, we get the equation

$$0.08a + 0.2b = 50.$$

Now, to find the solution to our problem, we solve the system

$$4a + 6b = 2,200$$
$$0.08a + 0.2b = 50 \quad .$$

First, we multiply the second equation by 100 to eliminate decimal coefficients, giving us

$$4a + 6b = 2,200$$
$$8a + 20b = 5,000.$$

Now multiply the top equation by -2 to get

$$-8a - 12b = -4,400$$
$$8a + 20b = 5,000 \quad .$$

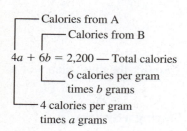

FIGURE 7.3 Equation describing calories consumed per day.

Highlight: *Is It Reasonable to Use Linear Equations as Models?*

Because linear equations are easy to work with, mathematicians often use them instead of more complex equations. Many times, assuming that quantities are related by linear equations is good enough. Remember that when building a model, we decide what goes into it and what to omit. There is always a trade off—the accuracy of the model versus dealing with lengthy computations.

You have used approximations in other situations. For example, your calculator cannot compute $\frac{5}{7}$. Does it surprise you to hear that? Your calculator cannot calculate the *exact* decimal representation of $\frac{5}{7}$ because it would need to compute an infinite number of decimal places. (If you compute $\frac{5}{7}$ by hand to 15 or 20 decimal places, you will see why this is so.) This does not mean that we should throw our calculators away, because often a good approximation gives us an acceptable answer to a problem.

For another example, if you look very carefully at the screen of a video game, you may be able to see angular edges instead of the smooth face of a character. Curved surfaces take too long to compute, so game designers compromise by using tiny planar surfaces and sophisticated shading techniques instead of drawing the true curved surfaces.* Planes are the geometric representation of linear equations in three variables. So, the game designer is replacing nonlinear surfaces in three-dimensional space with linear approximations to speed up the mathematics that allows you to play the game at real-life speed.

Those who design models always face this trade-off—higher accuracy versus doing time-consuming calculations. For practical reasons, the designer must use approximations. For example, a highly accurate model of the weather that requires four days of computer time to make a prediction would not be useful even though it could accurately tell us what the weather was three days ago! If we find that a model does not give acceptable results, then we must refine it, or perhaps discard it.

Adding these equations eliminates the a to give us $8b = 600$, or $b = 75$. Since 75 is easy to work with, we substitute it for b in the top equation and solve for a. Making this substitution gives us

$$4a + 6(75) = 2{,}200.$$

Subtracting 450 from both sides and dividing by 4 yields $a = 437.5$. If Yuki consumes 437.5 grams of food A and 75 grams of food B each day, she meets the dietary requirements for calories and protein.

A system of equations describes the relationship between supply and demand.

The **law of demand** in economics states that as the price of an item increases, the consumer is less willing to purchase it. According to the **law of supply,** as the price of an item increases, the producer is willing to produce more of the item. If

---

* As video game technology improves, it is harder to see these edges; however, be assured that in creating three-dimensional figures, the common technique is to approximate curved surfaces by flat polygons.

the price is low, consumer demand increases, but since producers are not willing to produce much, there will be a shortage of the product. On the other hand, if the price is high, producers will produce more of the product, but because consumers are not willing to pay the high prices, there is a surplus of the item. Eventually the market for the item adjusts so that there will be a price at which the quantity demanded and the quantity supplied are equal. This point is called an **equilibrium point.** We illustrate this notion of equilibrium in Example 6.

EXAMPLE 6   The Supply and Demand for Hand-Crafted Jewelry

Malik sells hand-crafted turquoise, Native American jewelry at regional craft shows. At a price of $20 per necklace, he is willing to buy 30 necklaces from his suppliers. However, at this price, his suppliers will only provide him with 9 necklaces. On the other hand, if he will pay $60 per necklace, he can only sell 15 necklaces per show. At this higher price, his suppliers will provide him with 29 necklaces. Assume that the equations relating price to demand and to supply are both linear, what should the price per necklace be in order for supply to equal demand?

SOLUTION: We summarize the information we have in Table 7.1.

Supply		Demand	
Price	Necklaces Supplied	Price	Necklaces Demanded
20	9	20	30
60	29	60	15

TABLE 7.1   The supply and demand for turquoise necklaces.

Since the supply equation is linear, its graph is a line that passes through the points (20, 9) and (60, 29). The slope of this line is

$$\frac{29 - 9}{60 - 20} = \frac{20}{40} = \frac{1}{2}.$$

If we write the supply equation in slope-intercept form* as $y = mx + b$, we can substitute $\frac{1}{2}$ for b to rewrite this equation as

$$y = \frac{1}{2}x + b.$$

_____

* If you are rusty on this, you should review Section 6.1.

Because (20, 9) lies on this line, we can substitute 20 for x and 9 for y, to get the equation

$$9 = \frac{1}{2}(20) + b.$$

Solving this equation, we get $b = -1$. The supply equation is therefore,

$$y = \frac{1}{2}x - 1.$$

We rewrite this in standard form as

$$x - 2y = 2.$$

The points (20, 30) and (60, 15) lie on the graph of the demand equation, so the slope of this line is $\frac{15 - 30}{60 - 20} = \frac{-3}{8}$. Therefore, we can write the demand equation in slope-intercept form as

$$y = \frac{-3}{8}x + b.$$

The point (20, 30) is on this line, so we substitute 20 for x and 30 for y to get

$$30 = \frac{-3}{8} \cdot 20 + b.$$

Solving this equation, we find that $b = \frac{75}{2}$. The equation representing demand for necklaces is therefore

$$y = \frac{-3}{8}x + \frac{75}{2}.$$

Multiplying by 8 and rearranging terms, we rewrite this equation in standard form as

$$3x + 8y = 300.$$

Therefore, the following system describes this supply and demand situation.

$$\begin{array}{ll} x - 2y = 2 & \text{(supply)} \\ 3x + 8y = 300 & \text{(demand)} \end{array}$$

We will now solve this system to find an equilibrium point. Multiplying the top equation by 4 and then adding the equations results in $7x = 308$, or $x = 44$. Thus at a price of $44 per necklace, the quantity supplied and the quantity demanded will be equal. If we substitute $x = 44$ in either equation, we find that $y = 21$, which means, at a price of $44 per necklace, Malik will be willing to buy 21 necklaces from the supplier and that the supplier will be willing to sell 21 necklaces at this price.

Following the spirit of Descartes' analytic geometry, it is helpful to interpret the solution to Example 6 geometrically. We graph both the supply and demand equations in Figure 7.4. The region to the left of the equilibrium point shows where demand exceeds supply. The region to the right of the equilibrium point shows where supply exceeds demand.

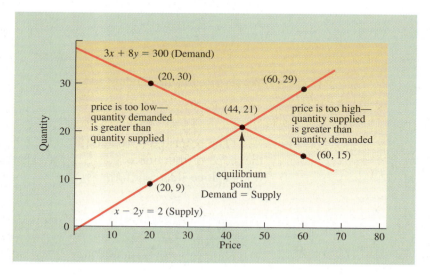

FIGURE 7.4 Supply and demand graphs intersect at the equilibrium point.

EXAMPLE 7 Solving a System of Equations in a Manufacturing Situation

Christian's Custom Woodworking Company builds cherry furniture. This week, they will manufacture only end tables and coffee tables. An end table requires 6 board feet and a coffee table requires 8 board feet. It takes 2 hours of labor to make an end table and it takes 4 hours of labor to make a coffee table. The company has 1,200 board feet of cherry wood available and also 480 hours of labor. Assume that we wish to use all the labor and wood. Describe the conditions on wood and labor as a system of two linear equations in two unknowns. Solve this system and interpret your answer.

SOLUTION: It is helpful to organize this information in Table 7.2, which we will call a **resource table.** We will represent the number of end tables by e and the number of coffee tables by c.

Resources	Needed for an End Table (e)	Needed for a Coffee Table (c)	Available
Wood	6 Board feet	8 Board feet	1,200 Board feet
Labor	2 Hours	4 Hours	480 Hours

TABLE 7.2 Resource table for furniture manufacturing problem.

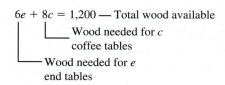

$6e + 8c = 1,200$ — Total wood available
└── Wood needed for c
coffee tables
└── Wood needed for e
end tables

FIGURE 7.5 Equation describing use of wood.

First let's consider the condition on the wood needed to manufacture the tables. Because we use 6 board feet for each end table and 8 board feet for each coffee table and have 1,200 board feet available, we get the equation as seen in Figure 7.5.

Let's now look at the restrictions on labor. We use 2 hours for each end table and 4 for each coffee table. Therefore, the equation $2e + 4c = 480$ describes how to use the 480 available hours to construct e end tables and c coffee tables. We will now solve the system:

$$6e + 8c = 1,200 \quad \text{(wood)}$$
$$2e + 4c = 480 \quad \text{(labor)}$$

We multiply the second equation by -2 to get:

$$6e + 8c = 1,200$$
$$-4e - 8c = -960$$

When we add these two equations, we get the equation $2e = 240$, whose solution is $e = 120$. If we substitute 120 for e in the second equation of the original system, we get $2(120) + 4c = 480$, which means that $c = 60$. Therefore, if the company manufacturers 120 end tables and 60 coffee tables, all 1,200 board feet of wood and 480 hours of labor will be used up. ◎

Exercises 7.1

The solution for each system in Exercises 1–12 is a pair of integers. Determine the solution by graphing each system and check your answer by substituting values in both equations.*

1. $\begin{aligned} 2x + 5y &= 18 \\ -3x + 4y &= -4 \end{aligned}$

2. $\begin{aligned} 3x + 4y &= 1 \\ 2x + 6y &= -6 \end{aligned}$

3. $\begin{aligned} 3x + 2y &= 9 \\ -x + 2y &= -3 \end{aligned}$

4. $\begin{aligned} -2x + 4y &= 8 \\ 3x + 8y &= 30 \end{aligned}$

5. $\begin{aligned} x + 5y &= 27 \\ -3x + 2y &= 4 \end{aligned}$

6. $\begin{aligned} -x + 2y &= -10 \\ 2x + y &= 5 \end{aligned}$

7. $\begin{aligned} 4x - y &= -1 \\ -4x + 3y &= 11 \end{aligned}$

8. $\begin{aligned} 5x - 2y &= -2 \\ 2x + 3y &= -16 \end{aligned}$

9. $\begin{aligned} 2x - 5y &= 18 \\ 3x + y &= 10 \end{aligned}$

10. $\begin{aligned} -x + 2y &= -8 \\ -2x + 3y &= -11 \end{aligned}$

11. $\begin{aligned} -2x + 3y &= -5 \\ -x + y &= -2 \end{aligned}$

12. $\begin{aligned} 5x + 3y &= -11 \\ x - y &= -7 \end{aligned}$

Use the elimination method to solve the following systems of linear equations.

13. $\begin{aligned} -3x + 2y &= 5 \\ 6x + 4y &= 22 \end{aligned}$

14. $\begin{aligned} 4x - y &= 13 \\ 5x + 8y &= -30 \end{aligned}$

15. $\begin{aligned} -4x + 2y &= -6 \\ 2x + y &= 13 \end{aligned}$

16. $\begin{aligned} 6x - 5y &= -12 \\ -4x + y &= -6 \end{aligned}$

17. $\begin{aligned} 3x - 2y &= -14 \\ -4x + y &= 22 \end{aligned}$

18. $\begin{aligned} -5x + 4y &= -44 \\ 2x + y &= 15 \end{aligned}$

19. $\begin{aligned} 8x - 2y &= -2 \\ -4x + y &= 3 \end{aligned}$

20. $\begin{aligned} 6x + 9y &= -4 \\ 2x + 3y &= -2 \end{aligned}$

21. $\begin{aligned} x - y &= -2 \\ -4x + 4y &= 8 \end{aligned}$

22. $\begin{aligned} 10x - 4y &= -4 \\ -5x + 2y &= 2 \end{aligned}$

23. $\begin{aligned} 6x - 9y &= 8 \\ -4x + 6y &= 10 \end{aligned}$

24. $\begin{aligned} 12x - 4y &= 2 \\ -9x + 3y &= 11 \end{aligned}$

25. $\begin{aligned} 12x - 8y &= -1 \\ -2x + 5y &= 2 \end{aligned}$

26. $\begin{aligned} 12x - 27y &= -1 \\ 5x + 2y &= 4 \end{aligned}$

---

* If you wish to use a graphing calculator or a computer algebra system to solve the following problems, go to the technology section of www.aw.com/pirnot for tutorials.

27. $\begin{aligned} 3x - 9y &= -3 \\ -4x + 2y &= 0 \end{aligned}$ **28.** $\begin{aligned} 4x - 5y &= -4 \\ -8x + 15y &= 13 \end{aligned}$

29. Communicating Mathematics Suppose that you are solving a system of two linear equations in two variables. If, as you are doing your calculations, the following situations arise, which case are you in?

a) You obtain the equation $0 + 0 = 7$?

b) You obtain the equation $0 + 0 = 0$?

c) You obtain the equation $x = 3$?

30. Communicating Mathematics Explain how you decide that a system of linear equations that you are trying to solve

a) has a unique solution.

b) has no solutions.

c) has an infinite number of solutions.

Use the elimination method in Exercises 31–40 to solve each system of two linear equations in two unknowns.

31. Sports. The 2001 Seattle Mariners had the best record in major league baseball. They played 162 games and won 70 more games than they lost. What was their record?

32. Sports. The U.S Olympic basketball team won a basketball game by 13 points over Germany. If the two teams together scored a total of 109 points, what was the final score?

33. Dietary requirements. An average bagel contains 30 mg of calcium and 2 mg of iron.* One ounce of cream cheese contains 25 mg of calcium and 0.4 mg of iron. If Leyla wishes to eat a combination of bagels and cream cheese that contains exactly 245 mg of calcium and 10 mg of iron, how much of each should she eat?

34. Dietary requirements. A slice of cheese pizza contains 40 g of carbohydrates and 220 mg of calcium. A 12-oz. cola contains 40 g of carbohydrates and 15 mg of calcium. If Karl eats several slices of cheese pizza and drinks cola, how much of each must Karl eat to give a nutritional benefit of exactly 200 g of carbohydrates and 690 mg of calcium?

35. Comparing job offers. Shaun can work at Best Deal Electronics for a base pay of $225 per week plus a $45 commission for each computer system that he sells. At Circuit Town his base pay would be $400 with a $20 commission for each computer system that he sells. In-terpret what the solution to this system tells you. How should Shaun decide which position to take?

36. Comparing job offers. Emily can work as freelance editor for Wild Adventure magazine for a base salary of $20 per hour plus 25 cents per page. At Travel World magazine, she can earn $22.10 per hour plus 18 cents per page. Interpret what the solution to this system tells you. How should Emily decide which position to take?

37. Women's basketball records. In the 2001 season, the New York Liberty in the Women's National Basketball Association played 32 games. The Liberty won ten more games than they lost. What was their win–loss record?

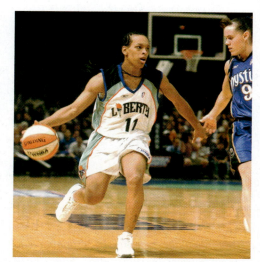

38. Women's basketball records. In the 2001 season, the Sacramento Monarchs in the Women's National Basketball Association played 32 games. The Monarchs won eight more games than they lost. What was their win–loss record?

39. Airport traffic. According to the Airports Association Council International, in 2000, the two busiest airports in the world were Atlanta's Hartsfield International and Chicago's O'Hare International. The two airports together handled 152 million passengers and Hartsfield handled 8 million more than O'Hare. How many passengers went through each airport?

40. Tourist travel. According to the World Tourism Organization, the two European countries that had the most tourists in 2000 were France and Spain. The two countries

---

* The nutritional content of the foods mentioned in these exercises is based on the *Home and Garden Bulletin No. 72, U.S. Department of Agriculture.*

together had 123 million visitors and France had 26 million more visitors than Spain. How many tourists visited each country?

Use the elimination method in Exercises 41–44 to find an equilibrium point for the described supply–demand situation. You may assume that the supply and demand equations are both linear.

41. Supply and demand—tutoring. A tutoring service provides tutoring for a fee. When the tutoring service charges $8/hour, there is a demand for 30 tutors per week. When the hourly rate rises to $15/hour, the demand drops to 9 tutors per week. On the other hand, at $8/hour, the service is only able to supply 9 tutors per week, while at the rate of $15/hour, they are able to provide 37 tutors per week. Find the price per hour at which the number of tutors demanded and the number of tutors supplied will be equal.

42. Supply and demand—housing. In doing a low-income housing survey, the local chamber of commerce found that if an apartment rents for $400 per month, there were only 275 available; however, when the rental price rose to $550 per month, the supply increased to 350. On the other hand, at a rental price of $400 per month, there was a demand for 450 units, but when the rental price increased to $550 per month, the demand dropped to only 200 units. Find the monthly rental price at which the number of apartments demanded and the number of apartments supplied will be equal.

43. Supply and demand—used books. A book store buys and sells used textbooks. For a current sociology text, they have found that if they pay an average price of $30 for each used book, they will be able to buy 60 texts for resale and will have 95 customers who are willing to buy the texts after markup. If they pay $42 per used text, they will be able to buy 120 texts but will only be able to sell 50 of them. What price should they offer for the used texts in order to sell all that they buy?

44. Supply and demand—selling bagels. The Coffee Shoppe makes fresh bagels each morning. The proprietors will make 4 dozen bagels if the price is $7/dozen and the demand will be for 10 dozen. At a price of $16/dozen, they will make 16 dozen but will only be able to sell 4 dozen. How should the bagels be priced so that all bagels made will be sold?

*The **substitution method** is another method for solving systems of linear equations which involves the following steps.*

Step 1. Solve either equation in the system for x or y. (You have four choices here: you can solve equation one for either x or y or you can solve equation two for either x or y. You make your decision as to which to do based upon what will give you the simplest expression to work with.)

Step 2. Substitute the expression you found in Step 1 for the variable in the other equation. This gives you a linear equation in a single variable.

Step 3. Solve the equation in Step 2 for the variable.

Step 4. Substitute the value for the variable that you found in Step 3 into the equation that you chose in Step 1. This gives you a linear equation in the other variable that you can solve easily.

We will use this method to solve the system

$$3x - 4y = 13$$
$$x - 2y = 5$$

Step 1. In considering our four choices, it is clear that our best choice is to solve x − 2y = 5 for x in equation 2. This gives us x = 2y + 5.

Step 2. We now substitute 2y + 5 for x in equation 1. This gives us the equation

$$3(2y + 5) - 4y = 13.$$

Step 3. We can rewrite this equation as

$$6y + 15 - 4y = 13,$$

and again as

$$2y = -2.$$

Thus, y = −1.

Step 4. Now we substitute −1 for y in equation 2 to get

$$3x - 4(-1) = 13.$$

Thus, 3x + 4 = 13, or x = 3. The solution to the system is therefore (3, −1).

Use the substitution method to solve the systems in Exercises 45–52. As with the elimination method, you may discover some systems to be either inconsistent or dependent.

45. $x - 2y = -7$
$3x + 5y = 12$

46. $x + 3y = 11$
$-x + 4y = 10$

47. $-2x + y = 6$
$-2x + 3y = 14$

48. $5x + 4y = 9$
$3x + y = 11$

49. $x + y = 6$
$2x = 8 - 2y$

50. $x + y = 3$
$x - y = -7$

51. $x - y = 4$
$3y = 3x - 12$

52. $2x + 6y = -36$
$2y - x = -7$

Further Exercises

We can use the elimination method to solve systems that have more than two equations and two unknowns. We outline this technique in Exercises 53 and 54.

53. Reduce this system to a system with two equations and two unknowns by completing the following steps:

$$2x + 3y + z = 10$$
$$x - y + 2z = 7$$
$$3x + y - z = 4$$

a) Consider the first and second equations:

$$2x + 3y + z = 10$$
$$x - y + 2z = 7$$

Multiply the first equation by -2 and then add this new equation to the second equation to get an equation containing only the variables x and y.

b) Consider the first and third equations:

$$2x + 3y + z = 10$$
$$3x + y - z = 4$$

If you add these equations together you will get another equation that contains only the variables x and y.

54. a) Now solve the system of equations obtained in Exercise 53. Since this system has two linear equations with two unknowns, we can solve for x and y as we did in this section.

b) Choose any of the three original equations and substitute the values for x and y that you found in part a). You are now able to solve for z.

Use the method that we outlined in Exercises 53 and 54 to solve the following systems.

55. $x - 3y + 2z = 6$
$x + y - 2z = 2$
$3x + 2y + z = 13$

56. $2x - 4y + 3z = 7$
$-2x + y + 2z = 5$
$x + 2y - z = 0$

57. $x - y - z = 2$
$4x + y + 2z = 24$
$x + y - z = 6$

58. $x + 4y + 3z = 9$
$-x + 2y + z = -3$
$2x - 2y - 2z = 4$

59. Communicating Mathematics Although it would be a tedious process, you could have solved some of Exercises 31–40 by trial and error. For example, in Exercise 31 you could try 90 wins for the Mariners, then 91 wins, and so on, until you found a solution by trial and error. If you found a solution this way, how can you be sure that it is the same solution that you would have found if you had done algebra? (Hint: Think about this geometrically and determine which case you are dealing with.)

*60. Communicating Mathematics How could you make up a word problem that has no solution? (Hint: Think about this geometrically and determine which case you are dealing with.) Take any one of the word problems given above and change one of the conditions so that the problem has no solution.

61. Communicating Mathematics Make up a system of two equations in two unknowns which has the solution (2, 3). Explain how you made this example.

62. Communicating Mathematics In making up exercises, mathematics instructors often begin by selecting an answer and then writing a question to give that answer. Think about how you would use this approach to create a supply and demand problem. For example, you could get the supply equation by finding the line through two points. Pick a real situation and write a complete problem using your method.

7.2 SYSTEMS OF LINEAR INEQUALITIES

When we model relationships, often we are not concerned with two quantities being exactly equal, so a linear equation is not appropriate. For example, a pregnant woman, concerned about getting enough calcium in her diet, may want to make sure that she consumes *at least* 1,500 mg of calcium daily. Or, an athlete who is taking dietary supplements might be careful not to consume *more than* the recommended daily requirement of 400 international units of vitamin D, because too much vitamin D can be toxic.

Because we cannot model these situations with linear equations, we need to develop a mathematical theory of inequalities. The techniques we use for working with inequalities are very similar to those that you have already learned for working with equations.

The solution to a linear inequality in two variables is a half-plane.

Like equations, inequalities can be linear or non-linear. In this section, we will be concerned with linear inequalities in two variables.

> **DEFINITION**
>
> A **linear inequality in two variables** is a statement we can write in one of the following forms:
>
> $$ax + by \geq c \text{ (read } ax + by \text{ is greater than or equal to } c),$$
>
> $$ax + by > c \text{ (read } ax + by \text{ is greater than } c),$$
>
> $$ax + by \leq c \text{ (read } ax + by \text{ is less than or equal to } c),$$
>
> $$ax + by < c \text{ (read } ax + by \text{ is less than } c),$$
>
> where $a, b,$ and c are real numbers with not both a and b equal to zero.

The following are examples of linear inequalities.*

$$2x + 3y \geq 6,$$
$$4x - 5y > 12,$$
$$y \leq 8,$$
$$x < 5.$$

As was the case with linear equations in two variables, solutions of linear inequalities are ordered pairs of numbers.

---

* In this section, we will only be dealing with linear inequalities in two variables; therefore, we will frequently omit the reference to the number of variables in the inequality.

EXAMPLE 1   Solutions to Inequalities

Which of the following ordered pairs are solutions of the inequality $2x + 3y \leq 6$?

a) $(2, 1)$ b) $(0, 2)$ c) $(-4, 2)$

SOLUTION: In order to determine if these ordered pairs are solutions, we substitute the first coordinate of the pair for x and the second coordinate for y. If we get a true statement, the pair is a solution of the inequality; if the statement is false, we do not have a solution.

a) $2 \cdot 2 + 3 \cdot 1 \leq 6$ is false; therefore $(2, 1)$ is not a solution.

b) $2 \cdot 0 + 3 \cdot 2 \leq 6$ is true, so $(0, 2)$ is a solution.

c) $2 \cdot (-4) + 3 \cdot 2 \leq 6$ is true, so $(-4, 2)$ is a solution.

 Quiz Yourself 4*

Which of the following ordered pairs are solutions of the inequality $3x - 4y \geq 5$?

a) $(4, 1)$ b) $(0, 2)$

c) $(1, 1)$

EXAMPLE 2   Solving a Linear Inequality Graphically

Solve the inequality $3x + 4y \geq 11$.

SOLUTION: We can rephrase this inequality as the pair of statements:

$$3x + 4y > 11 \quad \text{or} \quad 3x + 4y = 11$$

This means that the solutions to the original inequality include the solutions to the linear equation $3x + 4y = 11$. We can visualize *some* of the solutions to the original inequality by graphing this line, as in Figure 7.6.

We now want to convince you that all points lying above the line are also part of the solution set of $3x + 4y \geq 11$. Let's start by plotting the point $(1, 2)$,

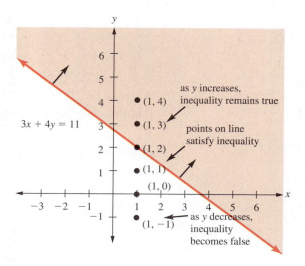

FIGURE 7.6   Points on and above the line $3x + 4y = 11$ are solutions; those below the line are not.

which satisfies the equation $3x + 4y = 11$ and is therefore on the line. As we see in Table 7.3, if we increase y from 2 to 3 and then to 4, we see that the expression $3x + 4y$ gets larger. Because all the points above $(1, 2)$ make $3x + 4y$ greater than 11, all of these points are solutions to the inequality $3x + 4y \geq 11$.

Point	$3x + 4y$	Comment
$(1, 4)$	$3 \cdot 1 + 4 \cdot 4 = 19$	Expression is greater than 11
$(1, 3)$	$3 \cdot 1 + 4 \cdot 3 = 15$	Expression is greater than 11
$(1, 2)$	$3 \cdot 1 + 4 \cdot 2 = 11$	Expression equals 11
$(1, 1)$	$3 \cdot 1 + 4 \cdot 1 = 7$	Expression is less than 11
$(1, 0)$	$3 \cdot 1 + 4 \cdot 0 = 3$	Expression is less than 11
$(1, -1)$	$3 \cdot 1 + 4 \cdot (-1) = -1$	Expression is less than 11

TABLE 7.3   Points above the line satisfy the inequality but points below the line do not.

Similarly, points such as $(1, 1)$, $(1, 0)$, and $(1, -1)$, which are below $(1, 2)$, make the expression $3x + 4y$ smaller than 11 and therefore are not solutions to the inequality $3x + 4y \geq 11$.

What you saw happen for the point $(1, 2)$ is true for any point lying on the line. Therefore, the solution to $3x + 4y \geq 11$ is all points *on* or *above* the line $3x + 4y = 11$. We indicated this region in Figure 7.6 by placing small arrows on the line and also shading the solution set.

If the inequality in Example 2 were a strict inequality, namely $3x + 4y > 11$, then the line $3x + 4y = 11$ would not be part of the solution set. In this case, we would draw the line as a dotted line rather than a solid line.

What happened in Example 2 is true for any linear inequality in two variables.

> **The Solution Set for a Linear Inequality in Two Variables**
>
> The solution set for a linear inequality in two variables is always a half plane with either a solid or dotted border.

The method that we used in Example 2 always works; however, there is a simpler way to solve linear inequalities.

The one-point test determines the half plane in solving a linear inequality.

The following method is a quick and reliable way for you to determine the correct half plane when solving a linear inequality.

Solving Linear Inequalities Using the One-Point Test

To solve a linear inequality in two variables, follow these steps:

1. Change the inequality to an equation and graph the line. If the inequality contains $\leq$ or $\geq$, draw a solid line. If the inequality contains $<$ or $>$, draw a dotted line.

2. Choose a point which is clearly either above or below the line you drew in step 1. The point $(0, 0)$ is usually the best point to use unless the line passes through or is very close to $(0, 0)$.

3. If the coordinates of the point you chose in step 2 satisfy the inequality, then the side of the line containing that point contains solutions for the inequality. Otherwise, the opposite side of the line contains solutions. We will refer to this test as **the one-point test.**

4. Use arrows or shading to indicate which side of the line contains the solutions to the inequality.

We will illustrate this method in Example 3.

EXAMPLE 3   Using the One-Point Test to Solve an Inequality

Use the steps we listed above to solve $4x - 3y \geq 9$.

SOLUTION: First we graph the equation $4x - 3y = 9$ with a solid line in Figure 7.7. Because $(0, 0)$ does not satisfy the inequality, we know that the $(0, 0)$ side of the line does not contain solutions. We choose the other side instead. The solution consists of all points on or below the line, as we show in Figure 7.7.

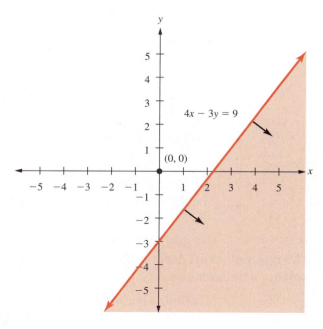

FIGURE 7.7   Using the one-point test to graph $4x - 3y \geq 9$.

SOME GOOD ADVICE

In working with inequalities, do not look for shortcuts to obtain the graphs more quickly. You cannot assume that if the symbols ≤ or < appear in the inequality that the solution to the inequality occurs below the line. Also, don't assume that the symbols ≥ or > mean that you should select the upper half plane. In order to choose the proper half plane, you should always use the one-point test.

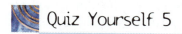

Quiz Yourself 5

Solve $2x - 3y > 6$.

The solution of a system of inequalities is the intersection of the solution sets for the individual inequalities.

Just as we can have systems of equations, we can construct systems of two or more inequalities. In solving a system of inequalities, we solve the inequalities separately and then find the intersection of the individual solution sets.

EXAMPLE 4   Solving a System of Inequalities

Solve the system

$$2x - 3y < -6$$
$$x + y \leq 7$$

SOLUTION: We first graph $2x - 3y = -6$ using a dotted line and $x + y = 7$ using a solid line, as in Figure 7.8.

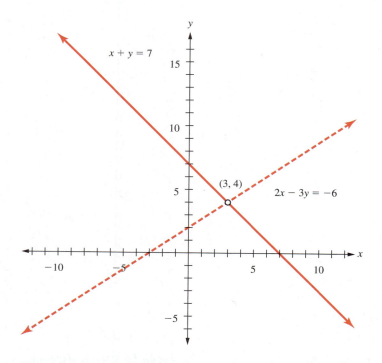

FIGURE 7.8   The dotted line contains points which are not part of the solution of $x + y < 7$.

Testing $(0, 0)$ with the first inequality, we get $2 \cdot 0 - 3 \cdot 0 = 0$, which is greater than -6, so $(0, 0)$ is not a solution. This means that the region above the line is the desired half plane to solve $2x - 3y < -6$.

When we test $(0, 0)$ with the second inequality, we get $0 + 0 = 0$, which is less than or equal to 7. Thus, the $(0, 0)$ side of the line $x + y = 7$ is the half plane we want.

We have indicated these regions with arrows in Figure 7.9 and also have shaded the intersection of the two half planes to show the solution to the system of inequalities.

Notice that the point $(3, 4)$ is a solution to the system of linear equations $2x - 3y = -6$ and $x + y = 7$. We found this solution using the elimination method from Section 7.1. The dot is open rather than solid because $(3, 4)$ lies on the line $2x - 3y = -6$, which is not part of the solution of $2x - 3y < -6$. If both boundary lines had been drawn solid, then we would also have drawn $(3, 4)$ as a solid dot. A point, such as $(3, 4)$, which is the intersection of two boundary lines of a solution set, is called a **corner point.**

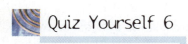

Quiz Yourself 6

Solve the system $\begin{array}{l} 3x - 4y \leq 8 \\ x + 2y \geq 6 \end{array}$.

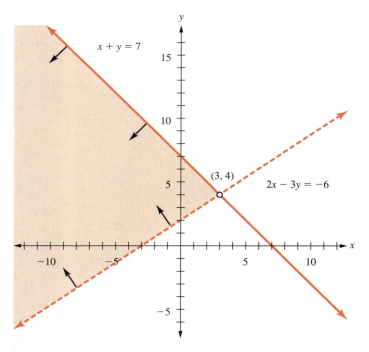

FIGURE 7.9   Graph of the solution of $\begin{array}{l} 2x - 3y \leq -6 \\ x + y < 7 \end{array}$.

Highlight: *Higher Dimensions, Science Fiction, and Mathematics*

A super hero passes through a secret portal and instantly becomes a distorted figure struggling to avoid destruction at the hands of grotesque villains inhabiting the fourth dimension. Such is the stuff of science fiction, where higher dimensions are portrayed as hidden and sometimes evil worlds, existing parallel to our three-dimensional world. Have you ever wondered, "What does the fourth dimension look like?" "What does the fourth dimension stand for?"

To the mathematician, the fourth dimension (and higher dimensions) are simply sets of ordered arrangements of numbers. In working with linear equations and inequalities in two dimensions, we represented points by ordered pairs of numbers such as (2, 3) and (6, − 8). If we were inclined, we could discuss linear equations and inequalities in three, four, or more dimensions. A point in the fourth dimension would simply be an ordered 4-tuple such as (2, − 5, 6, 9). Scientists solve real problems using spaces of much higher dimensions. One paper on a topic called wavelet theory contained mathematics done in a 64,000-dimensional space. Imagine how a cartoonist would draw superman if he were to accidentally fly through the portal to that space!

In order to work in a higher dimensional space, you do not have to be able to visualize it. In fact, mathematics progresses by creating mathematical representations of familiar objects and then generalizing these representations to more abstract situations. If you were to work with equations and inequalities in higher dimensions, you would have to learn techniques which rely less on geometry and more on other, abstract mathematical concepts.*

We use a system of linear inequalities in two variables to represent a set of conditions on two quantities.

We will now consider a situation involving several conditions imposed on a diet. Example 5 shows how to model these conditions with a system of linear inequalities.

EXAMPLE 5   Using Inequalities to Represent Nutritional Requirements

Khalid is a marathon runner who is interested in the amount of protein and calcium in his diet. Two of his favorite foods are fried shrimp and broccoli. A serving of fried shrimp contains approximately 15 g of protein and 60 mg of calcium. A spear of broccoli contains 5 g of protein and 80 mg of calcium. Assume that, as part of his diet, he wishes to get at least 60 g of protein and 600 mg of calcium from fried shrimp and broccoli. Express this pair of conditions as a system of inequalities and graph its solution set.

SOLUTION: Assume that Khalid eats s servings of shrimp and b spears of broccoli. Because each serving of shrimp has 15 g of protein, if Khalid eats s servings of shrimp, he will consume $15s$ g of protein. Each spear of broccoli has 5 g of protein, so eating b spears of broccoli provides $5b$ g of protein. He wants the amount of protein to be at least 60 g, which we can model with the inequality $15s + 5b \geq 60$.

---

* For a discussion of how to use the abstract notion of matrices to solve systems of linear equations, see www.aw.com/pirnot.

Also, s servings of shrimp provides $60s$ mg of calcium and b spears of broccoli have $80b$ mg of calcium. Because Khalid wants the amount of calcium to be at least 600 mg, we get the inequality $60s + 80b \geq 600$. Thus, the following system of inequalities describes the original pair of dietary requirements:

$$15s + 5b \geq 60$$
$$60s + 80b \geq 600$$

We solve this system as we did in Example 4 and show the solution in Figure 7.10. You should verify this.

In Figure 7.10, it does not make sense to shade the part of the solution set below the horizontal axis, since points in that region have a negative second coordinate and correspond to eating a negative number of broccoli spears. Similarly, we do not shade points to the left of the vertical axis.

In Section 7.1, when we used a system of linear equations in two variables as a model, a solution was usually* a unique ordered pair of numbers. The solution in Example 5 is very different because there are many points that satisfy the nutritional requirements stated in this problem. For example, because (6, 8) is in the solution set, this means that eating six servings of shrimp and eight spears of broccoli would provide at least 60 g of protein and 600 mg of calcium. Often, among all the possible solutions, we want to find a solution that is the best with respect to some other requirement. For example, in Figure 7.10, we may want a combination of shrimp and broccoli that minimizes the amount of fat in the diet. We will discuss these kind of considerations in the Of Further Interest section when we introduce the topic Linear Programming.

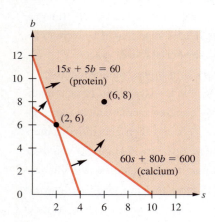

FIGURE 7.10 Shaded region represents the number of servings of shrimp and broccoli needed to satisfy the protein and calcium conditions.

Exercises 7.2

In Exercises 1–4, determine which points satisfy the given inequality.

1. $3x + 4y \geq 2$

   a) (3, 5)   b) (1, −2)   c) (0, 0)   d) (−4, 6)

2. $2x − 4y \geq 5$

   a) (5, 2)   b) (6, 0)   c) (0, −3)   d) (−2, 3)

3. $5y − 3x < 2$

   a) (2, 3)   b) (1, 1)   c) (0, 0)   d) (−3, −2)

4. $8x − y < 2$

   a) (4, 6)   b) (−1, −2)   c) (0, 0)   d) (−4, 3)

5. Communicating Mathematics   In the discussion immediately following Example 2, we stated that the solution of a linear inequality is a half plane. What did we see in Example 2 that makes that statement believable?

6. Communicating Mathematics   In some problems involving systems of linear inequalities, we may decide not to shade the region which lies below the x axis or to the left of the y axis. Explain why we might do this.

7. Communicating Mathematics   In graphing the solution to a linear inequality, we use either a solid line or a dotted line for the boundary. Explain when we use one versus the other.

---

* Cases 2 and 3 that we considered in solving systems in Section 7.1 are really extreme cases. Most times, when you are solving a system of equations, you are dealing with Case 1 in which there is one, unique solution.

8. Communicating Mathematics   Sometimes in graphing a corner point of the solution of an inequality we use an open dot and other times we use a solid dot. When do we use an open dot? When do we use a solid dot?

In Exercises 9–16, use the one-point test to solve the inequality. Shade the solution.

9. $3x + 4y \geq 12$

10. $2x - 4y \leq 6$

11. $2x - 4y < 12$

12. $5x + 4y > 10$

13. $x \geq 3y - 9$

14. $4y \leq 10 - 2x$

15. $4x - 8 < 2y$

16. $2x - 6 < 3y$

Solve each system. Indicate all corner points and shade the solution set.

17. $\begin{array}{l} 2x - 3y \leq -5 \\ x - 2y \leq -8 \end{array}$

18. $\begin{array}{l} 2x - 5y \leq 23 \\ 3x + y \leq 9 \end{array}$

19. $\begin{array}{l} 2x - y \geq 3 \\ x - y \leq -1 \end{array}$

20. $\begin{array}{l} 3y \geq 13 + x \\ 9y \leq 23 - x \end{array}$

21. $\begin{array}{l} -4x + 3y \leq 23 \\ 3x > 19 - 5y \end{array}$

22. $\begin{array}{l} 5x - 6y \geq 21 \\ 3y - 6 < 8x \end{array}$

23. $\begin{array}{l} 3x + 5y \leq 32 \\ y \geq 4 \end{array}$

24. $\begin{array}{l} 5x - 2y \geq 3 \\ x \leq 3 \end{array}$

25. $\begin{array}{l} 2x + 5y > 26 \\ x < 8 \end{array}$

26. $\begin{array}{l} -3x + 2y < 12 \\ y \geq 6 \end{array}$

In Exercises 27–30, each system of inequalities has no solutions. Solve each system graphically and then explain why the system has no solution.

27. $\begin{array}{l} 2x + 3y \geq 18 \\ 4x + 6y \leq 16 \end{array}$

28. $\begin{array}{l} 5x - 2y < 10 \\ -5x + 2y \leq -20 \end{array}$

29. $\begin{array}{l} -2x + 5y \geq 20 \\ 4x - 10y > -10 \end{array}$

30. $\begin{array}{l} -3x + 2y \leq 12 \\ -9x + 6y > 60 \end{array}$

Use a system of inequalities to represent the situations described in Exercises 31–42, then solve the system using the methods developed in this section. In drawing the graph of your solution set, you may want to take into consideration that certain quantities cannot be negative and therefore you may want to restrict the solution set appropriately as we did in Example 5.

31. Manufacturing. Scott owns a manufacturing company that produces two models of entertainment centers. The Athens requires 4 feet of fancy molding and on the average takes 4 hours to manufacture. The Barcelona needs 15 feet of molding and 3 hours to manufacture. In a given

week, there are 120 hours of labor available and the company has 360 feet of molding to use for the centers.

32. Investing. Gina is considering investing no more than $3,000 total in a pharmaceutical company and a communications company stock. She intends to invest at least three times as much money in the pharmaceutical stock as she puts into the communications stock.

33. Small business. The art coalition sells decorated hats and T-shirts at an arts festival. It takes 30 minutes to decorate a hat and 20 minutes to decorate a T-shirt. They have a total of 600 minutes to spend decorating the items.

34. Small business. Jaleel has a part-time business preparing slides for small businesses. It takes him 10 minutes to produce a plain text slide and 18 minutes to produce a slide with graphics. He intends to spend 300 minutes per week at his business.

35. Investing. Michael is planning to invest no more than $18,000 in a bond fund and a mutual fund. He plans to invest at least twice as much in the bond fund as in the mutual fund.

36. Diet. As part of her preparation for the World Cup in women's soccer, Cassandra is taking several nutritional supplements. Quantum contains 4 mg of niacin and 80 mg of calcium. NutraPlus contains 2 mg of niacin and 220 mg of calcium. She wants to supplement her diet by taking at least 12 mg of niacin and 960 mg of calcium.

37. Diet. Raphael is training for the Mr. Universe competition and is supplementing his diet with PowerUp and StressTabs. PowerUp contains 30 mg of niacin and 200 mg of vitamin C. StressTabs has 40 mg of niacin and 400 mg of vitamin C. Raphael wants to take at least 180 mg of niacin and 1,600 mg of vitamin C.

38. Small business. Joleen makes redware plates and cups to commemorate special occasions. It takes her 5 minutes to make a plate and 11 minutes to make a cup. A plate requires 1.5 pounds of red clay and a cup requires 1.1 pounds of clay. She has on hand 33 pounds of clay and plans to work 3 hours and 40 minutes today.

39. Diet. An average bagel contains 200 calories. One ounce of cream cheese contains 100 calories. Ailagh wishes to eat a combination of bagels and cream cheese and consume no more than 700 calories from these foods. She will eat at least 2 bagels.

40. Small business. Just Like New is a store that specializes in refinishing wooden furniture. It requires 3 hours and $4 worth of materials to refinish a table and one hour and $1 worth of materials to refinish a chair. There are 27 hours per week available for labor and $32 per week for materials.

41. Advertising. The newly opened Espresso Bar is deciding on advertising to generate business. The owners have $2,400 total to spend for ads in the local newspaper and on the local radio station. A radio ad costs $100 and a newspaper ad costs $300. They want to have at least three times as many radio ads as they have newspaper ads.

42. Diet. Caitlin is deciding how to feed her new dog. Premium dog food provides 180 calories and 200 mg of calcium per cup while the cheaper food provides only 150 calories and 50 mg per cup. She wants to provide her dog with at least 1,500 calories and 1,200 mg of calcium.

Further Exercises

We can have more than two inequalities in a system. Solve the following systems and find all corner points.

43.
$$2x + 3y \leq 25$$
$$5y \geq 20 + x$$
$$y \leq x + 5$$
$$x \geq 0, y \geq 0$$

44.
$$3x + 2y \leq 22$$
$$x + y \leq 8$$
$$x + 4y \leq 24$$
$$x \geq 0, y \geq 0$$

45.
$$x - 2y \leq 8$$
$$2x + y \leq 19$$
$$x - y \geq 5$$
$$x \geq 0, y \geq 0$$

46.
$$6y \leq 14 + 5x$$
$$2y \leq 23 - 2x$$
$$13y \geq 56 - 2x$$

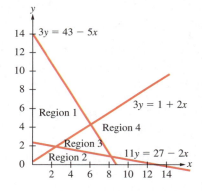

In Exercises 47–50, use the graph given to write a system of inequalities whose solution is the indicated region. Ignore the axes in describing these regions.

* **47.** Region 1

48. Region 2

49. Region 3

50. Region 4

---

* Exercise numbers circled in red can be used as group exercises.

CHAPTER SUMMARY

SECTION 7.1

system of linear equations, solution, graphic representation
A collection of linear equations is called a system of linear equations. A solution of the system is an ordered pair of numbers which makes all equations in the system true. We represent a system of two linear equations in two unknowns by a pair of lines.

elimination method
We replace the original system of equations with an equivalent system in which the coefficients of one of the unknowns are opposites of each other. Adding the two equations results in a single linear equation in one unknown which is easy to solve.

cases arising when solving a system of linear equations
Case 1—The unique solution of the system corresponds to the coordinates of the point of intersection of the lines represented by the equations in the system.
Case 2—There are no solutions to the system, which represents a pair of parallel lines.
Case 3—There are an infinite number of solutions to the system and the two equations of the system represent the same line.

systems of linear equations as models
A system of linear equations models a set of relationships between two quantities.

law of demand, law of supply, equilibrium point
The law of demand in economics states that as the cost of a item increases, the consumer is less willing to purchase it. The law of supply states that as the price of an item increases, the producer is willing to produce more of the item. An equilibrium point is a point at which supply and demand are equal.

SECTION 7.2

linear inequality in two variables
A linear inequality in two variables is a statement which can be written in one of the following forms:

$$ax + by \geq c, ax + by > c, ax + by \leq c, ax + by < c,$$

where a, b, and c are real numbers with not both a and b equal to zero.

solution of a linear inequality
A solution of a linear inequality is an ordered pair of numbers which, when substituted into the inequality, makes a true statement. The set of all solutions of a linear inequality is a half plane.

solving linear inequalities in two variables
To solve a linear inequality in two variables first, we change the inequality to an equation and graph the line. If the inequality contains $\leq$ or $\geq$, we draw a solid line. Otherwise, we draw a dotted line. Then, we use the one-point test to determine which half plane belongs to the solution set.

solution for a system of inequalities, corner point
The solution of a system of inequalities is the intersection of the solution sets for the individual inequalities. The intersection of two boundary lines of a solution set of a system of linear inequalities is called a corner point.

CHAPTER TEST

SECTION 7.1

1. Solve this system of equations.

$$4x - 3y = 27$$
$$x + 2y = -7$$

2. If you are solving a system of two linear equations in two variables and the following situations arise, which case are you in?

a) You obtain the equation $0 + 0 = 5$.

b) You obtain the equation $0 + 0 = 0$.

c) You obtain the equation $x = 5$.

3. The U.S Olympic basketball team won a basketball game by 17 points. If 123 total points were scored, what was the final score?

4. Aaron works in a fast-food restaurant which pays $5.85 per hour and as personal trainer for $15 per hour. In a given week, he worked 7 more hours in the restaurant than he did as a trainer and earned $145.20. How much did he work at each job?

SECTION 7.2

5. Use the one-point test to solve $6x + 5y > 20$. Shade the solution.

6. Solve the system $\begin{array}{l} 2x + 5y \geq 24 \\ 2y \leq 2x + 18 \end{array}$. Find all the corner points of the solution set.

7. The art club is selling custom-decorated T-shirts and hats to raise money for a homeless shelter. There is a profit of $4 on each T-shirt and $2 on each hat. The number of hats sold is at least 15 more than the number of T-shirts sold and the total profit is no more than $180. Use a system of inequalities to model this situation. Then solve the system using the methods developed in this chapter. In drawing your solution set, take into consideration that certain quantities cannot be negative and restrict the solution set appropriately.

Of Further Interest: LINEAR PROGRAMMING

A system of inequalities can have many solutions, so it is natural to ask, "Are some of these solutions better than others?" Recall that in Example 5 of Section 7.2, we were interested in the amount of shrimp and broccoli that a runner would eat to provide protein and calcium in his diet. The runner might also be interested in keeping the amount of sodium in his diet to a minimum. Or cost may be a consideration, and he may want to keep the cost of his diet low. When we start to look at situations this way, we realize that not all solutions to a system of inequalities are equally desirable.

As another example, consider Erin who makes antique copper chandeliers and sconces. In a given week, she may have on hand a certain amount of copper to use and a limited amount of time to spend in crafting the items. These two restrictions force her to make decisions as to how she will allocate her resources of copper and labor. There is always a trade-off. If she makes more chandeliers, then she has less material for making sconces. If Erin makes more sconces, then she has less time to devote to making chandeliers.

The question she would probably ask in deciding how to allocate her resources would be, "How many of each should I make in order to make the biggest profit?" To answer this question, we would first model the restrictions on materials and labor by linear inequalities. Then we would represent her profit by an algebraic expression and consider how to make this expression take on its largest value. A problem such as this is an example of what we call a **linear programming problem.**

All linear programming problems have the same three components.

There is a great similarity between all linear programming problems.

> **Components of a Linear Programming Problem**
>
> 1. The linear programming problems we study in this section will all have *two variables,* say *x* and *y*.*
> 2. We will represent *a set of conditions* on *x* and *y* by a system of linear inequalities. These conditions are called **(linear) constraints** and we will call the solution set for this system the **set of feasible solutions.**
> 3. There will be *some quantity that we wish to minimize or maximize.* We represent this quantity by a linear expression of the form *Ax* + *By* which we call the **objective function.**
>
> With these definitions, we can state the linear programming problem as:
>
> *Minimize (or maximize) a linear objective function in two variables subject to a set of linear constraints.*

---

* Linear programming problems can have many variables, but we will not consider such problems in this text.

■

We find the solutions to a linear programming problem at corner points.

The intersection of two boundary lines for a set of feasible solutions is called a **corner point.** There is a well-known theorem in linear programming that says that we always find the solutions to linear programming problems at the corner points of the set of feasible solutions. In order to keep our discussion of linear programming simple, we will only consider sets of feasible solutions that are bounded on all sides by lines.

THEOREM

Linear Programming Theorem

In a linear programming problem, the objective function attains its maximum and minimum values at corner points of the set of feasible solutions.

The linear programming theorem gives us a straightforward way to solve linear programming problems.

Method for Solving Linear Programming Problems

1. Graph the set of feasible solutions.
2. Find all corner points.
3. Evaluate the objective function for each corner point you find in step 2.
4. The largest value you find in step 3 is the maximum value for the objective function and the smallest value you find is the minimum.

We illustrate this method in Example 1.

EXAMPLE 1   **Solving a Linear Programming Problem by Using Corner Points**

Maximize the quantity $P = 2x - 5y$ subject to the constraints

$$x - 2y \geq -5,$$
$$2x + 3y \leq 18,$$
$$2x - y \leq 10,$$
$$x \geq 0, y \geq 0.$$

SOLUTION: We first graph the set of feasible solutions in Figure 7.11.

Next, we find the coordinates of each of the five corner points in Figure 7.11. We do this by using the elimination method to solve the system of linear equations that represents the pair of lines passing through each corner point. For example, the intersection of the lines $2x + 3y = 18$ and $2x - y = 10$ is the point $(6, 2)$ (see Table 7.4).

We finish the linear programming problem by evaluating the objective function $P = 2x - 5y$ for each of these corner points. From Table 7.4 we see that the maximum value of P is 10 and occurs at the corner point $(5, 0)$.   ◎

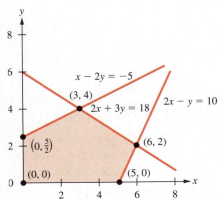

FIGURE 7.11   The set of feasible solutions for the system

$$x - 2y \geq -5,$$
$$2x + 3y \leq 18,$$
$$2x - y \leq 10,$$
$$x \geq 0, y \geq 0.$$

Historical Highlight: *George Dantzig—The Father of Linear Programming*

In 1947, George Dantzig invented linear programming and a method for solving linear programming called the simplex method. Although Dantzig has won numerous awards and prizes for his work, this was not always the case. As a child, Dantzig had little interest in school and in the ninth grade was actually failing his first algebra course. This made him angry with himself, and as a result, he dedicated himself to becoming an outstanding student. He credits his father, Tobias Dantzig, with developing his mathematical ability. Throughout high school, Dantzig's father challenged him with geometry problems, and by solving over 10,000 of these problems, young George developed the skills that later would make him a world-class mathematician.

As a graduate student at the University of Michigan, he once came late to class and copied two problems from the blackboard thinking that they were a homework assignment. After he had solved them, to his surprise, he learned that they were two famous unsolved problems in statistics. These solutions became the basis for his doctoral thesis and he earned a Ph.D. in statistics even though he had only a few courses in that area.

During World War II, Dantzig worked for the Air Force on ways to improve the deployment of troops and supplies. He felt that in order to decide on the best course of action it was necessary to consider all possibilities in a situation. However, to consider all possibilities in a situation can be an overwhelming task.

To put this problem in perspective, suppose that you wished to consider all the different ways you could assign 27 people to 27 different jobs in order to be sure that you had made the best assignment. Suppose further that you had a high-speed computer that could examine one billion different assignments per second. If you were to start today, the computer would take over 300 billion years to consider all the possible assignments of 27 people! Dantzig, through his remarkable theory of linear programming, was able to solve much larger scheduling problems in hours.

Many felt that Dantzig should have shared the 1975 Nobel prize for economics which was awarded for economic theories based on his work. However, because his work was in mathematics, rather than economics, he did not share the prize.*

System	Solution	Value of $P = 2x - 5y$
$x = 0$ $x - 2y = -5$	$(0, \frac{5}{2})$	$2 \cdot 0 - 5 \cdot (\frac{5}{2}) = -\frac{25}{2}$
$x - 2y = -5$ $2x + 3y = 18$	$(3, 4)$	$2 \cdot 3 - 5 \cdot 4 = -14$
$2x + 3y = 18$ $2x - y = 10$	$(6, 2)$	$2 \cdot 6 - 5 \cdot 2 = 2$
$2x - y = 10$ $y = 0$	$(5, 0)$	$2 \cdot 5 - 5 \cdot 0 = 10$
$x = 0$ $y = 0$	$(0, 0)$	$2 \cdot 0 - 5 \cdot 0 = 0$

TABLE 7.4   Evaluating the objective function at corner points.

---

* You might find it interesting to research why there is no Nobel prize in mathematics.

In many linear programming problems, there is only one point where the maximum or minimum of the objective function occurs. However, there are three other special situations that can arise in solving linear programming problems.

1. The set of feasible solutions can be empty.
2. There may be more than one point where the objective achieves its maximum and minimum values.
3. If the set of feasible solutions is unbounded, there may not be a maximum or minimum for the objective function.

We will examine these situations in the exercises.

Linear programming problems help us find the "best" way to allocate resources.

In Example 2, we return to the question of how Erin should decide how many sconces and chandeliers to make.

EXAMPLE 2   Using Linear Programming to Maximize Profit

Erin makes antique copper sconces and chandeliers. She needs 2 feet of copper to make a sconce and 16 feet of copper to make a chandelier. It takes her 1 hour to make a sconce and 3 hours to make a chandelier. She has 160 feet of copper available and plans to work 40 hours this week. Her profit on a sconce is $20 and she makes a profit of $75 on a chandelier. How many sconces and chandeliers should Erin make in order to achieve the greatest profit?

SOLUTION: Let us identify the components of this linear programming problem.

Variables—Let s be the number of sconces and c be the number of chandeliers

Constraints—(1) The number of feet of copper used in making the sconces + the number of feet of copper used in making the chandeliers must be less than or equal to 160 feet. (2) The amount of labor for making the sconces + the amount of labor for making the chandeliers must be less than or equal to 40 hours.

Objective Function—The artist wishes to maximize her profit. She makes $20 per sconce and $75 per chandelier.

The information that you are given in a linear programming problem such as this usually falls into a pattern. You can understand it better if you organize it in a **constraints-objective table** as in Table 7.5.

	Copper	Labor	Objective: Profit
Sconce s	2 Feet	1 Hour	$20s$
Chandelier c	16 Feet	3 Hours	$75c$
Constraints	160 Feet available	40 Hours available	

TABLE 7.5   Constraints-objective table for sconce–chandelier linear programming problem.

We can write the constraints in Table 7.5 algebraically as

$$2s + 16c \leq 160 \qquad \text{(copper)}$$
$$1s + 3c \leq 40 \qquad \text{(labor)}$$
$$s \geq 0, c \geq 0.$$

The objective function has the form

$$P = 20s + 75c.$$

We graph the set of feasible solutions in Figure 7.12 and determine its corner points in Table 7.6.

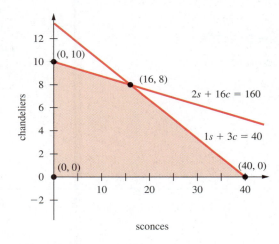

FIGURE 7.12 Set of feasible solutions for the sconce–chandelier problem.

System	Solution	Value of $P = 20s + 75c$
$s = 0, c = 0$	$(0, 0)$	$20 \cdot 0 + 75 \cdot 0 = 0$
$s = 0$ $2s + 16c = 160$	$(0, 10)$	$20 \cdot 0 + 75 \cdot 10 = 750$
$2s + 16c = 160$ $1s + 3c = 40$	$(16, 8)$	$20 \cdot 16 + 75 \cdot 8 = 920$
$1s + 3c = 40$ $c = 0$	$(40, 0)$	$20 \cdot 40 + 75 \cdot 0 = 800$

TABLE 7.6 Maximum profit occurs at (16, 8).

We see that the maximum profit occurs at the point (16, 8) which means Erin should make 16 sconces and 8 chandeliers. If she does this, she will make $920.

Highlight: *Using a Computer Algebra System to Solve a Linear Programming Problem**

It is very easy to use a computer algebra system to solve the linear programming problems we have been discussing. A typical session would look something like this, and would take only a few seconds to find a solution to Example 2. The bold type is what you would type in as commands to the system, and what follows is the computer's response.

**constraints:={2*s+16*c<=160, 1*s+3*c<=40,
s>=0,c>=0};**

objective:= 20*s+75*c;

maximize(objective,constraints);

{c = 8, s = 16}

You would first enter the constraints on the variables and also enter the objective function. Next, you would then tell the system to find the maximum value for the objective function subject to the constraints on the variables. The system will return the solution (16, 8) in less than a second.

When using such a computer algebra system, your real work is understanding the problem well enough to describe the constraints and the objective function algebraically so that the system can do the dirty work of the computations to find the answer.

Exercises

Find all the corner points for each of the following sets of feasible solutions

1.

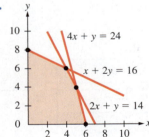

2.

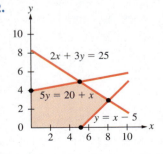

3.

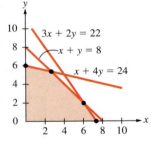

4.

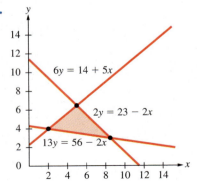

---

* For a tutorial on using technology to solve linear programming problems, see www.aw.com/pirnot.

Solve the given linear programming problems. Find the requested value and state at which corner point it occurs.

5. Maximize $P = 3x + 4y$ subject to the constraints

$$x + 2y \leq 16$$
$$x + y \leq 10$$
$$2x + y \leq 16$$
$$x \geq 0 \text{ and } y \geq 0$$

6. Maximize $P = x + y$ subject to the constraints

$$x - 3y \geq -9$$
$$x + 2y \leq 11$$
$$3x + y \leq 18$$
$$x \geq 0 \text{ and } y \geq 0$$

7. Minimize $P = 5x - y$ subject to the constraints

$$x - 6y \geq -18$$
$$3x + 2y \leq 26$$
$$x - 3y \leq 5$$
$$x \geq 0 \text{ and } y \geq 0$$

8. Minimize $P = x + 3y$ subject to the constraints

$$x + 5y \leq 40$$
$$3x + 2y \leq 29$$
$$4x + y \leq 32$$
$$x \geq 0 \text{ and } y \geq 0$$

9. Minimize $P = 20x - 12y$ subject to the constraints

$$x - y \geq -1$$
$$y \leq 5$$
$$y \geq 3$$
$$2x + y \leq 21$$

10. Maximize $P = 10x - 6y$ subject to the constraints

$$x - y \geq -2$$
$$x \leq 6$$
$$x \geq 2$$
$$x + 4y \geq 14$$

11. Maximize $P = 5x + 12y$ subject to the constraints

$$x - 2y \geq -2$$
$$x + y \leq 10$$
$$x + 4y \leq 16$$
$$x + y \leq 4$$

12. Maximize $P = x + y$ subject to the constraints

$$x - y \geq 2$$
$$3x - y \geq 12$$
$$3x + y \leq 30$$
$$x \geq 0, y \geq 0$$

13. Minimize $P = 4x + 2y$ subject to the constraints

$$6x - 7y \geq -6$$
$$3x + 4y \leq 42$$
$$6x - 22y \leq -51$$

14. Minimize $P = 3x + 6y$ subject to the constraints

$$x + y \geq 13$$
$$3x + 5y \leq 61$$
$$2x + y \leq 22$$

For each of the following linear programming problems, construct a constraints-objective table and then solve the problem by evaluating the objective function at the corner points of the set of feasible solutions as we did in Examples 2 and 3.

15. Manufacturing. Scott owns a manufacturing company that produces two models of entertainment centers. The Athens, requires 4 feet of fancy molding and, on the average takes 4 hours to manufacture. The Barcelona, needs 15 feet of molding and 3 hours to manufacture. In a given week, there are 120 hours of labor available and the company has 360 feet of molding to use for the centers. The company makes a profit of $9 on the Athens and $12 on the Barcelona. How many of each model should the company manufacture to maximize its profit?

16. Investing. Gina is considering investing no more than $3,000 total in a pharmaceutical company and a communications company stock. She intends to invest at least three times as much money in the pharmaceutical stock as she puts into the communications stock. The pharmaceutical stock is currently paying a return of 7.5 percent and the communications stock is paying 12.5 percent. How should she divide her investment in order to maximize her return?

17. Small business. Nicole sells decorated hats and T-shirts at an arts festival. It takes 10 minutes to decorate a hat and 9 minutes to decorate a T-shirt. She makes a profit of $3 on each T-shirt and $4 on each hat and has a total of 140 minutes to spend decorating the items. She intends to decorate at least twice as many T-shirts as hats. How many T-shirts and hats should Nicole decorate to obtain the largest profit?

18. **Small business.** Jaleel has a part-time business preparing slides for small businesses. It takes him 10 minutes to produce a plain text slide and 15 minutes to produce a slide with graphics. He intends to spend 350 minutes per week at his business. He also intends to make at least 5 more plain text slides than graphics slides. If he makes a profit of $1 on each text slide and $2.30 on each graphics slide, what combination of slides will earn the most profit for him?

19. **Diet.** Raphael is training for the Mr. Universe competition and is supplementing his diet with PowerUp and StressTabs. PowerUp contains 30 mg of niacin and 200 mg of vitamin C. StressTabs have 40 mg of niacin and 400 mg of vitamin C. Raphael wants to take at least 180 mg of niacin and 1,600 mg of vitamin C. He wants to take no more than eight tablets. If PowerUp costs 9 cents per tablet and StressTabs costs 13 cents per tablet, what combination of tablets will be the cheapest for him to take and meet his nutritional requirements?

20. **Small business.** Joleen makes redware plates and cups to commemorate special occasions. It takes her 10 minutes to make a plate and 20 minutes to make a cup. A plate requires 1 pound of red clay and a cup requires $\frac{1}{2}$ pound of clay. She has on hand 20 pounds of clay and plans to work 8 hours today. She expects to make no more than 28 pieces. Her profit on a plate is $12 and the profit on a cup is $8. How many plates and cups should she make today in order to maximize her profit?

21. **Small business.** The Reptile Farm has 400 square feet in which to house a collection of new lizards and frogs. A lizard requires 2 square feet of living space and costs $6 per month to feed. A frog also requires 2 square feet of

living space but costs only $1 per month to feed. The farm has budgeted $600 per month for food. The farm makes a profit of $17 on each lizard and $6 on each frog. What mixture of frogs and lizards will give the best profit?

22. **Real estate development.** A real estate developer wishes to construct an apartment building consisting of two- and three-bedroom apartments. She feels that for the project to be profitable to her, she must have at least 26,000 square feet devoted to the apartments, but no more than 30,000 square feet. Each two-bedroom apartment will contain 600 square feet of floor space and each three-bedroom apartment will contain 700 square feet. There will be at least as many three-bedroom apartments as there are two-bedroom apartments. The school board is concerned about the impact of this building on the school system. They believe that for every two-bedroom apartment there will be an increase of one child in the school system and for every three-bedroom apartment the increase will be two children. How many of each type of apartment should be constructed to minimize the increase in school children?

Further Exercises

*23. Communicating Mathematics   The following steps describe how we would create a linear programming problem to guarantee that it has an integer solution.

a) Select the corner points of the set of feasible solutions.

b) Find equations of the lines passing through consecutive pairs of these points.

c) State inequalities corresponding to the lines found in b).

d) Experiment by choosing different objective functions until you find one that has its maximal value at (3, 4).

---

* Exercise numbers circled in red can be used as group exercises.

Follow this procedure to construct a linear programming word problem so that the maximal value of the objective function occurs at the point (3, 4). Write a paragraph explaining how you constructed your example.

24. Communicating Mathematics The following linear programming problem has no solutions. Explain why this is the case. Maximize $A = 5x - 3y$ subject to the restrictions

$$x + 2y \le 8,$$
$$x + y \le 6,$$
$$x + y \ge 10$$
$$x \ge 0, y \ge 0.$$

25. Communicating Mathematics The following linear programming problem has several solutions. Pick several points along the line segment joining those solutions and see what happens when you substitute their coordi-

nates into the objective function. Describe the results of your investigations. Maximize $A = 4y - 2x$ subject to the restrictions

$$x + 2y \le 8,$$
$$x + y \le 6,$$
$$x + 2 \ge -4$$
$$x \ge 0, y \ge 0.$$

26. Communicating Mathematics The objective function for the following linear programming has no maximum. Explain why this is the case. (Hint: Think carefully about the set of feasible solutions.) Maximize $B = 3x + 5y$ subject to the restrictions

$$5x + 4y \ge 40,$$
$$3x + 4y \ge 32,$$
$$x \ge 0, y \ge 0.$$

8

Geometry:*
Ancient and Modern Mathematics Embrace

Geometry is one of the oldest areas of mathematics, and paradoxically it is also one of the most modern. Geometry is the study of points, lines, angles, surfaces, and solids and their measurement and relationships. Early mathematicians constructed geometric figures in two dimensions from simple objects such as lines and circles, as you might draw on a piece of paper. The discoveries of these ancient geometers formed the basis for surveying, mapmaking, architecture, and the use of perspective in painting.

As civilization advanced, this classical geometry became inadequate to describe accurately our physical world. In order to navigate on a spherical earth or understand the shape of the universe, mathematicians had to alter the meaning of lines. The straight lines of the ancient Greeks had to be replaced by lines that take on the contour of curved surfaces. Geometric figures in a curved space have different properties than the straight-line figures of a flat world.

More recently, mathematicians have invented yet another type of geometry called fractal geometry. In fractal geometry, lines are neither straight nor curved, but have an infinite number of microscopic "wiggles" such as we might find along a rough coastline, in the jagged peaks of a mountain range, or in the paths taken by streams as they merge to form a mighty river. Surprisingly, this fractal geometry also describes such things as the pattern of a beating heart and the way blood vessels and air passageways are arranged in our bodies.

---

* For more resources on geometry, see www.aw.com/pirnot.

We begin this chapter by introducing you to two-dimensional plane geometry consisting of a study of lines, angles, and circles. Using the properties of these objects, we then introduce other two- and three-dimensional geometric objects that we use to solve a wide range of practical problems. Next we apply the properties we have developed to investigate some modern topics in geometry such as symmetry and tiling the plane with repeated patterns. Finally, we will show you how to create, measure, and compare fractal objects.

8.1   LINES, ANGLES, AND CIRCLES

Over 2,300 years ago, Euclid of Alexandria wrote the most famous textbook in the history of the world. Euclid intended his text, called *Elements*, to be an introduction to elementary mathematics—including number theory, geometry, and a geometric version of what today we call algebra.

Euclid used points, lines, and planes to develop his geometric principles. In this section we will discuss some of the ideas that are based on these fundamental geometric objects, focusing particularly on the properties of lines, angles, and circles.

Euclid's book is remarkable because it set a standard for deductive reasoning that has lasted for over two thousand years. He began his discussion of geometry with intuitive descriptions of three undefined terms, saying that a *point* is "that which has no part," a *line* has "length but no breadth," and a *plane* has "length and breadth only."

In addition to these undefined terms, Euclid stated ten assumptions. He called five of these *postulates*; the other five he called common notions or *axioms*. Table 8.1 lists these postulates and axioms. Euclid used these assumptions to prove facts

Euclid's Postulates	Euclid's Common Notions or Axioms
1. It is possible to draw a straight line from any point to any point.	6. Things which are equal to the same thing are equal to each other.
2. It is possible to extend a straight line infinitely in either direction.	7. If equals are added to equals, the sums are equal.
3. It is possible to draw a circle with a given center and radius.	8. If equals are subtracted from equals, the remainders are equal.
4. All right angles are equal.	9. Figures which coincide with each other are equal.
5. Given a line and a point not on that line, there is one and only one line through the given point which is parallel to the given line.*	10. The whole is greater than any of its parts.

TABLE 8.1   Euclid's postulates and axioms.

---

* Euclid states his fifth postulate in a more complex way. This version of the fifth postulate is an equivalent form due to the eighteenth-century mathematician John Playfair.

about geometry that we call *theorems*. Later mathematicians used these postulates, axioms, and theorems to derive hundreds of other theorems in geometry.

Two lines in the plane are either parallel or intersecting.

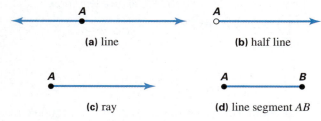

(a) line

(b) half line

(c) ray

(d) line segment *AB*

FIGURE 8.1 Some basic terminology regarding lines.

According to Euclid's geometry, lines either intersect or are parallel. In order to discuss some of the basic properties of lines and angles, we need to introduce some terminology and notation. We will label points with capital letters, such as A, B, and C, and lines with lowercase letters, such as l and m, or we may include subscripts, such as l_1 or l_2.

As we show in Figure 8.1, any point on a line divides the line into three parts—the point and two *half lines*. A **ray** is a half line with its endpoint included. In Figure 8.1(b), the open dot means that the point A is not included in the half line, whereas the solid dot in Figure 8.1(c) means that A is part of the ray. A piece of a line joining two points and including the points is called a **line segment.**

Although we do not try to define precisely what a plane is, you can think of it as being an infinite two-dimensional surface such as an infinitely large flat sheet of paper.

Parallel lines are lines that lie on the same plane and have no points in common. In Figure 8.3, lines l_1 and l_2 are parallel. We express this as $l_1 \| l_2$. If two different lines lying on the same plane lines are not parallel, then they have a single point in common and are called **intersecting lines.** In Figure 8.3, lines l_3 and l_4 are intersecting lines.

 Quiz Yourself 1*

Identify each object shown in Figure 8.2.

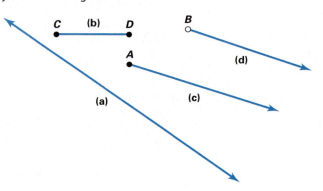

FIGURE 8.2 Line terminology.

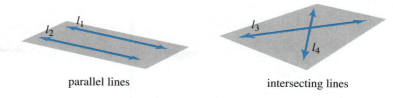

FIGURE 8.3   Parallel lines and intersecting lines.

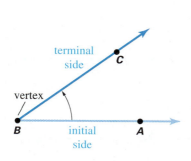

FIGURE 8.4   An angle formed by rotating a ray about point B.

Two rays having a common endpoint form an **angle.** In Figure 8.4, we form an angle by rotating ray AB, called the **initial side,** about the point B to finish in the position corresponding to ray BC, called the **terminal side.** We use the symbol $\angle$ to denote an angle; therefore we can call the angle in Figure 8.4 $\angle ABC$, or simply $\angle B$. Point B is called the **vertex** of the angle.

We measure angles in degrees.

We measure angles in units called degrees.* The symbol $°$ represents the word *degrees.* If you rotate the initial side of an angle one complete revolution about the vertex so that the terminal side coincides with the initial side, you will have formed a $360°$ angle, as we show in Figure 8.5(a). If you rotate the initial side only $\frac{1}{360}$ of the way around the vertex to reach the terminal side, that angle has size 1 degree. A $36°$ angle is shown in Figure 8.5(b).

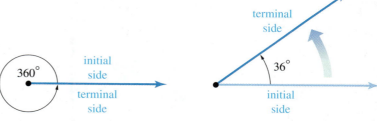

(a) one complete revolution   **(b)** $\frac{1}{10}$ of a complete revolution

FIGURE 8.5   We form a $36°$ angle by rotating the initial side $\frac{1}{10}$ of the way around the vertex.

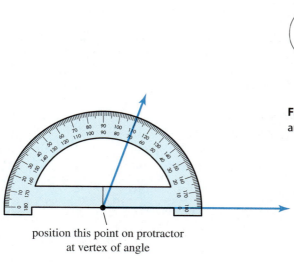

position this point on protractor at vertex of angle

FIGURE 8.6   A protractor measuring a 70° angle.

Figure 8.6 shows a tool called a *protractor* measuring a $70°$ angle. We write the measure (in degrees) of $\angle ABC$ as $m \angle ABC$.

Certain types of angles occur so often that we give them special names (see Figure 8.7). An angle whose measure is between $0°$ and $90°$ is called an **acute angle.** A **right angle** has a measure of $90°$. We indicate a right angle by placing a square at the vertex of the angle, as shown in Figure 8.7. An **obtuse**

---

* There are other possible units that are used to measure angles, such as radians. We will not discuss these other measures in this book.

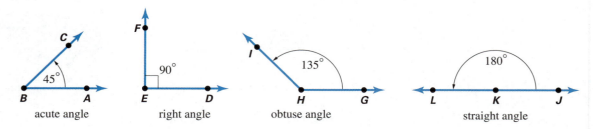

FIGURE 8.7   Some special types of angles.

angle has a measure between 90° and 180°, and a **straight angle** has a measure of 180°.

Two intersecting lines form two pairs of angles called **vertical angles.** Figure 8.8 shows one pair of vertical angles *ABC* and *EBD**. Vertical angles have the following important property.

> **Property of Vertical Angles**
>
> Vertical angles have equal measures.

We call a pair of angles **complementary** if the sum of their measures is 90°. Two angles having an angle sum of 180° are called **supplementary** angles. In Figure 8.8, angles *PQR* and *RQS* are complementary, and angles *WXY* and *YXZ* are supplementary angles. Two lines that intersect forming right angles are called **perpendicular lines.**

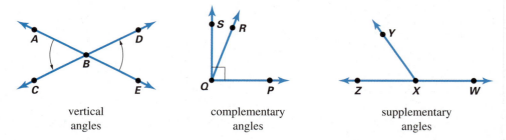

FIGURE 8.8   Some special pairs of angles.

Quiz Yourself 2

Identify the angles shown in Figure 8.9.

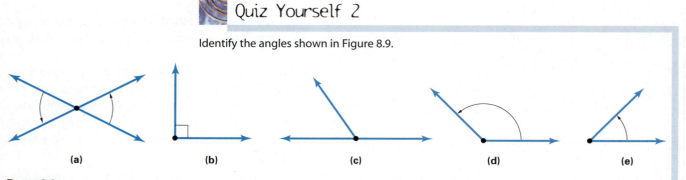

FIGURE 8.9

---

* *DBA* and *CBE* would be another pair of vertical angles.

Parallel lines cut by a transversal form several pairs of equal angles.

If we intersect a pair of parallel lines with a third line, called a **transversal,** we form eight angles as shown in Figure 8.10. Certain pairs of these angles have special names and also special properties regarding their angle measures.

We summarize some of these special properties in Table 8.2.

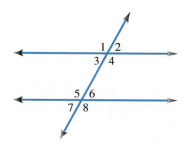

FIGURE 8.10   Special pairs of angles are formed when a transversal cuts parallel lines.

Type of Angles	Examples	Property
Corresponding angles	Angles 1 and 5 Angles 4 and 8	Corresponding angles have equal measures.
Alternate interior angles	Angles 3 and 6 Angles 4 and 5	Alternate interior angles have equal measures.
Alternate exterior angles	Angles 1 and 8 Angles 2 and 7	Alternate exterior angles have equal measures.
Interior angles on the same side of the transversal	Angles 3 and 5 Angles 4 and 6	Interior angles on the same side of the transversal are supplementary angles (sum of measures is 180°).

TABLE 8.2   Properties of pairs of angles formed by cutting parallel lines with a transversal.

PROBLEM SOLVING

It is easier to remember new mathematical terminology if you think about what the words mean in English. For example, when you use the term *alternate interior angles,* you are talking about "alternate" angles — one angle is on each side of the transversal — that are "interior" — inside, or between the parallel lines.

EXAMPLE 1   Finding Measures of Angles

In Figure 8.11, assume that lines *l* and *m* are parallel. Also assume that $m\angle A = 51°$ and $m\angle B = 76°$.

a) Find the measure of angle 9.   b) Find the measure of angle 2.

SOLUTION: a) Angles 8 and *B* are equal because they are corresponding angles. Thus $m\angle 8 = 76°$. Angles *A*, 8, and 9 form a straight angle, so

$$m\angle A + m\angle 8 + m\angle 9 = 180°.$$

Substituting the measures of angles *A* and 8 gives the equation

$$51° + 76° + m\angle 9 = 180°.$$

Solving this equation, we get $m\angle 9 = 53°$.

b) Because interior angles on the same side of the transversal are supplementary, we have $m\angle A + m\angle 2 = 180°$. Substituting 51° for $m\angle A$, we get $51° + m\angle 2 = 180°$. Solving this equation for $m\angle 2$, we get $m\angle 2 = 129°$.   ◎

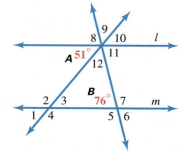

FIGURE 8.11   Finding measures of angles.

Quiz Yourself 3

Use Figure 8.11 to find (a) the measure of angle 7; (b) the measure of angle 6.

There is a correspondence between the measure of central angles and the length of arcs of circles.

Circles are common geometric objects that have many useful applications.

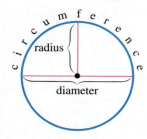

FIGURE 8.12   Radius (red), diameter (red), and circumference (blue) of a circle.

> **DEFINITIONS**
>
> A **circle** is the set of all points lying on a plane that are located at a fixed distance, called the **radius,*** from a given point called the **center.** A **diameter** of a circle is a line segment passing through the center with both endpoints lying on the circle. The **circumference** is the distance around the circle. (See Figure 8.12.)

An angle that has its vertex at the center of a circle is called a **central angle** (see Figure 8.13).

In Figure 8.13 the length of the arc between A and B is proportional to the central angle ACB. By this we mean that if we compare the measure of $\angle ACB$ with $360°$, the quotient $\frac{m\angle ACB}{360°}$ will tell us what fractional part the length of the arc from A to B is of the circumference of the circle. For example, if $m\angle ACB = 60°$, then the length of the arc from A to B is $\frac{60}{360} = \frac{1}{6}$ of the circumference of the circle.

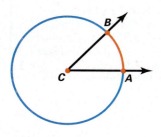

FIGURE 8.13   Angle ACB is a central angle.

EXAMPLE 2   Using a Central Angle to Measure the Length of an Arc of a Circle

Assume that a circle has a circumference of 12 meters. If central angle ACB has measure of $120°$, then what is the length of the arc from A to B?

SOLUTION: A good problem-solving strategy is to draw a diagram, as shown in Figure 8.14.

Because $\frac{120}{360} = \frac{1}{3}$, this means that the arc from A to B is one-third of the circumference. Therefore the length of arc AB is $(\frac{1}{3})12 = 4$ meters.

Surprisingly, we can use the small amount of geometry we have developed so far to solve meaningful problems.

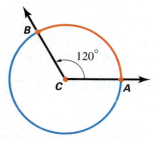

circumference = 12 meters

FIGURE 8.14   The measure of $\angle ACB$ determines the length of arc AB.

EXAMPLE 3   Finding the Circumference of the Earth

How can you use elementary geometry to estimate the circumference of the Earth?†

SOLUTION: Consider Figure 8.15 on the following page. Assume that lines l and m are parallel and cut by the transversal t. The point C is the center of the circle. Therefore angles α and β are equal.

---

* We will use the word *radius* in two ways. (1) It is the *distance* between the center and any point on the circle. (2) It is a *line segment* joining the center to any point on the circle.
† Although Earth is not a perfect sphere, we will assume that it is in order to simplify our calculations.

Quiz Yourself 4

Assume that a circle has a circumference of 40 inches. If a central angle has measure of 45°, what is the length of the arc of the circle determined by the sides of the angle?

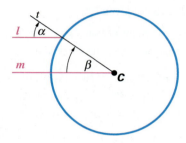

FIGURE 8.15 $\alpha = \beta$ because $l \parallel m$.

From our earlier discussion of central angles, we see that we have the following ratio:

$$\frac{\text{measure of angle } \beta}{360 \text{ degrees}} = \frac{\text{length of arc cut by sides of } \beta}{\text{total circumference of circle}}.$$

With this observation, it is easy to measure the circumference of the Earth. Begin by placing a vertical pole in the ground and waiting until high noon when the rays of the sun and the pole form an angle of 0°. Suppose at that very moment, you are talking to a friend who lives one thousand miles away and who also has a similar vertical pole. Your friend tells you that the sun's rays make an angle of 15° with his pole. If you redraw Figure 8.15 as in Figure 8.16, you will see how to find your answer.

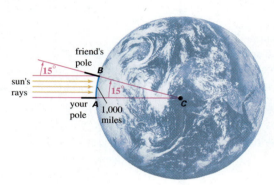

FIGURE 8.16   The measure of angle *ACB*, which is 15°, is the same fractional part of 360° as the length of the arc between *A* and *B* is of the Earth's circumference.

Denote the circumference of the Earth by c and set up the following ratio:

$$\frac{15°}{360°} = \frac{1,000}{c}. \tag{1}$$

Cross-multiplying equation (1) results in the equation

$$15c = 360(1,000). \tag{2}$$

Dividing both sides of equation (2) by 15 and simplifying gives us $c = 24,000$. From this calculation, you can estimate the circumference of the Earth to be 24,000 miles.

Historical Highlight: *Non-Euclidean Geometry (Part 1)*

What kind of geometry would the Greeks have invented if they had been able to view the earth from a space shuttle as twentieth-century astronauts? In ancient times, people believed the world was flat, so it is natural that

Euclid drew geometric figures on a flat plane. The shortest distance between two points on a plane is a straight line segment; however, on a sphere the shortest distance between two points is an arc of a *great circle*.* A great

continued

---

* Arcs of circles that are not great circles will not give the shortest distance between two points on the surface of a sphere.

circle, as shown in Figure 8.17, is a circle on a sphere that has the same center as the sphere. Therefore, in spherical geometry, we think of a "line" as a great circle.

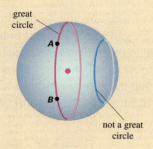

FIGURE 8.17 The distance between points A and B on a sphere is an arc of a great circle.

There can be no parallel lines in spherical geometry because any two great circles on a sphere intersect in two points. In Figure 8.18, the two great circles intersect at points P and Q.

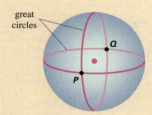

FIGURE 8.18 Two great circles intersect at points P and Q.

Notice that what we consider to be lines and what properties these lines have depends on the surface on which we are drawing the lines.

Mathematicians eventually developed non-Euclidean geometry as the result of centuries of speculation regarding Euclid's fifth postulate. Because the fifth postulate is more complicated than the other four, many felt that it should be a theorem rather than a postulate. After many fruitless attempts to prove the fifth postulate from the other postulates, several mathematicians decided to approach the problem in another way. They asked: "What if we assume that the fifth postulate is false? What would our geometry be like?"

There are two ways to deny the fifth postulate:

1) Assume that for a line and a point not on that line, there are no lines parallel to the given line.

2) Assume that for a line and a point not on that line, there are at least two lines parallel to the given line.

The first approach led to geometries such as spherical geometry that we mentioned above. Around 1850, the great German mathematician Bernhard Riemann invented such a geometry.

In the early part of the nineteenth century, Karl Friedrich Gauss from Germany, Janos Bolyai from Hungary, and Nicolai Lobachevsky from Russia each independently developed a geometry using the second approach. We can visualize this type of geometry on a surface called a *pseudosphere*. To form a pseudosphere, we rotate a curve called a *tractrix* about a line, as shown in Figure 8.19.

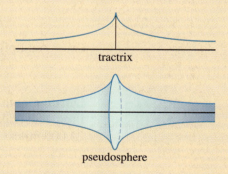

FIGURE 8.19 A pseudosphere is a model of a non-Euclidean geometry.

The surface looks somewhat like the bells of two trumpets joined together. We will discuss these non-Euclidean geometries further in Section 8.2.

Exercises 8.1

In Exercises 1–8, match each term with the numbered angles in the given figure. There may be several correct answers. We will state only one in the answer key. Lines l and m are parallel.

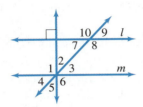

FIGURE FOR EX. 1–8

1. vertical angles

2. complementary angles

3. alternate interior angles

4. right angle

5. obtuse angle

6. corresponding angles

7. supplementary angles

8. acute angle

In Exercises 9–18, determine whether each statement is true or false. Remember by the Always Principle that if a statement is true, it must always be true without exception. If you think that a statement is false, you should try to find a counterexample.

9. Two lines lying on the same plane that do not intersect are parallel.

10. At least two of the angles formed by two intersecting lines are equal.

11. If two angles are complementary, then they must be equal.

12. If two angles (with measure greater than 0°) are complementary, then each must be an acute angle.

13. If two equal angles are supplementary, then each is a right angle.

14. An obtuse angle cannot be complementary to another angle.

15. An angle cannot be the complement of one angle and the supplement of another angle at the same time.

16. Four of the angles formed when parallel lines are cut by a transversal are equal.

17. The supplement of an acute angle must be an acute angle.

18. Alternate interior angles must be acute angles.

Use the given figure to answer Exercises 19–22. There may be several correct answers. We will state only one in the answer key.

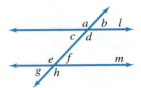

FIGURE FOR EX. 19–22

19. Find a pair of obtuse, alternate interior angles.

20. Find a pair of acute, alternate exterior angles.

21. Find a pair of acute, corresponding angles.

22. Find a pair of obtuse, corresponding angles.

23. Communicating Mathematics Explain how you remember the meaning of *alternate exterior angles*.

24. Communicating Mathematics Explain how you remember the meaning of *interior angles on the same side of the transversal*.

In Exercises 25–30, find the measure of a complementary angle and a supplementary angle for each angle.

25. 30°

26. 108°

27. 120°

28. 45°

29. 51.2°

30. 110.4°

In Exercises 31–36, find the measures of angles a, b, and c in each figure. In Exercises 33–36, lines l and m are parallel.

31.

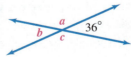

32.

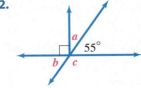

33.

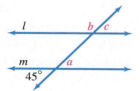

34.

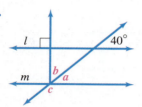

35.

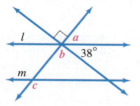

36.

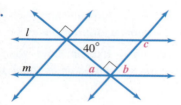

In Exercises 37–40, use the fact that two intersecting lines form four angles.

37. Communicating Mathematics Can all of the four angles be equal? Explain your answer.

38. Communicating Mathematics Can all of the four angles be acute? Explain your answer.

39. Communicating Mathematics What is the largest number of angles that can be obtuse? Explain your answer.

40. Communicating Mathematics Could three of these angles be acute and one of them be obtuse? Explain your answer.

41. If an angle measure is $x°$, represent its complement algebraically.

42. If an angle measure is $x°$, represent its supplement algebraically.

43. If an angle measure is $x°$, represent three times its complement algebraically.

44. If an angle measure is $x°$, represent two times its supplement algebraically.

Find angle x in each of the following situations.

45. The supplement of angle x is 30° more than twice its complement.

46. The supplement of angle x is 40° less than three times its complement.

47. The supplement plus the complement of the angle x is equal to 120°.

48. The supplement minus the complement of the angle x is equal to 90°.

49. Two lines in a plane can intersect forming four angles (some may have the same measure). What is the greatest number of angles we can form using three lines?

50. Solve Exercise 49 for four lines.

In Exercises 51–56, you are given two of the following three pieces of information: the circumference of the circle, the measure of the central angle ACB, and the length of arc AB. Find the third piece of information.

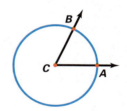

FIGURE FOR EX. 51–56

51. circumference = 24 feet; $m\angle ACB = 90°$

52. circumference = 150 centimeters; $m\angle ACB = 72°$

53. circumference = 12 meters; length of arc $AB = 4$ meters

54. circumference = 240 inches; length of arc $AB = 40$ inches

55. $m\angle ACB = 30°$; length of arc $AB = 100$ millimeters

56. $m\angle ACB = 120°$; length of arc $AB = 9$ feet

57. In Example 2, your friend has erred in measuring the angle the vertical pole makes with the sun's rays because the correct circumference of the Earth is closer to 25,000 miles than 24,000 miles. If we use the circumference of 25,000 miles, in Example 2, what should the angle measurement taken by your friend have been?

58. Reconsider Example 2. Assume that you do not know how far your friend is from you. At high noon, your time, your friend tells you that the sun's rays make an 18° angle with the vertical pole he has in the ground. Assume that the true circumference of the Earth is 25,000 miles. How far away is your friend?

Further Exercises

*(59.) Communicating Mathematics   Make up a question like Exercise 45. Explain how you made your example.

(60.) Communicating Mathematics   Make up a question like Exercise 47. Explain how you made your example.

61. Communicating Mathematics   Can alternate interior angles be complementary? Explain.

62. Communicating Mathematics   Can alternate interior angles be supplementary? Explain

(63.) Draw a diagram that has four lines and six points, and each line passes through exactly three of the six points.

(64.) Draw a diagram that has five lines and ten points, and each line passes through exactly four of the ten points.

65. Consider your solutions to Exercises 49 and 50. Without drawing a diagram, find the largest number of angles that can be formed by ten lines. (*Hint:* Consider the number of intersections that can be formed by five lines, six lines, and so on.)

66. A boat lost at sea has a radio that transmits a distress signal. Anyone receiving the signal can determine the direction of the signal, but not the distance. Explain why if one ship receives the signal, it cannot notify a search plane of the exact location of the boat. If two different ships receive the signal, the exact location of the boat can now be determined. Explain how to find the boat's position with the information from both ships.

8.2 POLYGONS

One of the many remarkable things about Euclidean geometry is that by starting with a few undefined terms and postulates, we can derive many important properties of geometric figures. In this section we build on the ideas we have discussed regarding lines and angles to define a fundamental class of geometric objects called polygons.

Polygons are special types of plane figures made up of line segments.

We begin with a few definitions (see Figures 8.20 and 8.21).

> **DEFINITIONS**
> A plane figure is **closed** if we can draw it without lifting the pencil and the starting and ending points are the same. A plane figure is **simple** if we can draw it without lifting the pencil and in drawing it we never pass through the same point twice, with the possible exception of the starting and ending points.

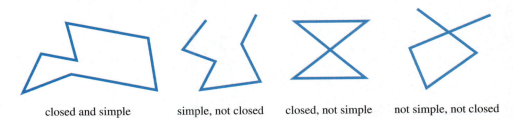

closed and simple   simple, not closed   closed, not simple   not simple, not closed

FIGURE 8.20   Closed and simple plane figures.

---

* Exercise numbers circled in red can be used as group exercises.

not closed
(a)

not simple;
not closed
(a)

not made of
line segments
(a)

nonregular polygons
(b)

regular polygons
(b)

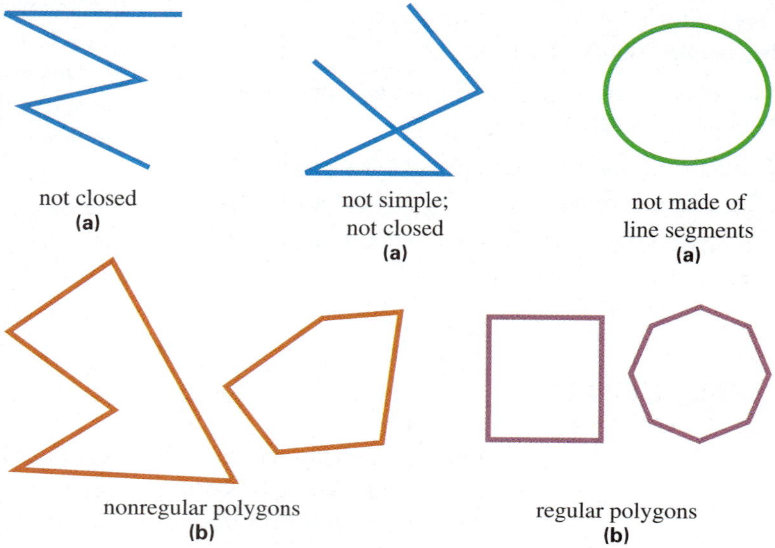

FIGURE 8.21 (a) Non-polygons. (b) Polygons.

Number of Sides	Name of Polygon
3	Triangle
4	Quadrilateral
5	Pentagon
6	Hexagon
7	Heptagon
8	Octagon
9	Nonagon
10	Decagon

TABLE 8.3 Names of polygons.

We classify polygons according to the number of their sides. Table 8.3 lists the names of polygons having up to ten sides.

There is another property that a polygon may have regarding its shape.

Figure 8.22 shows a convex and a nonconvex polygon.

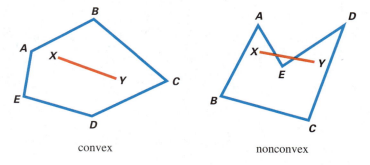

convex

nonconvex

FIGURE 8.22 A convex and a nonconvex polygon.

Highlight: *Geometry in Everyday Things*

If your floor is not perfectly even, then you perhaps should buy a table with three legs instead of four. The reason is that any three points lie on a plane, which means that a three-legged table will not wobble and spill your drinks, whereas a four-legged table probably will. This is also the reason why surveyors, such as those you often see along a highway, rest their transits on three legs and photographers steady their cameras on three-legged tripods.

When carpenters frame a house, they often nail a board diagonally across the vertical beams (see Figure 8.23). They do this so that the rectangles cannot shift into nonrectangular parallelograms, causing the building to collapse. By introducing extra lines that are not parallel to any existing lines in the geometric structure, they add rigidity to the structure.

A driver in a car equipped with a navigation system might want to locate the nearest Italian restaurant. By taking a reading of the car's position from different locations, the system determines that the car lies on

two nonparallel lines. Because these lines intersect in a unique point, the navigation system then determines the car's exact position and directs the driver to the nearest restaurant.

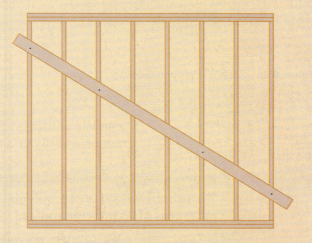

FIGURE 8.23 The diagonal board gives stability to the frame.

Some triangles and quadrilaterals may have special properties, in which case we use certain adjectives to describe them, as shown in Figures 8.24 and 8.25. In these figures, an object following an arrow has all the properties of the objects preceding the arrow. For example, in Figure 8.24 we see that every

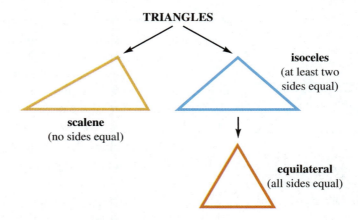

FIGURE 8.24 Classification of triangles.

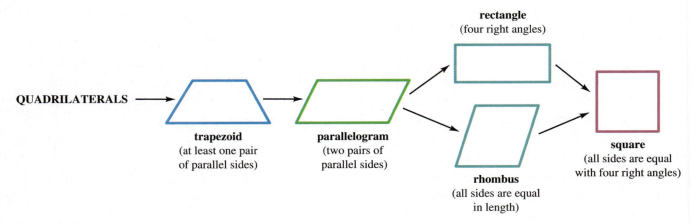

rectangle
(four right angles)

QUADRILATERALS

trapezoid
(at least one pair
of parallel sides)

parallelogram
(two pairs of
parallel sides)

rhombus
(all sides are equal
in length)

square
(all sides are equal
with four right angles)

FIGURE 8.25 Classification of quadrilaterals.*

equilateral triangle is also an isoceles triangle. Note that an isoceles triangle may not necessarily be an equilateral triangle.

The number of sides of a polygon determines the sum of the measures of its interior angles.

We can use polygons to solve real-life problems such as designing a deck or building a house. However, in order to do this, we often need to know the sum of measures of the interior angles of the polygons. For example, imagine how difficult it would be to design a hexagon-shaped gazebo without knowing the measure of the interior angles of the floor.

EXAMPLE 1 **The Angle Sum of a Triangle Is 180°**

Find the sum of the measures of the interior angles in $\triangle ABC$. (The notation $\triangle ABC$ is read "triangle ABC.")

SOLUTION: We begin by constructing a line m that contains line segment AC and a second line l through point B that is parallel to m, as shown in Figure 8.26.

Because angles 1 and 4 are alternate interior angles, they are equal. Similarly, angles 3 and 5 are equal. Therefore, the sum of the measures of angles 1, 2, and 3 equals the sum of the measures of angles 4, 2, and 5, which is 180°.

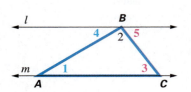

FIGURE 8.26 Lines l and m are parallel lines cut by transversals that form equal alternate interior angles.

Now that we know the sum of the angles of a triangle, we can use this information to find the interior angle sum of other polygons.

---

* Some define a trapezoid to be a quadrilateral that has *exactly one* pair of parallel sides. According to this alternate definition, a parallelogram would not be a trapezoid.

EXAMPLE 2   The Angle Sum of a Pentagon

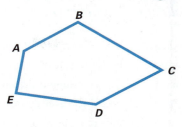

Find the interior angle sum of the convex pentagon *ABCDE*.

SOLUTION: Because we already know the interior angle sum for a triangle, it makes sense to divide the pentagon into a collection of triangles, as shown in Figure 8.27.

Notice in Figure 8.27 that ∠*A* is made up of three smaller angles 1, 2, and 3. Also, ∠*C* consists of angles 5 and 6 and ∠*D* is made up of angles 7 and 8.

The sum of the angles of the pentagon is

$$m\angle A + m\angle B + m\angle C + m\angle D + m\angle E.$$

If we replace $m\angle A$ by $m\angle 1 + m\angle 2 + m\angle 3$, $m\angle C$ by $m\angle 5 + m\angle 6$, and $m\angle D$ by $m\angle 7 + m\angle 8$, we see that the angle sum of the pentagon is

$$m\angle 1 + m\angle 2 + m\angle 3 + \cdots + m\angle 9.$$

This is the same as the angle sum of the three triangles Δ*AED*, Δ*ADC*, and Δ*ACB*, which equals $3 \times 180° = 540°$.

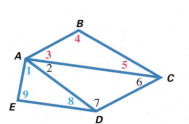

FIGURE 8.27 A pentagon divided into a collection of triangles.

 Quiz Yourself 5*

Apply the method used in Example 2 to find the interior angle sum of quadrilateral *ABCD*.

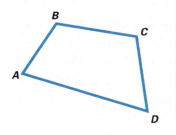

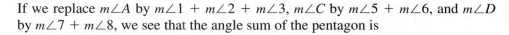

PROBLEM SOLVING

In Example 2, to find the angle sum of a pentagon, we used the strategy of converting a new problem to an older one. Often you can solve a new problem by relating it to a problem that you have solved earlier.

In Examples 1 and 2 and Quiz Yourself 5, you saw that a three-sided polygon has an angle sum of $180° = 1 \times 180°$, a four-sided convex polygon has an angle sum of $360° = 2 \times 180°$, and a five-sided convex polygon has an angle sum of $540° = 3 \times 180°$. We can generalize this pattern to *n*-sided convex polygons.

Angle Sum of a Polygon

The sum of the measures of the interior angles of a convex polygon having *n* sides is $(n - 2) \times 180°$.

---

* Quiz Yourself answers begin on page 849.

Historical Highlight: *Non-Euclidean Geometry (Part 2)*

Because the fundamental axioms of non-Euclidean geometry are different from Euclid's axioms, it is not surprising that when we rephrase well-known Euclidean geometry theorems in non-Euclidean terms, they sound somewhat strange. The Euclidean theorem that the interior angle sum of a triangle is 180° in Riemannian geometry becomes:

The interior angle sum of a triangle is greater than 180°.

If we consider the surface of a sphere, it is not hard to understand why this theorem should be true. Figure 8.28 shows a triangle whose interior angle sum is greater than 180°. Imagine beginning at the top of the sphere, the North Pole, so to speak (call this point A), and draw an arc of a great circle to meet another great circle that is horizontal, comparable to the equator. Call this point B. Make a right angle, draw an arc along this "equator" to point C, and then make a right angle and draw an arc of a great circle from C back to A. The angles at B and C are each 90°, so that when we add in the measure of angle A, we have an angle sum that exceeds 180°.

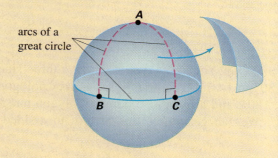

FIGURE 8.28 Triangle with angle sum greater than 180°.

On a pseudosphere, which has a shape like two bells of a trumpet, triangles look something like the curved triangle drawn in Figure 8.29. In this type of non-Euclidean geometry, triangles have an interior angle sum that is less than 180°.

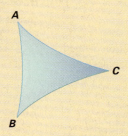

FIGURE 8.29 Triangle with an angle sum less than 180°.

At this point, you might be wondering which of the three geometries (Euclid's, Riemann's, Lobachevsky's) is the "correct" geometry. The answer is that there is no one "correct" geometry. Mathematicians have proved that all three types of geometries are perfectly consistent mathematical systems. Even more remarkable, they have also proved that in order for any one of these non-Euclidean geometries to be consistent, the other two types must be also be consistent. So in this sense, no one of the geometries is better. As to which geometry we use, it all depends on what application we have in mind. For surveying land and building buildings, Euclidean geometry is appropriate; however, Einstein found that in order to understand the vastness of the universe, non-Euclidean geometry is a better tool.

Although many word-processing and computer illustration programs have graphics capabilities, you often need to understand basic geometry in order to draw a figure exactly the way you want it to look.

EXAMPLE 3   Designing a Logo with Computer Graphics Software

a) Explain how you would use geometry to draw this star as part of a logo for your company. Assume that the graphics software you are using can draw circles and line segments and measure angles.

b) What is the angle measure of each point of the star?

SOLUTION: a) We begin by placing five points on a circle to divide it into five equal arcs. Because there is an angle measure of 360° around the center of the circle, we can mark off five equal angles measuring $\frac{360°}{5} = 72°$ to determine points A, B, C, D, and E, as shown in Figure 8.30.

Next we draw line segments from B to E to C to A to D, and back to B, as shown in Figure 8.31.

We can now trace the outline of the star by drawing line segments from B to Z to A to V to E, . . . , to Y, and back to B.

b) To find the measure of the angle for a point of the star, consider the pentagon ZVWXY shown in Figure 8.31. Because this pentagon is regular and has an interior angle sum of 540°, each interior angle

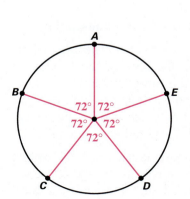

FIGURE 8.30   Five equally spaced points on a circle.

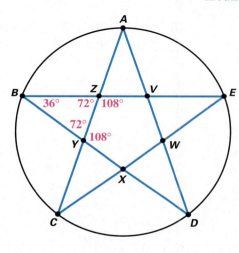

FIGURE 8.31   Constructing the five-point star.

measures $\frac{540°}{5} = 108°$. Because a straight angle equals 180°, we know that $\angle BZY$ and $\angle ZYB$ each measure 72°. The point of the star formed by $\angle YBZ$ therefore measures $180° - 72° - 72° = 36°$.

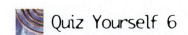

Quiz Yourself 6

Find the measure of each interior angle of a regular octagon.

Similar polygons have proportional sides and equal angles.

Industrial designers use scale models to create three-dimensional prototypes before devoting time and materials to building the final product. Similarly, a graphic artist creates a small, preliminary version of an advertisement that will later be enlarged to billboard size. Often clothing designers make sample garments in small sizes that are later resized after they have sold the design to stores. In each case, we are working with objects that have the same shape but a different size. In geometry, we also work with similar figures.

> **DEFINITION**
>
> Two polygons are **similar** if their corresponding sides are proportional and their corresponding angles are equal.

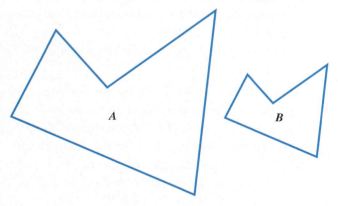

FIGURE 8.32   A and B are similar polygons.

Polygons A and B in Figure 8.32 are similar.

EXAMPLE 4   Using Geometry in Landscaping

Tall hedges are shading a rose garden. The edge of the garden is fifteen feet from the base of the hedges, as shown in Figure 8.33. At the time the gardener wishes the garden to be in sunlight, a four-foot vertical pole casts a ten-foot shadow. How much should the hedges be cut back so that they will not shade the garden?

SOLUTION: We will let x be the height of the hedges after trimming. Because line segments AC and DF represent the sun's rays, they are parallel lines cut by a transversal (the ground). This means that $\angle ACB$ and $\angle DFE$

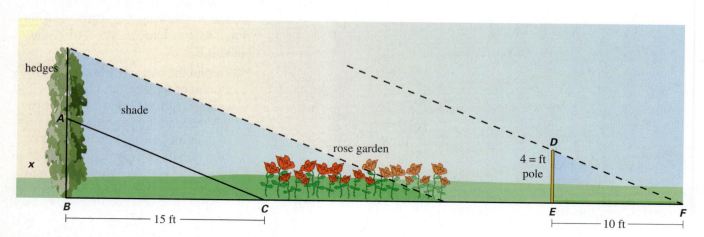

FIGURE 8.33   Landscaping a rose garden.

have equal measures. Angles B and E are right angles, making angles A and D equal. Therefore $\triangle ABC$ is similar to $\triangle DEF$. This means that corresponding sides of the two triangles are proportional. We can set up the following ratio:

$$\frac{x}{15} = \frac{4}{10}$$

Cross-multiplying, we get $10x = 4(15)$, which means that $x = 6$. Therefore, we should cut the hedges back to a height of six feet.

Exercises 8.2

In Exercises 1–10, determine whether each statement is true or false. Remember by the Always Principle that if a statement is true, it must always be true without exception.

1. An equilateral triangle is an isoceles triangle.

2. An isoceles triangle is not a scalene triangle.

3. A trapezoid is a parallelogram.

4. A rhombus is a parallelogram.

5. If a regular convex polygon has ten sides, then each interior angle measures 144°.

6. A regular polygon is convex.

7. If two polygons have corresponding angles equal, then the polygons are similar.

8. If two polygons have corresponding sides equal, then the polygons are similar.

9. A convex polygon can have an interior angle sum of 400°.

10. If a polygon has an interior angle sum of 180°, then the polygon is a triangle.

In Exercises 11–14, state whether each figure is a polygon. For those that are not polygons, state what part of the definition fails.

11.

12.

13.

14.

15. **Communicating Mathematics** Can an isoceles triangle also be a scalene triangle? Explain.

16. **Communicating Mathematics** Can an isoceles triangle also be an equilateral triangle? Explain.

17. **Communicating Mathematics** What is the difference between a rhombus and a square?

18. **Communicating Mathematics** What is the difference between a trapezoid and a parallelogram?

In Exercises 19–22, use the information that we give you to find the measures of angles A, B, and C in △ABC. All measures are in degrees.

19. $m\angle B = 2m\angle A, m\angle C = 3m\angle A$

20. $m\angle B = m\angle A, m\angle C = 5 + 3m\angle A$

21. $m\angle B = m\angle A + 10, m\angle C = 2m\angle A - 10$

22. $m\angle B = 3m\angle A, m\angle C = 2m\angle A + m\angle B$

23. Use the method of Example 2 to find the angle sum of a regular hexagon without using the formula $(n - 2) \times 180°$.

24. Use the method of Example 2 to find the angle sum of a regular octagon without using the formula $(n - 2) \times 180°$.

25. What is the measure of an interior angle of a regular 20-sided polygon?

26. What is the measure of an interior angle of a regular twelve-sided polygon?

27. If each interior angle of a regular polygon measures 160°, how many sides does the polygon have?

28. If each interior angle of a regular polygon measures 135°, how many sides does the polygon have?

In Exercises 29–32, assume that in each pair the figures are similar. Given the lengths of sides and measures of angles in the left figure, what information do you know about the right figure?

29.
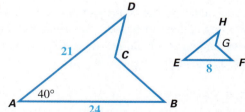

30.
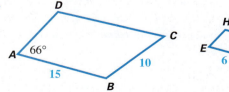

31.

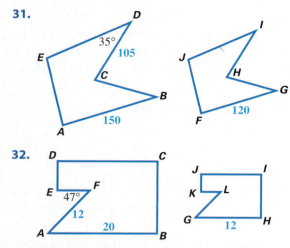

32.

33. If triangles *ABC* and *ADE* are similar in this diagram, what is the length of the pond?

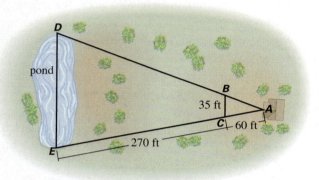

FIGURE FOR EX. 33

34. A private investigator wishes to place a surveillance camera at location *C* to observe activity at location *X*. The camera cannot be seen directly from position *X* because of a large clump of bushes. The camera will take pictures by aiming at a building, *M*, with shiny, mirror-like sides, as shown in the diagram. How far should the

camera be positioned from point *I* in order for this setup to work?

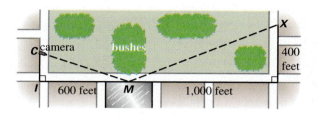

FIGURE FOR EX. 34

Some Japanese furniture makers construct elegant wooden furniture without using any glue or mechanical fasteners. In order to assemble the furniture, the pieces of wood must be cut precisely so that they fit together tightly.

35. A furniture maker is constructing a chair using a horizontal beam with a cross-section in the shape of a regular pentagon. The beam must fit into a notch cut into a support post so that an angle of the beam fits exactly in a notch cut in the post as shown in the given diagram. What should the measure of the indicated angles be so that the beam fits exactly in the notch?

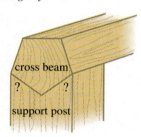

FIGURE FOR EX. 35

36. Repeat Exercise 35, but now assume that the horizontal beam has a cross-section in the shape of a regular hexagon. The beam will rest in the support post as shown in the given diagram.

FIGURE FOR EX. 36

Further Exercises

In Exercises 37–40, if it is possible to construct a triangle of the type described, then explain how you might draw one. If it is impossible to construct the triangle described, explain why it is impossible.

* **37.** A scalene triangle with two acute angles

38. A right triangle that also has an obtuse angle

39. An equilateral triangle that has all obtuse angles

---

* Exercise numbers circled in red can be used as group exercises.

40. An equilateral triangle that has all acute angles

41. Make up a description of a triangle as we did in Exercises 37–40 which is impossible to construct.

42. Make up a description of a triangle as we did in Exercises 37–40 which is possible to construct.

43. Communicating Mathematics What happens to the measure of the interior angles of a regular *n*-sided polygon as *n* gets larger and larger? Explain your answer.

44. Communicating Mathematics The Russians erected the world's largest full-figure statue, *Motherland*, as a World War II memorial. From her bare feet to the tip of her raised sword, *Motherland* stands 270 feet tall. If you are not allowed to touch the statue, how could you use a square piece of paper to determine her height? (*Hint:* Fold the paper into an isoceles triangle.)

45. Communicating Mathematics The English surveyors Mason and Dixon surveyed the famous Mason-Dixon line between Maryland and Pennsylvania using a level made from wood in the shape of the letter A,* as shown in the given diagram. A string with a weight hangs from the tip of the level.

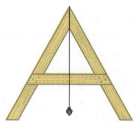

FIGURE FOR EX. 45

Explain how you would build such a level and also explain how it would work. Be sure to discuss lengths of line segments and measures of angles.

46. Communicating Mathematics Generalize Exercises 35 and 36. If the cross beam has a cross-section in the shape of a regular polygon with *n* sides, what would be the measure of the indicated angles for the support beam? Assume that one vertex of the end of the cross beam points directly down as was the case in Exercises 35 and 36.

47. In building scaffolding, often the scaffolding has many triangles, as shown in figure (a). What advantage do you see in using scaffolding of type (a) versus type (b)? (*Hint:* There is a fundamental property that triangles have that rectangles do not have.)

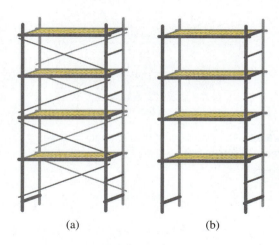

(a)　　　　　(b)

FIGURE FOR EX. 47

48. Suppose that you have a table with one leg, *AE*, five feet long and the other, *BD*, six feet long, as shown in the given diagram. Assume that you are going to join the legs at point *C* on leg *BD*, as shown. How far should *C* be from *A* so that the table will be level?

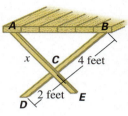

FIGURE FOR EX. 48

* J. E. Thompson, *Geometry for the Practical Worker* (Van Nostrand Reinhold, 1982), p. 82.

8.3 PERIMETER AND AREA

In using geometry to solve problems, we often need to know the length of the border of a plane figure or the size of its interior. If we want to install fencing around a game preserve, we must calculate the length of its border. In order to find the most efficient shape for a solar cell, we would be interested in the interior. In geometric terms, we need to measure perimeter and area.

> **DEFINITIONS**
>
> The **perimeter** of a polygon is the sum of the lengths of the sides of the polygon. The **area** is a measure of the amount of surface that the polygon covers.

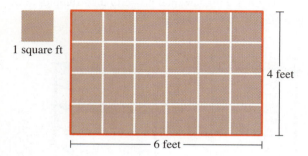

1 square ft

4 feet

6 feet

FIGURE 8.34 The perimeter of the rectangle is 20 feet; the area is 24 square feet.

We measure the perimeter of a figure with the same unit of measurement as we measure the sides, and we measure the area of the figure in square units. For example, as we see in Figure 8.34, a rectangle that has length 6 feet and width 4 feet has a perimeter of $6 + 4 + 6 + 4 = 20$ feet, and the area is $6 \times 4 = 24$ square feet.

The following are general formulas for calculating the perimeter and area of rectangles.

> **Perimeter and Area of a Rectangle**
>
> If a rectangle has length l and width w, then the perimeter of the rectangle is $P = 2l + 2w$ and the area is $A = l \cdot w$.

We can use the formula for the area of a rectangle to derive formulas for the area of other polygons.

Knowing how to do a calculation for one type of geometric figure often leads us to a method for doing that same computation for another type of figure. For example, once we know the formula for the area of a rectangle, we can easily derive the formula for the area of a parallelogram.

Consider the parallelogram in Figure 8.35(a), which has height h and base b. If we cut off the green triangle on the left, slide it over, and attach it to the right side of the parallelogram, we get the rectangle shown in Figure 8.35(b). We know that the area of this rectangle is $h \cdot b$, so therefore the area of the parallelogram is also $h \cdot b$.

> **Area of a Parallelogram**
>
> The area of a parallelogram with height h and base b is $A = h \cdot b$.

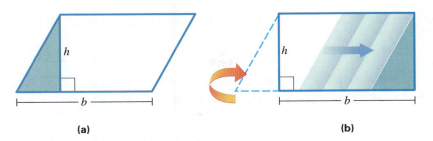

FIGURE 8.35 (a) Parallelogram. (b) Rectangle.

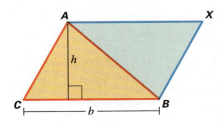

FIGURE 8.36 The area of $\triangle ABC$ is one-half the area of the parallelogram.

We next use the formula for the area of a parallelogram to derive the formula for the area of a triangle. In Figure 8.36, triangle ABC has height h and base b. It is easy to see that the area of $\triangle ABC$ is exactly one-half of the area of the parallelogram $AXBC$, which has area $h \cdot b$. Therefore $\triangle ABC$ has area $\frac{1}{2} h \cdot b$.

Area of a Triangle

A triangle with height h and base b has area $A = \frac{1}{2} h \cdot b$.

 SOME GOOD ADVICE

If you practice deriving the area formulas as we have developed them, it will be easier for you to remember the details of the formulas. For example, remembering that a triangle is one-half of a parallelogram helps you recall that there is a factor of $\frac{1}{2}$ in the formula for the area of a triangle.

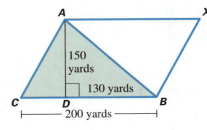

FIGURE 8.37 Playground shaped like a parallelogram.

X EXAMPLE 1   The Area of a Playground

a) Figure 8.37 shows a playground in the shape of a parallelogram. If a pound of grass seed covers 100 square yards, how much grass seed is needed to seed the entire playground?

b) Suppose that we want to seed only the triangular area ACD. How much grass seed will be needed then?

SOLUTION: a) The area of the parallelogram is $A = h \cdot b = 150 \cdot 200 = 30{,}000$ square yards. Dividing this area by 100, we see that 300 pounds of grass seed are needed for the entire playground.

b) Triangle ACD is a triangle with base 70 yards and height 150 yards. The area of this triangle is $\frac{1}{2} h \cdot b = \frac{1}{2} \cdot 150 \cdot 70 = 5{,}250$ square yards. Dividing

 Quiz Yourself 7*

Find the area of the given triangle.

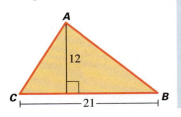

---

this by 100 we get 52.5, which is the number of pounds of grass seed required for this area.

Heron's formula uses the lengths of the sides of a triangle to find its area.

Sometimes we don't know the height of a triangle, but we do know the length of all three sides. In this case we can use Heron's formula to find the area of the triangle. Because this formula is not as intuitively clear as the formula we developed earlier, we will state it without proof.

> **Heron's Formula for the Area of a Triangle**
>
> Suppose that a triangle has sides $a, b,$ and c. We define the quantity $s = \frac{1}{2}(a + b + c)$. Then the area of the triangle is
>
> $$A = \sqrt{s(s - a)(s - b)(s - c)}.$$

EXAMPLE 2 **Finding the Area of a Triangular Flower Bed Using Heron's Formula**

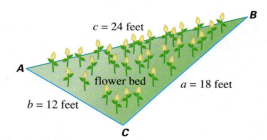

FIGURE 8.38 Triangular flower bed.

A gardener wishes to fill a triangular flower bed outside the city museum with yellow tulips. Figure 8.38 shows the dimensions of the flower bed. The gardener estimates that he will need four bulbs for each square foot of the flower bed. How many dozen bulbs should he buy?

SOLUTION: Because we know the length of the sides of the flower bed, we can use Heron's formula. First we find

$$s = \frac{1}{2}(a + b + c) = \frac{1}{2}(18 + 12 + 24) = 27.$$

Then the area of the flower bed is

$$A = \sqrt{s(s - a)(s - b)(s - c)} = \sqrt{27(9)(15)(3)} = \sqrt{10,935} \approx 105 \text{ square feet.}$$

Therefore, the gardener needs about $105 \times 4 = 420 = 35$ dozen bulbs.

We can use the area formulas that we already know to find the areas of other geometric figures.

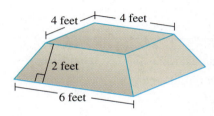

FIGURE 8.39 Base for statue formed by one square and four trapezoids.

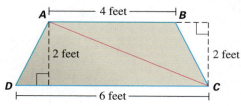

FIGURE 8.40 Trapezoid divided into two triangles.

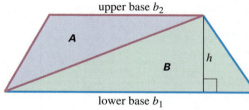

FIGURE 8.41 The area of the trapezoid is the sum of the areas of triangles A and B.

EXAMPLE 3 **Finding the Area of Trapezoids in the Base of a Statue**

A sculptor has created a statue for the lobby of the town's historical society building. Because the building is old and the town engineer wants to reduce the weight of the platform the statue will rest upon, he recommends a hollow base instead of a solid one. The platform will be formed by joining four congruent trapezoids and one square, as shown in Figure 8.39. In order to determine which material will be best for the platform, the sculptor needs to know the surface area of the platform. Use the formulas we have developed so far to determine this surface area.

SOLUTION: We know that the area of the top of the platform is $4 \times 4 = 16$ square feet. Each side of the platform is a trapezoid with one base 6 feet, the other base 4 feet, and height 2 feet, as shown in Figure 8.40. If we divide the trapezoid into two (noncongruent) triangles—ABC and ACD—then we can use the formula for finding the area of a triangle to solve the problem. Triangle ABC has a base of 4 feet and a height of 2 feet, so its area is $\frac{1}{2} \times 4 \times 2 = 4$ square feet. Triangle ACD has a base of 6 feet and a height of 2 feet, so its area is 6 square feet. The total area of the trapezoid is therefore 10 square feet. The area of the four trapezoidal sides plus the area of the rectangular top is $4 \times 10 + 16 = 56$ square feet.

From this example, we can see that the area of the trapezoid shown in Figure 8.41 equals the area of triangle A plus the area of triangle B. The area of the trapezoid is therefore $\frac{1}{2} \times b_1 \times h$ plus $\frac{1}{2} \times b_2 \times h$.

This leads us to the general formula for finding the area of a trapezoid.

Area of a Trapezoid

A trapezoid with lower base b_1, upper base b_2, and height h has area

$$A = \frac{1}{2} \times (b_1 + b_2) \times h.$$

 Quiz Yourself 8

Find the area of the given trapezoid.

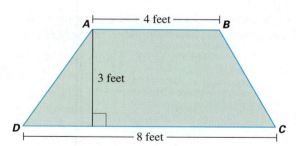

Highlight: *Geometry and Fashion Design**

One of the fundamental skills required in designing clothing is an understanding of the basic plane geometry that you are studying in this chapter. There are three basic methods of pattern making: the flat-pattern method, draping, and drafting.

A designer uses the flat-pattern method to create a new piece of clothing by beginning with a generic pattern called a sloper. Figure 8.42 shows a basic bodice sloper. As you can see, this sloper begins as a rectangle from which various triangles and curved pieces are removed. A blouse or jacket is then created by removing other geometric-shaped pieces from this fundamental shape. For women's clothes, there are slopers for skirts, pants, and sleeves as well.

Draping is working with fabric on a dress form or a live model. One way to drape garments is to cut shapes such as rectangles, squares, circles, half circles, quarter circles, and eighth circles and then experiment with these shapes. A talented designer who uses draping to create new fashions must have a good intuitive understanding of geometry.

Other designers draft patterns by drawing geometric shapes on paper. They create perfectly fitted patterns that are then scaled to fit different sizes. A drafter needs a good mind for mathematics, especially fractions and geometry.

We see that a serious clothing designer must understand the basic geometric notions of square, rectangle, circle, radius, and diameter and their relationships. So not only is geometry "all around us," but we wear it every day.

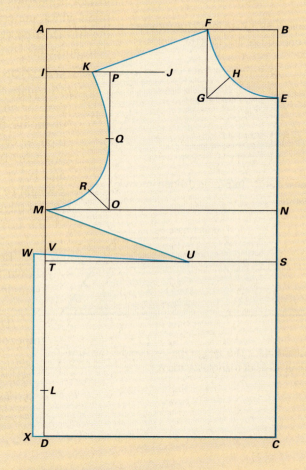

FIGURE 8.42 A bodice sloper.

We can use the Pythagorean theorem to find measurements involving right triangles.

In the sixth century B.C., the Greek mathematician Pythagoras proved one of the most famous theorems in the history of mathematics. The Pythagorean theorem states that the sum of the squares of the lengths of the legs of a right triangle equals the square of the length of the hypotenuse.

---

* This Highlight is based on the article "The Drafter, the Draper, the Flat Pattern," *Threads Magazine*, No. 11, June/July 1987, pp. 33–37.

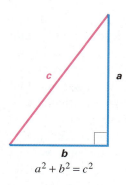

FIGURE 8.43 The Pythagorean theorem.

$$a^2 + b^2 = c^2$$

> **The Pythagorean Theorem**
>
> If a and b are the legs of a right triangle and c is the hypotenuse (the side opposite the right angle, shown in red) then $a^2 + b^2 = c^2$ (see Figure 8.43).

If we know the lengths of any two sides of a right triangle, we can use the Pythagorean theorem to find the length of the third side.

EXAMPLE 4 Using the Pythagorean Theorem

Use the lengths of the two given sides to find the length of the third side in the triangles in Figure 8.44.

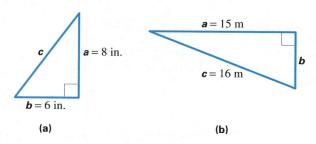

FIGURE 8.44 Using the Pythagorean theorem.

SOLUTION: a) We are given $a = 8$ and $b = 6$, so we must find c. If we substitute the values for a and b in the Pythagorean theorem, we get

$$8^2 + 6^2 = c^2.$$

Therefore $c^2 = 100$, which means that $c = 10$.

b) In this case, we are given $a = 15$ and $c = 16$ and we must find b. Again substituting these values in the Pythagorean formula, we get $15^2 + b^2 = 16^2$, or $225 + b^2 = 256$. Subtracting 225 from both sides of this equation, we get $b^2 = 31$. This means that $b = \sqrt{31} \approx 5.57$.

 Quiz Yourself 9

Assume that the hypotenuse of a right triangle measures 20 inches and one leg measures 10 inches. What is the length of the other leg?

In order to solve a problem, it may be necessary to use the Pythagorean theorem several times, as we demonstrate in Example 5.

EXAMPLE 5 Using the Pythagorean Theorem to Find the Height of a Pyramid

The Great Pyramid, near Cairo, Egypt, is the tomb of the Egyptian pharaoh Khufu (also called Cheops by the Greeks) and was considered to be one of the seven wonders of the ancient world. An archaeologist wishes to determine the height of this pyramid. This pyramid has a square base measuring 230 meters on each side, and she has found the distance from one corner of the base to the tip of the pyramid to be 219 meters. What is the height of the pyramid?

SOLUTION: If we imagine this pyramid to be hollow, we could drop a string with a weight from the tip of the pyramid, point T, to point M, which is the middle of the

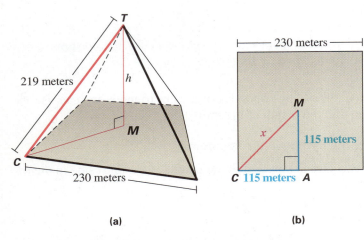

(a)

(b)

FIGURE 8.45 (a) The Great Pyramid. (b) The base of the pyramid. *M* is the center of the base, *C* is the corner of the base, and *A* is the midpoint of a side of the base.

base. If we then drew a line segment from *M* to a corner of the base, calling this point *C*, we would form a right triangle $\triangle TMC$, as we see in Figure 8.45(a). Our goal now is to find the length of line segment *TM*. In order to do this, we must first find the length of line segment *CM*. We draw the base of the pyramid in Figure 8.45(b), showing segment *CM*.

Because each edge of the base is 230 meters long, *MA* and *CA* are both 115 meters long. We can apply the Pythagorean theorem to triangle $\triangle MAC$ to get $115^2 + 115^2 = x^2$. Solving $26{,}450 = x^2$ gives us $x = 115\sqrt{2}$. (If you prefer, you can write $x \approx 162.6$.)

Now that we have found the length of *CM*, we can find the length of *TM* in $\triangle TMC$. Using the Pythagorean theorem again, we get

$$(\text{length of } CM)^2 + (\text{length of } TM)^2 = (\text{length of } TC)^2.$$

Substituting, for the various lengths in $\triangle TMC$ gives the equation

$$\left(115\sqrt{2}\right)^2 + h^2 = 219^2,$$

or

$$h^2 = 219^2 - \left(115\sqrt{2}\right)^2 = 47{,}961 - 26{,}450 = 21{,}511.$$

Taking the square root of both sides of this equation, we find that $h \approx 146.7$ meters.

Historical Highlight: *Hypatia*

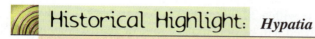

Throughout the history of mathematics, fewer women than men are mentioned for their contributions because traditionally women were discouraged from studying mathematics. One notable exception is Hypatia, who was born in Greece in 370 A.D. Her father, who was a professor of mathematics at the University of Alexandria, gave her a classical education, which included mathematics. She lectured on both philosophy and mathematics at Alexandria and wrote major papers on geometry, including the work of Eu-clid. She also did work in philosophy and astronomy and is believed to have invented several astronomical devices.

Unfortunately, her work in science caused her problems with the Christian church and, moreover, as a Greek she was seen as a pagan. In 415 a mob attacked and brutally murdered her, which caused other scholars to flee Alexandria. This tragic event marked the end of the golden age of Greek mathematics and, some feel, the beginning of the Dark Ages in Europe.

We use simple formulas to compute the circumference and area of circles.

As with polygons, we often need to calculate the circumference and area of circles. The derivation of these formulas is not as intuitive as those for the polygons we have been discussing, so we will state them without proof. The number π, which appears in these formulas, is the Greek letter pi. It stands for the measure of the circumference of a circle divided by the length of its diameter. We will use 3.14 to approximate pi.

> **Circumference and Area of a Circle**
>
> A circle with radius r has circumference $C = 2\pi r$ and area $A = \pi r^2$.

These formulas tell us that a circle with radius 10 feet has a circumference of $C = 2\pi r = 2\pi(10) \approx 62.8$ feet and area $A = \pi r^2 = \pi(10)^2 \approx 314$ square feet.

EXAMPLE 6   Laying Out a Basketball Court

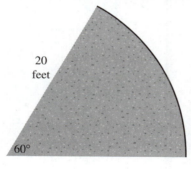

20 feet

60°

FIGURE 8.46   Basketball court.

A man is designing a basketball court for his children. The court is in the form of a segment of a circle, as shown in Figure 8.46. There will be a fence at the rounded end of the court and he will paint the court with a concrete sealer.

a) How much fencing is required?

b) What is the area of the surface of the court?

SOLUTION: Since a circle contains 360°, the court is one-sixth of the interior of a circle with a radius of 20 feet.

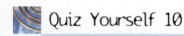

Quiz Yourself 10

Find the circumference and area of a circle that has radius 8.

a) The circumference of a circle with a 20 foot radius is $2\pi r = 2\pi(20) = 40\pi \approx$ 125.6 feet. The fencing we need will be one-sixth of this, or approximately 20.93 feet. The man should buy about 21 feet of fencing.

b) The area of a circle with a 20-foot radius is $\pi(20)^2 = 400\pi \approx 1{,}256$ square feet. One-sixth of this is about 209.3 square feet.

Exercises 8.3

In Exercises 1–12, find the area of each figure. *

1.

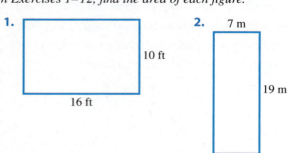

10 ft

16 ft

2.

7 m

19 m

3.

7 in.

20 in.

4.

14 ft

10 ft

5.

14 cm

6 cm

22 cm

6.

12 m

5 m

18 m

7.

6 yd

24 yd

8.

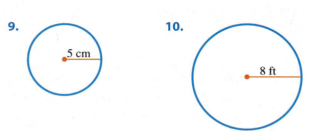

6 in.

15 in.

9.

5 cm

10.

8 ft

11.

8 m

12.

18 yd

In Exercises 13–20, find the area of the shaded regions. Recall that to solve a new problem, it is helpful to relate it to a problem that you have seen before.

13.

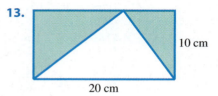

10 cm

20 cm

14.

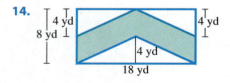

4 yd

8 yd

4 yd

4 yd

18 yd

———

* Throughout this exercise set approximate π by 3.14.

15.

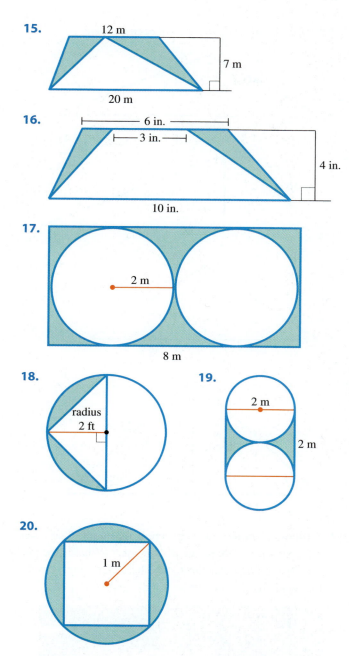

16.

17.

18.   **19.**

20.

Use the following figure to do Exercises 21–24.

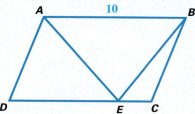

Assume that the area of the parallelogram ABCD is 60 square inches and the area of triangle BEC is 6 square inches.

21. What is the area of triangle *ABE*?

22. What is the area of triangle *ADE*?

23. What is the area of the trapezoid *ABED*?

24. What is the area of the trapezoid *ABCE*?

Use the following figure to do Exercises 25–28.

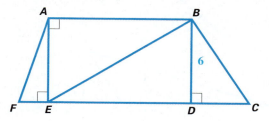

Assume that the area of triangle BAE is 30 square yards, The area of the trapezoid ABDF is 66 square yards, and the area of triangle BDC is twice the area of triangle AEF.

25. Find the area of triangle *BDC*.

26. Find the area of triangle *EBC*.

27. Find the area of trapezoid *ABCE*.

28. Find the area of trapezoid *ABCF*.

A geoboard is a board with rows of nails spaced one inch apart in both the vertical and horizontal directions. In Exercises 29–32, we stretched a rubber band around some of the nails. Find the area of the enclosed figures.

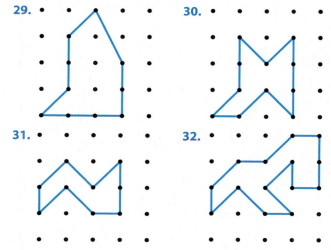

29.   **30.**

31.   **32.**

In Exercises 33–36, use Heron's formula to find the area of each triangle.

33.
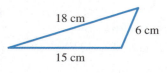
18 cm
6 cm
15 cm

34.
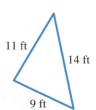
11 ft
14 ft
9 ft

35.
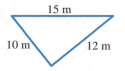
15 m
10 m
12 m

36.
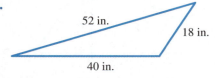
52 in.
18 in.
40 in.

In Exercises 37–38, find the height h for each triangle.

37.
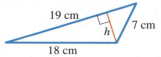
19 cm
7 cm
h
18 cm

38.
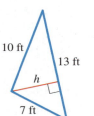
10 ft
13 ft
h
7 ft

In Exercises 39–42, find the length of side x for each triangle.

39.
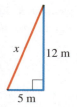
x
12 m
5 m

40.
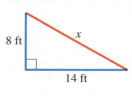
8 ft
x
14 ft

41.
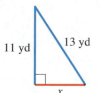
11 yd
13 yd
x

42.
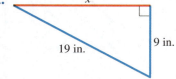
x
9 in.
19 in.

In Exercises 43–46, find the area of each triangle.

43.

13 in.
11 in.
12 in.

44.
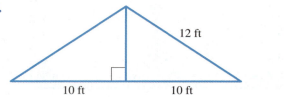
10 m
8 m
15 m

45.
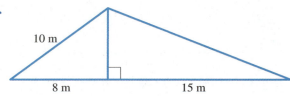
12 ft
10 ft
10 ft

46.
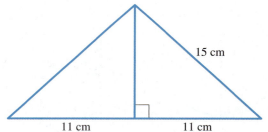
15 cm
11 cm
11 cm

47. Finding distances on a baseball diamond. The bases on a baseball diamond are 90 feet apart. What is the distance from home plate to second base? (All angles in the diamond are right angles.)

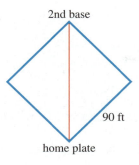

2nd base

90 ft

home plate

FIGURE FOR EX. 47

48. In baseball, home plate is shaped as shown ($m\angle FDE = 90°$).

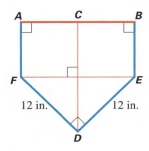

A   C   B

F   E

12 in.   12 in.

D

a) What is the length of line segment *AB*?

b) If we assume that line segment *CD* has the same length as segment *AB*, what is the area of home plate?

49. Find the length of line segment *AB* in the given figure.

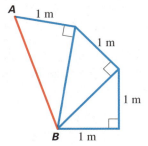

A   1 m

1 m

1 m

B   1 m

50. Continuing the pattern in Exercise 49, construct a line segment having length $\sqrt{6}$.

In Exercises 51–54, state whether perimeter or area would be the more appropriate quantity to measure.

51. You are covering an archeological plot with a tarpaulin.

52. As a zookeeper, you wish to buy fencing for a paddock to graze llamas.

53. You are putting a decorative stencil at the top of the walls in your bedroom.

54. You are surfacing your rooftop Japanese garden with slate tiles.

55. Making a stained glass window. A stained-glass window in a museum is in the shape of a rectangle with a semicircle on top. The height of the rectangular part of the window is twice the base. If the base of the window measures 6 feet, what is the area of the window?

├—6 ft—┤

FIGURE FOR Ex. 55

56. Making a stained glass window. Consider a window like the one in Exercise 55. Assume that the semicircular top has a radius of 2 feet. We want the rectangular part of the window to have the same area as the semicircular top. What should the dimensions of the rectangle be?

57. Making a flower bed. A gardener has enough tiger lilies to fill a circular flower bed having an area of 50 square feet. Find the radius of the flower bed to the nearest foot.

58. Comparing pizzas. At Camillo's Italian Ristorante, the medium pizza has a diameter of 12 inches and sells for $5.99. The large pizza has a diameter of 16 inches and sells for $8.99. Which pizza is the better buy? Explain.

59. True or false? If we double the radius of a circle, the area also doubles.

60. True or false? If we double the radius of a circle, the circumference also doubles.

61. Measuring a running track. A running track, 4 meters wide, has the dimensions shown in the following diagram. The ends of the track are semicircles with diameter 20 meters. What is the surface area of the track?

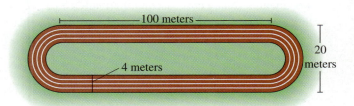

├————100 meters————┤

20 meters

4 meters

62. **Measuring a running track.** In Exercise 61, if the 100-meter dimension is increased to 120 meters, the 20-meter dimension is increased to 40 meters, and the width of the track is increased to 6 meters, what is the surface area of the track?

63. Redo Example 6, but now assume that the sides of the basketball court are 18 feet long and the angle measures 72°.

64. Redo Example 6. However, now the tip of the court has been removed, as shown in the figure.

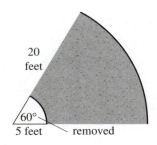

FIGURE FOR EX. 64

Further Exercises

65. In the figure below, *W*, *X*, *Y*, and *Z* are the midpoints of the line segments on which they lie. How does the area of *WXYZ* compare with the area of rectangle *ABCD*? Explain how you got your answer.

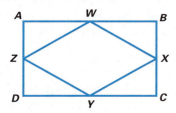

66. If *W* and *Y* were not the midpoints of the line segments on which they lie, would it change your answer in Exercise 65? Explain your answer.

* 67. You have 200 feet of fencing and wish to enclose a rectangular area. What shape will enclose the most area? You might solve this problem by experimenting with different lengths and widths until you can make a reasonable guess.

68. A furniture maker is shaping round and square posts. In figure (a) he is cutting a circular post from a piece of wood with a square cross-section. In figure (b) he is cutting a square post from a piece of wood with a circular cross-section. In which situation is there the smaller *percentage* of waste? (*Hint:* Answer this question by making up numerical examples.)

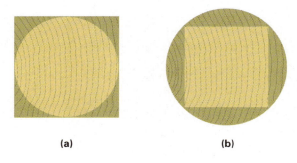

FIGURE FOR EX. 68

69. In a housing development, lots are clustered in circles with an open area in the center of the circle, as shown in the diagram. The side of a lot is 180 feet and the central common area is 11,304 square feet. Each lot in the cluster is the same size. What is the area of each lot? (We are using 3.14 for π.)

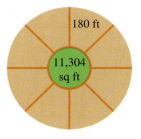

FIGURE FOR EX. 69

---

* Exercise numbers circled in red can be used as group exercises.

70. A property owner in Exercise 69 wishes to put a fence entirely around her lot, except for the curved section away from the common area. How much fencing is required?

In Exercises 71–72, express the shaded areas in terms of the radius r of the circle.

71.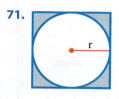

72.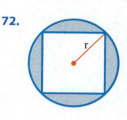

8.4 VOLUME AND SURFACE AREA

In this section we extend the ideas of two-dimensional Euclidean geometry to three dimensions. Instead of squares and rectangles, we will discuss boxlike figures called parallelepipeds; instead of circles we will talk about cylinders, spheres, and cones; and instead of measuring perimeters and areas, we will calculate the volumes and surface areas of three-dimensional geometric figures.

Many applications require a knowledge of three-dimensional geometry. When designing a skyscraper, an architect needs to know the volume of concrete to pour for a foundation. A manufacturer must be able to compute the amount of materials required to fabricate a cylindrical water tank. A state highway department engineer may wish to know how much road salt is contained in a conical storage shed.

When we measure length in one dimension, we use units such as inches, feet, centimeters, and meters. In measuring area in two dimensions, we use square units such as square inches and square centimeters. We measure the **volume** of a three-dimensional figure using *cubic units*. Even though we will study three-dimensional geometric objects, *we will still measure surface area in square units*.

Figure 8.47 shows a cube whose edges are each one inch long, so the cube has a volume of one cubic inch.

In determining the volume of a three-dimensional figure whose measurements are given in inches, we are really asking how many of these one-inch cubes will fit inside the figure. Because it might be impossible to fit these cubes inside the figure exactly, we can imagine filling one-cubic-inch containers with water and asking, "How many of these containers of water does it take to fill the figure exactly?"

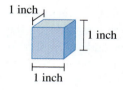

FIGURE 8.47 A cube whose volume is one cubic inch.

The volume of a rectangular parallelepiped is the product of its length, width, and height.

It is easy to compute the volume of the box-shaped rectangular solid, called a *rectangular parallelepiped*,* shown in Figure 8.48. If we slice along the lines drawn at one-inch intervals on the solid, we get four layers each containing fifteen

---

* A parallelepiped is a solid with six faces, each of which is a parallelogram. In this case the parallelepiped is rectangular, which means the faces are rectangles.

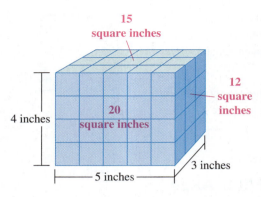

FIGURE 8.48 Rectangular solid sliced into 60 one-inch cubes.

one-inch cubes. The volume of this solid is therefore $5 \times 3 \times 4 = 60$ cubic inches.

It is also straightforward to find the surface area of the solid in Figure 8.48. Clearly the areas of the front and the back of the solid are $5 \times 4 = 20$ square inches, the areas of the top and bottom are each 15 square inches, and the areas of the two sides are 12 square inches. The total surface area is therefore $(2 \times 20) + (2 \times 15) + (2 \times 12) = 94$ square inches. Figure 8.48 helps us remember the formulas for volume and surface area.

 Quiz Yourself 11*

Find the volume and surface area of a rectangular solid with length 8 centimeters, width 6 centimeters, and height 3 centimeters.

> **Volume and Surface Area of a Rectangular Solid (Rectangular Parallelepiped)**
>
> If a rectangular solid has length l, width w, and height h, the volume of the solid is $V = lwh$ and the surface area of the solid is
>
> $$S = 2lw + 2lh + 2wh.$$

For many solids, volume equals base area times height.

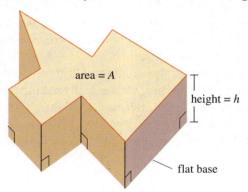

FIGURE 8.49 Volume = area of base times height.

Another way to look at the way we derived the formula for a rectangular solid is that we took the area of the base, which was lw, and multiplied it by the height h. In general, assume that we have a three-dimensional figure with a flat top and base and *sides perpendicular to the base*, as shown in Figure 8.49. If the area of the top and base is A and the height of the figure is h, then the volume of the figure is $V = Ah$.

EXAMPLE 1 Comparing Volumes

Figure 8.50 shows two blocks of cheese that are selling for the same price. Which block contains the greater volume?

---

* Quiz Yourself answers begin on page 849.

SOLUTION: We calculate the volume of each block by multiplying the area of its base times its height. The area of the base of the trapezoidal block is $(\frac{1}{2}) \times (9 + 5) \times 3 = 21$ square inches. This block has a volume of $21 \times 4 = 84$ cubic inches.

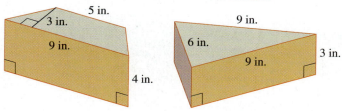

We use Heron's formula to find the area of the base of the triangular block of cheese. We first must compute $s = \frac{1}{2} \times (9 + 9 + 6) = 12$. Then the area is

$$\sqrt{12(12 - 9)(12 - 9)(12 - 6)} = \sqrt{648} \approx 25.5.$$

Multiplying this area by the height of 3 gives us $25.5 \times 3 = 76.5$ cubic inches. Therefore the trapezoidal block of cheese has slightly more volume and is the better deal.

FIGURE 8.50 Which block has greater volume?

Simple diagrams illustrate the formulas for finding the volume and surface area of a cylinder.

We can use this method of multiplying the base area times the height to find the volume of some common three-dimensional solids. A *right circular cylinder* is a solid that is shaped like a soup or tuna fish can (see Figure 8.51). We use the adjective *right* for these cylinders because the sides are perpendicular to the base. The base of a right circular cylinder is a circle with area $A = \pi r^2$.* Thus the volume of the cylinder is $V = \pi r^2 h$.

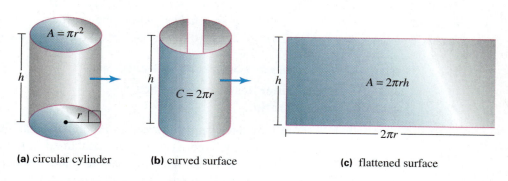

(a) circular cylinder   **(b)** curved surface   **(c)** flattened surface

FIGURE 8.51 (a) Right circular cylinder. (b) Remove top and bottom and cut side of cylinder. (c) Flatten side of cylinder.

To find the surface area of the cylinder in Figure 8.51(a), imagine that we remove the top and bottom of the cylinder and then cut the side of the cylinder and open it up, as in Figure 8.51(b). If we flatten this curved surface, we get a rectangle with length $2\pi r$ and height h, shown in Figure 8.51(c). Therefore the surface area of the side of the cylinder is $2\pi rh$. Adding to this the areas of the top

---

* Throughout this section, we again approximate π by 3.14.

and bottom of the cylinder, which are each πr^2, we get the total surface area of the cylinder.

If you recall the diagrams in Figure 8.51, it will be easy for you to remember the following formulas.

> **Volume and Surface Area of a Right Circular Cylinder**
>
> A right circular cylinder with radius r and height h has volume $V = \pi r^2 h$ and surface area $S = 2\pi rh + 2\pi r^2$.

We can use these formulas to compare purchases.

EXAMPLE 2   Comparing Volumes of Commercial Products

An art supply store sells paint solvent in a cylindrical can that has a diameter of four inches and a height of six inches. The giant economy size can of the same solvent has a diameter twice as large (the height of the can is the same) and costs three times as much as the smaller can.

a) Which can is the better deal?

b) Compare the surface areas of the two cans.

SOLUTION: a) The radius of the first can is 2, so its volume is $\pi r^2 h = \pi(2)^2(6) \approx 75.4$ cubic inches. The radius of the second can is 4, so it has a volume of $\pi r^2 h = \pi(4)^2(6) \approx 301.4$ cubic inches. Since the larger can contains four times as much solvent but costs only three times as much, the larger can is a better deal.

b) The surface area of the smaller can is $2\pi rh + 2\pi r^2 = 2\pi(2)(6) + 2\pi(2)^2 \approx 100.5$ square inches, while the surface area of the larger can is $2\pi rh + 2\pi r^2 = 2\pi(4)(6) + 2\pi(4)^2 \approx 251.2$ square inches.

Quiz Yourself 12

What is the volume and surface area of a right circular cylinder with a radius of five centimeters and height of ten centimeters?

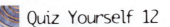

An efficient container has a small surface area – to – volume ratio.

You may have been surprised in Example 2 that although the larger can contains four times the amount of solvent contained in the smaller can, it takes only two and a half times as much material to make the larger can. Manufacturers are interested in such relationships so they can minimize the amount of materials they need to package a product and, by doing so, reduce their cost. What is the most efficient shape of a container to contain the largest amount of a product? For example, what is the most efficient shape of a soup can?

EXAMPLE 3   The Most Efficient Shape of a Can

What are the dimensions of a can that will contain one cubic foot of liquid and that will have the smallest amount of surface area?

SOLUTION: We must find the radius, r, and the height, h, of the can. Our intuition tells us that if the radius is small, as in Figure 8.52(a), then the height

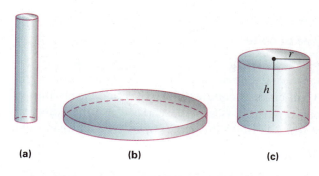

FIGURE 8.52   A can with a small radius requires a large height, whereas a can with a large radius requires a small height.

(a)   (b)   (c)

must be large. On the other hand, if the radius is large, as in Figure 8.52(b), then the height must be small. Our problem then is to find the ideal radius that gives the minimum surface area (Figure 8.52(c)).

Because the volume of the can is to be 1, we can set $\pi r^2 h = 1$. Dividing both sides of this equation by πr^2, we get $h = \frac{1}{\pi r^2}$. Thus, as our intuition tells us, the height depends on the choice of radius. Recall that the formula for finding the surface area of a cylinder is $S = 2\pi rh + 2\pi r^2$. We can substitute $\frac{1}{\pi r^2}$ for h in this equation to get

$$S = 2\pi r \left(\frac{1}{\pi r^2} \right) + 2\pi r^2 = \frac{2\pi r}{\pi r^2} + 2\pi r^2. \tag{1}$$

Canceling the π and an r from the quotient on the right side of equation (1), we get

$$S = \frac{2}{r} + 2\pi r^2. \tag{2}$$

The benefit of equation (2) is that we now have a formula for the surface area of a can (containing one cubic foot of liquid) expressed totally in terms of the variable r. We can now vary r to find which value gives us the smallest surface area.

We expect that a cylinder with a very small radius will not be very efficient and will need a lot of surface area to contain one cubic foot. As we increase the size of the radius, the can should become more efficient and require less surface area. If we keep increasing the size of the radius, eventually we will get a can that is too flat and again is not efficient. We see this happen in Table 8.4, where we use equation (2) to find the surface area for different values of r.

It appears in Table 8.4 that as we increase r, the surface area starts to increase somewhere between 0.5 and 0.6. To get a better estimate of the ideal radius, we could use a spreadsheet† to calculate the surface area for radii of size 0.51, 0.52, 0.53, and so on. If we did these calculations, we would find that $r \approx 0.54$ gives us the smallest surface area for a can with volume 1.   ◎

Radius, r	Surface Area, $S = \frac{2}{r} + 2\pi r^2$	
0.1*	20.063	
0.2	10.251	decreasing
0.3	7.232	decreasing
0.4	6.005	decreasing
0.5	5.570	decreasing
0.6	5.594	increasing
0.7	5.934	increasing
0.8	6.519	increasing

TABLE 8.4   Surface area first decreases, then increases as r increases.

---

* Because the volume is expressed in cubic feet, the radius of the can is measured in feet, so 0.1 is one-tenth of a foot. The surface area is then measured in square feet, so 20.063 is slightly over 20 square feet.
† Some graphing calculators have a table command that will allow you to generate these values automatically. See www.aw.com/pirnot.

The formulas for the volume and surface area of cones and spheres are related to the formulas for cylinders.

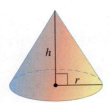

We now work with formulas for the volume of two other solid figures: cones and spheres. Figure 8.53 shows a right circular cone with height h and base radius r. We call the cone circular because its base is a circle and we use the adjective *right* because the line segment from its tip to the center of its base forms a right angle with the base.

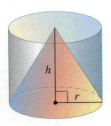

FIGURE 8.53 Right circular cone.

Volume and Surface Area of a Right Circular Cone

A right circular cone with height h and base radius r has volume $V = \frac{1}{3}\pi r^2 h$ and surface area $S = \pi r\sqrt{r^2 + h^2}$.

In the formula for the surface area of a cone, if we want to include the area of the base we must add πr^2.

FIGURE 8.54 The cone has a volume $\frac{1}{3}$ that of the cylinder.

PROBLEM SOLVING

We can use what we know about the volume of a cylinder to understand the formula for the volume of a cone. Figure 8.54 shows that a cone with height h and radius r easily fits inside a cylinder with height h and radius r. The cone does not appear to occupy even one-half of the volume of the cylinder, which is $V = \pi r^2 h$. Therefore it is easy to remember that the volume of the cone is $V = \frac{1}{3}\pi r^2 h$.

EXAMPLE 4   Finding the Volume and Surface Area of a Cone

Quiz Yourself 13

Find the volume and surface area of a right circular cone with a radius of four yards and a height of five yards.

What is the volume and surface area of a cone with a height of ten meters and a base radius of eight meters?

SOLUTION: The volume is $V = \frac{1}{3}\pi r^2 h = \frac{1}{3}\pi(8)^2(10) = \frac{640\pi}{3} \approx 669.87$ cubic meters. The surface area is

$$S = \pi r\sqrt{r^2 + h^2} = \pi(8)\sqrt{8^2 + 10^2} = 8\sqrt{164}\,\pi \approx 321.69 \text{ square meters.}$$

EXAMPLE 5   The Volume of Conical Storage Buildings

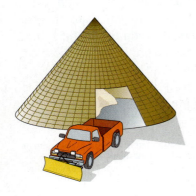

Many northern states stockpile salt to use for melting road ice. In one such state, the highway department builds sheds in the shape of right circular cones to store the salt. Current sheds have a diameter of 30 feet and a height of 10 feet. The department plans to build larger conical sheds to hold more salt. The larger shed will either have a ten-foot-longer diameter and the same height, or the same diameter and a ten-foot-greater height.

a) Calculate the volume for each of the proposed sheds.

b) Calculate the surface area for each of the proposed sheds.

c) Which design seems to be the more economical design?

SOLUTION: a) First we will compute the volume if we increase the diameter by ten feet. The radius of the cone will be $\frac{(30 + 10)}{2} = 20$ feet and the height will remain at 10 feet. The volume is therefore $V = \frac{1}{3}\pi(20)^2(10) \approx 4{,}187$ cubic feet.

If we increase the height by 10 feet, the radius will remain at 15 feet and the height will be 20 feet. In this case the volume is $V = \frac{1}{3}\pi(15)^2(20) \approx 4{,}710$ cubic feet.

b) The surface area for the shed with radius 20 and height 10 is $S = \pi(20)\sqrt{20^2 + 10^2} \approx 1{,}404$ square feet. A shed with radius 15 and height 20 has $S = \pi(15)\sqrt{15^2 + 20^2} \approx 1{,}178$ square feet.

c) Increasing the height by 10 feet and keeping the diameter at 30 feet gives us more volume and smaller surface area than if we increase the diameter to 40 feet and keep the height at 10 feet. Therefore, increasing the height to 20 feet seems to be the better design.

The last three-dimensional figure we consider is a sphere (see Figure 8.55).

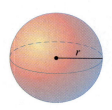

Figure 8.55 A sphere with radius *r*.

> **Volume and Surface Area of a Sphere**
>
> A sphere with radius *r* has volume $V = \frac{4}{3}\pi r^3$ and surface area $S = 4\pi r^2$.

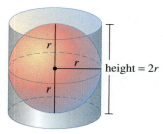

area of base $= \pi r^2$

Figure 8.56 A sphere with radius *r* inside a cylinder with radius *r* and height 2*r*.

Figure 8.56 will help you remember the formula for the volume of a sphere. A sphere with radius *r* fits exactly inside a cylinder having radius *r* and height 2*r*. The volume of this cylinder is $\pi r^2 \times 2r = 2\pi r^3$. Because the sphere does not completely fill the cylinder, you should remember that the volume of the sphere is *less than* $2\pi r^3$, or $\frac{4}{3}\pi r^3$.

Many containers, such as water tanks, are shaped like spheres.

EXAMPLE 6 Measuring the Capacity of Water Tanks

Metrodelphia is replacing existing spherical water tanks with larger spherical water tanks. The water commission insists that to provide for future expansion of the city, the new capacity of the tanks should be at least five times the capacity of the old water tanks. If the city purchases tanks that have a radius that is twice the radius of the old tanks, will these tanks satisfy the water commission?

Quiz Yourself 14

Find the volume of a sphere with a radius of six centimeters.

SOLUTION: Let the radius of the old water tanks be *r* feet, so the volume of each of these tanks will be $V = \frac{4}{3}\pi r^3$ cubic feet. The new tanks will each have a radius of 2*r* feet, so the volume of these new tanks will be $V = \frac{4}{3}\pi(2r)^3 = \frac{4}{3}\pi 8r^3 = 8\left(\frac{4}{3}\pi r^3\right)$ cubic feet, which is eight times the volume of the original tanks. Therefore, the new tanks will exceed the specifications of the water commission.

Exercises 8.4

In Exercises 1–8, find (a) the surface area and (b) the volume of each figure. Approximate π by 3.14.

1.

5 cm
6 cm
4 cm

2.

⊢3 ft⊣
6 ft

3.

8 in.
3 in.

4.

5 yd

5.

⊢5 ft⊣
8 ft

6.

18 m
⊢5 m⊣

7.

20 cm

8.

15 in.
40 in.   10 in.

In Exercises 13–20, find the volume of each figure.

13.

area = 25 sq ft
3 ft

14.

area = 40 sq ft
3 ft

15.

7 in.
4 in.
5 in.
11 in.

16.

6 ft
8 ft
8 ft
3 ft

17.

4 m
5 m
5 m
2 m

18.

circular arc
16 m
70°
16 m
4 m

19.

circular arc
8 in.
45°
8 in.
3 in.

20.

12 cm
6 cm
6 cm
15 cm

In Exercises 9–12, if you were tutoring a friend for a geometry test, how would you explain intuitively how to remember the formula for each figure?

9. Communicating Mathematics the volume of a cylinder

10. Communicating Mathematics the surface area of a cylinder

11. Communicating Mathematics the volume of a cone

12. Communicating Mathematics the surface area of a cone (*Hint:* Try to use the Pythagorean theorem in Figure 8.43.)

21. A punch bowl is in the shape of a hemisphere (half of a sphere) with a radius of 9 inches. The cup part of a ladle is also in the shape of a hemisphere with a radius of 2 inches. If the bowl is full, how many ladles of punch are there in the bowl?

22. Assume that we are filling cylindrical glasses that have a diameter of 3 inches and a height of 4 inches from the punch bowl in Exercise 21. If the bowl is completely full, how many glasses can we fill?

23. How many cubic inches are in a cubic foot?

24. How many cubic inches are in a cubic yard?

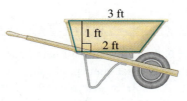

25. Building a Japanese garden. Mariko needs about sixteen wheelbarrows full of stone for her Japanese garden. Her wheelbarrow has vertical sides with the given shape and is $2\frac{1}{2}$ feet wide. How many cubic yards of stone should she order?

26. A gallon contains 231 cubic inches. A cylindrical barrel full of juice concentrate has a diameter of 2 feet and is 3 feet high. How many gallons of concentrate are in the barrel?

27. One cake is rectangular with a length of 14 inches, a width of 9 inches, and a height of 3 inches. The other cake is a round cake with a radius of 5 inches and a height of 4 inches. Which cake has more volume?

28. Which has more volume: a single scoop of ice cream made with a 3-inch scoop, or two scoops of ice cream made with a $2\frac{1}{2}$-inch scoop? (Assume that both ice cream scoops form perfectly filled spheres of ice cream.)

In Exercises 29–32, we describe a geometric object. Compute the volume you obtain if (a) you increase the height of the object by 2 inches and keep the radius the same and (b) you increase the radius of the object by 2 inches and keep the height the same.

29. A cylinder with radius 10 inches and height 5 inches

30. A cylinder with radius 4 inches and height 8 inches

31. A cone with radius 6 inches and height 3 inches

32. A cone with radius 5 inches and height 12 inches

33. Communicating Mathematics Reconsidering Exercises 29 and 30, which causes the volume of the cylinder to increase more: increasing its radius or its height? Explain why this is so.

34. Communicating Mathematics Reconsidering Exercises 31 and 32, which causes the volume of the cone to increase more: increasing its radius or its height? Explain why this is so.

35. Communicating Mathematics Assume that three tennis balls fit in a can with no room to spare. Which do you think is larger—the height of the can or the circumference of the can? Explain your answer.

36. Communicating Mathematics A spherical tank has a certain volume. If another spherical tank has twice the volume of the first tank, how do the two radii compare? Explain why this is so.

*37. Following the approach in Example 3, find the radius of a tin can that contains 2 cubic feet of liquid and has the smallest surface area.

38. Repeat Exercise 37, but now the can must contain 8 cubic feet.

39. Building an aviary. Luis plans to build a hummingbird aviary with a concrete floor shaped like a regular hexagon. Each side will measure 8 feet, as shown in the diagram, and the floor will be 6 inches deep. How many cubic feet of concrete are needed? Assume that Luis cannot buy a fraction of a cubic foot. (*Hint:* Divide the hexagon into equilateral triangles and use Heron's formula to find the area of the top surface of the patio.)

40. Constructing a patio. A homeowner is constructing a concrete patio that is 20 feet wide, 30 feet long, and 6 inches thick. Mixed concrete is sold by the cubic yard. How many cubic yards are needed for the project? Assume that the homeowner cannot buy a fraction of a cubic yard of mixed concrete. (*Hint:* You must convert cubic feet into cubic yards.)

41. Earth has a diameter of approximately 7,920 miles and a volume that is roughly 49 times the volume of the moon. Find the diameter of the moon.

42. The volume of Earth is roughly 6.7 times the volume of Mars. Find the diameter of Mars.

43. Communicating Mathematics If we double the radius of a right circular cone, what effect does that have on the volume? Explain your answer.

44. Communicating Mathematics If we double the radius of a sphere, what effect does that have on the volume? Explain your answer.

Further Exercises

45. Cooling drinks. Assume that ice cools a liquid proportionally to the amount of surface area of the ice that is in contact with the liquid. Determine whether a large ice cube or a number of smaller ice cubes whose total volume is equal to the larger cube will cool a drink faster. Give specific examples to support your claim.

46. Comparing hamburgers. A popular fast-food restaurant serves square hamburgers that are $4 \times 4 \times \frac{1}{4}$ inches. What should the radius of a round hamburger that is also $\frac{1}{4}$ inches thick be so that it has the same volume?

47. Comparing hamburgers. Another restaurant is serving a hamburger called the "tri-burger," which is shaped like an equilateral triangle and is $\frac{1}{4}$ inch thick. What should the length of its side be so that it has the same volume as the hamburgers in Exercise 46?

48. If we cut off the top of a cone by making a horizontal slice we get a figure called a *frustum* of a cone, as shown in the accompanying figure. Suppose that we cut off the top half of a cone that has radius r and height h. What is the volume of the remaining frustum? (*Hint:* The radius of the top of the frustum is $\frac{r}{2}$.)

FIGURE FOR Ex. 48

49. In Exercise 48, what is the surface area of the side of the frustum?

50. In Example 3, we determined the best shape for a cylinder that contains 1 cubic foot of liquid. Follow that example to determine the shape of a right circular cone with smallest surface area that contains 1 cubic foot. Estimate the radius to within one-tenth of a foot.

In this section we have found the volume of certain solids by multiplying the area of the base of the solid by the height of the solid. Imagine that you have a stack of thin sheets of paper that form a rectangular solid (see figure (a)). If you then gently push the stack so that its sides no longer form right angles with the base, it will look like figure (b).

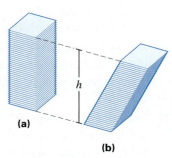

(a)

(b)

"Pushing" the solid to look like figure (b) has not changed the volume nor the height of figure (a). According to Cavalieri's principle, we can calculate the volume of figure (b) by multiplying the area of its base by its height.

51. Communicating Mathematics In what way is Cavalieri's principle similar to finding the area of a parallelogram?

8.5 THE METRIC SYSTEM AND DIMENSIONAL ANALYSIS

Would it surprise you if we were to tell you that the United States is an island? We don't mean to say that the United States is entirely surrounded by water, but we are surrounded by a world that uses a very different method for measuring length, volume, weight, and temperature. You are used to measuring length in inches, feet, yards, and miles, whereas the rest of the world uses centimeters, meters, and kilometers. We buy quarts of orange juice and gallons of gasoline but other people buy liters of these liquids. You may buy two pounds of hamburger or four ounces of Cajun seasoning, but outside of the United States, people would buy kilograms and grams of these products.

In this section, you will learn about the **metric** system of measurement that is officially known as the **Systèm International d'Unités** or the **SI system** and compare it with the U.S. system of measurement that is called the **U.S. customary system.**

In order to work in the metric system, you need to understand two things.

1. What are the basic units that we use to measure length, weight, and volume?
2. How do these different measures of length, volume, and weight relate to each other. As you will see, the relationship between different metric measures is much simpler than the relationship between different customary measures.

In the metric system, length measurement is based on the *meter*, which is slightly more than a yard. A tall man would be about two meters high. Volume is based on the *liter* which is a little more than a quart. A half gallon of milk would be roughly two liters. The basic unit of weight is the *gram*. A pound is just a little less than 500 grams. A 120-pound ice skater would weigh less than 60,000 grams or, 60 kilograms (a kilogram is 1,000 grams).

EXAMPLE 1   Estimating Metric Measurements

Give a rough estimate of each of the following using metric measurements.

a) the height of a door frame   b) a 10 pound bag of sugar

c) the length of a football field   d) 5 gallons of gasoline.

SOLUTION: a) The door frame is probably a little over 6 feet, which would be about 2 meters.

b) A pound is about 500 grams, so the sugar would weigh roughly 5,000 grams, or 5 kilograms.

c) Think of a yard as roughly a meter, so the length of the football field would be about 100 meters.

d) Five gallons of gasoline would be 20 quarts. A liter is approximately a quart, so the gasoline would be approximately 20 liters.

> **Quiz Yourself 15***
>
> Estimate each of the following using metric measurements.
> a) 1.5 gallons of lemonade
> b) length of your arm
> c) the weight of a 20-pound turkey

Units of measure in the metric system are based on powers of 10.

Although measurement in the metric system is based on meters, liters, and grams, we often use variations of these basic units to measure different quantities. For example, we might measure length in centimeters or kilometers, volume in milliliters, or weight in decigrams or kilograms. Table 8.5 explains the meaning of prefixes such as milli-, deka-, kilo-, centi-, and so on.

kilo- (k)	hecto- (h)	deka- (da)	base unit	deci- (d)	centi- (c)	milli- (m)
× 1,000	× 100	× 10		× 1/10 or × 0.1	× 1/100 or × 0.01	× 1/1,000 or × 0.001

TABLE 8.5   Some common metric prefixes.

---

* Quiz Yourself answers begin on page 849.

For example, a kilometer is 1,000 meters, a centigram is $\frac{1}{100}$ of a gram or 0.01 grams, and a milliliter is $\frac{1}{1,000}$ of a liter, or 0.001 liters. Conversion of units in the metric system is quite simple. From Table 8.5, we see that 1,000 of something is always a "kilo" and $\frac{1}{100}$ of something is always a "centi." The same pattern of prefixes works the same for length, volume, and weight. In the U.S. customary system, this is certainly not the case. Three feet equal 1 yard, but 4 quarts equal 1 gallon. There are 16 ounces in a pound, but only 2 cups in a pint. There is no consistency or uniformity.

We have listed the abbreviations for the metric prefixes in Table 8.5. Also, we use m for meter, g for gram, and l for liter. For example, to represent kilograms, we could use k for kilo and g for grams, thus we could write 10 kilograms as 10 kg. Similarly, we could write 25 milliliters as 25 ml.

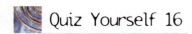 Quiz Yourself 16

a) One kiloliter is equal to how many liters?

b) One gram is equal to how many centigrams?

 PROBLEM SOLVING

It helps to remember the meaning of some metric prefixes if you connect them with some common words. For example, "centi" reminds you of cent which is one-one hundredth of a dollar. "Milli" might remind you of millennium which is 1,000 years. "Deci" is found in the word decimal and a decimal system is based on powers of 10.

If you remember the meaning of the prefixes in Table 8.5, it is easy to convert from one unit to the other in the metric system. For example, in measuring length, a kilometer is 10 times as long as a hectometer and a meter is 100 or 10^2 times as long as a centimeter. A millimeter is $\frac{1}{1,000}$ or 10^{-3} of a meter.

EXAMPLE 2   Converting Units of Measurement in the Metric System

Convert each of the following quantities to the unit of measurement that we specify.

a) 5 dekameters to centimeters

b) 2,300 milliliters to hectoliters

c) 2.5 kilograms to decigrams

SOLUTION: a) We can think of 5 dekameters as 5×10 meters, or 50 meters. Next, think of 50 meters as 50×10 decimeters, or 500 decimeters. Finally, represent each decimeter as 10 centimeters to get 500 decimeters = 500×10 centimeters = 5,000 centimeters.

We can do this more quickly if we recognize that each time we move one place to the right in Table 8.5, we require ten times the number of objects. That is,

1 dekameter = 10 meters = 10×10 decimeters = $10 \times 10 \times 10$ centimeters.

In other words, moving three places to the right in Table 8.5 requires us to multiply the amount of units by $10^3 = 1,000$. Note that because centimeters are much smaller than dekameters, we need many more centimeters than dekameters.

You can easily perform this calculation by moving the decimal point in 5.0 dekameters three places to the right to get 5,000 centimeters.

b) A quick way to solve this problem is to realize that because the prefix hecto is five places to the right of milli in Table 8.5, all we have to do to make the conversion is to move the decimal point in 2,300 milliliters five places to the left to get

$$0\underset{\frown}{.}2\underset{\frown}{3}\underset{\frown}{0}\underset{\frown}{0}\underset{\frown}{0}.$$

or, 0.023.

Quiz Yourself 17

a) 5.63 kiloliters is equal to how many deciliters?

b) 4,850 milligrams is equal to how many hectograms?

c) We will solve this problem using the information in Table 8.5 to move the decimal point. The prefix deci is four places to the right of kilo in Table 8.5; therefore, we will move the decimal point in 2.5 kilograms four places to the right to get 25,000 decigrams. It is good to check if this makes sense to you. Remember that decigrams are much smaller than kilograms, so therefore it requires many decigrams to equal 2.5 kilograms.

> **SOME GOOD ADVICE**
>
> Although it is tempting to simply memorize how to move the decimal point in doing conversions such as those we did in Example 2, it is important to *understand* why you are moving the decimal point and in which direction. If you want to make a number larger, you are multiplying by powers of 10, which means that you move the decimal point to the right. To make the number smaller, you move the decimal point to the left. In checking your answer, remember that it takes many small things to make one large thing and vice versa.

The meter is the basic unit of length in the metric system.

In 1790, the French Academy of Science defined the **meter** to be one ten-millionth of the distance from the North Pole to the equator—a distance of about 39.37 inches. Since that time, scientists have redefined the meter several times and currently the meter is the distance that light travels in a vacuum in $\frac{1}{299,792,458}$ of a second. A good way to visualize a meter is that it is slightly longer than a yard stick. Figure 8.57 shows the relationship between millimeters ($\frac{1}{1,000}$ of a meter), centimeters ($\frac{1}{100}$ of a meter), and inches.

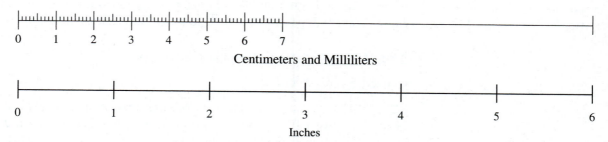

FIGURE 8.57   Comparing millimeters, centimeters, and inches.

From Figure 8.57, you can see that a millimeter is very small, about the thickness of a dime, and the centimeter is about the width of the body of a felt tip pen. One inch is about 2.5 cm, and a centimeter is roughly 0.4 inches. We often measure large distances in kilometers (1,000 meters). A kilometer is about 0.6 miles and a mile is approximately 1.6 km.

EXAMPLE 3   Estimating Metric Lengths

Match each of the following items with one of these metric measurements: 1 mm, 2 m, 3 cm, 10 m, 16 km.

a) The length of a 30-foot yacht

b) The thickness of the wire that is bent to make a standard paper clip

c) The length of a 10-mile race

d) The length of your bed

e) The thickness of this book

SOLUTION: a) Thirty feet equals 10 yards, which is about 10 m.

b) If you straighten the paper clip and place it in Figure 8.57, you will see that the paper clip is about 1 mm in diameter.

c) One mile is about 1.6 km, so a 10-mile race would be about 16 km long.

d) Your bed is probably a little more than 6 feet long, so it would be approximately 2 m long.

e) If you place the edge of your book on the diagram in Figure 8.57, you will see that the book is around 3 or 4 cm thick.   ◎

We use dimensional analysis to convert from one system to the other.

| 1 foot = 12 inches |
| 1 yard = 3 feet = 36 inches |
| 1 mile = 5,280 feet |

TABLE 8.6   Some relationships among units in the customary system.

Often we need to convert from customary units to metric units or vice versa. Before we explain how to convert from one system to the other, let us examine how we convert one unit of length to another within our customary system. Recall the relationships among various units of length, which we give in Table 8.6.

In order to convert one quantity to another, we will use unit fractions. A *unit fraction* is a quotient such as $\frac{3\ feet}{1\ yard}$ or $\frac{1\ foot}{12\ inches}$ that has different units of measurement in the numerator and denominator and whose value is equal to one. We will show you how to use unit fractions to make conversions in Example 4.

EXAMPLE 4   Converting Yards to Inches

Convert 5 yards to inches.

SOLUTION: We wish to have an answer that is in terms of inches instead of yards, so we will multiply by a unit fraction that has inches in the numerator and yards in the denominator, for example, $\frac{36\ inches}{1\ yard}$. Thus, $5\ yards \times \frac{36\ inches}{1\ yard} =$

1 inch = 2.54 centimeters
1 foot = 30.48 centimeters
1 yard = 0.9144 meters
1 mile = 1.6 kilometers

TABLE 8.7   Some basic relationships between customary and metric units of length.

5 ~~yards~~ $\times \frac{36 \text{ } inches}{1 \text{ } yard} = 5 \times 36 \text{ } inches = 180 \text{ } inches$. As you see, the unit "yard" cancels from the numerator and denominator leaving the answer expressed in inches.

You can use dimensional analysis to make conversions between the customary and metric systems, but first you need to know some basic relationships* between units of length in the two systems, which we give in Table 8.7.

From Table 8.7, we see that there are a number of unit fractions that we can use to make conversions such as $\frac{1 \text{ } inch}{2.54 \text{ } centimeters}$ and $\frac{0.9144 \text{ } meters}{1 \text{ } yard}$.

EXAMPLE 5   Converting Between the Metric and Customary Systems

Make each of the following conversions.

a) 420 miles to kilometers

b) 5 yards to centimeters

c) 0.3 kilometers to feet

SOLUTION: a) To make this conversion, we can use the unit fraction $\frac{1.6 \text{ } kilometers}{1 \text{ } mile}$. Therefore,

$$420 \text{ } miles \times \frac{1.6 \text{ } kilometers}{1 \text{ } mile} = 420 \text{ } miles \times \frac{1.6 \text{ } kilometers}{1 \text{ } mile}$$
$$= 420 \times 1.6 \text{ } kilometers = 672 \text{ } kilometers.$$

b) To make this conversion, we need a unit fraction that will convert yards to meters and then another to convert meters to centimeters.† Multiplying by the unit fraction $\frac{0.9144 \text{ } meters}{1 \text{ } yard}$ will convert yards to meters and because 1 meter = 100 centimeters, we can use the unit fraction $\frac{100 \text{ } centimeters}{1 \text{ } meter}$ to finish the conversion. Thus,

$$5 \text{ } yards = 5 \text{ } yards \times \frac{0.9144 \text{ } meters}{1 \text{ } yard} \times \frac{100 \text{ } centimeters}{1 \text{ } meter}$$
$$= 5 \times 0.9144 \times 100 \text{ } centimeters = 457.2 \text{ } centimeters.$$

c) We will solve this problem similar to the way we solved part b). We need one unit fraction, $\frac{1 \text{ } mile}{1.6 \text{ } kilometers}$, to convert kilometers to miles and then another, $\frac{5,280 \text{ } feet}{1 \text{ } mile}$, to convert miles to feet. Therefore,

$$0.3 \text{ } kilometers = 0.3 \text{ } kilometers \times \frac{1 \text{ } mile}{1.6 \text{ } kilometers} \times \frac{5,280 \text{ } feet}{1 \text{ } mile}$$
$$= \frac{0.3 \times 5,280}{1.6} = 990 \text{ } feet.$$

Quiz Yourself 18

Convert 0.65 kilometers to feet.

---

* You can find more accurate approximations for these relationships in many science books and encyclopedias.
† Of course, we could convert meters to centimeters by moving the decimal point as we did in Example 2. Either approach will give you the same answer.

The liter is the basic unit of volume in the metric system.

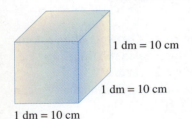

FIGURE 8.58 One liter = 1 cubic decimeter = 1,000 cubic centimeters.

The **liter** is defined to be one cubic decimeter (see Figure 8.58). That is, a liter is the volume of a cube that measures one decimeter (or 10 centimeters) on each side. From Figure 8.58, you can see that a liter is $10 \times 10 \times 10 = 1,000$ cubic centimeters.

As we mentioned earlier, a liter is slightly more than a quart. We can use the same prefixes (kilo, hecto, deka, and so on) that we used in measuring length. We use a milliliter, $\frac{1}{1,000}$ of a liter, to measure small quantities such as a dose of medicine. If you go to a clinic to get a flu shot, you might receive 2 milliliters of vaccine. We measure large volumes in kiloliters (1,000 liters). Figure 8.59 shows that a kiloliter is the same as a cubic meter. Each layer in Figure 8.59 contains $10 \times 10 = 100$ cubic decimeters, and because there are 10 layers, we have 1,000 cubic decimeters.

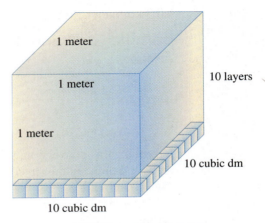

FIGURE 8.59 Representing a cubic meter in terms of liters.

In Canada, you might order 20 cubic meters of cement to build a patio floor. We also use cubic centimeters, cm^3 or cc, to measure dry materials.

EXAMPLE 6 Estimating Metric Volumes

Estimate the volume of each of the following items with one of these metric measurements: 3 ml (milliliters), 250 ml (milliliters), 100 l (liters), 500 m³ (cubic meters), 1 cubic decimeter (dm³).

a) The gas tank of car

b) A large pile of gravel at a construction site

c) A glass of soda

d) An injection of flu vaccine

e) A quart of orange juice

SOLUTION: a) 100 liters is about 100 quarts, or 25 gallons, which is the size of the gas tank of a medium-sized car.

b) 500 cubic meters is about 500 cubic yards which is the size of a fairly good-sized pile of gravel.

c) A glass of soda would be about $\frac{1}{4}$ of a quart. Since a quart is roughly equal to a liter, and a liter is 1,000 milliliters, the glass of soda has a volume of about $\frac{1,000}{4} = 250$ milliliters.

d) As we mentioned earlier, an injection is a small amount of liquid, so 3 milliliters would be a good estimate.

e) A quart is roughly a liter and a liter is 1 cubic decimeter, so the volume of the orange juice is about 1 cubic decimeter.

We make conversions among units of volume in the metric system in exactly the same way we made conversions among units of length.

EXAMPLE 7   Estimating the Amount of Vaccinations

The East Side Community Health Clinic has on hand four bottles that each contain 0.35 l of flu vaccine. Each vaccination requires 2 ml of vaccine. How many patients can be treated with the amount of vaccine that the clinic has on hand?

SOLUTION: The total amount of vaccine is $4 \times 0.35 = 1.4$ l. From Table 8.5, we see that to convert liters to millimeters, we can move the decimal point three places to the right. Therefore the clinic has 1,400 ml vaccine on hand, so the clinic can give $\frac{1,400}{2} = 700$ vaccinations.

You can use dimensional analysis to make conversions between units of volume in the metric and customary systems, but first you need to know some basic relationships, which we give in Table 8.8.

2 cups = 1 pint	1 cup = 0.2366 liters
2 pints = 1 quart	1 quart = 0.9464 liters
32 fluid ounces = 1 quart	1 cubic foot = 0.03 cubic meters
4 quarts = 1 gallon	1 cubic yard = 0.765 cubic meters

TABLE 8.8 Relationships between units of volume in the metric and customary system.

You can now use these relationships to define unit fractions to make conversions from one system to the other as we did earlier in this section.

EXAMPLE 8   Calculating the Volume of a Car's Gas Tank

Assume that an American car's gas tank has a capacity of 23 gallons. Suppose that this same car were in France, how many liters of gasoline would it take to fill the tank?

SOLUTION: To solve this problem, we will find two unit fractions—one to convert gallons to quarts and a second to convert quarts to liters. The unit fraction $\frac{4\ quarts}{1\ gallon}$ will convert gallons to quarts and the fraction $\frac{0.9464\ liters}{1\ quart}$ converts quarts to liters. Thus

$$23\ gallons = 23\ \cancel{gallons} \times \frac{4\ \cancel{quarts}}{1\ \cancel{gallon}} \times \frac{0.9464\ liters}{1\ \cancel{quart}}$$
$$= 23 \times 4 \times 0.9464 = 87.07\ liters.$$

So you would need to purchase 87.07 l of gasoline.

The gram is the basic unit of mass (weight) in the metric system.

The last type of measurement we will consider is mass, which, for simplicity sake, you might think of as weight. Strictly speaking, mass and weight are not the same thing. The mass of an object depends on its molecular makeup and does not change. The weight of an object, however, depends on the gravitational pull on that object. Gravity is stronger on a large planet and weaker on a small planet. Because the moon's gravity is only $\frac{1}{6}$ of Earth's gravity, an elephant that weighs 2,400 pounds on Earth would weigh only 400 pounds on the moon. However, the mass of the elephant would be the same on both Earth and the moon.

The basic unit of mass in the metric system is the **gram,** which is defined to be the mass of one cubic centimeter (one milliliter) of water at a certain specified temperature and pressure. A liter (roughly one quart) is 1,000 ml, so a liter of water has a mass of 1,000 g or 1 kg. It is common in the metric system to measure the mass of large object in kilograms, which are approximately 2.2 pounds. So, if you were having a barbecue in Spain and wanted to broil about 8 or 9 quarter-pounders, you would go to the market and buy 1 kg of ground beef.

The pattern of prefixes, and abbreviations, and also the rules for making conversions are the same for mass as they were for length and volume.

EXAMPLE 9 Estimating Mass in the Metric System

Estimate the mass of each of the following items with one of these metric measurements: 1 g, 50 cg, 5 kg, 1,000 kg

a) A decent-sized steak

b) A large bag of sugar

c) A medium-sized car

d) A large paper clip

SOLUTION: a) The steak might weigh a pound or more. A kilogram is 2.2 pounds, so the steak would be about one half of a kilogram, or 50 cg.

b) Typically, we buy sugar in 10-pound bags. Five kg would be equal to 5 × 2.2 = 11 pounds, so the sugar would weigh a little less than 5 kg.

c) A car might weigh 2,200 pounds, which is 1,000 kg.

d) A paper clip is pretty light, so its weight might be about 1 g.

In order to make mass conversions between the metric and customary systems, you can use the information in Table 8.9.

16 ounces = 1 pound	1 ounce = 28 grams
2,000 pounds = 1 ton	2.2 pounds = 1 kilogram
	1 metric tonne = 1,000 kilograms
	1.1 tons = 1 metric tonne

TABLE 8.9   Relationships between units of mass in the metric and customary systems.

Again, we can use dimensional analysis to convert units in one system to units in the other.

EXAMPLE 10   Refurbishing a Church with Italian Marble

A contractor, who is refurbishing a church, has ordered slabs of Italian marble that weigh 1.8 metric tonnes each. If he wishes to rent a crane to lift the slabs of marble, in pounds, how much should the cranes be able to lift?

SOLUTION: We will convert metric tonnes to tons and then tons to pounds. The unit fraction $\frac{1.1\ tons}{1\ metric\ tonne}$ converts metric tonnes to tons and the unit fraction $\frac{2,000\ pounds}{1\ ton}$ will convert tons to pounds. Therefore,

$$1.8\ metric\ tonnes = 1.8\ \cancel{metric\ tonnes} \times \frac{1.1\ \cancel{tons}}{1\ \cancel{metric\ tonne}} \times \frac{2,000\ pounds}{1\ \cancel{ton}}$$
$$= 1.8 \times 1.1 \times 2,000 = 3,960\ pounds$$

A crane that can lift 4,000 pounds would just barely do the job.

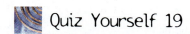

Quiz Yourself 19

Convert 3.8 kg to ounces.

You will find the following equivalents between the Metric and Customary systems useful in doing the exercises.

1 meter = 1.0936 yards
1 mile = 1.609 kilometers
1 pound = 454 grams
1 metric tonne = 1.1 tons
1 liter = 1.0567 quarts*

* Note that we are using this equivalence in the exercises, rather than using the equivalence 1 quart = 0.9464 liters that we stated earlier.

Exercises 8.5

In Exercises 1–10, match the highlighted words in the column in the left with the metric measurement in the column on the right that corresponds to it. Before calculating your final answer, it would be wise to first convert the given measurement to basic metric units such as meters, grams, or liters.

1. A *quarter-pound* hamburger   a) 0.02159 dam

2. A *gallon* of milk   b) 946.4 ml

3. A *15-foot* tall giraffe   c) 378.5 cl

4. *Five pounds* of potatoes   d) 1.77 dl

5. A *6-inch*-long ruler   e) 4,572 mm

6. A *quart* of motor oil   f) 0.0061 km

7. A book *eight and one-half inches* wide g) 15.24 cm

8. A *12-ounce* gerbil   h) 1,135 dg

9. *Six ounces* of orange juice   i) 3.405 hg

10. A *20-foot*-long swimming pool   j) 2.27 kg

Make the indicated conversions in Exercises 11–22.

11. 2.4 kiloliters to deciliters

12. 240 centigrams to dekagrams

13. 3.45 kilometers to hectometers

14. 4,300 milliliters to liters

15. 85 dekagrams to decigrams

16. 28 decimeters to millimeters

17. 1.8 kilograms to grams

18. 3 kiloliters to centiliters

19. 4,500 millimeters to meters

20. 5.6 hectograms to centigrams

21. 3.5 dekaliters to deciliters

22. 7,600 centimeters to meters

23. Communicating Mathematics   In converting milligrams to dekagrams, we would move the decimal point either four places to the right or to the left. Which is it? Explain how you would help a fellow classmate remember which to do.

24. Communicating Mathematics   If *a* kiloliters equals *b* dekaliters, which is larger, *a* or *b*? Explain how you arrived at your answer.

25. Communicating Mathematics   Explain how you might remember the difference between the meanings of the prefixes "centi" and "deci."

26. Communicating Mathematics   Explain why making conversions in the metric system is easier than making conversion in the U.S. customary system. Give several concrete examples.

Pick the most appropriate measurement for each of the following items. Explain your answer.

27. The length of a wooden pencil;

   a) 0.5 m   b) 18 cm   c) 20 mm

28. The volume of a small juice glass

   a) 125 ml   b) 500 ml   c) 2.5 l

29. The weight of books in your back pack

   a) 500 g   b) 80 hg   c) 6 kg

30. The length of your nose

   a) 4 dm   b) 3 mm   c) 5 cm

31. The volume of a bottle of wine

   a) 0.25 l   b) 0.2 kl   c) 750 ml

32. A quarter pound of hard salami

   a) 200 g   b) 0.11 kg   c) 5 dg

33. The height of a medium-sized dog

   a) 100 mm   b) 0.06 dam   c) 0.6 hm

34. The weight of an NFL lineman

   a) 136 kg   b) 13,000 g   c) 20 dag

Rewrite each statement, replacing the metric measure by the corresponding customary measure.

35. Hold your ground. Don't give him 2.54 centimeters.

36. Tex was wearing a 9.5 liter hat.

37. It is first down and 9.14 meters to go.

38. 28 grams of prevention is worth 0.45 kilograms of cure.

Use dimensional analysis* to make each of the following conversions. You may have to define several unit fractions to make the conversions. Also, because the constants that we give in the table are only approximations, depending on the

---

* In some cases, you may find it quicker not to use dimensional analysis to get your answers.

method with which you do your computations, your answers may differ slightly from ours.

39. 1,460 grams to pounds   **40.** 18 meters to feet

41. 27 gallons to liters   **42.** 3 kilograms to ounces

43. 4 yards to centimeters   **44.** 240 ounces to grams

45. 10,000 milliliters to quarts

46. 2,176 inches to hectometers

47. 514 decimeters to yards   **48.** 2.1 kiloliters to gallons

49. 47 pounds to kilograms

50. 10,000 deciliters to quarts

51. 507,820 milligrams to pounds

52. 176 centimeters to inches

53. 480 dekagrams to pounds

54. 3 yards to millimeters

55. 3,500 cm to yards   **56.** 4,200 ml to gallons

57. 45,000 kg to tons   **58.** 0.65 tonnes to pounds

59. 0.24 gallons to liters   **60.** 2.6 feet to decimeters

61. 400 yards to dekameters   **62.** 10 tons to tonnes

63. A rectangular tank is 5 m wide, 8 m long, and 4 m deep.

a) What is the volume of the tank in cubic meters?

b) How many liters of water does the tank contain?

c) What is the weight of the water in kilograms?

64. If a rectangular wall of a gymnasium that measures 2.5 m by 30.8 m requires 8 liters of paint, then how much paint would be needed to cover a wall that measures 3.2 m by 26.4 m?

In Exercises 65–68, use the fact that a hectare (pronounced "HEKtaire") is a square that measures 100 meters on each side.

65. Communicating Mathematics   What is the relationship between hectares and square meters?

66. Communicating Mathematics   What is the relationship between hectares and square kilometers?

67. Thiep purchased a rectangular piece of land that measures 0.75 km by 1.2 km. How many hectares of land is that?

68. If Anna purchased a rectangular piece of land that contains 40 hectares and is 0.65 km long, how long is the other dimension?

69. Finding the volume of a swimming pool. Adrian's rectangular swimming pool is 20 feet wide, 40 feet long, and averages 6 feet in depth. How many kiloliters of water are needed to fill the pool?

70. Buying canned food. Marco purchased a large can of chili that has a diameter of 10 centimeters and a height of 14 centimeters. Determine the volume of the chili in the can

a) in liters

b) in ounces.

71. Purchasing fruit in Europe. While visiting Poland, Anthony purchased some red plums that cost $2.75 per kilogram. What is the cost of the plums per pound?

72. Measuring a medication. Justin is taking the anti-inflammatory drug Niamoxin and the instructions say that the patient must receive 10 mg for each 15 kilograms of body weight. If Justin weighs 275 pounds, what dosage should he receive?

73. Fencing a dog pen. Serina wishes to fence in a rectangular exercise pen for her golden labrador retriever. The pen is 35 feet wide and 62 feet long. The fencing is sold in whole meters. How much fencing should she buy?

74. In Exercise 73, what is the area of Serina's pen in square meters?

75. Calculating gas mileage. If a car averages 30 miles per gallon, how many kilometers per liter would that be?

76. Calculating gas mileage. If a car averages 15 kilometers per liter, how many miles per gallon would that be?

In the metric system, temperatures are measured using the Celsius (also called centigrade) thermometer instead of the Fahrenheit that we commonly use. On the Celsius scale, water freezes at 0 degrees and boils at 100 degrees. In order to convert a temperature using one thermometer to a temperature using the other, you can use the following equation*

$$F = \frac{9}{5}C + 32,$$

---

* This equation converts Celsius degrees to Fahrenheit degrees. You can also use it (plus algebra) to convert Fahrenheit to Celsius, which we recommend that you do. If you really want another equation to convert Fahrenheit to Celsius, you can use the equation $C = \frac{5}{9}(F - 32)$. However, we strongly recommend that you memorize only one equation and use algebra to do the second conversion.

where F is the Fahrenheit temperature and C is the Celsius temperature. Use this equation in Exercises 77–84 to convert each temperature to a corresponding measurement in the other system.

77. 149 degrees Fahrenheit   **78.** 95 degrees Fahrenheit

79. 60 degrees Celsius   **80.** 85 degrees Celsius

81. 20 degrees Celsius   **82.** 131 degrees Fahrenheit

83. 113 degrees Fahrenheit   **84.** 50 degrees Celsius

Further Exercises

85. Communicating Mathematics Explain the advantages that you see in using the metric system over the customary system. Be specific. Give concrete examples.

86. Communicating Mathematics Why are conversions easy to do in the metric system? Give specific examples.

87. Communicating Mathematics Look up the definition of a cubit. Explain what it means.

88. Communicating Mathematics Look up the definition of a furlong. Explain what it means.

89. Communicating Mathematics Research the history of the definition of an inch. Explain the results of your research.

90. Communicating Mathematics Research the history of the definition of a yard. Explain the results of your research.

91. Communicating Mathematics Research the definition of avoirdupois and troy weight. Use your research to explain why a pound of silver is not the same as a pound of steak.

92. Communicating Mathematics In 1870, Jules Verne wrote his famous novel, *20,000 Leagues Under the Sea.* What is a league? If the submarine in Jules Verne's book traveled a distance of 20,000 leagues under water, how far did it travel? Give your answer in feet.

8.6 GEOMETRIC SYMMETRY AND TESSELLATIONS

Symmetry is a property that is often found in art, music, architecture, and literature. An artist may be enthralled by the beautiful reflection of a swan swimming on the smooth surface of a lake or by the brilliant patterns of a stained-glass window. In analyzing music, the musician notices how the recurrence of a theme balances its introduction earlier in the work. A writer may use symmetrical imagery and sound patterns in creating a poem. Biologists describe the symmetry of a butterfly's wings or the symmetry of the markings on a hawk. Psychologists have found that perceptions of beauty are influenced by the symmetry of our faces with respect to the placement of our eyes and the shape of our mouth.

Although you no doubt have an intuitive idea of what we mean by symmetry, it is not an easy concept to describe. In mathematics, we begin with an intuitive concept, such as symmetry, and define it precisely so that we can measure it and calculate with it. For example, in Figure 8.60, it is easy to see that the star has more symmetry than the arrowhead. But exactly what do we mean when we say that?

To understand symmetry, imagine that the arrowhead is made of a very stiff wire and is resting in shallow grooves cut into a piece of wood. Now suppose that you close your eyes and a friend picks up the arrowhead, flips it over, and returns

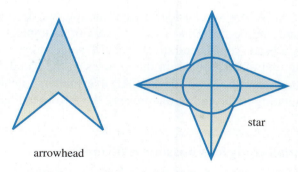

FIGURE 8.60   Objects possessing different amounts of symmetry.

it to its resting place. Figure 8.61(a) shows the arrowhead before the flip, and Figure 8.61(b) shows the arrowhead after the flip. Although many points on the arrowhead have been moved, the overall appearance of the arrowhead remains unchanged. When you open your eyes, you would not know whether the arrowhead was flipped. Other than this flip, there is really nothing else that we can do to the arrowhead that will move individual points, yet keep the overall appearance of the object the same.

On the other hand, there are many different ways to move the star that will change the position of individual points, but leave the overall appearance of the star the same. Figure 8.62 shows two possibilities. In Figure 8.62(b), we have rotated the star in a counterclockwise direction, so that A moves to the bottom of the star, B moves to what was A's position, C moves to what was B's position, and so on. In Figure 8.62(c), we have flipped the original star over, interchanging vertices A and C.

Figures 8.61 and 8.62 show examples of rigid motions.

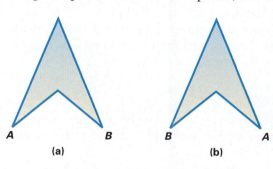

FIGURE 8.61   The arrowhead has been flipped over and returned to its resting place.

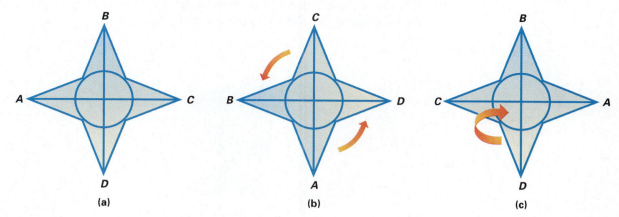

FIGURE 8.62   Two transformations of the star.

When we consider a rigid motion, we are interested only in the beginning and ending position of the object. For example, suppose that in flipping the arrowhead, your friend drops it and it rolls across the floor. If your friend retrieves it, and places it back in the grooves in the wood, most of what has happened to the arrowhead is irrelevant. All that matters mathematically is that from the beginning to the end of the process, the arrowhead has been flipped over.

There are essentially only four possible rigid motions.

You might think that there are many different ways that we could perform rigid motions; however, this is not the case.

> Every rigid motion is essentially a reflection, a translation, a glide reflection, or a rotation.

By the word *essentially*, we mean that if we consider only the beginning and ending positions of the object, and ignore the intermediate motions, the rigid motion can be accomplished in only one of four ways.

We first discuss reflections.

> **DEFINITION**
>
> A **reflection** is a rigid motion in which we move an object so that the ending position is a mirror image of the object in its starting position.

In a reflection, there is a line, called the *axis of reflection*, that acts as a mirror to transform the figure from its original position to its final position. We say that the original figure has been *reflected about the axis of reflection* to produce the final figure. We show two reflections in Figure 8.63.

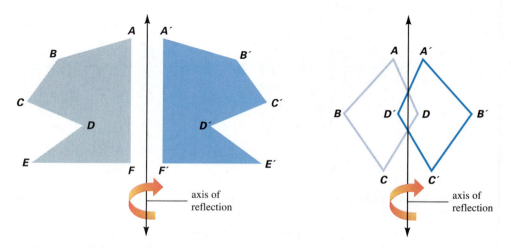

FIGURE 8.63 Original figures *ABCDEF* and *ABCD*; reflected figures *A′B′C′D′E′F′* and *A′B′C′D′*.

As in Figure 8.63, we label the vertices of the original polygons with *A, B, C,* and so on, and the vertices of the reflected polygons with *A′, B′, C′,* and so on.

EXAMPLE 1   Reflecting a Geometric Object

Reflect polygon *ABCDE* about axis of reflection *l* in Figure 8.64.

SOLUTION: In order to reflect the polygon, we must reflect each vertex *A, B, C, D,* and *E* of the polygon about *l*. We do this by drawing a line segment from *A* to point *A′* so that segment *AA′* is perpendicular* to *l* and also so that the distance from *A* to *l* is equal to the distance from *A′* to *l*. Then draw segments *BB′*, *CC′*, *DD′*, and *EE′* in a similar way, as shown in Figure 8.65.

We draw the reflected polygon by connecting vertices *A′*, *B′*, *C′*, *D′*, and *E′*, as shown in Figure 8.66.

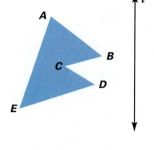

FIGURE 8.64   Polygon *ABCDE* is to be reflected about line *l*.

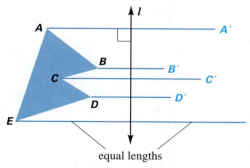

FIGURE 8.65   Reflecting points *A, B, C, D,* and *E* about *l*.

Quiz Yourself 20†

Reflect the quadrilateral *ABCD* about line *l*.

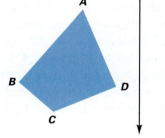

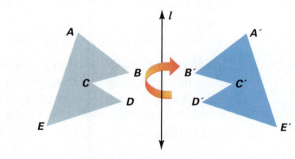

FIGURE 8.66   Reflecting polygon *ABCDE* about line *l*.

DEFINITION

A **translation** is a rigid motion in which we move a geometric object by sliding it along a line segment in the plane. The direction and length of the line segment completely determine the translation. We represent the distance and direction of a translation by a line segment with an arrow on it, called the *translation vector*.

---

* We have not explained how to construct a line that is perpendicular to another line. You can construct reasonably good perpendicular lines freehand that will give you an idea of what a reflection looks like.
† Quiz Yourself answers begin on page 849.

Figure 8.67 shows a translation.

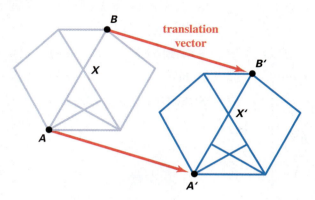

FIGURE 8.67 A translation vector determines a translation of an object.

DEFINITION

A **glide reflection** is a rigid motion formed by performing a translation (the glide) followed by a reflection.

EXAMPLE 2   **Producing a Glide Reflection of a Geometric Object**

Use the translation vector and the axis of reflection to produce a glide reflection of the object shown in Figure 8.68.

SOLUTION: First we must slide the object in the direction of the translation vector and for the length of the translation vector. To do this we place a copy of the translation vector at point Y on the object (Figure 8.69(a)). Next we slide the object along the translation vector so that point Y coincides with the tip of the translation vector (Figure 8.69(b)). Finally, we reflect the object about the axis of reflection to get the final object (Figure 8.69(c)). In Figure 8.69(d) we see the effect that the glide reflection has in moving the original polygon to the polygon containing the vertices X'', Y'', and Z''.

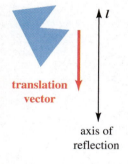

FIGURE 8.68 Forming a glide reflection of an object.

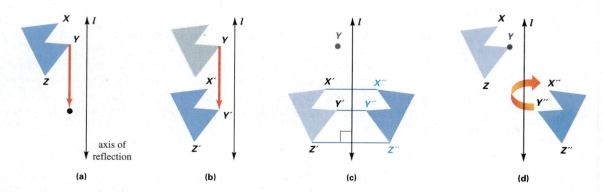

FIGURE 8.69 A glide reflection.

Quiz Yourself 21

Perform a glide reflection by first translating the triangle using the given translation vector, then reflecting the translated triangle about the axis of reflection.

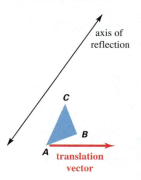

PROBLEM SOLVING

The Order Principle in Section 1.1 tells you to be careful about the order in which you perform a glide reflection. Performing a translation and then a reflection is not the same as performing a reflection followed by a translation.

The last rigid motion we will study is a rotation.

DEFINITION

We perform a **rotation** by first selecting a point, called the *center of the rotation*, and then, while holding this point fixed, we rotate the plane about this point through an angle called the *angle of rotation*.

A good way to think of a rotation is to imagine that the plane is a piece of paper. If you stick a pin in the plane at the center of rotation and then rotate the plane, the plane will turn about the pin and all points in the plane will move except the point where the pin is placed. We show a rotation in Figure 8.70.

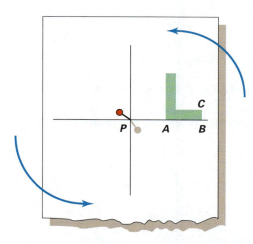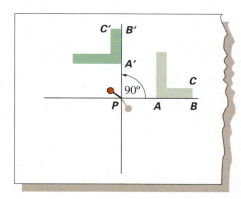

FIGURE 8.70   A rotation of 90° about point *P*.

A symmetry is a rigid motion that leaves the overall appearance of an object unchanged.

We can now define the concept of symmetry using the idea of a rigid motion.

DEFINITION

A **symmetry** of a geometric object is a rigid motion such that the beginning position and the ending position of the object by the motion are exactly the same.

Looking back at Figure 8.62, we can say that the rotation and flip are two examples of symmetries of the star. The rotation is of course a rotation as we defined earlier, while the flip is a reflection.

EXAMPLE 3   Symmetries of a Star

Find two symmetries (other than those shown in Figure 8.62) of the star in Figure 8.71.

SOLUTION: There are clearly many ways to reflect and rotate the star to produce symmetries of the star. For example, in Figure 8.72, we reflect the star about the line l. We call l a *line of symmetry* for the star.

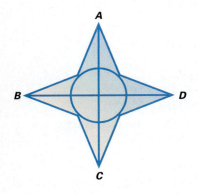

FIGURE 8.71   The star has many symmetries.

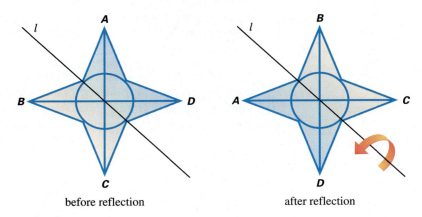

before reflection

after reflection

FIGURE 8.72   Reflection of the star about line l.

We can also rotate the star about its center, as shown in Figure 8.73.

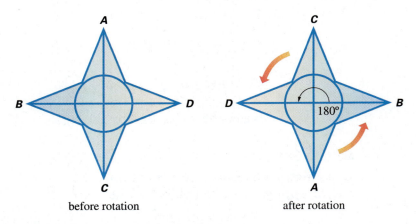

before rotation

after rotation

FIGURE 8.73   Rotation of the star about its center through an angle of 180°.

Let's return to our question about measuring the symmetry in the arrowhead and the star. In Figure 8.74, we see that the arrowhead has only two symmetries: a rotation about A through an angle of zero degrees (this symmetry leaves every point in the arrowhead unchanged) and a reflection about the vertical line through points *A* and *C*. The star also has these two symmetries as well as many more, as we saw in Example 3.

It can be shown that there are eight symmetries for the star. Using the language of set theory, we say that the set of symmetries of the arrowhead is a proper subset* of the set of symmetries of the star. It is in this sense that the star has greater symmetry than the arrowhead.

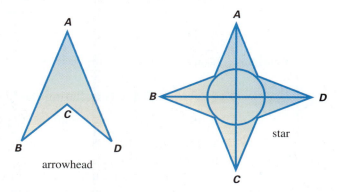

FIGURE 8.74   The star has more symmetries than the arrowhead.

Tessellations cover the plane with copies of polygons.

We can use rigid motions to place copies of a geometric figure at different positions in the plane. An interesting mathematical question is, "Can we begin with a set of polygons and then completely cover the plane with copies of these polygons?"

> **DEFINITIONS**
>
> A **tessellation** (or *tiling*) of the plane is a pattern made up entirely of polygons that completely covers the plane. The pattern must have no holes or gaps, and polygons cannot overlap except at their edges. A **regular tessellation** consists of regular polygons of the same size and shape such that all vertices of the polygons touch other polygons only at their vertices.

A designer of wallpaper, a fabric pattern, or a company logo should know that equilateral triangles, squares, and regular hexagons tessellate the plane, as we see in Figure 8.75 on the following page.

It is natural to ask, "Are there any other regular polygons that tessellate the plane?" To answer this question, recall that in Section 8.2 we stated that the sum of the measures of the interior angles of an *n*-sided convex polygon is $(n - 2) \times 180°$. Thus, for a regular *n*-sided polygon, the measure of each interior angle is $\frac{(n - 2) \times 180°}{n}$. For example, the measure of each interior angle of a regular 12-sided polygon is $\frac{(12 - 2) \times 180°}{12} = \frac{1,800°}{12} = 150°$. We can use this fact to determine which other regular polygons tessellate the plane.

---

* We discussed subsets in Section 1.4.

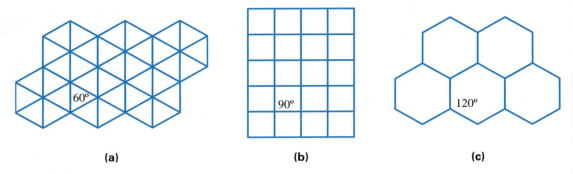

(a)   (b)   (c)

FIGURE 8.75   Regular tessellations of the plane using (a) triangles, (b) squares, and (c) hexagons.

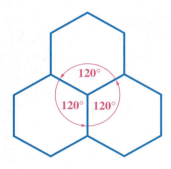

FIGURE 8.76   The angle sum around a vertex in a tessellation must add up to 360°.

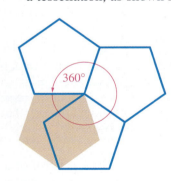

FIGURE 8.77   There is no regular tessellation of the plane using pentagons.

EXAMPLE 4   Tessellating the Plane with Regular Polygons

Which regular polygons can tessellate the plane?

SOLUTION: To understand the problem, let us look at a part of a tessellation using hexagons (Figure 8.76).

Notice that around a vertex of any regular tessellation, we must have an angle sum of 360°, and we also must have three or more polygons of the same shape and size. In Figure 8.75(a), we see that each vertex is surrounded by six 60° angles; in Figure 8.75(b), each vertex is surrounded by four 90° angles.

Let us now consider if it is possible to have a regular tessellation of the plane using pentagons. Interior angles of a regular pentagon measure $\frac{(5 - 2) \times 180°}{5} = \frac{540°}{5} = 108°$. If we have three pentagons surrounding the vertex of a tessellation, as shown in Figure 8.77, the sum of the angles around the vertex is $3 \times 108° = 324° < 360°$. We do not have enough pentagons to completely surround the vertex; on the other hand, if we include a fourth pentagon, the sum of the angles around the vertex exceeds 360°, which causes the pentagons to overlap. Thus there cannot be a regular tessellation using pentagons.

If a regular polygon has more than six sides, the measure of its interior angles exceeds 120°; therefore, we cannot have three such polygons surrounding a vertex of a tessellation.

Thus we conclude that we can construct only three regular tessellations. They use equilateral triangles, squares, or regular hexagons.   ◎

Although there are only three regular tessellations, many tessellations are not regular. Is it possible to construct a tessellation using two different types of polygons that have edges of the same length?

EXAMPLE 5   Decorating with Nonregular Tessellations

An artist specializes in producing decorative tessellations using ceramic tiles. At present she has a number of tiles shaped like equilateral triangles and squares. All tiles have sides of the same length. Is it possible to produce a tessellation using a combination that contains both types of tiles?

SOLUTION: We can solve this problem by considering the different possibilities systematically. We will investigate tessellations that have one square at a vertex, two squares at a vertex, and so on.

The angle sum around each vertex of a tessellation is 360°. Because the corners of the squares occupy some of the 360°, the angles of the triangles will occupy the remainder. After we have subtracted the amount of the 360° used by the squares, the remainder of the 360° must be evenly divisible by 60°, which is the measure of each angle of an equilateral triangle. We organize our calculations in Table 8.10.

Number of Squares at a Vertex	Amount of 360° Used by the Squares	Amount of 360° Remaining for the Triangles	Does 60° Divide This Number?	Is This Configuration Possible?
1	90°	270°	No	No
2	180°	180°	Yes	Yes
3	270°	90°	No	No
4	360°	0°	Yes	No (there will be no triangles)

TABLE 8.10   Number of possible equilateral triangle square combinations possible at the vertex of a tessellation.

We see in Table 8.10 that the only combination of equilateral triangles and squares resulting in a total angle sum at each vertex of 360° occurs when we have two squares and three triangles. Such a configuration at a vertex is shown in Figure 8.78(a). We show a larger part of the tessellation in Figure 8.78(b).

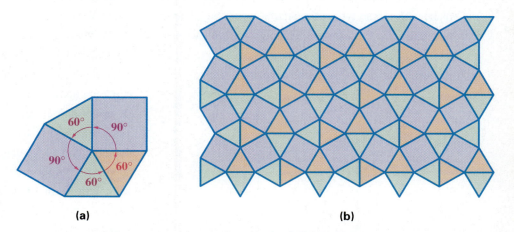

(a)   (b)

FIGURE 8.78   Two squares and three equilateral triangles around the vertex of a tessellation.

Quiz Yourself 22

Explain why the following tessellation is possible by considering the angles at each vertex in the tessellation.

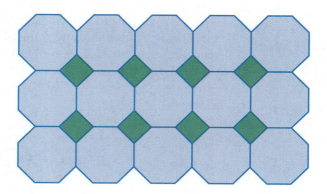

 Highlight: *Has All of Mathematics Been Discovered?*

It may surprise you to know that mathematicians are still discovering new facts about the topics you studied in this section. You have seen that it is possible to tessellate the plane with certain types of regular polygons and nonregular polygons. For example, although we cannot tessellate the plane with regular pentagons, we can tessellate the plane with the following nonregular pentagon.

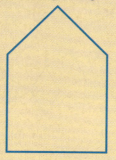

A simple question that a mathematician then asks is, "Exactly what types of pentagons can be used to tessellate the plane?" At one time it was thought that there were only eight such pentagons. However, in 1975, Marjorie Rice, a woman with only a high school background in mathematics, found a ninth pentagon that would tessellate the plane. In subsequent years, by applying her new methods to this problem, she went on to discover four more pentagonal tessellations. At this time, mathematicians do not know how many different types of convex pentagons will tile the plane. The example of Marjorie Rice is inspiring in that a person with little formal training in mathematics, but having interest and talent in mathematics, can make a significant contribution.

Exercises 8.6

Use the following figures for Exercises 1–4. You may want to use graph paper to do these exercises.

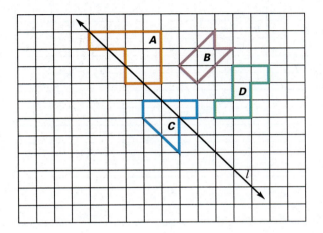

1. Reflect figure *A* about line *l*.
2. Reflect figure *B* about line *l*.
3. Reflect figure *C* about line *l*.
4. Reflect figure *D* about line *l*.

Use the following figure for Exercises 5–6.

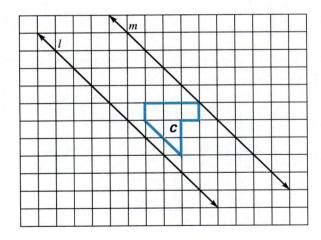

5. Reflect figure *C* about line *l*. Call the resulting figure *C'*. Then reflect figure *C'* about line *m*. Call the resulting figure *C"*.

6. Reflect figure *C* about line *m*. Call the resulting figure *C'*. Then reflect figure *C'* about line *l*. Call the resulting figure *C"*.

Use this figure for Exercises 7–8.

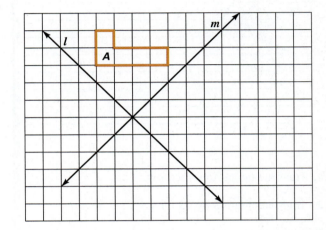

7. Reflect figure *A* about line *l*. Call the resulting figure *A'*. Then reflect figure *A'* about line *m*. Call the resulting figure *A"*.

8. Reflect figure *A* about line *m*. Call the resulting figure *A'*. Then reflect figure *A'* about line *l*. Call the resulting figure *A"*.

★ 9. Communicating Mathematics In Exercises 5–6, you were reflecting an object about one line and then another line that was parallel to the first line.

 a) Did the order in which you did the reflections make a difference? Explain.

 b) What is the effect of performing the two reflections on the object?

10. Communicating Mathematics In Exercises 7–8, you were reflecting an object about one line and then another line that was perpendicular to the first line. Did the order in which you did the reflections make a difference? Explain.

---

* Exercise numbers circled in red can be used as group exercises.

Use the figure for Exercises 11–14.

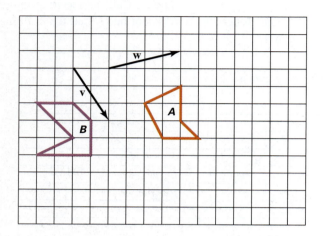

11. Translate figure *A* by vector **w**.

12. Translate figure *B* by vector **v**.

13. Translate figure *A* by vector **v**. Call the resulting object *A′*. Then translate figure *A′* by vector **w**. Call the resulting object *A″*.

14. Communicating Mathematics If you reverse the order in which you do the translations in Exercise 13, does it make a difference in the resulting object *A″*? Explain.

15. Perform the indicated glide reflection on figure *A*.

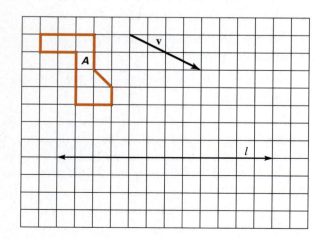

16. Perform the indicated glide reflection on figure *B*.

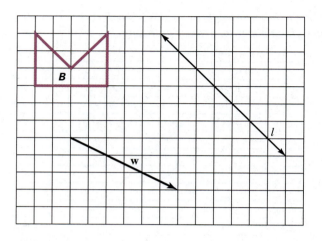

In Exercises 17–20, first rotate the object 45° about point P. Then rotate the figure 90° about P. Finally, rotate the object 180° about P.

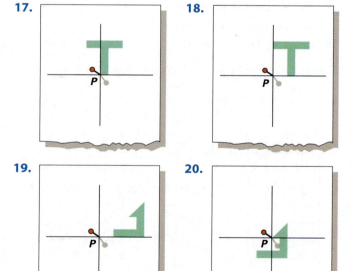

The following figure shows eight arrangements of tiles. Some of the arrangements can be obtained by applying a rigid motion to other arrangements. For example, we can obtain arrangement (e) by reflecting arrangement (a) about a vertical axis. Use these arrangements to solve Exercises 21–24.

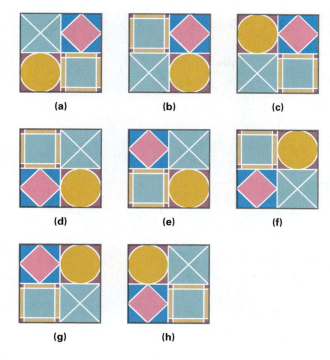

(a)	(b)	(c)
(d)	(e)	(f)
(g)	(h)	

FIGURE FOR EX. 21–24

21. List all arrangements that we obtain by reflecting arrangement (b) about a line. (Don't forget reflections about diagonal lines.)

22. List all arrangements that we obtain by rotating arrangement (g) about its center.

23. List all arrangements that we obtain by applying a rigid motion to arrangement (f).

24. Explain why it is impossible to find a rigid motion that we can apply to arrangement (a) to obtain arrangement (b).

In Exercises 25–30, list all the reflectional symmetries for each object. Also find all rotational symmetries of the object using angles between 1° and 359°.

25.

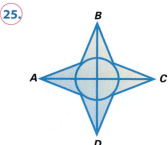

26.

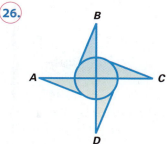

27.

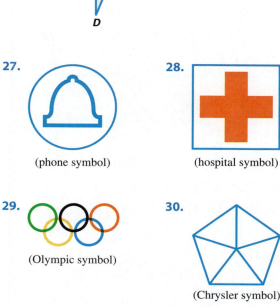

(phone symbol)

28.

(hospital symbol)

29.

(Olympic symbol)

30.

(Chrysler symbol)

31. What is the interior angle sum of a regular twelve-sided polygon? What is the measure of its interior angles?

32. What is the interior angle sum of a regular fifteen-sided polygon? What is the measure of its interior angles?

33. Communicating Mathematics   Explain why we can tessellate the plane with a regular hexagon, but not with a regular pentagon.

In Exercises 34–36, tessellate the plane with the given figure.

34.

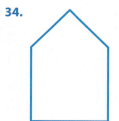

35.

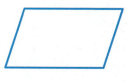

36.

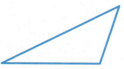

Communicating Mathematics *In Exercises 37–40, explain why each tessellation is possible by considering the angles at each vertex in the tessellation, as we did in Quiz Yourself 22.*

37.

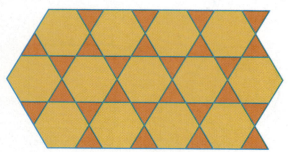

38.

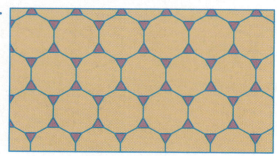

39.

40.

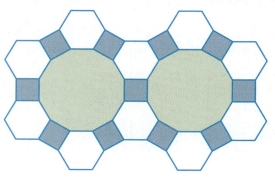

Further Exercises

41. Communicating Mathematics   Use the given figure to explain why any convex quadrilateral will tessellate the plane.

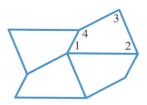

FIGURE FOR EX. 41

42. Communicating Mathematics   Use a drawing similar to the one in Exercise 41 to show that the given quadrilateral tessellates the plane.

FIGURE FOR EX. 42

A perpendicular bisector of a chord* of a circle passes through the center of a circle. Because the line l *divides the segment AB into two equal parts and is also perpendicular to the segment AB, we know that the center of the circle lies somewhere on* l. *Use this information to estimate the center of rotation for the rotations shown in Exercises 43–44.*

43.

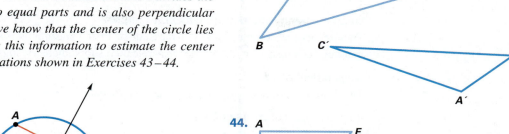

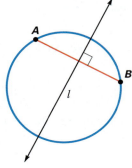

FIGURE FOR EX. 43 – 44

44.

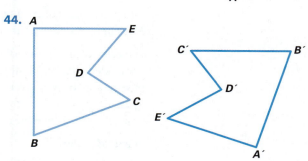

CHAPTER SUMMARY

SECTION 8.1

ray, line segment
A ray is a half line with its endpoint included. A piece of a line joining two points and including the points is called a line segment.

parallel lines, intersecting lines
Parallel lines are lines that lie on the same plane and have no points in common. Two lines that have a single point in common are called intersecting lines.

angle
An angle is formed by two rays that have a common endpoint. We measure an angle in degrees.

acute angle, right angle, obtuse angle, straight angle
An angle whose measure is between 0 and 90 degrees is called an acute angle. A right angle has a measure of 90 degrees. An obtuse angle has a measure between 90 and 180 degrees. A straight angle has a measure of 180 degrees.

* A *chord* of a circle is a line segment with both endpoints on the circle.

vertical angles

Two intersecting lines form two pairs of vertical angles.

complementary angles, supplementary angles

Two angles are complementary if the sum of their measures is 90 degrees. Two angles that have an angle sum of 180 degrees are called supplementary angles.

perpendicular lines

Two lines that intersect forming right angles are called perpendicular lines.

parallel lines cut by a transversal

Parallel lines cut by a transversal form several pairs of equal angles.

circle, radius, center, diameter, circumference

A circle is the set of all points lying on a plane that are located at a fixed distance, called the radius, from a given point called the center. A diameter of a circle is a line segment passing through the center, with both endpoints lying on the circle. The circumference is the distance around the circle.

central angle

An angle that has its vertex at the center of a circle is called a central angle.

SECTION 8.2

closed figure, simple figure

A plane figure is closed if we can draw it without lifting the pencil and the starting and ending points are the same. A plane figure is simple if we can draw it without lifting the pencil and in drawing it we never pass through the same point twice, with the possible exception of the starting and ending points.

polygon

A polygon is a simple closed figure consisting only of line segments, called edges, such that no two consecutive edges lie on the same line.

convex polygon

A polygon is convex if for any two points X and Y inside the polygon, the entire line segment XY also lies inside the polygon.

angle sum of a convex polygon

The sum of the measures of the interior angles of a convex polygon that has n sides is $(n - 2) \times 180°$.

similar polygons

Two polygons are similar if their corresponding sides are proportional and their corresponding angles are equal.

SECTION 8.3

perimeter and area of a rectangle

If a rectangle has length l and width w, then the perimeter of the rectangle is $P = 2l + 2w$ and the area is $A = l \cdot w$.

area of a parallelogram

The area of a parallelogram with height h and base b is $A = h \cdot b$.

area of a triangle
A triangle with height h and base b has area $A = \frac{1}{2} h \cdot b$.

Heron's formula
For a triangle with sides a, b, and c, we define the quantity $s = \frac{1}{2}(a + b + c)$. Then the area of the triangle is

$$A = \sqrt{s(s - a)(s - b)(s - c)}.$$

area of a trapezoid
A trapezoid with lower base b_1, upper base b_2, and height h has area

$$A = \frac{1}{2} \times (b_1 + b_2) \times h.$$

Pythagorean theorem
If a and b are the legs of a right triangle and c is the hypotenuse (the side opposite the right angle), then $a^2 + b^2 = c^2$.

circumference, area of a circle
A circle with radius r has circumference $C = 2\pi r$ and area $A = \pi r^2$.

SECTION 8.4

volume, surface area of a rectangular solid
If a rectangular solid has length l, width w, and height h, then the volume of the solid is $V = lwh$ and the surface area of the solid is $S = 2lw + 2lh + 2wh$.

volume, surface area of a cylinder
A right circular cylinder with radius r and height h has volume $V = \pi r^2 h$ and surface area $S = 2\pi rh + 2\pi r^2$.

volume, surface area of a cone
A right circular cone with height h and base radius r has volume $V = \frac{1}{3}\pi r^2 h$ and surface area $S = \pi r \sqrt{r^2 + h^2}$.

volume, surface area of a sphere
A sphere with radius r has volume $V = \frac{4}{3}\pi r^3$ and surface area $S = 4\pi r^2$.

SECTION 8.5

basic metric units
In the metric system, length measurement is based on the meter which is slightly more than a yard. Volume is based on the liter which is a little more than a quart. The basic unit of weight is the gram. A pound is 454 grams.

metric prefixes
The following table explains the prefixes that we use in the metric system.

kilo- (k)	hecto- (h)	deka- (da)	base unit	deci- (d)	centi- (c)	milli- (m)
× 1,000	× 100	× 10		× 1/10 or × 0.1	× 1/100 or × 0.01	× 1/1,000 or × 0.001

metric and customary equivalents

The following equivalents between the Metric and Customary systems are useful in doing dimensional analysis.

1 meter = 1.0936 yards
1 mile = 1.609 kilometers
1 pound = 454 grams
1 metric tonne = 1.1 tons
1 liter = 1.0567 quarts

SECTION 8.6

rigid motion

A rigid motion is the action of taking a geometric object in the plane and moving it in some fashion to another position in the plane without changing its shape or size. Every rigid motion is essentially a reflection, a translation, a glide reflection, or a rotation.

reflection

A reflection is a rigid motion in which we move an object so that the ending position is a mirror image of the object in its starting position.

translation

A translation is a rigid motion in which we move a geometric object by sliding it along some line segment in the plane. We represent the distance and direction of a translation by a translation vector.

glide reflection

A glide reflection is a rigid motion formed by performing a translation (the glide) followed by a reflection.

rotation

We perform a rotation by first selecting a point, called the center of the rotation, and then, while holding this point fixed, rotating the plane about this point through an angle called the angle of rotation.

symmetry

A symmetry of a geometric object is a rigid motion such that the beginning position and the ending position of the object are exactly the same.

tessellation

A tessellation (or tiling) of the plane is a pattern made up entirely of polygons that completely covers the plane with no gaps or overlapping of polygons.

regular tessellation

The only regular polygons that tessellate the plane are triangles, squares, and hexagons.

CHAPTER TEST

SECTION 8.1

1. In the given figure:

a) Find a pair of acute, alternate exterior angles.

b) Find a pair of obtuse, alternate interior angles.

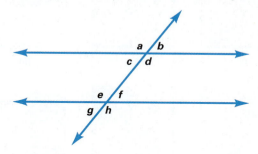

2. Find the measures of angles *a*, *b*, and *c* in the given diagram.

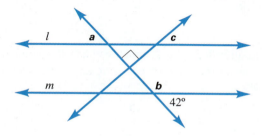

SECTION 8.2

3. What is the measure of an interior angle of a regular eighteen-sided polygon?

4. The given pair of figures are similar. What information do you know about the figure on the right?

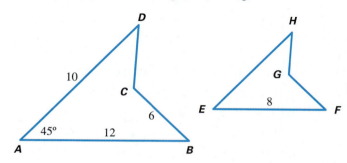

SECTION 8.3

5. Find the area of each figure.

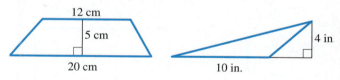

6. Find the shaded area of each figure.

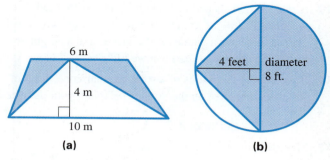

(a)   (b)

7. a) Find the area of the triangle.

b) Find the height *h* of the triangle.

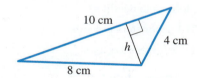

8. A running track, 5 meters wide, has the dimensions shown in the diagram. The ends of the track are semicircles with diameter 20 meters. What is the surface area of the track?

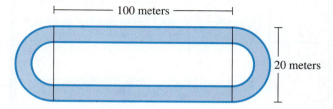

SECTION 8.4

9. Find the volume of each solid.

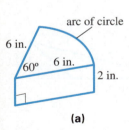
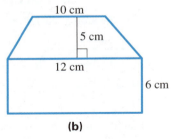

(a) (b)

10. A punch bowl shaped like a hemisphere with a radius of 9 inches is full of punch. If we are filling cylindrical glasses that have a diameter of 3 inches and a height of 3 inches, how many glasses can we fill?

11. If we double the radius of a right circular cone, what effect does that have on the volume? Explain your answer.

SECTION 8.5

12. Make the following conversions:

   a) 3,500 millimeters to meters

   b) 4.315 hectograms to centigrams

   c) 3.86 kiloliters to deciliters

13. Convert 514 decimeters to yards

14. Convert 2.1 kiloliters to quarts

SECTION 8.6

15. Perform the indicated glide reflection on figure *B*.

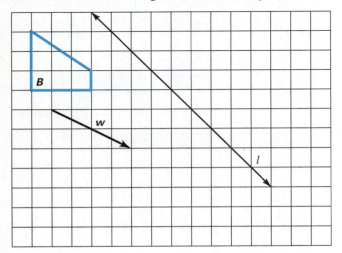

16. a) List all patterns that can be obtained by reflecting pattern (a) about a single line.

   b) List all patterns that can be obtained by rotating pattern (a) about its center.

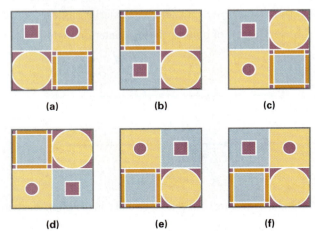

(a) (b) (c)

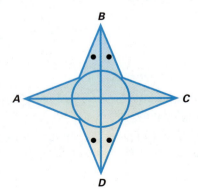

(d) (e) (f)

17. Find all reflectional symmetries and all rotational symmetries of the given object using angles between 1° and 359°.

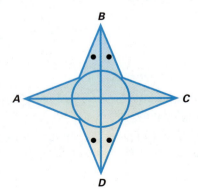

18. Tessellate the plane with the given quadrilateral.

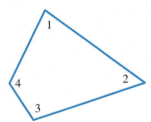

Of Further Interest: **FRACTALS**

The smooth lines, curves, and surfaces of Euclidean geometry are not adequate to describe many of the structures in the world around us. Because a lake's shoreline, a mountain, or a cloud all have rough, jagged edges, we need a different type of geometry to describe these irregularly shaped figures. Scientists use this geometry, called *fractal geometry*, to describe real-life objects more accurately than they can by using traditional Euclidean geometry.

Fractal objects are self-similar.

To understand how fractal geometry differs from Euclidean geometry, imagine a photograph of the edge of a large cloud taken from a weather satellite hundreds of miles away. The edge would not be a smooth curve, as clouds are often drawn in children's books; rather, it would be extremely jagged, as shown in Figure 8.79(a). If we enlarged a small portion of this edge, we would see a jagged curve that looks something like Figure 8.79(b). If we further enlarged a tiny portion of this smaller piece, we would still see an edge containing the same type of jaggedness that was present in the original photograph. With a fractal object, no matter how much we magnify the object, we still see patterns that are very similar to the patterns that were present in the original object. We say that an object with this property is *self-similar*.

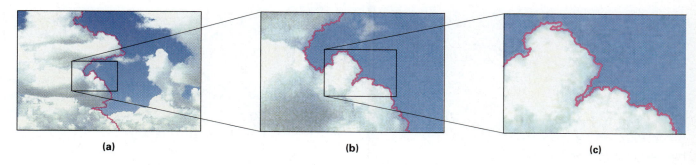

(a)   (b)   (c)

FIGURE 8.79   A cloud is self-similar.

Benoit Mandelbrot developed the theory of fractal geometry while working as a mathematician at IBM during the early 1960s. In order to get a better understanding of his geometry, we will construct a curve called the Koch curve.

EXAMPLE 1   The Koch Curve Is a Fractal

We begin the Koch curve by drawing a line segment* AB (step 0), which we divide into three equal parts. At segment CD, we construct an equilateral triangle

---

* To avoid cluttering fractal drawings, we will not show the endpoints of line segments.

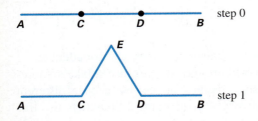

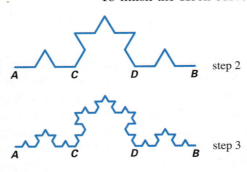

FIGURE 8.80 Beginning the Koch curve.

ΔCED, and then remove segment CD, giving the object shown as step 1 in Figure 8.80.

To continue the construction of the curve, we divide each of the four line segments in step 1 into three parts and replace the middle segment by a triangular bump, as we did in going from step 0 to step 1. The resulting curve is shown in step 2 of Figure 8.81. If we repeat this process again for the sixteen line segments in step 2, we get the curve shown in step 3 of Figure 8.81.

To finish the Koch curve, we must repeat the process indefinitely of subdividing each line segment and replacing it with a line segment having a triangular bump.

In Figure 8.82, we see why the Koch curve is a fractal. If we enlarge a small portion of the curve, we see that the enlargement has the exact same structure as the original curve. No matter how many times we magnify this curve, we always see the same repeated pattern.*

FIGURE 8.81 Steps 2 and 3 of the Koch curve.

Quiz Yourself 23†

How many line segments would there be in step 4 of the Koch curve?

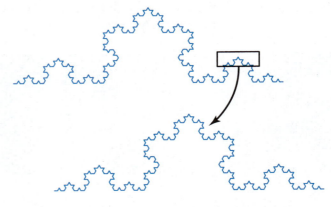

FIGURE 8.82 Small portions of the Koch curve repeat the same pattern as the original curve.

Beautiful fractal art, such as in Figure 8.83, has the same self-similarity property as the Koch curve. If we were able to put the fractal in Figure 8.83 under a microscope, we would see the same beautiful patterns at every level of magnification.

We can begin with a two-dimensional object and, by applying some rule repeatedly, create a fractal, as we see in Example 2.

---

* In order to learn how to use Geometer's Sketchpad to draw fractals such as the Koch curve, see www.aw.com/pirnot.
† Quiz Yourself answers begin on page 849.

Figure 8.83 Fractal art.

EXAMPLE 2   The Sierpinski Gasket Is a Fractal

We construct a fractal called the Sierpinski gasket by first constructing an equilateral triangle, as in Figure 8.84(a). We then divide this triangle into four smaller equilateral triangles and remove the middle triangle, as in Figure 8.84(b). Next, we apply this same rule to each of the three remaining triangles; that is, we divide each triangle into four smaller triangles and remove the center one. We show the results of this step in Figure 8.84(c).

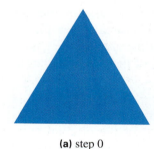

(a) step 0   **(b)** step 1   **(c)** step 2   step 3

Figure 8.84 The first two steps in forming the Sierpinski gasket.

Figure 8.85 Step 3 in forming the Sierpinski gasket.

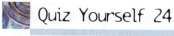
Quiz Yourself 24

How many dark triangles would appear in step 5 of forming the Sierpinski gasket?

As with the Koch curve, we must continue this process of removing the center of each solid equilateral triangle indefinitely in order to complete this fractal. Figure 8.85 shows the next step in forming the Sierpinski gasket.

We cannot draw the entire gasket, because in order to do this, we would have to draw smaller and smaller triangles that eventually become so small that their size would be finer than the resolution of any printing device.

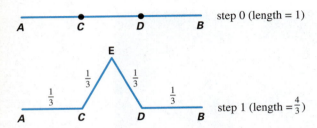

step 0 (length = 1)

step 1 (length = $\frac{4}{3}$)

FIGURE 8.86 In step 1, the Koch curve is $\frac{4}{3}$ as long as the original line segment.

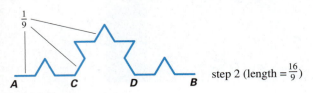

step 2 (length = $\frac{16}{9}$)

FIGURE 8.87 Each segment now has length $\frac{1}{9}$.

 Quiz Yourself 25

What is the length of the Koch curve at step 3 of the construction process?

As in Euclidean geometry, we are interested in the length, area, and volume of fractal objects. When we investigate the length of the Koch curve, we find a surprising result.

EXAMPLE 3   The Length of the Koch Curve Is Infinite

Find the length of the Koch curve.

SOLUTION: We began the Koch curve with a line segment that has length 1 (step 0). In step 1 we replaced this curve with a curve $\frac{4}{3}$ as long, as shown in Figure 8.86.

Figure 8.87 shows that in step 2, the curve consists of sixteen small line segments with length $\frac{1}{9}$, so the length is now $\frac{16}{9}$.

At each successive step, the curve is $\frac{4}{3}$ as long as the curve in the previous step. This means that the curve's length keeps growing larger and larger as we construct further steps of the curve. At step 40, we would find that the length of the Koch curve is $(\frac{4}{3})^{40}$, which is slightly over 99,437 units long. At stage 100, the curve is over three trillion units long! Of course, we do not stop at stage 100. Since we must go through an infinite number of stages to construct the whole curve, the total length of the Koch curve is therefore infinite.

EXAMPLE 4   The Area of the Sierpinski Gasket Is Zero

What is the area of the Sierpinski gasket?

SOLUTION: To make the computations easier to follow, let us assume that we begin the Sierpinski gasket with an equilateral triangle with area 1 (see Figure 8.88). In step 1 of the construction, we have removed $\frac{1}{4}$ of the area, so the gasket now consists of three triangles, each with area $\frac{1}{4}$. The area of the dark triangles at step 1 is therefore $\frac{3}{4}$.

Now consider subtriangle 1. In step 2, we remove $\frac{1}{4}$ of its area, which is $(\frac{1}{4})(\frac{1}{4}) = \frac{1}{16}$ of the original area, leaving three smaller triangles also with area $\frac{1}{16}$. The area that remains in subtriangle 1 is therefore $\frac{3}{16}$ of the original area. Likewise, $\frac{3}{16}$ of the original area remains in subtriangles 2 and 3 after we remove their centers. Therefore, in step 2 of constructing the gasket, the remaining area is

$$\frac{3}{16} + \frac{3}{16} + \frac{3}{16} = \frac{9}{16} = \left(\frac{3}{4}\right)\left(\frac{3}{4}\right).$$

 Quiz Yourself 26

What is the area of the Sierpinski gasket at step 3 of the construction?

We see that at each successive step in constructing the gasket, we get an area that is $\frac{3}{4}$ the area of the previous step. Therefore, the area of the gasket keeps getting smaller and smaller with each successive step in the construction. For example, at step 20 the area is $(\frac{3}{4})^{20} \approx 0.0032$ square units. We conclude that the area of the gasket is 0.

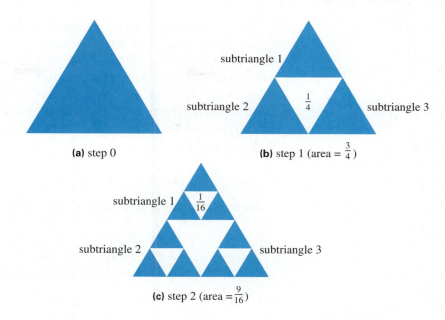

(a) step 0

(b) step 1 (area = $\frac{3}{4}$)

subtriangle 1

subtriangle 2 $\frac{1}{4}$ subtriangle 3

subtriangle 1 $\frac{1}{16}$

subtriangle 2   subtriangle 3

(c) step 2 (area = $\frac{9}{16}$)

FIGURE 8.88   Step 1 of the Sierpinski gasket has area $\frac{3}{4}$; step 2 has area $\frac{9}{16}$.

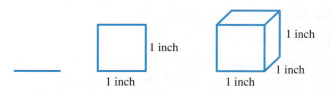

FIGURE 8.89   A unit of measurement in one, two, and three dimensions.

You may have a feeling that the Koch curve, because of all of its "wiggling around," is somewhat thicker than the kinds of curves we draw in Euclidean geometry. Therefore we might say that the Koch curve, in some sense, has a larger dimension than a smooth curve in Euclidean geometry. On the other hand, all the "wiggling" is not enough for the curve to fill entirely some region of the plane, which would make it a two-dimensional object. In order to make this concept of dimension more clear, consider the line segment, the square, and the cube in Figure 8.89.

Imagine that we place the line, the square, and the cube into a three-dimensional copying machine. This copying machine will increase or decrease the size of any object that we place into it. If we set the size of our copies to two times the original, the copies of the line, square, and cube would come out looking as in Figure 8.90.

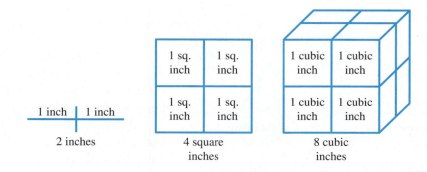

FIGURE 8.90   The line, square, and cube have been enlarged by a factor of 2.

A general way of looking at this is to say we used a scaling factor of s ($s = 2$ in this case). In the one-dimensional case, the copier returned a copy that had length equal to $s^1 = 2$ times the length of the original. In the two-dimensional case, the copier returned a copy containing an area equal to $s^2 = 4$ times the original area. For the cube, the copier returned a copy with $s^3 = 8$ times the volume of the original. It seems then that we can think of the dimension of an object as an exponent D that satisfies these types of equations. We now define the notion of dimension for fractals.

DEFINITION

The **fractal dimension** of an object is a number D that satisfies the equation

$$n = s^D,$$

where s is a scaling factor and n is the amount by which the quantity we are measuring (length, area, volume) of the object changes when we apply the scaling factor to the object.

To understand the notion of fractal dimension, let's look again at the Koch curve.

EXAMPLE 5   The Fractal Dimension of the Koch Curve

What is the dimension of the Koch curve?

SOLUTION: We need to understand what it means to magnify the curve by a factor. Consider Figure 8.91(a), where we show a picture of the *entire* Koch curve but because the picture is so small, we cannot see much detail. We next enlarge Figure 8.91(a) by a factor of 3 and display this enlargement in Figure 8.91(b).

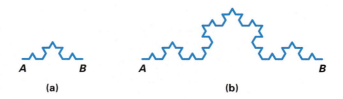

(a)   (b)

FIGURE 8.91   The Koch curve enlarged by a factor of 3.

We see that when each of the sixteen tiny line segments in Figure 8.91(a) is enlarged, it appears as a line segment with a "bump" on it, as shown in Figure 8.92. The line segment with the bump is $\frac{4}{3}$ times as long as the line segment before enlargement.

What we are saying is that enlarging any portion of the curve by a scale factor of 3 appears to increase its length by 4. If we call the dimension of the curve D, this means that

$$3^D = 4. \tag{1}$$

line segment   line segment enlarged
                      by a factor of 3

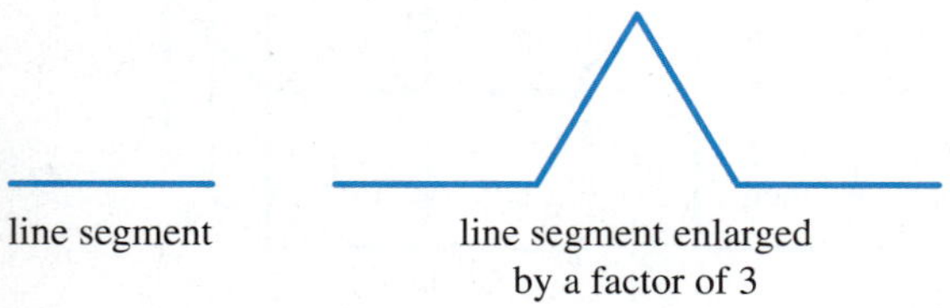

FIGURE 8.92   Each enlarged segment of the curve appears $\frac{4}{3}$ as long as the original.

Highlight: *Applications of Fractals*

It is surprising that the fractal geometry that produces such strange and beautiful images can also describe many important natural phenomena. Scientists use fractal geometry to explain the similarities and differences between the way a tree branch divides repeatedly to form finer subbranches and the way the bronchi of the lungs subdivide to form a tree of airways inside our lungs. This same sort of branching occurs in a reverse fashion as small tributaries join together to form streams and eventually a river.

Geographers classify the roughness of coastlines according to their fractal dimension. The South African coast is relatively smooth, with a fractal dimension close to 1, while the extreme irregularity of Norway's coast has a fractal dimension of 1.52.

Understanding the fractal pattern seen in a lightning bolt also helps explain how electrical insulators break down when subjected to high voltages. This same pattern occurs at a microscopic level when crystals form.

Economists studying the stock market have found that if they graph the market over hundreds of days, then over hundreds of hours, and finally over hundreds of thirty-second intervals, the graphs look remarkably similar. Cardiologists have learned that the healthy heart beats according to a fractal rhythm rather than a steady, regular rhythm, as we used to think.

Using percolation theory, mathematicians apply fractal geometry to describe the way coffee percolates in a coffeepot and the way groundwater seeps into the soil. It is surprising that this same area of fractal mathematics also explains how a fire "percolates" through a forest, how an epidemic "percolates" through a population, and how galaxies "percolate" throughout the universe.

We now solve for D. In order to do this, we need to use the log key on our calculator. The log function has the property that $\log a^x = x \log a$; we will use this property to solve equation (1).

Taking the log of both sides of equation (1) gives us

$$\log 3^D = \log 4. \tag{2}$$

Now using the property we just stated for log, we get

$$D \log 3 = \log 4. \tag{3}$$

Dividing both sides of equation (3) by log 3 and using a calculator to evaluate the result, we find that

$$D = \frac{\log 4}{\log 3} \approx 1.26.$$

The fractal dimension of the Koch curve is therefore approximately 1.26. ◎

The idea that the Koch curve has dimension 1.26 means that in a certain sense it is thicker, or fills space better, than a one-dimensional object such as a line segment. On the other hand, because this dimension is less than 2, the Koch curve does not fill space as well as a two-dimensional object such as a solid square.

Quiz Yourself 27

Assume that for a fractal curve, enlarging the curve by a factor of 4 increases its length by a factor of 8. What is the fractal dimension of the curve?

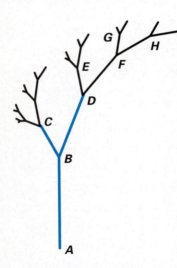

FIGURE 8.93 A fractal tree.

Artists use fractal geometry in movies to create beautiful mountains, clouds, and other natural-looking objects. We will show you a simple example of how to create a tree using fractals.

EXAMPLE 6   Drawing a Fractal Tree

Explain why the "tree" in Figure 8.93 is a fractal.

SOLUTION: The basic pattern in the tree is determined by the line segments joining points A, B, C, and D. At the end of segment AB is a branching represented by segments BC and BD. Smaller versions of this Y-shaped motif occur repeatedly throughout the tree. The fractal pattern is clear. To add finer branches to the tree, we choose a branch such as DF and replace it with a small Y-shaped figure.

Natural-looking scenes such as those shown in Figure 8.94 are generated using techniques that are similar to the method we described in Example 6.

Anne Burns, 2002

FIGURE 8.94   Fractal landscape.

Exercises

Exercises 1–4 pertain to the Koch curve in Example 1.

1. How many line segments will the curve have in step 5?

2. What is the length of the curve at step 5?

3. What is the length of the curve at step 10?

4. True or false: Doubling the number of steps doubles the length of the curve.

In Exercises 5–6, you are given steps 0 and 1 for constructing a fractal.

    a) *Construct step 2 of the fractal.*

    b) *Assume that the length of the line segment in step 0 is 1. Find the length of the curve in step 5.*

5.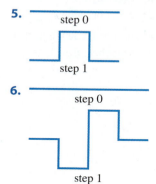

6.

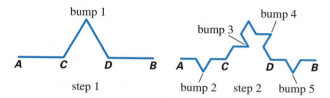

* **7.** We can get a realistic "coastline" effect if we slightly vary the construction of the Koch curve. In the construction of the Koch curve in Example 1, whenever we added a bump to a line segment, we always added it above the curve. Now when we add a bump, we will use the following list of random numbers.

   87127   03570   73103   16946   81852   94819
   33108   72734   43411   31078

We will add bumps above the curve for even digits and below the curve for odd digits. The first random digit is even, so we will add the first bump above the curve; the second and third digits are odd, so we add the second and third bumps below the curve. The fourth digit is even, so we draw the fourth bump above the curve, and so on. We show steps 1 and 2 in the following diagram. Redraw steps 1 and 2 of this fractal; however, now use random digits beginning with 16946.

FIGURE FOR EX. 7

8. What is the dimension of the curve described in Exercise 7?

9. Solve $4^D = 6$.

10. Solve $4^D = 12$.

11. Find the fractal dimension of the curve described in Exercise 5.

12. Find the fractal dimension of the curve described in Exercise 6.

In Exercises 13–14, construct step 2 for each fractal.

13. the Sierpinski carpet

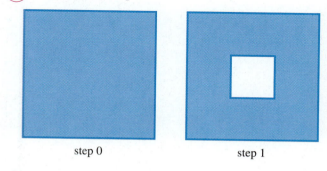

          step 0   step 1

14.

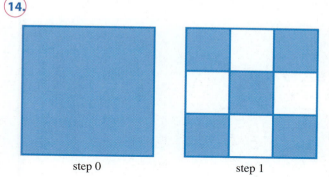

          step 0   step 1

Further Exercises

15. Find a formula for the number of line segments at step n for the fractal in Exercise 5.

16. Find a formula for the number of line segments at step n for the fractal in Exercise 6.

17. Find a formula for the area of the Sierpinski gasket at step 10; at step n.

18. Find a formula for the area of the Sierpinski carpet in Exercise 13 at step 10; at step n.

19. Draw a fractal tree using the method of Example 6; however, vary the length and angles of the branches.

20. Draw a fractal tree using the method of Example 6; however, now use a motif with three branches.

CHAPTER 9

Apportionment:* How Do We Measure Fairness?

Because the founding fathers wanted to ensure that each citizen would be fairly represented in the new government, they placed the requirements for state representatives at the very beginning of the United States Constitution. Article I, Section 2, states:

> *Representatives and direct taxes shall be apportioned among the several states which may be included within the Union, according to their respective numbers. . . . The actual Enumeration shall be made within three years after the first Meeting of the Congress of the United States, and, within every subsequent Term of ten years, in such Manner as they shall by Law direct.*

Apportioning representatives is not as easy as it might seem. The problem is that once we decide on the total membership of the house, one state might deserve 13.465 representatives while another deserves 11.702. Just like dividing ten party favors among three small children, the question is, "How do we decide who gets the extra whole representatives?"

Daniel Webster believed that it was not possible to solve the apportionment problem perfectly. Speaking before the House of Representatives in 1832, he said,

> *The Constitution . . . must be understood, not as enjoining an absolute relative equality, because that would be demanding an impossibility, but as requiring of Congress to make the apportionment of Representatives among the several States*

---

* You can find further resources on apportionment at www.aw.com/pirnot.

according to their respective numbers, as near as may be. *That which cannot be done perfectly must be done in a manner as near perfection as can be.**

We will first explain a method that was used to apportion early Congresses and show how this method can lead to serious problems. Next, we will use inequalities to develop the method that is currently used to apportion the U.S. House of Representatives. We will then use both apportionment methods to show you how to allocate resources other than political representatives. In Section 9.4, we will discuss several other apportionment methods suggested by Thomas Jefferson, John Quincy Adams, and Daniel Webster.

9.1 UNDERSTANDING APPORTIONMENT

In 1881, Congress made a surprising discovery when reapportioning the House of Representatives. Using an apportionment method developed by Alexander Hamilton, Alabama would be entitled to 8 representatives in a House having 299 members, but it would receive only 7 representatives in a 300-member House. Under Hamilton's method, Alabama would receive fewer representatives in a larger House *although no state had a change in population.* This strange situation is known as the **Alabama paradox** and was not an isolated incident. Using the same Hamilton method following the 1890 census, Arkansas, with no change in any state's population, lost a representative when the House increased from 359 members to 360.

Following the 1890 census, a different apportionment plan caused the same paradoxical situation for Maine. As the number of representatives proposed for Maine increased and decreased, Representative Littlefield remarked:

Now you see it and now you don't. In Maine comes and out Maine goes. The House increases in size and still she is out. It increases a little more in size, and then, forsooth, in she comes. A little further increase, and out she goes, and then a little further increase and in she comes. God help the State of Maine when mathematics reach for her and undertake to strike her down in this manner in connection with her representation on this floor. . . . †

In order to develop an apportionment method that avoids an Alabama paradox, we must first understand why the problem occurs. The following example will help explain it.

Naxxon, Aroco, and Eurobile have formed a consortium to develop an oil drilling platform off the coast of Africa. The companies have agreed to form a nine-member board with executives from the three companies to oversee the project. Each company will have at least one representative on the board, and

---

* *The Works of Daniel Webster*, Vol. III, 16th ed. (Little, Brown & Company, 1872).
† You can read the full text of Representative Littlefield's remarks in the *Congressional Record*, 56th Congress, 2nd session, Vol. 34, beginning on p. 590.

additional board members will be assigned in proportion to the number of stockholders in each company. Naxxon has 4,700 stockholders, Aroco has 3,700 stockholders, and Eurobile has 1,600 stockholders.

Because the total number of stockholders in the three companies is 10,000, this means that Naxxon is entitled to $4,700/10,000 = 47\%$ of the board's members, Aroco 37%, and Eurobile 16%. Naxxon is therefore entitled to exactly $47\% \times 9 = 0.47 \times 9 = 4.23$ members on the board. We show similar calculations for the other partners in Table 9.1.

A company cannot have a part of a board member. Therefore, because Naxxon is entitled to exactly 4.23 members, it will be given either 4 or 5. We will call 4 the *integer part* of 4.23 and .23 the *fractional part* of 4.23.

Company	Percent of Stockholders	Board Members Deserved
Naxxon	47	$0.47 \times 9 = 4.23$
Aroco	37	$0.37 \times 9 = 3.33$
Eurobile	16	$0.16 \times 9 = 1.44$
Total	100	9.0

TABLE 9.1   The exact number of representatives allotted to each company on the board.

The Hamilton method uses fractional parts to apportion representatives.

If we give each company its integer part as its number of board members, then Naxxon will have four members, Aroco will have three, and Eurobile will have one. Thus there will be only eight members on the board. In order to have the required nine, we must now decide which company gets the additional member. It seems reasonable to assign the member to Eurobile, which has the highest fractional part—namely 0.44. In fact, this is exactly how the **Hamilton apportionment method*** would allocate the last board member.

---

* This method, which is attributed to Alexander Hamilton, was used to apportion the first U.S. Congress.

Hamilton Apportionment Method (Applied to the Consortium Board)

Follow these steps for each company.

1. Determine the exact number of board members due to the company by computing

$$\text{percent of stockholders} \times \text{size of board.}$$

2. Assign the integer part of the exact number of board members to each company.

If there are more members to be allocated, then go to step 3.

3. Assign additional members according to the fractional parts of the exact number of members for each company. The first additional member goes to the company having the largest fractional part; the second additional member, if any, goes to the company with the second largest fractional part. Continue in this manner until you have assigned all additional members.

To illustrate step 3, we add two columns to Table 9.1 to get Table 9.2.

Company	Percent of Stockholders	Board Members Deserved	Step 2: Assign Integer Parts	Examine Fractional Parts	Step 3: Assign Additional Member
Naxxon	47	$0.47 \times 9 = 4.23$	4	0.23	4
Aroco	37	$0.37 \times 9 = 3.33$	3	0.33	3
Eurobile	16	$0.16 \times 9 = 1.44$	1	0.44	2
Total	100	9.0	8		9

TABLE 9.2   Using the Hamilton method to allocate nine board members.

EXAMPLE 1   Using the Hamilton Apportionment Method

Suppose the consortium decides to increase the size of the board to ten members. Use the Hamilton apportionment method to apportion the ten-member board.

SOLUTION: We show the steps in making this apportionment in Table 9.3.

The last column of Table 9.3 shows the assignment of members according to the Hamilton method. We assigned the first eight members using the integer parts

Company	Percent of Stockholders	Board Members Deserved	Step 2: Assign Integer Parts	Examine Fractional Parts	Step 3: Assign Additional Members
Naxxon	47	$0.47 \times 10 = 4.7$	4	0.7	5
Aroco	37	$0.37 \times 10 = 3.7$	3	0.7	4
Eurobile	16	$0.16 \times 10 = 1.6$	1	0.6	1
Total	100	10.0	8		10

TABLE 9.3   Using the Hamilton method to allocate ten board members.

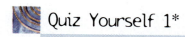

of the exact amounts deserved by each company. Then we gave the two additional members to Naxxon and Aroco since they have the highest fractional parts.

Notice in Example 1 that Eurobile lost one member when the board increased in size, even though the number of stockholders in each company remained the same. This Alabama paradox occurred because with the Hamilton method, we assigned a block of members on the board to each company. When the board size increased to ten, we reassigned the original nine seats although they had been apportioned earlier. As we saw, the second apportionment allowed for the possibility that some companies lose members in the larger board.

Although the Hamilton method is not a good one to use to apportion representatives to states, there are some situations where in assigning objects (other than representatives) the total will not increase once the apportionment has been done, so the Hamilton method is appropriate.

We can use the Hamilton method to present survey results.

One such situation occurs when summarizing the results of a survey in a table of statistics in a newspaper article. The tables often have footnotes stating that the percentages do not add up to 100% due to roundoff error. Therefore, an interesting question is, how can we reasonably adjust the individual numbers in a table so that their sum is exactly 100%? We will need to "apportion" a quantity to answer this question. However, the situation is different from apportioning representatives and we will see that it is appropriate to use the Hamilton method in solving this problem. Before discussing this further, we introduce some terminology.

With the Hamilton method, it was useful when dealing with a number such as 12.3756 to consider its integer part, which is 12, and its fractional part, which is 0.3756. Sometimes, in working with this number, we will need to discuss a "piece" of the number that is neither the integer nor the fractional part. For example, we may be interested in a number such as 12.375. If we drop all decimal places beyond the tenths place in a number, we say we are *truncating to tenths*. In the preceding case, we would get 12.3. Similarly, if we drop all decimal places beyond the hundredths place, to get 12.37, we say we are *truncating to hundredths*. Without formally defining this further, we extend this terminology to "truncating to thousandths," "truncating to millionths," and so on.

EXAMPLE 2 Truncating the Fractional Part of a Number

Truncate the number 183.6574893 to:

a) tenths.

b) thousandths.

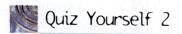

SOLUTION: a) Truncating the number to tenths means we drop all decimal places beyond the 6 in the tenths place. This gives us 183.6. Notice that the result is different from the one we get by rounding.

b) Truncating to thousandths means dropping all decimal places beyond the 7 in the thousandths place to get 183.657.

EXAMPLE 3   Presenting Survey Results

The results of a survey of voters on key issues are listed in column A of Table 9.4. If we wish to show these percentages using only one decimal place, then we might round these numbers to get the values in column B. The total is not 100 percent, which is unsatisfactory.

	Exact Percentage (column A)	Percentage Rounded to Tenths (column B)
Taxes	34.4235	34.4%
Education	13.456	13.5
Crime	14.75	14.8
Health care	37.3705	37.4
Total	100.0%	100.1%

TABLE 9.4   Survey of issues important to voters.

Present a table with numbers expressed to the tenths decimal place, but adding to exactly 100.0%.

SOLUTION: First truncate the numbers in column A of Table 9.4 to tenths to get column A of Table 9.5.

	Original Data Truncated to Tenths (column A)	Part of Number That Is Discarded (column B)
Taxes	34.4%	0.0235
Education	13.4	**0.056**
Crime	14.7	0.05
Health care	37.3	**0.0705**
Total	99.8%	

TABLE 9.5   Truncating survey data to tenths.

We have discarded all decimal places beyond the tenths place, so it is not surprising that our total falls short of 100.0%. We can use the Hamilton method to "apportion" the missing two-tenths of a percent needed to make the total 100.0%. In particular, we need to decide which two of the four percentages should get an additional tenth of a percent.

To do this, we look at the portions remaining after truncating the numbers to the tenths place. In column B of Table 9.5, we see that the fourth number is the largest and the second number is the second largest from the hundredths place on. Thus we give additional tenths to the fourth number and second number. Our result appears as Table 9.6.

We will call the technique that we used in Example 3 *adjusting a list of numbers*. Quiz Yourself 3 illustrates this technique.

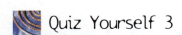

Quiz Yourself 3

A group of consumers was asked how they expected their spending to change in the next six months. Adjust the percentages in the following table so that they are shown to the tenths place and their sum is 100.0%.

Spending will increase	32.03
Spending will decrease	24.74
Spending will stay the same	22.75
Unsure	20.48
Total	100.00%

Taxes	34.4%
Education	13.5
Crime	14.7
Health care	37.4
Total	100.0%

TABLE 9.6   Using the Hamilton method to adjust survey percentages.

Historical Highlight: *Apportionment in U.S. History*

Although the Constitution originally set the number of representatives at 65, Congress does have the power to vary the size of the House. Of course, there can be no more than one representative for every 30,000 people. After the country's first census in 1790, Congress passed an apportionment act in 1792 that allotted 120 seats to the then fifteen states. Secretary of State Thomas Jefferson believed that the bill was not in keeping with the spirit of the founding fathers and wrote in opposition to the bill, "No invasions of the Constitution are fundamentally so dangerous as the tricks played on their own numbers . . ."* George Washington, acting on Jefferson's advice, turned down the bill in what became the very first presidential veto.

Another apportionment bill was drafted that fixed the House at 103 members for a population of then roughly four million. After each succeeding census a further apportionment was enacted, usually with a corresponding increase in the size of the House. In 1910, the number of representatives was set at 435, which is the size of the present House. It is interesting to note that if the principle of approximately one representative for every 30,000 people were applied with our present population, there would be over 6,000 members in the House of Representatives!

Certainly one way to prevent an Alabama paradox is to use an apportionment method so that once seats are assigned, they are never reassigned at a later date if the size of the governing body increases. In order to make assignments on a "once and done" basis, at each stage of the apportionment process, we must have a clear measure of who is most deserving of the next representative.

Average constituency measures the fairness of an apportionment.

To begin the development of such a measure, consider the representation of two hypothetical states A and B in the House of Representatives. Suppose state A has a population of 2,000,000 people and eight representatives and state B has 800,000 people and four representatives. Each of state A's representatives has an average of 2,000,000/8 = 250,000 constituents, whereas each representative from state B averages 800,000/4 = 200,000 constituents. Because a representative from state A serves more constituents than a representative from state B, it is fair to say that state A is more poorly represented in the House than state B. This leads to the following definitions.

> **DEFINITIONS**
>
> The **average constituency** of a state† is the quotient
>
> $$\frac{\text{population of the state}}{\text{number of representatives from the state}}.$$
>
> Comparing the representation of two states A and B, we say that state A is **more poorly represented** than state B if the average constituency of A is greater than the average constituency of B.

---

* Paul Leicester Ford, editor, *The Works of Thomas Jefferson*, Vol. VI (G. P. Putnam's Sons, 1904), pp. 460–471.
† Although we will discuss apportionment in terms of assigning representatives to the U.S. House of Representatives, we can apply these ideas to many other situations.

EXAMPLE 4   Determining Which State Is More Poorly Represented

According to a recent census, Wisconsin had a population of 4,891,770 and was allocated 9 representatives; Texas had a population of 16,865,510 and was allocated 30 representatives. Calculate the average constituency of each state and determine which state is more poorly represented.

SOLUTION: The average constituency of Wisconsin is

$$\frac{\text{the population of Wisconsin}}{\text{the number of representatives assigned to Wisconsin}} = \frac{4{,}891{,}770}{9} = 543{,}530,$$

whereas the average constituency of Texas is

$$\frac{\text{the population of Texas}}{\text{the number of representatives assigned to Texas}} = \frac{16{,}865{,}510}{30} \approx 562{,}184.$$

Thus Texas is more poorly represented.

Quiz Yourself 4

a) If the 420-member electricians' union has three representatives on the United Labor Council, what is the average constituency of this group?

b) If the 440-member plumbers' union has four representatives on the council, are the electricians or the plumbers more poorly represented?

Absolute unfairness is the difference in average constituencies.

It would be ideal, of course, to have an apportionment in which the average constituencies are the same for all states because it would be consistent with the "one man, one vote"* concept. However, it is usually not possible to achieve this ideal when making an actual apportionment. If we cannot have equality, then we should try assigning the representatives to make the average constituencies as equal as possible. One measure of how close we come to this goal is called absolute unfairness.

> **DEFINITIONS**
>
> Suppose that representatives are apportioned between two states A and B. We define the **absolute unfairness** of this apportionment as the difference between the larger average constituency and the smaller one. If state A has the larger average constituency, then the absolute unfairness is
>
> (average constituency of state A) − (average constituency of state B).
>
> If the two states have the same average constituencies, then we say that the two states are *equally well represented.*

Notice that in computing the absolute unfairness we subtract the smaller average constituency from the larger. Consequently, the absolute unfairness of an apportionment *cannot* be negative.

---

* We have retained this terminology for historical reasons. See the highlight box in Section 9.4.

Quiz Yourself 5

Assume that state X has a population of 974,116 with four representatives and state Y has a population of 730,779 with three representatives. Compute the absolute unfairness for this apportionment.

EXAMPLE 5 Finding Absolute Unfairness

Suppose the Weavers' Guild, with 1,542 members, has six delegates on the National Arts Commission and the Artists' Alliance, with 1,445 members, has five delegates. Calculate the absolute unfairness for this assignment of delegates.

SOLUTION: The average constituency of the Weavers' Guild is $\frac{1,542}{6} = 257$ and the average constituency of the Artists' Alliance is $\frac{1,445}{5} = 289$. The Artists' Alliance is more poorly represented than the Weavers' Guild; the absolute unfairness of this apportionment is $289 - 257 = 32$.

Relative unfairness considers the size of constituencies in calculating absolute unfairness.

Although absolute unfairness measures the imbalance of an apportionment between two states, it is inadequate to compare the unfairness of two apportionments. You may think that a larger absolute unfairness indicates a greater imbalance.

This conclusion is wrong!

When measuring the unfairness of an apportionment, it is important to take into consideration the size of the states involved. The following examples will help you understand why we need a different measure of unfairness.

In the 1960 presidential election, John Kennedy and Richard Nixon each received slightly over 64 million votes. Because Kennedy received only 120,000 more votes than Nixon, newspapers might have proclaimed:

KENNEDY AND NIXON
EVEN IN POPULAR VOTE

On the other hand, if Kennedy and Nixon were competing for a borough council seat in a small town and Kennedy got 115 votes while Nixon received 35, the town paper might read:

KENNEDY BURIES NIXON
IN LANDSLIDE

Although the difference in the number of votes is important, the significance of that difference depends on the number of votes they received. A difference of 120,000 votes is considered small when a candidate receives 64 million, whereas an 80-vote difference seems large if a candidate receives only 35 votes.

Similarly, the absolute unfairness of $562{,}184 - 543{,}530 = 18{,}654$ in Example 4 is *relatively* small due to the sizes of the average constituencies for Wisconsin and Texas. On the other hand, the absolute unfairness 32 that we computed in Example 5 is *relatively* large when we compare it to the average constituencies of 289 and 257. From these examples, we see that we must consider the sizes of the average constituencies when measuring the unfairness of an apportionment.

DEFINITION

When apportioning the representatives for two states, the **relative unfairness** of the apportionment is defined as

$$\frac{\text{the absolute unfairness of the apportionment}}{\text{the smaller average constituency of the two states}}.$$

EXAMPLE 6   **Determining Relative Unfairness**

Compute the relative unfairness for the apportionment of representatives to Wisconsin and Texas in Example 4 and the apportionment of delegates for the Weavers' Guild and the Artists' Alliance in Example 5.

SOLUTION: Recall that the absolute unfairness of the Wisconsin-Texas apportionment was 18,654. Since Wisconsin had the smaller average constituency of 543,530, the relative unfairness of this apportionment is therefore $\frac{18,654}{543,530} \approx 0.034$.

The absolute unfairness of the apportionment in Example 5 is 32. Since the Weavers' Guild had the smaller average constituency, namely 257, the relative unfairness is $\frac{32}{257} \approx 0.125$.

The smaller relative unfairness for the apportionment for Wisconsin and Texas means that apportionment is fairer than the apportionment for the Weavers' Guild and the Artists' Alliance.   ◎

We now have the necessary background concepts on unfairness to develop an apportionment criterion and principle that avoids an Alabama paradox. We will do this in Section 9.2.

> **Quiz Yourself 6**
>
> If state A has a population of 11,710 and five representatives and state B has a population of 16,457 and seven representatives, calculate the relative unfairness of the apportionment.

Exercises 9.1

In Exercises 1–2, identify the integer and fractional parts of each number.

1. a) 24.098   b) 5.02   **2.** a) 0.843   b) 1.720

In Exercises 3–8, use the Hamilton method to make the assignment.

3. Apportioning a water authority. Assume that California, Arizona, and Nevada are cooperating to build a dam to

provide water to communities currently lacking adequate water supplies. Seats on the eleven-member Southwest Water Authority, which governs the project, are assigned according to the number of customers in each state who use the water from the project. There are 56,000 customers in California, 52,000 in Arizona, and 41,000 in Nevada. Assign the seats on this authority.

4. **Apportioning a negotiations committee.** The employees of the Jungle World theme park are negotiating a new contract. There are 213 performers, 273 food workers, and 178 maintenance workers. The nine-person negotiations committee has members in proportion to the number of employees in each of the three groups. Assign members to the negotiations committee.

5. **Assigning booths at an art show.** The Civic Arts Guild is having a show. There is room for 31 booths and the guild has decided that the booths will be assigned in proportion to the type of members in the guild. The guild has 87 painters, 46 sculptors, and 53 weavers. Assign the booths to the three groups.

6. **Allocating a resident council.** An apartment complex has three buildings. Building A has 78 units, building B has 36 units, and building C has 51 units. A twelve-person resident council will set rules governing the complex. Membership in the council is to be proportional to the number of units in each building. Assign representatives to this council.

7. **Apportioning representatives.** In a recent census, Alabama's population in thousands was 4,041, Mississippi's was 2,573, and Louisiana's was 4,220. Allocate nineteen members of the U.S. House of Representatives to these three states.

8. **Apportioning representatives.** In a recent census, Michigan's population in thousands was 9,295, Minnesota's was 4,375, and Wisconsin's was 4,892. Allocate 33 members of the U.S. House of Representatives to these three states.

9. Communicating Mathematics Discuss what must occur to cause an Alabama paradox when apportioning a representative council. Specifically, why does the Hamilton method cause such a paradox to occur?

10. Communicating Mathematics Refer back to the example of apportioning the nine-member oil consortium.

Give an argument that Eurobile might use in opposition to using the Hamilton method for assigning the extra representative.

11. Suppose an electronics company has three divisions: (V)ideo, (C)omputers, and (B)usiness products. Division V has 140 employees, C has 85, and B has 30. Assume that a twelve-member quality-improvement council has membership on the council proportional to the number of employees in the three divisions.

   a) Apportion this council using the Hamilton method.

   b) Now increase the council's size to thirteen, and then to fourteen, and so on, redoing the apportionment each time until an Alabama paradox occurs.

   c) What is the first size above twelve when the paradox occurs, and what division loses a seat when the size of the council increases?

12. There are 47 local police, 13 federal agents, and 40 state police involved with drug enforcement in Metro City. A special nine-person task force will be formed to investigate a particular case. Officers will be assigned to this task force with membership proportional to the number of the three types of law enforcement officers.

   a) Apportion this task force using the Hamilton method.

   b) Now increase the council's size to ten, and then to eleven, and so on, redoing the apportionment each time until an Alabama paradox occurs.

   c) What is the first size above ten when the paradox occurs, and what group loses an officer when the size of the task force increases?

13. Truncate the number 353.785378 to:

   a) tenths.   b) thousandths.

14. Truncate the number 1,453.1535821 to:

   a) hundredths.   b) millionths.

15. **Presenting survey results.** The sum of the numbers

$$12.56, 34.237, 25.40, 11.723, 16.08$$

is 100.000. Adjust the list so that the numbers are shown to the tenths place and their sum is 100.0. The following table shows the numbers truncated to the tenths place

Original Number	Number Truncated to Tenths	Discarded Portion
12.56	12.5	0.060
34.237	34.2	0.037
25.40	25.4	0.000
11.723	11.7	0.023
16.08	16.0	0.080
100.000	99.8	

and the discarded portions. Apportion the additional two tenths so that the sum of the resulting numbers is 100.0.

16. Presenting survey results. The sum of the numbers

$$2.356, 14.587, 25.62, 31.85, 12.60, 12.987$$

is 100.000. Adjust the list so that the numbers are shown to the tenths place and their sum is 100.0. The following table shows the numbers truncated to the tenths place and the discarded portions. Apportion the additional three tenths so that the sum of the resulting numbers is 100.0.

Original Number	Number Truncated to Tenths	Discarded Portion
2.356	2.3	0.056
14.587	14.5	0.087
25.62	25.6	0.020
31.85	31.8	0.050
12.60	12.6	0.000
12.987	12.9	0.087
100.000	99.7	

17. Adjusting lists of numbers. Adjust the list of numbers

$$34.789, 11.2, 23.897, 12.034, 2.987, 15.093$$

so that they are displayed to the hundredths place and their sum is 100.00.

18. Adjusting lists of numbers. Adjust the list of numbers

$$11.047, 27.333, 36.901, 3.006, 21.713$$

so that they are displayed to the hundredths place and their sum is 100.00.

19. Presenting survey results. You have collected the following data for an article you are writing on public opinion regarding legalized gambling in your state. Your editor thinks your data are appropriate but should be published with the percentages shown to the tenths place. Adjust the numbers to accomplish this.

Opinion	Percentages
Allow gambling with no restrictions	12.45%
Allow gambling, but with limited stakes	31.11
Allow gambling, but only in western counties	36.02
Do not allow gambling anywhere in the state	15.73
No opinion	4.69
Total	100.00%

20. Presenting survey results. Suppose the newspaper editor in Exercise 19 asks to have the data shown as whole percentages—that is, integers with no fractional parts. Adjust the list of percentages to achieve this objective.

Some of the following lists of data contain numbers that are not percentages and therefore their sum is not 100.

21. Presenting data. Adjust the list of numbers

$$101.45328, 96.00059, 33.98023, 67.1329$$

so that they are displayed to the thousandths place and their sum is 298.567.

22. Presenting data. Adjust the list of numbers

$$6.323, 11.046, 32.7839, 3.8053, 21.004, 0.4578$$

so that they are displayed to the hundredths place and their sum is 75.42.

23. If the American Nurses Association has 177,408 members and three representatives on the National Health Board, what is the association's average constituency?

24. If the International Brotherhood of Electrical Workers has 819,000 members and thirteen representatives on the Building Trades Council, what is its average constituency?

25. Which state is more poorly represented: state A with a population of 27,600 and sixteen representatives, or state B with a population of 23,100 and fourteen representatives? What is the absolute unfairness of this apportionment? What is the relative unfairness of this apportionment?

26. Which state is more poorly represented: state C with a population of 85,800 and eleven representatives, or state D with a population of 86,880 and twelve representatives? What is the absolute unfairness of this apportionment? What is the relative unfairness of this apportionment?

27. Recall that on a ten-member board, Naxxon, with 4,700 stockholders, received five members and Aroco, with 3,700 stockholders, received four members. Calculate the absolute and relative unfairness of this apportionment.

28. Redo Exercise 27 for Aroco and Eurobile. Recall that Eurobile had 1,600 stockholders and one board member.

29. According to a recent census, Colorado had a population of approximately 2.2 million people, while Delaware had 0.6 million people. Colorado was allotted five seats in the House and Delaware was allotted one. Calculate the absolute and relative unfairness of this apportionment.

30. According to a recent census, Iowa's population was approximately 2.8 million people while Utah had 1 million people. Iowa was allotted six seats and Utah was allotted two. Calculate the absolute and relative unfairness for this apportionment.

31. Assume that on the oil consortium board, Naxxon currently receives five representatives and Aroco receives four. Suppose one additional representative can be given to either Naxxon or Aroco.

a) Calculate the relative unfairness of the apportionment if the additional representative is given to Naxxon.

b) Calculate the relative unfairness of the apportionment if the additional representative is given to Aroco.

c) Based on your answers to parts (a) and (b), which company should get the additional representative? Why?

32. Assume that on the oil consortium board, Naxxon currently receives six representatives and Eurobile receives two. Suppose one additional representative can be given to either Naxxon or Eurobile.

a) Calculate the relative unfairness of the apportionment if the additional representative is given to Naxxon.

b) Calculate the relative unfairness of the apportionment if the additional representative is given to Eurobile.

c) Based on your answers to parts (a) and (b), which company should get the additional representative? Why?

33. Communicating Mathematics   Explain what we mean by an Alabama paradox.

34. Communicating Mathematics   Why can an Alabama paradox occur when using the Hamilton method?

35. Communicating Mathematics   What must we do to avoid an Alabama paradox?

36. Communicating Mathematics   Why was the notion of relative unfairness necessary?

9.2   THE HUNTINGTON–HILL APPORTIONMENT PRINCIPLE

Recall that an Alabama paradox may occur if we reapportion seats that have already been assigned. Therefore we will avoid this paradox if we apportion *only* the new seats when the representative body increases in size. Thus, we want a method that tells us at each stage in the apportionment process who gets the next seat. We use relative unfairness and the following criterion in making the decision.

Apportionment Criterion

When assigning a representative among several parties, make the assignment so as to give the smallest relative unfairness.

EXAMPLE 1 Using the Apportionment Criterion

Suppose that state A has a population of 13,680 and five representatives, and state B has a population of 6,180 and two representatives. Use the apportionment criterion to determine which state is more deserving of one additional representative.

SOLUTION: Recall from Section 9.1 that to determine relative unfairness, we first compute the average constituency for each state after an assignment of seats has been made. Then the relative unfairness is

$$\frac{\text{the absolute unfairness of the apportionment}}{\text{the smaller average constituency of the two states}} =$$

$$\frac{\text{larger average constituency} - \text{smaller average constituency}}{\text{smaller average constituency}}$$

Now let us see what happens if we first assign the additional representative to state A. If state A gets the additional representative, then it will have six, whereas state B has two. The average constituencies are then as follows:

$$\text{state A: } \frac{13,680}{6} = 2,280 \qquad \text{state B: } \frac{6,180}{2} = 3,090$$

Because state A has the smaller average constituency, the relative unfairness of this apportionment is

$$\frac{3,090 - 2,280}{2,280} = \frac{810}{2,280} \approx 0.355.$$

Next we compute the relative unfairness if the representative is given instead to state B. In this case, state B will have three representatives and state A will have five. The average constituencies are as follows:

$$\text{state A: } \frac{13,680}{5} = 2,736 \qquad \text{state B: } \frac{6,180}{3} = 2,060$$

B now has the smaller average constituency, so the relative unfairness of this apportionment is

$$\frac{2,736 - 2,060}{2,060} = \frac{676}{2,060} = 0.328.$$

Because a smaller relative unfairness occurs if state B is given the additional representative, state B should be assigned a third representative before state A receives a sixth one. ◎

Quiz Yourself 7*

State A, with a population of 41,440, presently has seven representatives. State B, with a population of 25,200, has four representatives.

a) Determine the relative unfairness of the apportionment if we give an additional representative to state A.

b) Determine the relative unfairness of the apportionment if this representative is given instead to state B.

c) Use the apportionment criterion to decide which state should receive the additional representative.

Note that our solution in Example 1 violates the Hamilton method of apportionment. To see how, assume that we are apportioning eight representatives between states A and B. The exact number of representatives due state A is

$$8 \times \frac{13,680}{13,680 + 6,180} = 8 \times \frac{13,680}{19,860} \approx 5.511,$$

whereas the exact number due state B is

$$8 \times \frac{6,180}{19,860} \approx 2.489.$$

Thus, using the Hamilton method, state A would get the extra representative.

Comparing the relative unfairness of two apportionments as we did in Example 1 is tedious. There is an easier way to determine which of two states is more deserving of an additional representative. To understand this method, let us return to the problem in Example 1 again. This time we will use a and b for the populations of states A and B, respectively. If we give five representatives to state A and three to state B, we can write the relative unfairness of this assignment as

$$\frac{\frac{a}{5} - \frac{b}{3}}{\frac{b}{3}}.$$

Next we simplify this as follows:

$$\frac{\frac{a}{5} - \frac{b}{3}}{\frac{b}{3}} = \frac{\frac{a}{5}}{\frac{b}{3}} - \frac{\frac{b}{3}}{\frac{b}{3}} = \frac{\frac{a}{5}}{\frac{b}{3}} - 1 = \frac{3a}{5b} - 1 \qquad (1)$$

If instead we give the additional representative to state A rather than state B, the relative unfairness is

$$\frac{\frac{b}{2} - \frac{a}{6}}{\frac{a}{6}} = \frac{\frac{b}{2}}{\frac{a}{6}} - \frac{\frac{a}{6}}{\frac{a}{6}} = \frac{\frac{b}{2}}{\frac{a}{6}} - 1 = \frac{6b}{2a} - 1. \qquad (2)$$

In Example 1, we saw that state B should receive the additional representative because that resulted in a smaller relative unfairness. We can rewrite this as follows:

(relative unfairness if extra representative is given to B) $<$

(relative unfairness if extra representative is given to A)

Using equations (1) and (2), we can express this algebraically as

$$\frac{a}{5} \cdot \frac{3}{b} - 1 < \frac{b}{2} \cdot \frac{6}{a} - 1.$$

If we add 1 to both sides of this inequality, we get

$$\frac{a}{5} \cdot \frac{3}{b} < \frac{b}{2} \cdot \frac{6}{a}. \qquad (3)$$

Since a and b represent populations and are therefore positive, we see that $\frac{a}{6} \cdot \frac{b}{3}$ is also positive. Multiplying both sides of inequality (3) by this number we get

$$\left(\frac{a}{6} \cdot \frac{b}{3}\right)\frac{a}{5} \cdot \frac{3}{b} < \left(\frac{a}{6} \cdot \frac{b}{3}\right)\frac{b}{2} \cdot \frac{6}{a}.$$

Canceling common factors from the numerators and denominators of both sides of this inequality gives us the equivalent inequality

$$\frac{a^2}{5 \cdot 6} < \frac{b^2}{2 \cdot 3}. \qquad (4)$$

The point we are making is that the word inequality

(relative unfairness if extra representative is given to B) $<$

(relative unfairness if extra representative is given to A)

is equivalent to the algebraic inequality*

$$\frac{a^2}{5 \cdot 6} < \frac{b^2}{2 \cdot 3}.$$

Therefore to determine whether state A or state B deserves the additional representative, we could compute for each state a number of the form

$$\frac{(\text{population of the state})^2}{(\text{number of representatives}) \cdot [(\text{number of representatives}) + 1]}$$

and compare the two. The larger number indicates which state should receive the additional representative. These observations for states A and B lead us to the following principle.

The Huntington–Hill Apportionment Principle

If states X and Y have already been allotted x and y representatives, respectively, then state X should be given an additional representative in preference to state Y provided that

$$\frac{(\text{population of Y})^2}{y \cdot (y + 1)} < \frac{(\text{population of X})^2}{x \cdot (x + 1)}.$$

Otherwise, state Y should be given the additional representative. We will often refer to a number of the form $\dfrac{(\text{population of X})^2}{x \cdot (x + 1)}$ as a **Huntington–Hill number**.

---

* Although we began by comparing the relative unfairness of two apportionments, the quantities $\dfrac{a^2}{5 \cdot 6}$ and $\dfrac{b^2}{2 \cdot 3}$ are simply algebraic expressions. *They are not measures of relative unfairness.*

Historical Highlight: *Is There a Perfect Apportionment Method?*

The Huntington–Hill apportionment principle was developed by Edward Huntington and Joseph Hill, two mathematicians who were classmates at Harvard. This method was signed into law by Franklin D. Roosevelt in 1941 and is currently used to apportion the U.S. House of Representatives after each census.

Although the Huntington–Hill method avoids the Alabama paradox, we should ask, is the Huntington–Hill method perfect? In fact, since there have been many apportionment methods proposed over the course of U.S. history, we might ask, does a perfect apportionment method exist?

To answer this question, we must specify criteria that an apportionment method should meet. The following two conditions are often expected:

- An apportionment method should not be subject to the Alabama paradox or other similar types of paradoxes.*
- An apportionment should satisfy the *quota rule*. That is, if the exact number of representatives due to a state

is 31.462, then we expect the number apportioned to be either 31 or 32. It is not acceptable for the state to be given 33 or even 30 representatives.

In 1980, Michael Balinski and H. Peyton Young proved a surprising theorem, known as **Balinski and Young's Impossibility Theorem,** which states:

There is no apportionment method that avoids all paradoxes and at the same time satisfies the quota rule.

In other words, if an apportionment method satisfies the quota rule, then it is always possible to produce a paradox using that method. Conversely, if a method is paradox-free, then we can find examples in which the quota rule is not satisfied. Thus, any apportionment method must be flawed. Because of this, the politics of which states benefit and which suffer often plays as large a role as mathematics when Congress discusses an apportionment method.

Quiz Yourself 8

According to a recent census, Iowa had a population of approximately 2.8 million people and six representatives to the U.S. House of Representatives; and Nebraska had a population of 1.6 million people and 3 representatives. Use the Huntington–Hill apportionment principle to determine which state is most deserving of an additional representative.

EXAMPLE 2   Using the Huntington–Hill Apportionment Principle

There are 320 full-time nurses and 148 part-time nurses at Community General Hospital. The nursing supervisor has chosen four full-time nurses and two part-time nurses to serve on a committee to evaluate proposed nursing guidelines. Use the Huntington–Hill apportionment principle to decide whether the seventh nurse on the committee should be full-time or part-time.

SOLUTION: We compute the Huntington–Hill numbers for the full-time nurses and for the part-time nurses:

$$\frac{(\text{number of full-time nurses})^2}{(\text{current representation}) \cdot (\text{current representation} + 1)} = \frac{(320)^2}{4 \cdot 5} = 5{,}120$$

$$\frac{(\text{number of part-time nurses})^2}{(\text{current representation}) \cdot (\text{current representation} + 1)} = \frac{(148)^2}{2 \cdot 3} \approx 3{,}651$$

Comparing these numbers, we find that the next nurse selected for the committee should be a full-time nurse.

---

* There are other types of paradoxes that we will discuss in Section 9.4. The interested reader may want to read M. L. Balinski and H. P. Young, *Fair Representation: Meeting the Ideal of One Man, One Vote* (Yale University Press, 1982).

Example 3 shows how we can use the Huntington–Hill apportionment principle to apportion representatives among more than two parties.

EXAMPLE 3 **Apportioning Representatives among Three States**

Assume that the oil consortium board currently has two members from Naxxon, two from Aroco, and one from Eurobile. Use the Huntington–Hill apportionment principle to decide which company should receive the next member on the board.

SOLUTION: Recall that Naxxon had 4,700 stockholders, Aroco had 3,700, and Eurobile had 1,600. We compute the Huntington–Hill numbers for each company.

$$\text{Naxxon:} \frac{47^2}{2 \cdot 3} \approx \mathbf{368} \qquad \text{Aroco:} \frac{37^2}{2 \cdot 3} \approx 228 \qquad \text{Eurobile:} \frac{16^2}{1 \cdot 2} = 128$$

Thus Naxxon should get the next representative, because it has the largest Huntington–Hill number.

The Huntington–Hill apportionment principle tells us at any stage in an apportionment process which party most deserves the next seat. As we saw in Example 3, we can also use this principle when allocating representatives to more than two parties. Because this method meets the apportionment criterion we stated earlier, we will use it to apportion the entire oil consortium board in Section 9.3.

Exercises 9.2

1. The Musicians' Guild, with 908 members, has four representatives on the National Arts Advisory Board (NAAB); the Artists' Alliance, with 633 members, has three representatives. If either the musicians or the artists can be given one additional representative to the NAAB, determine which group should get it by doing the following steps:

a) Determine the relative unfairness of the apportionment if we give an additional representative to the musicians.

b) Determine the relative unfairness of the apportionment if this representative is given to the artists instead.

c) Use the apportionment criterion to decide which group deserves the additional representative more.

2. The 1,218-member carpenters' union has six representatives on the state labor council, and the 720-member plumbers' union has four representatives. If either the carpenters or the plumbers can have another representative on the council, which union is more deserving of this representative? Use the method outlined in Exercise 1.

3. The Metro City Transit System (MCTS) is made up of the red line and the blue line. The red line has nine cars and averages 405 passengers per run. The blue line has seven cars and averages 287 passengers per run. Use the method outlined in Exercise 1 to determine which line is more deserving of an additional car.

4. The Family Services Agency has two offices. The Grand Lakes office has seven caseworkers and a caseload of 595 clients. The Plains City office has thirteen caseworkers and 819 clients. Use the method outlined in Exercise 1 to determine which office is more deserving of an additional caseworker.

In Exercises 5–8, we provide the populations* of two states and their representation in the U.S. House of Representatives. Use the apportionment criterion to decide which state is more deserving of an additional representative in the House.

5. Alabama: population 4,221,826, 7 representatives

   New York: population 17,990,778, 31 representatives

6. Michigan: population 9,295,287, 16 representatives

   Pennsylvania: population 11,882,842, 21 representatives

7. Colorado: population 3,294,473, 6 representatives

   Delaware: population 666,168, 1 representative

8. Florida: population 12,938,071, 23 representatives

   California: population 29,785,857, 52 representatives

In Exercises 9–16, calculate the Huntington–Hill number for each party. To keep the calculations manageable, round the populations to the nearest tenth of a million. For example, for Florida use 12.9 million.

9. The Musicians' Guild in Exercise 1

10. The carpenters' union in Exercise 2

11. The blue line in Exercise 3

12. The Grand Lakes office in Exercise 4

13. Alabama (see Exercise 5)

14. Michigan (see Exercise 6)

15. Colorado (see Exercise 7)

16. Florida (see Exercise 8)

In Exercises 17–20, use the Huntington–Hill apportionment principle to decide which state is most deserving of an additional representative in the U.S. House of Representatives.

17. The populations of Indiana and Illinois are 5.5 million and 11.4 million, respectively. Indiana has 10 representatives and Illinois has 20 representatives.

18. The populations of Maine and New Hampshire are 1.2 million and 1.1 million, respectively. The two states each have 2 representatives.

19. The populations of Alaska, New Hampshire, and Wyoming are 0.6 million, 1.1 million, and 0.5 million, respectively. Alaska has 1 representative, New Hampshire has 2, and Wyoming has 1.

20. Tennessee has a population of 4.9 million with 9 representatives; West Virginia has a population of 1.8 million with 3 representatives; and Georgia has a population of 6.5 million with 11 representatives.

Exercises 21–24 refer to Exercises 1–4, respectively. In each case we add one more party to the situation described earlier. Compute Huntington–Hill numbers to decide which of the three parties is most deserving of an additional object.

21. The Actors' Coalition has 420 members and two members on the NAAB.

22. The electricians' union has 681 members and three representatives on the labor council.

23. The MCTS adds a yellow line with three cars that average 156 passengers per run.

24. The new Family Services Agency office in Great Mountain has five caseworkers with 405 clients.

25. Reconsider Exercise 21. Suppose that instead of one member, we can add two members to the NAAB. How should these two members be allocated? (*Hint:* Do not assign both additional members at the same time. Use the Huntington–Hill apportionment principle to assign one member and then use the principle again to assign the second.)

26. Reconsider Exercise 23. Suppose that instead of one car, we can add two cars to the MCTS. How should these two cars be assigned?

27. Is the assignment of the 11 representatives to the NAAB in Exercise 25 consistent with the Hamilton apportionment method? Explain.

28. Is the assignment of the 21 cars to the MCTS in Exercise 26 consistent with the Hamilton apportionment method? Explain.

29. Communicating Mathematics   Explain the thinking behind the apportionment principle.

---

* All state populations in these exercises are based on a recent census.

30. Communicating Mathematics In using the Huntington–Hill apportionment principle, we use the inequality

$$\frac{(\text{population of } Y)^2}{y \cdot (y + 1)} < \frac{(\text{population of } X)^2}{x \cdot (x + 1)}.$$

How did we get that inequality? What is its purpose?

31. Communicating Mathematics How is the Huntington–Hill apportionment method different from Hamilton's method?

32. Communicating Mathematics Why must the Huntington–Hill method avoid the possibility of an Alabama paradox occurring?

Further Exercises

*(33.) Michigan has a population of approximately 9.3 million and 16 representatives, while Wisconsin has a population of approximately 4.9 million and 9 representatives. Suppose Wisconsin's population starts growing while Michigan's remains constant. Further, suppose with Wisconsin's growth the 25 representatives between the two states would be redistributed using the Huntington–Hill method so that Wisconsin will have 10 seats in the House. Now, how large must Wisconsin's population become to take a representative away from Michigan? Explain how you arrived at your answer.

(34.) New Jersey has a population of 7.7 million and 13 representatives, while Washington has a population of 4.9 million and 9 representatives. Suppose Washington's population starts growing while New Jersey's remains constant. Further, suppose with Washington's growth the 22 representatives between the two states would be redistributed using the Huntington–Hill method so that Washington will have 10 seats in the House. Now, how large must Washington's population become to take a representative away from New Jersey? Explain how you arrived at your answer.

35. Are the situations that we described in Exercises 33 and 34 examples of an Alabama paradox? Explain your answer.

9.3 APPLICATIONS OF THE APPORTIONMENT PRINCIPLE

We will now use the Huntington–Hill principle to apportion the oil consortium board. Recall that Naxxon has 4,700 stockholders, Aroco has 3,700, and Eurobile has 1,600. We will assign the representatives one at a time until we have filled all nine seats on the council.

It would seem unjust if any company had no representatives, so we begin by giving one seat to each company. This is consistent with a provision in the U.S. Constitution that says each state must have at least one representative. Now that three seats have been filled, let us assign the remaining six seats.

We proceed as follows: Each time we allocate a representative, we will calculate the Huntington–Hill number

$$\frac{(\text{number of stockholders in the company})^2}{(\text{current representation}) \cdot (\text{current representation} + 1)}$$

---

* Exercise numbers circled in red can be used as group exercises.

for all three companies. The largest of these numbers determines which company is most deserving of the next representative. At present each company has one representative, so we compute the numbers:

Naxxon	Aroco	Eurobile
$\dfrac{(47)^2}{1 \times 2} = 1104.5$	$\dfrac{(37)^2}{1 \times 2} = 684.5$	$\dfrac{(16)^2}{1 \times 2} = 128$

Since Naxxon has the largest Huntington–Hill number, it should be given its second representative in preference to Aroco and Eurobile.

Seat Number	Goes To	Number of Representatives Naxxon Has	Number of Representatives Aroco Has	Number of Representatives Eurobile Has
1	N	1	0	0
2	A	1	1	0
3	E	1	1	1
4	N	2	1	1
5	?			

To decide on the fifth seat, we only need to recompute the Huntington–Hill number for Naxxon, because its representation is the only one that has changed. The Huntington–Hill numbers for the three companies are now as follows:

Naxxon	Aroco	Eurobile
$\dfrac{(47)^2}{2 \times 3} \approx 368.2$	$\dfrac{(37)^2}{1 \times 2} = 684.5$	$\dfrac{(16)^2}{1 \times 2} = 128$

These calculations show that Aroco should receive the fifth seat, so we add the following line to the table.

5	A	2	2	1
6	?			

We continue in this fashion to build Table 9.7.

We did not compute the entries marked with an asterisk (*) because, for example, it is unlikely that Aroco will be allotted five or more representatives. If our intuition is wrong, we can easily calculate the necessary entries. We write the entries from Table 9.7 in decimal form in Table 9.8.

Quiz Yourself 9

Calculate Huntington–Hill numbers to determine which company gets the sixth seat on the consortium board. We are assuming that Naxxon and Aroco have two representatives each and Eurobile has one.

Current Representation	Naxxon	Aroco	Eurobile
1	$\dfrac{(47)^2}{1 \times 2}$	$\dfrac{(37)^2}{1 \times 2}$	$\dfrac{(16)^2}{1 \times 2}$
2	$\dfrac{(47)^2}{2 \times 3}$	$\dfrac{(37)^2}{2 \times 3}$	$\dfrac{(16)^2}{2 \times 3}$
3	$\dfrac{(47)^2}{3 \times 4}$	$\dfrac{(37)^2}{3 \times 4}$	$\dfrac{(16)^2}{3 \times 4}$
4	$\dfrac{(47)^2}{4 \times 5}$	$\dfrac{(37)^2}{4 \times 5}$	*
5	$\dfrac{(47)^2}{5 \times 6}$	*	*
6	$\dfrac{(47)^2}{6 \times 7}$	*	*

TABLE 9.7 Huntington–Hill numbers for the oil consortium board.

EXAMPLE 1 **Using Huntington–Hill Numbers to Allocate a Seat on the Council**

Use Table 9.8 to decide which company receives the seventh seat.

Current Representation	Naxxon	Aroco	Eurobile
1	1,104.5	684.5	**128.0**
2	368.2	**228.2**	42.7
3	**184.1**	114.1	21.3
4	110.5	68.5	*
5	73.6	*	*
6	52.6	*	*

TABLE 9.8 Huntington–Hill numbers in decimal form for the oil consortium board.

Quiz Yourself 10

Use Table 9.8 to determine which company should get the eighth seat on the oil consortium board.

SOLUTION: Naxxon has three members, Aroco has two, and Eurobile has one, so we compare the highlighted entries of Table 9.8 to make our decision. Because 228.2 is greater than either 184.1 or 128.0, Aroco receives the seventh seat. ✺

SOME GOOD ADVICE

Once you have used the largest of three Huntington–Hill numbers to apportion a representative, cross that number off so that you do not use it again.

We can add the following line to the table, which summarizes how the seats have been assigned so far.

Seat Number	Goes To	Number of Representatives Naxxon Has	Number of Representatives Aroco Has	Number of Representatives Eurobile Has
6	N	3	2	1
7	A	3	3	1

Table 9.9 shows how to allocate all nine members to the oil consortium board.

Seat Number	Goes To	Number of Representatives Naxxon Has	Number of Representatives Aroco Has	Number of Representatives Eurobile Has
1	N	1	0	0
2	A	1	1	0
3	E	1	1	1
4	N	2	1	1
5	A	2	2	1
6	N	3	2	1
7	A	3	3	1
8	N	4	3	1
9	E	4	3	2
10	A	4	4	2

TABLE 9.9 Apportionment of the oil consortium board.

Thus in a nine-member board, Naxxon receives four, Aroco receives three, and Eurobile two. We also see in Table 9.9 that for a ten-member board, Aroco should be given the additional representative, and the representation for the other companies remains the same.

The procedure we used in this example satisfies two important criteria. First, it prevents the Alabama paradox from occurring; second, using Table 9.8 to determine which company is to receive the next seat, the assignment can be made so that the relative unfairness between any two companies is minimal. Furthermore, it can be proved that the relative unfairness between any two states cannot be improved, even by transferring a representative from one state to another when the apportionment is done as we described.* Although the Huntington–Hill method is not flawless because it can violate the quota rule, it is the method currently used to apportion Congress. It is still the subject of lively debate by mathematicians and politicians alike.

---

* E. V. Huntington, "The Apportionment of Representatives in Congress," *Trans. Am. Math. Soc.* 30 (1928), pp. 85–110.

Highlight: *Using a Spreadsheet to Compute Huntington–Hill Numbers*

Generating many Huntington–Hill numbers is a long and tedious process and, of course, subject to error. A spreadsheet* can both avoid the tedium and ensure accuracy. The top spreadsheet shown here calculates the Huntington–Hill numbers listed in Table 9.8. (In fact, we used it to compute all the Huntington–Hill numbers in this section.) Spreadsheet 1 shows the formulas used to calculate the Huntington–Hill numbers, whereas Spreadsheet 2 shows the numbers after they have been computed.

	A	B	C
1	47	37	16
2	(A1*A1)/(1*2)	(B1*B1)/(1*2)	(C1*C1)/(1*2)
3	(A1*A1)/(2*3)	(B1*B1)/(2*3)	(C1*C1)/(2*3)
4	(A1*A1)/(3*4)	(B1*B1)/(3*4)	(C1*C1)/(3*4)
5	(A1*A1)/(4*5)	(B1*B1)/(4*5)	(C1*C1)/(4*5)
6	(A1*A1)/(5*6)	(B1*B1)/(5*6)	(C1*C1)/(5*6)
7	(A1*A1)/(6*7)	(B1*B1)/(6*7)	(C1*C1)/(6*7)

SPREADSHEET 1 Formulas used to calculate Huntington–Hill numbers for apportioning the oil consortium board.

	A	B	C
1	47	37	16
2	1104.5	684.5	128.0
3	368.2	228.2	42.7
4	184.1	114.1	21.3
5	110.5	68.5	12.8
6	73.6	45.6	8.5
7	52.6	32.6	6.1

SPREADSHEET 2 Huntington–Hill numbers for apportioning the oil consortium board.

If we were apportioning a large body such as the U.S. House of Representatives, a tool such as a spreadsheet helps to keep the calculations manageable.

---

* See www.aw.com/pirnot for this spreadsheet.

■

We can use the Huntington–Hill method to allocate objects other than representatives.

If we view apportionment as a process by which we assign objects (namely, the representatives) to interested parties (namely, the states), then we can recognize other situations in which the same problem arises. In Example 2, we discuss a different situation that has some similarities to it.

EXAMPLE 2   **Using the Huntington–Hill Method to Assign Police Officers**

Suppose a small town has seven police officers. Furthermore, assume that the town is divided rather naturally into three regions. The police chief has gathered data regarding the number of incidents (crimes, traffic accidents, and so on) that have occurred in each region over the past several months, as listed in Table 9.10. If the chief decides to assign officers for duty in proportion to the number of incidents per region, how should he make these assignments?

SOLUTION: We can look at this as an apportionment problem in which the objects being apportioned are the officers and the parties receiving the objects are the regions. We do not want any region to lose a police officer if there are future increases in the size of the police force. Consequently, we will use the Huntington–Hill method to apportion the seven officers among the three regions.

We first compute the Huntington–Hill numbers in Table 9.11. Instead of using populations as we did earlier, we used the number of incidents per region to calculate these numbers. As usual, we start by giving one officer to each region. For example, $\frac{(107)^2}{1 \times 2} = 5724.5$ is the Huntington-Hill number for region 1 if it already has one officer assigned.

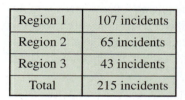

Region 1	107 incidents
Region 2	65 incidents
Region 3	43 incidents
Total	215 incidents

TABLE 9.10   Number of police incidents.

If Given the Next Officer, the Number a Region Would Have Is:	Region 1	Region 2	Region 3
2	(a) 5724.5	(b) 2112.5	(e) 924.5
3	(c) 1908.2	(f) 704.2	308.2
4	(d) 954.1	352.1	154.1
5	572.5	211.3	92.5

TABLE 9.11   Huntington–Hill numbers for assigning police officers.

Considering the entries in the table labeled (a) through (d), we see that after one officer has been assigned to each of the three regions, the remaining four officers should be assigned as follows:

(a) Officer 4 goes to region 1
(b) Officer 5 goes to region 2
(c) Officer 6 goes to region 1
(d) Officer 7 goes to region 1

If the town hires an eighth officer, looking at entries (e) and (f) we see that the next assignment should be made to region 3.

Exercises 9.3

A labor council is being formed from the members of three unions. The electricians' union has 25 members, the plumbers have 18 members, and the carpenters have 31 members. The council will have seven representatives, with each union having at least one representative. Use the Huntington–Hill numbers listed in Table 9.12 to answer Exercises 1–4.

1. Knowing that each union currently has one representative, which entries in Table 9.12 do we use to assign the fourth seat on the council and which union gets that seat?

2. Assume that the carpenters have two representatives, whereas the electricians and plumbers each have one. Which entries in Table 9.12 do we use to assign the fifth seat on the council and which union gets that seat?

3. Suppose the current council has five seats, apportioned so that the electricians and carpenters have two seats each and the plumbers have one. Which entries in Table 9.12 do we use to assign the sixth seat on the council and which union gets that seat?

4. If the current council has six seats, each union has two representatives, which entries in Table 9.12 do we use to assign the seventh seat on the council and which union gets that seat?

We want to apportion ten seats on a condominium association board to buildings Allen, Baker, and Carney. There are 23 units in Allen, 33 in Baker, and 43 in Carney. The board will have ten seats, with each building having at least one seat. Use the Huntington–Hill numbers listed in Table 9.13 to answer Exercises 5–8.

5. Knowing that each building currently has one representative, which entries in Table 9.13 do we use to assign the fourth seat on the board and which building gets that seat?

6. Assume Carney has two representatives, whereas Allen and Baker each have one. Which entries in Table 9.13 do we use to assign the fifth seat on the board and which building gets that seat?

7. Suppose the current assignment of representatives to five seats on the council is Baker and Carney each with two and Allen with one. Which entries in Table 9.13 do we use to assign the sixth seat on the board and which building gets that seat?

8. Suppose the current assignment of representatives to six seats on the board is Carney with three, Baker with two, and Allen with one. Which entries in Table 9.13 do we use to assign the seventh seat on the board and which building gets that seat?

Current Representation	Electricians	Plumbers	Carpenters
1	312.5	162.0	480.5
2	104.2	54.0	160.2
3	52.1	27.0	80.1
4	31.3	*	48.1

TABLE 9.12 Huntington–Hill numbers for apportioning a labor council.

Current Representation	Allen	Baker	Carney
1	264.5	544.5	924.5
2	88.2	181.5	308.2
3	44.1	90.8	154.1
4	26.5	54.5	92.5
5	17.6	36.3	61.6
6		25.9	44.0
7			33.0

TABLE 9.13 Huntington–Hill numbers for apportioning a condominium board.

Current Representation	(T)heater	(M)usic	(D)ance
1	88,200.0	94,612.5	10,512.5
2	29,400.0	31,537.5	3,504.2
3	14,700.0	15,768.8	1,752.1
4	8,820.0	9,461.3	*
5	5,880.0	6,307.5	*
6	4,200.0	4,505.4	*
7	3,150.0	3,379.0	*

TABLE 9.14 Huntington–Hill numbers for apportioning a performing arts board.

9. **Apportioning an arts board.** Ten seats on a state performing arts board are to be apportioned. In the state, there are 420 participating in theater, 435 in music, and 145 in dance. Use Table 9.14, which contains the Huntington–Hill numbers, to apportion the ten seats to this board. Begin by giving one representative to each area and then assign the remaining seven seats. List the order in which the representatives are apportioned.

10. **Allocating fellowships.** Eight fellowships are to be apportioned among the schools of humanities, science, and business at a small  university. The apportionment is based on the number of full-time graduate students in each school, which is 30 for humanities, 20 for science, and 40 for business. Use the Huntington–Hill numbers in Table 9.15 to carry out the apportionment.

Current Representation	(H)umanities	(S)cience	(B)usiness
1	450.0	200.0	800.0
2	150.0	66.7	266.7
3	75.0	33.3	133.3
4	45.0	*	80.0
5	*	*	53.3

TABLE 9.15 Huntington–Hill numbers for allocating fellowships.

Begin by giving one fellowship to each school. List the order in which the fellowships are apportioned.

11. **Apportioning a city council.** A city is made up of three boroughs—Alsace, Bradford, and Cambria. Representation on a ten-member city council is allocated in proportion to the population in each of the three boroughs. Alsace has a population of 23,000, Bradford has 34,000, and Cambria has 14,000. Apportion the ten council seats

a) using the Hamilton method.

b) using the Huntington–Hill apportionment principle.

12. **Assigning members to a labor council.** The Unified Labor Council has eleven members representing the carpenters', electricians', plumbers', and painters' unions. The four trades are represented on the council in proportion to the size of their membership. If there are 84 carpenters, 48 electricians, 40 plumbers, and 28 painters, how many from each trade should be on the council

a) using the Hamilton method?

b) using the Huntington–Hill apportionment principle?

13. **Apportioning representatives.** Use the Huntington–Hill method to apportion ten representatives among Utah, Idaho, and Oregon. The recent populations of the states are Utah, 1.7 million; Idaho, 1.0 million; and Oregon, 2.8 million. Begin by giving one representative to each state. List the order in which the representatives are apportioned.

14. **Apportioning representatives.** Use the Huntington–Hill method to apportion eleven representatives among Arizona, New Mexico, and Nevada. The recent populations of the states are Arizona, 3.7 million; New Mexico, 1.5 million; and Nevada, 1.2 million. Begin by giving one representative to each state. List the order in which the representatives are apportioned.

15. **Apportioning representatives.** Use the Huntington–Hill method to apportion eleven representatives among Arkansas, Kansas, and Nebraska. The recent populations of the states are Arkansas, 2.4 million; Kansas, 2.5 million; and Nebraska,

1.6 million. Begin by giving one representative to each state. List the order in which the representatives are apportioned.

16. Refer to your answer to Exercise 9 and Table 9.14. Suppose the arts board is increased to twelve representatives. How should the two new seats be distributed and how many representatives would each area have on the twelve-member council?

17. Refer to your answer to Exercise 10 and Table 9.15. Suppose that three additional fellowships are to be awarded. How should the three new fellowships be distributed and which schools will get them?

18. Assigning medical personnel. A hospital administrator wishes to assign seven emergency room medical teams to three community outreach centers. The number of patients treated last week at each center is listed in the given table. Begin by giving each center one team. Use the Huntington–Hill method to decide how the administrator should assign the medical teams to these three centers.

Center 1	98 patients
Center 2	34 patients
Center 3	57 patients
Total	189 patients

19. Assigning police officers. Consider the problem of apportioning police officers given in Example 2. Suppose we still want to apportion the seven officers among the three regions, but the number of incidents per region has changed. The number of related incidents for regions 1, 2, and 3 are listed in the table.

Region 1	123 incidents
Region 2	44 incidents
Region 3	79 incidents
Total	246 incidents

How should the seven officers be assigned to the regions, given this new data?

20. Scheduling fitness classes. A health club instructor has a course load that allows her to teach six two-credit-hour classes. A pre-registration survey indicates the following interests.

> 56 want to take high-impact aerobics
> 29 want to take low-impact aerobics
> 11 want to take Jazzercise
> 4 want to take step exercise

Assume that the instructor will teach at least one class for each activity.

a) Use the Huntington–Hill method to apportion the instructor's remaining two classes among the four activities.

b) Does your solution in part (a) agree with your intuition? Explain.

21. Scheduling fitness classes. Consider again the problem described in Exercise 20. But now do the apportionment when the pre-registration numbers are as follows:

> 58 want to take high-impact aerobics
> 32 want to take low-impact aerobics
> 11 want to take Jazzercise
> 39 want to take step exercise

Further Exercises

*(22.) **Assigning police officers.** Consider again the problem of assigning officers as described in Example 2. As stated, it would not be appropriate to apply the Hamilton method, since an Alabama paradox could arise if the number of officers was increased. Verify that this is indeed the case.

a) Change the number of incidents in regions 1, 2, and 3.

b) Use the Hamilton method to apportion officers. Increase the number of officers until a region has a decrease in the number of officers assigned to it.

* Exercise numbers circled in red can be used as group exercises.

You may have to do step (a) several times until you get a reasonable example of an Alabama paradox.

23. **Assigning police officers.** Consider again the problem of assigning officers as described in Example 2. Table 9.11 contains the Huntington–Hill numbers used in solving this problem. Explain why the table is not adequate if the number of officers to be apportioned increased to nine—that is to say, why would we have to compute additional Huntington–Hill numbers to do the apportionment?

24. When constructing a table of Huntington–Hill numbers, it is not always necessary to continue the computation after a certain point. For example, consider the apportionment problem in Exercise 9 and Table 9.14. We did not compute the Current Representation entries 4 through 7 for dance. Explain why computing these entries is unnecessary. (*Hint:* Look at the last entries in the table under theater and music.)

25. Consider the Huntington–Hill numbers in Table 9.16, which we might use to apportion seats on a representative council. As usual, assume that we start the apportionment by giving each party one representative. Now, what would the smallest number of representatives on the council be so as to guarantee that B has three of them? Explain how to determine this number without actually carrying out an apportionment calculation.

Current Representation	A	B	C
1	32.00	60.50	112.50
2	10.67	20.17	37.50
3	5.33	10.08	18.75
4	3.20	6.05	11.25
5	2.13	4.03	7.50
6	1.52	2.88	5.36

TABLE 9.16   Huntington–Hill numbers for a representative council.

26. Make up an example of an apportionment situation in which the Hamilton and the Huntington–Hill methods give you the same apportionment.

27. Make up an example of an apportionment situation in which the Hamilton and the Huntington–Hill methods give you different apportionments.

9.4   OTHER PARADOXES AND APPORTIONMENT METHODS

In Section 9.1, we showed that although the Hamilton method was a straightforward way to apportion representatives, it was flawed because it was subject to the Alabama paradox. As you will see shortly, it also has some other serious problems. After discussing these other paradoxes, we will investigate several apportionment methods proposed by Thomas Jefferson, Daniel Webster, and John Quincy Adams.

We can explain apportionment using the standard divisor.

In order to understand the paradoxes and these other apportionment methods, we will have to look at apportionment in a slightly different way. Recall that to use the Hamilton method to compute the number of representatives* deserved by a state, we calculated the expression

$$\frac{\text{State's Population}}{\text{Total Population}} \times \text{Number of Representatives Being Allocated.}$$

If we represent this expression by the algebraic expression $\frac{s}{t} \times n$, we can then rewrite it as

$$\frac{s}{t} \times n = s \times \frac{n}{t} = \frac{s}{\frac{t}{n}}.$$

That is, in calculating the exact number of representatives that a state deserves, we could have first calculated

$$\frac{\text{Total Population}}{\text{Number of Representatives Being Allocated}},$$

and then divided this number into the state's population. Although this may not seem as intuitively clear as our original approach, this idea of first computing a divisor and then dividing into the state's population will be the central theme of our calculations for the rest of this section.

> ### DEFINITIONS
>
> In allocating a group of representatives among several states, we define the **standard divisor** by
>
> $$\text{Standard Divisor} = \frac{\text{Total Population}}{\text{Number of Representatives Being Allocated}},$$
>
> and a state's **standard quota** is defined to be
>
> $$\text{Standard Quota} = \frac{\text{State's Population}}{\text{Standard Divisor}}.$$

Intuitively, we can think of the standard divisor as the number of constituents that each representative must represent and the standard quota as the number of representatives that a state deserves. For example, if we wish to allocate eight representatives among several states that have a total population of 4,000,000, then the standard divisor is $\frac{4,000,000}{8} = 500,000$. So, each representative must represent 500,000 constituents. If one of the states has a population of

---

* In this section, we will continue to use the language of states, populations, and representatives, even though you have seen in earlier sections that the apportionment principles we discuss apply to many other situations.

1,500,000, then its standard quota is standard quota $= \frac{\text{state's population}}{\text{standard divisor}} = \frac{1,500,000}{500,000} = 3$, so the state deserves three representatives.

Example 1 shows that when we calculate the standard quotas, we get exactly the same results that we obtained in calculating the exact members deserved by Naxxon, Aroco, and Eurobile on the oil consortium board that we discussed in Section 9.1.

EXAMPLE 1   Calculating Standard Quotas for the Oil Consortium Board

Recall that we were apportioning a nine-member board to oversee the oil consortium formed by Naxxon, Aroco, and Eurobile. The members of the board would be apportioned according the number of stockholders of each company. Naxxon had 4,700 stockholders, Aroco had 3,700, and Eurobile had 1,600.

a) Calculate the standard divisor for this apportionment.

b) Calculate the standard quota for Naxxon, Aroco, and Eurobile.

SOLUTION: a) The total (population) of stockholders is

$$4,700 + 3,700 + 1,600 = 10,000.$$

To find the standard divisor, we divide this by the number of items being apportioned, which is nine, to get, $\frac{10,000}{9} \approx 1,111.11$.

b) To find the standard quota for each company, we divide the number of stockholders for the company by the standard divisor, as in Table 9.17.

Quiz Yourself 11*

Assume we are apportioning eight representatives among three states A, B, and C, which have populations of 3 million, 4 million, and 5 million, respectively.

a) Calculate the standard divisor for this apportionment.

b) Calculate each state's standard quota.

	Number of Stockholders	Standard Quota
Naxxon	4,700	$\frac{4,700}{1111.11} = 4.23$
Aroco	3,700	$\frac{3,700}{1111.11} = 3.33$
Eurobile	1,600	$\frac{1,600}{1111.11} = 1.44$

TABLE 9.17   Calculating the standard quotas for each oil company.

Notice that the standard quotas we found in Table 9.17 are the exact number of representatives that were due to each company as we calculated in Section 9.1.

SOME GOOD ADVICE

Keep in mind that the standard divisor is a single number that we calculate once and then use for the entire apportionment process. However, we must compute the standard quota individually for each state.

* Quiz Yourself answers begin on page 849.

Recall that in using the Hamilton apportionment method, we always gave each state a number of representatives that was immediately below or immediately above its standard quota (the exact number of representatives that it was due). For example, if a state's standard quota was 4.375 representatives, we always gave it either 4 or 5 representatives. We will now make this notion more precise.

> **DEFINITIONS**
>
> If we round the standard quota down, we call that number the **lower quota;** if we round the standard quota up, then we call that number the **upper quota.** If in making an apportionment, each state is allocated a number of representatives that is between its lower quota and upper quota, then we say the apportionment satisfies the **quota rule.**

For example, if in doing an apportionment, we find that a state's standard quota is 4.683, then 4 is the lower quota and 5 is the upper quota. We can use this new language to restate Hamilton's apportionment in slightly different terms.

> **Hamilton's Apportionment Method**
>
> a) Find the standard divisor for the apportionment (total population/total number of representatives).
> b) Find the standard quota (state's population/standard divisor) for each state and round it down to its lower quota. Assign that number of representatives to each state.
> c) If there are any representatives left over, assign them to states in order according to the size of the fractional parts of the states' standard quotas.

The method that we just described is exactly what we did in apportioning the oil consortium board in Example 1 of Section 9.1.

Perhaps the Hamilton method would be the apportionment method used today if its only problem was the Alabama paradox. However, there are other paradoxes that we will discuss now.

The Hamilton method can have the population paradox and the new states paradox.

In the early 1900s, it was discovered that Hamilton's method had another serious flaw in that a state with a faster growing population could lose a representative to a state whose population was growing more slowly.

> **DEFINITION**
>
> The **population paradox** occurs when state A's population is growing faster than state B's population, yet A loses a representative to state B. (We are assuming that the total number of representatives in the legislature is not changing.)

EXAMPLE 2   **The Population Paradox Can Occur with the Hamilton Method**

The Graduate school at Great Eastern University used the Hamilton method to apportion 15 graduate assistantships among the colleges of education, liberal arts, and business based on their undergraduate enrollments, as shown in Table 9.18.

a) Use Hamilton's method to allocate the graduate assistantships to the three colleges.

b) Assume that after the allocation was made in part a) that education gains 30 students, liberal arts gains 46, and the business enrollment stays the same. Reapportion the graduate assistantships again using Hamilton's method.

c) Explain how this illustrates the population paradox.

SOLUTION: a) The standard divisor for this apportionment is the total number of undergraduate students divided by the number of assistantships being awarded, or $\frac{4010}{15} = 267.33$. In Table 9.18, we compute the standard quota and then determine the allocation of graduate assistantships for each college.

College	Number of Students	Standard Quota	Integer Parts	Fractional Parts	Assign 2 Additional Assistantships
Education	940	$\frac{940}{267.33} = 3.52$	3	0.52	4
Liberal arts	1,470	$\frac{1,470}{267.33} = 5.5$	5	0.50	5
Business	1,600	$\frac{1,600}{267.33} = 5.99$	5	0.99	6
Total	4,010		13		15

TABLE 9.18   Apportioning the 15 graduate assistantships before the enrollments increase.

b) Table 9.19 shows how the apportionment would be done after the enrollment increases. The standard divisor is now $\frac{4086}{15} = 272.4$.

College	Number of Students	Standard Quota	Integer Parts	Fractional Parts	Assign 2 Additional Assistantships
Education	940 + 30 = 970	$\frac{970}{272.4} = 3.56$	3	0.56	3
Liberal arts	1,470 + 46 = 1,516	$\frac{1,516}{272.4} = 5.57$	5	0.57	6
Business	1,600	$\frac{1,600}{272.4} = 5.87$	5	0.87	6
Total	4,086		13		15

TABLE 9.19   Apportioning the 15 graduate assistantships after the enrollments increase.

c) Notice how the college of education lost an assistantship to the college of liberal arts. However, the percent in the increase in the number of education students was $\frac{30}{940} = 0.0319$ or 3.19 percent but the percent increase in the number of liberal arts students was $\frac{46}{1470} = 0.0313 = 3.13$ percent. So, even though the college of liberal arts grew more slowly than the college of education, it was able to take one of its assistantships away.

Admittedly, it was a close call in Example 2, and to be truthful it can be a lot of work to find such a counterexample (a spreadsheet* helps!); but the point is, Hamilton's method can allow the population paradox to occur.

When Oklahoma joined the union in 1907, the House of Representatives had to be reapportioned. Congress decided to increase the size of the house by five and give those five representatives to Oklahoma. However, when the house was reapportioned, New York was required to give one of its seats to Maine. This is called the new-states paradox.

> **DEFINITION**
>
> The **new-states paradox** occurs when a new state is added, and its share of seats is added to the legislature causing a change in the allocation of seats previously given to another state.

EXAMPLE 3 **The New-States Paradox Can Occur with the Hamilton Method**

A small country, Namania, consists of three states A, B, and C with populations given in Table 9.20. Namania's legislature has 37 representatives that are to be apportioned to these states using the Hamilton method.

a) Apportion these representatives using the Hamilton method.

b) Assume that Namania annexes the country Darelia whose population is 3,000 (thousands). Give Darelia its current share of representatives using the current standard divisor and add that number to the total number of representatives of Namania. Reapportion Namania again using the Hamilton method.

c) Explain how the new-states paradox has occurred.

SOLUTION: a) The standard divisor for this apportionment is the total population, 12,140 (thousand) divided by the number of representatives, 37. So the standard divisor is $\frac{12,140}{37} = 328.11$. We will use this to find each states' standard quota in Table 9.20.

State	Population (thousands)	Standard Quota	Integer Parts	Fractional Parts	Assign Additional Representative
A	2,750	$\frac{2,750}{328.11} = 8.38$	8	0.38	8
B	6,040	$\frac{6,040}{328.11} = 18.41$	18	0.41	19
C	3,350	$\frac{3,350}{328.11} = 10.21$	10	0.21	10
Total	12,140		36		37

TABLE 9.20 Apportioning representatives to states A, B, and C before the annexation of Darelia.

---

* For a spreadsheet to do these calculations, see the web site www.aw.com/pirnot.

Historical Highlight: *Apportionment and Political Inequalities*

U.S. Code requires that the apportionment counts are delivered by December 31 of the census year. Within a week of the opening of the next Congress, the president must give the census counts to the Clerk of the House of Representatives and within 15 days more, state governors are informed of the number of representatives to which they are entitled.

Until the 1960s, state legislatures had great power in how the representatives were apportioned within a state. Customarily, representation favored rural districts with small populations. For example, in the mid 1950s, Speaker of the House, Sam Rayburn, of Bonham, Texas, presided over a district of 227,735 constituents, whereas Albert Thomas, of Houston, had 806,701 constituents. Several lawsuits were filed before the Supreme Court and the court ruled that representation must be fairly distributed among the people within a state. This decision is sometimes called the "one man, one vote" rule. After these landmark decisions, representatives within a state were assigned more fairly. In 1973, again in Texas, the smallest district had 461,216 constituents, whereas the largest had 468,148.

b) According to the apportionment in part a), using the standard divisor of 328.11, Darelia's standard quota is $\frac{3,000}{328.11} = 9.14$. This means that Darelia deserves nine representatives under the current apportionment, so we will add nine representatives to Namania's legislature and reapportion it. Now, because Namania's population is $12,140 + 3,000 = 15,140$, the new standard divisor is $\frac{15,140}{46} = 329.13$. We will use this to compute the new standard quotas in Table 9.21.

State	Population (thousands)	Standard Quota	Integer Parts	Fractional Parts	Assign Additional Representative
A	2,750	$\frac{2,750}{329.13} = 8.36$	8	0.36	9
B	6,040	$\frac{6,040}{329.13} = 18.35$	18	0.35	18
C	3,350	$\frac{3,350}{329.13} = 10.18$	10	0.18	10
(D)arelia	3,000	$\frac{3,000}{329.13} = 9.11$	9	0.11	9
Total	15,140		45		46

TABLE 9.21   Apportioning representatives to states A, B, C, and D after the annexation of Darelia.

c) We saw in part b) that by adding Darelia, and increasing the size of the legislature according to the number of representatives that it deserved, B had to give one of its representatives to A. This illustrates the new-states paradox.

For the remainder of this section we will discuss some alternative apportionment methods proposed by Jefferson, Webster, and Adams. These three methods are similar in that each uses a divisor that is different from the standard divisor which is called a **modified divisor.** (As you will see, we find this modified divisor by trial and error.) When we divide a state's population by the modified divisor, we get the state's **modified quota.**

Once we find the modified quotas for each state, the question is what do we do with them? Jefferson says round them down to get the required number of representatives, Adams says to round up, and Webster says to round in the usual way.

Jefferson's method rounds quotas down.

In Jefferson's method, we are going to round the modified quotas down, so we will need modified quotas which are *larger** than the standard quotas. This means that the modified divisor has to be *smaller* than the standard divisor to give us the larger quotas.

> **Jefferson's Apportionment Method**
>
> a) Use trial and error to find a modified divisor which is smaller than the standard divisor for the apportionment.
> b) Calculate the **modified quota** (state's population/modified divisor) for each state and round it down. Assign that number of representatives to each state. (Keep varying the modified divisor until the sum of these assignments is equal to the total number being apportioned.)

We will now reapportion the oil consortium board using Jefferson's method.

EXAMPLE 4   Using Jefferson's Apportionment Method

Recall in Example 1, we were apportioning a nine-member board to oversee the oil consortium formed by Naxxon, Aroco, and Eurobile. We used Hamilton's method to apportion the board according to the number of stockholders of each company. Naxxon had 4,700 stockholders, Aroco had 3,700, and Eurobile had 1,600. Now use Jefferson's method to apportion the board.

SOLUTION: In Example 1, we found that the standard divisor for this apportionment was $\frac{10,000}{9} \approx 1,111.11$, and the standard quotas for the three companies are listed in Table 9.17. When we used Hamilton's method and rounded the standard quotas down, the total was eight and we had to assign the ninth representative.

In applying Jefferson's method, we want to find modified quotas so that when we round these quotas down, the total is nine. This means that the modified quotas must be larger than the standard quotas. In order to get larger modified quotas, we must have a modified divisor that is smaller than the standard divisor of 1,111.11.

Let's try a modified divisor of 1,050 to see what happens. We show the calculations in Table 9.22.

Because the total number of board members assigned is too small, we need *larger* modified quotas. In order to do this, we must try *smaller* modified divisors. In making up this example, we used a spreadsheet and tried 1,000, and then 950, but neither of these modified divisors worked. Finally when we used the modified

---

* Sometimes the standard divisor and quotas will work when using the Jefferson method.

Modified Divisor = 1,050			
	Naxxon	**Aroco**	**Eurobile**
Number of Stockholders	4,700	3,700	1,600
Standard Quota	4.23	3.33	1.44
Modified Quota	$\frac{4,700}{1,050} = 4.48$	$\frac{3,700}{1,050} = 3.52$	$\frac{1,600}{1,050} = 1.52$
Round Modified Quota Down	4	3	1
			Total = 8

TABLE 9.22   Using the Jefferson method to apportion the oil consortium board using a modified divisor of 1,050.

Quiz Yourself 12

Assume we are apportioning nine representatives among three groups A, B, and C which have populations of 1,276, 1,427, and 2,697, respectively.

a) Calculate the standard divisor for this apportionment.

b) Use the modified divisor of 550 to calculate group C's modified quota.

c) Using Jefferson's method, how many representatives would C receive if we use the modified divisor 550?

divisor of 935, we found the results in Table 9.23. Of course there are numbers other than 935 that would be acceptable modified divisors. We will ask you to investigate which other modified divisors would be acceptable in the exercises.

Modified Divisor = 935			
	Naxxon	**Aroco**	**Eurobile**
Number of Stockholders	4,700	3,700	1,600
Standard Quota	4.23	3.33	1.44
Modified Quota	$\frac{4,700}{935} = 5.03$	$\frac{3,700}{935} = 3.96$	$\frac{1,600}{935} = 1.71$
Round Modified Quota Down	5	3	1
			Total = 9

TABLE 9.23   Using the Jefferson method to apportion the oil consortium board using a modified divisor of 935.

We will next illustrate how to use Adam's method to apportion the oil consortium board.

Adams' method rounds quotas up.

In the Adams method, we are going to round the modified quotas up, so we will need modified quotas which are *smaller* than the standard quotas. This means that the modified divisor has to be *larger* than the standard divisor to give us the smaller quotas.

> **Adams' Apportionment Method**
>
> a) Use trial and error to find a modified divisor which is larger than the standard divisor for the apportionment.
> b) Calculate the modified quota (state's population/modified divisor) for each state and round it up. Assign that number of representatives to each state. (Keep varying the modified divisor until the sum of these assignments is equal to the total number being apportioned.)

We will now reapportion the oil consortium board using the Adams method.

EXAMPLE 5 Using the Adams Apportionment Method

Recall Naxxon had 4,700 stockholders, Aroco had 3,700, and Eurobile had 1,600. Now use Adams' method to apportion the nine-member board.

SOLUTION: Because the standard divisor for this apportionment was $\frac{10,000}{9} \approx$ 1,111.11, we need to try a larger modified divisor.

We will first try a modified divisor of 1,200, as we show in Table 9.24.

Modified Divisor = 1,200			
	Naxxon	**Aroco**	**Eurobile**
Number of Stockholders	4,700	3,700	1,600
Standard Quota	4.23	3.33	1.44
Modified Quota	$\frac{4,700}{1,200} = 3.92$	$\frac{3,700}{1,200} = 3.08$	$\frac{1,600}{1,200} = 1.33$
Round Modified Quota Up	4	4	2
			Total = 10

TABLE 9.24 Using the Adams method to apportion the oil consortium board using a modified divisor of 1,050.

Quiz Yourself 13

Assume we are apportioning ten representatives among three groups A, B, and C which have populations of 1,300, 950, and 2,550, respectively.

a) Calculate the standard divisor for this apportionment.

b) Use the modified divisor of 500 to calculate group A's modified quota.

c) Using Adams' method, how many representatives would A receive if we use the modified divisor 500?

Because the total board members assigned is too large, we need *smaller* modified quotas. In order to do this, we must try *larger* modified divisors. When we try a modified divisor of 1,250 we get the results in Table 9.25.

Modified Divisor = 1,250			
	Naxxon	**Aroco**	**Eurobile**
Number of Stockholders	4,700	3,700	1,600
Standard Quota	4.23	3.33	1.44
Modified Quota	$\frac{4,700}{1,250} = 3.76$	$\frac{3,700}{1,250} = 2.96$	$\frac{1,600}{1,250} = 1.28$
Round Modified Quota Up	4	3	2
			Total = 9

TABLE 9.25 Using the Adams method to apportion the oil consortium board using a modified divisor of 1,250.

As in the Jefferson method, there will be acceptable modified divisors other than 1,250 that we could have used in Example 5. We will pursue this question further in the exercises.

We will finally discuss Webster's apportionment method.

Webster's method rounds quotas in the usual way.

In Webster's method, we round the modified quotas in the usual way. That is, if the quota has a decimal part of 0.5 or greater, we round up, otherwise, we round down. Again, as in the Jefferson and Adams method, we find the modified divisor by trial and error.

> **Webster's Apportionment Method**
>
> a) Use trial and error to find a modified divisor.
> b) Calculate the modified quota (state's population/modified divisor) for each state and round it in the usual way. Assign that number of representatives to each state. (Keep varying the modified divisor until the sum of these assignments is equal to the total number being apportioned.)

We will now reapportion the oil consortium board using Webster's method.

EXAMPLE 6   Using Webster's Apportionment Method

Naxxon has 4,700 stockholders, Aroco has 3,700, and Eurobile has 1,600. Use Webster's method to apportion the nine member board.

SOLUTION: We need a starting place in using Webster's method so as a first step, let's try a modified divisor of 1,111.11 which is the standard divisor (see Table 9.26).

Modified Divisor = 1,111.11			
	Naxxon	**Aroco**	**Eurobile**
Number of Stockholders	4,700	3,700	1,600
Standard Quota	4.23	3.33	1.44
Modified Quota	$\frac{4,700}{1,111.11} = 4.23$	$\frac{3,700}{1,111.11} = 3.33$	$\frac{1,600}{1,111.11} = 1.44$
Round Modified Quota in Usual Way	4	3	1
			Total = 8

TABLE 9.26   Using the Webster method to apportion the oil consortium board using a modified divisor of 1,111.11.

Because the total board members assigned is too small, we need *larger* modified quotas. In order to do this, we must try *smaller* modified divisors. After trying several modified divisors, we tried the modified divisor of 1,060 to get the results in Table 9.27.

Modified Divisor = 1,060	Naxxon	Aroco	Eurobile
Number of Stockholders	4,700	3,700	1,600
Standard Quota	4.23	3.33	1.44
Modified Quota	$\frac{4,700}{1,060} = 4.43$	$\frac{3,700}{1,060} = 3.49$	$\frac{1,600}{1,060} = 1.51$
Round Modified Quota in Usual Way	4	3	2
			Total = 9

TABLE 9.27 Using Webster's method to apportion the oil consortium board using a modified divisor of 1,060.

Quiz Yourself 14

Assume we are apportioning eight representatives among three groups A, B, and C which have populations of 2,700, 4,300, and 5,000, respectively.

a) Calculate the standard divisor for this apportionment.

b) Use the modified divisor of 1,540 to calculate group B's modified quota.

c) Using Webster's method, how many representatives would B receive if we use the modified divisor 1,540?

Again, there will be acceptable modified divisors other than 1,060. In the exercises, we will ask you to find the largest and the smallest integers that would be acceptable modified divisors for Example 6.

> **PROBLEM SOLVING**
>
> In learning these apportionment methods, try to relate them to each other. In each case, first you must find a modified divisor and then round in one of three ways: Jefferson (down), Adams (up), Webster (usual way).
>
> Don't bother memorizing how to select the modified divisor — you can always start with the standard divisor and remember that if your apportionment is too large, then you need smaller quotas and therefore you need a larger modified divisor. Likewise, if the apportionment is too small, you need a smaller modified divisor.

We summarize the four methods we have studied in Table 9.28. As you see from this table, the Jefferson, Adams, and Webster methods are not perfect because they all can fail to satisfy the Quota Rule. We will give examples of such violations in the exercises.

	Hamilton	Jefferson	Adams	Webster
Can have Alabama paradox	Yes	No	No	No
Can have population paradox	Yes	No	No	No
Can have new-states paradox	Yes	No	No	No
Can violate quota rule	No	Yes	Yes	Yes

TABLE 9.28 Summary of four apportionment methods.

Exercises 9.4*

In Exercises 1–12 we give you a total population, state A's population, and the total number of representatives that are to be apportioned. Find: a) the standard divisor, b) the standard quota for the state, and c) the number of representatives that would be given to the state if we use the standard divisor as the modified divisor with the specified method.

1. Total Population 100,000; A's population 27,000; Number of representatives being apportioned 8; Method: Jefferson

2. Total Population 14,000; A's population 2,850; Number of representatives being apportioned 56; Method: Jefferson

3. Total Population 220,000; A's population 92,000; Number of representatives being apportioned 11; Method: Jefferson

4. Total Population 512,000; A's population 84,160; Number of representatives being apportioned 16; Method: Jefferson

5. Total Population 120,000; A's population 36,800; Number of representatives being apportioned 15; Method: Adams

6. Total Population 135,000; A's population 48,000; Number of representatives being apportioned 9; Method: Adams

7. Total Population 134,400; A's population 56,160; Number of representatives being apportioned 28; Method: Adams

8. Total Population 50,400; A's population 17,520; Number of representatives being apportioned 21; Method: Adams

9. Total Population 504,000; A's population 74,200; Number of representatives being apportioned 18; Method: Webster

10. Total Population 28,800; A's population 13,920; Number of representatives being apportioned 9; Method: Webster

11. Total Population 102,000; A's population 19,975; Number of representatives being apportioned 12; Method: Webster

12. Total Population 204,000; A's population 45,832; Number of representatives being apportioned 30; Method: Webster

13. Communicating Mathematics   Explain the main difference between the Hamilton method and the three other apportionment methods that we discussed.

14. Communicating Mathematics   In what ways are the Jefferson and Adams methods similar? How are they different?

15. Communicating Mathematics   Explain why the modified divisor in the Jefferson method is usually smaller than the standard divisor.

16. Communicating Mathematics   Explain why the modified divisor in the Adams method is usually greater than the standard divisor.

17. Communicating Mathematics   In using the Adams method to apportion a legislature, Juanita got an assignment with too many representatives. What change must she make in her approach?

18. Communicating Mathematics   In using the Webster method to apportion a legislature, Joe got an assignment with too few representatives. What change must he make in his approach?

19. **Assigning seats on a water authority.** Assume that California, Nevada, and Arizona are cooperating to build a dam to provide water to communities currently lacking adequate water supplies. Seats on the 11-member Southwest Water Authority, which governs the project, are assigned according to the number of customers in each state who use the water from the project. There are 56,000 customers in California, 52,000 in Arizona, and 41,000 in Nevada. Find the standard divisor and each state's standard quota.

20. Use the Jefferson method to assign the seats on the Southwest Water Authority.

21. Use the Adams method to assign the seats on the Southwest Water Authority.

22. Use the Webster method to assign the seats on the Southwest Water Authority.

---

* For some of these exercises, particularly for the Further Exercises, you will find it useful to go to the Web site www.aw.com/pirnot and download spreadsheets and TI-83 programs to help you with the computations for these apportionment problems.

23. Choosing representatives on a negotiations committee. The employees of Jungle World theme park are negotiating a new contract. There are 213 performers, 273 food workers, and 178 maintenance workers. The 20-person negotiations committee has members in proportion to the number of employees in each of the three groups. Find the standard divisor and each group's standard quota.

24. Use the Jefferson method to assign the members of the negotiations committee.

25. Use the Adams method to assign the members of the negotiations committee.

26. Use the Webster method to assign the members of the negotiations committee.

27. Apportioning a labor council. A labor council is being formed from the members of four unions. The electricians union has 25 members, the plumbers have 18 members, the painters have 29, and the carpenters have 31 members. The council will have 20 representatives. Find the standard divisor and each union's standard quota.

28. Use the Jefferson method to assign the representatives to the council.

29. Use the Adams method to assign the representatives to the council.

30. Use the Webster method to assign the representatives to the council.

31. Awarding assistantships. Nineteen fellowships are to be apportioned to students in a writing program at a major university. The apportionment is to be based on the number of full-time graduate students in each area of the writing program. There are 30 fiction majors, 20 poetry majors, 17 majoring in technical writing, and 14 majoring in writing for the media. Find the standard divisor and each major's standard quota.

32. Use the Jefferson method to assign the fellowships.

33. Use the Adams method to assign the fellowships.

34. Use the Webster method to assign the fellowships.

35. Assigning police officers. Suppose a small town has 13 police officers. Furthermore, assume that the town is divided rather naturally into five regions. The police chief has gathered data regarding the number of incidents (crimes, traffic accidents, etc.) that have occurred in each region over the past several months, as listed in Table 9.29.

Region 1	107 incidents
Region 2	65 incidents
Region 3	43 incidents
Region 4	59 incidents
Region 5	27 incidents
Total	301 incidents

TABLE 9.29

The chief wishes to assign officers for duty in proportion to the number of incidents per region. Find the standard divisor and the standard quota for each region.

36. Use the Jefferson method to allocate the police officers.

37. Use the Adams method to allocate the police officers.

38. Use the Webster method to allocate the police officers.

39. A health club instructor has a course load that allows her to teach six two-credit hour classes. A preregistration survey indicates the following interests.

56 want to take high-impact aerobics

29 want to take low-impact aerobics

11 want to take jazzercise

4 want to take step exercise

Find the standard divisor and the standard quotas for each interest area.

40. Use the Jefferson method to assign the number of classes to each area.

41. Use the Adams method to assign the number of classes to each area.

42. Use the Webster method to assign the number of classes to each area.

In Exercises 43–50, we are using the Hamilton method to do the apportionment.

43. Ten representatives are apportioned among three states A, B, and C, which have the populations (in thousands) shown in Table 9.30.

Ten years later, the population of A has increased by 20 thousand and the population of B has increased by 60

State	Population
A	570
B	2,557
C	6,873

TABLE 9.30

thousand, while C's population remained the same. The ten representatives are reapportioned. Does this example illustrate the population paradox? Explain.

44. The Metrodelphia Regional Transportation Authority has three lines, the Arrow, the Blue Line, and the City division. The numbers of passengers that use each line per day are given in Table 9.31.

Line	Passengers per Day (now)	Passengers per Day (one year later)
Arrow	23,530	23,930
Blue	5,550	5,650
City	70,920	71,100

TABLE 9.31

Assume that 100 new cars are apportioned among the three lines according to the number of passengers using each line. One year later, the MRTA is able to purchase another 100 cars and again apportions them among the three divisions. We show this increase in ridership in Table 9.31. Does this example illustrate the population paradox? Explain.

45. A group of three regional airports has hired a company to increase security. The contract calls for 150 security personnel who are to be apportioned to the airports according to the number of passengers using each airport each

Airport	Passengers per Week (now)	Passengers per Week (one year later)
Allenport	120,920	121,420
Bakerstown	5,550	5,650
Columbia City	23,530	23,930

TABLE 9.32

week. Two years later, the number of passengers per week has increased at the three airports, so the 150 security personnel are reallocated. We show these data in Table 9.32. Does this example illustrate the population paradox? Explain.

46. The current enrollments of a university's three campuses are shown in Table 9.33.

Campus	Students (now)	Students (one year later)
Altus	7,200	7,480
Brightsfield	20,500	21,140
Center City	52,300	52,900

TABLE 9.33

The New State University's administration has decided to upgrade 18 computer labs according to the enrollments at the three campuses. One year later, the enrollments have increased at each campus and again it is decided to upgrade 18 labs. Does this example illustrate the population paradox? Explain.

47. Two states A and B have formed a 100-member commission to explore ways in which they can work jointly to improve the business climate in their states. The members will be apportioned to the commission according to the populations of the two states. After seeing the success of the commission, a third state asks to join and appoint members to the commission. States A and B agree and will allow state C to add as many members to the commission as they would deserve now using the current standard divisor. The populations of the three states are given in Table 9.34.

State	Population (in thousands)
A	2,000
B	7,770
C	600

TABLE 9.34

Does this example illustrate the new states paradox? Explain.

48. Redo Exercise 47, but now use the populations in Table 9.35. Also assume that the original commission has 40 members.

State	Population (in thousands)
A	12,550
B	4,450
C	8,250

TABLE 9.35

49. Redo Exercise 47, but now use the populations in Table 9.36. Also assume that the original commission has 56 members. Assume that A, B, and C are the original states and D is the new state.

State	Population (in thousands)
A	8,700
B	4,300
C	3,200
D	6,500

TABLE 9.36

50. Redo Exercise 49, but now use the populations in Table 9.37. Also assume that the original commission has 70 members.

State	Population (in thousands)
A	230
B	740
C	380
D	600

TABLE 9.37

Exercises 51–54 illustrate that the Jefferson and Adams apportionment methods can violate the quota rule. In each case, determine which method illustrates the violation and explain how the quota rule is not satisfied.

51. We are apportioning 200 representatives among the states A, B, C, D, E, and F. The populations of these states are given in Table 9.38.

State	Population
A	700
B	1,500
C	820
D	4,530
E	2,200
F	550

TABLE 9.38

52. We are apportioning 100 representatives among the states A, B, C, D, E, and F. The populations of these states are given in Table 9.39.

State	Population
A	700
B	8,000
C	820
D	1,500
E	4,000
F	550

TABLE 9.39

53. We are apportioning 500 representatives among the states A, B, C, D, E, and F. The populations of these states are given in Table 9.40.

State	Population
A	1,400
B	2,000
C	1,500
D	9,000
E	4,300
F	500

TABLE 9.40

54. We are apportioning 400 representatives among the states A, B, C, D, E, and F. The populations of these states are given in Table 9.41.

State	Population
A	700
B	1,000
C	750
D	200
E	2,100
F	4,500

TABLE 9.41

Further Exercises

***55.** In Example 4, we found that a modified divisor of 935 worked in using Jefferson's apportionment method. Use trial and error to find the smallest and the largest integers that we could have used in this example as modified divisors.

56. In Example 5, we found that a modified divisor of 1,250 worked in using Adams' apportionment method. Use trial and error to find the smallest and the largest integers that we could have used in this example as modified divisors.

57. In Example 6, we found that a modified divisor of 1,060 worked in using Webster's apportionment method. Use trial and error to find the smallest and the largest integers that we could have used in this example as modified divisors.

58. Make up several hypothetical countries each consisting of several large and several small states. Experiment with different apportionments and see if the Jefferson method seems to favor large or small states.

59. Repeat Exercise 58 for the Adams method.

60. Repeat Exercise 58 for the Webster method.

CHAPTER SUMMARY

SECTION 9.1

Alabama paradox
While apportioning the U.S. House of Representatives in 1881, although the size of the legislature was increased, Alabama's representation was decreased.

Hamilton apportionment method
The Hamilton method apportions representatives to states by giving each state the integer part of the exact representation due to it. If more representatives remain to be apportioned, then they are allocated to the states that have the largest fractional part of their exact representation.

---

* Exercise numbers circled in red can be used as group exercises.

average constituency

The average constituency of a state is the following quotient:

$$\frac{\text{population of the state}}{\text{number of representatives from the state}}$$

more poorly represented

State A is more poorly represented than state B if the average constituency of A is greater than the average constituency of B.

absolute unfairness

If state A is more poorly represented than state B, then the absolute unfairness is

(average constituency of state A) − (average constituency of state B).

relative unfairness

The relative unfairness of an apportionment between two states is the absolute unfairness divided by the smaller average constituency.

SECTION 9.2   **apportionment criterion**

When assigning a representative to one of two states or divisions, make the assignment so as to give the smallest relative unfairness.

Huntington–Hill apportionment principle

If states X and Y have been allotted x and y representatives, respectively, then state X should be given an additional representative in preference to state Y provided that

$$\frac{(\text{population of Y})^2}{y \cdot (y + 1)} < \frac{(\text{population of X})^2}{x \cdot (x + 1)}.$$

Otherwise, state Y should be given the additional representative.

Huntington–Hill number

The number

$$\frac{(\text{population of X})^2}{x \cdot (x + 1)}$$

is called a Huntington–Hill number.

SECTION 9.3   **apportioning the oil consortium board**

Each time we allocate a representative, we will calculate the Huntington–Hill number

$$\frac{(\text{number of stockholders in the company})^2}{(\text{current representation}) \cdot (\text{current representation} + 1)}$$

for all three companies. The largest of these numbers determines which company is most deserving of the next representative.

SECTION 9.4   **standard divisor, standard quota**

In allocating a group of representatives among several states, we define the standard divisor by

$$\text{standard divisor} = \frac{\text{total population}}{\text{number of representatives being allocated}},$$

and a state's standard quota is defined to be

$$\text{standard quota} = \frac{\text{state's population}}{\text{standard divisor}}.$$

lower quota, upper quota, quota rule

If we round the standard quota down, we call that number the lower quota, if we round the standard quota up, then we call that number the upper quota. If in making an apportionment, each state is allocated a number of representatives that is between its lower quota and upper quota, then we say the apportionment satisfies the quota rule.

Hamilton's apportionment method

a) Find the standard divisor for the apportionment (total population/total number of representatives).

b) Find the standard quota (state's population/standard divisor) for each state and round it down to its lower quota. Assign that number of representatives to each state.

c) If there are any representatives left over, assign them to states in order according to the size of the fractional parts of the states standard quotas.

population paradox, new-states paradox

The population paradox occurs when state A's population is growing faster than state B's population, yet A loses a representative to state B. The new-states paradox occurs when a new state is added, and its share of seats is added to the legislature causing a change in the allocation of seats previously given to another state.

Jefferson's apportionment method

a) Use trial and error to find a modified divisor which is smaller than the standard divisor for the apportionment.

b) Calculate the modified quota for each state and round it down. Assign that number of representatives to each state.

Adams' apportionment method

a) Use trial and error to find a modified divisor which is larger than the standard divisor for the apportionment.

b) Calculate the modified quota for each state and round it up. Assign that number of representatives to each state.

Webster's apportionment method

a) Use trial and error to find a modified divisor.

b) Calculate the modified quota for each state and round it in the usual way. Assign that number of representatives to each state.

summary of apportionment methods

	Hamilton	Jefferson	Adams	Webster
Can have Alabama paradox	Yes	No	No	No
Can have population paradox	Yes	No	No	No
Can have new-states paradox	Yes	No	No	No
Can violate quota rule	No	Yes	Yes	Yes

CHAPTER TEST

SECTION 9.1

1. What is the Alabama paradox?

2. The governing board of the American History Roundtable Society has eleven members. The seats are apportioned among the four divisions of the society—Revolutionary War (560 members), Civil War (524), World War I (431), and World War II (485)—according to the number of members in each division. Use the Hamilton method to apportion seats on the board.

3. Explain why an Alabama paradox can occur when using the Hamilton apportionment method.

4. Suppose state A has a population of 935,000 and five representatives, whereas state B has a population of 2,343,000 and eleven representatives.

   a) Determine which state is more poorly represented and calculate the absolute unfairness for this assignment of representatives.

   b) Determine the relative unfairness for this apportionment.

5. Adjust the list of percents

   11.045, 27.333, 36.901, 3.006, 21.715

   so that they are displayed to the tenths place and their sum is 100.00.

SECTION 9.2

6. Suppose that South Carolina has a population of approximately 3.5 million and has six representatives, and that Maryland, with a population of 4.1 million, has eight.

   a) Calculate the Huntington–Hill numbers for both South Carolina and Maryland.

   b) According to the Huntington–Hill apportionment principle, which of these states is most deserving to receive an additional representative?

7. Explain why the Huntington–Hill apportionment principle will avoid an Alabama paradox.

SECTION 9.3

8. You are to allocate thirteen buses to four routes of a regional transportation authority. The routes are called the red, blue, yellow, and green routes. Each route is guaranteed at least one bus and you are to use the Huntington–Hill method to apportion the remaining nine buses. Use the given table of Huntington–Hill numbers to allocate the buses to each route.

Current Representation	Red	Blue	Yellow	Green
1	21,012.5	13,448.0	3,362.0	27,378.0
2	7,004.2	4,482.7	1,120.7	9,126.0
3	3,502.1	2,241.3	560.3	4,563.0
4	2,101.3	1,344.8	336.2	2,737.8
5	1,400.8	896.5	224.1	1,825.2
6	1,000.6	640.4		1,303.7

9. An instructor at the Physical Fitness Institute can teach eight classes. A preregistration survey indicates the following interests.

   66 want to take tae-bo
   39 want to take karate
   18 want to take weight training
   23 want to take tai chi

   Assume that the instructor will teach at least one class for each area. Use the Huntington–Hill method to apportion the instructor's remaining four classes among the four areas.

SECTION 9.4

10. A health club instructor has a course load that allows her to teach eight classes. A preregistration survey indicates the following interests.

   8 want to take high-impact aerobics
   64 want to take low-impact aerobics
   11 want to take jazzercise
   31 want to take step exercise

   Apportion her classes using the Jefferson Method.

11. Apportion the classes in question 10 using the Adams Method.

12. Apportion the classes in question 10 using the Webster Method.

13. A group of three regional airports has hired a company to increase security. The contract calls for 150 security personnel who are to be apportioned to the airports according to the number of passengers using each airport each week. One year later, the number of passengers per week has increased at the three airports, so the 150 security personnel are reallocated. We show these data in the given table. (We are using the Hamilton method.)

Airport	Passengers per Week (now)	Passengers per Week (one year later)
A	80,500	87,700
B	6,800	7,410
C	93,300	101,300

Does this example illustrate the population paradox? Explain.

14. What do we mean by the new-states paradox?

15. Which apportionment method(s) do not satisfy the quota rule?

Of Further Interest: FAIR DIVISION

There are situations that are similar to apportioning representatives in which we need to distribute items but in which we cannot compute Huntington–Hill numbers to determine the "fair portion." This would be the case in dividing an inheritance among the heirs of an estate, splitting a pile of different types of candy among children at a party, or awarding property between a couple getting divorced. Problems such as these fall within the area of mathematics called *fair division*. Many different mathematicians have developed various methods for dealing with fair division problems. In this section we will discuss one method for handling a *discrete fair division* problem as opposed to a *continuous fair division* problem.

> ### DEFINITIONS
>
> **Discrete fair division** means that the items being distributed cannot be subdivided into fractional parts. **Continuous fair division** means that the items being distributed can be subdivided into parts and the division can be done as finely as we wish.

For example, we could not subdivide an antique car into parts to distribute it among heirs to an estate. On the other hand, we can subdivide a cake and distribute pieces of it among children at a birthday party.

We will now discuss the *method of sealed bids*, which is based on work done by the Polish mathematician Hugo Steinhaus.† We start by looking at the fair division of an estate consisting of one item being shared equally by several people. Our problem is to determine how each person, in his or her opinion, can get a fair share when the item cannot be physically divided. But first, we must explain what we mean by a person's *fair share* of an estate.

> ### DEFINITION
>
> A person's **fair share** of an estate is the person's estimate of the estate's value divided by the number of people sharing equally in the estate.

Dear Uncle Charlie has died and left a rare painting that has been in the family for generations to three cousins—Martha, George, and Sam. The three cousins agree that the painting should not be sold, but each is entitled to an equal share of the bequest. Obviously the painting cannot be divided, so they agree that one of them will get the painting and pay cash to the others.

Martha, George, and Sam contact several appraisers to establish the painting's value. However, they receive such a wide range of estimates that it only

Quiz Yourself 15*

Identify each situation as dealing with either discrete or continuous division of items.

a) Apportioning the seats in the U.S. House of Representatives.

b) Dividing a piece of land among three people.

c) Dividing a mixture of M&Ms, hard candy, and pieces of bubble gum among five children.

d) Sharing a can of soda between three people.

---

† Hugo Steinhaus, *Mathematical Snapshots*, 3rd ed. (Oxford University Press, 1969), pp. 65–74.

causes confusion. Martha, who has taken a mathematics course in which she studied fair division, suggests that they proceed in the following way.

Each of them secretly writes down an estimate of the painting's value and places the bid in a sealed envelope. By doing this, the three of them will establish an opinion on the estate's value and, in turn, their fair share of the estate. For the process to work, each must promise to give an honest estimate and be content to receive a fair share. Next, they will open the bids and the person making the highest bid will get the painting.

Upon opening the envelopes, they find that Martha has valued the painting at $84,000, George feels the painting is worth $60,000, and Sam thinks it is worth $72,000. As agreed, Martha gets the painting because she had the highest bid. But now she must pay the estate an amount of money. The amount is her estimate of the estate's value less her fair share of the estimate.

Since Martha had valued the estate at $84,000 and since the cousins are to share equally in the estate, Martha's estimate of her fair share is

$$\frac{1}{3} \times 84{,}000 = \$28{,}000.$$

Therefore, she pays the estate the difference

$$\$84{,}000 - \$28{,}000 = \$56{,}000.$$

But should George and Sam split that money and get $28,000 apiece? It would be unfair if they did, because they each estimated the estate at a lower value. Further, they agreed to be content with receiving their fair share. Since George bid $60,000, he thinks that his fair share is

$$\frac{1}{3} \times 60{,}000 = \$20{,}000.$$

With Sam's bid of $72,000, he perceives that his fair share of the estate is

$$\frac{1}{3} \times 72{,}000 = \$24{,}000.$$

Giving George $20,000 and Sam $24,000 from the estate leaves a balance of

$$\$56{,}000 - \$20{,}000 - \$24{,}000 = \$12{,}000$$

in the estate. Now what do we do with the balance? Since the three cousins have equal interests in the estate, we will divide this balance equally among them. Therefore, George and Sam each get an additional $4,000, and the remaining $4,000 is returned to Martha.

Table 9.42 shows each person's estimate of the estate's value and his or her opinion of a fair share. Table 9.43 summarizes the accounting of cash based on our method of fair division. A plus sign indicates the person pays that amount to the estate, while a minus sign means the person receives that amount from the estate. Observe in Table 9.43 that the amount paid to the estate must equal the sum of the amounts received from the estate.

	Martha ($\frac{1}{3}$)	George ($\frac{1}{3}$)	Sam ($\frac{1}{3}$)
Bid on painting	$84,000	$60,000	$72,000
Fair share of estate	$\dfrac{\$84,000}{3} = \$28,000$	$\dfrac{\$60,000}{3} = \$20,000$	$\dfrac{\$72,000}{3} = \$24,000$
Item obtained with high bid	Painting		

TABLE 9.42   Estimates of value and fair share.

	Martha ($\frac{1}{3}$)	George ($\frac{1}{3}$)	Sam ($\frac{1}{3}$)
Pays to estate (+) or receives from estate (−)	$56,000 (+)	$20,000 (−)	$24,000 (−)
Division of estate balance ($12,000)	$4,000 (−)	$4,000 (−)	$4,000 (−)
Summary of cash	Pays $52,000	Receives $24,000	Receives $28,000

TABLE 9.43   Fair division accounting of estate cash.

It may appear at first that there is some unfairness in this division, because Sam received more money than George. However, keep in mind that each party agreed to be content with his or her estimate of a fair share when using this method. But what is interesting is that each party ends up with more than

 Quiz Yourself 16

Rosa, Juan, Carlos, and Luis have inherited a house that was left to them by their parents. Since they have equal interests in the house and wish to keep it in the family, they will use the method of sealed bids to decide who gets the house. That person will pay cash to the others. The following table shows the four bids on the house. Complete the table by showing each person's estimate of his or her fair share of the estate and who gets the house.

	Rosa ($\frac{1}{4}$)	Juan ($\frac{1}{4}$)	Carlos ($\frac{1}{4}$)	Luis ($\frac{1}{4}$)
Bid on house	$250,000	$240,000	$260,000	$200,000
Fair share of estate				
Item obtained with highest bid				

expected. First, Martha gets the painting and pays $4,000 less than her estimate of the others' fair share, namely $52,000 versus $56,000. On the other hand, George and Sam each received $4,000 more than their estimated fair share. This "win-win" outcome is due to Martha's higher assessment of the painting, which raised George and Sam's share, while their smaller estimates lowered the amount that Martha had to pay to the estate.

 Quiz Yourself 17

We continue the fair division problem discussed in Quiz Yourself 16. Because Carlos gets the house, he must pay an amount of cash to the estate, whereas the others will receive cash. Construct a table to show an accounting of the estate's cash transactions.

EXAMPLE 1   Dividing an Estate with Several Items

Ann, Tom, Karen, and Bob have inherited equal portions of an estate left by their grandfather. The estate consists of a house, an expensive car, and a boat. They decide to use the method of sealed bids in dividing the estate, and each of the four submits a sealed bid on each item in the estate. Their bids, and therefore their individual estimates of a fair share of the estate, are listed in Table 9.44. Determine how the estate should be divided using the method of sealed bids.

	Ann $(\frac{1}{4})$	Tom $(\frac{1}{4})$	Karen $(\frac{1}{4})$	Bob $(\frac{1}{4})$
Bid on house	$165,000	$180,000	$160,000	$190,000
Bid on car	$55,000	$50,000	$35,000	$40,000
Bid on boat	$40,000	$36,000	$25,000	$30,000
Total value	$260,000	$266,000	$220,000	$260,000
Estimate of fair share of estate	$\frac{1}{4} \cdot \$260,000 =$ $65,000	$\frac{1}{4} \cdot \$266,000 =$ $66,500	$\frac{1}{4} \cdot \$220,000 =$ $55,000	$\frac{1}{4} \cdot \$260,000 =$ $65,000

TABLE 9.44   Estimates of value and fair share.

SOLUTION: We see from the last line of Table 9.44 that Ann and Bob will be satisfied if they each receive a combination of items and cash worth at least $65,000, whereas Karen would be happy with $55,000, but Tom would not be satisfied with less than $66,500. Remember that no one knows the others' estimates of the estate's value until the bids are opened. When the bids are opened, Bob gets the house because he has the high bid of $190,000. Since Ann bid the highest on both the car and the boat, she gets those items. Of course, Karen and Tom get none of the items but will receive cash from the estate. However, it is interesting to note that even though Tom did not submit the highest bid on any item, he had the largest estimate of the total value of the estate. Let us now do the final accounting of the cash transactions for the estate.

We start with those who received items from the estate. Bob got the house, which is worth $190,000 to him. But he thinks that his fair share of the total estate is worth $65,000. Therefore, he received considerably more than his fair share, namely the difference

$$\$190,000 - \$65,000 = \$125,000.$$

He must pay this difference to the estate. Ann received the car and the boat from the estate. Based on her bids for these items, she received a total value of

$$\$55,000 + \$40,000 = \$95,000.$$

However, this is more than what she feels is her fair share of the estate. Therefore, she must also pay the estate the difference between the value of items received and her fair share, namely,

$$\$95,000 - \$65,000 = \$30,000.$$

The combined cash paid to the estate by Bob and Ann is

$$\$125,000 + \$30,000 = \$155,000.$$

Last, Karen and Tom receive their perceived fair share from these payments. The cash accounting thus far is tabulated in Table 9.45.

	Ann $(\frac{1}{4})$	Tom $(\frac{1}{4})$	Karen $(\frac{1}{4})$	Bob $(\frac{1}{4})$
Item obtained with high bid	Car, Boat			House
Pays to estate (+) or receives from estate (−)	$30,000 (+)	$66,500 (−)	$55,000 (−)	$125,000 (+)

TABLE 9.45 Fair division accounting of estate.

The last line of Table 9.45 shows that there is a cash balance left in the estate, namely,

$$\$30,000 + \$125,000 - \$66,500 - \$55,000 = \$155,000 - \$121,500 = \$33,500.$$

Since the four heirs each have one-quarter interest in the estate, they each receive

$$\frac{1}{4} \cdot \$33,500 = \$8,375$$

of the balance. As with any fair division based on sealed bids, the four heirs get more than they expected. Both Ann and Bob get items but pay less to the estate than they had expected. Although Karen and Tom do not receive items, they each get an amount of cash that is more than they think is their respective fair share. Table 9.46 summarizes the fair division of the estate. As before, the total paid to the estate equals the total received from the estate and the estate is now closed with a zero balance.

	Ann ($\frac{1}{4}$)	Tom ($\frac{1}{4}$)	Karen ($\frac{1}{4}$)	Bob ($\frac{1}{4}$)
Item obtained with high bid	Car, Boat			House
Pays to estate (+) or receives from estate (−)	$30,000 (+)	$66,500 (−)	$55,000 (−)	$125,000 (+)
Division of estate balance ($33,500)	$8,375 (−)	$8,375 (−)	$8,375 (−)	$8,375 (−)
Summary of cash	Pays $21,625	Receives $74,875	Receives $63,375	Pays $116,625

TABLE 9.46 Fair division accounting of estate.

Let us summarize fair division through the method of sealed bids. We explain the process in terms of dividing an estate; however, the method can be applied to other similar situations.

The Method of Sealed Bids (Equal Interests)

Each person having a stake in the fair division of the items makes a sealed bid on each item.

1. Each person's fair share is determined as follows:
   - Add the person's bids on the various items.
   - Divide this total by the number of people sharing in the estate.

   Because different people make different bids, what is a fair share for one person may not be the same as a fair share to another.

2. Open the bids. The person bidding the highest amount on an item gets that item. If several people have the same highest bid, then the decision of who gets the item can be based on random selection.

3. If the total value of the items awarded to a person is more than the person's fair share, then the person pays the difference in cash to the estate. If the total value is less than the person's fair share, then the person receives the difference in cash from the estate. If a person does not receive any items, then the person gets his or her fair share in cash from the estate.

4. If, after completing step 3, there is a cash balance in the estate, then it is divided equally among the heirs.

The method of sealed bids will work successfully only when

- the parties involved make an honest bid on each item;
- the person making the highest bid on an item is willing to accept the item and pay cash to the estate if the bid is larger than the fair share; and

- most important, each person makes a bid without knowledge of the others' bids.

We can also use this method when the parties have different percent interests in the estate. For example, three heirs to an estate may have, respectively, 50%, 30%, and 20% interests as stated in a will. In this case, a person's fair share is his or her estimate of the estate's value times his or her percent interest in the estate. The exercises have problems of this type.

Exercises

Identify each situation as dealing with either discrete or continuous division of items.

1. a) Dividing a collection of rare books among five people

b) Dividing a chocolate bar among several children

c) Apportioning the seats on a city council with representatives chosen from different neighborhoods of the city

2. a) Splitting candy collected at Halloween among three trick-or-treaters

b) Dividing an estate consisting of a house and a boat among five heirs

c) Dividing a piece of pie among among four people

Use the method of sealed bids to complete the tables in Exercises 3–14.

3. Dividing an inheritance. Greg and Dharma want to determine who should inherit their rich aunt's antique car.

	Greg ($\frac{1}{2}$)	Dharma ($\frac{1}{2}$)
Bid on car	$110,000	$85,000
Fair share of estate	a.	b.
Item obtained with highest bid	c.	d.

4. Dividing an inheritance. Karl and Friedrich have inherited their father's copy of a rare book and want to decide who should receive it.

	Karl ($\frac{1}{2}$)	Friedrich ($\frac{1}{2}$)
Bid on book	$27,000	$35,000
Fair share of estate	a.	b.
Item obtained with highest bid	c.	d.

5. Dissolving a partnership. Alex and Marianne have written a book together and now want to decide how to dissolve their partnership, with one of them obtaining sole rights to the copyright. Their current entitlement to the book is 40% belonging to Alex and 60% belonging to Marianne.

	Alex (40%)	Marianne (60%)
Bid on copyright	$20,000	$28,000
Fair share of copyright	a.	b.
Item obtained with highest bid	c.	d.

6. Dissolving a partnership. Matt and Tony own a small pizza stand at the shore, with Matt having 35% interest and Tony 65% interest. They want to dissolve their partnership, with one buying the other out.

	Matt (35%)	Tony (65%)
Bid on pizza stand	$120,000	$90,000
Fair share of pizza stand	a.	b.
Item obtained with highest bid	c.	d.

7. Dividing an inheritance. Three brothers—Ed, Al, and Jerry—want to decide who gets a valuable painting and statue that they have inherited from their grandfather.

	Ed ($\frac{1}{3}$)	Al ($\frac{1}{3}$)	Jerry ($\frac{1}{3}$)
Bid on painting	$13,000	$8,000	$9,000
Bid on statue	$14,000	$13,000	$15,000
Total value	$27,000	a.	b.
Fair share of estate	c.	d.	e.

8. Dissolving a partnership. Franz, Ida, Bill, and Monica are members of an investment club. Their portfolio consists of three stocks—a computer manufacturer, an oil company, and a pharmaceutical company. They wish to dissolve their group and determine who gets the stocks and who gets cash.

	Franz ($\frac{1}{4}$)	Ida ($\frac{1}{4}$)	Bill ($\frac{1}{4}$)	Monica ($\frac{1}{4}$)
Bid on computer stock	$75,000	$80,000	$70,000	$90,000
Bid on oil stock	$35,000	$40,000	$45,000	$40,000
Bid on pharmaceutical stock	$40,000	$30,000	$25,000	$35,000
Total value	a.	b.	c.	d.
Fair share of portfolio	e.	f.	g.	h.

9. Dividing an inheritance. Consider your answer to Exercise 3. Complete the settlement of the estate by completing the fair division accounting of the estate cash as indicated in the table.

	Greg ($\frac{1}{2}$)	Dharma ($\frac{1}{2}$)
Pays to estate (+) or receives from estate (−)		
Division of estate balance ($)		
Summary of cash		

10. Dividing an inheritance. Consider your answer to Exercise 4. Complete the settlement of the estate by completing the fair division accounting of the estate cash as indicated in the table.

	Karl ($\frac{1}{2}$)	Friedrich ($\frac{1}{2}$)
Pays to estate (+) or receives from estate (−)		
Division of estate balance ($)		
Summary of cash		

11. Dissolving a partnership. Consider your answer to Exercise 5. Complete the settlement of the copyright by completing the fair division accounting of the cash as indicated in the table.

	Alex (40%)	Marianne (60%)
Pays to pool (+) or receives from pool (−)		
Division of pool balance ($)		
Summary of cash		

12. Dissolving a partnership. Consider your answer to Exercise 6. Complete the settlement of the pizza stand by completing the fair division accounting of the cash as indicated in the table.

	Matt (35%)	Tony (65%)
Pays to pool (+) or receives from pool (−)		
Division of pool balance ($)		
Summary of cash		

13. **Dividing an inheritance.** Consider your answer to Exercise 7. Complete the settlement of the estate by completing the fair division accounting of the estate cash as indicated in the table.

	Ed ($\frac{1}{3}$)	Al ($\frac{1}{3}$)	Jerry ($\frac{1}{3}$)
Items obtained with highest bid			
Pays to estate (+) or receives from estate (−)			
Division of estate balance ($)			
Summary of cash			

14. **Dissolving a partnership.** Consider your answer to Exercise 8. Complete the settlement of the portfolio by completing the fair division accounting of the cash as indicated in the table.

	Franz ($\frac{1}{4}$)	Ida ($\frac{1}{4}$)	Bill ($\frac{1}{4}$)	Monica ($\frac{1}{4}$)
Stock obtained with highest bid				
Pays to pool (+) or receives from pool (−)				
Division of pool balance ($)				
Summary of cash				

In Exercises 15–16, use the method of sealed bids to determine how the items and cash are distributed.

15. **Dividing an inheritance.** Betty and Dennis wish to divide a diamond ring, an antique desk, and a collection of rare books that were bequeathed to them by their dear aunt who passed away. Use the method of sealed bids to divide the items and determine a fair division of the estate. Betty's and Dennis's estimates of these items are given in the table.

	Betty	Dennis
Ring	$16,000	$18,000
Desk	$4,500	$5,000
Books	$4,000	$3,000

16. **Dissolving a partnership.** Matt, Dani, and Christian have a partnership in which they own a chain of four fast-food restaurants (call them A, B, C, and D). They have decided to dissolve their partnership but keep ownership of the restaurants individually among themselves. Determine a fair division by using the method of sealed bids. The table shows the partners' estimates of the values of the individual restaurants.

	Matt	Dani	Christian
A	$170,000	$180,000	$160,000
B	$145,000	$150,000	$155,000
C	$200,000	$190,000	$210,000
D	$70,000	$50,000	$45,000

Further Exercises

In Exercises 17–20, we reconsider our opening example in which Martha, George, and Sam are dividing Uncle Charlie's estate. In Exercises 21–24, we take another look at Example 1, in which Ann, Tom, Karen, and Bob are dividing their grandfather's estate. These questions are intended to be open-ended, with no right or wrong answers. Just consider the implications of what might happen if the division is done when the rules of fairness are not being followed.

* (17.) What can happen if George has prior knowledge of Martha's bid?

(18.) What can happen if Martha has prior knowledge of George's bid?

(19.) What can happen if George bids unfairly by bidding too low? Bidding too high?

(20.) What can happen if Martha bids unfairly by bidding too low? Bidding too high?

(21.) What can happen if Ann has prior knowledge of others' bids?

(22.) What can happen if Bob has prior knowledge of others' bids?

(23.) What can happen if Tom bids unfairly by bidding too low? Bidding too high?

(24.) What can happen if Karen bids unfairly by bidding too low? Bidding too high?

---

* Exercise numbers circled in red can be used as group exercises.

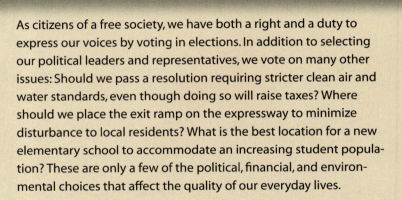

Voting:*
Using Mathematics to Make Choices

As citizens of a free society, we have both a right and a duty to express our voices by voting in elections. In addition to selecting our political leaders and representatives, we vote on many other issues: Should we pass a resolution requiring stricter clean air and water standards, even though doing so will raise taxes? Where should we place the exit ramp on the expressway to minimize disturbance to local residents? What is the best location for a new elementary school to accommodate an increasing student population? These are only a few of the political, financial, and environmental choices that affect the quality of our everyday lives.

If there are only two choices, it is easy to determine the winner; the choice receiving more votes wins. This voting system is known as *majority rule*. However, when there are more than two choices, then selecting the winner is more complicated. We encounter voting situations with more than two alternatives in conducting primary elections, choosing union leaders, selecting the rookie of the year, and awarding a Grammy or an Academy Award. In such situations we want to know, what is the best method for conducting an election?

The answer to that question is not simple. In this chapter, we will study various voting systems and learn how winners are chosen using each system. We will investigate the fairness of these systems and see that each system has curious quirks built into it that go against our intuition. For example, you will see a surprising example

* For further resources on voting theory, see www.aw.com/pirnot.

where, in order to win an election, a candidate asks, "Please vote against me so I can win."

Later in the chapter, we will also investigate different *weighted voting systems*, in which not every voter has the same number of votes; you will also learn several ways to measure voters' power in such systems.

10.1   VOTING METHODS

To see why we study voting theory, let's examine the 1992 presidential election results in Texas. When the final votes were counted, George Bush had 2,496,071 votes, Bill Clinton had 2,281,815 votes, and Ross Perot had 1,354,781 votes. If we do the arithmetic, we find that Bush won Texas and its 32 electoral votes even though he had only 41 percent of the popular vote. Another way of looking at this is that 59 percent of the voters voted against Bush, yet he won the election in Texas!*

To many people, this outcome is troublesome because it appears that we determine the winner of an election not only by the number of votes cast, but also by the way we agree to use those votes. Various methods have been proposed over the course of centuries as to how voting should be conducted. We will explain several voting methods in this section; in Section 10.2 we will investigate their weaknesses.

Using the plurality method, the person with the most votes wins an election.

The **plurality method** is the simplest way to determine the outcome of an election; the person earning the most votes is the winner.

> **DEFINITION**
>
> **The Plurality Method**
>
> Each person votes for his or her favorite candidate. The candidate receiving the most votes is declared the winner.

The plurality method is used in many state and local elections, including the 1992 presidential election in Texas. It is easy to determine the winner using this method, because all we need to do is tally the votes for each candidate.

---

* In U.S. presidential elections, each state is assigned a certain number of electoral votes. Without getting into details of exactly how the electoral college operates, it is fair to say that if a candidate wins a state, then that candidate is given the entire number of electoral votes allocated to that state. The winner of the presidential election is then the candidate who acquires the most electoral votes.

EXAMPLE 1 Determining the Winner Using the Plurality Method

In a closely fought election for president of the 33-member United Labor Council (ULC), the election results are as follows:

Ann	10
Ben	9
Carim	11
Doreen	3

Using the plurality method, who is the winner of this election?

SOLUTION: Since Carim has the most votes, he is declared the winner.

Note in Example 1 that using the plurality method, Carim won the election even though $\frac{22}{33} = \frac{2}{3}$ of the ULC members voted against him.

Using the Borda count method, the person earning the most points wins an election.

Although it was too late once the election had been conducted, Ann realized that even though she was not the first choice of most voters, she was the second choice of a great many voters. Unfortunately for Ann, the ballot did not allow voters to state their second, third, and fourth preferences. A voting method called the **Borda count method** permits a voter to "fine-tune" his or her vote in the sense that not only can the voter designate a first choice, but also a second choice, a third choice, and so on. In the ULC election, we could have used the Borda method by specifying that on a voter's ballot the first choice would be given four points, the second choice three points, the third choice two points, and the fourth choice one point.

> **DEFINITION**
>
> **The Borda Count Method**
>
> If there are k candidates in an election, each voter ranks all candidates on the ballot. Then the first choice is given k points, the second choice is given $k - 1$ points, the third choice $k - 2$ points, and so on.* The candidate who receives the most total points wins the election.

In order to use the Borda count method in the United Labor Council election in Example 1, voters must rank candidates on their ballots. For example, the given ballot would mean that a voter preferred Carim first, Ann second, Ben

---

* In some variations of the Borda count method, different numbers of points are awarded for first place, second place, third place, and so on. For example, we may give five points to first place, three to second, and one to third. We investigate these variations in the exercises.

1st	C
2nd	A
3rd	B
4th	D

third, and Doreen fourth. Such a ballot is called a **preference ballot**. With 33 voters, some preference ballots would be the same.* In tallying the votes, we group identical preference ballots together to speed the counting process, as we illustrate in Example 2. A table that summarizes preference ballots is called a **preference table**.

EXAMPLE 2 **Determining the Winner Using the Borda Count Method**

Assume that the United Labor Council used the Borda count method to determine its president. Table 10.1 summarizes the preference ballots cast in the election. Who is the winner of the election?

	Number of Ballots					
Preference	6	7	5	3	9	3
1st	C	A	C	A	B	D
2nd	A	C	D	D	A	A
3rd	B	B	B	B	D	C
4th	D	D	A	C	C	B

TABLE 10.1 Preference table for ULC election.

SOLUTION: The numbers at the top of the columns in Table 10.1 show how many voters submitted the particular preference ballot in that column. For example, the 6 means that six voters had preference ballots choosing Carim first, Ann second, Ben third, and Doreen fourth. Notice, as before, Carim has 11 first-place votes, Ann has 10, Ben 9, and Doreen 3. However, since we are using the Borda method, we now assign four points for each first place vote received by a candidate, three for each second place vote, etc. We summarize the point totals in Table 10.2.

Candidates	Number of Points				
	1st-Place Votes × 4 (Points)	2nd-Place Votes × 3 (Points)	3rd-Place Votes × 2 (Points)	4th-Place Votes × 1 (Points)	Total Points
A	10 × 4 = 40	18 × 3 = 54	0 × 2 = 0	5 × 1 = 5	99
B	9 × 4 = 36	0 × 3 = 0	21 × 2 = 42	3 × 1 = 3	81
C	11 × 4 = 44	7 × 3 = 21	3 × 2 = 6	12 × 1 = 12	83
D	3 × 4 = 12	8 × 3 = 24	9 × 2 = 18	13 × 1 = 13	67

TABLE 10.2 Points earned using the Borda method for the ULC election.

---

* When we discuss permutations in the Of Further Interest section, you will see that there are only 24 possible different preference ballots for A, B, C, and D.

When we examine the voting, we see that Ann now wins the election.

PROBLEM SOLVING

- -

After completing a problem, double-check your answer for accuracy. In Example 2, we found that 330 points were awarded. This happened because each of the 33 voters was awarding 10 points.

Many polls use the Borda method to rank sports teams. For example, since there is no playoff system to determine a national champion in college football, often there is controversy as to which team is number one. One of several polls used in determining this championship is conducted by the Associated Press. Each voter in this poll gives 25 points to his or her number one choice, 24 to the second choice, and so on, to select the top 25 teams each week.

Quiz Yourself 1*

Use the preference table to determine the winner of this election using the Borda count method. How many points does the winning candidate receive?

	Number of Ballots			
Preference	8	7	5	7
1st	C	D	C	A
2nd	A	A	B	D
3rd	B	B	D	B
4th	D	C	A	C

Because the outcomes of the plurality method and the Borda count method are different, you may wonder which is the "correct" method for determining the ULC president. We will not answer this question. Rather, knowing that these two methods yield different results, which do you feel is more appropriate?

The plurality-with-elimination method eliminates the weakest candidate before revoting.

When we used the plurality method in the ULC election in Example 1, we obtained the following results.

Ann	10
Ben	9
Carim	11
Doreen	3

----

* Quiz Yourself answers begin on page 849.

Historical Highlight: *Approval Voting*

The 1980 New Hampshire presidential primary was essentially a three-person race between Ronald Reagan, George Bush, and Howard Baker. In the vote, conducted using the plurality method, Reagan and Bush placed first and second with 50 percent and 23 percent, respectively. Baker, a distant third with 13 percent, also had a similarly poor showing in Massachusetts the following week and withdrew from the race. Third-party candidates such as George Wallace in 1968, Jesse Jackson in 1984, and Ross Perot in 1992 also suffered from a voting system in which the winner gains all the electoral votes and other candidates get none. Thus, particularly in presidential primaries, those who may seem to be viable candidates early in the race quickly fall by the wayside after several small primary elections.

Another method called **approval voting** allows voters to vote for as many candidates on the ballot as they wish. Candidates are not ranked, and the winner is the person who earns the most votes of approval. According to ABC exit polls in New Hampshire in 1980, if the approval voting method had been used, Ronald Reagan would have come in first with a 58 percent approval rating, Baker would have been second with 41 percent, and Bush would have been last with 39 percent. With a strong showing such as this, it is unlikely that Baker would have withdrawn from the race so early, and we can only speculate as to the long-term political implications had approval voting been used.

Some argue that approval voting tends to select the strongest candidate and to reduce infighting among candidates, particularly in primaries, because in order to gain support a candidate does not have to take it away from someone else. Although this method is rarely used in American political elections, it is used by numerous universities, the National Academy of Sciences, and the Mathematical Association of America, and is also used to elect the secretary-general of the United Nations.

One could argue that since Doreen is so far out of the running, perhaps we should eliminate her and conduct a new election with just Ann, Ben, and Carim as candidates. If no candidate earns a majority (more than 50 percent) in this second election, then we could eliminate the person receiving the smallest number of votes and then conduct a third election. This voting method is called the **plurality-with-elimination method**.

> **DEFINITION**
>
> **The Plurality-with-Elimination Method**
>
> Each voter votes for one candidate. A candidate receiving a majority of votes is declared the winner. If no candidate receives a majority of votes, then the candidate (or candidates) with the fewest votes is dropped from the ballot and a new election is held.* This process continues until a candidate receives a majority of votes.

---

* In order to reduce the number of rounds of voting, some variations of this method may drop more than one candidate from the second round of voting. For example, it may be specified that only the top two candidates are allowed to proceed to the second round of voting.

To avoid many rounds of voting, we can simplify the elimination process. If we assume that each voter has a ranking of all the candidates, we can construct a preference table as in the Borda count method. Of course, we will use this table differently than we do in the Borda method.

Throughout this section we will assume that once a voter has ranked the candidates, *this ranking does not change during subsequent rounds of voting.* That is, if a voter prefers Ann to Doreen and Doreen to Ben in round one, if Doreen is eliminated, then the voter will prefer Ann to Ben in round two.

EXAMPLE 3 **Determining the Winner Using the Plurality-with-Elimination Method**

Determine the winner of the United Labor Council election using the plurality-with-elimination method.

SOLUTION: The preference table for this election is shown in Table 10.3.

Preference	Number of Ballots					
	6	**7**	**5**	**3**	**9**	**3**
1st	C	A	C	A	B	D
2nd	A	C	D	D	A	A
3rd	B	B	B	B	D	C
4th	D	D	A	C	C	B

TABLE 10.3 Preference table for the ULC election.

From Table 10.3, we see that Doreen has only three first-place votes, so she is eliminated and a new election is held. Remember, *we are assuming that voters do not change their preferences from one round of voting to the next.* For example, because three voters had Doreen as their first choice, now that she is eliminated, those three voters now have Ann as their first choice. One way to think of this adjustment is to imagine that when we eliminate D from a column of the table, all candidates below D in that column move up one row. By eliminating Doreen, we get Table 10.4.

Preference	Number of Ballots					
	6	**7**	**5**	**3**	**9**	**3**
1st	C	A	C	A	B	A
2nd	A	C	B	B	A	C
3rd	B	B	A	C	C	B

TABLE 10.4 Preference table for the ULC election after Doreen is eliminated.

We combine identical columns in Table 10.4 to get Table 10.5.

Preference	Number of Ballots				
	6	10	5	3	9
1st	C	A	C	A	B
2nd	A	C	B	B	A
3rd	B	B	A	C	C

TABLE 10.5 Preference table for the ULC election with three candidates.

We see now that Ann has thirteen first-place votes, Ben has nine, and Carim has eleven. Thus Ben is eliminated and a third round of voting takes place. Table 10.6 summarizes the preferences at this point.

Preference	Number of Ballots				
	6	10	5	3	9
1st	C	A	C	A	A
2nd	A	C	A	C	C

TABLE 10.6 Preference table for the ULC election after removing Ben.

In Table 10.6 we see that Ann has 22 first-place votes and Carim has 11, so Ann wins the ULC election.

Quiz Yourself 2

Use the preference table to determine the winner of this election using the plurality-with-elimination method.

Preference	Number of Ballots				
	8	9	5	4	2
1st	C	E	B	A	A
2nd	A	D	C	D	C
3rd	B	B	E	B	B
4th	E	C	A	C	E
5th	D	A	D	E	D

In the pairwise comparison method, the candidate who can beat the most other candidates "head to head" is the winner.

It is natural to expect that the winner of an election should be capable of beating each of the other candidates "head to head." We show a variation of this in the next method, the **pairwise comparison method**.

> **DEFINITION**
>
> **The Pairwise Comparison Method**
>
> Voters first rank all candidates. If A and B are a pair of candidates, we count how many voters prefer A to B and vice versa. Whichever candidate is preferred the most receives 1 point. If A and B are tied, then each receives $\frac{1}{2}$ point. Do this comparison, assigning points, for each pair of candidates. At the end, the candidate receiving the most points is the winner.

As with the plurality-with-elimination method, we assume that if a voter prefers A to B and B to C, then the voter will prefer A to C. In Example 4, we see that we can use a voting method to choose among alternatives as well as decide elections.

EXAMPLE 4   **Determining a Preference Using the Pairwise Comparison Method**

In order to decide which new item to add to its menu, a fast-food chain did a market survey in which customers were asked to rank their preferences for (T)acos, (N)achos, and (B)urritos. The chain will use the pairwise comparison method to decide which item to add to its menu. The results of counting the ballots are shown in Table 10.7. Which item should the chain select?

Preference	Number of Ballots					
	2,108	864	1,156	1,461	1,587	1,080
1st	T	T	N	N	B	B
2nd	N	B	T	B	T	N
3rd	B	N	B	T	N	T

TABLE 10.7   Preference table for voting on menu items.

SOLUTION: We must compare (a) T with N, (b) T with B, and (c) N with B.

a) In comparing T with N, we ignore all references to B in Table 10.7. Thus we see that $2{,}108 + 864 + 1{,}587 = 4{,}559$ customers prefer T over N, and $1{,}156 + 1{,}461 + 1{,}080 = 3{,}697$ customers prefer N over T. Thus we award option T 1 point.

b) Now we compare T with B. From Table 10.7 we find that $2,108 + 864 + 1,156 = 4,128$ customers prefer T over B, and $1,461 + 1,587 + 1,080 = 4,128$ customers prefer B over T. Therefore T and B each receive $\frac{1}{2}$ point.

c) Finally we compare N with B. Table 10.7 tells us that $2,108 + 1,156 + 1,461 = 4,725$ customers prefer N over B. Also, $864 + 1,587 + 1,080 = 3,531$ customers prefer B over N. We award 1 point to N.

Since T has $1\frac{1}{2}$ points, N has 1 point, and B has $\frac{1}{2}$ point, option T is declared the winner.   ◎

 Quiz Yourself 3

Redo Example 4 using the following preference table to determine which item to add to the restaurant chain's menu. We are again using the pairwise comparison method. Also list how many points each item gets.

	Number of Ballots					
Preference	985	864	1,156	1,021	1,187	1,080
1st	T	T	N	N	B	B
2nd	N	B	T	B	T	N
3rd	B	N	B	T	N	T

Table 10.8 summarizes the voting methods that we have discussed in this section.

Method	How the Winning Candidate Is Determined
Plurality	The candidate receiving the most votes wins.
Borda count	Voters rank all candidates by assigning a set number of points to first choice, second choice, third choice, and so on; the candidate with the most points wins.
Plurality-with-elimination	Successive rounds of elections are held, with the candidate receiving the fewest votes being dropped from the ballot each time, until one candidate receives a majority of votes.
Pairwise comparison	Candidates are compared in pairs, with a point being assigned the voters' preference in each pair. (In the case of a tie, each candidate gets a half point.) After all candidates have been compared, the candidate receiving the most points wins.

TABLE 10.8   Summary of voting methods.

Which of these voting methods is the best? Unfortunately, each of the methods we have introduced has serious flaws, as you will see in Section 10.2.

Exercises 10.1

1. Four candidates running for a vacant seat on the town council receive votes as follows: Edelson, 2,156; Borowski, 1,462; Callow, 986; Levan, 428.

   a) Is there a candidate who earns a majority?

   b) Who wins the election using the plurality method?

2. Five candidates running for mayor receive votes as follows: Kinison, 415; Barron, 462; Lichman, 986; Devlin, 428; Saito, 381.

   a) Is there a candidate who earns a majority?

   b) Who wins the election using the plurality method?

The university administration has asked a group of student leaders to vote on the aspects of college life to target for improvement over the next year. The choices were (D)ining facilities, (A)thletic facilities, (C)ampus security, and the (S)tudent union building. The votes are summarized in the following preference table. Use this information to answer Exercises 3–6.

	Number of Ballots					
Preference	15	30	18	17	10	2
1st	A	D	A	C	C	S
2nd	C	C	S	D	S	A
3rd	S	S	D	S	A	C
4th	D	A	C	A	D	D

3. **Voting on improving college life.** What option is selected using the plurality method?

4. **Voting on improving college life.** What option is selected using the Borda count method?

5. **Voting on improving college life.** What option is selected using the plurality-with-elimination method?

6. **Voting on improving college life.** What option is selected using the pairwise comparison method?

The drama society members are voting for the type of play they will perform next season. The choices are a (D)rama, (C)omedy, (M)ystery, or (G)reek tragedy. The votes are summarized in the following preference table. Use this information to answer Exercises 7–10.

	Number of Ballots					
Preference	10	15	13	12	5	7
1st	C	D	C	M	M	G
2nd	M	M	G	D	G	C
3rd	G	G	D	G	C	M
4th	D	C	M	C	D	D

7. **Selecting a play.** What type of play is selected using the plurality method?

8. **Selecting a play.** What type of play is selected using the Borda count method?

9. **Selecting a play.** What type of play is selected using the plurality-with-elimination method?

10. **Selecting a play.** What type of play is selected using the pairwise comparison method?

Before a conference on "Trends in the Next Decade," a panel of participants voted on the topic for the keynote address. The choices were (T)echnology, (P)overty in Third World Countries, (E)nvironment, (G)ender Roles, and (F)amilies. The votes are summarized in the following preference table. Use this information to answer Exercises 11–14.

	Number of Ballots				
Preference	15	7	13	5	2
1st	T	E	G	P	E
2nd	P	P	F	G	F
3rd	E	G	E	F	G
4th	F	F	P	E	P
5th	G	T	T	T	T

11. **Choosing a speaker.** What topic is selected using the plurality method?

12. **Choosing a speaker.** What topic is selected using the Borda count method?

13. **Choosing a speaker.** What topic is selected using the plurality-with-elimination method?

14. Choosing a speaker. What topic is selected using the pairwise comparison method?

A small employee-owned Internet company is voting on a merger with one of its competitors. The choices are (E)-Mall, (F)irestorm, (S)ecurenet, (T)echenium, and (L)ondram. The employees' votes are summarized in the following preference table. Use this information to answer Exercises 15–18.

Preference	Number of Ballots				
	15	11	9	10	2
1st	L	F	S	T	L
2nd	F	S	L	E	S
3rd	S	L	F	F	E
4th	E	E	T	L	F
5th	T	T	E	S	T

15. Deciding on a merger. What option is their first choice using the plurality method?

16. Deciding on a merger. What option is their first choice using the Borda count method?

17. Deciding on a merger. What option is their first choice using the plurality-with-elimination method?

18. Deciding on a merger. What option is their first choice using the pairwise comparison method?

Variations of the Borda count method are often used to determine winners of sports awards. In these methods, voters assign candidates a certain number of points for first place, fewer for second place, and so on. In Exercises 19–20, first determine the voting scheme being used and then complete the table.

19. Awarding the Heisman Trophy. The voting for the 2001 Heisman Trophy, which was awarded to Eric Crouch, is summarized in the following table. Determine the voting method that is being used and then complete the table.

Player, Team	1st	2nd	3rd	Total
Eric Crouch, Nebraska	162	a.__	88	770
Rex Grossman, Florida	b.__	105	87	708
Ken Dorsey, Miami	109	122	c.__	638
Joey Harrington, Oregon	54	68	66	364
David Carr, Fresno State	34	60	d.__	280

20. Awarding the Cy Young Award. The voting for the 2001 American League Cy Young Award, which was given to Roger Clemens, is summarized in the following table. Determine the voting method that is being used and then complete the table.

Player, Team	1st	2nd	3rd	Total
Roger Clemens, New York	21	a.__	2	122
Mark Mulder, Oakland	b.__	13	11	60
Freddy Garcia, Seattle	4	8	11	55
Jamie Moyer, Seattle	1	c.__	1	12
Mike Mussina, New York	0	0	d.__	2

We can use each of the voting methods described in this section to rank candidates in addition to determining the winner. The plurality method and the Borda count method can rank candidates in order of the most first-place votes and the points earned, respectively. Use this information to answer Exercises 21–22.

21. Devise a method for ranking candidates using the plurality-with-elimination method.

22. Devise a method for ranking candidates using the pairwise comparison method.

In Exercises 23–26, refer to the preference table that was given prior to Exercise 3 regarding choices for improvement of college life. Use that table to rank the choices using the specified method.

23. the plurality method   **24.** the Borda count method

25. the plurality-with-elimination method

26. the pairwise comparison method

In Exercises 27–30, refer to the preference table that was given prior to Exercise 11 regarding choices for the conference keynote address. Use that table to rank the choices using the specified method.

27. the plurality method   **28.** the Borda count method

29. the plurality-with-elimination method

30. the pairwise comparison method

In Exercises 31–34, assume that only two candidates are running in the election.

31. Communicating Mathematics   Explain why the winner using the plurality method will have a majority of the votes.

32. Communicating Mathematics   Explain why the winner using the Borda count method will have a majority of the votes.

33. Communicating Mathematics   Explain why the winner using the plurality-with-elimination method will have a majority of the votes.

34. Communicating Mathematics   Explain why the winner using the pairwise comparison method will have a majority of the votes.

35. What is the total number of points that all candidates can earn in an election using the Borda count method if there are four candidates and 20 voters?

36. What is the total number of points that all candidates can earn in an election using the Borda count method if there are five candidates and 18 voters?

37. What is the maximum number of points that a candidate can earn in an election using the Borda count method if there are five candidates and 22 voters?

38. What is the minimum number of points that a candidate can earn in an election using the Borda count method if there are six candidates and 20 voters?

39. What is the total number of points that all candidates can earn in an election using the pairwise comparison method if there are four candidates?

40. What is the total number of points that all candidates can earn in an election using the pairwise comparison method if there are five candidates?

41. What is the maximum number of points that a candidate can earn in an election using the pairwise comparison method if there are five candidates?

42. What is the minimum number of points that a candidate can earn in an election using the pairwise comparison method if there are five candidates?

Further Exercises

*43. Construct a preference table for an election with candidates A, B, C, and D such that using that table

    A wins using the plurality method
    B wins using the Borda count method
    C wins using the plurality-with-elimination method
    D wins using the pairwise comparison method.

44. **The Pareto condition.** The Pareto condition states that in an election with several choices, including *x* and *y*, if everyone prefers alternative *x* to alternative *y*, then *y* is not chosen.

    Determine which of the voting methods we have discussed satisfy the Pareto condition. You might begin by looking at specific preference tables and from those investigations try to state a general reason why the Pareto condition is satisfied by a particular method. Of course, by the Always Principle of Section 1.1, if you can find a single failure for a particular voting method, then that method will not satisfy the Pareto condition.

45. Make a voting example using a variation of the Borda count method. Perhaps 10 points is given to first place, 5 to second, 3 to third, and 1 to fourth. Construct a preference table and use this method to determine the winner. Then using the same preference table, change the number of points assigned to each place, so that a different winner results from this new assignment of points.

46. Research the topic of voting theory. Briefly describe a voting method that we have not discussed in this chapter.

---

* Exercise numbers circled in red can be used as group exercises.

47. Choose a topic you think will be of interest to your class and have them use preference ballots to rank their preferences in choosing among four or five alternatives regarding that topic. Use the different voting methods we presented in this section to determine the winner. It would be good if, when you present the alternatives, there is not an obvious favorite, so that one alternative might win using one voting method, but lose using another. Discuss why you think that one option wins using one method, but loses using another.

48. This question is the same as 47; however, now use the approval method of voting to make your choice. Compare the results with what you obtained in Exercise 47. Do you see an advantage to using approval voting, or not?

10.2   DEFECTS IN VOTING METHODS

In Section 10.1 you saw that the voting method used to select the winner of an election is as important as the number of votes each candidate receives. In this section, we will investigate whether there is a perfect method for conducting an election.

We want an election to satisfy several criteria:

- the majority criterion
- Condorcet's criterion
- the independence-of-irrelevant-alternatives criterion
- the monotonicity criterion

We will discuss which methods satisfy these criteria.

According to the majority criterion, a candidate with more than half the first-place votes wins an election.

The first criterion that we will consider is the majority criterion.

> **DEFINITION**
>
> **The Majority Criterion**
>
> If a majority* of the voters rank a candidate as their first choice, then that candidate should win the election.

It is clear that the plurality method and the pairwise comparison method satisfy the majority criterion. (Verify this.) However, this is not the case with the Borda count method, as we show in Example 1.

-----
* Majority means *more than* half.

EXAMPLE 1   The Borda Count Method Violates the Majority Criterion

Authors Alvarez, Byron, Cawley, and Dickson have been nominated for a literary award for the best short story of the year. A three-person panel using the Borda count method will make the final selection. Preference ballots for the panel members are as shown at left. Determine the winner using the Borda count method, and show that this outcome violates the majority criterion.

1st	C	C	D
2nd	D	D	A
3rd	A	B	B
4th	B	A	C

SOLUTION: According to the Borda method, first-place votes are worth four points, second-place votes are worth three points, and so on. The following table summarizes the election results.

	Number of Points				
	1st-Place Votes (× 4 Points)	2nd-Place Votes (× 3 Points)	3rd-Place Votes (× 2 Points)	4th-Place Votes (× 1 Point)	Total
A	$0 \times 4 = 0$	$1 \times 3 = 3$	$1 \times 2 = 2$	$1 \times 1 = 1$	6
B	$0 \times 4 = 0$	$0 \times 3 = 0$	$2 \times 2 = 4$	$1 \times 1 = 1$	5
C	$2 \times 4 = 8$	$0 \times 3 = 0$	$0 \times 2 = 0$	$1 \times 1 = 1$	9
D	$1 \times 4 = 4$	$2 \times 3 = 6$	$0 \times 2 = 0$	$0 \times 1 = 0$	10

Although the Borda count winner is Dickson, the majority of the first-place votes were cast for Cawley.

Quiz Yourself 4*

Use the Borda count method and the following preference table to determine whether the election satisfies the majority criterion.

	Number of Ballots			
Preference	8	10	5	2
1st	C	D	C	A
2nd	A	A	B	B
3rd	B	B	D	D
4th	D	C	A	C

---

* Quiz Yourself answers begin on page 849.

Highlight: *The Marquis de Condorcet*

Marie-Jean-Antoine-Nicolas Caritat, the Marquis de Condorcet, was an eighteenth-century French nobleman who was also a highly regarded philosopher and mathematician. He became a member of the French Academy of Sciences at age 26; during the French Revolution, he was active both in politics and the reform of France's educational system. At the core of Condorcet's philosophy was his belief that the human race was continuously progressing toward a state of physical, intellectual, and moral perfection. He believed that social scientists would be able to develop social and moral laws, similar to those that had been recently discovered in physics and chemistry, that would allow society to regulate its citizens for the public good.

In 1785, Condorcet wrote a paper, *Essay on the Applications of Mathematics to the Theory of Decision Making,* in which he intended to prove that mathematics could be used in the social sciences to discover laws as precise as those in the physical sciences. By applying calculus, social scientists would free society from its reliance on greedy capitalists for industrial progress. Mathematicians using probability theory would develop insurance programs to protect the poor. And once social scientists discovered moral laws, crime and war would disappear.

Although Condorcet's ideas may seem naive by twentieth-century standards, he was a forward thinker whose liberal and humanitarian views are still relevant today. He believed in abolishing slavery, opposed capital punishment, advocated free speech, and supported early feminists lobbying for equal rights. As far as Condorcet was concerned, human rights were derived from people's ability to use reason in forming moral concepts.

*Women, having these same qualities, must necessarily possess equal rights. Either no individual of the human species has any true rights, or all have the same. And he who votes against the rights of another, of whatever religion, color, or sex, has thereby abjured his own.**

Condorcet's criterion says that a candidate who defeats everyone else head-to-head is the winner.

The next condition, which was proposed by the Marquis de Condorcet, is a desirable property of a voting method.

> **DEFINITION**
>
> **Condorcet's Criterion**
>
> If candidate X can defeat each of the other candidates in a head-to-head vote, then X is the winner of the election.

Clearly a candidate with a majority of first-place votes can defeat every other candidate in a head-to-head election. However, Example 2 shows that the plurality method may violate Condorcet's criterion.

---

* This quote is taken from an article written by Condorcet in 1790 entitled "On the Admission of Women to the Rights of Citizenship."

EXAMPLE 2   The Plurality Method Violates Condorcet's Criterion

A seven-member judicial committee is voting using preference ballots for three candidates A, B, and C for committee leader. The partially completed preference ballots are listed below.

1st	A	A	A	B	B	C	C
2nd							
3rd							

Using the plurality voting method, A is the winner. Complete the ballots so that in head-to-head voting, B defeats candidates A and C.

SOLUTION: Since A defeats B in the first three ballots, we must force B to defeat A on the last four ballots. We could do this as follows.

1st	A	A	A	B	B	C	C
2nd				A	A	B	B
3rd						A	A

At this point, B defeats A by 4 to 3. We wish B to also defeat C in head-to-head voting. Since C is currently leading B by 2 to 0, if we make C the last choice on the uncompleted ballots, we force B to defeat C. The final ballots are then as follows.

1st	A	A	A	B	B	C	C
2nd	B	B	B	A	A	B	B
3rd	C	C	C	C	C	A	A

We have shown that although A wins the election using the plurality method, B defeats both A and C in head-to-head contests, so Condorcet's criterion is not satisfied.   ◎

You must be careful to understand what we are saying in Example 2 when we state that the plurality method violates Condorcet's criterion. We are not claiming that *every* vote using the plurality method violates Condorcet's criterion.* Example 2 simply provides a counterexample to the statement, "The plurality method satisfies Condorcet's criterion."

---

* Recall the Always Principle and the Counterexample Principle in Section 1.1.

Quiz Yourself 5

Complete the following ballots so that not only does B defeat A and C in head-to-head voting, but also C defeats A in head-to-head voting.

1st	A	A	A	B	B	C	C
2nd						B	B
3rd						A	A

In Quiz Yourself 5, we see that A wins the election using the plurality method, even though most voters prefer both B and C to A!

The independence-of-irrelevant-alternatives criterion says that removing losers from the ballot does not affect the outcome of the election.

Suppose that after an election has been conducted using preference ballots we discover that one of the losing candidates should never have been listed on the ballot. Rather than have a new election we could decide to strike that candidate's name from each ballot and do a recount. It is a desirable condition of voting that the winner using the original count should also be the winner using the modified ballots.

> **DEFINITION**
>
> **Independence-of-Irrelevant-Alternatives Criterion**
>
> If candidate X wins an election, some nonwinners are removed from the ballot, and a recount is done, then X still wins the election.

EXAMPLE 3   **The Plurality Method Violates the Independence-of-Irrelevant-Alternatives Criterion**

The county board of supervisors is voting on methods to raise taxes to finance a new sports stadium. The options are levying a tax on (H)otel rooms, increasing the tax on (A)lcohol, and increasing the tax on (G)asoline. The board will use the plurality method to make its decision. The following preference table shows the results of the vote.

	Number of Ballots		
Preference	8	6	6
1st	A	H	G
2nd	G	G	A
3rd	H	A	H

Highlight: *Using Spreadsheets* to Find Counterexamples*

It is tedious to generate preference tables by hand to find flaws in a voting method. Instead, you can use spreadsheets to do the repetitive computations necessary to find a particular counterexample. We used the following spreadsheet to generate a preference table showing that the Borda count method fails Condorcet's criterion.

C7		=4*(E1+F1)+3*(B1+C1)+2*(G1)+1*(D1)					
	A	**B**	**C**	**D**	**E**	**F**	**G**
1		15	4	28	10	8	30
2	1st	C	D	C	A	A	B
3	2nd	A	A	B	D	B	C
4	3rd	B	B	D	B	C	A
5	4th	D	C	A	C	D	D
6							
7	BORDA	A pts	217				
8		B pts	286				
9		C pts	292	(C is Borda Winner)			
10		D pts	155				
11							
12		Total	950				
13		NVoters	95				
14							
15	CONDORCET						
16	A vs B	A=	37	B=	58		
17	A vs C	A=	22	C=	73		
18	A vs D	A=	63	D=	32		
19	B vs C	B=	52	C=	43		
20	B vs D	B=	81	D=	14		
21	C vs D	C=	81	D=	14		

Notice that although C is the Borda winner, B defeats every other candidate in a head-to-head contest. Thus the Borda count method does not satisfy Condorcet's criterion.

We have highlighted cell C7, which contains a formula (shown at the top of the spreadsheet) for computing A's points using the Borda count method. By creating formulas such as this, we can simply change the numbers in row 1 to recalculate Borda point counts as well as tallies of who beats whom head to head. By adjusting the numbers in trial-and-error fashion, we quickly found the desired counterexample.

---

* You can find sample spreadsheets to do these calculations at www.aw.com/pirnot.

Does this vote satisfy the independence-of-irrelevant-alternatives criterion?

SOLUTION: What we are really asking is whether the removal of one of the losing options changes the outcome of the election. If, due to lobbying pressure, the board removes the tax on hotel rooms as an option, we get the following.

Preference	Number of Ballots		
	8	6	6
1st	A	G	G
2nd	G	A	A

We see that the gasoline tax now wins by a vote of 12 to 8; thus, the plurality method does not satisfy the independence-of-irrelevant-alternatives criterion.   ◎

Example 4 shows that the pairwise comparison method also violates the independence-of-irrelevant-alternatives criterion.

EXAMPLE 4   **The Pairwise Comparison Method Violates the Independence-of-Irrelevant-Alternatives Criterion**

The following table summarizes the preference ballots cast for candidates A, B, C, and D.

Preference	Number of Ballots			
	8	4	5	1
1st	A	D	C	D
2nd	B	A	B	A
3rd	C	C	D	B
4th	D	B	A	C

Using the pairwise comparison method, who is the winner of this election? If any of the losing candidates are removed, does this change the results of the election?

SOLUTION: For each pair of candidates, we must determine who is the winner in a head-to-head vote. The following table shows the results of these comparisons.

	Vote Results	Points Earned
A vs. B	A wins 13 to 5	A gets 1 point
A vs. C	A wins 13 to 5	A gets 1 point
A vs. D	D wins 10 to 8	D gets 1 point
B vs. C	tie — each has 9	B and C get $\frac{1}{2}$ point
B vs. D	B wins 13 to 5	B gets 1 point
C vs. D	C wins 13 to 5	C gets 1 point

Thus A gets 2 points, B and C each get $1\frac{1}{2}$ points, and D gets 1 point. If we remove candidates B and C, the preference table looks like this.

	Number of Ballots			
Preference	8	4	5	1
1st	A	D	D	D
2nd	D	A	A	A

We see that D defeats A by 10 votes to 8. Since the removal of some original losing candidates changes the outcome of the election, this method does not satisfy the independence-of-irrelevant-alternatives criterion.

Quiz Yourself 6

Use the pairwise comparison method to determine the winner of the election summarized in the following preference table. Is the independence-of-irrelevant-alternatives criterion satisfied?

	Number of Ballots			
Preference	5	3	3	1
1st	W	Z	Y	Z
2nd	X	W	X	W
3rd	Y	Y	Z	X
4th	Z	X	W	Y

The monotonicity criterion states that if a candidate wins an election and then gains more support, then that candidate will win a reelection.

Often before an election takes place, pollsters are hired to determine the preferences of the voters. If candidate X, the current front-runner, draws away support from one of his opponents, then it would seem likely that X has improved his chances of winning the election.* This idea motivates our final voting criterion, the monotonicity criterion.

> **DEFINITION**
>
> **The Monotonicity Criterion**
>
> If X wins an election and in a reelection all voters who change their votes only change their votes to favor X, then X also wins the reelection.

---

* Of course, our intuition tells us that it depends on the voting method we are using.

If we are using the plurality method or the Borda count method, a candidate who wins an election and then gains more support will win any reelection.

Example 5 is the example that we described at the start of the chapter opener explaining why it may be an advantage for voters to vote against you in order for you to win an election.

EXAMPLE 5   **The Plurality-with-Elimination Method Violates the Monotonicity Criterion**

Table 10.9 summarizes the preference ballots cast for the president of the International Students' Organization (ISO), in which Michael Chang, Rukevwe Kwami, and Anna Woytek are candidates. The plurality-with-elimination method is being used to determine the winner. The day before the election, three supporters of Woytek, who had preferred Chang over her, tell her that since they know she is going to win the election, they are going to throw their support to her in the election tomorrow. Later that day, after talking with her uncle who is an expert in voting theory, Woytek calls the three new supporters and asks them to vote for Chang instead.

	Number of Ballots			
Preference	12	9	3	8
1st	W	K	C	C
2nd	C	W	W	K
3rd	K	C	K	W

TABLE 10.9   Preference table for the ISO president election.

If the three voters indicated in the highlighted column in Table 10.9 change their votes so that Woytek is first, Chang is second, and Kwami is third, why should this cause Woytek concern?

SOLUTION: Let us first determine the winner of the election using the ballots in Table 10.9. We see that Woytek gets twelve first-place votes, Chang eleven, and Kwami nine. Thus Kwami is eliminated and a runoff election is held. Since voters do not change their preferences in the runoff, we can simply eliminate all references to Kwami on the original ballots and recount the votes. The ballots will look like this.

	Number of Ballots			
Preference	12	9	3	8
1st	W		C	C
2nd	C	W	W	
3rd		C		W

With Kwami removed from the ballots, we see that Woytek defeats Chang by a vote of 21 to 11.

Now let us see what would have happened if the three voters who preferred Chang to Woytek to Kwami had switched their votes. The ballots before the first elimination would have been as follows.

	Number of Ballots			
Preference	15	9	0	8
1st	W	K	C	C
2nd	C	W	W	K
3rd	K	C	K	W

Since Chang has the fewest first-place votes, he is eliminated and the ballots now look like this.

	Number of Ballots			
Preference	15	9	0	8
1st	W	K		
2nd		W	W	K
3rd	K		K	W

It is now clear why Woytek did not want the votes changed; in the runoff election between Woytek and Kwami, she loses 17 to 15.

We have seen that each of the voting methods discussed in Section 10.1 violates an important voting criterion. Table 10.10 summarizes what we know about voting methods' defects at this point and indicates where we address this issue in the examples and exercises. In Table 10.10, a "Yes" entry means the criterion stated in the row is always satisfied by the method listed in the column. Otherwise we reference a counterexample from this section showing that the method need not satisfy the stated criterion.

	Plurality	Borda Count	Plurality with Elimination	Pairwise Comparison
Majority	Yes	No—Example 1	Yes	Yes
Condorcet's	No—Example 2	No—Exercise 5	No—Example 2	Yes
Independence-of-irrelevant-alternatives	No—Example 3	No—Exercise 7	No—Exercise 22	No—Example 4
Monotonicity	Yes	Yes	No—Example 5	Yes

TABLE 10.10 Flaws in voting methods.

Because all of the voting methods we have studied have serious flaws, it is natural to wonder whether a perfect voting method exists. This question was answered in 1951 by Kenneth Arrow, a researcher at a government think tank called the RAND Corporation, who was studying decision making as it applied to governmental affairs. He posed a set of simple criteria* that we would expect any reasonable voting system to have.

1. The voting method must allow at least three voters to rank at least three alternatives.
2. If every voter prefers choice X to choice Y, then the voting method must rank X above Y.
3. No one voter can force his or her choice on the other voters.
4. The method must satisfy the independence-of-irrelevant-alternatives criterion.

Arrow proved the remarkable result that *no voting method can simultaneously satisfy these four simple criteria*. This theorem is often referred to as **Arrow's impossibility theorem.**

Exercises 10.2

Some of these exercises have no fixed solution method. Often constructing a preference table and adjusting the entries by trial and error will eventually lead to a solution.

1. In the preference table, A has the majority of first-place votes. Who wins the election if we use the Borda count method?

	Number of Ballots			
Preference	12	15	9	13
1st	A	B	C	A
2nd	B	C	B	D
3rd	C	A	D	B
4th	D	D	A	C

2. In the preference table, D has the majority of first-place votes. Who wins the election if we use the Borda count method?

	Number of Ballots			
Preference	4	10	3	2
1st	C	D	C	A
2nd	A	A	A	D
3rd	B	B	D	B
4th	D	C	B	C

3. Voters are choosing among three choices. Give a preference table in which the Borda count winner violates the majority criterion.

4. Voters are choosing among five choices. Give a preference table in which the Borda count winner violates the majority criterion.

5. **Determining the legal drinking age.** A state commission is voting on changing the legal drinking age. The options are A, lower the age to 18; B, lower the age to 19; C, lower the age to 20; and D, keep the age at 21. Use the preference table to determine the winner using the Borda count method. Show that Condorcet's criterion is not satisfied.

	Number of Ballots				
Preference	8	10	14	3	10
1st	C	D	C	A	B
2nd	A	A	B	D	C
3rd	B	B	D	B	A
4th	D	C	A	C	D

6. **Voting for the president of a club.** A chapter of the Sierra Club is voting for president. The candidates are (A)lvaro, (B)rown, (C)lark, and (D)ukevitch. Use the preference table to determine the winner using the Borda

---

* These conditions are sometimes stated in slightly different ways depending on which author you read. Our discussion is based on Alfred F. MacKay, *Arrow's Theorem: The Paradox of Social Choice* (Yale University Press, 1980).

count method. Show that Condorcet's criterion is not satisfied.

	Number of Ballots				
Preference	4	23	8	3	12
1st	C	D	C	A	A
2nd	A	A	B	D	B
3rd	B	B	D	B	C
4th	D	C	A	C	D

7. Choosing a location for a research facility. NASA is considering (A)tlanta, (B)oston, (C)hicago, and (D)enver for a new research facility. A group of senior managers voted to determine where the facility will be located. Use the preference table to determine the city that was chosen using the Borda count method. Show that the independence-of-irrelevant-alternatives criterion is not satisfied.

	Number of Ballots					
Preference	15	4	8	10	8	2
1st	C	D	C	B	B	A
2nd	B	B	A	D	A	C
3rd	A	A	D	A	C	B
4th	D	C	B	C	D	D

8. Locating a new factory. The Land Mover Tractor Company is going to build a new factory in either (A)labama, (C)alifornia, (O)regon, or (T)exas. The vote of the board

of directors is shown in the preference table. Determine the state that they chose using the Borda count method. Show that the independence-of-irrelevant-alternatives criterion is not satisfied.

	Number of Ballots					
Preference	9	2	4	5	4	1
1st	C	T	C	A	A	O
2nd	A	A	O	T	O	C
3rd	O	O	T	O	C	A
4th	T	C	A	C	T	T

9. Reducing a budget. Due to a budget problem, a citizens' committee is recommending to the school board ways to reduce expenses. The options are A, reduce sports programs; B, reduce expenditures on art and music programs; C, increase class size; and D, defer maintenance on buildings. Use the preference table to determine the choice that the committee recommends using the plurality-with-elimination method. Show that Condorcet's criterion is not satisfied.

	Number of Ballots					
Preference	9	12	4	5	4	1
1st	C	D	C	A	A	B
2nd	A	A	B	D	B	C
3rd	B	B	D	B	C	A
4th	D	C	A	C	D	D

10. Voting on an award for best restaurant. A group of columnists is voting on the restaurant of the year. The choices are The (A)lamo, The (B)ar-B-Q, (C)hez Nous, and (D)anny's Place. Use the preference table to determine the winner using the plurality-with-elimination method. Show that Condorcet's criterion is not satisfied.

Preference	Number of Ballots				
	8	11	3	4	3
1st	B	D	B	C	C
2nd	C	C	A	D	A
3rd	A	A	D	A	B
4th	D	B	C	B	D

Use the following preference table for Exercises 11–12.

	13	10	5
1st	A	B	C
2nd	B	C	B
3rd	C	A	A

11. Who wins this election using the pairwise comparison method? Why does this election not violate the majority criterion?

12. a) Who wins the election using the plurality-with-elimination method?

   b) If the last 5 voters change their ballots to

C
A
B

who now wins the election using the plurality-with-elimination method? Is this a violation of the monotonicity criterion? Explain.

13. Complete the preference table so that the Borda count winner violates Condorcet's criterion.

Preference	a.	b.
1st	B	A
2nd	A	C
3rd	C	B

14. Complete the preference table so that A is the Borda count winner, but when we remove C, then B is the Borda count winner, thus violating the independence-of-irrelevant-alternatives criterion.

Preference	a.	b.	c.
1st	A	B	C
2nd	C	A	B
3rd	B	C	A

15. Communicating Mathematics   Explain how to construct a preference table as in Example 2, showing an election using the plurality method that violates Condorcet's criterion.

16. Use the insight that you gained in answering Exercise 15 to construct a preference table in which voters vote on three choices using the plurality method and the vote violates Condorcet's criterion.

17. Communicating Mathematics   Explain how to construct a preference table as in Example 3, showing an election using the plurality method that violates the independence-of-irrelevant-alternatives criterion.

18. Use the insight that you gained in answering Exercise 17 to construct a preference table in which voters vote on three choices using the plurality method and the vote violates the independence-of-irrelevant-alternatives criterion.

19. Complete the preference table so that the plurality-with-elimination method violates Condorcet's criterion.

Preference	Number of Ballots				
	20	__	8	8	12
1st	C	D	A	A	B
2nd	A	A	D	B	C
3rd	B	B	B	C	A
4th	D	C	C	D	D

20. Communicating Mathematics   Explain why the plurality-with-elimination method satisfies the majority criterion.

21. Communicating Mathematics   Explain why the plurality method satisfies the majority criterion.

22. Voters are choosing among four choices. Construct a preference table in which the plurality-with-elimination method violates the independence-of-irrelevant-alternatives criterion.

23. Communicating Mathematics Explain how to construct an example such as Example 5 showing that the plurality-with-elimination method may violate the monotonicity criterion.

*⃝**24.** Use the insight that you gained in answering Exercise 25 to construct a preference table in which voters vote on three choices using the plurality-with-elimination method and the vote violates the monotonicity criterion.

25. Communicating Mathematics Explain why the pairwise comparison method satisfies the majority criterion.

26. Communicating Mathematics Explain why the pairwise comparison method satisfies Condorcet's criterion.

27. Communicating Mathematics Explain how to construct an example such as Example 4 to show how an election using the pairwise comparison method may violate the independence-of-irrelevant-alternatives criterion.

⃝**28.** Use the insight that you gained in answering Exercise 27 to construct a preference table in which voters vote on five choices using the pairwise comparison method and the vote violates the independence-of-irrelevant-alternatives criterion.

⃝**29.** Voters are choosing among five choices. Construct a preference table in which the plurality-with-elimination method violates the monotonicity criterion.

⃝**30.** Voters are choosing among four choices. Construct a preference table in which the plurality-with-elimination method violates the monotonicity criterion.

Further Exercises

31. A famous example called *Condorcet's paradox* is illustrated in the preference table.

	Number of Ballots		
Preference	1	1	1
1st	A	B	C
2nd	B	C	A
3rd	C	A	B

Notice that two-thirds of the voters prefer A over B, two-thirds prefer B over C, and two-thirds prefer C over A. Construct a preference table for five candidates A, B, C, D, and E such that 80 percent of the voters prefer A over B, 80 percent prefer B over C, 80 percent prefer C over D, 80 percent prefer D over E, and 80 percent prefer E over A.

32. In using the pairwise comparison method, as the number of candidates grows it is easy to see that the number of comparisons grows quite rapidly.

a) Complete the following table:

Candidates	Number of Comparisons
A, B	1
A, B, C	3
A, B, C, D	6
A, B, C, D, E	
A, B, C, D, E, F	

b) In Chapter 12 we will show that for k candidates there are $k(k - 1)/2$ comparisons necessary. How many comparisons are necessary for 10 candidates? For 20?

⃝**33.** Communicating Mathematics Can you construct a table of preference ballots for three candidates in which one candidate has a majority of first place votes but would lose to another candidate using the Borda count method? If you can, create such a table. If you cannot, explain why it is not possible.

34. Communicating Mathematics Can you construct a preference table with fewer than seven ballots showing that the winner using the plurality method may not satisfy Condorcet's criterion? If you can, create such a table. If you cannot, explain why it is not possible.

35. Research how the electoral college system is used to elect the president of the United States. How does this procedure conflict with the majority criterion?

36. Locate several books in your library on voting theory or social choice. Briefly describe a voting criterion that has not been discussed in this chapter.

37. Write a short essay on the reasonableness of Condorcet's expectations that the social sciences could be put on as firm a mathematical foundation as the physical sciences.

10.3 WEIGHTED VOTING SYSTEMS

Not all members of the United Nations Security Council have the same voting power. The present council consists of five permanent members (Great Britain, France, the United States, China, and Russia) and ten nonpermanent members. According to its rules, the council cannot pass a resolution unless all of the permanent members vote yes and, in addition, four of the nonpermanent members also vote yes. There are other situations in which voting is done very differently from the methods we have discussed in this chapter. For example, when a jury votes in a criminal trial, a vote of eleven to one is not enough to convict the accused. In corporate decision making, a large stockholder may possess 40 or 50 times the voting power of a smaller stockholder. In this section, we will develop a method to measure the power of voters in a system in which not everyone has the same strength.

We describe a weighted voting system by its quota and the weight of each voter.

In order to understand the concept of a weighted voting system, consider the following situation. A corporation that owns the Phoenix Flames, a professional football team, has six stockholders, each of whom owns different amounts of stock. Let's say that Alicia Mendez and her son Ben each own 26 percent of the stock and Carl, Dante, Emily, and Felix each own 12 percent. (We represent these stockholders as A, B, C, D, E, and F.) Assume that each stockholder possesses as many votes as the percentage of stock owned. Voting in this corporation clearly does not reflect the "one person, one vote" principle. In fact, A and B possess a lot of power compared to C, D, E, and F. If a resolution requires a vote of more than 50 percent in order to pass, we see that A and B together can pass any resolution they wish, whereas C, D, E, and F are considerably weaker in their ability to get their resolutions passed.

In this situation we see a number of characteristics of weighted voting systems that are present in the examples we will study. First, there is a number of votes, 51, required in order for a resolution to pass. This number is called a *quota*. Second, there are voters, each of whom controls a number of votes. We call this number the voter's *weight*. We will make these ideas more precise.

DEFINITIONS

A **weighted voting system** with *n* voters is described by a set of numbers which are listed in the following format:

[quota : weight of voter 1, weight of voter 2, . . . , weight of voter *n*].

The **quota** is the number of votes necessary in this system to get a resolution passed. The numbers that follow, called **weights**, are the amount of votes controlled by voter 1, voter 2, etc.

EXAMPLE 1   Weighted Voting Systems

Explain each of the following weighted voting systems.

a) [51 : 26, 26, 12, 12, 12, 12]   b) [4 : 1, 1, 1, 1, 1, 1, 1]

c) [14 : 15, 2, 3, 3, 5]   d) [10 : 4, 3, 2, 1]

e) [12 : 1, 1, 1, 1, 1, 1, 1, 1, 1, 1, 1, 1]   f) [12 : 1, 2, 3, 1, 1, 2]

g) [39 : 7, 7, 7, 7, 7, 1, 1, 1, 1, 1, 1, 1, 1, 1, 1]

SOLUTION: a) This is the stockholder situation that we described earlier. The quota is 51; there are six voters; the first two voters each have 26 votes and the last four voters each have 12 votes.

b) In this situation, there are seven voters, each of whom has one vote. Since the quota is 4, a simple majority suffices to pass a resolution. This is an example of a "one person, one vote" situation.

c) The quota is 14; because voter 1 has fifteen votes, he or she has total control. Since the other four voters have no power whatsoever in this system, we call voter 1 a **dictator**.

d) Notice that the sum of the votes is 10, which is also the quota. In this system, even though the first voter has greater weight than the others, in fact he or she has no more power, because the support of even the weakest voter is necessary in order to pass a resolution. A voter who can, by himself or herself, prevent a motion from passing has **veto power**.

e) This describes our jury system for trying criminal cases. Since the quota is 12, every voter must vote for a resolution in order for it to pass. Each voter has veto power.

f) In this system, the sum of all the possible votes is less than the quota, so no resolutions can be passed.

g) This system describes the voting in the United Nations Security Council. Notice that the quota cannot be achieved unless all of the first five voters vote for a resolution. In addition, four of the next ten voters must also vote in order for a resolution to be passed.

Groups of voters form coalitions to pass resolutions.

The weighted voting system in Example 1(d) is interesting in the sense that even though voter 1 seems to have more power in an election, this voter, in fact, does not. This system behaves the same as the system $[4 : 1, 1, 1, 1]$. Notice that the number of votes that a voter possesses is not the same thing as the power a voter has to pass resolutions. In order to describe the concept of power in a voting system we need to introduce some more definitions. We first describe the subsets of voters that have the power to pass resolutions.

> **DEFINITIONS**
>
> Any set of voters who vote the same way is called a **coalition**. The sum of the weights of the voters in a coalition is called the **weight of the coalition**. If a coalition has a weight that is greater than or equal to the quota, then that coalition is called a **winning coalition**.

In a weighted voting system $[4 : 1, 1, 1, 1, 1, 1, 1]$, any coalition of four or more voters is a winning coalition.

> **PROBLEM SOLVING**
>
> Recall from Chapter 1 that a k-element set has 2^k subsets. Since one of these is the empty set, we see that a set of k voters can form $2^k - 1$ possible coalitions. For example, a set of five voters can form $2^5 - 1 = 32 - 1 = 31$ different coalitions. It is useful to list coalitions systematically. That is, consider all the one-element sets, then all the two-element sets, followed by the three-element sets, and so on.

EXAMPLE 2   Finding Winning Coalitions

A town has two large political parties, (R)epublican and (D)emocrat, and one small party, (I)ndependent. Membership on the town council is proportional to the size of the parties. We will assume that R has nine members on the council, D has eight, and I only three. Traditionally each party votes as a single bloc, and resolutions are passed by a simple majority. List all possible coalitions and their weights, and identify the winning coalitions.

SOLUTION: A coalition is a non-empty subset of the set of all parties {R, D, I}. We list these subsets and their weights in the following table. Because there are 20 members on the council, any coalition with a weight of 11 or more is a winning coalition.

Coalition	Weight	
{R}	9	
{D}	8	
{I}	3	
{R, D}	17	Winning
{R, I}	12	Winning
{D, I}	11	Winning
{R, D, I}	20	Winning

Quiz Yourself 7*

Consider the weighted voting system [5 : 3, 2, 2, 1].

a) How many voters are there?

b) What is the quota?

c) List all winning coalitions.

It is interesting that even though party I has less representation on the council than either R or D, it still appears in as many winning coalitions as does R or D.

In Example 2, it seems that all three parties have the same amount of voting power. We will make this idea of power more precise shortly. The key to understanding a voter's power is knowing how many coalitions need the voter in order to win.

> **DEFINITION**
>
> A voter in a winning coalition is called **critical** if it is the case that if he or she were to leave the coalition, then the coalition would no longer be winning.

EXAMPLE 3 Identifying Critical Voters in a Coalition

Identify the critical voters in the winning coalitions in the town council in Example 2.

SOLUTION:

Coalition	Weight		Critical Voters
{R}	9		
{D}	8		
{I}	3		
{R, D}	17	Winning	R, D
{R, I}	12	Winning	R, I
{D, I}	11	Winning	D, I
{R, D, I}	20	Winning	none

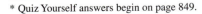

The Banzhaf power index measures voting power.

We now define how to measure voting power in a weighted voting system.

> **DEFINITION**
>
> In a weighted voting system, a voter's **Banzhaf power index*** is defined as
>
> $$\frac{\text{the number of times the voter is critical in winning coalitions}}{\text{the total number of times voters are critical in winning coalitions}}.$$

Quiz Yourself 8

Reconsider the weighted voting system [5:3, 2, 2, 1] from Quiz Yourself 7, which had winning coalitions {A, B}, {A, C}, {A, B, C}, {A, B, D}, {A, C, D}, {B, C, D}, and {A, B, C, D}.

a) Determine the critical voters in each winning coalition.

b) Compute the Banzhaf power index for voters in this system.

EXAMPLE 4   Computing the Banzhaf Power Index

In Example 3 we saw that R, D, and I each were critical voters twice. Thus R's Banzhaf power index is

$$\frac{\text{the number of times R is critical in winning coalitions}}{\text{the total number of times voters are critical in winning coalitions}} = \frac{2}{6} = \frac{1}{3}.$$

Similarly, D and I each have a Banzhaf power index of $\frac{1}{3}$.

EXAMPLE 5   Calculating the Banzhaf Power Index Indirectly

In the law firm of Krook, Cheatum, and Associates, there are two senior partners (Krook and Cheatum) and four associates (W, X, Y, and Z). In order to change any major policy of the firm, a vote must be taken in which Krook, Cheatum, and at least two associates agree to the change. Calculate the Banzhaf power index for each member of this firm.

SOLUTION: We can represent the members of the firm by {K, C, W, X, Y, Z}. At first it seems tempting to list all of the coalitions (subsets) of the firm and determine which are winning. However, this is a tedious job and is not necessary. Realize that every winning coalition includes {K, C}, so all we have to do is determine the subsets of {W, X, Y, Z} with two or more members and form the union of these with {K, C}.

2-Element Subsets of {W, X, Y, Z}	3-Element Subsets of {W, X, Y, Z}	4-Element Subsets of {W, X, Y, Z}
{W, X}, {W, Y}, {W, Z}, {X, Y}, {X, Z}, {Y, Z}	{W, X, Y}, {W, X, Z}, {W, Y, Z}, {X, Y, Z}	{W, X, Y, Z}

---

* There are other slightly different ways of defining this index, depending on which author you read.

The winning coalitions of the firm and their critical members are therefore:

	Winning Coalitions	Critical Members
1	{K, C, W, X}	K, C, W, X
2	{K, C, W, Y}	K, C, W, Y
3	{K, C, W, Z}	K, C, W, Z
4	{K, C, X, Y}	K, C, X, Y
5	{K, C, X, Z}	K, C, X, Z
6	{K, C, Y, Z}	K, C, Y, Z
7	{K, C, W, X, Y}	K, C
8	{K, C, W, X, Z}	K, C
9	{K, C, W, Y, Z}	K, C
10	{K, C, X, Y, Z}	K, C
11	{K, C, W, X, Y, Z}	K, C

From this table we see that K and C are critical members eleven times whereas W, X, Y, and Z are each critical members only three times. The Banzhaf power indices for the members of this firm are therefore:

Members	Banzhaf Power Index	
K, C	$\dfrac{11}{11 + 11 + 3 + 3 + 3 + 3}$	$= \dfrac{11}{34}$
W, X, Y, Z	$\dfrac{3}{11 + 11 + 3 + 3 + 3 + 3}$	$= \dfrac{3}{34}$

Notice that the sum of the Banzhaf power indices for the members of the firm is one. This is always the case when computing Banzhaf power indices in a weighted voting system.   ◎

Often the chair of a committee votes only in the case of breaking a tie. In fact, this is the way the U.S. Senate votes. The vice-president of the United States presides over the 100-member U.S. Senate and votes only to break a tie. The mathematics to determine the Banzhaf power index for the vice-president and the members of the senate is too lengthy to include here. We will, however, analyze a much simpler example based on the same voting principle.

EXAMPLE 6   Finding the Banzhaf Power Index of a "Tie Breaker"

A five-person air safety review board is developing in-flight safety procedures to deal with skyjackings. The board is chaired by a federal aviation administrator (A) and consists of two senior pilots (S and T) and two flight attendants (F and G).

Historical Highlight: *Blocking Coalitions, Banzhaf, and the Electoral College*

The Banzhaf power index is usually computed in a more complex way than we have explained it. In addition to counting the winning coalitions to which a voter belongs, we also count how many times the voter is in a blocking coalition. A **blocking coalition** is a set of voters that has enough votes to defeat a resolution. A critical voter in a blocking coalition is one who, by leaving the coalition, changes it from blocking to nonblocking. The Banzhaf index is then computed by counting critical voters in winning and blocking coalitions.

This method can be used to compute the Banzhaf power index of the states voting in the electoral college that elects the president of the United States. In the electoral college, each state has as many electoral votes as the total of its senators and representatives. For example, New York presently has 31 representatives and 2 senators; thus it has 33 votes in the electoral college. Similarly, Missouri has 11 electoral votes. We can then think of the electoral college as a weighted voting system with the 50 states (plus the District of Columbia). The state with the most weight is California (54 votes), and there are eight states with the fewest votes (3): Alaska,

Delaware, Montana, North Dakota, South Dakota, Vermont, Wyoming, and the District of Columbia.

Although there are more than nine trillion winning and blocking coalitions to be considered, using computers the Banzhaf index can be calculated for each state.* We list several of these here.

State	Banzhaf Power Index (%)
California	11.14
New York	6.20
Florida	4.63
Pennsylvania	4.25
Arkansas	1.09
New Mexico	0.91
Idaho	0.73
Delaware	0.55

From this table, we see that California has almost 3 times the Banzhaf power of Pennsylvania and over 20 times the power of Delaware.

It was the intent when the board was established that the administrator have considerably less power than the pilots and flight attendants. Therefore, the administrator votes only in the case of a tie; otherwise, cases are decided by a simple majority. How much less power does the administrator have than the other members of the board?

SOLUTION: To answer this question, we compute the Banzhaf power index for each member of the board. We first list the winning coalitions and the critical members of these coalitions. Clearly if three or four of {S, T, F, G} vote together, they form a winning coalition. We list these five coalitions first

---

* The Banzhaf indices presented here are from the paper by P. J. Affuso and S. J. Brams, "Power and Size: A New Paradox," *Theory and Decision* 7(1976): pp. 29–56.

	Winning Coalitions	Critical Members
1	{S, T, F}	S, T, F
2	{S, T, G}	S, T, G
3	{S, F, G}	S, F, G
4	{T, F, G}	T, F, G
5	{S, T, F, G}	none
6	{A, S, T}	A, S, T
7	{A, S, F}	A, S, F
8	{A, S, G}	A, S, G
9	{A, T, F}	A, T, F
10	{A, T, G}	A, T, G
11	{A, F, G}	A, F, G

and then list the six ways ties can be formed, which then require the addition of the administrator to break them.

If we count all of the number of times members A, S, T, F, and G are critical members of some coalition, we get a total of 30. We see that any single board member (including the chair) is a critical member of exactly six coalitions. Therefore each of the five board members has exactly the same Banzhaf power index, which is $\frac{6}{30}$. Even though it is not apparent at first, the chair of the board has exactly the same power as the other board members.   ◎

In Example 6, the administrator and each other member of the board were critical in exactly the same number of circumstances, by being one of a bare majority in favor of a motion. Similarly, in the U.S. Senate, the vice-president and each senator have exactly the same number of opportunities to be critical by being on the winning side of a 51-to-50 vote. Thus the vice-president has the same power as each senator. In Chapter 12, you will learn principles of counting that will enable you to calculate the number of cases that we have to consider to prove this.

Exercises 10.3

In Exercises 1–12, the weights represent voters A, B, C, and so on, in that order. Identify (a) the quota, (b) the number and weights of the voters, (c) dictators, and (d) those having veto power.

1. [5 : 1, 1, 1, 1, 1]   **2.** [15 : 5, 4, 3, 2, 1]

3. [11 : 10, 3, 4, 5]   **4.** [6 : 6, 1, 2, 2]

5. [15 : 1, 2, 3, 3, 4]   **6.** [11 : 1, 2, 3, 4]

7. [12 : 1, 3, 5, 7]   **8.** [16 : 1, 5, 7, 9]

9. [25 : 4, 4, 6, 7, 9]   **10.** [21 : 3, 5, 6, 8, 9]

11. [51 : 20, 20, 20, 10, 10]

12. [67 : 15, 15, 15, 15, 10, 10]

In Exercises 13–16, write out all of the winning coalitions in each voting system. Do not approach this randomly. Be systematic by considering coalitions from smallest to largest.

13. [12 : 1, 3, 5, 7]   **14.** [16 : 1, 5, 7, 9]

15. [25 : 4, 4, 6, 7, 9]   **16.** [23 : 3, 5, 6, 8, 9]

17. Find the critical voters in the winning coalitions that you found in Exercise 13.

18. Find the critical voters in the winning coalitions that you found in Exercise 14.

19. Find the critical voters in the winning coalitions that you found in Exercise 15.

20. Find the critical voters in the winning coalitions that you found in Exercise 16.

21. A theater guild. The Theater Guild consists of (P)erformers, (T)echnicians, and (S)upport staff. Representation on the guild is proportional to the number in each group, and the representatives of each group tend to vote in a bloc. A simple majority is required to pass a resolution.  Assume there are five performers, four technicians, and two members of the support staff on the guild. List all the coalitions of {P, T, S} and their weights. Which are the winning coalitions?

22. A theater guild. Repeat Exercise 21, but now assume that there are eight performers, six technicians, and two members of the support staff on the guild.

23. A state committee. The state athletics procedures committee makes policy decisions affecting high school athletics programs throughout the state. The committee consists of three (A)dministrators, four (F)aculty members in athletics departments, three a(T)hletes, and two (N)on-athlete students. Assume that each of these four groups votes in a bloc. In order to make a policy change, a vote of at least 8 is required. Find all the winning coalitions of {A, F, T, N} and state their weights.

24. A state committee. Repeat Exercise 23, but now assume that there are four administrators, five faculty members, four athletes, and three non-athletes. Also assume that a vote of 12 is required to make a policy change.

25. A theater guild. Determine the critical voters in the winning coalitions in Exercise 21.

26. A theater guild. Determine the critical voters in the winning coalitions in Exercise 22.

27. A state committee. Determine the critical voters in the winning coalitions in Exercise 23.

28. A state committee. Determine the critical voters in the winning coalitions in Exercise 24.

In Exercises 29–34, determine the Banzhaf power index for each voter in each weighted voting system.

29. the weighted voting system in Exercise 1

30. the weighted voting system in Exercise 3

31. the weighted voting system in Exercise 9

32. the weighted voting system in Exercise 11

33. the weighted voting system in Exercise 13

34. the weighted voting system in Exercise 15

Communicating Mathematics   *In Example 5, we analyzed the voting power of the law partners in the firm of Krook, Cheatum, and Associates. Recall that to change a policy required the votes of Krook, Cheatum, and two associates. For each scenario described in Exercises 35–38, do the following:*

(a) Give an intuitive explanation as to how you think the voting power of Krook, Cheatum and each associate is changed.

(b) Calculate the Banzhaf power index for each member of the firm to see how this calculation corresponds to your intuition.

35. A law firm. Another associate is added to the original firm, so the firm now has two senior partners and five associates.

36. A law firm. Another senior partner, named Fair, is added to the original firm, so the firm now has three senior partners and four associates. Assume that two senior partners and two associates must vote to change a policy.

37. A law firm. One of the original associates, named Howe, is promoted to senior partner. The firm now has three senior partners and three associates. Assume that two senior partners and two associates must vote to change a policy.

38. A law firm. Krook resigns, leaving Cheatum as the only senior partner with four associates. Assume that Cheatum and two associates must vote to change a policy.

39. Communicating Mathematics   The system [3 : 1, 1, 1, 1, 1] is an example of a "one person, one vote" situation.

a) Calculate the Banzhaf power index for each person in this system.

b) How does this conform with your intuition? Explain.

40. Communicating Mathematics   A twelve-person jury corresponds to the weighted voting system

$$[12 : 1, 1, 1, 1, 1, 1, 1, 1, 1, 1, 1, 1].$$

a) Calculate the Banzhaf power index for each person in this system.

b) How does this conform with your intuition? Explain.

41. Communicating Mathematics   Consider the system [14 : 15, 2, 3, 3, 5], in which A is a dictator.

a) Calculate the Banzhaf power index for each person in this system.

b) How does this conform with your intuition? Explain.

42. Communicating Mathematics   Consider the system [12 : 1, 2, 3, 1, 1, 2], in which no resolution can be passed because the quota is too high.

a) Explain why Banzhaf power indices cannot be calculated in this system.

b) How does this conform with your intuition? Explain.

Further Exercises

In Exercises 43–44, devise a voting system that behaves with specifications that are similar to the United Nations Security Council described in Example 1.

*43. A committee has three standing members and six temporary members. The three standing members and two of the temporary members must vote for a resolution in order for it to be passed.

44. A committee has four standing members and eight temporary members. The four standing members and three of the temporary members must vote for a resolution in order for it to be passed.

It is interesting to compare the Banzhaf power indices for states in the electoral college with the percentage of electoral votes and with the states' percentage of the total U.S. population. Use a world almanac or similar source to gather the information needed to do Exercises 45–47.

45. **U.S. Congress.** Make the suggested comparisons for California, New York, Florida, and Pennsylvania.

46. **U.S. Congress.** Make the suggested comparisons for Arkansas, New Mexico, Idaho, and Delaware.

47. Communicating Mathematics In Exercises 45–46, what patterns do you notice in comparing the larger states with the smaller states? Can you explain this?

CHAPTER SUMMARY

SECTION 10.1

plurality method
Each person votes for his or her favorite candidate. The candidate receiving the most votes is declared the winner.

Borda count method
If there are k candidates in an election, each voter ranks all candidates on the ballot. Then the first choice is given k points, the second choice is given $k - 1$ points, the third choice $k - 2$ points, and so on. The candidate who receives the most total points wins the election.

preference ballot, preference table
A preference ballot allows a voter to indicate a first choice, a second choice, and so on. A table that summarizes preference ballots is called a preference table.

approval voting
Approval voting allows voters to vote for as many candidates on the ballot as they wish. Candidates are not ranked, and the winner is the candidate who earns the most votes of approval.

plurality-with-elimination method
Each voter votes for one candidate. A candidate receiving a majority of votes is declared the winner. If no candidate receives a majority of votes, then the candidate or candidates with the fewest votes are dropped from the ballot and a new election is held. This process continues until a candidate receives a majority of votes.

---

* Exercise numbers circled in red can be used as group exercises.

pairwise comparison method
Voters first rank all candidates. If A and B are a pair of candidates, we count how many voters prefer A to B and vice versa. Whichever candidate, A or B, is preferred more receives 1 point. If A and B are tied, then each receives $\frac{1}{2}$ point. Do this comparison, assigning points, for each pair of candidates. At the end, the candidate receiving the most points is the winner.

SECTION 10.2

majority criterion
If a majority of the voters rank a candidate as their first choice, then that candidate wins the election.

Condorcet's criterion (head-to-head criterion)
If candidate X can defeat each of the other candidates in a head-to-head vote, then X is the winner of the election.

independence-of-irrelevant-alternatives criterion
If candidate X wins an election, some nonwinners are removed from the ballot, and a recount is done, then X still wins the election.

monotonicity criterion
If X wins an election and in a reelection all voters who change their votes only change their votes to favor X, then X also wins the reelection.

Arrow's impossibility theorem
This theorem, discovered by Kenneth Arrow, proves that no perfect voting method exists.

SECTION 10.3

weighted voting system, quota, weights
A weighted voting system with n voters is described by a set of numbers

$$[\text{quota}: \text{weight of voter 1, weight of voter 2}, \ldots, \text{weight of voter } n].$$

The quota is the number of votes necessary to get a resolution passed. The weights are the amount of votes controlled by voter 1, voter 2, and so on.

dictator, veto power
A dictator is a voter having total control of a voting system in the sense that resolutions can be approved even if only the dictator votes for the resolution. A voter who can, by himself or herself, prevent a motion from passing has veto power.

coalition, weight of a coalition, winning coalition
A set of voters who vote the same way is called a coalition. The sum of the weights of the voters in a coalition is called the weight of the coalition. If a coalition has a weight that is greater than or equal to the quota, then that coalition is called a winning coalition.

critical voter
A voter in a winning coalition is called critical if it is the case that if he or she were to leave the coalition, then the coalition would no longer be winning.

Banzhaf power index
In a weighted voting system, a voter's Banzhaf power index is defined as

$$\frac{\textit{the number of times the voter is critical in winning coalitions}}{\textit{the total number of times voters are critical in winning coalitions}}.$$

blocking coalition
A blocking coalition is a set of voters with enough votes to defeat a resolution.

CHAPTER TEST

SECTION 10.1

1. Four candidates running for town council receive votes as follows: Myers, 2,156; Pulaski, 1,462; Harris, 986; Martinez, 428.

   a) Is there a candidate who earns a majority?

   b) Who wins the election using the plurality method?

2. Use the preference table to determine the winner of the election using the Borda count method.

	Number of Ballots					
Preference	8	5	7	4	3	6
1st	A	D	A	B	B	C
2nd	B	B	C	D	C	A
3rd	C	C	D	C	A	B
4th	D	A	B	A	D	D

3. Members of the chamber of commerce have been asked to vote on their preference for a topic for a speaker for their convention. The choices are (S)ocial justice, the (R)ole of government in a free society, (E)ducation in the future, and (G)lobal issues. Their preferences are summarized in the table. What option is their first choice using the plurality-with-elimination method?

	Number of Ballots				
Preference	1,531	1,102	906	442	375
1st	G	R	S	S	G
2nd	R	S	G	E	S
3rd	S	G	R	R	E
4th	E	E	E	G	R

4. Using the preference table, who wins the election using the pairwise comparison method?

	Number of Ballots			
Preference	8	4	5	6
1st	A	D	B	C
2nd	B	B	D	B
3rd	C	C	C	A
4th	D	A	A	D

SECTION 10.2

5. Consider the following three preference ballots.

Preference			
1st	R	R	D
2nd	D	D	P
3rd	P	Q	Q
4th	Q	P	R

Who is the winner of this election using the Borda method? Does this election satisfy the majority criterion? Explain.

6. Use the preference table to determine the winner using the Borda count method. Is Condorcet's criterion satisfied in this election?

	Number of Ballots				
Preference	6	8	12	1	8
1st	R	D	R	A	M
2nd	A	A	M	D	R
3rd	M	M	D	M	A
4th	D	R	A	R	D

7. Use the preference table to determine the winner using the Borda count method. Is the independence-of-irrelevant-alternatives criterion satisfied? Explain.

Preference	Number of Ballots					
	11	3	6	9	4	3
1st	C	D	C	B	B	A
2nd	B	B	A	D	A	C
3rd	A	A	D	A	C	B
4th	D	C	B	C	D	D

8. Use the plurality-with-elimination method to determine the winner of the election using the preference table. Is the independence-of-irrelevant-alternatives criterion satisfied? Explain.

Preference	Number of Ballots			
	10	7	2	4
1st	A	B	C	C
2nd	C	A	A	B
3rd	B	C	B	A

SECTION 10.3

9. Find the quota, find the weights of the voters, determine whether there is a dictator, and find those having veto power in the weighted voting system [17:1, 5, 7, 8].

10. Write out all of the winning coalitions in the voting system [11:2, 3, 5, 7].

11. Determine the Banzhaf power index for each voter in the weighted voting system [11:2, 3, 5, 7].

12. Determine the Banzhaf power index for each voter in each weighted voting system. Explain intuitively why you would expect the calculations to come out as they did.

a) [10:1, 2, 3, 4]

b) [10:11, 1, 3, 3, 2]

Of Further Interest: THE SHAPLEY-SHUBIK INDEX

In her campaign flyers, your state legislator emphasizes that as a member of the Transportation Committee, she voted for a bill to improve public transportation in your city. Although she is being truthful, she has neglected to state that initially she was neutral on this matter and only threw in her support after it was clear that the bill had enough votes to pass. In fact, her only motive for supporting the bill was to win votes in your district, in which many people rely on public transportation.

From this example you can see that it may be important to know the order in which members join a coalition to make it a winner. In the discussion of the Banzhaf power index, we only considered the members of a coalition, specifically the critical members. The method we study here, which is based on the work of the mathematician Lloyd Shapley and the economist Martin Shubik, focuses not only the makeup of winning coalitions, but also the order in which winning coalitions are formed.

In order to understand the Shapley-Shubik index, we must make a clear distinction between sets whose elements have an order to them versus sets in which the order is unimportant. Recall from Chapter 1, when we use the notation {A, B, C}, order is not important; if we wish, we could write this set as {B, C, A} or {C, A, B} instead. If we want to emphasize that the order of the elements in a set is important, then we must use a different notation.

DEFINITION

An ordering of the elements of the set $\{x_1, x_2, x_3, \ldots, x_n\}$ in a straight line is called a **permutation*** of that set. We denote a permutation as follows

$$(x_{i_1}, x_{i_2}, x_{i_3}, \ldots, x_{i_n}),$$

where x_{i_1} is the first element in the permutation, x_{i_2} is the second element in the permutation, and so on.

Using this definition we see that (A, B, C), (B, C, A), and (C, A, B) are all different permutations of the set {A, B, C}, because the elements are listed in different orders.

From now on, we will assume the following:

A permutation of voters specifies a coalition in which the voters were added one at a time.

---

* We will study permutations at much greater length in Chapter 12.

PROBLEM SOLVING

The Splitting Hairs Principle in Section 1.1 stated that different notation usually means that you are dealing with different concepts. You can conclude that because {A, B, C} and (A, B, C) look different, they do not mean the same thing. Try to connect new notation with notation you have seen before. Recalling that the ordered pair of numbers (3, 4) does not mean the same thing as the ordered pair (4, 3) helps you understand the meaning of the notation (A, B, C).

It will be important to know how many different ways we can order the elements in a set. We will see how to generate different permutations of a set by looking at the problem graphically, as in the next example.

EXAMPLE 1   Finding Permutations of a Set

How many permutations are there of each set?

a) {A, B, C}

b) {A, B, C, D}

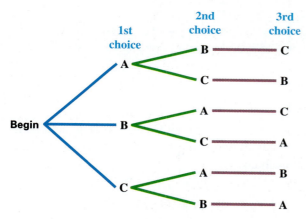

FIGURE 10.1   Tree diagram showing all permutations of {A, B, C}.

SOLUTION: a) If we think about forming the orderings by deciding which is the first element of the ordering, then which is the second, and finally which is the third element, we can visualize the permutations as shown in Figure 10.1, which is called a *tree diagram*.

Following the six branches of this tree starting with "Begin," we generate the permutations (A, B, C), (A, C, B), (B, A, C), (B, C, A), (C, A, B), and (C, B, A). In this case we found $3 \times 2 \times 1 = 6$ ways to order the three elements A, B, and C.

b) In this case, in Figure 10.2 (on the following page), we draw a tree beginning with four branches for our first choice, three branches at the next stage for our second choice, and so on.

From Figure 10.2 we see that there are $4 \times 3 \times 2 \times 1 = 24$ permutations of {A, B, C, D}. They are (A, B, C, D), (A, B, D, C), (A, C, B, D), . . . , (D, C, B, A).

Example 1 leads us to state the following principle.

The Number of Permutations of a Set

If we wish to order a set with n elements, following the method in Example 1, there are n ways to choose the first element, followed by $n - 1$ ways to choose the second element, followed by $n - 2$ ways to choose the third element, and so on. Thus there are $n \times (n - 1) \times (n - 2) \times . . . \times 2 \times 1$ permutations of these elements. This product is called n **factorial** and is written $n!$.

Quiz Yourself 9

How many permutations are there of a five-element set?

In forming a coalition by adding voters one at a time, at some point we add a voter who makes the coalition a winning coalition. That voter is particularly important, because he or she has changed a nonwinning coalition into a winning one.

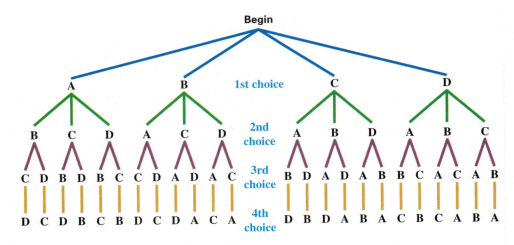

FIGURE 10.2   Tree diagram showing all permutations of {A, B, C, D}.

DEFINITION

As we add voters to a coalition one at a time, the first player added who makes the coalition a winning coalition is called a **pivotal voter** for the coalition.

EXAMPLE 2   Identifying the Pivotal Voter

Consider the weighted voting system [5 : 3, 2, 2, 1] with voters A, B, C, and D, in that order. Recall that A has three votes, B and C have two votes each, and D has one vote. Find the pivotal voter in each coalition.

a) (A, B, C, D)   b) (A, D, C, B)

c) (D, A, B, C)   d) (D, C, A, B)

SOLUTION: In each coalition we are looking for the first voter who makes the coalition a winning one by giving it a voting weight of 5 or more.

Coalition	Weight after 1st, 2nd, 3rd, 4th Voter Is Added	Pivotal Voter
a) (A, B, C, D)	3, <u>5</u>, 7, 8	B
b) (A, D, C, B)	3, 4, <u>6</u>, 8	C
c) (D, A, B, C)	1, 4, <u>6</u>, 8	B
d) (D, C, A, B)	1, 3, <u>6</u>, 8	A

We are now in a position to define the Shapley-Shubik index for measuring voter power.

In a weighted voting system, the **Shapley-Shubik index** for a voter is

$$\frac{\text{the number of times the voter is pivotal in some permutation of the voters}}{\text{the total number of permutations of the voters}}.$$

EXAMPLE 3   Finding the Shapley-Shubik Index

Compute the Shapley-Shubik index for each voter in the weighted voting system [4 : 3, 2, 1].

SOLUTION: Consider the voters to be A, B, and C, in that order. We first list all permutations of these voters and determine in each permutation which voter is pivotal.

Coalition	Weight after 1st, 2nd, and 3rd Voter Is Added	Pivotal Voter
a) (A, B, C)	3, 5, 6	B
b) (A, C, B)	3, 4, 6	C
c) (B, A, C)	2, 5, 6	A
d) (B, C, A)	2, 3, 6	A
e) (C, A, B)	1, 4, 6	A
f) (C, B, A)	1, 3, 6	A

A's Shapley-Shubik index is

$$\frac{\text{the number of times A is pivotal in some permutation of the voters}}{\text{the total number of permutations of the voters}} = \frac{4}{6} = \frac{2}{3}.$$

Since B is pivotal once out of six times, B's index is $\frac{1}{6}$. Similarly, C's index is also $\frac{1}{6}$.

Note in Example 3 that even though B has twice the votes of C, both B and C have the same voting power according to the Shapley-Shubik index. In general, the number of votes a person has is not the same as that person's voting power as measured according to the Shapley-Shubik index.

PROBLEM SOLVING

As we recommended in Section 1.1, it is always good to check to see if your answers are reasonable. Notice when computing the Shapley-Shubik index, the sum of the indices of all the voters is always 1.

As you might expect, the Banzhaf and Shapley-Shubik indices do not measure power the same way.

EXAMPLE 4   **The Banzhaf Power Index Differs from the Shapley-Shubik Index**

Let us compute the Banzhaf power index for the voters in Example 3.

SOLUTION: We summarize the computations in the following table.

Winning Coalition	Weight	Critical Voters
{A, B}	5	A, B
{A, C}	4	A, C
{A, B, C}	6	A

Since A is a critical voter three out of five times, her Banzhaf power index is $\frac{3}{5}$. Similarly B and C have Banzhaf power indices of $\frac{1}{5}$ each.

Notice that both the Shapley-Shubik index and the Banzhaf power index rate A as the most powerful member of the weighted voting system in Examples 3 and 4. However, also note that the power indices are not identical for both methods.

Often the chair of a committee has more power than the other members of the committee. It is possible to measure exactly how much greater this power is using the Shapley-Shubik index.

EXAMPLE 5   **Comparing the Power of Committee Members**

Assume your local newspaper has an editorial board consisting of the managing editor and four other members. In order to begin to develop a story, the managing editor and two other board members must agree that the story is newsworthy. Use the Shapley-Shubik index to compare the power of the managing editor with the other board members.

SOLUTION: Let us assume that the managing editor is A and the other board members are B, C, D, and E. We will solve this problem by first determining B's power. We can then deduce that C, D, and E also have this same power. From this information we can easily determine the power that A has.

In forming permutations of {A, B, C, D, E}, we can think of filling slots in a list as follows:

$$\underline{\quad}\ \ \underline{\quad}\ \ \underline{\quad}\ \ \underline{\quad}\ \ \underline{\quad}$$
1st   2nd   3rd   4th   5th

Clearly B cannot be a pivotal voter unless he is in the 3rd slot. So let us assume that B is occupying the 3rd slot.

$$\underline{\quad}\ \ \underline{\quad}\ \ \underset{\text{B}}{\underline{\quad}}\ \ \underline{\quad}\ \ \underline{\quad}$$
1st   2nd   3rd   4th   5th

Now in order for B to be pivotal, clearly the editor A must be in slot 1 or slot 2.

Case 1: Assume that A is in slot 1. The diagram now looks like this:

$$\underset{\text{1st}}{\underline{\text{A}}} \quad \underset{\text{2nd}}{\underline{}} \quad \underset{\text{3rd}}{\underline{\text{B}}} \quad \underset{\text{4th}}{\underline{}} \quad \underset{\text{5th}}{\underline{}}$$

The remaining three slots can be filled in with the remaining three committee members in 3! = 6 ways.

Case 2: Assume that A is in slot 2. The diagram now looks like this:

$$\underset{\text{1st}}{\underline{}} \quad \underset{\text{2nd}}{\underline{\text{A}}} \quad \underset{\text{3rd}}{\underline{\text{B}}} \quad \underset{\text{4th}}{\underline{}} \quad \underset{\text{5th}}{\underline{}}$$

Again the remaining three slots can be filled in six ways.

From Cases 1 and 2, we see that there are exactly twelve ways to form a permutation in which B is pivotal. Thus B's Shapley-Shubik index is

$$\frac{12}{5!} = \frac{12}{120} = \frac{1}{10}.$$

By a similar analysis we would find that C, D, and E also have a Shapley-Shubik index of $\frac{1}{10}$. Since B, C, D, and E each are pivotal in 12 permutations, it follows that A is pivotal in the other $120 - 12 - 12 - 12 - 12 = 72$ permutations. Thus A's index is

$$\frac{72}{5!} = \frac{72}{120} = \frac{6}{10}.$$

Therefore A is six times as powerful as the other editorial board members. ◎

Exercises

In Exercises 1–4, use tree diagrams to find all the permutations of each set.

1. {X, Y, Z}

2. {P, Q, R}

3. {W, X, Y, Z}

4. {P, Q, R, S}

5. How many permutations are there of a six-element set?

6. How many permutations are there of an eight-element set?

7. How many permutations are there of a twelve-element set?

8. How many permutations are there of a thirteen-element set?

9. For the weighted voting system [5 : 3, 2, 2], complete the following table, which is similar to the table given in Example 3.

Coalition	Weight after 1st, 2nd, and 3rd Voter Is Added	Pivotal Voter
a) (A, B, C)		
b) (A, C, B)		
c) (B, A, C)		
d) (B, C, A)		
e) (C, A, B)		
f) (C, B, A)		

10. For the weighted voting system [7 : 4, 3, 2], complete the following table, which is similar to the table given in Example 3.

Coalition	Weight after 1st, 2nd, and 3rd Voter Is Added	Pivotal Voter
a) (A, B, C)		
b) (A, C, B)		
c) (B, A, C)		
d) (B, C, A)		
e) (C, A, B)		
f) (C, B, A)		

In Exercises 11–16, determine the Shapley-Shubik index of each voter in each weighted voting system.

11. [6 : 3, 3, 2]

12. [8 : 4, 3, 3]

13. [8 : 3, 3, 2, 2]

14. [9 : 4, 3, 3, 1]

15. [4 : 2, 1, 1, 1]

16. [13 : 6, 6, 5, 3]

17. The system [3 : 1, 1, 1, 1, 1] is an example of a "one person, one vote" situation.

   a) What is the Shapley-Shubik index for each person in this system?

   b) Explain how you obtained your answer in part (a).

   c) How does your answer in part (a) conform with your intuition?

18. Measuring power on a jury. We can consider a twelve-person jury as the weighted system [12 : 1, 1, 1, 1, 1, 1, 1, 1, 1, 1, 1, 1].

   a) What is the Shapley-Shubik index for each person in this system?

   b) Explain how you obtained your answer in part (a).

   c) How does your answer in part (a) conform with your intuition?

19. Consider the system [14 : 15, 2, 3, 3, 5], in which A is a dictator.

   a) What is the Shapley-Shubik index for each person in this system?

   b) Explain how you obtained your answer in part (a).

   c) How does your answer in part (a) conform with your intuition?

20. Consider the system [12 : 1, 2, 3, 1, 1, 2], in which no resolution can be passed because the quota is too high.

   a) What is the Shapley-Shubik index for each person in this system?

   b) Explain how you obtained your answer in part (a).

   c) How does your answer in part (a) conform with your intuition?

21. Measuring power on a committee. Committee {A, B, C, D} has chairperson A. In order for an item to be placed on the committee's agenda, the chairperson and at least one other committee member must agree to that item. What is the Shapley-Shubik index for each person in this system?

22. Measuring power on a theater guild. The Theater Guild consists of (P)erformers, (T)echnicians, and (S)upport staff. Representation on the guild is proportional to the number in each group, and the representatives of each group vote in a bloc. A simple majority is required to pass a resolution. There are five performers, four technicians, and two members of the support staff on the guild.

   a) List all the permutations of {P, T, S}.

   b) Determine the pivotal voter in each permutation of {P, T, S}.

   c) Find the Shapley-Shubik index for each voter in this system.

23. Measuring power on a state committee. The state athletics procedures committee makes policy decisions af-

fecting high school athletics programs throughout the state. The committee consists of three (A)dministrators, four (F)aculty members in athletics departments, three a(T)hletes, and two (N)on-athlete students. Assume that each of these four groups votes in a bloc. In order to make a policy change, a vote of at least 8 is required.

a) List all the permutations of {A, F, T, N}.

b) Determine the pivotal voter in each permutation of {A, F, T, N}.

c) Find the Shapley-Shubik index for each voter in this system.

24. Measuring power in a law firm. In Example 5 of Section 10.3, we analyzed the voting power of the law partners in the firm of Krook, Cheatum, and Associates. Recall that to change a policy required the votes of Krook, Cheatum, and at least two of the associates: W, X, Y, and Z. Find the Shapley-Shubik index for each voter in this system.

CHAPTER 11

Consumer Mathematics:* The Mathematics of Everyday Life

According to Ben Franklin, a penny saved is a penny earned. However, as an informed and careful consumer, you can improve upon Franklin's advice and make those pennies add up quickly. If you understand how interest works, you can increase the value of your investments and minimize the cost of your debts. The topics that you study in this chapter will help you solve typical consumer finance problems such as these:

- How much must you invest each month in order to have enough to buy a house or a car at some future date?
- How much must you save to ensure that your child will be able to afford a college education?
- When should you begin saving to have $1 million when you retire?
- How long will it take for you to pay off your credit card balance?
- What will be your monthly payments if you decide to refinance a mortgage? What if there is a charge for refinancing? Will it benefit you to refinance?
- How much interest will you pay if you take out a loan for college?
- If you can afford a monthly payment of $200, what is the highest-priced car that you can presently afford? What if you can increase your payments to $250?

In your lifetime you will face dozens of questions such as these. As you study the mathematical theory of interest, annuities, loans, and mortgages presented in this chapter, you will learn how to make the most of your money. And soon you will be able to answer questions about your finances for yourself.

---

* For further resources on consumer mathematics see www.aw.com/pirnot.

11.1 PERCENT

In this section, you will study the notion of percent that you encounter so often in everyday life. Understanding the meaning of percent is critical to understanding the ideas that we will introduce later in this chapter.

Percent means "per hundred."

The word percent is derived from the Latin "per centum" which means "per hundred." Therefore, 17 percent means "seventeen per hundred." We can write 17 percent as $\frac{17}{100}$ or in decimal form as 0.17. In this chapter, we will usually write percents in decimal form.

EXAMPLE 1 Writing Percents as Decimals

a) Write each of the following percents in decimal form.

   36 percent   5 percent   19.32 percent   0.25 percent

b) Write each of the following decimals as percents.

   0.29   0.354   1.37   0.003

SOLUTION: a) Think of 36 percent as the decimal thirty-six hundredths, or 0.36, as we show in Figure 11.1(a).

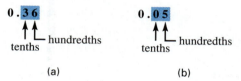

(a)   (b)

FIGURE 11.1   Interpreting percents in term of hundredths.

To write 5 percent, realize that we have zero tenths and only five hundredths, as indicated in Figure 11.1(b). Thus, 5 percent equals 0.05.

To write 19.32 percent, recall the problem solving advice from Section 1.1 that it is often helpful to solve a simpler problem instead. If we had asked you to write 19 percent as a decimal, you would write it as 0.19. Now, once you have positioned the one and the nine properly in the decimal, write the three and the two immediately to their right, as we show in Figure 11.2(a). Thus, **19**.32 percent is equal to 0.**19**32.

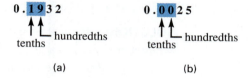

(a)   (b)

FIGURE 11.2   The key to converting percents to decimals is what you write in the hundredths place.

We handle 0.25 percent in a similar way. Realize that 0.25 percent is not even 1 percent, so therefore we have zeros in the first two decimal places after the decimal point, as you see in Figure 11.2(b).

b) To translate decimals to percents, we first examine what numbers are in the tenths and hundredths place. To rewrite 0.29, we see that we have 29 hundredths, so 0.29 equals 29 percent.

To rewrite 0.**35**4, recognize that 0.35 would be 35 percent, so 0.354 is 35.4 percent.

In the decimal 1.**37**, the key is to recognize that 0.37 is 37 percent. Therefore, in converting 1.37 to a percent, we place the one to the left of the three in the 37 percent. So we write the decimal number 1.37 as 137 percent.

Figure 11.3 shows you how to convert 0.**003** to a percent. We see that we have zero whole percent, so we begin by writing 0 percent and then writing the three to the right of the decimal point. Thus 0.003 equals 0.3 percent.

FIGURE 11.3   We have zero whole percent in the number 0.003.

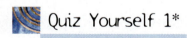

PROBLEM SOLVING

You may have been told that when converting a percent to a decimal, you move the decimal point two places to the left and that in converting a decimal to a percent, you move the decimal point two places to the right. It is all right to use such memory devices, provided that you understand where they come from.

Always remember that per cent means hundredths. If you understand how to rewrite 0.29 as 29 percent, then you also know how to rewrite 1.29. Similarly, if you know that 19 percent equals 0.19, then you also know how to rewrite 19.32 percent as a decimal.

If you wish, you can use the following rules:

To convert from a percent to a decimal, drop the percent sign and divide by 100.

To convert from a decimal to a percent, multiply by 100 and add a percent sign.

Quiz Yourself 1*

a) Write 17.45 percent as a decimal

b) Write 0.05 percent as a decimal

c) Write 2.45 as a percent

d) Write 0.025 as a percent

We often have to convert a fraction to a percent. Example 2 shows how to do this.

EXAMPLE 2   Converting a Fraction to a Percent

Write $\frac{3}{8}$ as a percent.

SOLUTION: Because we already know how to rewrite a decimal as a percent, we first convert $\frac{3}{8}$ to a decimal and then rewrite the decimal as a percent.

If we divide the denominator into the numerator, we get $\frac{3}{8} = 0.375$. Next, we see that 0.375 is equal to 37.5 percent. Thus, $\frac{3}{8} = 37.5$ percent.

---

* Quiz Yourself answers begin on page 849.

Example 3 shows how you can use the simple notion of percent to compare changes in data over long periods of time.

EXAMPLE 3   **Changes in Defense Spending as a Percent of the Federal Budget over Time**

According to the Office of Management and Budget, in 1970 the United States government spent 82 billion dollars for defense at a time when the Federal budget was 196 billion dollars. Thirty years later, in the year 2000, spending for defense was 294 billion dollars and the budget was 1,789 billion dollars. What percent of the Federal budget was spent for defense in 1970? In 2000?

SOLUTION: In 1970, 82 billion out of 196 billion dollars was spent for defense. We can write this as the fraction $\frac{82}{196} \approx 0.418 = 41.8$ percent. Thirty years later, this fraction was $\frac{294}{1,789} \approx 0.164 = 16.4$ percent.

Thus, you see, as a percentage of the Federal budget, defense spending was considerably less in 2000 than it was in 1970. ◎

The media often uses percentages to explain the change in some quantity. For example, you may hear that, on a particularly bleak day on Wall Street, the stock market is down 1.2 percent. Or on a good day, the evening news tells us that consumer confidence in the economy is up 13.5 percent over last month. In order to make such statements we have to understand several quantities.

The percent of change is always in relationship to a previous, or **base amount.** We then compare a **new amount,** with the base amount as follows

$$\text{percent of change} = \frac{\text{new amount} - \text{base amount} *}{\text{base amount}}$$

We illustrate this idea in Example 4.

EXAMPLE 4   **Finding the Percent of a Tuition Increase**

This year the tuition at Good Old State was $7,965 and for next year, the board of trustees has decided to raise the tuition to $8,435. What is the percent of increase in the tuition?

SOLUTION: In this example, the base amount is $7,965 and the new amount is $8,435. To find the percent of tuition increase, we calculate

$$\text{percent of change} = \frac{\text{new amount} - \text{base amount}}{\text{base amount}}$$

$$= \frac{8,435 - 7,965}{7,965} = \frac{470}{7,965} \approx 0.059 = 5.9 \text{ percent}$$

Thus, the tuition will increase almost 6 percent from this year to the next. ◎

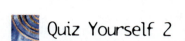 Quiz Yourself 2

The population of Florida increased from 12.9 million in 1990 to 16 million in 2000. What was the percent of increase of Florida's population over these ten years?

---

* If the new amount is less than the base amount then the percent of change will be negative.

Merchants often use percents to describe the deals that they are giving to the public when they have a sale. The increase that a merchant adds to his base price is called **markup.**

EXAMPLE 5 Investigating a Sale Price on a New Car

Monte's Autorama is having an End Of Year Clearance in which the TV ads proclaim that all cars are sold at 5 percent markup over the dealer's cost. Monte has a new Exfinity for sale for $18,970. On the Internet, you find out that this particular model has a dealer cost of $17,500. Is Monte being honest in his advertising?

SOLUTION: To find the percent markup on this particular Exfinity, you can calculate the percent of markup, which is the same as the percent of change in the base price (dealer's cost) of the car.

We calculate as in Example 3,

$$\text{percent of markup} = \frac{\text{selling price} - \text{dealer cost}}{\text{dealer cost}}$$

$$= \frac{18,970 - 17,500}{17,500} = \frac{1,470}{17,500} = 0.084 = 8.4 \text{ percent}$$

Thus Monte is not being quite truthful here since his markup on this model is 8.4 percent, which is well above the amount he stated in his TV ads.

Many percent problems are based on the same equation.

We will conclude this section with several examples of percent problems; however, it is important for you to recognize that these problems are all variations of the same equation. In each case, we will be taking some *percent* of a *base* quantity, and setting it equal to an *amount*. We can write this as the equation

$$\text{percent} \times \text{base} = \text{amount.}$$

We will call this equation, *the percent equation.*

You saw this pattern in Example 5, where the percent was 8.4 percent = 0.084, the base (dealer's price) was $17,500, and the amount (markup) was $1,470. Notice that $0.084 \times 17,500 = 1,470$.

In the remaining examples, we will give you two of the three quantities, percent, base, and amount, and ask you to find the third.

EXAMPLE 6 Using the Percent Equation

a) What is 35 percent of 140?

b) 63 is 18 percent of what number?

c) 288 is what percent of 640?

SOLUTION: a) Here the base is 140, the percent is 35 percent = 0.35, and we must find the amount. We begin by substituting for the percent and the base in the percent equation to get $0.35 \times 140 = amount$, so the amount is $0.35 \times 140 = 49$.

b) Now you are given the percent, 18 percent = 0.18, and the amount, 63. You can use the percent equation to find the base. Substituting in the percent equation, we find that

$$0.18 \times base = 63,$$

or, dividing both sides of the equation by 0.18, we get

$$base = \frac{63}{0.18} = 350.$$

Quiz Yourself 3

a) What is 15 percent of 60?

b) 18 is 24 percent of what number?

c) 96 is what percent of 320?

c) Here we are given the base, 640, and the amount, 288, and must find the percent. Therefore, percent $\times$ 640 = 288. Solving for percent, we get

$$percent = \frac{288}{640} = 0.45 = 45 \text{ percent}$$

EXAMPLE 7   New AIDS Cases

According to the National Center for Health Statistics, the number of new AIDS cases among 20 to 29 year olds in 1999 was only 47.3 percent of the number that was reported in 1995. If there were 8,404 new cases in 1995, how many new cases were there in 1999?

SOLUTION: We will use the percent equation *percent* $\times$ *base* = *amount*. We are given the percent, 47.3 percent = 0.473, and the base, 8,404. Therefore the amount of new AIDS cases reported in 1999 was 0.473 $\times$ 8,404 $\approx$ 3,975.

EXAMPLE 8   Calculating Sports Statistics

In the 2000–2001 season, the Philadelphia 76ers in the National Basketball Association had a record of 56 wins and 26 losses. What percent of their games did they win?

SOLUTION: Again you can use the percent equation to solve this problem; however, you have to be careful. The base is not 56, but rather the total number of games played, which is 56 + 26 = 82. The amount is the number of victories, 56. In order to find the winning percentage, you have to solve the equation percent $\times$ 82 = 56. Dividing both sides of the equation by 82 gives us

$$percent = \frac{56}{82} \approx 0.683 = 68.3 \text{ percent.}$$

Therefore, the sixers won 68.3 percent of their games.

EXAMPLE 9   Increase in Satellite Dishes

In 2000, there were 4,250 (thousand) satellite dish systems sold in the United States, an increase of 17.24 percent over 1999. How many satellite dish systems were sold in the United States in 1999?

SOLUTION: In this case, the base is unknown. In deciding what percent to use in the percent equation, you have to be careful. The 17.24 percent is only the increase. The amount 4,250 (thousand) systems sold in 2000 represents 100 percent of the systems sold in 1999 plus the 17.24 percent increase. Therefore the percent that we will use in the percent equation is 100 percent + 17.24 percent = 117.24 percent = 1.1724. Substituting the amount and the percent in the percent equation, we get

$$1.1724 \times base = 4,250.$$

Solving this equation for base, gives us

$$base = \frac{4,250}{1.1724} \approx 3,625.$$

Therefore, about 3,625 (thousand) satellite systems were sold in the United States in 1999.

Exercises 11.1

Change the following percents to decimals.

1. 78 percent

2. 65 percent

3. 8 percent

4. 3 percent

5. 27.35 percent

6. 83.75 percent

7. 0.35 percent

8. 0.08 percent

Write each of the following decimals as percents.

9. 0.43

10. 0.95

11. 0.365

12. 0.875

13. 1.45

14. 2.25

15. 0.002

16. 0.0035

Convert each of the following fractions to percents.

17. $\dfrac{3}{4}$

18. $\dfrac{7}{8}$

19. $\dfrac{5}{16}$

20. $\dfrac{9}{25}$

21. $\dfrac{5}{2}$

22. $\dfrac{11}{8}$

23. $\dfrac{4}{250}$

24. $\dfrac{3}{500}$

Assume that you are studying with a friend for a quiz on percents. Your friend insists on blindly memorizing which way to

"move the decimal point" in doing percent problems. Explain how you would help your friend understand how to do the conversions in Exercises 25–28 without relying solely on memorization.

25. Communicating Mathematics You are converting 28.35 percent to a decimal.

26. Communicating Mathematics You are converting 1.285 to a percent.

27. Communicating Mathematics You are converting 0.0375 to a percent.

28. Communicating Mathematics You are converting 1.375 percent to a decimal.

29. What is 28 percent of 350?

30. 20 is 16 percent of what number?

31. 12 is what percent of 80?

32. What is 24 percent of 125?

33. 77 is 22 percent of what number?

34. 33.6 is what percent of 96?

35. What is 12.25 percent of 160?

36. 47.74 is 38.5 percent of what number?

37. 8.4 is what percent of 48?

38. What is 23 percent of 140?

39. 29.76 is 23.25 percent of what number?

40. 149.5 is what percent of 130?

41. Comparing cookie sales. In 2001, the top-selling cookie in America was Nabisco's Oreos with sales of $606 million. The total cookie sales for that year was $3,124 million. What percent of the total cookie sales was due to Oreos? (Source: Information Resources, Inc.)

42. Comparing coffee sales. In 2001, Folgers sold $505 million dollars of nondecaf coffee. The total sales of nondecaf coffee that year was $2,012 million. What percent of the total nondecaf coffee sales was due to Folgers? (Source: Information Resources, Inc.)

43. Increase in the price of a new home. From 1999 to 2001, the average price of a new home in Boston increased by roughly 8.28 percent to $314,000. To the nearest $1,000, what was the average price of a new home in Boston in 1999?

44. Decline in radio stations. According to *The 2002 World Almanac,* from 1999 to 2001, the number of country music radio stations declined from 2,306 to 2,190. What was the percent of decline?

We summarize the circulation (in thousands) of the five leading U.S. newspapers in 2000 in the graph below. Use this information to do Exercises 45–48. (Source: The 2002 World Almanac.*)*

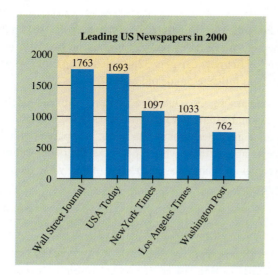

45. Comparing newspapers circulations. What percent of the circulation of these five papers was due to *The Wall Street Journal?*

46. Comparing newspapers circulations. What percent of the circulation of these five papers was due to *The New York Times* and the *Los Angeles Times* combined?

47. Comparing newspapers circulations. What percent greater was the circulation of *USA Today* than the circulation of *The New York Times?*

48. Comparing newspapers circulations. What percent smaller was the circulation of *The Washington Post* than the circulation of *The Wall Street Journal?*

The graph below shows the sales (in thousands) of the top five selling cars in 2000. Use this information to do Exercises 49–52. (Source: The NY Times 2002 Almanac.*)*

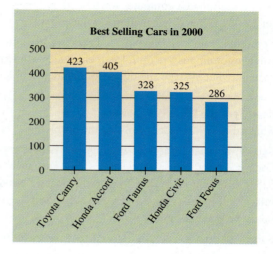

49. Comparing car sales. What percent of the sales of these five cars was due to the Toyota Camry?

50. Comparing car sales. What percent of the sales of these five cars was due to Fords?

51. Comparing car sales. What percent greater was sales of the Camry than the Civic?

52. Comparing car sales. What percent smaller was the sales of the Ford models than the Honda models?

53. Population growth. According to the latest census, from 1990 to 2000 the U.S. population had grown by 12.85 percent to 281 million people. To the nearest million, what was the U.S. population in 1990?

54. Population growth. From 1990 to 2000, California's population rose from 29.8 million to 33.9 million. What was the percent of increase?

55. Declining audio cassette sales. According to the Recording Industry Association of America, over a recent one year period, audio cassette sales dropped by 41 percent to 626 million. To the nearest million, what was the number of audio cassette sales before this decline?

56. Declining audio cassette sales. According to the Recording Industry Association of America, over a recent one-year period, music video sales dropped from 19.8 million to 18.2 million. What was the percent of decrease?

57. Dealer markup on a car. If a dealer buys a car from the manufacturer for $14,875 and then sells it for $16,065, what is his markup?

58. Dealer markup on a computer. A computer retailer buys a multimedia computer for $1,850 and then sells it for $2,081. What is her markup on the computer?

59. Communicating Mathematics In doing Exercise 57, Angela got an answer of 7.4 percent which is incorrect. Explain what mistake she did in solving that problem.

60. Communicating Mathematics In doing Exercise 58, Sanjiv got the incorrect answer of 11.1 percent. Explain what mistake he made in solving the problem.

61. Salary increase. Marcy is working in a two-year probationary period as a paralegal. She currently earns $28,000 a year and after her probationary period, her salary will increase by 35 percent. What will her yearly salary be at that time?

62. Salary decrease. Jed works for Metal Fabricators, Inc. which just lost a large contract and has asked employees to take a 12 percent pay cut. If Jed now makes $34,500 per year, what will his yearly pay be after the pay cut?

63. Buying a grill. Tomas is buying a new gas barbecue grill that has been reduced for an end of summer sale by 15 percent to $578. What was the original price of the grill.

64. Buying groceries. Ashley saw that lettuce that had sold for $1.75 last week now costs 16 percent more. What is the current price of the lettuce?

65. Increasing work load. Gisela is an insurance claims adjuster. Last quarter she settled 124 claims and this quarter she settled 155 claims. What is the percent increase in the claims that she settled this quarter over last quarter?

66. Reducing blood pressure. Joshua's diastolic blood pressure reading had been 160. After several weeks of medication, his reading is now 134. By what percent did he reduce his blood pressure?

67. Markup on a boat. Carlos is a boat dealer who bought a sailboat for $11,400 and then sold it for $12,711. What percent was his markup on the boat?

68. Markup on an appliance. Anna, who owns a small appliance store, bought a food processor for $524 and then sold it for $589.50. What percent was her markup on the food processor?

69. Change in the stock market. Due to a slump in the economy, Annalee's mutual fund has dropped by 12 percent from last quarter to this quarter. If her fund is now worth $11,264, how much was her fund worth last quarter?

70. Weight lifting. Mikal finds that he can bench press 15 percent more weight this month than he could last month. If he now bench presses 161 pounds, how much more is he able to bench press than he could last month?

71. Gas mileage. Renaldo bought a new car that gets 35 percent more miles per gallon than he got with his old car. If his new car gets 32.4 miles per gallon, what was the gas mileage on his old car?

72. Air passengers. Due to a concern with air safety, the Regional Municipal Airport noticed a 24 percent drop in passengers per month this year compared with three years ago. If there were 3,344 passengers per month this year, how many passengers per month were there three years ago?

Further Exercises

73. Depreciation on a new car. If a new car costs $18,000 and depreciates at a rate of 12 percent per year, what will be the value of the car in four years?

74. Compounding raises. If you are given a raise of 8 percent this year and a raise of 5 percent the following year, what single raise would give you the same yearly salary in the second year?

*(75.) Communicating Mathematics   If a merchant increases the price of a home entertainment system by x percent and then later reduces the price by x percent, is the price of the system the same as the original price? If not, what is the relationship between the two prices? Explain your answer by using appropriate examples.

(76.) Communicating Mathematics   If a merchant reduces the price of a luxury speedboat by x percent and then later increases the price by x percent, is the price of the boat the same as the original price? If not, what is the relationship between the two prices? Explain your answer by using appropriate examples.

(77.) Communicating Mathematics   If you increase an amount by 10 percent and then again by 20 percent, is that the same as increasing the amount in one step by 30 percent. Explain your answer by using appropriate examples.

(78.) Communicating Mathematics   Several years ago in an actual contract negotiation, the college faculty in a large eastern state system were encouraged by the state to "backload" their new contract, stating that it would be beneficial to them. That is, rather than take 3 percent, 2 percent, and 0 percent raises for the next three years, the state encouraged them to take 0 percent, 2 percent, and 3 percent. Which is the better deal? Explain your answer by using appropriate examples. (Hint: There is more involved here than just looking at the percentages.)

11.2   INTEREST

One way to save money for a house, a car, or a vacation is to deposit money in a bank account where it will earn interest. The bank uses your money to make loans to other customers and pays you interest in exchange for using your funds. On the other hand, if you borrow money from the bank, then you will pay interest to the bank. In essence, **interest** is the money that one person (a borrower) pays another (a lender) to use the lender's money. Savers earn interest; borrowers pay interest.

In this section, we will show you how to compute the interest you earn when you deposit your money in a bank account; in Section 11.3 we will discuss the cost of loans.

Simple interest is a straightforward way to compute interest.

The amount you deposit in a bank account is called the **principal.** The bank specifies an **interest rate** for that account as a percentage of your deposit. The rate is usually expressed as an annual rate. For example, a bank may offer an account that has an annual interest rate of 5 percent. To find the interest that you will earn in such an account, we also need to know how long the deposit remains in the account. The time is usually stated in years. There is a simple formula that relates principal, interest earned, interest rate, and time. In words,

$$\text{interest earned} = \text{principal} \times \text{interest rate} \times \text{time}$$

When we compute interest this way, it is called **simple interest.**

---

* Exercise numbers circled in red can be used as group exercises.

Formula for Computing Simple Interest

We calculate simple interest using the formula

$$I = Prt,$$

where I is the interest earned, P is the principal, r is the interest rate, and t is the time.

EXAMPLE 1   Calculating Simple Interest

If you deposit $500 in a bank account paying 6 percent annual interest, how much interest will the deposit earn in four years if the bank computes the interest using simple interest?

SOLUTION: In this example:

> P is the principal, which is $500
> r is the annual interest rate, which is 6 percent (written as 0.06)
> t is the time, which is 4 (years)

Thus the interest earned is

$$I = Prt = 500 \times 0.06 \times 4 = 120.$$

In four years, this account earns $120 in interest.

Future value equals principal plus interest.

If you are saving for a particular goal, you will want to know how much money you will have in your savings account after a specific time period. To answer this question, we have to find the future value of the account. The **future value** is the amount A that will be in the account after the interest is added to the principal. The future value is sometimes called the *future amount*, which explains why we often represent it by A. Symbolically, we can say that $A = P + I$.

If we replace I by Prt, we get $A = P + Prt = P(1 + rt)$.

Computing Future Value Using Simple Interest

To find the future value of an account that pays simple interest, use the formula

$$A = P(1 + rt),$$

where A is the future value, P is the principal, r is the annual interest rate, and t is the time in years.

EXAMPLE 2   Computing Future Value Using Simple Interest

Assume that you deposit $1,000 in a bank account paying 3 percent annual interest and leave the money there for six years. Use the simple interest formula to compute the future value of this account.

SOLUTION: We see that $P = 1,000$, $r = 0.03$, and $t = 6$. Therefore,

$$A = 1,000(1 + (0.03)(6)) = 1,000(1 + 0.18) = 1,000(1.18) = 1,180.$$

Thus, your bank account will have $1,180 at the end of six years.

In contrast to future value, the principal that you have to invest in an account now in order to have a specified amount in the account in the future is called the **present value** of the account. Notice that the formula for computing future value has four unknowns. If we wish, we can use this formula for finding the present value of an account provided we know the future value, interest rate, and time.

EXAMPLE 3   Finding the Present Value of an Account

Assume that you plan to save $2,500 to take a white-water rafting trip in two years. Your bank offers a certificate of deposit (CD) that pays 4 percent annual interest computed using simple interest. How much must you put in this CD now in order to have the necessary money in two years?

SOLUTION: We can use the formula

$$A = P(1 + rt).$$

We know that $A = 2,500$, $r = 4\% = 0.04$, and $t = 2$. Therefore,

$$2,500 = P(1 + (0.04)(2)).$$

We can rewrite this equation as

$$2,500 = P(1.08).$$

Dividing both sides of the equation by 1.08, we get

$$P = \frac{2,500}{1.08} \approx 2314.814815.$$

We will round this *up* to $2,314.82 to guarantee that if you put this amount in the certificate of deposit now, in two years you will have the $2,500 you need for your white-water rafting trip.†

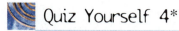

Quiz Yourself 4*

Redo Example 3, but now assume that you wish to save $2,400 in four years and the CD has an annual interest rate of 5 percent.

SOME GOOD ADVICE

In Example 3 we used the earlier formula for computing future value to find the present value rather than stating a new formula to solve this specific problem. You will find it easier to learn a few formulas well and use them, together with simple algebra, to solve new problems rather than trying to memorize separate formulas for every type of problem.

Compounding pays interest on previously earned interest.

It seems fair that if money in a bank account has earned interest, the bank should compute the interest due, add it to the principal, and then pay interest on this new, larger amount. This is in fact the way most bank accounts work. Interest that is paid on principal plus previously earned interest is called **compound interest.** If the interest is added yearly, we say that the interest is *compounded annually*. If the interest is added every three months, we say the interest is *compounded quarterly*. Interest can also be compounded monthly and daily.

EXAMPLE 4   Calculating Compound Interest the Long Way

Assume that you wish to replace your sailboat with a larger one in three years. In order to save for a down payment for this purchase, you deposit $2,000 for three years in a bank account that pays 10 percent annual interest, compounded annually. How much will be in the account at the end of three years?

SOLUTION: We will use the formula $A = P(1 + rt)$ for calculating future value. However, because interest is being added to the principal at the end of each year, we must recalculate the future value at the end of each year.

First year: $P = 2,000$, $r = 0.10$, $t = 1$. The future value at the end of the first year is $A = P(1 + rt) = 2,000(1 + (0.10)(1)) = 2,000(1.10) = 2,200$. This will be the principal for the second year.

Second year: $P = 2,200$, $r = 0.10$, $t = 1$. The future value at the end of the second year is $A = 2,200(1 + (0.10)(1)) = 2,200(1.10) = 2,420$. We use this amount for the principal for the third year.

Third year: $P = 2,420$, $r = 0.10$, $t = 1$. The future value is now $A = 2,420(1 + (0.10)(1)) = 2,420(1.10) = 2,662$.

PROBLEM SOLVING

You should always check answers to see whether they are reasonable. In Example 4, if we had used simple interest to find the future value, we would have obtained $A = 2,000(1 + (0.10)(3)) = 2,000(1.30) = 2,600$. The interest we found in Example 4 is a *little* larger because as the interest is added to the principal each year, the bank is paying interest on an increasingly larger principal.

We illustrate the growth of your money in the bank account with the time-line in Figure 11.4.

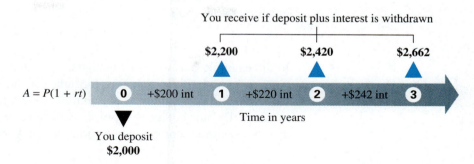

FIGURE 11.4 10 percent interest being compounded annually.

We could continue the process we began in Example 4 to calculate the future value of the account year by year to eventually find the future value of the account for any time period. Suppose, for example, that we left the $2,000 on deposit for 30 years. We could use this method, but the calculations would be very tedious. Fortunately there is a method that allows us to do the calculation more quickly regardless of the number of years. Let us look at the computations in Example 4 in a slightly different way.

In the first year, we began with $P = 2,000$, $r = 0.10$, $t = 1$ and saw that the future value at the end of the first year was

$$A = 2,000(1 + (0.10)(1)) = 2,000(1.10).$$

Without multiplying this out, we use $2,000(1.10)$ as the principal for the second year. The future value at the end of the second year is therefore

$$A = 2,000(1.10)(1 + 0.10) = 2,000(1.10)(1.10) = 2,000(1.10)^2.$$

We now use $2,000(1.10)^2$ as the principal for the third year. Therefore the future value at the end of the third year is

$$A = 2,000(1.10)^2(1 + 0.10) = 2,000(1.10)^2(1.10) = 2,000(1.10)^3.$$

If we were to continue this pattern to compute the future value of the account at the end of 30 years, we would see that

$$A = 2,000(1.10)^{30} \approx 2,000(17.44940227) \approx 34,898.80.*$$

This large amount shows how your money can grow if it is compounded over a long period of time.

---

* In order to ensure greater accuracy, we often show calculations with eight decimal places. If your calculations do not agree with ours, it may be due to the difference in the way we are rounding our calculations.

We can describe the general pattern that you just saw in this example by a relatively simple equation. That is, if we deposit a principal P in an account paying an annual interest rate r for t years, then the future value of the account is given by the formula

$$A = P(1 + r)^t.$$

In the example that we just calculated, $P = 2,000$, $r = 0.10$, and $t = 30$. It is important to understand that this formula for calculating compound interest only works for the case when r is the *annual interest rate* and t is time being measured in *years*. Do not bother to learn this formula, because in just a moment we will give you a similar compounding formula that works for more general situations.

Historical Highlight: *The Knights Templars**

Following the Crusades, a small group of twelfth-century knights devoted their service to the Catholic Church by providing safe passage to pilgrims visiting the Holy Lands. These knights swore vows of personal poverty and founded a military order in Jerusalem called the Knights of the Temple of Solomon. The Knights Templar, as they came to be called, lived by a strict moral and disciplinary code, and over the years their military wing grew in size and strength until they were counted among the most feared warriors in Europe and the Middle East.

Wealth flowed into the order through donations and military conquest and the Templars established and maintained almost nine hundred castles throughout the region. Eventually the knights' chain of treasure houses and their guarantee of safe transportation persuaded people to use them as bankers. Because of the Templars' reputation for honor and strength, people trusted them to protect their money and valuables. For example, a traveler could deposit gold in one castle and receive a note that could be cashed for currency in another country. The Templars charged fees for such transactions, eventually becoming bankers to kings and popes and providing a range of financial services to the nobility.

As the wealth of the Templars grew, the order attracted the attention of King Philip IV of France, who desperately needed money to reverse his mismanagement of his country's finances. He waged a successful campaign to discredit the Templars, and by the fourteenth century, the order was destroyed—its assets had been seized and many of its members executed.

Knowing the principal, the periodic interest rate, and the number of compounding periods, it is easy to determine future value.

All banks and most other financial institutions compound interest more frequently than once a year. For example, many banks send savings account customers a monthly statement showing the balance in their accounts. So far in our discussion of compounding, we have used a yearly interest rate. If com-

----

* This Highlight is based on a discussion of the Knights Templar in Jack Weatherford, *The History of Money* (Crown, 1997), pp. 64–72.

pounding takes place more frequently, then the interest rate must be adjusted accordingly. For example, a yearly interest rate of 12 percent $= 0.12$ corresponds to a monthly interest rate of $\frac{12\%}{12} = \frac{0.12}{12} = 0.01 = 1$ percent. If the interest is being compounded quarterly, the quarterly interest rate would then be $\frac{12\%}{4} = \frac{0.12}{4} = 0.03 = 3$ percent.

In order to handle situations such as these, we will modify the formula $A = P(1 + r)^t$ slightly.

The Compound Interest Formula

Assume that an account with principal P is paying an annual interest rate r and compounding is being done n time periods per year for t years. To find the future value, A, of the account, use the formula

$$A = P\left(1 + \frac{r}{n}\right)^{nt}.$$

Notice in this definition that we have replaced r, the annual interest rate, by the annual rate divided by the number of compounding periods per year and t, the number of years of compounding, by nt, which is the number of compounding periods.

You can use the compound interest formula for computing compound interest to compare investments.

EXAMPLE 5   Using the Compound Interest Formula

Your rich uncle has passed away and bequeathed you $10,000. Because you are cautious with your investments, you decide to buy a five-year certificate of deposit to save for the future. Prudent Savings and Loan has a certificate with a yearly interest rate of 5 percent compounded quarterly, whereas First Friendly National Bank has a certificate with a yearly interest rate of 4.8 percent compounded monthly. Which institution gives the better return on your money?

SOLUTION: We will use the formula for computing future value using compound interest stated prior to this example.

Prudent Savings and Loan: Here $P = 10,000$, $r = 0.05$, $n = 4$, and $t = 5$. Using the compounding formula to find the future value, we get

$$A = P\left(1 + \frac{r}{n}\right)^{nt} = 10,000\left(1 + \frac{0.05}{4}\right)^{4 \cdot 5} = 10,000(1.0125)^{20} \approx 12,820.37.$$

First Friendly National Bank: In this case, $P = 10,000$, $r = 0.048$, $n = 12$, and $t = 5$. Again, using the compounding formula, we find that

$$A = P\left(1 + \frac{r}{n}\right)^{n \cdot t} = 10,000\left(1 + \frac{0.048}{12}\right)^{12 \cdot 5} = 10,000(1.004)^{60} \approx 12,706.41.$$

From this you see that you will earn slightly more interest if you purchase the certificate of deposit from Prudent Savings and Loan.   ◎

Quiz Yourself 6

Sarah deposits $1,000 in a CD paying 6 percent annual interest for two years. What is the future value of her account if the interest is compounded quarterly?

Example 6 illustrates a different way to use the compound interest formula.

EXAMPLE 6 Finding the Present Value for a College Tuition Account

Upon the birth of a child, a parent wishes to make a deposit into a taxfree account to use later for the child's college education. Assume that the account has an annual interest rate of 8 percent and that the compounding is done yearly. How much must the parent deposit now so that the child will have $60,000 at age eighteen?

SOLUTION: We will use the compound interest formula $A = P(1 + \frac{r}{n})^{n \cdot t}$. Because we know that $A = 60,000$, $r = 0.08$, $n = 1$, and $t = 18$, we can find the present value by solving the equation

$$60,000 = P\left(1 + \frac{0.08}{1}\right)^{1 \cdot 18}$$

$$= P(1 + 0.08)^{18} \text{ for } P.$$

Therefore,

$$P = \frac{60,000}{(1.08)^{18}} \approx \frac{60,000}{3.99601950} \approx 15,014.94$$

A deposit of slightly more than $15,000 now will guarantee $60,000 for college in eighteen years.

Although $60,000 may seem like a lot of money, realize that *inflation*, the increase in the price of goods and services, also will cause the cost of a college education to increase. We will consider the effects of inflation in the exercises.

So far, we have used the formula $A = P(1 + \frac{r}{n})^{n \cdot t}$ to find A and P. Sometimes we want to find r or $n \cdot t$. In order to do this, we need to introduce some new techniques.

We use the log function to solve for *nt* in the formula $A = P(1 + \frac{r}{n})^{nt}$.

In order to solve for nt in the formula $A = P(1 + \frac{r}{n})^{nt}$, we need to be able to solve an equation of the form $a^x = b$, where a and b are fixed numbers. A property of logarithmic functions enables us to solve such equations. Many calculators have a key labeled either "log" or "log x," which stands for the **common logarithmic function.** Pressing this key reverses the operation of raising 10 to a power. For example, suppose that you compute $10^5 = 100,000$ on your calculator. If you next press the log key, the display will show 5. If you enter 1,000, which is 10 raised to the third power, and press the log key, the display will show 3. Practice finding the log of powers of 10 such as 100 and 1,000,000. If you enter 23 and then press the log key, the display will show 1.361727836. The interpretation of this result is that $10^{1.361727836} = 23$.* The log function has an important property that will help us to solve equations of the form $a^x = b$.

Exponent Property of the Log Function

$$\log y^x = x \log y$$

In order to understand this property, you should use your calculator to verify the following:

$$\log 4^5 = 5 \log 4$$
$$\log 6^3 = 3 \log 6$$

Example 7 illustrates how to use the exponent property to solve equations.

EXAMPLE 7   Solving an Equation Using the Exponent Property of the Log Function

Solve $3^x = 20$.

SOLUTION: We illustrate the steps required to solve this equation.

Step 1	Take the log of both sides of the equation.	$\log 3^x = \log 20$
Step 2	Use the exponent property of the log function.	$x \log 3 = \log 20$
Step 3	Divide both sides by log 3.	$x = \dfrac{\log 20}{\log 3}$
Step 4	Use a calculator to evaluate the right side of the equation (your calculator may give a slightly different answer).	$x = 2.726833028$

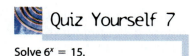

Quiz Yourself 7

Solve $6^x = 15$.

In Example 8, we use the exponent property of the log function to find the time it takes an investment to grow to a certain amount.

EXAMPLE 8   Saving for Equipment for a Business

Mara wants to buy lighting equipment from her cousin to start a dance studio. He will sell his equipment for $2,800. She presently has $2,500 and found an investment that will pay her 9 percent annual interest, compounded monthly. In how many months will Mara be able to pay her cousin for the equipment?

---

* We will not discuss what it means to raise 10 to a power such as 1.361727836.

SOLUTION: We know that the future value that Mara must pay her cousin is $A = 2,800$, the amount she has now is $P = 2,500$, and the monthly interest rate is $\frac{r}{n} = \frac{0.09}{12} = 0.0075$. We must solve the compound interest formula $A = P(1 + \frac{r}{n})^{nt}$ for nt. Because we are dealing with monthly payments and an annual interest rate, nt represents the number of months of the compounding. Substituting for A, P, and r, we get the equation

$$2,800 = 2,500\left(1 + \frac{0.09}{12}\right)^{nt}.$$

We solve this equation by the following steps:

$1.12 = (1.0075)^{nt}$ (Divide both sides of the equation by 2,500 and simplify.)

$\log(1.12) = \log(1.0075)^{nt}$ (Take the log of both sides.)
$\log(1.12) = nt\log(1.0075)$ (Use the exponent property of the log function.)

Solving for nt, we get the equation

$$nt = \frac{\log(1.12)}{\log(1.0075)} \approx 15.17.$$

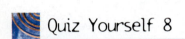

Quiz Yourself 8

Do Example 8 again, but now assume that the interest rate is 6 percent.

This means that Mara will have the money she needs by the end of the sixteenth month.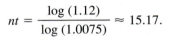

The last situation that we will consider is how to solve the compound interest equation $A = P(1 + \frac{r}{n})^{nt}$ for r. In order to do this, we have to be able to solve an equation of the form $x^a = b$, where a and b are fixed numbers. We show how to solve such an equation in Example 9.

EXAMPLE 9 Negotiating a Basketball Contract

Rasheed is negotiating a new basketball contract with the Lakers and expects to retire after playing one more year. In order to reduce his current taxes, his agent has agreed to defer a bonus of $1.4 million to be paid as $1.68 million in two years. If the Lakers invest the $1.4 million now, what rate of investment would they need in order to have $1.68 million to pay Rasheed in two years? Assume that you want to find an annual interest rate that is compounded monthly.

SOLUTION: To solve this compound interest problem we again use the formula $A = P(1 + \frac{r}{n})^{nt}$. We know $A = 1.68$, $P = 1.4$, $t = 2$ years, and $n = 12$. We must find the annual interest rate r.

Substituting for A, P, t, and n, we get the equation

$$1.68 = 1.4\left(1 + \frac{r}{12}\right)^{12 \cdot 2}.$$

Dividing both sides of the equation by 1.4 gives us $1.2 = (1 + \frac{r}{12})^{24}$. We can get rid of the exponent 24 if we raise both sides of the equation to the $\frac{1}{24}$ power. This gives us the equation

$$(1.2)^{1/24} = \left(\left(1 + \frac{r}{12}\right)^{24}\right)^{1/24} = 1 + \frac{r}{12}.^{*}$$

Subtracting 1 from both sides of the equation, we get

$$\frac{r}{12} = (1.2)^{1/24} - 1 = 1.00762566 - 1 = 0.00762566.$$

Now, multiplying this equation by 12, we find the annual interest rate, r, to be $12(0.00762566) \approx 0.0915$. Thus the Lakers need to find an investment that pays an annual interest rate of about 9.15 percent compounded monthly.

SOME GOOD ADVICE

Be careful to distinguish between the situations in Examples 8 and 9. In Example 8, we used the log function to solve an equation of the form $a^x = b$. In Example 9, we solved an equation of the form $x^a = b$ by raising both sides of the equation to the $\frac{1}{a}$ power.

Exercises 11.2

Communicating Mathematics *In Exercises 1–3, explain each variable in the formula.*

1. (simple interest) $I = Prt$

2. (future value using simple interest) $A = P(1 + rt)$

3. (future value using compound interest) $A = P(1 + \frac{r}{n})^{nt}$

4. Solve the equation in Exercise 3 for P.

In Exercises 5–8, use the formula for simple interest, $I = Prt$, and elementary algebra to find the missing quantity.

5. $P = \$1,000$; $r = 8$ percent; $t = 3$ years

6. $I = \$196$; $r = 7$ percent; $t = 2$ years

7. $I = \$700$; $P = \$3,500$; $t = 4$ years

8. $I = \$1,920$; $P = \$8,000$; $r = 6$ percent

In Exercises 9–14, use the formula for future value, $A = P(1 + rt)$, and elementary algebra to find the missing quantity.

9. $P = \$2,500$; $r = 8$ percent; $t = 3$ years

10. $P = \$1,600$; $r = 4$ percent; $t = 5$ years

11. $A = \$1,770$; $r = 6$ percent; $t = 3$ years

12. $A = \$2,332$; $r = 3$ percent; $t = 2$ years

13. $A = \$1,400$; $P = \$1,250$; $t = 2$ years

14. $A = \$966$; $P = \$840$; $r = 5$ percent

15. Buying an entertainment system. You have purchased a home entertainment system for \$3,600 and have agreed to pay off the system in 36 monthly payments of \$136 each.

a) What will be the total sum of your payments?

b) What will be the total amount of interest that you have paid?

16. Buying a car. You have purchased a used car for \$6,000 and have agreed to pay off the car in 24 monthly payments of \$325 each.

---

* In algebra, $(a^x)^y = a^{xy}$. That is why $((1 + \frac{r}{12})^{24})^{1/24} = (1 + \frac{r}{12})^{(24)(1/24)} = (1 + \frac{r}{12})^1 = 1 + \frac{r}{12}$.

a) What will be the total sum of your payments?

b) What will be the total amount of interest that you have paid?

Often, through government-supported programs, students may obtain "bargain" interest rates such as 6 percent or 8 percent in order to attend college. Frequently payments are not due and interest does not accumulate until you stop attending college. In Exercises 17–18, calculate the amount of interest due one month after you must begin payments.

17. You have borrowed $10,000 at an annual interest rate of 8 percent.

18. You have borrowed $15,000 at an annual interest rate of 6 percent.

19. How much must you deposit in an account paying 8 percent annual interest computed using the simple interest formula if you are to earn $800 in interest in two years?

20. How much must you deposit in an account paying 5 percent annual interest computed using the simple interest formula if you are to earn $400 in interest in four years?

In Exercises 21–28, we will assume that the lender is using simple interest to compute the interest on the loan.

21. Borrowing for a trip. You plan to take a trip to the Grand Canyon in two years. You wish to buy a certificate of deposit for $1,200 that you will cash in for your trip. What annual interest rate must you obtain on the certificate if you need $1,500 for your trip?

22. Borrowing for a trip. This question is the same as Exercise 21, except that now you are going to take the trip in four years.

23. Paying interest on late taxes. Jonathan wishes to defer payment of his $4,500 tax bill for four months. If he must pay an annual interest rate of 15 percent for doing this, what will his total payment be?

24. Borrowing money for a business. To start up a new photography business, Taylor is borrowing $18,000, which she intends to pay back in sixteen months. Her bank is charging her 9 percent annual interest on the loan. How much will she pay back?

25. Taking a bank loan. Linda's union is preparing to go on strike. To help members meet expenses, the union has arranged with a bank to provide loans at an annual interest rate of 12 percent. If Linda borrows $2,000, how much will she have to repay the bank in three months?

26. Borrowing money for a business. Nikolas borrowed $14,000 to purchase new equipment for his landscaping business. The bank has loaned him this money at an annual interest rate of 7.5 percent. How much will Nikolas owe the bank when he repays his loan in eighteen months if the bank is using simple interest?

27. Borrowing from a pawn shop. Sanjay has borrowed $400 on his father's watch from the Main Street Pawn Shop. He has agreed to pay off the loan with $425 one month later. What is the annual interest rate that he is being charged?

28. Borrowing from a bail bondsman. If a person accused of a crime does not have sufficient resources, he may have a bail bondsman post bail in order to be released until a trial is held. Assume that a bondsman charges a $50 fee plus 8 percent of the amount of the bail. If a bondsman posts $20,000 for a trial that takes place in two months, what is the interest rate being charged by the bondsman? (Treat the $50 fee plus the 8 percent as interest on a $20,000 loan for two months.)

In Exercises 29–30, assume that the amount P is deposited in an account paying the specified annual interest rate. The interest is being compounded yearly. Follow the procedure demonstrated in Example 4 to complete the table. Find the amount in the account at the end of the last year in the table.

29. $P = \$2,000; r = 8\%$

	P	I (Interest)	A (Value of Account)
Beginning of year 1	$2,000	$0	$2,000
End of year 1			
End of year 2			
End of year 3			

30. $P = \$500$; $r = 4\%$

	P	I (Interest)	A (Value of Account)
Beginning of year 1	$500	$0	$500
End of year 1			
End of year 2			
End of year 3			
End of year 4			

In Exercises 31–34, you are given an annual interest rate and the compounding period. Find the interest rate per compounding period.

31. 18 percent; monthly

32. 8 percent; quarterly

33. 12 percent; daily*

34. 10 percent; daily

In Exercises 35–42, you are given the principal, the annual interest rate, and the compounding period. Use the formula for computing future value using compound interest to determine the value of the account at the end of the specified time period.

35. $5,000, 5 percent, yearly; five years

36. $7,500, 7 percent, yearly; six years

37. $4,000, 8 percent, quarterly; two years

38. $8,000, 4 percent, quarterly; three years

39. $20,000, 8 percent, monthly; two years

40. $10,000, 6 percent, monthly; five years

41. $4,000, 10 percent, daily; two years

42. $6,000, 4 percent, daily; three years

In Exercises 43–44, you are given an annual interest rate and the compounding period for two investments. Decide which is the better investment.

43. 5 percent compounded yearly; 4.95 percent compounded quarterly

44. 4.75 percent compounded monthly; 4.70 percent compounded daily

In Exercises 45–46, Ann and Tom wish to establish a fund for their grandson's college education. What lump sum must they deposit in each account in order to have $30,000 in the fund at the end of fifteen years?

*We will assume there are 365 days in a year.

45. Saving for college. 6 percent annual interest rate, compounded quarterly

46. Saving for college. 7.5 percent annual interest rate, compounded monthly

In Exercises 47–54, solve each equation.

47. $3^x = 10$

48. $2^x = 12$

49. $(1.05)^x = 2$

50. $(1.15)^x = 3$

51. $x^3 = 10$

52. $x^2 = 10$

53. $x^4 = 10$

54. $x^4 = 25$

In Exercises 55–62, use the compound interest formula $A = P(1 + r)^t$ and the given information to solve for either t or r.

55. $A = \$2,500$, $P = \$2,000$, $t = 5$

56. $A = \$3,000$, $P = \$1,500$, $t = 7$

57. $A = \$1,000$, $P = \$100$, $t = 25$

58. $A = \$400$, $P = \$20$, $t = 35$

59. $A = \$1,500$, $P = \$1,000$, $r = 4$ percent

60. $A = \$2,500$, $P = \$1,000$, $r = 6$ percent

61. $A = \$800$, $P = \$200$, $r = 4.5$ percent

62. $A = \$1,000$, $P = \$100$, $r = 8$ percent

63. Comparing investments. Jocelyn purchased 100 shares of airline stock for $23.75 per share. Eight months later she sold the stock at $24.50 per share.

a) What annual rate, calculated using simple interest, did she earn on this transaction?

b) What annual rate would she have to earn in a savings account compounded monthly to earn the same money on her investment?

64. Comparing investments. Dominick purchased a bond to preserve a wildlife sanctuary for $2,400 and ten months later he sold it for $2,580.

a) What annual rate, calculated using simple interest, did he earn on this transaction?

b) What annual rate would he have to earn in a savings account compounded monthly to earn the same money on his investment?

65. Investment earnings. Emily purchased a bond for highway construction valued at $20,000 for $9,420. If the bond pays 7.5 percent annual interest compounded monthly, how long must she hold it until it reaches its full face value?

66. Investment earnings. Lucas purchased a bond to build a new sports stadium with a face value of $10,000 for $4,200. If the bond pays 6.5 percent annual interest compounded monthly, how long must he hold it until it reaches its full face value?

The computations for dealing with inflation are the same as for determining future value. If an item sells for $100 today and there is an annual inflation rate of 4 percent for ten years, the same item would then cost $100(1.04)^{10} = $148.02. In Exercises 67–70, determine what each item's cost would be in 20 years using the given inflation rate.

67. Inflation. fast-food meal, $4.65; inflation rate, 3 percent

68. Inflation. automobile, $15,000; inflation rate, 5 percent

69. Inflation. salary, $35,000 per year; inflation rate, 4 percent

70. Inflation. athletic shoes, $80; inflation rate, 3.5 percent

71. Inflation. From 1992 to 1995, Albania experienced a yearly inflation rate of 226 percent. Determine the price of the fast-food meal in Exercise 67 after five years at a 226 percent inflation rate.

72. Inflation. The inflation rate in Hungary during the mid-1990s was about 28 percent. Determine the price of the athletic shoes in Exercise 70 after ten years at a 28 percent inflation rate.

Further Exercises

Savings institutions often state two rates in their advertising. One is the nominal yield, *which you can think of as an annual simple interest rate. The other is called the* effective annual yield, *which is the actual interest rate that the account earns due to the compounding. If $1,000 is in an account that pays a nominal yield of 9 percent and if the compounding is done monthly, then after one year, the account would contain $1,093.80, which corresponds to a simple interest rate of 9.38 percent. We would say that this account has an effective annual yield of 9.38 percent. In Exercises 73–76, find the effective annual yield for each account.*

73. nominal yield, 7.5 percent; compounded monthly

74. nominal yield, 10 percent; compounded twice a year

75. nominal yield, 6 percent; compounded quarterly

76. nominal yield, 8 percent; compounded daily

Some banks advertise that money in their accounts is compounded continuously. To get an understanding of what this means, apply the compound interest formula using a very large number of compounding periods per year. In Exercises 77–78, divide the year into 100,000 compounding periods per year. Apply the compound interest formula for finding future value to approximate what the effective annual yield would be if the compounding were done continuously for the stated nominal yield.

*(77.) nominal yield, 10 percent

78. nominal yield, 12 percent

If the principal P is invested in an account that pays an annual interest rate of r percent and the compounding is done continuously, then the future value, A, that will be in the account after t years is given by the formula

$$A = Pe^{rt}.$$

The number e is approximately 2.718281828.

79. Use the formula for continuous compounding to find the effective annual yield if the compounding in Exercise 77 is done continuously.

80. Use the formula for continuous compounding to find the effective annual yield if the compounding in Exercise 78 is done continuously.

81. Communicating Mathematics Explain the relationship between the formulas $A = P(1 + r)^t$ and $A = P(1 + \frac{r}{n})^{nt}$.

82. Communicating Mathematics Under what circumstances will $A = P(1 + r)^t$ and $A = P(1 + \frac{r}{n})^{nt}$ give you the same answers to a compound interest problem?

83. Communicating Mathematics Explain the difference in the techniques that you have to use to solve a problem like Example 8 versus a problem like Example 9.

84. Communicating Mathematics Explain why money grows faster when you are using compound interest versus simple interest.

---

* Exercise numbers circled in red can be used as group exercises.

11.3 CONSUMER LOANS

Imagine that you have just signed the lease for your first apartment and now all you have to do is furnish it. If you buy living-room furniture for $1,100, which you pay for in payments, you are taking out an installment loan. Loans having a fixed number of payments are called *closed-ended credit* agreements (or *installment loans*). Each payment is called an *installment*. The size of your payments is determined by the amount of your purchase and also by the interest rate that the seller is charging. The interest charged on a loan is often called a *finance charge*.

■

The add-on interest method is a simple way to compute payments on an installment loan.

We use the simple interest formula from Section 11.2 to calculate the finance charge for an installment loan. To determine the payments for an installment loan, we add the simple interest due on the loan to the loan amount and then divide this sum by the number of monthly payments.

> **Formula for Determining the Monthly Payment of an Installment Loan**
>
> $$\text{monthly payment} = \frac{P + I}{n},$$
>
> where P is the amount of the loan, I is the amount of interest due on the loan, and n is the number of monthly payments.

This method is sometimes called the **add-on interest method**, because we are adding on the interest due on the loan before determining the payments.

EXAMPLE 1 Determining Payments for an Add-On Interest Loan

You have decided to purchase a new audio system, which has a list price of $720. You intend to take out an installment loan for two years at an annual interest rate of 18 percent. If the store is using the add-on interest method, what will be your monthly payments?

SOLUTION: We first use the simple interest formula to calculate the interest:

$$I = Prt = 720(0.18)2 = 259.20.$$

Next, we add the interest to the purchase price:

$$720 + 259.20 = 979.20.$$

Quiz Yourself 9*

Suppose that you take an installment loan for $360 for one year at an annual interest rate of 21 percent. What are your monthly payments?

To find the monthly payments, we divide this amount by 24:

$$\frac{979.20}{24} = 40.80$$

Monthly payments are therefore $40.80.

In Example 1, the annual interest rate of 18 percent is quite misleading. If we think about it, the purchase price was $720, so it would be fair to say that you are paying off $720/24 = $30 of the loan amount each month; the other $10.80 is interest. When you reach the last month, although you only owe $30 on the purchase, you are still paying $10.80 in interest. Simple arithmetic shows that 10.80/30 = 0.36. So in a certain sense, the interest rate for the last month of the loan is actually 36 percent. Because you are paying 36 percent interest for one month, this is equivalent to an annual interest rate of 12 × 36 percent = 432 percent! What we want to point out here is that although simple interest is easy to compute, as you pay off the loan amount, the actual interest you are paying on the outstanding balance is higher than the stated interest rate.

When you use your credit card to pay for gas at a gas station, you are using *open-ended credit*. With open-ended credit, the calculation of finance charges can be more complicated than with closed-ended credit. Although you may be making monthly payments on your loan, you may also be increasing the loan by making further purchases.

There are several ways that credit card companies compute finance charges. We will look at two methods and compare them at the end of this section. You will see that if you understand the method being used to compute your finance charges, you can use this information to reduce the cost of borrowing money.

The unpaid balance method computes finance charges on the balance at the end of the previous month.

The first method that we will discuss for computing finance charges is called the **unpaid balance method**. With this method, the interest is based on the previous month's balance.

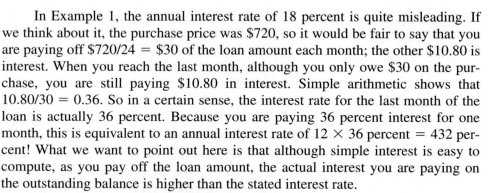

> **The Unpaid Balance Method for Computing the Finance Charge on a Credit Card Loan**
>
> This method also uses the simple interest formula $I = Prt$; however,
>
> P = previous month's balance + finance charge + purchases made − returns − payments
>
> The variable r is the annual interest rate, and $t = \frac{1}{12}$.

EXAMPLE 2   Using the Unpaid Balance Method for Finding Finance Charges

Assume that the annual interest rate on your credit card is 18 percent and your unpaid balance at the beginning of last month was $600. Since then, you purchased ski boots for $130 and sent in a payment of $170.

a) Using the unpaid balance method, what is your credit card bill this month?

b) What is your finance charge next month?

SOLUTION: a) The finance charge is 18 percent per year, or 1.5 percent = 0.015 per month. The balance on your credit card last month was $600, so your finance charge is $600(0.015) = \$9$. Therefore you owe $600 + 9 + 130 - 170 = \$569$.

b) Thus, the finance charge next month will be $569(0.015) = \$8.54$.

Note that you can use the unpaid balance method to your advantage by making a large purchase early in the billing period and then paying it off just before the billing date. This is not to the credit card company's advantage, because you can use the credit card company's money for free for almost a whole month.

It is easy to use credit cards for purchases, and so tempting to pay only the minimum payment that appears on your credit card bill, that you may find your debt increasing even though you are sending in a payment every month. Example 3 illustrates how credit card debt can get out of hand.

EXAMPLE 3   Paying Off a Credit Card Debt

Assume that you wish to pay off your credit card debt of $6,589 by making the minimum payment of $100 a month. What will your balance be at the end of one month? Assume that the annual interest rate on your card is 18 percent and that the credit card company is using the unpaid balance method to compute your finance charges.

SOLUTION: When you send in your $100, it is credited to your account, so the next month you have an unpaid balance of $6,589 - 100 = \$6,489$. The annual interest rate is 18 percent, so your monthly interest rate is $\frac{18}{12} = 1.5$ percent.

Therefore, at the end of the month you still owe

$$6,489 + (0.015)(6,489) = 6,489 + 97.34 = \$6,586.34.$$

Thus your $100 payment has reduced your debt by only $6,589 - 6,586.34 = \$2.66$!

Example 3 illustrates how difficult it is to pay off a large credit card bill. If your debt is large enough, the minimum payment shown on a credit card statement may be insufficient to pay *any* of the principal. This means that the interest you owe is being compounded and that your debt is growing rapidly. The best practice is to pay off as much of your outstanding balance as you can to avoid paying a large amount of interest. Although it is possible to borrow cash from one company to pay off debts at another, doing this is not reducing the size of your debt. Instead, the debt may still grow at a rate of 18 percent or more per year.

 Quiz Yourself 10

Assume that the annual interest rate on your credit card is 21 percent. Your outstanding balance last month was $300. Since then you have charged a purchase for $84 and made a payment of $100. What is the outstanding balance on your card at the end of this month? What is next month's finance charge on this balance?

When you use credit, always look at the annual interest rate. In some cases, if you read the fine print in the agreement, you will find rates as high as 24 or 25 percent. If you are unfortunate enough to borrow from a loan shark, it is possible that for a loan of $1,000, you would be asked to pay $1,025 a week later. This may not seem like much, but realize that the interest rate is 2.5 percent for a single week—or an annual rate of $2.5 \times 52 = 130$ percent. Also, when you take a cash advance on your credit card, the interest rate is usually much higher than the rate charged for making purchases on your card.

■ ▬▬▬▬▬▬▬▬▬▬▬▬▬▬▬▬▬▬▬▬▬▬▬▬

The average daily balance method computes finance charges based upon the balance in the account for each day of the month.

A more complicated method for determining the finance charge on a credit card is called the **average daily balance method**, which is one of the most common methods used by credit card companies. With this method, the balance is the average of all the daily balances for the previous month.

> **The Average Daily Balance Method for Computing the Finance Charge on a Credit Card Loan**
>
> 1. Add the outstanding balance for your account for each day of the previous month.
> 2. Divide the total in step 1 by the number of days in the previous month to find the average daily balance.
> 3. To find the finance charge, use the formula $I = Prt$, where P is the average daily balance found in step 2, r is the annual interest rate, and t is the number of days in the previous month divided by 365.

EXAMPLE 4   **Using the Average Daily Balance Method for Finding Finance Charges**

Suppose that you begin the month of September (which has 30 days) with a credit card balance of $240. Assume that your card has an annual interest rate of 18 percent and that during September the following adjustments are made on your account:

> September 11: A payment of $60 is credited to your account.
> September 18: You charge a dozen roses for $24.
> September 23: You charge $12 for gasoline.

Use the average daily balance method to compute the finance charge that will appear on your October credit card statement.

SOLUTION: To answer this question, we must first find the average daily balance for September. The easiest way to calculate the balance is to keep a day-by-day record of what you owe the credit card company for each day in September, as we do in Table 11.1.

Day	Balance	Number of Days × Balance
1, 2, 3, 4, 5, 6, 7, 8, 9, 10	$240	10 × 240 = 2,400
11, 12, 13, 14, 15, 16, 17	$180	7 × 180 = 1,260
18, 19, 20, 21, 22	$204	5 × 204 = 1,020
23, 24, 25, 26, 27, 28, 29, 30	$216	8 × 216 = 1,728

TABLE 11.1   Daily balances for September.

The average daily balance is therefore

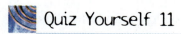

Quiz Yourself 11

Recalculate the average daily balance in Example 4, except now assume you bought the roses on September 3 instead of September 18. Make a table similar to Table 11.1.

$$\frac{(10 \times 240) + (7 \times 180) + (5 \times 204) + (8 \times 216)}{30} =$$

$$\frac{2{,}400 + 1{,}260 + 1{,}020 + 1{,}728}{30} = \frac{6{,}408}{30} = 213.6.$$

We next apply the simple interest formula, where $P = \$213.60$, $r = 0.18$, and $t = \frac{30}{365}$.* Thus $I = Prt = 213.6(0.18)(\frac{30}{365}) = 3.16$. Your finance charge on the October statement will be $3.16.

As you will see in Example 5, the amount of finance charges you pay on a loan will vary depending on the method used to compute the charges.

EXAMPLE 5   Comparing Methods for Finding Finance Charges

Suppose that you begin the month of May (which has 31 days) with a credit card balance of $500. The annual interest rate is 21 percent. On May 11, you use your credit card to pay for a $400 car repair and on May 29 you make a payment of $500. Calculate the finance charge that will appear on the statement for next month using the two methods we have discussed.

SOLUTION:

Method	P	r	t	Finance Charge = I = Prt
Unpaid balance	last month's balance + finance charge − payment + charge for car repair = 500 + 8.75 + 400 − 500 = 408.75	21%	$\frac{1}{12}$	$(408.75)(0.21)\left(\dfrac{1}{12}\right) = \$7.15$
Average daily balance	$\dfrac{10 \times 500 + 18 \times 900 + 3 \times 400}{31}$ $= \dfrac{22{,}400}{31}$ $= 722.58$	21%	$\frac{31}{365}$	$(722.58)(0.21)\left(\dfrac{31}{365}\right) = \$12.89$

———

*$t = \frac{30}{365}$ because we are using the credit card for 30 days out of a 365-day year.

With the unpaid balance method, the finance charge is $7.15; with the average daily balance method the finance charge is $12.89.

Example 5 shows that the exact same charges on two different credit cards having the same annual interest rate can result in very different finance charges. If you understand the method your credit card company is using, you may be able to schedule your purchases and payments to minimize your finance charges.

In deciding how to use credit, you must consider many other issues that we have not discussed in this section. Some credit card companies charge an annual fee; others return part of your interest payments. For some credit cards there is a grace period. If you reduce your balance to zero during this grace period, then you pay no finance charges. A credit card may have a low introductory rate that changes to a much higher rate at a later time.

One common enticement is that you can make a purchase and pay no interest payments until several months later. With such deals, however, you must be careful. Often, if you do not pay off the purchase completely at the end of the interest-free period, then all the interest that would have accumulated is added to your balance. It is difficult to understand all the pros and cons of the many different types of credit contracts. However, if you read credit agreements carefully

Historical Highlight: *Credit and Interest**

Credit cards appeared in the United States in the early 1900s but did not become widely used until the 1950s. Early cards such as Diners Club, Carte Blanche, and American Express made the use of plastic money more popular. Today Americans charge about one trillion dollars per year on their cards.

Credit is neither a modern idea nor exclusively American. Surprisingly, there are ancient Sumerian documents dating back to about 3000 B.C. that show the regular use of credit in borrowing grain and metal. Interest was charged on loans, often in the 20 to 30 percent range—not unlike the 18 to 21 percent charged on many of today's credit cards. As the use of credit increased, so did its misuse. Many societies wrote laws to prevent its abuse—particularly the charging of high interest rates, which is called *usury*. Interest rates were

either limited, or the charging of interest on a loan was forbidden. The Israelites did not allow a lender to charge interest, and the ancient Iranians felt that charging interest dishonored a person.

Credit and interest appear in the twentieth century in many diverse forms. The Ifugao tribe of the Philippines have the rule that 100 percent interest must be charged on a loan. If rice is borrowed, then at the next harvest the loan must be paid in double. If one pig is borrowed, two must be returned. In Vancouver, Canada, the Kwakiutl have a system of credit based on blankets. The rules of interest state that if five blankets are borrowed, in six months they become seven. If the five blankets are borrowed for a year, then ten must be returned. In Northern Siberia, loans are made in reindeer usually at a 100 percent interest rate.

* This Highlight is based on S. Homer and R. Sylla, *A History of Interest Rates*, 3rd ed. (Rutgers University Press, 1991), pp. 21–30.

and remember the principles that you learned in this section, you will be an intelligent consumer who will be able to use credit wisely.

Exercises 11.3

In Exercises 1–4, compute the monthly payments for each add-on interest loan. The amount of the loan, the annual interest rate, and the term of the loan are given.

1. $900; 12 percent; two years

2. $840; 10 percent; three years

3. $1,360; 8 percent; four years

4. $1,710; 9 percent; three years

5. Paying off a computer. Luis took out an add-on interest loan for $1,280 to buy a new laptop computer. The loan will be paid back in two years and the annual interest rate is 9.5 percent. How much interest will he pay? What are his monthly payments?

6. Paying off furniture. Mandy bought furniture for her new apartment costing $1,460. To pay for it, her bank gave her a five-year add-on interest loan at an annual interest rate of 10.4 percent. How much interest will she pay? What are her monthly payments?

7. Financing equipment. Angela's bank gave her a four-year add-on interest loan for $6,480 to pay for new equipment for her antiques restoration business. The annual interest rate is 11.65 percent. How much interest will she pay? What are her monthly payments?

8. Paying for a sculpture. Mikeal purchased an antique sculpture from a gallery for $1,320. The gallery offered him a three-year add-on loan at an annual rate of 9.75 percent. How much interest will he pay? What are his monthly payments?

In Exercises 9–12, use the add-on method for determining interest on the loan. Determine the annual interest rate during the last month of the loan.

9. $900; 12 percent; two years

10. $840; 10 percent; three years

11. $1,360; 8 percent; four years

12. $1,710; 9 percent; four years

13. Financing a boat. Paul is buying a new boat for $11,000. The dealer is charging him an annual interest rate of 9.2 percent and is using the add-on method to compute his monthly payments.

a) If Paul pays off the boat in 48 months, what are his monthly payments?

b) If he makes a down payment of $2,000, how much will this reduce his monthly payments?

c) If he wishes to have monthly payments of $200, how large should his down payment be?

14. Financing a swimming pool. Joline is buying a new swimming pool for $14,000. The dealer is charging her an annual interest rate of 8.5 percent and is using the add-on method to compute her monthly payments.

a) If Joline pays off the pool in 48 months, what are her monthly payments?

b) If she makes a down payment of $3,000, how much will this reduce her monthly payments?

c) If she wishes to have monthly payments of $250, how large should her down payment be?

15. Financing rare coins. Anna is buying $15,000 worth of rare coins as an investment. The dealer is charging her an annual interest rate of 9.6 percent and is using the add-on method to compute her monthly payments.

a) If Anna pays off the coins in 36 months, what are her monthly payments?

b) If she makes a down payment of $3,000, how much will this reduce her monthly payments?

c) If she wishes to have monthly payments of $300, how large should her down payment be?

16. Financing a musical instrument. Walt is buying a music synthesizer for his rock band for $6,500. The music store is charging him an annual interest rate of 8.5 percent and is using the add-on method to compute his monthly payments.

a) If Walt pays off the synthesizer in 24 months, what are his monthly payments?

b) If he makes a down payment of $1,500, how much will this reduce his monthly payments?

c) If he wishes to have monthly payments of $150, how large should his down payment be?

In Exercises 17–22, use the unpaid balance method to find the finance charge on the credit card account. Last month's balance, the payment, the annual interest rate, and any other transactions are given.

17. last month's balance, $475; payment, $225; interest rate, 18 percent; bought ski jacket, $180; returned camera, $145

18. last month's balance, $510; payment, $360; interest rate, 21 percent; bought exercise equipment, $470; bought fish tank, $85

19. last month's balance, $640; payment: $320; interest rate, 16.5 percent; bought dog, $140; bought pet supplies, $35; paid veterinarian bill, $75

20. last month's balance, $340; payment, $180; interest rate, 17.5 percent; bought coat, $210; bought hat, $28; returned boots, $130

21. last month's balance, $460; payment, $300; interest rate, 18.8 percent; bought plane ticket, $140; bought luggage, $135; paid hotel bill, $175

22. last month's balance, $700; payment, $480; interest rate, 21 percent; bought ring, $210; bought theater tickets, $142; returned vase, $128

In Exercises 23–26, use the average daily balance method to compute the finance charge on the credit card account for the previous month. The starting balance and transactions on the

account for the month are given. Assume an annual interest rate of 21 percent in each case.

23. Computing a finance charge. Month: August (31 days); previous month's balance: $280

Date	Transaction
August 5	Made payment of $75
August 15	Charged $135 for hiking boots
August 21	Charged $16 for gasoline
August 24	Charged $26 for restaurant meal

24. Computing a finance charge. Month: October (31 days); previous month's balance: $190

Date	Transaction
October 9	Charged $35 for a book
October 11	Charged $20 for gasoline
October 20	Made payment of $110
October 26	Charged $13 for lunch

25. Computing a finance charge. Month: April (30 days); previous month's balance: $240

Date	Transaction
April 3	Charged $135 for a coat
April 13	Made payment of $150
April 23	Charged $30 for CDs
April 28	Charged $28 for groceries

26. Computing a finance charge. Month: June (30 days); previous month's balance: $350

Date	Transaction
June 9	Made payment of $200
June 15	Charged $15 for gasoline
June 20	Charged $180 for skis
June 26	Made payment of $130

In Exercises 27–30, redo the specified exercise using the unpaid balance method to calculate the finance charges.

27. Exercise 23   **28.** Exercise 24

29. Exercise 25   **30.** Exercise 26

31. Mayesha purchased a large-screen TV for $1,000 and can pay it off in ten months with an add-on interest loan at an annual rate of 10.5 percent, or she can use her credit card that has an annual rate of 18 percent. If she uses her credit card, she will pay $100 per month (beginning next month) plus the finance charges for the month. Assume that Mayesha's credit card company is using the unpaid balance method to compute her finance charges and that she is making no other transactions on her credit card. Which option will have the smaller total finance charges on her loan?

32. Repeat Exercise 31, but now assume that Mayesha purchased an entertainment center for $2,000, the rate for the add-on loan is 9.6 percent, and she is paying off the loan in 20 months.

33. An appliance store advertises 0 percent financing for three months for purchases made before the new year. The fine print in the advertisement states that if the purchase is not paid off within three months, the purchaser must pay interest that has accumulated at a monthly rate of 1.75 percent. Assume that you buy a refrigerator for $1,150 and make no payments during the next three months. How much interest has accumulated on your purchase during this time?

34. Repeat Exercise 33, but now assume that the purchase is for $1,450 and the annual interest rate is 24 percent.

Further Exercises

35. Communicating Mathematics   In Example 5, we found that the average daily balance method gave the highest finance charges. Explain why this happened.

*__36.__ Communicating Mathematics   Make up transactions on a hypothetical credit card so that the unpaid balance method will give you lower finance charges than the average daily balance method does. Assume an annual interest rate of 18 percent.

__37.__ Communicating Mathematics   Make up transactions on a hypothetical credit card so that the average daily balance method will give you lower finance charges than the unpaid balance method does. Assume an annual interest rate of 21 percent.

__38.__ Communicating Mathematics   Explain a buying strategy that would enable you to benefit from each of the finance charge methods.

__39.__ Communicating Mathematics   In our discussions about credit, we have ignored the fact that whatever money is not used to pay off a loan can be invested. Assume that you can earn 5 percent on any money that you do not use for paying off a loan. However, any money you earn as interest is subject to federal, state, and local taxes. Assume that these taxes total 20 percent. Discuss how that might affect your decision to pay off your credit card debt.

__40.__ Communicating Mathematics   Examine some of the features of a credit card agreement. Discuss how these features can be to your advantage or disadvantage.

11.4   ANNUITIES

Over the course of your lifetime, you will probably save for large expenses such as a college education, a new home, or retirement. It would be nice to be able to put a large sum of money in an interest-bearing account now so that you will have the money you need at some future date. However, most people do not have enough money to make such a large single deposit. It is more reasonable to assume that you will make a series of payments over many years to accumulate the money you need. In this section, we will discuss this type of investment, which is called an annuity.

---

* Exercise numbers circled in red can be used as group exercises.

We make regular payments into an annuity.

An **annuity** is an interest-bearing account into which we make a series of payments of the same size. If one payment is made at the *end* of every compounding period, the annuity is called an **ordinary annuity**. The **future value of an annuity** is the amount in the account, including interest, after making all payments.

To illustrate the future value of an annuity, suppose that in January you begin making payments of $100 at the end of each month into an account paying 12 percent yearly interest compounded monthly. How much money will be in this account for a summer vacation beginning on July 1?

This problem is different from those in Section 11.2. In the earlier problems, we deposited a lump sum that earned a stated interest rate for an entire period. In this problem we are depositing a *series* of payments, and each payment earns interest for a different number of periods.

The January deposit earns interest for February, March, April, May, and June. Using the formula for compound interest, this deposit will grow to

$$100(1 + 0.01)^5 = \$105.10.$$

However, the May deposit earns interest for only one month and therefore grows to only

$$100(1 + 0.01)^1 = \$101.$$

The June deposit earns no interest at all. We illustrate this pattern with the timeline in Figure 11.5.

If we compute how much each deposit contributes to the account and sum these amounts, we will have the value of the annuity on July 1.

January	$100(1.01)^5 = \$105.10$
February	$100(1.01)^4 = \$104.06$
March	$100(1.01)^3 = \$103.03$
April	$100(1.01)^2 = \$102.01$
May	$100(1.01)^1 = \$101.00$
June	$100(1.01)^0 = \$100.00$
	Total = $615.20

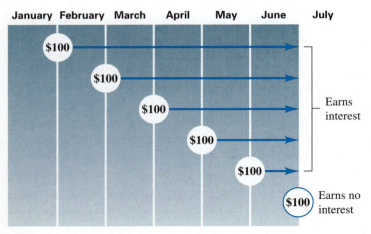

FIGURE 11.5 Timeline for ordinary annuity with deposits at end of January, . . . , June.

We can express the value of this annuity as

$$100(1.01)^5 + 100(1.01)^4 + 100(1.01)^3 + 100(1.01)^2 + 100(1.01)^1 + 100.$$

By factoring out the 100, we can write the value of the annuity in the form

$$100[(1.01)^5 + (1.01)^4 + (1.01)^3 + (1.01)^2 + (1.01)^1 + 1]. \tag{1}$$

Notice that although you deposited $100 at the end of each month for *six* months, the first deposit earned interest for *five* months, the second deposit earned interest for *four* months, and so on.

Following this pattern, it is clear that if you had deposited the $100 in a vacation savings account at the end of each month for ten months, the value of the annuity would have been

$$100[(1.01)^9 + (1.01)^8 + \cdots + (1.01)^2 + (1.01)^1 + 1]. \qquad (2)$$

Notice that in equation (1) we have an expression of the form $x^5 + x^4 + x^3 + x^2 + x^1 + 1$, and in equation (2) we have an expression of the form $x^9 + x^8 + \cdots + x^2 + x^1 + 1$. Fortunately, there is a way to write these lengthy expressions in a simpler form.

EXAMPLE 1   Simplifying Annuity Computations

Show that $x^5 + x^4 + x^3 + x^2 + x^1 + 1 = \dfrac{x^6 - 1}{x - 1}$.

SOLUTION: To show this relationship, we multiply the polynomials $x^5 + x^4 + x^3 + x^2 + x^1 + 1$ and $x - 1$ in the usual way.

$$
\begin{array}{r}
x^5 + x^4 + x^3 + x^2 + x^1 + 1 \\
\times\ x - 1 \\
\hline
x^6 + x^5 + x^4 + x^3 + x^2 + x^1 \\
-\ x^5 - x^4 - x^3 - x^2 - x^1 - 1 \\
\hline
x^6 \qquad\qquad\qquad\qquad - 1
\end{array}
$$

 Quiz Yourself 12*

Write $x^3 + x^2 + x^1 + 1$ as a quotient of two polynomials. (Do computations that are similar to those we did in Example 1.)

This result shows that $(x^5 + x^4 + x^3 + x^2 + x^1 + 1)(x - 1) = x^6 - 1$. Dividing this equation by $x - 1$, we obtain our desired relationship.

Following the pattern of Example 1, we can prove that

$$x^n + x^{n-1} + x^{n-2} + \cdots + x^2 + x^1 + 1 = \frac{x^{n+1} - 1}{x - 1}. \qquad (3)$$

Returning to our vacation savings account example, we can think of 1.01 as x in equation (1). Then we can use equation (3) to simplify our calculations. Because

$$(1.01)^5 + (1.01)^4 + (1.01)^3 + (1.01)^2 + (1.01)^1 + 1 =$$

$$\frac{(1.01)^6 - 1}{1.01 - 1} = \frac{1.061520151 - 1}{1.01 - 1} = \frac{0.061520151}{0.01} \approx 6.1520,$$

we can write

$$100[(1.01)^5 + (1.01)^4 + (1.01)^3 + (1.01)^2 + (1.01)^1 + 1] \approx 100(6.1520) = \$615.20.$$

This is the same amount that we found earlier.

---

* Quiz Yourself answers begin on page 849.

Doing similar computations, we find that the ten-month vacation savings account annuity has a value of

$$100[(1.01)^9 + (1.01)^8 + \cdots + (1.01)^2 + (1.01)^1 + 1] =$$

$$100\left[\frac{(1.01)^{10} - 1}{1.01 - 1}\right] \approx 100(10.4622) = \$1,046.22.$$

■

The future value of an annuity depends on the size of the payment, the interest rate, and the number of payments.

We can generalize the patterns that we have just seen in a formula for finding the future value of an annuity.

> **Formula for Finding the Future Value of an Ordinary Annuity**
>
> Assume that we are making a regular payment, R, at the end of each compounding period for an annuity that has an annual interest rate, r, which is being compounded n times per year. Then the value, A, of the annuity after t years is
>
> $$A = R\frac{\left(1 + \dfrac{r}{n}\right)^{nt} - 1}{\dfrac{r}{n}}.$$

In order to calculate this expression you should do the following steps:

1st: Find $\frac{r}{n}$ and then add 1.

2nd: Raise $1 + \frac{r}{n}$ to the nt power and then subtract one.

3rd: Divide the amount you found in step 2 by $\frac{r}{n}$.

4th: Multiply the quantity that you found in step 3 by R.

EXAMPLE 2 Finding the Future Value of an Ordinary Annuity

Assume that we make a payment of $50 at the end of each month into an account paying a 6 percent annual interest rate, compounded monthly. How much will be in that account after three years?

SOLUTION: This account is an ordinary annuity. The payment R is 50, $\frac{r}{n} = \frac{0.06}{12} = 0.005$, and the number of payments nt is $12 \times 3 = 36$. Using the formula for finding the future value of an ordinary annuity, we get

$$A = 50\left[\frac{(1.005)^{36} - 1}{0.005}\right] = 50\left(\frac{1.19668053 - 1}{0.005}\right)$$

$$= 50\left(\frac{0.19668053}{0.005}\right)$$

$$= 50(39.336105) \approx \$1,966.81.^*$$

Quiz Yourself 13

Redo Example 2, except now assume that you are depositing $75 per month for two years.

---

* When using your calculator, you should hold off on rounding your answers as long as possible. If you round off too soon, your answers will differ slightly from the answers in this text.

If you make regular payments into an annuity for many years, the value of the annuity can become enormous due to the compounding of your interest. Figure 11.6 shows that if the annual interest rate is 6.6 percent, then in roughly nineteen years, the amount of interest that the annuity has earned exceeds the amount of the deposits. The fact that the interest curve is rising so rapidly indicates that the future value of your account is also growing rapidly.

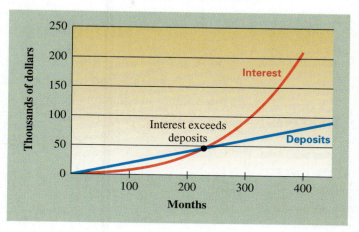

FIGURE 11.6 At an annual rate of 6.6 percent, the amount of interest earned in an ordinary annuity exceeds the amount of deposits in about 230 months.

With a sinking fund, we make payments to save a specified amount.

You may wish to save regularly to have a fixed amount available in the future. For example, you may want to save $1,800 to travel to Jamaica in two years. The question is, how much should you save each month to accomplish this? The account that you establish for your deposits is called a **sinking fund.** You could estimate the amount to save each month by simply dividing 1,800 by 24 months to get $\frac{1,800}{24} = \$75$ per month. Because your estimate ignores the interest that your deposits will generate, the actual amount you would need to put aside each month is somewhat less. Knowing exactly how much you need to save each month could be important if you were on a tight budget.

Because a sinking fund is a special type of annuity, it is not necessary to find a new formula to answer this question. We can use the formula for calculating the future value of an ordinary annuity that we have stated earlier. In this case we know the value of A and we want to find R.*

EXAMPLE 3 Calculating Payments for a Sinking Fund

In the sinking fund that you have set up to save $1,800 in two years, assume that you will make payments at the end of each month and that the account pays

---

* In making payments into a sinking fund, we will always round the payment *up* to the next cent.

6 percent yearly interest compounded monthly. What should be your monthly payment?

SOLUTION: Recall the formula for finding the future value of an ordinary annuity:

$$A = R \frac{\left(1 + \dfrac{r}{n}\right)^{nt} - 1}{\dfrac{r}{n}} \qquad (4)$$

We want the value A of the annuity to be \$1,800, the (monthly) rate $\frac{r}{n}$ is $\frac{6\%}{12} = \frac{0.06}{12} = 0.005$, and the number of payments nt is $12 \times 2 = 24$. Substituting these values in equation (4), we get:

$$1,800 = R\left[\frac{(1 + 0.005)^{24} - 1}{0.005}\right] = R(25.43195524).$$

Dividing both sides of this equation by 25.43195524, we get the monthly payment:

$$R = \frac{1,800}{25.43195524} \approx \$70.78$$

Quiz Yourself 14

What payments must you make to a sinking fund that pays 9 percent yearly interest compounded monthly if you wish to save \$2,500 in two years?

SOME GOOD ADVICE

You may be tempted to memorize a new formula to solve sinking fund problems. This is not necessary, because once you have learned to solve annuity problems, you can use the same formula (and a little bit of algebra) to solve sinking fund problems.

We use the log function to find how long it takes for an annuity to accumulate a specified value.

Sometimes in working with annuities, we want to know how long it will take to save a certain amount. That is, in the annuity formula we want to find nt. This problem is a little more complicated than those we have solved so far. In order to solve such problems, we will use the exponent property for the log function, which we introduced in Section 11.3. We will show you how to use this property in Example 4.

EXAMPLE 4   Finding the Time Required to Accumulate \$1,000,000

Suppose you have decided to retire as soon as you have saved \$1,000,000. Your plan is to put \$200 each month into an ordinary annuity that pays an annual interest rate of 8 percent. In how many years will you be able to retire?

SOLUTION: We can use the future value formula for an ordinary annuity to solve this problem:

$$A = R \frac{\left(1 + \dfrac{r}{n}\right)^{nt} - 1}{\dfrac{r}{n}}$$

Here A is the future value of $1,000,000, $\frac{r}{n}$ is the monthly interest rate of $\frac{0.08}{12} \approx 0.00667$, and R is 200. We must find nt, the number of months for which you will be making deposits. Therefore we must solve for nt in the following equation:

$$1,000,000 = 200\left[\frac{(1 + 0.00667)^{nt} - 1}{0.00667}\right]$$

We begin by multiplying both sides of the equation by 0.00667:

$$6,670 = 200[(1 + 0.00667)^{nt} - 1]$$

Next, we divide both sides of the equation by 200 and then add 1 to both sides of the equation:

$$34.35 = (1.00667)^{nt}$$

Then we take the log of both sides:

$$\log 34.35 = \log (1.00667)^{nt}$$

Now we use the exponent property of the log function to simplify the equation:

$$\log 34.35 = nt \log 1.00667$$

We divide by $\log 1.00667$ and use a calculator to find nt.

$$nt = \frac{\log 34.35}{\log 1.00667} = 531.991532 \approx 532$$

This tells how many months it will take you to save $1,000,000. Dividing 532 by 12 gives us $\frac{532}{12} = 44.33$ years until your retirement. ⊚

Quiz Yourself 15

If you put $150 per month into an ordinary annuity that pays an annual interest rate of 9 percent, how long will it take for the annuity to have a value of $100,000?

Exercises 11.4

In Exercises 1–4, simplify each algebraic expression, as in Example 1.

1. $x^4 + x^3 + x^2 + x^1 + 1$
2. $x^6 + x^5 + x^4 + x^3 + x^2 + x^1 + 1$
3. $x^7 + x^6 + \cdots + x^2 + x^1 + 1$
4. $x^8 + x^7 + \cdots + x^2 + x^1 + 1$

Exercises 5–8 are based on the vacation account example at the beginning of this section. Assume that, beginning in January, you make payments at the end of each month into an account paying the specified yearly interest. Interest is compounded monthly. How much will you have available for your vacation by the specified date?

5. monthly payment, $150; yearly interest rate, 12 percent; July 1
6. monthly payment, $100; yearly interest rate, 6 percent; August 1

7. monthly payment, $200; yearly interest rate, 3 percent; May 1

8. monthly payment, $75; yearly interest rate, 8 percent; June 1

In Exercises 9–12 the monthly payment R, the annual interest rate r, and the number of monthly payments nt, are given. Use the formula for finding the future value of an ordinary annuity to find A.

9. $R = \$150$; $r = 6\%$; $nt = 8$

10. $R = \$200$; $r = 7.5\%$; $nt = 24$

11. $R = \$400$; $r = 9.5\%$; $nt = 48$

12. $R = \$800$; $r = 6.5\%$; $nt = 30$

In Exercises 13–18, find the value of each ordinary annuity at the end of the indicated time period. The amount, frequency of deposits (which is the same as the frequency of compounding), annual interest rate, and time period are given.

13. amount, $500; quarterly; 8 percent; five years

14. amount, $750; quarterly; 9 percent; three years

15. amount, $400; monthly; 9 percent; four years

16. amount, $350; monthly; 10 percent; ten years

17. amount, $600; monthly; 9.5 percent; eight years

18. amount, $500; monthly; 7.5 percent; twelve years

In Exercises 19–24, assume that the compounding is being done monthly.

19. **Saving for a motorcycle.** Matt is saving to buy a new motorcycle. If he deposits $75 at the end of each month in an account that pays an annual interest rate of 6.5 percent, how much will he have saved in 30 months?

20. **Saving for a trip.** Glenna wants to save for an African safari. She is putting $200 each month in an ordinary

Governors Camp

annuity that pays an annual interest rate of 9 percent. If she makes payments for two years, how much will she have saved for her trip?

21. **Saving for a car.** Monica deposits $150 each month in an ordinary annuity to save for a new car. If the annuity pays a monthly interest rate of 0.85 percent, how much will she be able to save in three years?

22. **Saving for retirement.** Alton is saving for his retirement in ten years by putting $500 each month into an ordinary annuity. If the annuity has an annual interest rate of 9.35 percent, how much will he have when he retires?

23. **Saving for a vacation home.** Judy has set up an ordinary annuity to save for a vacation home in Florida in fifteen years. If her monthly payments are $400 and the annuity has an annual interest rate of 6.5 percent, what will be the value of the annuity when she retires?

24. **Saving for retirement.** Thiep has set up an ordinary annuity to save for his retirement in 20 years. If his monthly payments are $350 and the annuity has an annual interest rate of 7.5 percent, what will be the value of the annuity when he retires?

In Exercises 25–28, find the monthly payment R needed to have a sinking fund accumulate the future value A. The yearly interest rate r and the number of payments nt are given. Interest is compounded monthly. Use the formula for finding the future value of an ordinary annuity. Round your answer up to the next cent.

25. $A = \$2,000$; $r = 6\%$; $nt = 12$

26. $A = \$10,000$; $r = 12\%$; $nt = 60$

27. $A = \$5,000$; $r = 7.5\%$; $nt = 24$

28. $A = \$8,000$; $r = 4.5\%$; $nt = 36$

29. **Saving for a condominium.** Antawn wants to save $14,000 in eight years by making monthly payments into an ordinary annuity for a down payment on a condominium at the shore. If the annuity pays 0.7 percent monthly interest, what will his monthly payment be?

30. **Saving to start a business.** Vanessa is making monthly payments into an annuity that pays 0.8 percent monthly interest to save enough for a down payment to start her own business. If she wishes to save $10,000 in five years, what should her monthly payments be?

31. **Saving for consumer goods.** Sarafina is making monthly payments into an annuity. She wishes to have $600 in the fund to buy a new convection range in six months, and the account pays 8.2 percent annual interest. What are her monthly payments to the account?

32. **Saving for consumer goods.** Bernie is making monthly payments into an annuity. He wishes to have $1,150 in ten months to buy exercise equipment. His account pays 9 percent annual interest. What are his monthly payments to the account?

33. Communicating Mathematics   Explain the meaning of each symbol in the formula for finding the future value of an ordinary annuity:

$$A = R \frac{\left(1 + \dfrac{r}{n}\right)^{nt} - 1}{\dfrac{r}{n}}.$$

34. Communicating Mathematics   Explain why

$$x^n + x^{n-1} + x^{n-2} + \cdots + x^2 + x^1 + 1 = \frac{x^{n+1} - 1}{x - 1}.$$

Use examples that are different than the ones we used in this section. Explain how we used this fact in this section?

35. Communicating Mathematics   If you were tutoring a friend the night before a test, how would you explain how we derived the formula for finding the future value of an ordinary annuity?

36. Communicating Mathematics   Do you see any similarities in the formula for finding the future value of an ordinary annuity and the formula we gave in a previous section for calculating compound interest?

In Exercises 37–44 solve each equation for x.

37. $3^x = 20$

38. $5^x = 15$

39. $8^x = 10$

40. $20^x = 100$

41. $\dfrac{8^x - 5}{6} = 20$

42. $\dfrac{4^x + 6}{3} = 10$

43. $\dfrac{8^x + 2}{5} = 12$

44. $\dfrac{5^x - 8}{10} = 14$

In Exercises 45–50, use the formula for finding the future value of an ordinary annuity,

$$A = R \frac{\left(1 + \dfrac{r}{n}\right)^{nt} - 1}{\dfrac{r}{n}},$$

to solve for nt. You are given A, R, and r. Assume that payments are made monthly and that the interest rate is an annual rate.

45. $A = \$10,000$; $R = 200$; $r = 9$ percent

46. $A = \$12,000$; $R = 400$; $r = 8$ percent

47. $A = \$5,000$; $R = 150$; $r = 6$ percent

48. $A = \$8,000$; $R = 400$; $r = 5$ percent

49. $A = \$6,000$; $R = 250$; $r = 7.5$ percent

50. $A = \$7,500$; $R = 100$; $r = 8.5$ percent

In Exercises 51–54, assume that monthly deposits are being placed in an ordinary annuity and interest is compounded monthly.

51. **Saving for a fire truck.** The Reliance Volunteer Fire Company wants to take advantage of a state program to save money to purchase a new fire truck. The truck will cost $400,000, and members of the finance committee estimate that with community and state contributions, they can save $5,000 per month in an account paying 10.8 percent annual interest. How long will it take to save for the truck?

52. **Saving for new equipment.** BioCon, a bioengineering company, must replace its water treatment equipment within two years. The new equipment will cost $80,000 and the company will transfer $3,800 per month into a special account that pays 9.2 percent interest. How long will it take to save for this new equipment?

53. **Saving for a condominium.** Karen wants to save $30,000 for a down payment on a condominium at a ski resort. She feels that she can save $550 per month in an account that has a 7.8 percent annual interest rate. How long will it take to acquire her down payment?

54. **Saving for a business.** Bob needs to save a $25,000 down payment to start a photo restoration business. He intends to save $300 per month in an account that pays an annual interest rate of 6 percent. How long will it take for him to save his down payment?

Further Exercises

55. Saving for retirement. Reconsider Exercise 22. Suppose that Alton had started his annuity ten years earlier. What would his monthly payments have been to accumulate the same future value in his annuity as you found in Exercise 22?

56. Saving for retirement. Reconsider Exercise 24. Suppose that Thiep had started his annuity ten years earlier. What would his monthly payments have been to accumulate the same future value in his annuity as you found in Exercise 24?

*(57.) Saving for retirement. Carlos began to save for his retirement at age 25, and for ten years he put $200 per month into an ordinary annuity at an annual interest rate of 6 percent. After the ten years, he could no longer make payments, so he placed the value of the annuity into another account that paid 6 percent annual interest compounded monthly. He left the money in this account for 30 years until he was ready to retire. How much did he have for retirement?

(58.) Saving for retirement. If Carlos had waited until age 45 to think about retirement and then decided to put money into an ordinary annuity for 20 years, what would his monthly payments have to be to accumulate the same amount for retirement as we found in Exercise 57? (We assume that the interest rate for the annuity is the same.)

(59.) Communicating Mathematics   Examine your solutions to Exercises 57 and 58. What do you notice? Explain why this is so.

(60.) Communicating Mathematics   The difference between an *annuity due* and an ordinary annuity is that with an annuity due, the payment is made at the *beginning* of the month rather than at the end of the month. This means that each payment generates one more month of interest than with an ordinary annuity.

a) How does this change the formula for finding the future value of an annuity?

b) Use this formula to find the value of the annuity in Example 2, assuming that the annuity is an annuity due.

11.5   AMORTIZATION

When people borrow to make a large purchase such as a house or car, they take out a loan that they repay in monthly payments. The process of paying off a loan (plus interest) by making a series of regular, equal payments is called **amortization**, and such a loan is called an **amortized loan**. One of the first questions most consumers ask when taking out such a loan is, "What are my monthly payments?" Although the lender will be able to answer this question, we will show you how to do the mathematics to answer this question for yourself.

Paying off a loan with regular payments is called amortization.

Assume that you have purchased a new car and after your down payment, you borrowed $10,000 from a bank to pay for the car. Also assume that you have agreed to pay off this loan by making equal monthly payments for four years. Let's look at this transaction from two points of view:

---

* Exercise numbers circled in red can be used as group exercises.

Banker's point of view: Instead of thinking about your payments, the banker might think of this transaction as a future value problem in which she is making a $10,000 loan to you now and compounding the interest monthly for four years. At the end of four years she expects to be paid the full amount due. Recall from Section 11.2 that this future value is

$$A = P\left(1 + \frac{r}{n}\right)^{nt}.$$

Your point of view: For the time being, you could also ignore the question of monthly payments and choose to pay the banker in full with one payment at the end of four years. In order to have this money available you could make monthly payments into a sinking fund to have the amount A available in four years. As you saw in Section 11.3, the formula for doing this is

$$A = R\,\frac{\left(1 + \dfrac{r}{n}\right)^{nt} - 1}{\dfrac{r}{n}}.$$

Thus to find your monthly payment, we will set the amount the banker expects to receive equal to the amount that you will save in the sinking fund and then solve for R.

> **Formula for Finding Payments on an Amortized Loan**
>
> Assume that you borrow an amount P which you will repay by taking out an amortized loan. You will make n periodic payments per year for t years and the annual interest rate is r. Then, you can find your payment by solving for R in the equation
>
> $$P\left(1 + \frac{r}{n}\right)^{nt} = R\left(\frac{\left(1 + \dfrac{r}{n}\right)^{nt} - 1}{\dfrac{r}{n}}\right).^{*}$$

Do not let this equation intimidate you. You have done the calculation on the left side many times in Section 11.2, and the computation on the right side in Section 11.4. Once you find these two numbers, you do a simple division to solve for R, as you will see in Example 1.

---

* Certainly we could do the necessary algebra to solve this equation for R. Then we could use this new formula for solving problems to find the monthly payments for amortized loans. We chose not to do this because our philosophy is to minimize the number of formulas that you have to memorize in order to do the problems in this chapter. We will round payments on a loan *up* to the next cent.

EXAMPLE 1   Determining the Payments on an Amortized Loan

Assume that you have taken out an amortized loan for $10,000 to buy a new car. The yearly interest rate is 18 percent and you have agreed to pay off the loan in four years. What is your monthly payment?

SOLUTION: We will use the preceding equation. The values of the variables in this equation are

$$P = 10,000$$
$$nt = 12 \times 4 = 48$$
$$\frac{r}{n} = \frac{18\%}{12} = 0.015.$$

We must solve for R in the equation

$$10,000(1 + 0.015)^{48} = R\left[\frac{(1 + 0.015)^{48} - 1}{0.015}\right].$$

If we calculate the numerical expressions on both sides of this equation as we did in Sections 11.2 and 11.4, we get

$$20,434.78289 = R(69.56521929)$$

Therefore your monthly payment is

$$R = \frac{20,434.78289}{69.56521929} \approx \$293.75.$$

Quiz Yourself 16

What would your payments be in Example 1 if you agree to pay off the loan in five years?

Payments that a borrower makes on an amortized loan partly pay off the principal and partly pay interest on the outstanding principal. As the principal is reduced, each successive payment pays more toward principal and less toward interest. A list showing payment-by-payment how much is going to principal and interest is called an **amortization schedule**. We illustrate such a schedule in Example 2.

EXAMPLE 2   Constructing an Amortization Schedule

In order to expand your business selling collectibles on the Internet, you need a loan of $5,000. Your banker loans you the money at a 12 percent annual interest rate, which you agree to pay back in three equal monthly installments of $1,700.12.* Construct an amortization schedule for this loan.

SOLUTION: At the end of the first month, you have borrowed $5,000 for one month at 1 percent monthly interest; using the formula for simple interest $I = Prt$, the interest you owe the bank is $5,000 \times 0.01 \times 1 = \$50$. For the payment of $1,700.12, we see that $50 pays the interest and the rest, $1,700.12 - 50 = \$1,650.12$, is applied to the principal. For the second month you are now borrowing only $5,000 - 1,650.12 = \$3,349.88$ at 1 percent monthly interest. We complete the computations for the second and third payments in Table 11.2.

_____
* We used the method of Example 1 to calculate the exact payment to be $1,700.110557. Because we increase this ever so slightly to $1,700.12, after the third payment we have overpaid by $0.03.

Payment Number	Amount of Payment	Interest Payment	Applied to Principal	Balance
				$5,000.00
1	$1,700.12	$50.00	$1,650.12	$3,349.88
2	$1,700.12	$33.50	$1,666.62	$1,683.26
3	$1,700.12	$16.83	$1,683.29	−$0.03

TABLE 11.2   An amortization schedule.

As expected, we ended with a negative balance, because the payment of $1,700.12 is a fraction of a cent larger than it needs to be. In an actual banking situation, the bank would adjust the final payment so that the final balance is exactly $0.00.

Example 3 illustrates how discouraging it can be when you make your first payment on a mortgage for a house, and realize how little of your payment goes toward paying the principal.

EXAMPLE 3   Constructing an Amortization Schedule

Assume that you have saved money for a down payment on your dream house, but you still need to borrow $120,000 from your bank to complete the deal. The bank offers you a 30-year mortgage at an annual rate of 7 percent. The monthly payment is $798.37. Construct an amortization schedule for the first three payments on this loan.

SOLUTION: We compute Table 11.3* as we did Table 11.2 in Example 2.

Payment Number	Amount of Payment	Interest Payment	Applied to Principal	Balance
				$120,000.00
1	$798.37	$700.00	$98.37	$119,901.63
2	$798.37	$699.43	$98.94	$119,802.69
3	$798.37	$698.85	$99.52	$119,703.17

TABLE 11.3   Making an amortization schedule for a lengthy mortgage.

Quiz Yourself 17

Compute the fourth line of Table 11.3.

You see that for such a lengthy amortized loan, the early payments are mostly interest. Fortunately, since the debt is being reduced, each month a little more of the payment goes toward principal and a little less toward interest.

---

* If you verify these computations by hand, your answers may differ slightly from ours due to a difference in the way we are rounding off our intermediate calculations.

Highlight: *Using a Spreadsheet to Make an Amortization Schedule**

A spreadsheet can create an amortization schedule in the blink of an eye. The following is a spreadsheet that calculates the schedule for an amortized loan for $10,000 with 60 monthly payments of $202.77. We first show the spreadsheet displaying the formulas in the cells of the spreadsheet.

	A	B	C	D	E	F
1	End of Month	Payment	Interest	Principal	Balance	
2	0	$202.77			$10,000	
3	1	$202.77	E2*0.08/12	B3 − C3	E2 − D3	
4	2	$202.77	E3*0.08/12	B4 − C4	E3 − D4	
5	3	$202.77	E4*0.08/12	B5 − C5	E4 − D5	
6	4	$202.77	E5*0.08/12	B6 − C6	E5 − D6	
7	5	$202.77	·	·	·	
8	6	$202.77	·	·	·	
9	7	$202.77	·	·	·	

Here is the same spreadsheet when the formulas in the spreadsheet are evaluated.

	A	B	C	D	E	F
1	End of Month	Payment	Interest	Principal	Balance	
2	0	$202.77			$10,000.00	
3	1	$202.77	$66.67	$136.10	$9,863.90	
4	2	$202.77	$65.76	$137.01	$9,726.89	
5	3	$202.77	$64.85	$137.92	$9,588.96	
6	4	$202.77	$63.93	$138.84	$9,450.12	
7	5	$202.77	·	·		·
8	6	$202.77	·	·		·
9	7	$202.77	·	·		·

To generate a new schedule for a mortgage, all we have to do is change the formulas in several cells and the entire spreadsheet will be recalculated.

---

* To see how to do amortization with a graphing calculator, see www.aw.com/pirnot.

We use the formula for finding the size of monthly payments to determine the present value of an annuity.

When buying a car, your budget determines what size monthly payments you can afford, and that determines what price car you can buy. Assume that you can afford car payments of $200 per month for four years and your bank will grant you a car loan at an annual rate of 12 percent. We can think of this as a future value of an annuity problem where R is 200, $\frac{r}{n}$ is 1 percent, and nt is 48 months. We know from Section 11.4 that the future value of this annuity is

$$A = 200 \left[\frac{(1 + 0.01)^{48} - 1}{0.01} \right] = \$12{,}244.52.$$

This result does not mean that now you can afford a $12,000 car!

This amount is the *future value* of your annuity, not what that amount of money would be worth in the *present*.

> **DEFINITION**
>
> If we know the monthly payment, the interest rate, and the number of payments, then the amount we can borrow is called the **present value of the annuity**.

We can find the present value of an annuity by setting the expression for the future value of an account using compound interest equal to the expression for finding the future value of an annuity and solving for the present value P.

> **Finding the Present Value of an Annuity**
>
> Assume that you are making n periodic payments per year for t years into an annuity that pays an annual interest rate of r. Also assume that each of your payments is R. Then to find the present value of your annuity, solve for P in the equation
>
> $$P\left(1 + \frac{r}{n}\right)^{nt} = R\left(\frac{\left(1 + \frac{r}{n}\right)^{nt} - 1}{\frac{r}{n}} \right).$$

Again, you have done the computations on both sides of this equation many times in Sections 11.2 and 11.4.

EXAMPLE 4   Determining What Price Car You Can Afford

If you can afford to spend $200 each month on car payments and the bank offers you a four-year car loan with an annual rate of 12 percent, what is the present value of this annuity?

SOLUTION: To solve this problem, we can use the formula for finding payments on an amortized loan:

$$P\left(1 + \frac{r}{n}\right)^{nt} = R\left(\frac{\left(1 + \frac{r}{n}\right)^{nt} - 1}{\frac{r}{n}}\right) \tag{1}$$

We know $R = 200$, $\frac{r}{n} = 1$ percent $= 0.01$, and $nt = 48$ months. If we substitute these for the variables in equation (1), we get

$$P(1 + 0.01)^{48} = 200\left[\frac{(1 + 0.01)^{48} - 1}{0.01}\right]. \tag{2}$$

Calculating the numerical expressions on both sides of (2) gives us

$$P(1.612226078) = 12{,}244.52155.$$

Now dividing both sides of this equation by 1.61222608 we find

$$P = \left[\frac{12{,}244.52155}{1.61222608}\right] \approx \$7{,}594.79.$$

You may find this answer surprising, but the mathematics of this problem are clear. If you can only afford payments of $200 per month, then you can only afford to finance a car loan for about $7,600!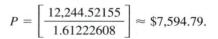

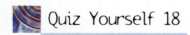
Quiz Yourself 18

Redo Example 4, but now assume that you can afford payments of $250 per month.

In order to refinance a loan, we must know the unpaid balance on the loan.

During times when interest rates are high, people are forced to borrow money at these high rates if they want to buy a car or a house on credit. If interest rates decline, then it is wise to consider paying off the remaining debt on the first loan by taking out a second loan at a lower interest rate. This procedure is called **refinancing** the loan. In order to understand refinancing, we must be able to compute how much debt remains on a loan after a certain number of payments have been made.

EXAMPLE 5   Finding the Unpaid Balance on a Loan

a) Assume that you take out a 30-year mortgage for $100,000 at an annual interest rate of 9 percent. If, after ten years, interest rates drop and you wish to refinance, how much remains to be paid on your mortgage?

b) If you can refinance your mortgage for the remaining 20 years at an annual interest rate of 7.2 percent, what will your monthly payments be?

c) How much will you save in interest in 20 years by paying the lower rate?

SOLUTION: a) Recall that in determining the monthly payment on an amortized loan, we looked at the situation from the banker's point of view and also from your point of view. The banker expects to be paid $P(1 + \frac{r}{n})^{nt}$ after 360 months. You want to accumulate

$$R\left(\frac{\left(1 + \dfrac{r}{n}\right)^{nt} - 1}{\dfrac{r}{n}}\right)$$

in an annuity in 360 months. Doing the same kind of calculations as we did in Example 1, we find that the monthly payment is $804.63.

Your monthly payment was based on the assumption that you would be paying the loan for 30 years. Therefore, after 10 years, you will not have accumulated enough in your annuity to pay off the banker. This means that the amount you owe, $P(1 + \frac{r}{n})^{nt}$, must be larger than the amount that you have accumulated in your sinking fund,

$$R\left(\frac{\left(1 + \dfrac{r}{n}\right)^{nt} - 1}{\dfrac{r}{n}}\right).$$

The unpaid balance U on the loan is therefore

$$U = P\left(1 + \frac{r}{n}\right)^{nt} - R\left(\frac{\left(1 + \dfrac{r}{n}\right)^{nt} - 1}{\dfrac{r}{n}}\right)^{*}. \tag{3}$$

It is important to recognize in equation (3) that only ten years have elapsed, so $nt = 12 \times 10 = 120$, not 360.

We can now substitute the values $P = \$100,000$, $\frac{r}{n} = 0.09/12 = 0.0075$, $nt = 120$, and $R = \$804.63$ in equation (3).

$$U = 100,000(1 + 0.0075)^{120} - 804.63\left[\frac{(1 + 0.0075)^{120} - 1}{0.0075}\right] = \$89,428.32.$$

Therefore, you still owe $89,428.32 on this mortgage.

b) Because you have a balance of $89,428.32 on your mortgage, in effect you are now taking out a new mortgage for this amount at an annual interest rate

---

* The unpaid balance appears in the balance column of an amortization table.

of 7.2 percent for the remaining 20 years. In this case, $P = \$89{,}428.32$, $nt = 12 \times 20 = 240$, and $\frac{r}{n} = .072/12 = 0.006$.

We now solve for the monthly payment R in the equation

$$89{,}428.32(1 + 0.006)^{240} = R\left[\frac{(1 + 0.006)^{240} - 1}{0.006}\right].$$

If we calculate the numerical expressions on both sides of this equation as we did in Sections 11.2 and 11.4, we get

$$375{,}829.1355 = R(533.7623389).$$

Therefore your new monthly payment is

$$R = \frac{375{,}829.1355}{533.7623389} \approx \$704.12.$$

Therefore, by refinancing, you are able to reduce your monthly mortgage payment by $\$804.63 - \$704.12 = \$100.51$ per month.

c) If you make monthly payments on the unpaid balance for 20 years at the old interest rate, your total payments will be $240 \times \$804.63 = \$193{,}111.20$. If you make monthly payments for 20 years at the new interest rate, you will pay $240 \times \$704.12 = \$168{,}988.80$. The difference

$$\$193{,}111.20 - \$168{,}988.80 = \$24{,}122.40^*$$

is the amount that you save in interest over 20 years at the reduced rate.

Often, when refinancing a mortgage you must pay a refinancing fee. The refinancing fee is often calculated as a percentage of the balance remaining on the mortgage. Suppose in Example 5 that you had to pay a two percent refinancing fee. Two percent of $\$89{,}428.32$ is $\$1{,}788.57$. You would gain this fee back in eighteen months with the reduced payments on the loan. In this case it is clear that after eighteen months, you would benefit from the refinancing.

Exercises 11.5

In Exercises 1–10 solve the equation

$$P\left(1 + \frac{r}{n}\right)^{nt} = R\left(\frac{\left(1 + \frac{r}{n}\right)^{nt} - 1}{\frac{r}{n}}\right)$$

for R *to find the monthly payment necessary to pay off the* loan. *You are given the loan amount, the annual interest rate, and the length of the loan.*

1. amount, $5,000; rate, 10 percent; time, four years

2. amount, $6,000; rate, 8 percent; time, three years

3. amount, $8,000; rate, 7.5 percent; time, four years

4. amount, $10,000; rate, 8.4 percent; time, four years

---

* Note that you can find this amount more quickly by multiplying the difference in mortgage payments, $100.51, by 240 months.

5. amount, $12,500; rate, 8.25 percent; time, four years

6. amount, $10,500; rate, 9.75 percent; time, four years

7. amount, $1,900; rate, 8 percent; time, 18 months

8. amount, $1,050; rate, 6.5 percent; time, 15 months

9. amount, $1,250; rate, 10.8 percent; time, 20 months

10. amount, $1,175; rate, 10.8 percent; time, 14 months

In Exercises 11–14, complete the first three lines of an amortization schedule for each loan. Your answer should look like Table 11.2.

11. the loan described in Exercise 1

12. the loan described in Exercise 3

13. the loan described in Exercise 5

14. the loan described in Exercise 7

15. **Paying off a mortgage.** Assume that you have taken out a 30-year mortgage for $100,000 at an annual rate of 7 percent.

   a) Construct the first three lines of an amortization schedule for this mortgage.

   b) Assume that you have decided to pay an extra $100 per month to pay off the mortgage more quickly. Find the first three lines of your payment schedule under this assumption.

   c) What is the difference in interest that you will pay on the mortgage during the fourth month if you pay the extra $100 per month versus paying only the required payment?

16. **Paying off a mortgage.** Repeat Exercise 15, but assume that the mortgage is a 20-year mortgage for $80,000 and the annual rate is 8 percent.

In Exercises 17–24, find (a) the monthly payment for each amortized loan and (b) the total interest paid on the loan. Assume that all interest rates are annual rates.

17. **Paying off a boat.** Andrew bought a new boat for $13,500. He paid $2,000 for the down payment and financed the rest for four years at an interest rate of 7.2 percent.

18. **Paying off a car.** Beatrice bought a new car for $14,800. She received $3,500 as a trade-in on her old car and took out a four-year loan at 8.4 percent to pay the rest.

19. **Paying off a consumer debt.** Danielle bought new furniture for $1,900. She took out a loan at 15 percent, which she will repay in eighteen months.

20. **Paying off a consumer debt.** Derrick bought a new refrigerator for $970. He made a down payment of $100 and the appliance store gave him a twelve-month loan at 16.4 percent to pay off the rest.

21. **Paying off a consumer debt.** Franklin's new skis cost $350. After his down payment of $75, he financed the remainder at 18 percent for five months.

22. **Paying off a consumer debt.** Sheree bought exercise equipment for $1,150. She paid $150 down and financed the rest at 15 percent for 20 months.

23. **Paying off a consumer debt.** Maryrose's new entertainment center cost $1,775. After her down payment of $200, she financed the rest at 14 percent for fifteen months.

24. **Paying off a consumer debt.** Richard's used motorcycle cost $3,500. He paid $1,100 down and financed the rest at 8.5 percent for two years.

In Exercises 25–32, find the present value of each annuity. Assume all interest rates are annual rates.

25. **Paying charitable contributions.** Ann has been asked to contribute $30 per month for the next three years toward the construction of a new town library. She has decided instead to make her contribution in one lump sum. Assume an interest rate of 6 percent, and find the amount she must contribute now.

26. **Paying charitable contributions.** Tom has agreed to contribute to his alumni association campaign to construct new classrooms. Rather than contribute $1,000 per year for the next five years, he wishes instead to contribute in one lump sum. Assume an interest rate of 8 percent, and determine what lump sum he should contribute.

27. The value of a lottery prize. Marcus has won a $1,000,000 state lottery. He can take his prize as either 20 yearly payments of $50,000 or a lump sum of $425,000. Which is the better deal? Assume an interest rate of 10 percent.

28. The value of a lottery prize. Belinda has won a $3,400,000 lottery. She can take her prize as either 20 yearly payments of $170,000 or a lump sum of $1,500,000. Which is the better deal? Assume an interest rate of 10 percent.

29. Present value of a car. If Bea can afford car payments of $350 per month for four years, what is the price of a car that she can afford now? Assume an interest rate of 10.8 percent.

30. Present value of a car. If Edmund can afford car payments of $250 per month for five years, what is the price of a car that he can afford now? Assume an interest rate of 9.6 percent.

31. Planning for retirement. Donyelle has a retirement plan with an insurance company. He can choose to be paid either $350 per month for 20 years, or he can receive a lump sum of $40,000. Which is the better deal? Assume an interest rate of 9 percent.

32. Planning for retirement. Valerie has a retirement plan with an investment company. She can choose to be paid either $400 per month for ten years, or she can receive a lump sum of $30,000. Which is the better deal? Assume an interest rate of 9 percent.

In Exercises 33–38, find the unpaid balance on each loan.

33. Paying off a loan early. You have taken an amortized loan at 8.5 percent for five years to pay off your new car, which cost $12,000. After three years, you decide to pay off the loan.

34. Paying off a loan early. In order to pay for new scuba equipment, you took an amortized loan for $1,800 at 11 percent, which you agreed to repay in three years. After eighteen payments, you decide to pay off the loan.

35. Paying off a mortgage early. The Morgans took out a 30-year mortgage for $120,000 on their vacation home at an annual interest rate of 7 percent. They decide to refinance the mortgage after eight years.

36. Paying off a mortgage early. In order to modernize their restaurant, the Alvarezes took out a 25-year mortgage for $135,000 at an annual interest rate of 6 percent. They decide to refinance the mortgage after ten years.

37. Paying off a loan early. Garrett took out an amortized loan for $8,000 for two years at 8 percent to finish culinary school. After twelve payments he decided to pay off the loan.

38. Paying off a loan early. Sheila borrowed $14,000 to invest in her floral shop. She took out an amortized loan at 6 percent for five years. After making payments for one year, she decided to pay off the loan.

In Exercises 39–44, you are given the amount of an amortized loan P, the annual interest rate r, the number of payments of the loan nt, and the monthly payment R. After the specified number of months, the borrower decides to refinance at the new interest rate for the remaining length of the loan.

a) What is the new monthly payment?

b) How much does the borrower save on interest?

39. $P = \$10,000$; $r = 8\%$; $nt = 48$; $R = \$244.13$; refinance after 24 months at an interest rate of 6.5 percent.

40. $P = \$20,000$; $r = 9\%$; $nt = 48$; $R = \$497.71$; refinance after 12 months at an interest rate of 7 percent.

41. $P = \$100,000$; $r = 8\%$; $nt = 240$; $R = \$836.45$; refinance after 60 months at an interest rate of 7 percent.

42. $P = \$100,000$; $r = 9.5\%$; $nt = 360$; $R = \$840.86$; refinance after 120 months at an interest rate of 7.5 percent.

43. $P = \$40,000$; $r = 8\%$; $nt = 48$; $R = \$976.52$; refinance after 24 months at an interest rate of 7.5 percent.

44. $P = \$50,000$; $r = 10\%$; $nt = 60$; $R = \$1,062.36$; refinance after 24 months at an interest rate of 9.5 percent.

Further Exercises

45. Communicating Mathematics What do the words *amortize* and *mortality* have in common? In light of the Three-Way Principle, how does this help you to understand the word *amortization*?

* **46. Communicating Mathematics** In deciding to refinance at a lower interest rate, how does it affect your payments if you decide to refinance early in the loan period versus later in the loan period? For example, for a 60-month loan, would the new payments be larger, smaller, or the same if you refinance after 12 months instead of 36 months? Make up numerical examples to answer to this question. Explain your answer.

47. Some mortgage agreements allow the borrower to make payments that are larger than what is required. Because this extra money goes toward reducing principal, increasing your payments may allow you to pay off the mortgage many years earlier, thus saving a great amount of interest. Assume you take out a 30-year amortized loan at 8 percent for $100,000 and your monthly payments will be $733.77. Suppose that instead of making the specified payment, you increase it by $100 to $833.77. How much do you save on interest over the life of the loan if you make the larger payment?

CHAPTER SUMMARY

SECTION 11.1

percent
The word percent is derived from the Latin "per centum" which means "per hundred."

converting decimals to percents
Examining what numbers are in the tenths and in the hundredths place of the decimal tells you how to write the decimal properly as a percent.

converting percents to decimals
Deciding which digits to put in the tenths and in the hundredths place in the decimal will allow you to convert the percent to a decimal properly.

the percent equation
Most percent problems are based on the equation *percent* $\times$ *base* = *amount*. You are usually given two of these quantities and must solve for the third.

SECTION 11.2

interest
Interest is the money that a borrower pays to use a lender's money.

principal
The amount borrowed is called the principal.

interest rate
The interest rate is specified as a percentage of the principal.

---

* Exercise numbers circled in red can be used as group exercises.

simple interest

We compute simple interest, I, using the formula $I = Prt$, where P is the principal, r is the interest rate, and t is the time.

future value

The equation $A = P(1 + rt)$ computes the future value of an account using simple interest. A is the future value, P is the principal, r is the interest rate, and t is the time.

present value

We find the present value of an account by solving the equation $A = P(1 + rt)$ for P.

future value of an account using compound interest

To find A, the future value of an account using compound interest, we use the formula $A = P(1 + \frac{r}{n})^{nt}$. P is the principal, $\frac{r}{n}$ is the interest rate per time period, and nt is the number of time periods.

common logarithmic function

The common logarithmic function reverses the operation of raising 10 to a power.

solving for nt and $\frac{r}{n}$ in the compound interest formula

To solve for nt in the compound interest formula, take the log of both sides of the equation and use the exponent property of the log function. To solve for $\frac{r}{n}$, divide both sides of the equation $A = P(1 + \frac{r}{n})^{nt}$ by P and then raise both sides of the resulting equation to the $\frac{1}{nt}$ power.

SECTION 11.3 **add-on interest method**

In using the add-on interest method for computing the amount to be repaid on a loan, we add the simple interest due on the loan to the loan. The monthly payments for the loan are given by the expression $\frac{P + I}{n}$, where P is the amount of the loan, I is the amount of interest due on the loan, and n is the number of monthly payments.

unpaid balance method for computing finance charges

This method also uses the simple interest formula $I = Prt$, where P is

previous month's balance + finance charge + purchases made − returns − payments.

The variable r is the annual interest rate, and $t = \frac{1}{12}$.

average daily balance method for computing finance charges

To compute finance charges using this method, do the following:

1. Add the outstanding balance for your account for each day of the previous month.
2. Divide the total in step 1 by the number of days in the previous month to find the average daily balance.
3. To find the finance charge, use the formula $I = Prt$, where P is the average daily balance found in step 2, r is the annual interest rate, and t is the number of days in the previous month divided by 365.

SECTION 11.4 **annuity, ordinary annuity, future value of an annuity**

When we make a series of payments into an interest bearing account, this account is called an annuity. If one payment is made at the *end* of every compounding period, the annuity is called an ordinary annuity. The sum of the deposits plus all the interest earned is called the future value of the annuity. The interest in an annuity is compounded with the same frequency as the payments.

finding the future value of an ordinary annuity
Assume that we are making a regular payment, R, at the end of each compounding period for an annuity that has an annual interest rate, r, which is being compounded n times per year. Then the value, A, of the annuity after t years is

$$A = R \frac{\left(1 + \dfrac{r}{n}\right)^{nt} - 1}{\dfrac{r}{n}}.$$

sinking fund
A sinking fund is an account into which we make regular payments for the purpose of saving some specified amount in the future. The interest in the account is compounded with the same frequency as the payments.

finding payments for a sinking fund
To find the regular payments that must be made into a sinking fund to save the amount A, solve the equation $A = R \dfrac{\left(1 + \dfrac{r}{n}\right)^{nt} - 1}{\dfrac{r}{n}}$, for R.

finding the number of payments for an annuity to accumulate a certain value
To find the number of payments required for an annuity to accumulate a value A, use the log function to solve the equation $A = R \dfrac{\left(1 + \dfrac{r}{n}\right)^{nt} - 1}{\dfrac{r}{n}}$ for nt.

SECTION 11.5 **amortization, amortized loan**
The process of paying off a loan (plus interest) by making a series of regular equal payments is called amortization, and such a loan is called an amortized loan.

finding payments on an amortized loan
Assume that you borrow an amount P which you will repay by taking out an amortized loan. You will make n periodic payments per year for t years, and the annual interest rate is r. Then, you can find your payment by solving for R in the equation

$$P\left(1 + \frac{r}{n}\right)^{nt} = R\left(\frac{\left(1 + \dfrac{r}{n}\right)^{nt} - 1}{\dfrac{r}{n}}\right).$$

amortization schedule
A list showing payment by payment how much applies to principal and interest on an amortized loan is called an amortization schedule.

present value of an annuity
If we know the monthly payment, the interest rate, and the number of payments, then the amount we can borrow is called the present value of the annuity. Assume that you are making n periodic payments per year for t years into an annuity that pays an annual interest rate of r. Also assume that each of your payments is R. Then to find the present value of your annuity, solve for P in the equation

$$P\left(1 + \frac{r}{n}\right)^{nt} = R\left(\frac{\left(1 + \frac{r}{n}\right)^{nt} - 1}{\frac{r}{n}}\right).$$

unpaid balance on a loan
The unpaid balance U on a loan after nt payments is

$$U = P\left(1 + \frac{r}{n}\right)^{nt} - R\left(\frac{\left(1 + \frac{r}{n}\right)^{nt} - 1}{\frac{r}{n}}\right).$$

P is the amount borrowed, r is the annual interest rate, t is the number of years that you have been making payments, and R is the monthly payment.

CHAPTER TEST

SECTION 11.1

1. Convert 0.1245 to a percent.

2. Convert 1.365 percent to a decimal.

3. Convert $\frac{11}{16}$ to a percent.

4. 2,890 is what percent of 3,400?

5. Great Buy Consumer Electronics buys a laptop computer for $1,380 and then marks it up 12 percent. What is the selling price of the computer?

SECTION 11.2

6. Find the future value of an account paying simple interest if $P = \$1,500$, $r = 9$ percent, and $t = 2$ years.

7. You have agreed to pay off an $8,000 car loan with 24 monthly payments of $400 each. Use the simple interest formula to determine the interest rate that you are being charged.

8. If Julia deposits $5,000 in an account that has an annual interest rate of 6 percent, compounded monthly, what is the value of the account in five years?

9. Palma wishes to establish a fund for her granddaughter's college education. What lump sum must she deposit in an account that pays a simple annual interest rate of 6 percent, compounded monthly, if she wishes to have $10,000 in ten years?

10. If you invest $1,000 in an account that pays an annual interest rate of 6.4 percent, compounded monthly, how long will it take for your money to double?

11. If $A = \$1,400$, $P = \$1,200$, and $t = 5$, solve $A = P(1 + r)^t$ for r.

SECTION 11.3

12. Bernie purchased a riding lawn mower for $1,320. The store offered him a three-year add-on loan at an annual rate of 8.25 percent. How much interest will he pay? What are his monthly payments?

13. Calculate the finance charges for the following credit card account for August (which has 31 days) using the average daily balance method. July's balance was $275 and the annual interest rate is 18 percent.

Date	Transaction
August 6	Made payment of $75
August 12	Charged $115 for clothes
August 19	Charged $20 for gasoline
August 24	Charged $16 for lunch

SECTION 11.4

14. Will is saving for his retirement by putting $175 each month into an ordinary annuity. If the annuity pays an annual interest rate of 9.35 percent, how much will he save for his retirement in ten years?

15. Find the monthly payment in order to have a sinking fund accumulate $2,000 in 36 months if the annual interest rate is 6 percent.

16. Solve $\dfrac{3^x - 4}{2} = 10$ for x.

17. You are making monthly payments of $300 into an annuity that pays 9 percent annual interest. How long will it take to accumulate $10,000?

SECTION 11.5

18. Find the monthly payment necessary to pay off a four-year amortized loan of $5,000 if the annual interest rate is 10 percent.

19. Complete the first two lines of an amortization schedule for a 20-year, $100,000 mortgage if the annual interest rate is 8 percent.

20. Jesse has won a $1,000,000 state lottery. She can take her prize as either 20 yearly payments of $50,000 or a lump sum of $500,000. Which is the better deal? Assume an interest rate of 8 percent.

21. You have taken an amortized loan at 7.5 percent for five years to pay off your new car, which cost $13,000. After two years, you decide to pay off the loan. What is your unpaid balance?

Of Further Interest: THE ANNUAL PERCENTAGE RATE

Because the mathematics of borrowing money is complicated, money lenders can easily take advantage of uninformed consumers. In 1968, Congress passed a law requiring lenders to inform consumers of the true cost of borrowing money.

To illustrate the problem, assume that you agree to repay a loan for $3,000 (plus the interest) in three yearly payments using an add-on interest rate of 10 percent. What is your true interest rate? It depends on how you look at this agreement. Using the add-on method described in Section 11.2, we compute the interest using the formula $I = Prt = (3,000)(0.10)(3) = \$900$. Thus the amount to be repaid in three equal installments is $3,000 + 900 = \$3,900$. Each payment is therefore $\frac{3,900}{3} = \$1,300$, of which $1,000 is being paid on the principal and $300 is interest. Therefore, for the first year of your loan, you have borrowed $3,000 and paid $300 in interest. Solving the equation $300 = (3,000)(r)(1)$, we see that your interest rate is actually 10 percent.

At the end of the second year you make another payment of $1,300, of which $1,000 goes to reduce the principal and $300 is interest. Thus for the second year you have paid $300 interest on a $2,000 loan. Solving the equation $300 = (2,000)(r)(1)$ for r, we find $r = 0.15$. So, in reality, the interest rate on your loan for the second year is 15 percent.

But, it gets worse! At the end of the third year, you make the final payment of $1,300. For this last year you have paid $300 interest on the $1,000 remaining on the balance of the loan. Solving the equation $300 = (1,000)(r)(1)$ for r, we find $r = 0.30$. Now the interest rate is 30 percent.

What is the true interest rate?

The "true" interest rate we are looking for is called the *annual percentage rate* or *APR*, which we will denote by a. Looking at our previous calculations, we see that

interest for the first year + interest for the second year +

interest for the third year = $900.

Using the interest formula $I = Prt$, we can rewrite this as

$$(3,000)(a)(1) + (2,000)(a)(1) + (1,000)(a)(1) = 900.$$

Collecting like terms, we get $(6,000)(a)(1) = 900$. Solving this equation gives us $a = 0.15$. Thus the annual percentage rate is 15 percent. You can verify that if you borrow $3,000 for one year at 15 percent and then $2,000 for one year at 15 percent and then $1,000 for one year at 15 percent, the total interest for the three years is $900.

SOME GOOD ADVICE

In doing these calculations, it is easy to make a minor computational error that will throw the answer off by a large amount. To detect such errors, always ask yourself, does the answer seem reasonable?

Suppose that we borrowed $6,000 at a simple add-on interest rate of 10 percent and agreed to repay it by making 60 monthly payments. In calculating the APR we should remember that we are repaying $100 per month, so we have really borrowed $6,000 for one month, $5,900 for one month, $5,800 for one month, etc. Instead of having three terms on the left side of the equation as we did above, we would have 60 terms. In order to avoid such lengthy computations, lenders use tables similar to Table 11.4 to determine the APR. Since most loans are repaid with monthly payments, for the remainder of this section on APR, in order to keep the discussion simple, we will consider only payment plans having monthly payments.

	APR						
	10%	**11%**	**12%**	**13%**	**14%**	**15%**	**16%**
Number of Payments	Finance Charge per $100						
6	$2.94	$3.23	$3.53	$3.83	$4.12	$4.42	$4.72
12	$5.50	$6.06	$6.62	$7.18	$7.74	$8.31	$8.88
24	$10.75	$11.86	$12.98	$14.10	$15.23	$16.37	$17.51
36	$16.16	$17.86	$19.57	$21.30	$23.04	$24.80	$26.57
48	$21.74	$24.06	$26.40	$28.77	$31.17	$33.59	$36.03

TABLE 11.4   Finding the annual percentage rate.*

In order to use Table 11.4, you must first know the *finance charge* on a loan, which is the total amount the borrower pays to use the money. This amount may include interest and fees. Then you must find the finance charge per $100 of the amount financed. To find this, divide the finance charge by the amount financed and multiply by 100. For example if you borrow $780 and pay a finance charge of $148.20, then the finance charge per $100 of the amount financed is

$$\frac{\text{finance charge}}{\text{amount borrowed}} \times 100 = \frac{148.20}{780} \times 100 = 0.19 \times 100 = \$19.$$

---

* We have kept this table simple in order to emphasize how it is used. A real table would have more columns for the APR, such as 14.5% and 14.25%.

> **Using Table 11.4 to Find the APR on a Loan**
>
> 1. Find the finance charge on the loan if it is not already given to you.
> 2. Determine the finance charge per $100 on the loan.
> 3. Use the line of Table 11.4 that corresponds to the number of payments to find the number closest to the amount found in step 2.
> 4. The top of the column containing the number found in step 3 is the APR.

EXAMPLE 1   Using the APR Table

Hector has agreed to pay off a $3,500 loan by making 24 monthly payments. If the total finance charge on his loan is $460, what is the APR he is being charged?

SOLUTION: The finance charge per $100 financed is

$$\frac{\text{finance charge}}{\text{amount borrowed}} \times 100 = \frac{460}{3,500} \times 100 \approx 0.13143 \times 100 \approx \$13.14.$$

Because he is making 24 monthly payments, we use the row in Table 11.4 for 24 payments. Read across this line from left to right to find the number closest to $13.14, which is $12.98. Now look at the top of this column to find the APR, which is approximately 12 percent.

We can find the APR if we know the number and size of payments on a loan.

EXAMPLE 2   Finding the APR Using Table 11.4

Jessica is considering buying a car costing $11,850. The terms of the sale require a down payment of $2,000 and the rest to be paid off by making 48 monthly payments of $250 each. What APR will she be paying on the car financing?

SOLUTION: The amount being financed is the purchase price minus the down payment, which is $11,850 - 2,000 = 9,850$. Because her payments amount to $48 \times 250 = \$12,000$, this makes the finance charge equal to $12,000 - 9,850 = \$2,150$. The finance charge per $100 financed is therefore

$$\frac{\text{finance charge}}{\text{amount borrowed}} \times 100 = \frac{2,150}{9,850} \times 100 \approx 0.2183 \times 100 = \$21.83.$$

We now look at the row for 48 payments in Table 11.4. The number in this row that is closest to $21.83 is $21.74. Looking at the top of this column, we find that the APR for this car financing is about 10 percent.

If we know the add-on interest rate and the length of the loan, we can determine the APR.

Quiz Yourself 19*

Assume that Jason will repay a loan for $11,250 by making 36 payments. Assume that the finance charge is $1,998.

a) What is the finance charge per $100 financed?

b) What is the APR?

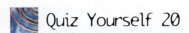

Quiz Yourself 20

In Example 2, assume that Jessica's payments are $265 instead of $250. What is the APR that she is being charged?

---

* Quiz Yourself answers begin on page 849.

EXAMPLE 3 Finding the APR on an Add-On Interest Loan

To furnish his apartment, Nehemiah borrowed $1,800 from a finance company at an add-on interest rate of 8 percent. If his loan is for three years, what is the APR on this loan?

SOLUTION: The finance charge, which is the interest on this loan, is $I = Prt = (1,800)(0.08)(3) = \$432$. The finance charge per $100 financed is therefore

$$\frac{\text{finance charge}}{\text{amount borrowed}} \times 100 = \frac{432}{1,800} \times 100 = 0.24 \times 100 = \$24.$$

Using the row for 36 payments in Table 11.4, we see that the closest number to $24 is $24.80. Therefore, the APR that is being charged to Nehemiah is a little less than 15 percent.

Knowing the APR enables consumers to determine the best deal when shopping for a loan.

EXAMPLE 4 Comparing Two APRs

Minxia must borrow $4,000 to pay for tuition for her last year in college. Her bank will give her an add-on interest loan at a rate of 7.7 percent for three years. The financial aid office can arrange for a loan requiring 24 payments of $190 each. Which loan has the better APR?

SOLUTION: We will calculate the APR for each loan.
Bank: The interest is $I = Prt = (4,000)(0.077)(3) = \$924$. The finance charge per $100 financed is therefore

$$\frac{\text{finance charge}}{\text{amount borrowed}} \times 100 = \frac{924}{4,000} \times 100 = 0.231 \times 100 = \$23.10.$$

From Table 11.4 we see that for a 36-payment loan, this corresponds to an APR of roughly 14 percent.
Financial aid office: With 24 payments of $190, Minxia will repay $24 \times 190 = \$4,560$, so her finance charge is $4,560 - 4,000 = \$560$. The finance charge per $100 financed is therefore

$$\frac{\text{finance charge}}{\text{amount borrowed}} \times 100 = \frac{560}{4,000} \times 100 = 0.14 \times 100 = \$14.$$

For a 24-payment loan, this represents an APR of slightly less than 13 percent.
The financial aid office is giving the better deal.

Since the Consumer Credit Protection Act was passed in 1968, it is not as common for lenders to offer add-on interest loans because they must reveal the APR. One way merchants can avoid having to state the APR is by offering the consumer the opportunity to rent rather than purchase the product outright. With a "rent-to-own" contract, since there is no loan, the merchant does not have to reveal the APR. If we consider these rental agreements as loans, we would often find that the APRs are outrageously high.

EXAMPLE 5 Computing the Cost of "Renting to Own"

Jeff is considering renting a TV from a nearby rent-to-own store. He can rent the TV, which he saw priced at $479 in a local store, for $19.95 a month. If he rents the TV for 36 months, then the set is his to keep. Analyze this rental agreement to determine whether Jeff is making a wise decision.

SOLUTION: There is really not much difference here between what Jeff is considering and purchasing the set at another store with an agreement to make monthly payments of $19.95. Of course, with the rent-to-own agreement, Jeff can stop renting before 36 months. If Jeff were to rent the TV until he owned it, he would make 36 payments of $19.95, so he would pay a total of $36 \times 19.95 = 718.20. His finance charge would be $718.20 - 479 = 239.20. The finance charge per $100 financed is therefore

$$\frac{\text{finance charge}}{\text{amount borrowed}} \times 100 = \frac{239.20}{479} \times 100 \approx 0.4994 \times 100 = \$49.94.$$

If we consider his rental as a 36-payment loan, we can try to use Table 11.4. Unfortunately, 49.94 is so large that we cannot find it in Table 11.4, which implies that the interest rate is quite high. Using other methods, which we explain in the Highlight, we find the APR to be about 28.5 percent. Jeff should carefully consider whether or not this rental is worth the cost.

Highlight: *Using a Computer Algebra System* to Find an APR*

If we consider Jeff's rental as a loan with payments of $19.95 per month, then we can reason as we did in Section 11.5 when we were determining the payments on a mortgage. Recall that we thought of the banker offering a loan where interest was computed monthly and we thought of the borrower as paying it off by making monthly payments into an annuity. The equation we used was

$$P\left(1 + \frac{r}{n}\right)^{nt} = R\left(\frac{\left(1 + \frac{r}{n}\right)^{nt} - 1}{\frac{r}{n}}\right) \quad (1)$$

where P was the amount borrowed, nt was the number of payments, and r was the annual interest rate. Since we knew these three quantities, we were able to solve for R.

The situation here is slightly different. We know P, nt, and R, and wish to find r. This is a very difficult equation to solve by hand. However, a computer algebra system can solve this equation with one command. The boldface is the command we typed in and what follows is the computer algebra system's response.

solve ((479*(1 + $\frac{r}{12}$)^36) =
                (19.95*((1 + $\frac{r}{12}$)^36 − 1)/($\frac{r}{12}$));
    .2852782388

Thus the APR is roughly 28.5 percent.

---

* To see a tutorial on using a graphing calculator to find an APR, see www.aw.com/pirnot.

Exercises

In Exercises 1–2, use an approach similar to the discussion preceding Example 1 to find the APR of the loan. Realize that Table 11.4 does not apply to these situations because the payments are not monthly. You are given the amount of the loan, the number and type of payments, and the add-on interest rate.

1. loan amount, $6,000; three yearly payments; rate = 8 percent

2. loan amount, $8,000; four yearly payments; rate = 12 percent

In Exercises 3–6, find the finance charge per $100 for each loan.

3. loan, $1,800; finance charge, $270

4. loan, $3,000; finance charge, $840

5. loan, $2,000; finance charge, $260

6. loan, $5,000; finance charge, $1,125

Use Table 11.4 to solve Exercises 7–18. Find the APR to the nearest whole percent. Assume that interest rates are annual rates.

7. **Finding the APR on a loan.** Michael has agreed to pay off a $3,000 loan by making 24 monthly payments. The total finance charge on his loan is $420.

8. **Finding the APR on a loan.** Joanna has agreed to pay off a $4,500 loan by making 24 monthly payments. The total finance charge on her loan is $600.

9. **Finding the APR on a loan.** Luisa pays a finance charge of $165 on a six-month $4,000 loan.

10. **Finding the APR on a loan.** Wesley pays a finance charge of $310 on a twelve-month $5,000 loan.

11. **Finding the APR on a remodeling loan.** Thiep took out a $10,000 add-on loan to remodel his house and will repay it by making 24 payments of $485.

12. **Finding the APR on a car loan.** Diana took out an $8,000 add-on loan to buy a car and will repay it by making 36 payments of $270.

13. **Finding the APR on a consumer loan.** Pete took out a $4,500 add-on loan to buy a sound system for his band and will repay it by making 48 payments of $116.50.

14. **Finding the APR on a consumer loan.** Amanda took out a $1,500 add-on loan to buy a new computer and will repay it by making 24 payments of $71.25.

15. **Finding the APR on a travel loan.** Emily took out a 24-month $2,000 add-on loan at an interest rate of 8 percent to go to China.

16. **Finding the APR on a vehicle loan.** John took out a 48-month $26,000 add-on loan at an interest rate of 7.9 percent to pay off his truck.

17. What is the APR of a 36-month add-on loan with an interest rate of 8.2 percent?

18. What is the APR of a 48-month add-on loan with an interest rate of 8.75 percent?

In Exercises 19–22, decide which has the better APR to repay a $5,000 loan. Assume that you are making monthly payments.

19. a) an add-on interest loan at 8.4 percent for three years

    b) 24 payments of $230

20. a) an add-on interest loan at 7.2 percent for two years

    b) 36 payments of $165

21. a) an add-on interest loan at 8.9 percent for four years

    b) 36 payments of $165

22. a) an add-on interest loan at 8.4 percent for one year

    b) 24 payments of $240

23. Communicating Mathematics We began this section with an example of an add-on interest loan in which you borrowed $3,000 and agreed to pay back the loan (plus

interest) in three equal payments of $1,300 each. What was the point of that example?

24. Communicating Mathematics   In what way is a rent-to-own agreement different than an add-on interest loan?

In Exercises 25–26, think of the rent-to-own agreement as though it were an add-on loan. If the consumer rents until the item is paid for, find the finance charge per $100 financed.

Although Table 11.4 does not contain enough columns to estimate the APR, guess as to what you think it might be.

25. **Evaluating a rent-to-own agreement**. Marcus rents a TV worth $375 for monthly payments of $18.75. After two years he will own the TV.

26. **Evaluating a rent-to-own agreement**. Maria rents furniture worth $1,375 for monthly payments of $49. After three years she will own the furniture.

Further Exercises

27. Communicating Mathematics   We often see advertisements stating that we can consolidate our loans and have more manageable payments. The advertiser may do this for us by extending the length of our loans. Consider an add-on loan for $1,000 at an interest rate of 10 percent for three years and then for four years. How does this affect the APR? Is it more financially sound to pay off loans in a short period of time or a longer period of time? (Here we are only thinking about the APR—of course there may be other considerations.)

28. Communicating Mathematics   Is the APR affected by the size of the loan?

29. Communicating Mathematics   Some approximate the APR by multiplying the add-on rate by 1.8. How does using this compare with using Table 11.4 to compute the APR?

CHAPTER 12

Counting:*
Just How Many Are There?

In the early 1990s an Australian-based syndicate developed a scheme to win lotteries around the world. They planned to sell shares priced at $3,000 to investors, and after raising $15 million, they would play small lotteries in which they could buy tickets to cover every possible combination of numbers. The proceeds from these small wins would then be used to buy tickets for more lucrative lotteries.

Suppose this syndicate offered you a chance to purchase a share. Would you buy it? I received just such a proposal in April 1991. My immediate question was, "How many tickets must the syndicate buy to guarantee that it had covered all possible ticket combinations?" I watched a videotape in which the syndicate's founder explained that he had a new mathematical method for covering all the possibilities. Although he would not reveal his method, he assured investors that he could improve on the well-known counting principles to guarantee success.

Because the videotape was not very convincing, I chose not to invest in the syndicate. Several months later, I was stunned to read the following headline.

Australian Syndicate Wins Virginia Lottery

Did I make a mistake? After learning the basic principles of counting in this chapter, you can evaluate my decision.

---

* For further resources on counting see www.aw.com/pirnot.

12.1 INTRODUCTION TO COUNTING METHODS

What is the largest set that you have ever counted? Was it the number of cars ahead of you while you were stuck in a traffic jam? The number of days until your birthday? The number of fans at a football game? Have you ever counted a set that had 1,000 members? What about 10,000? In our everyday life we deal mostly with relatively small sets that do not require sophisticated techniques to count them.

In situations where we need to count a complex set of objects, we can easily underestimate the number of ways something can happen. Most lottery players don't realize just how many ticket combinations there are and how slim the chances are of selecting the winning combination. A newspaper recounted the sad story of how a college student used his savings to buy lottery tickets in hopes of increasing his nest egg. Instead he lost all of his money. My young son confronted a similar counting problem, but with less dire consequences. He thought it would be easy to open a combination lock he had found. The lock had 40 numbers and required a combination of three numbers to open it. He expected that he had to try 120 combinations (3 × 40), but soon gave up, exclaiming, "There are too many ways too do this!"

To count small sets, you will probably count sequentially in the familiar fashion, 1, 2, 3, . . . , until you are done. If you have to count a larger number of items, such as the people seated in an auditorium, you may try to streamline the process by counting the number of people in a row and then multiplying by the total number of rows. In this section, you will learn several techniques to count large, complex sets of items effectively.

We can count a set by listing its elements systematically.

One of the simplest ways to count a set is to list its elements.

EXAMPLE 1   Counting Sets by Listing

How many ways can we do each of the following?

a) Flip a coin

b) Roll a single die (singular of *dice*)

c) Pick a card from a standard deck of cards (see Figure 12.1)

d) Choose a features editor from a five-person newspaper staff

SOLUTION: a) The coin can come up either heads or tails, so there are two ways to flip a coin.

b) The die has six faces numbered 1, 2, 3, 4, 5, and 6, so there are six ways the die can be rolled.

c)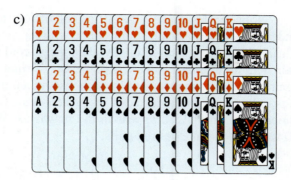

FIGURE 12.1 A standard deck of cards.

There are 52 different ways to choose a card from a standard deck.

d) There are five ways to choose one of the five staff members to be editor.

For more complex sets, it is useful to list the elements systematically as we suggested in Section 1.1 on problem solving.

EXAMPLE 2   Counting Animal Pairs

An environmental group plans to develop a fund-raising campaign featuring two endangered species. The list of candidates includes the (C)heetah, the (O)tter, the black-footed (F)erret, and the Bengal (T)iger. One animal will appear in a series of TV commercials and a different animal will appear in magazine advertisements. In how many ways can we choose the two animals for the campaign?

SOLUTION: In order not to overlook any possibilities, we will list all pairs of animals systematically. We denote the animals by the letters C, O, F, and T. We begin by assuming that C is the animal selected for the TV commercials and then consider each of the other animals for the magazine campaign. That gives us

CO, CF, CT.

If we next list O as the animal used on TV with each of the other animals appearing in the magazines, we get

OC, OF, OT.

Continuing in this fashion, the complete list is:

CO, CF, CT
OC, OF, OT
FC, FO, FT
TC, TO, TF

Thus there are twelve ways to select animals for the TV and magazine ads.

 Quiz Yourself 1*

A, B, C, D, E, F, and G are finalists in a downhill ski race. Medals will be awarded for first and second place. In how many different ways can we award these two medals? Do this either by listing all possible pairs, or by reasoning abstractly as we do in the discussion following the Solution to Example 2.

---

* Quiz Yourself answers begin on page 849.

Notice in Example 2 that if instead of four animals there were ten animals under consideration, we could count without actually writing down the pairs. Clearly if C were featured on TV, then there would be nine ways to complete the pair with an animal for the magazine ads. Similarly, if we used O on TV, there again would be nine ways to then choose an animal for the magazine campaign. It is easy to see that with each of the ten choices for the TV animal, we could choose nine animals to appear in the magazine. We would then have $10 \times 9 = 90$ ways to form the desired pairs.

Tree diagrams help visualize counting situations that take place in stages.

The Three-Way Principle in Section 1.1 recommended viewing situations graphically. It is good to keep this principle in mind when solving counting problems.

EXAMPLE 3   A Tree Diagram Shows How Three Coins Are Flipped

How many ways can three coins be flipped?

SOLUTION: To emphasize that the three coins are different, let's assume we are flipping a penny, a nickel, and a dime. A **tree diagram** is a handy way to illustrate the possibilities. First, in Figure 12.2(a) we draw a tree with two branches to show the possibility of a head or a tail for the penny. Next, in Figure 12.2(b), we see that if the penny shows either a head or a tail, the nickel also can show a head or a tail.

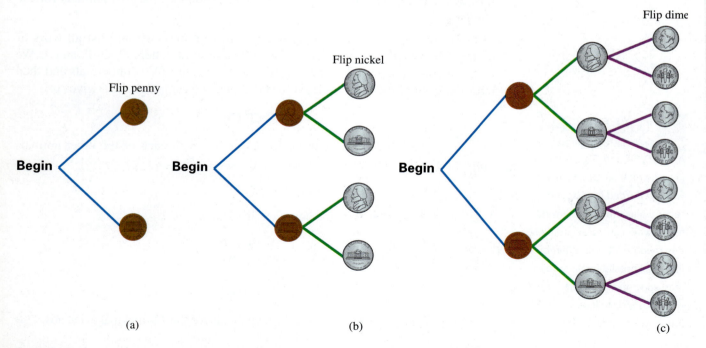

(a)   (b)   (c)

FIGURE 12.2   Tree diagrams representing all possible ways to flip (a) a penny; (b) a penny and then a nickel; (c) a penny, nickel, and a dime.

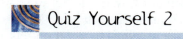

Quiz Yourself 2

How many ways are there to flip four coins?

So far there are four ways to flip the two coins. Finally, with each of these four possibilities, the dime can show a head or a tail, as we see in Figure 12.2(c).

Moving through the tree from left to right, we can trace eight branches corresponding to the ways the three coins can be flipped:

HHH, HHT, HTH, HTT, THH, THT, TTH, TTT

The situation shown in Example 4 will occur frequently in Chapter 13 when we discuss probability.

EXAMPLE 4   Rolling Two Dice

If we roll two dice, how many different pairs of numbers can appear on the upturned faces?

SOLUTION: To emphasize that the dice are different, we assume one is red and the other green. Clearly, a red 2 and a green 3 is not the same as a red 3 and a green 2. We will use ordered pairs of the form (red number, green number) to represent the pairs showing on the dice. For example, (4, 5) represents a red 4 and a green 5. Figure 12.3 illustrates this situation.

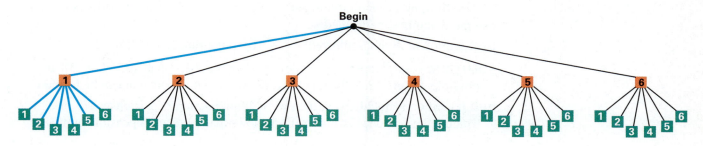

FIGURE 12.3   Tree diagram showing how many ways to roll two dice.

The leftmost set of six branches in Figure 12.3 shows pairs corresponding to 1 on the red die and either a 1, 2, 3, 4, 5, or 6 on the green. These branches correspond to the pairs (1, 1), (1, 2), (1, 3), (1, 4), (1, 5), and (1, 6). Similarly the second set of six branches corresponds to the pairs (2, 1), (2, 2), (2, 3), (2, 4), (2, 5), and (2, 6). Continuing this pattern, we get the following 36 pairs:

$$
\begin{array}{cccccc}
(1,1), & (1,2), & (1,3), & (1,4), & (1,5), & (1,6), \\
(2,1), & (2,2), & (2,3), & (2,4), & (2,5), & (2,6), \\
(3,1), & (3,2), & (3,3), & (3,4), & (3,5), & (3,6), \\
(4,1), & (4,2), & (4,3), & (4,4), & (4,5), & (4,6), \\
(5,1), & (5,2), & (5,3), & (5,4), & (5,5), & (5,6), \\
(6,1), & (6,2), & (6,3), & (6,4), & (6,5), & (6,6).
\end{array}
$$

In order to be able to solve a wide class of counting problems, including the lottery question we posed at the beginning of the chapter, we need to develop some more counting techniques—but first we will introduce some more terminology. In some counting problems, objects can be repeated; in others they cannot. To open a combination lock we can use the same number for each turn of the dial. For example, 23-23-23 could be a valid combination. In other situations, such as choosing the pairs of animals in Example 2, we cannot use the same animal twice. If objects are allowed to be used more than once in a counting problem, we will use the phrase *with repetition*. If we do not want objects to be used more than once, we will say *without repetition*.

SOME GOOD ADVICE

A good mathematician often takes a complex situation and breaks it into simpler components to solve a problem. We used this approach in Examples 2, 3, and 4. Frequently in counting problems it is useful to think of a situation not as occurring all at once but rather in distinct stages. Think of a first thing happening, then a second, then a third, and so on. If we count the possibilities at each stage, we then can combine this information to arrive at a final answer.

EXAMPLE 5   Constructing Three-Digit Numbers

We are forming three-digit numbers using the digits 5, 7, and 8. How many ways can these numbers be formed if

a) repetition of digits is allowed?   b) repetition is not allowed?

SOLUTION: a) If repetition is allowed, then numbers such as 557 or 858 are possible. Figure 12.4 illustrates the 27 possible numbers that can be formed.

b) If repetition is not allowed, then we will not list numbers such as 577 or 878. Tracing the branches in Figure 12.5 shows that there are 6 possible numbers.

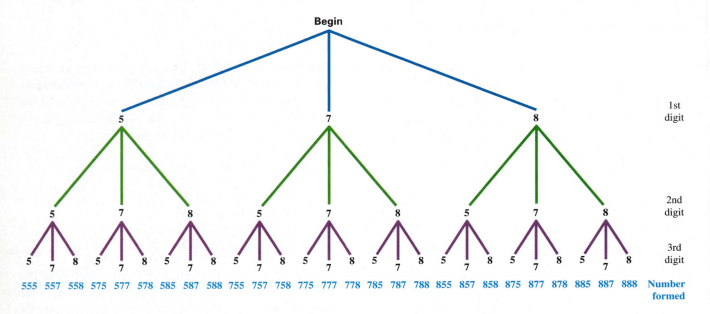

FIGURE 12.4   Tree diagram representing all different ways to form a three-digit number using 5, 7, and 8 with repetition.

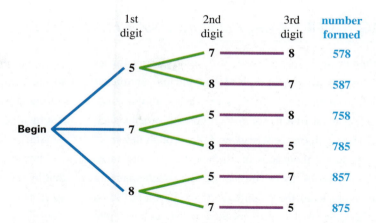

FIGURE 12.5 Tree diagram representing all different ways to form a three-digit number using 5, 7, and 8 without repetition.

We can solve counting problems by imagining a tree instead of drawing it.

Drawing large trees is tedious. After you have drawn several trees, you should be able to "see" the tree in your mind without actually drawing it.

EXAMPLE 6 Counting the Ways to Choose an Outfit

Ramona has designed a new collection for her part in the spring fashion competition. She has created five tops, four pairs of pants, and three jackets. If we consider an outfit to be a top, pants, and a jacket, how many different combinations can Ramona's models wear without repeating the exact same outfit?

SOLUTION: We can imagine a tree where the leftmost branches represent the choices of tops, the middle branches depict the choices of pants, and the rightmost branches show the choices for jackets.

We begin the tree with 5 branches, one for each top. Then to each of these branches we attach 4 branches representing the pants. So far we would have a tree with 20 branches. Finally, to each of these branches we attach 3 more representing the jackets, to give us a total of 60 different branches. Therefore Ramona's models can wear 60 different outfits.

Quiz Yourself 3

How many outfits could Ramona have in Example 6 if she had four tops, two pairs of pants, and three jackets?

Highlight: *Using Counting to Solve Practical Problems*

Although mathematics is often developed only for its own worth, it frequently ends up having significant practical applications. Such is the case with counting, which plays a role in solving complex problems such as routing millions of telephone calls through networks or scheduling thousands of airplane flights efficiently. Before solving these problems, the mathematician must understand the magnitude of the problem. Some problems grow large so rapidly as we increase the number of objects under investigation that it is not feasible to use certain methods to try to solve them. It does us no good to devise a solution that would take an unreasonable amount of time to carry out.

An example of a real problem whose solution is so complex that it is not practical to implement it is cutting hot steel as it comes out of a rolling mill. There is a method that can tell us what is the *best* way to cut the steel to minimize waste. The solution is impractical, however, because it requires so many calculations that by the time we would have found the solution, the steel would have cooled. In this case, the mathematician abandons hope to determine the *best* way to cut the steel and searches instead for another approach which provides an *acceptable*, although not perfect, solution.

Exercises 12.1

*In Exercises 1–4, you are selecting letters from the set S =
{A, B, C, D, E}.*

1. List all the ways you can select two different members from *S*. The order in which you select the members is not important. For example, AB is the same selection as BA.

2. List all the ways you can select two members from *S* with repetition. That is, selections such as AA and BB are allowed. Also, the order in which you select the members is not important. For example, AB is the same selection as BA.

3. List all the ways you can select two different members from *S*. The order in which you select the members *is important*. For example, AB is not the same selection as BA.

4. List all the ways you can select two members from *S* with repetition. The order in which you select the members *is important*. For example, AB is not the same selection as BA.

5. Communicating Mathematics   If you were studying for a test on counting, explain what the important difference is between Exercises 1 and 2. How would this difference affect your solution to each exercise?

6. Communicating Mathematics   If you were studying for a test on counting, explain what the important difference is between Exercises 1 and 3. How would this difference affect your solution to each exercise?

7. **Assigning tasks.**   Susan's friends Vicki, Luis, Maddy, and Todd have volunteered to help with the preparations for a party. How many ways can Susan assign someone to buy beverages, someone to arrange for food, and someone to send out invitations? Assume that no person does two jobs.

8. **Making staff assignments.** Suppose that the staff of a weekly newspaper, the *Southern California Sentinel*, consists of (A)drian, (B)rian, (C)armen, (D)avid, and (E)mily. The editor will choose a features editor and a sports editor from these five people. If the sports editor must be different from the features editor, in how many ways can the editor make this selection?

In Exercises 9–12, draw the tree only if you must. Try to picture the tree mentally without actually putting it on paper.

9. If you drew a tree diagram showing how many ways five coins could be flipped, how many branches would it have?

10. If you drew a tree diagram showing how many ways six coins could be flipped, how many branches would it have?

11. The game Dungeons and Dragons uses a tetrahedral die that has four congruent triangular sides numbered 1, 2, 3, and 4. How many branches would a tree diagram have that shows the way two tetrahedral dice could be rolled?

12. Dungeons and Dragons also uses twelve-sided dice. How many branches would a tree diagram have that shows the way two twelve-sided dice could be rolled?

Draw a tree diagram that illustrates the different ways to flip a penny, nickel, dime, and quarter. Use this diagram to solve Exercises 13–16.

13. In how many ways can you get exactly one head?

14. In how many ways can you get no tails?

15. In how many ways can you get exactly two tails?

16. In how many ways can you get exactly three heads?

17. How many different two-digit numbers can you form using the digits 1, 2, 5, 7, 8, and 9 without repetition? For example, 55 is not allowed.

18. How many different two-digit numbers can you form using the digits 1, 2, 5, 7, 8, and 9 with repetition? For example, 55 is allowed.

19. How many different three-digit numbers can you form using the digits 1, 2, 5, 7, 8, and 9 without repetition?

20. How many different three-digit numbers can you form using the digits 1, 2, 5, 7, 8, and 9 with repetition?

Use the given diagram to solve Exercises 21–24.

Squares such as ABCD and EFGH are called 1 × 1 squares. A square such as CXGY is called a 2 × 2 square.

21. How many 1 × 1 squares can be formed?

22. How many 4 × 4 squares can be formed?

23. How many 2 × 2 squares can be formed?

24. How many 3 × 3 squares can be formed?

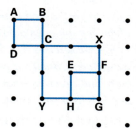

25. Answer Exercise 23 assuming the diagram has six rows with six dots in each row.

26. Answer Exercise 24 assuming the diagram has six rows with six dots in each row.

In Exercises 27–32, assume you are rolling two dice; the first one is red, and the second one is green. Use a systematic listing to determine the number of ways you can roll each of the following. For example, a total of three can be rolled in two ways: (1, 2) and (2, 1).

27. Roll a total of five.

28. Roll a total of seven.

29. Roll both numbers the same.

30. Roll a three on the red die.

31. Roll a total less than six.

32. Roll a total greater than nine.

33. Communicating Mathematics   How would you explain to a friend how to be systematic in solving Exercise 31?

34. Communicating Mathematics   How would you explain to a friend how to be systematic in solving Exercise 32?

Recall in Example 6 that Ramona has designed different tops, pants, and jackets to create outfits. How many different outfits can her models wear if she has designed the following:

35. **Counting outfits.** six tops, five pants, four jackets

36. **Counting outfits.** seven tops, six pants, three jackets

37. **Counting license plates.** An eyewitness to a crime said that the license plate of the getaway car began with the letters T, X, and L, but he could not remember the order. The rest of the plate had the numbers 8, 3, and 4, but again he did not recall the order. How many license plates must the police investigate to find this car?

38. **Counting license plates.** In a small state, the license plate for a car begins with two letters, which may be repeated and ends with three digits which also may be repeated. How many license plates are there possible in that state?

Two couples, Al and Monica and Ed and Vicki, have bought tickets for four adjacent seats to see the play 42nd Street. *Use this information to do Exercises 39–42. You may want to solve the problems by systematic listing, drawing a tree*

diagram, or by imagining a tree diagram. Also, drawing a diagram of the seats may help you.

39. In how many ways can the couples occupy their four seats?

40. In how many ways can the couples be seated if the women are to sit together?

41. In how many ways can the couples be seated if the women do not sit together?

42. In how many ways can the couples be seated if the couples are to sit together?

43. The base of a stack of oranges in a supermarket consists of five rows with five oranges in each row. The oranges in the next row will be placed as shown by the highlighted orange in the figure. How many oranges will there be in the stack?

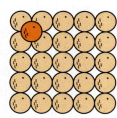

44. If the bottom row of oranges consisted of seven rows of seven oranges each, how many oranges would there be in the stack?

In Exercises 45–48, you are buying a triple-deck ice cream cone with vanilla, strawberry, and chocolate as possible flavors. The flavors can be repeated or not, and we will consider two cones to be different if the flavors are the same but occur in different order.

45. How many different cones are possible?

46. How many cones have only vanilla and strawberry?

47. How many possible cones have only two different flavored scoops?

48. How many possible cones have all different flavors?

Further Exercises

49. Building class schedules. Anthony is building his class schedule for next semester. He has decided to take four classes: mathematics, English, sociology, and music. He intends to schedule classes only on Monday, Wednesday, and Friday mornings. The possible classes that he can take are:

Mathematics: MWF—9:00, 11:00, 12:00
English: MWF—9:00, 10:00, 12:00
Sociology: MWF—10:00, 11:00, 12:00
Music: MWF—9:00, 10:00, 11:00

Use a tree to determine the possible schedules he can have and then list all those schedules.

50. Building appointment schedules. A pharmaceutical salesperson wishes to schedule appointments with three doctors to introduce a new line of antibiotics. The times that the doctors have available are as follows:

Dr. Antawn: Monday at 10:00, Wednesday at 11:00, Thursday at 11:00
Dr. Berrios: Monday at 9:00, Monday at 10:00
Dr. Calabria: Monday at 10:00, Wednesday at 11:00, Thursday at 11:00

Use a tree to determine all possible schedules that would allow the salesperson to meet with each doctor. List those schedules.

Assume that you are a contestant on a game show called Wheel of Destiny. *The final round of this game, called the Gem Round, is* *played as follows. There are four treasure chests each containing a gem. Two of the gems are red, one is blue, and one is green. The game show host will ask you to open the chests one at a time in any order that you choose. If the fourth chest you open contains "Your Final Gem," then you win the grand prize.*

You must spin the Wheel of Destiny to learn what your final gem will be. For example, when you spin the wheel, it may stop at "a green gem." This means that the round ends whenever you open a box containing the green gem.

In Exercises 51–53, you are given the final gem. List the different orders in which you can select the colors red, blue, and green in order to win the grand prize. Drawing a tree will help you see the possibilities.

51. Enumerating game show possibilities. A green gem.

52. Enumerating game show possibilities. A blue gem.

53. Enumerating game show possibilities. A red gem.

54. Communicating Mathematics   How are the coin-flipping problem in Example 3 and the dice-rolling problem in Example 4 similar?

55. Communicating Mathematics   If you were to draw tree diagrams to illustrate the problems in Exercises 51–53, how would these trees be different from those you would draw to illustrate flipping coins and rolling dice?

12.2   THE FUNDAMENTAL COUNTING PRINCIPLE

If you glance back over the situations discussed in Section 12.1, you will notice a recurring pattern. Whether we were flipping coins, rolling dice, or selecting items in a wardrobe, the problems all had the same basic pattern:

A first thing happens in *a* number of ways,
then a second thing happens in *b* number of ways,
then a third thing happens in *c* number of ways,
and so on.

For example, suppose that you wish to work out your abs, your arms, and your cardiovascular system *in that order*. Assume your gym has six machines to strengthen your abs, four to strengthen your arms, and eight to improve your cardiovascular system. If you want to count the different ways that you can do your workout, you can imagine a tree diagram. The tree would have six branches depicting the ways to choose the machine for your abs. Each of these six branches would then have four branches showing the ways you could exercise your arms. At this point the tree would have 6×4 branches. Each of these 24 branches would then have eight branches showing the way you could choose your cardiovascular workout. The final tree would have $6 \times 4 \times 8 = 192$ branches. If you wanted to add a fourth or fifth type of exercise to your workout, we would add a fourth collection of branches and then a fifth collection of branches to the tree, and so on. This thinking leads us to the following principle.

> **The Fundamental Counting Principle (FCP)**
>
> If we wish to perform a series of tasks and the first task can be done in *a* ways, the second can be done in *b* ways, the third can be done in *c* ways, and so on, then all the tasks can be done in $a \times b \times c \times \cdots$ ways.

The fundamental counting principle solves problems without listing elements or drawing tree diagrams.

We will revisit several situations from Section 12.1 to illustrate the power of the fundamental counting principle.

EXAMPLE 1   Using the FCP to Count Animal Pairs

Recall that an environmental group plans to launch an advertising campaign featuring two endangered species from among the following: the (C)heetah, the (O)tter, the black-footed (F)erret, and the Bengal (T)iger. One animal will appear

in a series of TV commercials and a different animal will be featured in magazine advertisements. In how many ways can this be done?

SOLUTION: The first task we have is determining the animal for the TV campaign, which can be done in four ways. The second task is to select a different animal for the magazine ads. There are three ways to perform this task, so the total number of ways to choose both animals is $4 \times 3 = 12$.

>
>
> **PROBLEM SOLVING**
>
> If you get confused when using counting formulas as to when to multiply and when to add, you may find that a picture often helps you recall how we derived the formula. For example, by imagining a tree diagram, you will find the fundamental counting principle easier to remember.

EXAMPLE 2   Using the FCP in Coin and Dice Problems

a) How many ways can four coins be flipped?

b) How many ways can three dice (red, green, blue) be rolled?

SOLUTION: a) The first task, flipping the first coin, can be done in two ways. The second, third, and fourth tasks also each consist of flipping a coin in one of two ways. Therefore, by the fundamental counting principle, the four coins can be flipped in $2 \times 2 \times 2 \times 2 = 16$ ways.

b) The first task is rolling the red die, which can be done in six ways. The second and third tasks also each can be done in six ways. Thus the three dice can be rolled in $6 \times 6 \times 6 = 216$ ways.

EXAMPLE 3   Counting Outfits Using the FCP

Matthew is working as a sales trainee and wishes to vary his outfit by wearing different combinations of coats, pants, shirts, and ties. If he has three sports coats, five pairs of pants, seven shirts, and four ties, how many different ways can he select an outfit consisting of a coat, pants, shirt, and tie?

SOLUTION: In selecting an outfit we consider the following tasks:

Task	Number of Ways to Perform Task
Select coat	3
Select pants	5
Select shirt	7
Select tie	4

By the fundamental counting principle, Matthew has $3 \times 5 \times 7 \times 4 = 420$ outfits.

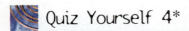

Quiz Yourself 4*

Assume that you are shopping for a new car. There are two different models to choose from, there are eight different colors available, each model has three different interior packages, and each model comes with either the plain or sport exterior trim package. How many different cars do you have to choose from?

Slot diagrams help organize information before applying the fundamental counting principle.

Sometimes special conditions affect the number of ways we can perform the various tasks. A useful technique for solving problems such as these is to draw a series of blank spaces, as shown in Figure 12.6, to keep track of the number of ways to do each task. We will call such a figure a **slot diagram**.

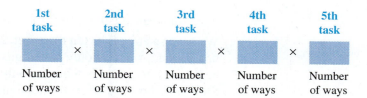

FIGURE 12.6 A slot diagram helps organize counting problems.

EXAMPLE 4 Counting Keypad Patterns

In order to open a personal locker, a person must enter five digits in order from the set 0, 1, 2, . . . , 9. How many different keypad patterns are possible if:

a) Any digits can be used in any position and repetition of digits is allowed?

b) The digit 0 cannot be used as the first digit, but otherwise any digit can be used in any position and repetition is allowed?

c) Any digits can be used in any position, but repetition is not allowed?

SOLUTION: a) We can use any of the ten digits for each of the five tasks, as shown in the slot diagram in Figure 12.7.

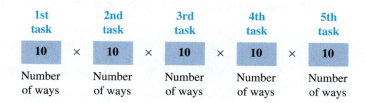

FIGURE 12.7 The slot diagram shows five tasks, each of which can be done in ten ways.

Thus we see that there are $10 \times 10 \times 10 \times 10 \times 10 = 100{,}000$ possible keypad patterns.

b) In this situation, we have only nine possible ways to select the first digit. The slot diagram is shown in Figure 12.8.

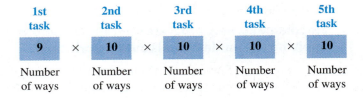

FIGURE 12.8 The first task can be done in nine ways; the remaining tasks each can be done in ten ways.

Therefore there are $9 \times 10 \times 10 \times 10 \times 10 = 90{,}000$ possible keypad patterns.

c) We can use any of the ten digits for the first number. But, since repetitions are not allowed, we have only nine digits for the second number. After two numbers have been chosen, there are only eight possibilities for the third number. Similarly, there are seven for the fourth number and six for the fifth number, as we see in Figure 12.9.

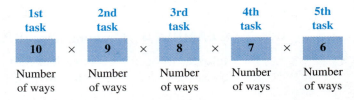

FIGURE 12.9 Since repetition of digits is not allowed, there are fewer possibilities for each keypad combination.

So we see that there are $10 \times 9 \times 8 \times 7 \times 6 = 30{,}240$ possible keypad patterns.

In solving counting problems, consider special conditions first.

We will now look at a counting situation requiring a slightly more complex analysis than we needed in the previous examples.

EXAMPLE 5 Counting Seating Patterns with Special Conditions

Professor Frantz teaches an advanced psychology class of ten students. She has a visually handicapped student, Louise, who must sit in the front row next to her tutor, who is also a member of the class. If there are six chairs in the first row of her classroom, how many different ways can Professor Frantz assign students to sit in the first row?

SOLUTION: To solve this problem, we must identify the separate tasks involved in determining the seating.

We begin by considering the special condition that Louise and her tutor must sit together.

Highlight: *Do You Want to Be a Millionaire?**

Over one-quarter of Americans think that they have a better chance of acquiring money for retirement by playing the lottery than by systematically investing their money over a long period of time. The Consumer Federation of America sponsored a survey in July 1999 that found that many Americans do not understand how remote their chances are of winning a lottery. They are also unaware that by saving and investing regularly, they can accumulate hundreds of thousands or even a million dollars. The survey found that younger and lower-income people were the least likely to understand how systematic saving can accumulate a substantial sum if done over many years. For example, if a person were to invest $25 per week for 40 years at a 7 percent interest rate, the amount of the investment would be almost $300,000. Yet in the survey, less than a third thought that the investment would be worth more than $150,000.

Among those earning $35,000 or less, 40 percent believed that lotteries rather than saving were the path to wealth. An understanding of the counting principles that we present in this chapter, together with a knowledge of how investments grow (as we discussed in Chapter 11), should convince you that mutual funds rather than lotteries are the surer road to financial security.

We must first decide which two seats (L)ouise and the (T)utor will occupy. The second task will be to decide where L and T will sit in these two seats. The remaining tasks are to decide who sits in the four other open seats.

As Figure 12.10 shows, the first task can be done in five different ways.

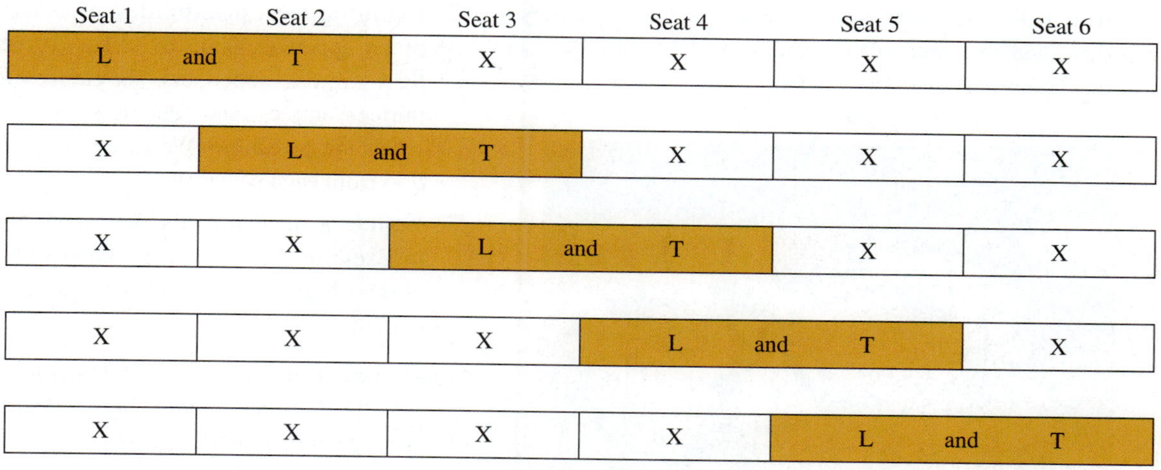

Seat 1	Seat 2	Seat 3	Seat 4	Seat 5	Seat 6
L and T		X	X	X	X
X	L and T		X	X	X
X	X	L and T		X	X
X	X	X	L and T		X
X	X	X	X	L and T	

FIGURE 12.10   There are five ways to choose two seats for Louise and her tutor.

Once we have determined which seats L and T occupy, there are two ways of deciding how L and T will sit within those two seats—L sits either on the right or on the left.

---

* This Highlight is based on an Associated Press article that appeared in the *New York Journal News,* October 29, 1999.

Quiz Yourself 5

Redo Example 5, but now assume that the class has twelve students and the first row has eight seats.

After Louise and her tutor have been seated, we fill the remaining four seats from left to right. There are eight students left; thus we have eight students for the first remaining seat, seven for the second seat, six for the third seat, and finally five for the last seat. Therefore, there are $5 \times 2 \times 8 \times 7 \times 6 \times 5 = 16,800$ ways for the students to be seated in the first row.

We will use the fundamental counting principle in Section 12.3 to develop further counting tools that will enable us to answer the lottery question we posed at the beginning of the chapter.

Exercises 12.2

1. **Assigning responsibilities.** The board of an Internet startup company has seven members. If one person is to be in charge of marketing and another in charge of research, in how many ways can these two positions be filled?

2. **Assigning positions.** If there are twelve members on the staff of the *Sandpiper*, a newspaper covering shore news, in how many ways can we choose an entertainment editor and a sales manager?

3. **Assigning officers.** The Equestrian Club has eight members. If the club wishes to select a president, vice-president, and treasurer (all of whom must be different), in how many ways can this be done?

4. **Assigning officers.** If the chamber of commerce has 20 members, in how many ways can they elect a different president, vice-president, and treasurer?

5. **Building music systems.** In how many ways can Elaine build a music system consisting of amplifier, speakers, tape deck, and CD player, if she can select from four amplifiers, eight speakers, three tape decks, and five CD players?

6. **Counting schedules.** Jorge is using his educational benefits from the navy to enroll in four courses to prepare him for a career as an architect: design, science, math, and social science. If there are seven design courses, five science courses, four math courses, and six social science courses that fit his schedule, in how many ways can he select his courses?

7. **Counting meal possibilities.** A restaurant meal consists of an appetizer, soup or salad, entrée, and dessert. If there are five appetizers, six choices for soup or salad, thirteen entrées, and four desserts, how many different meals are possible? (We assume that you make a selection from each category.)

8. **Counting meal possibilities.** What would your answer to Exercise 7 be if you are allowed to skip some categories? For example, you may choose to skip the appetizer and dessert.

In games such as Dungeons and Dragons, in addition to the common 6-sided dice, dice having 4, 8, 12, or 20 sides are also used, as shown in the figure.

In Exercises 9–12, determine the number of possibilities for each situation.

9. The 8-sided die is rolled twice.

10. The 12-sided die is rolled twice.

11. The 8-sided die is rolled and then the 12-sided die is rolled.

12. The 20-sided die is rolled and then the 6-sided die is rolled.

13. Enumerating call letters. Radio stations in the United States that are east of the Mississippi River have call letters consisting of a W followed by three letters (W can be used again.) How many different call letters of this type are possible?

14. Enumerating zip codes. Postal zip codes currently consist of five digits. For example, the zip code for Honolulu, Hawaii, is 96820. If we used letters instead of digits, how many letters would be required to make as many zip codes as five digits?

15. Counting musical patterns. In the early part of the twentieth century, the Austrian composers Arnold Schoenberg, Alban Berg, and Anton Webern created a system of music called *atonal music*, in which a note cannot be repeated in a composition until all other notes are used. In atonal music, a melody, called a *tone row*, consists of a sequence of twelve different notes. How many different tone rows are possible in this system?

16. Counting license plates. In a certain state, license plates currently consist of two letters followed by three digits. How many such license plates are possible? If the state department of transportation decides to change the plates to have three letters followed by two digits, how many plates will now be possible?

In Exercises 17–22, using the digits 0, 1, 2, . . . , 8, 9, determine how many of each type of four-digit number can be constructed.

17. Zero cannot be used for the first digit; digits may be repeated.

18. Zero cannot be used for the first digit; digits may not be repeated.

19. The number must begin and end with an odd digit; digits may not be repeated.

20. The number must begin and end with an odd digit; digits may be repeated.

21. The number must be odd and greater than 5,000; digits may be repeated.

22. The number must be between 5,001 and 8,000; digits may not be repeated.

A pin tumbler lock has a series of pins, each of which must be positioned at the proper height in order to allow the lock to open. The security of the lock is determined by the number of pins in the lock as well as the number of different possible lengths for the pins. Inserting the proper key in the lock aligns the pins properly and allows the interior of the lock to be turned.

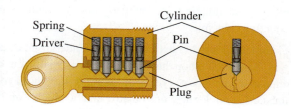

23. Counting keys. If a pin tumbler lock has five pins, each of three different lengths, how many different keys would you have to try in order to guarantee that you could open the lock?

24. Counting keys. If a pin tumbler lock has seven pins, each of four different lengths, how many different keys would you have to try in order to guarantee that you could open the lock?

25. Counting computer settings.

a) A controller for a computer disk drive has eight switches that can be set in either the "on" or "off" position. These switches must be set properly in order for the disk drive to work. In how many different ways can the group of switches be set?

b) If it takes two minutes to set the switches and test to see whether the disk drive is working properly, what is the longest possible amount of time it would take you to find the proper settings using trial and error?

26. Counting computer settings. In installing a computer modem, it is necessary to specify each of the following items: baud rate (five choices), parity (three choices), number of bits (two choices), handshaking (three choices), and type of modem (eight choices).

a) In how many different ways can you specify these five items?

b) If you can eliminate three of the baud rates but you still are unsure about the other settings, what is the longest possible amount of time it would take you to find the proper settings by using trial and error? Assume that it takes three minutes to change the settings and test the modem.

27. **Communicating Mathematics** Mamma's Pizza advertises a special in which you can choose a thin crust, thick crust, or cheese crust pizza with any combination of several different toppings. The ad says that there are almost 200 different ways that you can order the pizza. What is the smallest number of toppings available? Explain your thinking by referring to the fundamental counting principle.

28. **Communicating Mathematics** The Underground Railway Sandwich Shop has a sandwich special. You can choose whole wheat, rye, or an onion roll, with mayo or mustard, and lettuce or tomato, and a choice of one meat. What is the minimum number of meats available if there are at least 300 possible sandwich combinations? In your counting don't forget the possibility of "both" or "neither" in the mayo–mustard choice and the lettuce–tomato choice. Explain your thinking by referring to the fundamental counting principle.

29. **Communicating Mathematics** What does the fundamental counting principle have to do with the tree diagrams that we discussed earlier in this chapter?

30. **Communicating Mathematics** Why would the number 29 be an unlikely answer for a problem that used the fundamental counting principle?

Exercises 31–32 are alternative versions of the question posed in Example 5 about the seating of Louise and her tutor in Professor Frantz's class.

31. **Counting seating arrangements.** Assume that there are twelve students in the class, the front row has eight chairs, and the tutor must be seated to the right of Louise.

32. **Counting seating arrangements.** Assume that there are ten students in the class, the front row has six chairs, and Louise will sit between her tutor and her friend Jonathan. It does not matter on which side of Louise the tutor sits.

Assume that we wish to seat six people—Alex, Bonnie, Carl, Daria, Edith, and Frank—in a row of six chairs. Answer Exercises 33–36 using the fundamental counting principle. In order to apply this principle, first identify the separate tasks involved in making the seating arrangements.

33. **Counting seating arrangements.** In how many ways can we seat these people with no restrictions?

34. **Counting seating arrangements.** In how many ways can we seat these people if men and women must alternate?

35. **Counting seating arrangements.** In how many ways can we seat these people if all the men must sit together and all the women must sit together?

36. **Counting seating arrangements.** In how many ways can we seat these people if all the women must sit together but no woman sits on an end seat?

Further Exercises

In Exercises 37–40, create a counting problem whose answer is the given number.

(37.) 5 × 4

38. 7 × 8

(39.) 6 × 5 × 4

40. 9 × 8 × 7

(41.) Communicating Mathematics   Explain in your own words why the fundamental counting principle works.

42. Communicating Mathematics   In Exercises 51–53 of Section 12.1, we described the Gem Round of the game show *Wheel of Destiny*. In this round a contestant had to open a series of treasure chests containing a colored gem in a certain order to win the grand prize. There were two red gems, one blue gem, and one green gem. Explain why we cannot obtain the number of possible ways to select the gems in this round by using the fundamental counting principle.

43. Communicating Mathematics   Suppose that Yuki is selecting her schedule for next semester and has identified four English courses, three art courses, three psychology courses, and four history courses as possibilities to take. Assume that no two of the courses are offered at conflicting times.

a) Can the fundamental counting principle be used in solving this problem? Explain.

b) If any of the courses are offered at conflicting times, can you still use the fundamental counting principle? Explain.

44. Communicating Mathematics   Explain what must be present in a counting problem to make it possible to solve the problem using the fundamental counting principle.

12.3   PERMUTATIONS AND COMBINATIONS

If you look closely at the examples in Section 12.2, you will notice that certain patterns of numbers keep reoccurring. In Example 4, in which we constructed five-digit keypad patterns, we encountered the product 10 × 9 × 8 × 7 × 6. And in Example 5, after seating Louise and her tutor, we saw the pattern 8 × 7 × 6 × 5 when filling the remaining seats. In this section we will explore these patterns in more detail.

Permutations are arrangements of objects in a straight line.

In both of these problems we were selecting objects from a set and arranging them in order in a straight line. This notion occurs so often that we give it a name.

> **DEFINITION**
>
> A **permutation** is an ordering of distinct objects in a straight line. If we select r different objects from a set of n objects and arrange them in a straight line, this is called a permutation of n objects taken r at a time. The number of permutations of n objects taken r at a time is denoted by $P(n, r)$.†

---

* Exercise numbers circled in red can be used as group exercises.
† $_nP_r$ is another common notation for $P(n, r)$.

If you think carefully about the notation $P(n, r)$, it is easy to remember what it means. The letter P reminds you of the word *permutation*. The n tells you the number of objects from which you may select, and the r specifies how many you are selecting. For example, $P(5, 3)$ indicates that you are counting permutations (straight-line arrangements) formed by selecting three different objects from a set of five available objects.

EXAMPLE 1   Counting Permutations

a) How many permutations are there of the letters a, b, c, and d? Write the answer using $P(n, r)$ notation.

b) How many permutations are there of the letters a, b, c, d, e, f, and g if we take the letters three at a time? Write the answer using $P(n, r)$ notation.

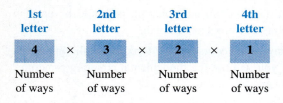

1st letter		2nd letter		3rd letter		4th letter
4	×	3	×	2	×	1
Number of ways		Number of ways		Number of ways		Number of ways

Figure 12.11   Slot diagram showing the number of ways to arrange four different objects in a straight line.

1st number		2nd number		3rd number
7	×	6	×	5
Number of ways		Number of ways		Number of ways

Figure 12.12   Slot diagram showing the number of ways to select three different objects from seven possible objects and arrange them in a straight line.

SOLUTION: a) In this problem, we are arranging the letters a, b, c, and d in a straight line without repetition. For example, *abcd* would be one permutation, and *bacd* would be another. There are four letters that we can use for the first position, three for the second, and so on. Figure 12.11 is a slot diagram for this problem. From Figure 12.11, we see that there are $4 \times 3 \times 2 \times 1 = 24$ permutations of these four objects. We can write this number more succinctly as $P(4, 4) = 24$.

b) Figure 12.12 shows a slot diagram for this question. In this diagram, we see that because we have seven letters, we can use any one of the seven for the first position, one of the remaining six for the second position, and one of the remaining five for the third position.

   We see that there are $7 \times 6 \times 5 = 210$ permutations of the seven objects taken three at a time. We write this as $P(7, 3) = 210$.

 Quiz Yourself 6*

a) Explain in your own words the meaning of $P(8, 3)$.

b) Find $P(8, 3)$.

 SOME GOOD ADVICE

It is important to remember that an equation in mathematics is a symbolic form of an English sentence. In Example 1, when we write $P(7, 3) = 210$, we read this as, "The number of permutations of seven objects taken three at a time is 210."

We use factorial notation to express $P(n, r)$.

Suppose that we want to arrange 100 objects in a straight line. The number of ways to do this is $P(100, 100) = 100 \cdot 99 \cdot 98 \cdot \cdots \cdot 3 \cdot 2 \cdot 1$. Since such a product

---

* Quiz Yourself answers begin on page 849.

is very tedious to write out, we will introduce a notation to write such products more concisely.

DEFINITION

If n is a counting number, the symbol $n!$, called n *factorial*, stands for the product $n \cdot (n - 1) \cdot (n - 2) \cdot (n - 3) \cdot \cdots \cdot 2 \cdot 1$. We define $0! = 1$.

EXAMPLE 2   Using Factorial Notation

Compute each of the following:

a) $6!$ b) $(8 - 3)!$ c) $\dfrac{9!}{5!}$ d) $\dfrac{8!}{5!3!}$

SOLUTION: a) $6! = 6 \cdot 5 \cdot 4 \cdot 3 \cdot 2 \cdot 1 = 720$.

b) Remembering the Order Principle from Section 1.1, we work in parentheses first and do the subtraction *before* computing the factorial. Therefore,

$$(8 - 3)! = 5! = 5 \cdot 4 \cdot 3 \cdot 2 \cdot 1 = 120.$$

c) $\dfrac{9!}{5!} = \dfrac{9 \cdot 8 \cdot 7 \cdot 6 \cdot \cancel{5} \cdot \cancel{4} \cdot \cancel{3} \cdot \cancel{2} \cdot \cancel{1}}{\cancel{5} \cdot \cancel{4} \cdot \cancel{3} \cdot \cancel{2} \cdot \cancel{1}} = 9 \cdot 8 \cdot 7 \cdot 6 = 3{,}024$.

Notice how we canceled common factors from the numerator and denominator to make our computations simpler.

d) $\dfrac{8!}{5!3!} = \dfrac{8 \cdot 7 \cdot \cancel{6} \cdot \cancel{5} \cdot \cancel{4} \cdot \cancel{3} \cdot \cancel{2} \cdot \cancel{1}}{\cancel{5} \cdot \cancel{4} \cdot \cancel{3} \cdot \cancel{2} \cdot \cancel{1} \cdot \cancel{3} \cdot \cancel{2} \cdot \cancel{1}} = 8 \cdot 7 = 56$.

In Example 1(b), we first wrote $P(7, 3)$ in the form $7 \times 6 \times 5$. This product started as though we were going to compute $7!$; however, the product $4 \cdot 3 \cdot 2 \cdot 1$ is missing. Another way of saying this is that $P(7, 3) = \frac{7!}{4!} = \frac{7!}{(7 - 3)!}$. We state this as a general rule.

Formula for Computing $P(n, r)$

$$P(n, r) = \frac{n!}{(n - r)!}$$

We will use this formula in the computations in Example 3.

EXAMPLE 3   Counting Ways to Fill Positions at a Community Theater

The twelve-person community theater group produces a yearly musical. Club members select one person from the group to direct the play, a second to supervise the music, and a third to handle publicity, tickets, and other administrative details. In how many ways can the group fill these positions?

SOLUTION: This can be viewed as a permutation problem if we consider that we are selecting three people from twelve and then arranging those names in a straight line—director, music, administration. The answer to this question then is

$$P(12, 3) = \frac{12!}{(12 - 3)!} = \frac{12 \cdot 11 \cdot 10 \cdot \cancel{9} \cdot \cancel{8} \cdot \cancel{7} \cdot \cancel{6} \cdot \cancel{5} \cdot \cancel{4} \cdot \cancel{3} \cdot \cancel{2} \cdot \cancel{1}}{\cancel{9} \cdot \cancel{8} \cdot \cancel{7} \cdot \cancel{6} \cdot \cancel{5} \cdot \cancel{4} \cdot \cancel{3} \cdot \cancel{2} \cdot \cancel{1}} = 1,320$$

ways to select three people from the twelve available for the three positions.   ◎

In forming combinations, order is not important.

In order to introduce a new idea, let's change the conditions in Example 3 slightly. Suppose that instead of selecting three people to carry out three different responsibilities, we choose a three-person committee that will work jointly to see that the job gets done. In Example 3, if A, B, and C were selected to be in charge of directing, music, and administration, respectively, that would be different than if B directed, C supervised the music, and A handled administrative details. However, with the new plan, it makes no difference whether we say A, B, C or B, C, A. From Table 12.1, we see that all of the six different assignments of A, B, and C under the old plan are equivalent to a single committee under the new plan.

Directing	Music	Administration
A	B	C
A	C	B
B	A	C
B	C	A
C	A	B
C	B	A

TABLE 12.1   Six different assignments of responsibilities for the musical are now handled by one committee.

What we are saying for A, B, and C applies to any three people in the theater group. Thus the answer 1,320 we found in Example 3 is too large and must be reduced by a factor of six under the new plan. Therefore the number of three-person committees we could form would be $\frac{1,320}{6} = 220$. Remember that the factor of six we divided out is really 3! (the number of ways we can arrange three people in a straight line). This means that the number of three-person committees we can select can be written in the form

$$\frac{1,320}{6} = \frac{P(12, 3)}{3!}.$$

The reason this number is smaller is that now we are concerned only with *choosing a set* of people to produce the play, but the *order* of the people chosen *is not important.*

We can generalize what we just saw. If we are choosing r objects from a set of n objects and are not interested in the order of the objects, then to count the number of choices we must divide $P(n, r)$* by $r!$. We now state this formally.

> **Formula for Computing $C(n, r)$**
>
> If we choose r objects from a set of n objects, we say that we are forming a **combination** of n objects taken r at a time. The notation $C(n, r)$ denotes the number of such combinations.† Also,
>
> $$C(n, r) = \frac{P(n, r)}{r!} = \frac{n!}{r! \cdot (n - r)!}.$$

---

* Recall that $P(n, r) = \frac{n!}{(n - r)!}$.

† $_nC_r$ is another common notation for $C(n, r)$.

Historical Highlight: *The Origins of Permutations and Combinations*

As is the case with so many ideas in mathematics, some of the fundamental concepts we have discussed in this chapter have origins that can be traced back to ancient times. The notion of permutation can be found in a Hebrew work called *The Book of Creation,* which may have been written as early as 200 A.D.

Even more surprising, it is believed that Chinese mathematicians tried to solve problems involving permutations and combinations in the fourth century B.C. Other early counting formulas can be traced back to Euclid in the third century B.C. and the Hindu mathematician Brahmagupta in the seventh century.

SOME GOOD ADVICE

In working with permutations and combinations we are choosing *r* different objects from a set of *n* objects. The big difference is whether the order of the objects is important. If the order of the objects matters, we are dealing with a permutation. If the order does not matter, then we are working with a combination.

Do not try to use the theory of permutations or combinations when that theory is not relevant. If a problem involves something other than simply choosing *different* objects (and maybe ordering them), then perhaps the fundamental counting principle may be more appropriate.

EXAMPLE 4   Using the Combination Formula to Count Committees

a) How many three-element sets can be chosen from a set of five objects?

b) How many four-person committees can be formed from a set of ten people?

SOLUTION: a) Since order is not important in considering the elements of a set, it is clear that this is a combination problem rather than a permutation problem. The number of ways to choose three elements from a set of five is

$$C(5, 3) = \frac{5!}{3! \cdot (5 - 3)!} = \frac{5 \cdot 4 \cdot \cancel{3} \cdot \cancel{2} \cdot \cancel{1}}{\cancel{3} \cdot \cancel{2} \cdot \cancel{1} \cdot 2 \cdot 1} = \frac{20}{2} = 10 \,.$$

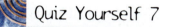

Quiz Yourself 7

How many five-person committees can be formed from a set of twelve people?

b) The number of different ways to choose four elements from a set of ten is

$$C(10, 4) = \frac{10!}{4! \cdot (10 - 4)!} = \frac{10 \cdot \overset{3}{\cancel{9}} \cdot 8 \cdot 7 \cdot \cancel{6} \cdot \cancel{5} \cdot \cancel{4} \cdot \cancel{3} \cdot \cancel{2} \cdot \cancel{1}}{\cancel{4} \cdot \cancel{3} \cdot \cancel{2} \cdot \cancel{1} \cdot \cancel{6} \cdot \cancel{5} \cdot \cancel{4} \cdot \cancel{3} \cdot \cancel{2} \cdot \cancel{1}} = 210.$$

We can now address the question about the lottery syndicate that we posed at the beginning of this chapter.

EXAMPLE 5   **How Many Tickets Are Necessary to Cover All Possibilities in a Lottery?**

Recall that a syndicate intended to raise $15 million to buy all the tickets for certain lotteries. The plan was that if the prize was larger than the amount spent on the tick-

ets, then the syndicate would be guaranteed a profit. In February 1992, the jackpot for the Virginia lottery reached $27 million. To play this lottery, the player buys a ticket for $1 containing a combination of six numbers from 1 to 44. Assuming that the syndicate has raised the $15 million, does it have enough money to buy enough tickets to be guaranteed a winner?

SOLUTION: Since we are choosing a combination of 6 numbers from the 44 possible, the number of different tickets possible is $C(44, 6) = 7,059,052$. Thus, the syndicate has more than enough money to buy enough tickets to be guaranteed a winner.

In addition to wondering whether $15 million would cover all combinations in a lottery, there were some other, practical concerns that caused me not to invest in the syndicate.

1) Suppose the syndicate did not raise all $15 million. What would be done with the money then? Would they play lotteries in which not all of the tickets can be covered? Doing this could lose all the money we invested.

2) What if there are multiple winners? In this case, the jackpot is shared and it is possible that the money won will not cover the cost of the tickets. This case would result in a loss to the syndicate.

3) Is it physically possible to actually purchase 7,059,052 tickets? Since there are $60 \times 60 \times 24 \times 7 = 604,800$ seconds in a week, it would take almost twelve weeks for one person buying one ticket every second to purchase enough tickets to cover all the combinations. In Virginia the syndicate bought only about five million tickets, thus opening up the possibility that all of the money could have been lost!

4) How much of the money raised will be used by the syndicate for its own administrative expenses?

5) Since the jackpot is to be shared with thousands of members of the syndicate and the prize money will be paid over a 20-year period, is the amount of return better than if the money had been placed in a high-interest-bearing account?

In light of these concerns, was I too cautious? What do you think? As a final note on this matter, since the Virginia lottery episode, some states have made it illegal to play a lottery by covering all ticket combinations.

As Example 6 shows, applications of counting appear in many gambling situations.*

---

* We cover counting as it applies to gambling in the "Of Further Interest" section at the end of this chapter.

Highlight: *Counting and Technology*

We rarely do the kind of calculations shown in Example 6 with pencil and paper. Not only are these computations tedious, but, more important, it is very easy to make errors when doing lengthy pencil-and-paper calculations.

Computer algebra systems have packages of routines that allow us to work painlessly with permutations and combinations as well as other complex counting formulas. For example, if we wished to use a computer algebra system to solve Example 7(a), we might type something like

numbercombinations(52,5)

and the computer would instantly respond with

2598960

One benefit of computer algebra systems is that they are not afraid of very large numbers. For example, if we were to type

numbercombinations(1000,20)

the computer would respond with

339,482,811,302,457,603,895,512,614,
793,686,020,778,700

If you have time on your hands, and a lot of paper, you can calculate $\frac{1,000!}{980!\,20!}$ by hand to see whether the computer is correct.

You can also calculate the number of permutations and combinations on a graphing calculator. In order to find $C(1000,20)$ on a TI-83, you would

1st:　Type 1000 (but do not press ENTER)

2nd:　Press the MATH key

3rd:　Move the cursor to the right to PRB and then down to nCr (see Screen 1)

4th:　Press ENTER

5th:　Type 20 and press ENTER (see Screen 2)

Screen 1

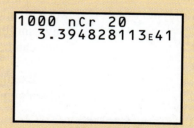

Screen 2

Notice that we get the same answer as we found with the computer algebra system; however, we cannot display as many significant digits.

EXAMPLE 6 **Counting Card Hands**

a) In the game of poker, five cards are drawn from a standard 52-card deck.* How many different poker hands are possible?

b) In the game of bridge, a hand consists of thirteen cards drawn from a standard 52-card deck. How many different bridge hands are there?

* Figure 12.1 shows that a standard 52-card deck contains 13 hearts, 13 spades, 13 clubs, and 13 diamonds. Each of these suits consists of A, K, Q, 10, 9, . . . , 3, 2.

SOLUTION: a) $C(52, 5) = \dfrac{52!}{5!\,47!} = \dfrac{\overset{13}{52} \cdot \overset{17}{51} \cdot \overset{10}{50} \cdot 49 \cdot \overset{24}{48}}{5 \cdot 4 \cdot 3 \cdot 2 \cdot 1} = 2{,}598{,}960.$

b) $C(52, 13) = \dfrac{52!}{13!\,39!} = 635{,}013{,}559{,}600.$

In some situations it is necessary to combine several counting methods to arrive at an answer. We see this technique in the next example.

EXAMPLE 7   Combining Counting Methods

A firm must select a group of two men and two women to represent them at a national Equal Opportunity in the Workplace conference. If ten men and nine women are qualified for the conference, in how many different ways can management make its decision?

SOLUTION: It is tempting to say that since we are selecting four people from a possible nineteen, the answer is $C(19, 4)$. But this is wrong. If we think about it, certain choices are unacceptable. For example, we could not choose all men, or three women and one man.

We can use the fundamental counting principle once we recognize that the group can be formed in two stages. First we will select the two women and then we will select the two men. The women can be chosen in $C(9, 2) = \frac{9!}{2!\,7!} = 36$ ways and the men can be chosen in $C(10, 2) = \frac{10!}{2!\,8!} = 45$ ways. Thus both decisions can be made in $36 \cdot 45 = 1{,}620$ ways.

Example 8 shows another situation in which you have to use several counting techniques at the same time.

EXAMPLE 8   Forming a Committee of Singers

Ye Olde Madrigal Singers has 16 vocalists and is electing a governing committee that consists of a president, a vice-president, and three executive members. In how many ways can this committee be formed?

SOLUTION: We can form the committee in two stages:

A. Choose the president and vice-president

B. Select the remaining three executive members

Thus we can apply the fundamental counting principle.

Since order is important in stage A, we recognize that we can choose the president and vice-president in $P(16, 2)$ ways.

Order is not important in choosing the remaining 3 committee members, so we can select the rest of the committee from the 14 remaining singers in $C(14, 3)$ ways.

Thus you see that we can do stage A followed by stage B in $P(16, 2) \times C(14, 3)$ ways.

The rows of Pascal's triangle count the subsets of a set.

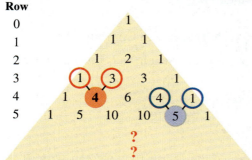

Row

FIGURE 12.13 Pascal's triangle.

It is surprising how the same mathematical idea appears again and again in different cultures over a period of centuries. For example, Pascal's triangle in Figure 12.13 occurs in the writings of fourteenth-century Chinese mathematicians, which some historians believe are based on twelfth-century manuscripts. Pascal's triangle is also found in the work of eighteenth-century Japanese mathematicians.

In Pascal's triangle, we begin numbering the rows of the triangle with zero and also begin numbering the entries of each row with zero. For example, in Figure 12.13, the number 6 is the second entry in the fourth row. Each entry in this triangle after row 1 is the sum of the two numbers immediately above it, which are a little to the right and a little to the left.

Pascal's triangle has an interesting application in set theory. Suppose that we wish to find all subsets of a particular set. Recall that the Be Systematic strategy from Section 1.1 encourages us to list possibilities systematically to solve problems. We can find all subsets of $\{1, 2, 3, 4\}$ by first listing sets of size 0, then sets of size 1, 2, 3, and 4, as in the following list:

$$\varnothing, \{1\}, \{2\}, \{3\}, \{4\}, \{1, 2\}, \{1, 3\}, \{1, 4\}, \{2, 3\}, \{2, 4\}, \{3, 4\}$$
$$\{1, 2, 3\}, \{1, 2, 4\}, \{1, 3, 4\}, \{2, 3, 4\}, \{1, 2, 3, 4\}$$

We see that there is one subset of size 0, four of size 1, six of size 2, four of size 3, and one of size 4. Note that this pattern,

$$1 \quad 4 \quad 6 \quad 4 \quad 1,$$

occurs as the fourth row in Pascal's triangle.

Similarly, the pattern

$$1 \quad 5 \quad 10 \quad 10 \quad 5 \quad 1,$$

which is the fifth row of Pascal's triangle, counts the subsets of size 0, 1, 2, and so on of a five-element set. In fact, the following result is true.

Quiz Yourself 8

a) What is the seventh row of Pascal's triangle?

b) How many subsets of size 3 are there in a seven-element set?

> **Pascal's Triangle Counts the Subsets of a Set**
>
> The nth row of Pascal's triangle counts the subsets of various sizes of an n-element set.

The entries of Pascal's triangle are numbers of the form C(n, r).

Let us look at the subsets of the set $S = \{1, 2, 3, 4\}$ again. The sets $\{1\}$, $\{2\}$, $\{3\}$, and $\{4\}$ are the four different ways we can choose a one-element set from S. The sets $\{1, 2\}$, $\{1, 3\}$, $\{1, 4\}$, $\{2, 3\}$, $\{2, 4\}$, and $\{3, 4\}$ are all the different ways that we can choose a two-element set from S. Because combinations are sets, this means that we can also use the entries in Pascal's triangle to count combinations.

> **Entries of Pascal's Triangle as *C(n, r)***
>
> The *r*th entry of the *n*th row of Pascal's triangle is *C(n, r)*.

EXAMPLE 9   Relating Entries in Pascal's Triangle to Combinations

Interpret the numbers in the fourth row of Pascal's triangle as counting combinations.

SOLUTION: The fourth row of Pascal's triangle is

$$1 \quad 4 \quad 6 \quad 4 \quad 1.$$

Because the leftmost 1 is the zeroth entry in the fourth row, we can write it as $C(4, 0)$. That is to say, $C(4, 0) = 1$. The first entry in the fourth row is 4, which means that $C(4, 1) = 4$. Similarly, $C(4, 2) = 6$, $C(4, 3) = 4$, and $C(4, 4) = 1$. ◎

Quiz Yourself 9

Write the numbers in the third row of Pascal's triangle using $C(n, r)$ notation.

For counting problems that are not too large, it is quicker to use entries in Pascal's triangle to count combinations rather than to use the formula for $C(n, r)$ that we stated earlier as you will see in Example 10.

EXAMPLE 10   Using Pascal's Triangle to Count Drug Combinations

Frequently doctors experiment with combinations of new drugs to combat hard-to-treat illnesses such as AIDS and hepatitis. Assume that a pharmaceutical company has developed five antibiotics and four immune-system stimulators. In how many ways can we choose a treatment program consisting of three antibiotics and two immune-system stimulators to treat a disease? Use Pascal's triangle to speed your computations.

SOLUTION: We can select the drugs in two stages: first select the antibiotics and then the immune-system stimulators. Therefore we can apply the fundamental counting principle, introduced in Section 12.2.

In selecting the antibiotics we are choosing three drugs from five, so this can be done in $C(5, 3)$ ways. Looking at the fifth row of Pascal's triangle, which is

$$1 \quad 5 \quad 10 \quad \mathbf{10} \quad 5 \quad 1,$$

we see that the third entry (remember, we start counting row entries with 0) is 10. Thus $C(5, 3) = 10$.

Choosing two immune-system stimulators from four can be done in $C(4, 2)$ ways. The fourth row in Pascal's triangle is

$$1 \quad 4 \quad \mathbf{6} \quad 4 \quad 1.$$

This means that $C(4, 2) = 6$. It follows that we can choose the antibiotics and then the immune-system stimulators in $C(5, 3) \times C(4, 2) = 10 \times 6 = 60$ ways.

◎

Historical Highlight: *Blaise Pascal*

Little was done to develop the theory of counting until the sixteenth century, when mathematicians began to analyze games of chance. While answering questions about throwing dice and drawing cards, a group of European mathematicians began to organize their results into a formal theory of counting. One of the most prominent figures in this development was the Frenchman Blaise Pascal, who wrote a paper in 1654 dealing with the theory of combinations.

Pascal was a child prodigy who became interested in Euclid's *Elements* at age twelve. Within four years he was doing original research and wrote a paper of such quality that some of the leading mathematicians of the time refused to believe that it had been written by a sixteen-year-old boy.

Pascal later abandoned mathematics to devote himself completely to philosophy and religion. In 1658, however, while unable to sleep because of a toothache, he decided to think about geometry to take his mind off the pain and, surprisingly, the pain stopped. Pascal took this as a sign from heaven that he should return to mathematics. For a short while he returned to his research, but soon became seriously ill with dyspepsia, a digestive disorder. Pascal spent the remaining years of his life in excruciating pain, doing little work until his death at age 39 in 1662.

Exercises 12.3

In Exercises 1–12, calculate each value.

1. $4!$

2. $3!$

3. $(8 - 5)!$

4. $(10 - 5)!$

5. $\dfrac{10!}{7!}$

6. $\dfrac{11!}{9!}$

7. $\dfrac{10!}{7!3!}$

8. $\dfrac{11!}{9!2!}$

9. $P(6, 2)$

10. $P(5, 3)$

11. $C(10, 3)$

12. $C(10, 4)$

In Exercises 13–16, find the number of permutations.

13. Eight objects taken three at a time

14. Seven objects taken five at a time

15. Ten objects taken eight at a time

16. Nine objects taken six at a time

In Exercises 17–20, find the number of combinations.

17. Eight objects taken three at a time

18. Seven objects taken five at a time

19. Ten objects taken eight at a time

20. Nine objects taken six at a time

21. Communicating Mathematics   Explain the main difference between a permutation and a combination.

22. Communicating Mathematics   Explain the meaning of each symbol in the following notation.

a) $P(10, 3)$ b) $C(6, 2)$

In Exercises 23–34, specify the number of ways to perform the task described. Give your answers using $P(n, r)$ *or* $C(n, r)$ *notation. The key in recognizing whether a problem involves permutations or combinations is deciding whether order is important.*

23. Quiz possibilities. On a biology quiz, a student must match eight terms with their definitions. Assume that the same term cannot be used twice.

24. Interview schedule. Bob wishes to schedule interviews in six available time slots with six job candidates.

25. Choosing tasks. Mulder will read the files on three unsolved cases from a list of seventeen. He does not care about the order in which he reads the files.

26. Selecting roommates. Annette has rented a summer house for next semester. She wishes to select four roommates from a group of six friends.

27. Baseball lineups. The manager of a baseball team has already chosen the first, second, fourth, and ninth batters in the batting lineup. He has selected five others to play and wishes to write their names in the batting lineup.

28. Race results. There are seven boats that will finish the America's Cup yacht race.

29. Journalism awards. Ten magazines are competing for three identical awards for excellence in journalism. No magazine can receive more than one award.

30. Journalism awards. Redo Exercise 29, but now assume the awards are all different. No magazine can receive more than one award.

31. Lock patterns. In order to unlock a door, five different buttons must be pressed down on the panel shown. When the door opens, the depressed buttons pop back up. The order in which the buttons are pressed is not important.

32. Lock patterns. A bicycle lock has three rings with the letters *A* through *K* on each ring. To unlock the lock, a letter must be selected on each ring. Duplicate letters are not allowed, and the order in which the letters are selected on the rings does not matter.

33. Choosing articles for a literary magazine. The editorial staff of *Inca*, a literary magazine, has received seventeen articles. The staff wishes to choose eight articles for the next issue of the magazine.

34. Choosing articles for a literary magazine. Redo Exercise 33, except that in addition to choosing the stories, the editorial staff must also decide in what order the stories will appear.

35. Six players are to be selected from a 25-player major league baseball team to visit schools to support a summer reading program. In how many ways can this selection be made?

36. If in Exercise 35 we not only want to select the players, but also want to assign each player to visit one of six schools, how many ways can that assignment be done?

37. In Exercise 34, if it takes one minute to write a list of eight articles selected for the magazine, how many years would it take to write all possible lists of eight articles? Assume 60 minutes in an hour, 24 hours in a day, and 365 days in a year.

38. In Exercise 36, if it takes one minute to write a list of six players with their assignments, how many years would it take to write all possible lists of assignments?

A typical bingo card is shown in the figure. The numbers 1–15 are found under the letter B, *16–30 under the letter* I, *31–45 under the letter* N, *46–60 under the letter* G, *and 61–75 under the letter* O. *The center space on the card is labeled "FREE."*

39. Bingo cards. How many different columns are possible under the letter *B*?

40. Bingo cards. How many different columns are possible under the letter *N*?

41. Bingo cards. How many different bingo cards are possible? *Hint:* Use the fundamental counting principle.

42. Bingo cards. Why is the answer to Exercise 37 not *P*(75, 24)?

B	I	N	G	O
5	28	33	51	75
7	17	41	59	63
2	22	FREE	48	61
11	29	44	46	72
9	30	38	52	68

In Exercises 43–52, use the fundamental counting principle.

43. Selecting astronauts. NASA wishes to appoint two men and three women to become residents of an orbiting space station. The finalists for these positions consist of six men and eight women. In how many ways can NASA make this selection?

44. Computer passwords. A password for a computer consists of three different letters of the alphabet followed by four different digits from 0 to 9. How many different passwords are possible?

45. Forming a public safety committee. Nicetown is forming a committee to investigate ways to improve public safety in the town. The committee will consist of three representatives from the seven-member town council, two members of a five-person citizens advisory board, and three of the eleven police officers on the force. How many ways can that committee be formed?

46. Choosing a team for a seminar. HazMat, Inc. is sending a group of eight people to attend a seminar on toxic waste disposal. They will send two of eight engineers, three of nine crew supervisors, and the remainder will be chosen from among the five senior managers. In how many ways can these choices be made?

47. Designing a fitness program. In order to lose weight and shape up, Julio is considering doing two types of exercises to improve his cardiovascular fitness from among running, bicycling, swimming, stair stepping, and Tae Bo. He is also going to take two from among the nutritional supplements AllFit, Energize, ProTime, and DynaBlend. In how many ways can he choose his cardio-vascular exercises and nutritional supplements?

48. Counting assignment possibilities. Gina must write evaluation reports on three hospitals and two health clinics as part of her degree program in community health services. If there are six hospitals and five clinics in her vicinity, in how many ways can she complete her assignment?

49. Picking a team for a competition. The students in the twelve-member advanced communications design class at Center City Community College are submitting a project to a national competition. They must select a four-member team to attend the competition. The team must have a team leader and a main presenter; the other two members have no particularly defined roles. In how many different ways can this team be formed?

50. Communicating Mathematics Why is Exercise 49 neither a strictly permutations nor a strictly combinations problem?

51. Choosing an evaluation team. The academic computing committee at Sweet Valley College is in the process of evaluating different computer systems. The committee consists of five administrators, seven faculty, and four students. A five-person subcommittee is to be formed. The chair and vice chair of the committee must be administrators; the remainder of the committee will consist of faculty and students. In how many ways can this subcommittee be formed?

52. Communicating Mathematics Why is Exercise 51 neither a strictly permutations nor a strictly combinations problem?

An armored car must pick up receipts at a shopping mall and at a jewelry store and deliver them to a bank. For security purposes, the driver will vary the route that the car takes each day. Use the map to answer Exercises 53 and 54.

53. Armored car routes. How many different direct routes can the driver take from the shopping mall to the bank?

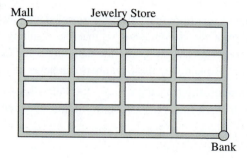

54. Armored car routes. How many different direct routes can the driver take from the jewelry store to the bank?

55. Playing cards. How many five-card hands chosen from a standard deck contain two diamonds and three hearts?

56. Playing cards. How many five-card hands chosen from a standard deck contain two kings and three aces?

57. Find the eighth row in Pascal's triangle.

58. Find the tenth row in Pascal's triangle.

Use the seventh row of Pascal's triangle to answer Exercises 59–60.

59. How many two-element subsets can we choose from a seven-element set?

60. How many four-element subsets can we choose from a seven-element set?

In Exercises 61–64, describe where each number would be found in Pascal's triangle.

61. $C(18, 2)$ **62.** $C(19, 5)$

63. $C(20, 6)$ **64.** $C(21, 0)$

Further Exercises

65. Communicating Mathematics Consider the question: How many four-digit numbers can be formed that are odd and greater than 5,000? Why is this not a permutation problem?

66. Communicating Mathematics Consider the question: How many four-digit numbers can be formed using the digits 1, 3, 5, 7, and 9? Repetition of digits is not allowed. Why is this not a combination problem?

Communicating Mathematics *In Exercises 67–70, recall that combinations are subsets of a given set and* $C(n, r)$ *is the number of subsets of size* r *in an* n-*element set.*

a) *Explain the meaning of the numbers given as numbers of subsets.*

b) *Use your intuition to explain the number of combinations in each case.*

c) *Verify your guess in part (b) by using the formula for computing the number of combinations.*

67. $C(5, 0)$ **68.** $C(5, 1)$

69. $C(5, 4)$ **70.** $C(5, 5)$

71. Communicating Mathematics Whenever we used the formulas for computing $P(n, r)$ or $C(n, r)$, we saw that many factors from the numerator canceled with all the factors from the denominator so that the resulting number was never a fraction. Using the definitions of $P(n, r)$ and $C(n, r)$, explain why these numbers can never be fractions.

72. Communicating Mathematics Explain why $C(n, r) = C(n, n - r)$. In order to understand this statement, recall

the Three-Way Principle from Section 1.1, which states that making numerical examples is useful in gaining understanding of a situation.

73. Communicating Mathematics We numbered the rows of Pascal's triangle beginning with 0 instead of 1. Explain why we did this.

74. Communicating Mathematics We numbered the entries of each row in Pascal's triangle beginning with 0 instead of 1. Explain why we did this.

The following diagram shows a portion of Pascal's triangle found in a book called The Precious Mirror, *written in China in 1303. Use this diagram to answer Exercises 75–78.*

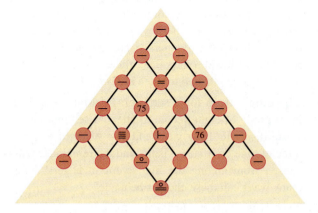

*75. What symbol do you expect to find in the circle labeled 75?

---

* Exercise numbers circled in red can be used as group exercises.

76. What symbol do you expect to find in the circle labeled 76?

77. What number does the symbol ⊢ represent?

78. What number does the symbol ≜ represent?

Consider rows n *and* n − 1 *of Pascal's triangle, shown in the following diagram. Use this diagram to answer Exercises 79–83.*

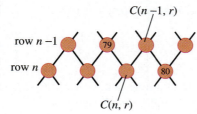

$C(n-1, r)$

row n −1 79

row n 80

$C(n, r)$

79. What expression will be in the circle labeled 79?

80. What expression will be in the circle labeled 80?

81. Complete the following equation: ☐ + C(n − 1, r) = ☐

82. Complete the following equation: $C(n, r)$ + ☐ = ☐

83. Communicating Mathematics   Notice that the arrangement of numbers in each row of Pascal's triangle is symmetric. See how the fours are balanced in the fourth row, the fives are balanced in the fifth row, and so on. Can you give a set-theory explanation to account for this?

CHAPTER SUMMARY

SECTION 12.1

systematic listing
A small set can be counted by listing its elements systematically.

tree diagram
Tree diagrams help visualize counting situations that happen in stages. It is often enough to imagine the tree without actually drawing it to solve a counting problem.

SECTION 12.2

fundamental counting principle
If we wish to perform a series of tasks and the first task can be done in *a* ways, the second can be done in *b* ways, the third can be done in *c* ways, and so on, then all the tasks can be done in $a \times b \times c \times \cdots$ ways.

slot diagram
Slot diagrams help organize information before applying the fundamental counting principle.

counting problems with special conditions
Any special conditions in a counting problem should be considered first.

SECTION 12.3

permutation
A permutation is an ordering of distinct objects in a straight line. The number of permutations of *n* objects taken *r* at a time is denoted by $P(n, r)$.

formula for computing $P(n, r)$

$$P(n, r) = \frac{n!}{(n - r)!}$$

combination
If we choose *r* objects from a set of *n* objects, we are forming a combination of *n* objects taken *r* at a time. The symbol $C(n, r)$ denotes the number of such combinations.

formula for computing $C(n, r)$

$$C(n, r) = \frac{P(n, r)}{r!} = \frac{n!}{r!\,(n - r)!}$$

Pascal's triangle counts subsets of a set
The nth line of Pascal's triangle counts the subsets of various sizes of a set having n elements.

Pascal's triangle is related to combinations
The rth entry of the nth row of Pascal's triangle is $C(n, r)$.

CHAPTER TEST

SECTION 12.1

1. List all the ways you can select two different members from $S = \{P, Q, R, S\}$. The order in which you select the members is important. For example, PQ is not the same selection as QP.

2. The game Dungeons and Dragons uses a tetrahedral die that has four congruent triangular sides numbered 1, 2, 3, and 4. If you were to draw a tree to show the number of ways that three tetrahedral dice could be rolled, how many branches would it have?

3. An outfit consists of a shirt, pants, tie, and jacket. If you have five shirts, four pairs of pants, six ties, and two jackets, how many different outfits can you make?

SECTION 12.2

4. The rock-climbing club has fourteen members. If the club wishes to select a president, vice-president, and treasurer (all of whom must be different), in how many ways can this be done?

5. A restaurant meal consists of an appetizer, entrée, and dessert. If there are four appetizers, twelve entrées, and six desserts, how many different meals are possible? (Assume that you have to make a selection from each category.)

6. In installing a computer modem, it is necessary to specify each of the following: baud rate (five choices), parity (three choices), number of bits (two choices), and type of modem (three choices). How many different specifications are there?

7. Assume that we wish to seat Alex, Bonnie, Carl, Daria, Edith, and Frank in a row of six chairs. In how many ways can the seating be done if Alex and Bonnie must sit together?

SECTION 12.3

8. Explain the main difference between a permutation and a combination.

9. On a psychological test, a patient must match eight terms with their pictures. We are assuming that the same term cannot be used twice. In how many ways can this be done?

10. The research director of NASA must choose three experiments for the next space shuttle. If seventeen experiments are proposed, in how many ways can this decision be made?

11. A password for a computer consists of three different letters of the alphabet followed by two different digits from 0 to 9. How many passwords are possible?

12. How is the formula for $P(n, r)$ related to the formula for $C(n, r)$?

13. Use the seventh row of Pascal's triangle to find the number of two-element subsets that can be chosen from a seven-element set.

14. Describe where $C(18, 2)$ would be found in Pascal's triangle.

Of Further Interest: COUNTING AND GAMBLING

A neon sign in a casino proclaims:

$100,000 jackpot !!

The jackpot was formed by putting one nickel of every dollar gambled on slot machines into the jackpot. Periodically, a lucky player wins the jackpot. At that time the jackpot is emptied and must start building all over again. If we look at this jackpot another way, the neon sign should perhaps proclaim:

$2,000,000 gambled on slot machines since the last jackpot !!

Operators of gambling establishments earn huge fortunes using the simple mathematical principle that the number of ways they can win is greater than the number of ways that you can win. In this section, we will use the theory of counting to explain why people win (and lose) at games such as slot machines and poker.

Early mechanical slot machines had three wheels with 20 positions on each wheel containing symbols such as oranges, bells, lemons, cherries, and bars. A gambler put a coin in the machine and pulled a lever, causing the wheels to spin. When the wheels stopped, the machine displayed a symbol from each wheel in a window. For certain arrangements of symbols, the player received a payoff. Arrangements of symbols that occur infrequently pay larger payoffs than those that occur more frequently. Figure 12.14 shows a typical arrangement of the symbols that might appear on each of the three wheels.

EXAMPLE 1   Using the FCP to Analyze a Slot Machine

a) In how many different ways can the three wheels of the slot machine described in Figure 12.14 stop?

b) One payoff is for cherries on wheel 1 and no cherries on the other wheels. In how many different ways can this happen?

SOLUTION: a) It is useful to think of the first wheel stopping, then the second, and finally the third. The solution becomes apparent if we draw a slot diagram. Using the fundamental counting principle, we see that the total number of different ways the three wheels can stop is $20 \times 20 \times 20 = 8,000$.

b) As before, think of the wheels stopping in stages. Again, a slot diagram helps with the counting. By the fundamental counting principle, we see that the total number of ways we can get cherries only on the first wheel is $2 \times 15 \times 12 = 360$.

1st wheel		2nd wheel		3rd wheel
20	×	**20**	×	**20**
Number of ways first wheel can stop		Number of ways second wheel can stop		Number of ways third wheel can stop

1st wheel		2nd wheel		3rd wheel
2	×	**15**	×	**12**
Number of ways first wheel can stop with cherries		Number of ways second wheel can stop with no cherries		Number of ways third wheel can stop with no cherries

Wheel 1   Wheel 2   Wheel 3

FIGURE 12.14   Three wheels of a typical mechanical slot machine.

EXAMPLE 2 **Counting the Number of Ways to Get Three Bars**

For the slot machine in Figure 12.14, the largest payoff is for three bars. In how many ways can three bars be obtained?

SOLUTION: There are two ways a bar can come up on the first wheel, three ways a bar can come up on the second wheel, and only one way a bar can come up on the third wheel. By the fundamental counting principle, a bar on the first, second, and third wheel can occur in $2 \times 3 \times 1 = 6$ ways.

From Examples 1 and 2 we would expect that the payoff for three bars is much greater than the payoff for cherries on the first wheel only.

Counting principles also explain why one poker hand beats another. In poker, a hand that is less likely to occur beats a hand that occurs more commonly. We first determine the total number of possible poker hands.

EXAMPLE 3 **Counting the Number of Poker Hands**

How many ways can a five-card poker hand be drawn from a standard 52-card deck?

SOLUTION: In drawing a five-card poker hand, we are choosing 5 cards from 52, which can be done in

$$C(52, 5) = \frac{52!}{5!47!} = \frac{52 \cdot 51 \cdot 50 \cdot 49 \cdot 48}{5 \cdot 4 \cdot 3 \cdot 2 \cdot 1} = 2{,}598{,}960 \text{ ways.}$$

Royal flush	10, J, Q, K, and A, all of the same suit
Straight flush	Five cards in sequence,* all of the same suit
Four of a kind	Four cards of one rank, plus another card
Full house	Three cards of one rank and two of another
Flush	All five cards are from the same suit
Straight	Any five cards in sequence
Three of a kind	Three cards of one rank and two other cards that are different from each other
Two pair	Two different pairs of cards and a fifth card that is different from the first four
Pair	Two cards of the same rank and three other cards, all of which are different from each other
Nothing	None of the above hands

TABLE 12.2 Poker hands in order of strength.

As we will show in Examples 4 and 6, the reason four of a kind beats a full house is that there are fewer ways to obtain four of a kind than there are to obtain a full house. In these examples, we will make heavy use of the fundamental counting principle.

---

* In choosing a sequence of five cards, we allow the ace to be considered higher than the king and also as a one. For example, the sequence A, 2, 3, 4, 5 is allowable as well as the sequence 10, J, Q, K, A.

EXAMPLE 4   Drawing Four of a Kind

Determine the number of different ways we can obtain four of a kind when drawing five cards from a standard 52-card deck.

SOLUTION: We can construct a four-of-a-kind hand in two stages.

Stage 1: Pick the rank of the card of which we will have four of a kind. For example, the four-of-a-kind cards could be aces, kings, threes, and so on. This decision can be made in thirteen ways.

Stage 2: Now that we have chosen four cards, we select the fifth card. For example, if we have selected four kings, then we could pick a jack for the fifth card. We can choose any one of the remaining 48 cards for the fifth card.

Therefore, by the fundamental counting principle, we can build a four-of-a-kind hand in $13 \times 48 = 624$ ways. ◎

Recall that a full house is a hand that has three cards of one rank and two of another rank. For example, three kings and two jacks make a full house. Before determining how many different ways we can draw a full house, we will attack a simpler question.

EXAMPLE 5   Drawing Three Cards of the Same Rank

Determine the number of different ways to draw three cards of the same rank when drawing three cards from a standard 52-card deck.

SOLUTION: This problem can be divided into two stages. First, choose the rank of the three cards that will be the same. This decision can be made in thirteen ways. Second, choose three cards from the four cards in the deck with this rank. This choice can be made in $C(4, 3) = $ four ways. Using the fundamental counting principle, we see that the three cards of the same rank can be selected in $13 \times 4 = 52$ ways. ◎

Quiz Yourself 10

Determine the number of different ways to draw two cards of the same rank when drawing two cards from a standard 52-card deck.

EXAMPLE 6   Drawing a Full House

Determine the number of different ways we can obtain a full house when drawing five cards from a standard 52-card deck.

SOLUTION: We can think of this problem as consisting of two simpler subproblems.

Subproblem A: Determine the number of ways we can choose the three cards having the same rank. We solved this subproblem in Example 5, where we found that this selection can be made in 52 ways.

Subproblem B: Determine the number of ways we can choose the remaining two cards having the same rank. It is tempting to use the answer found in Quiz Yourself 10, but this would be a mistake. Realize that at this stage, we are selecting the last two cards *after* we have already decided on the three-of-a-kind cards.

Since we have already used one rank for the three-of-a-kind cards, we have only twelve ranks remaining. For example, if we have chosen three kings, then kings can not be used again for the remaining two cards. Thus the rank of the last two cards can be chosen in only twelve ways. After we have decided the rank of the last two cards, we must select two of the four cards with this rank. This can be done in $C(4, 2) = 6$ ways. Using the fundamental counting principle, we see that the two-of-a-kind cards can be selected in $12 \times 6 = 72$ ways.

Now we are ready to answer the original question. In constructing a full house we can first choose the three-of-a-kind cards in 52 ways. Then we can choose the two-of-a-kind cards in 72 ways. Using the fundamental counting principle a final time, we see that a full house can be constructed in $52 \times 72 = 3,744$ ways.

EXAMPLE 7 Drawing Three of a Kind

If we select five cards from a standard 52-card deck, in how many ways can we draw a hand with three of a kind?

SOLUTION: We first must count the number of ways that we can select three cards of the same rank. In Example 5, we found the answer to this question to be 52.

Next we must select the remaining two cards. The fourth card can be chosen among the twelve remaining ranks in 48 ways, and then the fifth card (which must be a different value than the first four) can be chosen in 44 ways. It would seem then that the fourth and fifth cards could be selected in 48×44 ways.

However, there is a slight flaw in our thinking! Suppose, for example, we began with three kings and then for our fourth and fifth cards chose a queen and an eight *in that order*. We could then rearrange the queen and eight in our hand and the hand would be unchanged. Thus we recognize that in counting the ways to complete the hand with a fourth and fifth card, we have counted every possibility twice. Therefore there are only $\frac{48 \times 44}{2} = 1,056$ ways to select the last two cards.

Since the first three cards (of the same rank) can be selected in 52 ways and the last two cards can be selected in 1,056 ways, a hand with three of a kind can be chosen in $52 \times 1,056 = 54,912$ ways.

Exercises

Exercises 1–8 are based on the slot machine described in Figure 12.14.

1. **Slot machines.** In how many ways can we obtain cherries on only the first two wheels?

2. **Slot machines.** In how many ways can we obtain cherries on all three wheels?

3. **Slot machines.** In how many ways can we obtain oranges on all three wheels?

4. **Slot machines.** In how many ways can we obtain plums on all three wheels?

5. **Slot machines.** In how many ways can we obtain bells on all three wheels?

6. **Slot machines.** In how many ways can we obtain bars on all three wheels?

7. Communicating Mathematics Would you expect the payoff for three cherries or three oranges to be larger? Explain.

8. Communicating Mathematics Would you expect the payoff for three plums or three bells to be larger? Explain.

Exercises 9–15 deal with the poker hands described in Table 12.2. We assume that we are drawing five cards from a standard 52-card deck.

9. **Playing poker.** How many ways can we obtain a royal flush (10, J, Q, K, and A, all of the same suit)?

10. **Playing poker.** In constructing a straight flush, we first choose a suit and then choose a sequence of five cards within the suit.

    a) How many ways can we choose the suit?

b) How many ways can we choose the sequence of five cards within the suit?

c) How many ways can we construct a straight flush?

11. **Playing poker.** To construct a flush, we first select a suit and then choose five cards without regard to order from that suit.

   a) How many ways can we choose the suit?

   b) How many ways can we choose the five cards?

   c) How many ways can we construct a flush?

12. **Playing poker.** To construct a straight, we must choose five cards in sequence (not all of the same suit).

   a) How many ways can we choose a sequence of five cards?

b) For each sequence in part (a) we must select a suit for the first card, a suit for the second card, and so on. In how many ways can we do this? *Hint:* Draw a slot diagram for the five cards.

c) Multiply the results from parts (a) and (b).

d) Subtract the number of royal flushes and straight flushes that are not royal, found in Exercises 9 and 10.

13. **Playing poker.** How many ways can we construct a five-card poker hand that contains two pair?

14. **Playing poker.** How many ways can we construct a poker hand that contains only one pair and no other cards of any value?

Further Exercise

15. **Playing poker.** How many poker hands contain not even one pair? To solve this:

   a) Calculate the number of hands of each type in Table 12.2.

   b) Subtract the total of the hands found in part (a) from the total of all possible hands.

13

Probability:*
What Are the Chances?

On a hot summer evening, the National Weather Service issues a warning about the *possibility* of a tornado until 11:00 P.M. Las Vegas bookmakers establish the *odds* against the Dallas Cowboys winning next year's Super Bowl as being 7 to 2. Of patients using a certain hair growth product, 45 percent experience a reduction in their baldness. In the game Monopoly, to get out of jail you must roll a double. It is *not likely* that you will do this on your first try. A tire on your car that is *guaranteed* for 40,000 miles has to be replaced after only 17,000 miles.

In each situation, there is a *chance*, *likelihood*, or *possibility* that some event will happen. The lure of predicting the future has fascinated people for centuries. Those in ancient cultures relied on oracles, astrologers, and fortune-tellers to predict the outcomes of wars, harvests, and individual destinies. Mathematicians have replaced these intuitive and unreliable attempts to foresee the future with a formal, mathematical way of reasoning called probability theory. In this chapter, you will learn the basic principles and rules for computing probabilities. You will also understand how probability is applied to situations such as testing pharmaceuticals, determining the profit an insurance company can expect on its policies, and evaluating the best way to make business decisions.

As you will see, the results of probability computations can go against our intuition and, in fact, may be quite surprising. For example, who

---

* For further resources on probability, see www.aw.com/pirnot.

would argue against testing applicants for a marriage license for the HIV virus? Reliable tests exist to identify those infected with AIDS. It seems that testing would be a good way to prevent the spread of AIDS. Yet later in the chapter we use probability theory to make a case that such testing may not only be a waste of resources, but may do more harm than good.

13.1   THE BASICS OF PROBABILITY THEORY

Tomorrow's weather, the hand you are dealt in a card game, whether you will contract the latest flu virus, and the numbers that come up in the jackpot lottery are all examples of random phenomena. *Random phenomena* are occurrences that vary from day to day, case to case, and situation to situation. The hand that you receive in a card game is an example of a random phenomenon because it varies from deal to deal. The situations that we discuss in this chapter are all examples of random phenomena.

Although we never know for certain how a random phenomenon will turn out, often we can calculate a numerical measure, or a *probability*, that the phenomenon will occur in a certain way. For example, a weather forecaster might say that there is a 70 percent chance that a fast-moving hurricane will strike Florida before midnight. Typically we express a probability as a percentage, fraction, or decimal between 0 and 1.

In this section we introduce some basic terminology and then explain several simple methods for calculating the probabilities of random phenomena.

Knowing the sample space helps us compute probabilities.

Our first step in calculating the probability of a random phenomenon is to determine the sample space of an experiment.

> **DEFINITIONS**
>
> An **experiment** is any observation of a random phenomenon. The different possible results of the experiment are called **outcomes**. The set of all possible outcomes for an experiment is called a **sample space**.

If we observe the results of flipping a single coin, we have an example of an experiment. The possible outcomes are head and tail, so a sample space for the experiment would be the set {head, tail}.

EXAMPLE 1   Finding Sample Spaces

Determine a sample space for each experiment.

a) We select an item from a production line and determine whether it is defective or not.

b) Three children are born to a family and we note the birth order with respect to sex.

c) We select one card from a standard 52-card deck,* and then without returning the card to the deck, we select a second card.

d) We roll two dice and observe the total showing on the top faces.

SOLUTION: In each case, we find the sample space by collecting the outcomes of the experiment into a set.

a) This sample space is {defective, nondefective}.

b) In this experiment, we wish to know not only how many boys and girls are born, but also the birth order. For example, a boy followed by two girls is not the same as two girls followed by a boy. The tree diagram† in Figure 13.1 helps us find the sample space.

There are two ways that the first child can be born, followed by two ways for the second child, and finally two ways for the third. If we abbreviate "boy" by "b" and "girl" by "g," following the branches of the tree diagram gives us the following sample space:

{bbb, bbg, bgb, bgg, gbb, gbg, ggb, ggg}

c) We will assume that the order in which we select the cards is important. This sample space is too large to list; however, we can use the fundamental counting principle from Section 12.2 to count its members. Because we can choose the first card in 52 ways and the second card in 51 ways (we are not replacing the first card), we can select the two cards in 52 × 51 = 2,652 ways.

d) We discussed this situation in Example 4 of Section 12.1. We think of the dice as having different colors, say red and green, so that the outcome (1, 3) is considered different from the outcome (3, 1). With this in mind, the sample space for this experiment consists of the following 36 pairs:

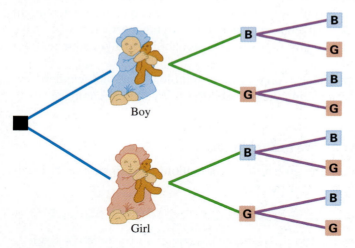

FIGURE 13.1 Tree diagram showing the sex of three children.

{(1, 1), (1, 2), (1, 3), (1, 4), (1, 5), (1, 6),
(2, 1), (2, 2), (2, 3), (2, 4), (2, 5), (2, 6),
(3, 1), (3, 2), (3, 3), (3, 4), (3, 5), (3, 6),
(4, 1), (4, 2), (4, 3), (4, 4), (4, 5), (4, 6),
(5, 1), (5, 2), (5, 3), (5, 4), (5, 5), (5, 6),
(6, 1), (6, 2), (6, 3), (6, 4), (6, 5), (6, 6)}

* See Section 12.1 for a picture of a standard 52-card deck.
† We introduced tree diagrams in Chapter 12.

An event is a subset of a sample space.

As you study probability further, you will find that often you will need to know the probability that a particular collection of outcomes occurs in an experiment. For instance, in the sample space consisting of 36 pairs in Example 1 d), you might be interested in the subset consisting of pairs such as (1, 6) and (2, 5) whose total is seven. In a test of a new psoriasis medication, a researcher might be interested in studying further those patients who were substantially improved by the treatment. A meteorologist would be concerned with the probability of outcomes that would be likely to foreshadow a violent thunderstorm. This concern with subsets of the sample space leads to the following definition.

> **DEFINITION**
>
> In probability theory, an **event** is a subset of the sample space.

Keep in mind that *any* subset of the sample space is an event, including such extreme subsets as the empty set, one-element sets, and the whole sample space.

> **SOME GOOD ADVICE**
>
> Although we usually describe events verbally, you should remember that *an event is always a subset of the sample space.* You can use the verbal description to identify the set of outcomes that make up the event.

Example 2 illustrates some events from the sample spaces in Example 1.

EXAMPLE 2   Describing Events as Subsets

Write each event as a subset of the sample space.

a) A head occurs when we flip a single coin.

b) Two girls and one boy are born to a family.

c) A total of five occurs when we roll a pair of dice.

SOLUTION: a) The set {head} is the event.

b) Noting that the boy can be the first, second, or third child, the event is
{bgg, gbg, ggb}.

c) The following set shows how we can roll a total of five on two dice: {(1, 4), (2, 3), (3, 2), (4, 1)}.

Quiz Yourself 1*

Write each event as a set of outcomes.

a) A total of six occurs when we roll two dice.

b) In a family with three children, there are more boys than girls.

■ ████████████████████████████████

The probability of an event is the sum of the probabilities of the outcomes in that event.

We will use the notions of outcome, sample space, and event to compute probabilities. Intuitively, you may expect that when rolling a fair die, each number has the same chance, namely $\frac{1}{6}$, of showing. In predicting weather, a forecaster may state that there is a 30 percent chance of rain tomorrow. You may believe that you have a 50-50 chance (that is, a 50 percent chance) of getting a job offer in your field. In each of these examples we have assigned a number between 0 and 1 to represent the likelihood that the outcome will occur.

> **DEFINITIONS**
>
> The **probability of an outcome** in a sample space is a number between 0 and 1 inclusive. The sum of the probabilities of all the outcomes in the sample space must be 1. The **probability of an event** E, written $P(E)$,* is defined as the sum of the probabilities of the outcomes that make up E.

One way to determine probabilities is to use *empirical information*. That is, we make observations and assign probabilities based on those observations.

> **Empirical Assignment of Probabilities**
>
> If E is an event and we perform an experiment several times, then we estimate the probability of E as follows:
>
> $$P(E) = \frac{\text{the number of times } E \text{ occurs}}{\text{the number of times the experiment is performed}}.$$
>
> This ratio is sometimes called the *relative frequency* of E.

Side Effects	Number of Times
None	67
Mild	25
Severe	8

TABLE 13.1   Summary of side effects of the flu vaccine.

 Quiz Yourself 2

Use Table 13.1 to find the probability that a patient receiving the flu vaccine will experience no side effects.

EXAMPLE 3   Using Empirical Information to Assign Probabilities

A pharmaceutical company is testing a new flu vaccine. The experiment is to inject a patient with the vaccine and observe the occurrence of side effects. Assume that we perform this experiment 100 times and obtain the information in Table 13.1.

Based on Table 13.1, if a physician injects a patient with this vaccine, what is the probability that the patient will develop severe side effects?

SOLUTION: In this case we base our probability assignment of the event

---

* Do not confuse $P(E)$, which is the notation for the probability of an event, with $P(n, r)$, which is the notation for the number of permutations of n objects taken r at a time (see Section 12.3).

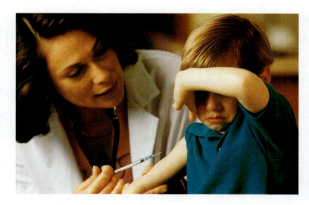

that severe side effects occur on observations. We use the formula for the relative frequency of an event as follows:

$$P \text{ (severe side effects)} =$$

$$\frac{\text{the number of times severe side effects occurred}}{\text{the number of times the experiment was performed}} =$$

$$\frac{8}{100} = 0.08$$

Thus, from this empirical information, we can expect an 8 percent chance that there will be severe side effects from this vaccine. ◎

EXAMPLE 4   Investigating Wage Data

Table 13.2 summarizes the hourly wage rates paid to U.S. workers in 2001.* If we randomly pick a person who is between 20 and 24 years old, inclusive, what is the probability that the person earns $10.00 or more per hour?

	All Workers (thousands)		
	$5.15 or less	**$5.16 to $9.99**	**$10.00 or more**
16–19 years	640	5,743	618
20–24 years	660	6,729	3,387
25 and older	1,015	20,117	35,860

TABLE 13.2   U.S. hourly wages in 2001.

SOLUTION: It is important to recognize that not all the data entries in Table 13.2 are relevant to the question that you were asked. We are only interested in 20 to 24 year olds, so the line that we have highlighted in the table contains all the information that we need to solve the problem. Therefore, we can consider the sample space, S, to consist of the

$$660 + 6{,}729 + 3{,}387 = 10{,}776$$

thousand people in this age group.

The event, call it A, is the set of 3,387 (thousand) people who earn at least ten dollars an hour. Therefore, the probability that we would select a person earning at least ten dollars per hour from the 20 to 24 year olds is

$$P(A) = \frac{n(A)}{n(S)} = \frac{3{,}387{,}000}{10{,}776{,}000} \approx 0.314. \qquad ◎$$

* Data provided by the U.S. Department of Labor as stated in the *World Almanac* (2002).

We can use counting formulas to compute probabilities.

Another way we can determine probabilities is to use *theoretical information*, such as the counting formulas we discussed in Chapter 12.

In order to understand the difference between using theoretical and empirical information, compare these two experiments: Suppose that without looking you reach into a box containing red and blue balls, draw a ball out, note its color, and return the ball to the box. If you repeat this experiment 100 times and get 60 red balls and 40 blue balls, based on this empirical evidence, you would expect that the probability of getting a red ball on your next draw is 60 percent. On the other hand, if you draw a five-card hand from a standard 52-card deck, you can use a combination formula from Chapter 12 to determine how many different hands consist of cards from the same suit.

EXAMPLE 5   Using Counting Formulas to Calculate Probabilities

Assign probabilities to the outcomes in the following sample spaces.

a) We flip three fair coins. What is the probability of each outcome in this sample space?

b) We draw a five-card hand randomly from a standard 52-card deck. What is the probability that we draw one particular hand?

SOLUTION: a) This sample space has eight outcomes, as we show in Figure 13.2.

FIGURE 13.2   Eight theoretically possible outcomes for flipping three coins.

Since the coins are fair, we expect that heads and tails are equally likely to occur. Therefore it is reasonable to assign a probability of $\frac{1}{8}$ to each outcome in this sample space.

b) In Chapter 12 we found that there are $C(52, 5) = 2{,}598{,}960$ different ways to choose five cards from a deck of 52. Since we are drawing the cards randomly, each hand has the same chance of being drawn. Therefore, the probability of drawing any one hand is $\frac{1}{2{,}598{,}960}$.

In a sample space with equally likely outcomes, it is easy to calculate the probability of any event by using the following formula.

Highlight: *Why Does It Always Happen to You?**

When you wait at a toll booth, why does it seem that the line of cars next to you moves faster than your line? If you look into a sock drawer, do you notice how many unmatched socks there are? Why does the buttered side of a slice of bread almost always land face down if you drop the bread while making a sandwich? Do such annoyances affect only you? Or is there a mathematical explanation?

First let's consider the socks. Suppose you have ten pairs of socks in a drawer and lose one sock, destroying a pair. Of the nineteen remaining socks, there is only one unpaired sock. Therefore, if you lose a second sock, the probability of losing a paired sock is $\frac{18}{19}$. Now you have two unpaired socks and sixteen paired socks. The probability that the third sock you lose will be part of a pair is still overwhelming. Continuing this line of thought, you see that it will be quite a while before probability theory predicts that you can expect to lose an unpaired sock.

The problem with the buttered bread is simple to explain. In fact you might conduct an experiment to simulate this situation with an object such as a computer mouse pad. If you slide the object off the edge of a table, as it begins to fall it rotates halfway around so that the top side is facing down. However, there is usually not enough time for the object to rotate back to the upward position before it hits the floor. You could conduct this experiment 100 times and determine the empirical probability that an object you slide off the table will land face down. So again, it's not bad luck—it's probability.

The slow-moving line at the toll booth is the easiest to explain. If we assume that delays in any line occur randomly, then one of the lines—yours, the one to the left, or the one to the right—will move fastest. Therefore, the probability that the fastest line will be yours is only $\frac{1}{3}$.

Calculating Probability When Outcomes Are Equally Likely

If *S* is a sample space with all *equally likely outcomes*, then each outcome has a probability equal to

$$\frac{1}{\text{number of outcomes in } S} = \frac{1}{n(S)}.$$

For an event *E* in this sample space,

$$P(E) = \frac{n(E)}{n(S)}.^{\dagger}$$

EXAMPLE 6   Computing Probability of Events

a) What is the probability in a family with three children that two of the children are girls?

b) What is the probability that a total of four shows when we roll two fair dice?

---

* This Highlight is based on Robert Matthews, "Murphy's Law or Coincidence." *Reader's Digest*, March 1998, pp. 25–30.
† Recall that $n(E)$ is the cardinal number of set *E*.

c) If we draw a five-card hand from a standard 52-card deck, what is the probability that all five cards are hearts?

d) Alexis, Julio, Brett, and Suresh are friends who belong to their college's 10-person international relations club. Two people from the club will be selected randomly to attend a conference at the United Nations building. What is the probability that two of these four friends will be selected?

SOLUTION: In each situation we assume that the outcomes are equally likely.

a) We saw in Example 1 that there are eight outcomes in this sample space. We denote the event that two of the children are girls by the set $G = \{$bgg, gbg, ggb$\}$. Thus

$$P(G) = \frac{n(G)}{n(S)} = \frac{3}{8}.$$

b) The sample space for rolling two dice has 36 ordered pairs of numbers. We will represent the event "rolling a four" by F. Then $F = \{(1, 3), (2, 2), (3, 1)\}$. Thus

$$P(F) = \frac{n(F)}{n(S)} = \frac{3}{36} = \frac{1}{12}.$$

c) From Example 5, we know that there are $C(52, 5)$ ways to select a five-card hand from a 52-card deck. If we wish to draw only hearts, then we are selecting five hearts from the thirteen available, which can be done in $C(13, 5)$ ways. Thus the probability of selecting all five cards to be hearts is

$$\frac{C(13, 5)}{C(52, 5)} = \frac{1{,}287}{2{,}598{,}960} \approx 0.000495.$$

 Quiz Yourself 3

a) If we roll two fair dice, what is the probability of rolling a total of eight?

b) If we select two cards randomly from a standard 52-card deck, what is the probability that both are face cards?

d) The sample space, S, consists of all the ways we can select two people from the ten members in the club. As you know from Chapter 12, we can choose two people from ten in $C(10, 2) = \frac{10!}{8! \cdot 2!} = \frac{10 \cdot 9}{2} = 45$ ways. The event, call it E, consists of the ways we can choose two of the four friends. This can be done in $C(4, 2) = 6$ ways. Because all of elements in S (choices of two people) are equally likely, the probability of E is

$$P(E) = \frac{n(E)}{n(S)} = \frac{C(4, 2)}{C(10, 2)} = \frac{6}{45} = \frac{2}{15}.$$

SOME GOOD ADVICE

If the outcomes in a sample space are not equally likely, then you cannot use the formula $P(E) = \frac{n(E)}{n(S)}$. In that case, to find $P(E)$, you must add the probabilities of all of the individual outcomes in E.

Suppose that we have a sample space with equally likely outcomes. Since $0 \le n(E) \le n(S)$, if we divide this inequality by the positive quantity $n(S)$, we get the first probability property listed below. The other properties are easy to see.

Basic Properties of Probability

Assume that S is a sample space for some experiment and E is an event in S.

1. $0 \le P(E) \le 1$
2. $P(\emptyset) = 0$
3. $P(S) = 1$

Probability theory helps explain genetic theory.

In the nineteenth century the Austrian monk Gregor Mendel, while cross-breeding plants, noticed that often a characteristic of the plants would disappear in the first-generation offspring but reappear in the second generation. He theorized that the first-generation plants contained a hidden factor (which we now call a *gene*) that was somehow transmitted to the second generation to enable the characteristic to reappear.

To check his hypothesis, he devised a series of experiments with pea plants that had different characteristics. Some plants were short and bushy and others were tall and climbing. Some had purple flowers; others had white. Other pairs of characteristics he considered were yellow seeds versus green seeds and smooth seeds versus wrinkled seeds.

After crossing many pea plants, when Mendel was sure that he had bred plants having a pure strain of a particular characteristic—say, having smooth seeds but having none of the "hidden" wrinkled characteristic—he was ready to begin his experiment. Table 13.3* shows Mendel's results of cross-breeding plants for three pairs of opposite characteristics.

Characteristics That Were Cross-Bred	First-Generation Plants	Second-Generation Plants
Tall vs. short	All tall	787 tall 277 short
Yellow seed vs. green seed	All yellow	6,022 yellow 2,001 green
Smooth seeds vs. Wrinkled seeds	All smooth seeds	5,474 smooth 1,850 wrinkled

TABLE 13.3   Results of Mendel's cross-breeding studies.

---

* Table 13.3 is taken from Albert Shulte, editor, *Teaching Statistics and Probability*, 1981 Yearbook of the National Council of Teachers of Mathematics, p. 176..

When Mendel crossed large numbers of his purebred plants, in each case, one of the paired characteristics disappeared in the first generation. Mendel concluded that in these pairs one of the characteristics was *dominant*. He called the other trait *recessive*. When Mendel next bred plants from the first generation, he found that in all cases the recessive trait now reappeared in the second generation and that the ratio of the dominant trait to the recessive trait was approximately 3 to 1, as you can see in Table 13.3. Because Mendel had been schooled in mathematics as well as biology, he gave the following explanation for the phenomena he had observed.

Consider the pair of traits "yellow seed versus green seed." We will represent the yellow seed gene by Y and the green seed gene by g. The uppercase Y indicates that yellow is dominant and the lowercase g indicates that green is recessive.*

Figure 13.3 shows the possible genetic makeup of the offspring from crossing a plant with pure yellow seeds and a plant with pure green seeds. Every one of these offspring has a Yg pair of genes. Because "yellow seed" is dominant over "green seed," every plant in the first generation will have yellow seeds.

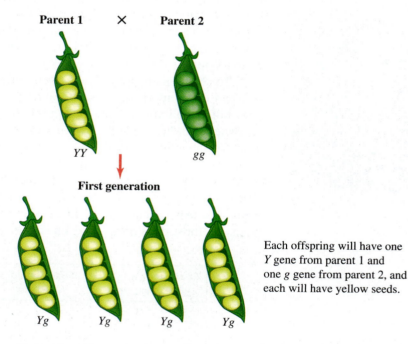

Each offspring will have one Y gene from parent 1 and one g gene from parent 2, and each will have yellow seeds.

FIGURE 13.3   The possible first-generation offspring we obtain by crossing pure yellow and pure green parents.

Figure 13.4 shows the possible genetic outcomes that can occur if we cross two first-generation pea plants.

---

* In biology books it is customary to use slightly different notation to indicate genes. For example, to indicate a pair of genes in which one is yellow and one green, where yellow is dominant, you may see the notation Yy. The Y indicates the dominant yellow gene and the y indicates the recessive green gene.

First generation   X   First generation

Yg   *Yg*

Second generation

Each offspring will have either a *Y* gene or a *g* gene from each parent. Only the seeds of the offspring with the *gg* pair of genes will be green.

YY   *Yg*   *gY*   *gg*

FIGURE 13.4   The possible second-generation offspring we obtain by crossing the first-generation offspring of pure yellow and pure green parents.

We summarize Figure 13.4 in Table 13.4, which is called a *Punnett square.*

We see that in the second generation it is possible to get plants with three different types of genetic makeup. A plant with the *YY* pair of genes will have yellow seeds. A plant with the *Yg* or *gY* pair of genes will also have yellow seeds. A plant with the *gg* pair of genes will have green seeds. Thus from the four possible ways that the genes can be paired, three result in yellow seeds and one results in green seeds. This explains why Mendel saw the recessive characteristic returned in roughly one-quarter of the plants in the second generation.

		First-Generation Plant	
		Y	**g**
First-Generation Plant	**Y**	*YY*	*Yg*
	g	*gY*	*gg*

TABLE 13.4   The genetic possibilities of crossing two plants that each have one yellow-seed and one green-seed gene.

EXAMPLE 7   Using Probability to Explain Genetic Diseases

Sickle-cell anemia is a serious inherited disease that is about 30 times more likely to occur in African American babies than in non–African American babies. A person with two sickle-cell genes will have the disease, but a person with only one sickle-cell gene will be a carrier of the disease.

If two parents who are carriers of sickle-cell anemia have a child, what is the probability of each of the following:

a) The child has sickle-cell anemia?

b) The child is a carrier?

c) The child is disease-free?

SOLUTION: Table 13.5 shows the genetic possibilities when two people who are carriers of sickle-cell anemia have a child. We will denote the sickle-cell gene by s and the normal gene by n. We use lowercase letters to indicate that neither s nor n is dominant.

From Table 13.5 we see that there are four equally likely outcomes for the child.

- The child receives two sickle-cell genes and therefore has the disease.

		Second Parent	
		s	n
First	s	ss	sn
Parent	n	ns	nn

TABLE 13.5   The genetic possibilities for children of two parents with sickle-cell trait.

- The child receives a sickle-cell gene from the first parent and a normal gene from the second parent and therefore is a carrier.

- The child receives a normal gene from the first parent and a sickle-cell gene from the second parent and therefore is a carrier.

- The child receives two normal genes and therefore is disease-free.

From this analysis it is clear that:

a) P(the child has sickle-cell anemia) $= \frac{1}{4}$.

b) P(the child is a carrier) $= \frac{1}{2}$.

c) P(the child is normal) $= \frac{1}{4}$.

In computing odds remember "against" versus "for."

We often use the word *odds* to express the notion of probability. When we do this, we will usually state the odds against something happening. For example, before the 1999 Kentucky derby, the odds against Charismatic, the eventual winner, were 31 to 1. When you calculate odds, it is helpful to think of what is against the event as compared with what is in favor of the event.

> **DEFINITION**
>
> If the outcomes in the sample space are *equally likely*, then the **odds** against event E are simply the number of outcomes against E compared to the number of outcomes in favor of E occurring. The odds against E are therefore the ratio $\frac{n(E')}{n(E)}$, where the set E' is the complement of E.* We often write this ratio as $n(E') : n(E)$.

Note that if the odds against an event are a to b, then the odds in favor of the event are b to a.

_____

* We discussed the complement of a set in Section 1.5.

Historical Highlight: *Early Probability Theory*

We can trace probability theory back to prehistoric times. Archaeologists have found numerous small bones, called astrogali, that are believed to have been used in playing various games of chance. Similar bones were later ground into the shape of cubes and decorated in various ways. They eventually evolved into modern dice. Games of chance were played for thousands of years, but it was not until the sixteenth century that a serious attempt was made to study them using mathematics.

Among the first to study the mathematics of games of chance was the Italian Girolamo Cardano. In addition to his interest in probability, Cardano was an astrologer. This caused him frequent misfortune. In one instance, Cardano cast the horoscope of Jesus and for this blasphemy he was sent to prison. Shortly thereafter, he predicted the exact date of his own death and, when the day arrived, Cardano made his prediction come true by taking his own life.

PROBLEM SOLVING

Making an analogy with a real-life situation helps you to remember the meaning of mathematical terminology. For example, when General Custer fought Chief Sitting Bull at the battle of Little Big Horn, Custer had about 650 soldiers and Sitting Bull had 2,500 braves. If we think of how many were for Custer and how many were against him, we might say that the odds against Custer winning were 2,500 to 650. Similarly, the odds against Sitting Bull winning were 650 to 2,500.

EXAMPLE 8   Calculating Odds on a Roulette Wheel

A common type of roulette wheel has 38 equal-size compartments. Thirty-six of the compartments are numbered 1 to 36 with half of them colored red and the other half black. The remaining two compartments are green and numbered 0 and 00. A small ball is placed on the spinning wheel and when the wheel stops, the ball rests in one of the compartments. What are the odds against the ball landing on red?

SOLUTION: This is an experiment with 38 equally likely outcomes. Since 18 of these are in favor of the event "the ball lands on red" and 20 are against the event, the odds against red are 20 to 18. We can write this as 20:18 or $\frac{20}{18}$, which we may reduce to $\frac{10}{9}$.

Although we have defined odds in terms of counting, we also can think of odds in terms of probability.

Probability Formula for Computing Odds

If E' is the complement of the event E, then the odds against E are

$$\frac{P(E')}{P(E)}.$$

EXAMPLE 9 **Using the Probability Formula for Computing the Odds in a Football Game**

Suppose that the probability of Green Bay winning the Super Bowl is 0.35. What are the odds against Green Bay winning the Super Bowl?

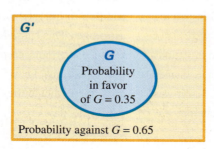

FIGURE 13.5 The odds against Green Bay winning are the probability of Green Bay losing versus the probability of Green Bay winning.

SOLUTION: Recall from Section 1.1 on problem solving that a diagram often helps us remember what to do. Call the event "Green Bay wins the Super Bowl" G. We illustrate this in Figure 13.5.

Thus the odds against Green Bay winning the Super Bowl are

$$\frac{P(G')}{P(G)} = \frac{0.65}{0.35} = \frac{0.65 \times 100}{0.35 \times 100} = \frac{65}{35} = \frac{13}{7}.$$

In this case we would say the odds against Green Bay winning the Super Bowl are 13 to 7. ◉

Now that we have introduced some of the basic terminology and properties of probability, in the next section we will learn rules for calculating probabilities.

Exercises 13.1

In Exercises 1–8, write the sample space for the given experiment as a set. If the set is large, you may want to describe the set without listing its members. Often a tree diagram will help you list the members of a sample space.

1. We flip three coins.

2. A family has four children and we note the birth order with respect to sex.

3. We roll two four-sided dice having the numbers 1, 2, 3, and 4 on their faces.

4. A jar contains one red, one blue, and one green ball. You select a ball at random, return it, and then select another ball. We will represent these outcomes by ordered pairs. For example, the pair (r, b) indicates that a red ball was selected first and then a blue ball was selected.

5. You and your best friend are looking for a job. Your possibilities for employment are a (b)ank, an (i)nvestment firm, and a (h)ealth provider. Your friend's possibilities are a (f)itness center and a (r)ehabilitation hospital. Represent the outcomes by ordered pairs.

6. We select three items from a production line and examine them to determine whether they are defective.

7. You buy two stocks as an investment and each stock can either (i)ncrease in value, (d)ecrease in value, or (s)tay the same. Represent the outcomes as ordered pairs.

8. An experimenter testing for extrasensory perception has five cards with a picture of a (s)tar, a (c)ircle, (w)iggly lines, a (d)ollar sign, and a (h)eart. She selects two cards without replacement. Represent the outcomes as ordered pairs.

In Exercises 9–16, write each event as a set of outcomes. If the event is large, you may describe the event without writing it out.

9. When we roll two dice, the total showing is seven.

10. When we roll two dice, the total showing is five.

11. We flip three coins and obtain more (h)eads than (t)ails.

12. We select a red face card from a standard 52-card deck.

13. In Exercise 5, your friend takes the job in the fitness center.

14. In Exercise 6, we find at least two defective items.

15. In Exercise 7, exactly one of the stocks increases in value.

16. In Exercise 8, a star is on the second card.

In Exercises 17–20, use the given spinner to write the event as a set of outcomes. Abbreviate "red" as "r," "blue" as "b" and "yellow" as "y".

17. We obtain red exactly once in two spins.

18. Red appears at least once in two spins.

19. Red appears exactly twice in three spins.

20. Red does not appear on any of three spins.

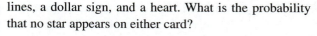

21. Communicating Mathematics   Explain the difference between an outcome and an event.

22. Communicating Mathematics   When rolling two dice, describe in words an event that is the empty set.

23. Communicating Mathematics   When rolling two dice, describe in words an event that is the whole sample space.

24. Communicating Mathematics   Can an event have just one single outcome in it? Give an example.

In Exercises 25–26, construct a Punnett square as we did in Table 13.5 to show the probabilities for the offspring.

25. One parent who has sickle-cell anemia and one parent who is a carrier have a child. Find the probability that the child is a carrier of sickle-cell anemia.

26. One parent who has sickle-cell anemia and one parent who is a carrier have a child. Find the probability that the child has sickle-cell anemia.

27. Two cards are selected without replacement from a set of five cards having a picture of a star, a circle, wiggly lines, a dollar sign, and a heart. What is the probability that a star is on one of the cards?

28. Two cards are selected without replacement from a set of five cards having a picture of a star, a circle, wiggly

lines, a dollar sign, and a heart. What is the probability that no star appears on either card?

29. A ball is randomly selected from a jar containing 70 red balls, 20 green balls, and 10 blue balls. What is the probability that a non-red ball is chosen?

30. The residents of a small town and its surrounding area are divided over the proposed construction of a sprint car racetrack in the town. The following table shows how opinion is split on this issue.

	Support Racetrack	Oppose Racetrack
Live in Town	1,745	2,345
Live in Surrounding Area	1,621	1,289

If a newspaper reporter randomly selects a person to interview from among these people, what is the probability that the person selected supports the racetrack?

31. Calculate the odds against the event in Exercise 27.

32. Calculate the odds against the event in Exercise 28.

33. Calculate the odds against the event in Exercise 29.

34. Calculate the odds against the event in Exercise 30.

35. Communicating Mathematics   Explain the difference between the probability of an event and the odds *in favor* of the event.

36. Communicating Mathematics   You know that the probability of an event can never be greater than one. Can the odds in favor of an event ever exceed one. Explain

37. In a given year, 2,048,861 males and 1,951,379 females were born in the United States. If a child is selected randomly from this group, what is the probability that it is a female?

38. Over a given period of time, the Department of Justice reported that there were 5,852 hate crimes in the United States. Of these, 3,505 were based on race, 619 on ethnicity, 1,051 on religion, and 677 on sexual orientation. (Assume that no crime was reported in two categories.) If you selected one of these incidents randomly, what is the probability that it would be a crime not based on race?

39. The following table lists some of the empirical results that Mendel obtained in his experiments in cross-breeding pea plants.

Characteristics That Were Crossbred	First-Generation Plants	Second-Generation Plants
Tall vs. short	All tall	787 tall 277 short
Smooth seeds vs. wrinkled seeds	All smooth seeds	5,474 smooth 1,850 wrinkled

Assume that we are cross-breeding genetically pure tall plants with genetically pure short plants. Use this information to assign the probability that a second-generation plant will be short. How consistent is this with the theoretical results that Mendel derived?

40. Assume that we are cross-breeding genetically pure smooth-seed plants with genetically pure wrinkled-seed plants. Use the information provided in Exercise 39 to assign the probability that a second-generation plant will have smooth seeds. How consistent is this with the theoretical results that Mendel derived?

In cross-breeding snapdragons, Mendel found that flower color does not dominate, as happens with peas. For example, a snapdragon with one red and one white gene will have pink flowers. In Exercises 41–42, analyze the cross-breeding experiment as we did in the discussion prior to Example 7.

41. a) Construct a Punnett square showing the results of crossing a purebred white snapdragon with a purebred red one.

b) What is the probability of getting red flowers in the first-generation plants? What is the probability of getting white? Of getting pink?

42. a) If we cross two pink snapdragons, draw a Punnett square that shows the results of crossing two of these first-generation plants.

b) What is the probability of getting red flowers in the second-generation plants? What about white? Pink?

43. Cystic fibrosis is a serious inherited lung disorder that often causes death in victims during early childhood. Because the gene for this disease is recessive, two apparently healthy adults, called carriers, can have a child with the disease. We will denote the normal gene by N and the cystic fibrosis gene by c to indicate its recessive nature.

a) Construct a Punnett square as we did in Example 7 to describe the genetic possibilities for a child whose two parents are carriers of cystic fibrosis.

b) What is the probability that this child will have the disease?

44. From the Punnett square in Exercise 43, what is the probability that a child of two carriers will

a) be normal?

b) be a carrier?

45. Assume that the following table summarizes a survey involving the relationship between living arrangements and grade point average for a group of students.

	On Campus	At Home	Apartment	Totals
Below 2.5	98	40	44	182
2.5 to 3.5	64	25	20	109
Over 3.5	17	4	8	29
Totals	179	69	72	320

If we select a student randomly from this group, what is the probability that the student has a grade point average of at least 2.5?

46. Using the data in Exercise 45, if a student is selected randomly, what is the probability that the student lives off campus?

In Exercises 47–52, use theoretical information such as counting formulas to determine the probability.

47. If we roll two fair dice, what is the probability that a total of nine shows?

48. If we roll two fair dice, what is the probability that a total of three shows?

49. If we select a card randomly from a standard 52-card deck, what is the probability that we draw a heart?

50. If we select a card randomly from a standard 52-card deck, what is the probability that we draw a face card?

51. If we toss four fair coins, what is the probability that we get exactly two heads?

52. If we toss four fair coins, what is the probability that we get more heads than tails?

Use this replica of the Monopoly game board to answer Exercises 53–56.

53. Assume that your game piece is on the Electric Company. If you land on either St. James Place, Tennessee Avenue, or New York Avenue, you will go bankrupt. What is the probability that you avoid these properties?

54. Assume that your game piece is on Pacific Avenue. If you land on either Park Place or Boardwalk, you will go bankrupt. What is the probability that you avoid these properties?

55. Your game piece is on Virginia Avenue. What is the probability that you will land on a railroad on your next move?

56. Your game piece is on Pennsylvania Avenue. What is the probability that you will have to pay a tax on your next move?

In Exercises 57–60, assume that we are drawing a five-card hand from a standard 52-card deck.

57. What is the probability that all cards are diamonds?

58. What is the probability that all cards are face cards?

59. What is the probability that all cards are of the same suit?

60. What is the probability that all cards are red?

Use Table 13.2 in Example 4 to do Exercises 61–64.

61. If we randomly pick a person 25 or older, what is the probability that that person earns less than $5.16 per hour?

62. If we randomly pick a person 19 or younger, what is the probability that that person earns more than $9.99 per hour?

63. If we randomly pick a person younger than 25, what is the probability that that person earns less than $10.00 per hour?

64. If we randomly pick a person older than 19, what is the probability that that person earns more than $5.15 per hour?

In horse racing, a trifecta is a race in which you must pick the first-, second-, and third-place winners in their proper order to win a payoff.

65. If eight horses are racing, and you randomly select three as your bet in the trifecta, what is the probability that you will win? (Assume that all horses have the same chance to win.)

66. If ten horses are racing, and you randomly select three as your bet in the trifecta, what is the probability that you will win? (Assume that all horses have the same chance to win.)

In Exercises 67–68, assume that to log in to a computer network you must enter a password. Assume that a hacker who is trying to break into the system randomly types one password every ten seconds. If the hacker does not enter a valid password within three minutes, the system will not allow any further attempts to log in. What is the probability that the hacker will be successful in discovering a valid password for each type of password?

67. The password consists of two letters followed by three digits. Case does not matter for the letters. Thus Ca154 and CA154 would be considered the same password.

68. The password consists of any sequence of two letters and three digits. Case does not matter for the letters.

Thus B12q5 and b12Q5 would be considered the same password.

69. If the odds against event E are 5 to 2, what is the probability of E?

70. If $P(E) = 0.45$, then what are the odds against E?

71. If the odds against the United States women's soccer team winning the World Cup are 7 to 5, what is the probability that they will win the World Cup?

72. If the odds against the United States men's hockey team defeating Russia in the winter Olympic Games is 9 to 3, what is the probability that the United States will defeat Russia?

73. Suppose the probability that the Yankees will win the World Series is 0.30.

    a) What are the odds in favor of the Yankees winning the World Series?

    b) What are the odds against the Yankees winning the World Series?

74. Suppose the probability that Magic Touch will win the Triple Crown in horse racing is 0.15.

    a) What are the odds in favor of Magic Touch winning the Triple Crown?

    b) What are the odds against Magic Touch winning the Triple Crown?

75. Communicating Mathematics Find some examples of advertising claims in the media. In what way are these claims probabilities?

76. Frequently in state lotteries it is possible to play a daily number. You usually do this by picking a three-digit number. If your number is drawn you win. What is the probability of winning such a lottery?

Further Exercises

*(77.) a) Flip a coin 1,000 times. How do your empirical results compare with the theoretical probabilities for obtaining heads and tails?

    b) Roll a pair of dice 1,000 times. How do your empirical results compare with the theoretical probabilities for rolling a total of two, three, four, and so on?

    c) Toss an irregular object such as a thumbtack 1,000 times. After doing this, what probability would you assign to the thumbtack landing point up? What about point down? Does it matter what kind of thumbtack you use? Explain.

(78.) Communicating Mathematics Investigate other genetic diseases, such as Tay-Sachs or Huntington's disease. Explain how the mathematics of the genetics of these dis-

eases is similar or dissimilar to the examples we studied in this section.

(79.) Experiment with an object such as a mouse pad. Slide it off a table 100 times and record how often it lands face down. If the object slips off the table, what is the probability that the object will land face down?

(80.) Simulate the lost socks example discussed in the Highlight by doing the following. Take 20 three-by-five cards. Label two of them "pair one," two of them "pair two," two of them "pair three," and so on. Put the 20 cards into a box and draw cards randomly, without replacement, until you have drawn two cards that have the same label. Do this 30 times. On the average, how many cards do you draw before you have drawn two cards with the same label?

---

* Exercise numbers circled in red can be used as group exercises.

13.2 COMPLEMENTS AND UNIONS OF EVENTS

We will now build upon the method we introduced in Section 13.1 for calculating probability that used the formula $P(E) = \frac{n(E)}{n(S)}$. Using results from set theory, that we introduced in Chapter 1, we will develop rules for computing the probability of the complement and the union of events.

We can compute the probability of an event by finding the probability of its complement.

A mathematician frequently solves a problem by rephrasing it as an equivalent question that is easier to answer. For example, a person who wants to count the number of people present in a room that has 60 seats can do so quickly by subtracting the number of empty chairs from 60. We often use this same indirect approach to find the probability of an event. Recall that $P(S) = 1$.* Therefore if it is complicated to compute $P(E)$, we may find it simpler instead to compute $P(E')$ and then subtract the result from 1.

> **Computing the Probability of the Complement of an Event**
>
> If E is an event, then $P(E') = 1 - P(E)$.

Of course we can state this result differently as $P(E) = 1 - P(E')$ or $P(E) + P(E') = 1$. We illustrate this formula in Figure 13.6.

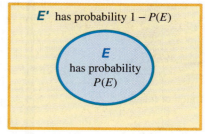

Sample space ($\boldsymbol{S}$)

$\boldsymbol{E'}$ has probability $1 - P(E)$

$\boldsymbol{E}$ has probability $P(E)$

FIGURE 13.6 $P(E) + P(E') = 1$.

PROBLEM SOLVING

The Analogies Principle in Section 1.1 says that you can learn a new area of mathematics more easily if you connect it with an area that you already know. By drawing Venn diagrams, you can visualize many of the rules of probability theory that will help you to remember them and use them properly.

---

* Throughout the rest of this chapter, unless we state otherwise, E, F, and so on will represent events in the sample space S.

Party	Percent
Democrat	42.4
Republican	26.3
Libertarian	1.3
Other	6.3
None	23.7

TABLE 13.6 Party affiliation of young voters.

 Quiz Yourself 4*

If a pair of dice are rolled, what is the probability of rolling a total that is less than 11? (Use the complement rule.)

EXAMPLE 1   Using the Complement Formula

Table 13.6 classifies a certain group of first-time voters according to political party affiliation.

If we randomly select a person from this group, what is the probability that the person has a party affiliation?

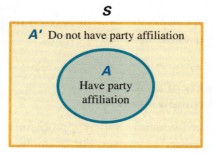

FIGURE 13.7   Calculate $P(A)$ by first finding $P(A')$.

SOLUTION: Let A be the event that the person we select has some party affiliation. Rather than compute the probability of this event, it is simpler to calculate the probability of A', which is the event that the person we select has no party affiliation. We illustrate this situation in Figure 13.7.

Because 23.7 percent have no party affiliation, the probability of selecting such a person is 0.237. Thus $P(A) = 1 - P(A') = 1 - 0.237 = 0.763$.

We often describe the union of events using the word *or*.

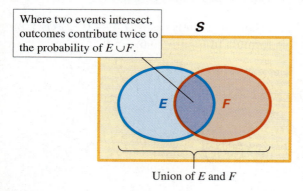

Where two events intersect, outcomes contribute twice to the probability of $E \cup F$.

Union of E and F

FIGURE 13.8   The union of two events.

We often combine objects in a mathematical system to obtain other objects in the system. For example, in number systems, we add and subtract pairs of numbers to get other numbers. In probability, we join events using the set operations union and intersection, which we discussed in Section 1.5. We will investigate the union of events in this section and the intersection of events in Section 13.3. We often describe the union of events using the word *or*.

Figure 13.8 shows the union of events E and F. In computing $P(E \cup F)$ it is a *common mistake* to simply add $P(E)$ and $P(F)$. Because some outcomes may be in both E and F, you will count their probabilities twice.

In order to compute $P(E \cup F)$ properly, you must subtract $P(E \cap F)$.

Rule for Computing the Probability of a Union of Two Events

If E and F are events, then

$$P(E \cup F) = P(E) + P(F) - P(E \cap F).$$

If E and F have no outcomes in common they are called *mutually exclusive events*. In this case, since $E \cap F = \varnothing$, the preceding formula simplifies to

$$P(E \cup F) = P(E) + P(F).$$

EXAMPLE 2   **Finding the Probability of the Union of Two Events**

If we select a single card from a standard 52-card deck, what is the probability that we draw either a heart or a face card?

SOLUTION: Let H be the event "draw a heart" and F be the event "draw a face card." We are looking for $P(H \cup F)$. There are thirteen hearts, twelve face cards, and three cards that are both hearts and face cards. Figure 13.9 helps us remember the formula to use.

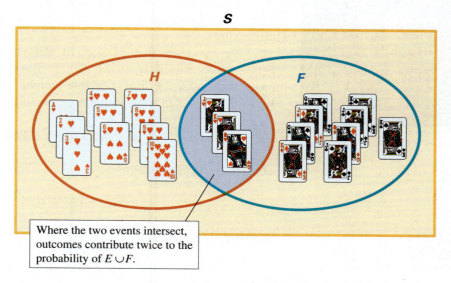

Where the two events intersect, outcomes contribute twice to the probability of $E \cup F$.

FIGURE 13.9   The union of the events "draw a heart" and "draw a face card."

Therefore,

$$P(H \cup F) = P(H) + P(F) - P(H \cap F) = \frac{13}{52} + \frac{12}{52} - \frac{3}{52} = \frac{22}{52} = \frac{11}{26}.$$

SOME GOOD ADVICE

If you are given any three of the four quantities in the formula

$$P(E \cup F) = P(E) + P(F) - P(E \cap F),$$

you can use algebra to solve for the other. See Example 3.

EXAMPLE 3   **Using Algebra to Find a Missing Probability**

A magazine conducted a survey of readers age 18 to 25 regarding their health concerns. The editors will use this information to choose topics that are relevant to their readers. The survey found that 35 percent of the readers were concerned

with improving their cardiovascular fitness and 55 percent wanted to lose weight. Also, the survey found that 70 percent are concerned with either improving their cardiovascular fitness or losing weight. If the editors randomly select one of those surveyed to profile in a feature article, what is the probability that the person is concerned with both improving cardiovascular fitness *and* losing weight?

SOLUTION: Let C be the event "the person wants to improve cardiovascular fitness," and let W be the event "the person selected wishes to lose weight." We want to find $P(C \cap W)$. We will use a method* similar to the one in Example 2; however, we will assign probabilities based on the given empirical information.

   We are given that 35 percent want to improve cardiovascular fitness, so $P(C) = 0.35$. Similarly, $P(W) = 0.55$. The event "a person selected wants to improve cardiovascular fitness or lose weight" is the set $C \cup W$. Therefore, $P(C \cup W) = 0.70$. We can substitute these probabilities in the following equation:

$$P(C \cup W) = P(C) + P(W) - P(C \cap W)$$

This gives us the equation

$$0.70 = 0.35 + 0.55 - P(C \cap W).$$

Rewriting this equation as

$$P(C \cap W) = 0.35 + 0.55 - 0.70,$$

we find that $P(C \cap W) = 0.20$.

   This means that there is a 20 percent chance that the person chosen will be interested in both improving cardiovascular fitness and losing weight.   ◎

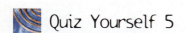

Quiz Yourself 5

Suppose that A and B are events such that $P(A) = 0.35$, $P(A \cap B) = 0.15$, and $P(A \cup B)$ = 0.65. Find $P(B)$.

We may use several formulas to calculate an event's probability.

In Example 4, we use both the complement formula and the union formula to compute an event's probability.

EXAMPLE 4   **Finding the Probability of the Complement of the Union of Two Events**

A survey of consumers comparing the amount of time they spend shopping on the Internet per month with their annual income produced the results in Table 13.7.†

---

* We will not always draw diagrams to illustrate the rule for calculating the probability of the union of two events; however, we do encourage you to do so to aid you in your computations.
† We will assume that all times are rounded to the nearest hour.

Annual Income	10 + Hours	3–9 Hours	0–2 Hours	Totals
Above $60,000	192	176	128	496
$40,000–$60,000	160	208	144	512
Below $40,000	128	192	272	592
Totals	480	576	544	1,600

TABLE 13.7   Survey results on Internet shopping.

Assume that these results are representative of all consumers. If we select a consumer randomly, what is the probability that the consumer neither shops on the Internet ten or more hours per month nor has an annual income above $60,000?

SOLUTION: Although we could use the probability techniques that we introduced in Section 13.1 to answer this question, we will use instead the formulas for the complement and union of events.

We will let T be the event "the consumer selected spends ten or more hours per month shopping on the Internet." Event T corresponds to the first column in Table 13.7. Also, let A be the event "the consumer selected has an annual income above $60,000," which is described by the first line in Table 13.7.

In Figure 13.10, you can see that the event "the consumer selected neither shops on the Internet ten or more hours per month nor has an annual income above $60,000" is the region outside of $T \cup A$, which is the complement of $T \cup A$.

From Table 13.7, you see that the number of outcomes in T is $192 + 160 + 128 = 480$. The total number of outcomes in the sample space, S, is the total number of people surveyed, which is 1,600. Therefore $P(T) = \frac{480}{1,600} = 0.30$. Similarly, because there are 496 outcomes in event A, we find that $P(A) = \frac{496}{1,600} = 0.31$. In order to use the complement formula, we also need to know the probability of the event $T \cap A$. If you look at the intersection of column one and row one in Table 13.7, you can see that the number of outcomes in $T \cap A$ is 192. Therefore, $P(T \cap A) = \frac{192}{1,600} = 0.12$.

With this information, we can now finish the solution to the problem. We calculate the probability of the complement of $T \cup A$ as follows:

$$P((T \cup A)') = 1 - P(T \cup A) = 1 - [P(T) + P(A) - P(T \cap A)]$$
$$= 1 - [0.30 + 0.31 - 0.12] = 1 - 0.49 = 0.51.$$

This means that if we select a consumer randomly, there is a 51 percent chance that the consumer neither spends ten or more hours per month shopping on the Internet nor has a yearly income above $60,000.

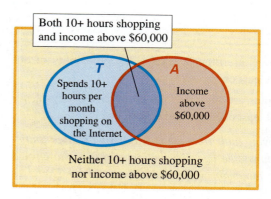

Both 10+ hours shopping and income above $60,000

T A

Spends 10+ hours per month shopping on the Internet

Income above $60,000

Neither 10+ hours shopping nor income above $60,000

FIGURE 13.10   The event "neither T nor A" corresponds to the complement of $T \cup A$.

Historical Highlight: *Modern Probability Theory*

In the seventeenth century, Antoine Gombauld, the Chevalier de Mere, asked Blaise Pascal several questions about gambling:

"How many times should a single die be thrown before we could reasonably expect two sixes?"

"How should prize money in a contest be fairly divided in the case that the contest, for some reason, cannot be completed?"

In responding to these questions, Pascal and his friend Pierre Fermat began to investigate the mathematical principles of games of chance, which many now recognize as the beginning of modern probability theory. In 1812, Pierre-Simon de Laplace wrote *Theorie Analytique des Probabilities*, in which he presented classical probability theory. Laplace boldly affirmed that all knowledge could be obtained by using the principles he set forth.

During the nineteenth and twentieth centuries, scientists have found many important applications of probability theory. Physicists use it to study radiation and atomic physics, while biologists apply it in studying genetics and mathematical learning theory. In addition, probability provides a theoretical basis for statistics, which is used in scientific, industrial, and social research.

Exercises 13.2

In Exercises 1–8, use the complement formula to find the probability of each event.

1. If the probability that your DVD player breaks down before the extended warranty expires is 0.015, what is the probability that the player will not break down before the warranty expires?

2. If the probability that a vaccine you took will protect you from getting the flu is 0.965, what is the probability that you will get the flu?

3. If there is a one in one thousand chance that you will pick the numbers correctly in tonight's lottery, what is the probability that you will not pick the numbers correctly?

4. If there is a one in four chance that it will rain for your Fourth of July barbecue, what is the probability that it won't rain?

5. If two dice are rolled, find the probability that neither die shows a five on it.

6. If two dice are rolled, find the probability that the total showing is less than ten.

7. If five coins are flipped, what is the probability of obtaining at least one head?

8. If five coins are flipped, what is the probability of obtaining at least one head and at least one tail?

Use the following table from the U.S. Bureau of Labor Statistics, which shows the age distribution of those who earned less than the minimum wage in 2001, to answer Exercises 9–10.

Age	Working Below Minimum Wage (in thousands)
16–19	640
20–24	660
25–34	372
35–44	276
45–54	171
55–64	111
65 and older	84

9. If we select a worker randomly from those surveyed, what is the probability that the person is younger than 55?

10. If we select a worker randomly from those surveyed, what is the probability that the person is older than 19?

11. If a single card is drawn from a standard 52-card deck, what is the probability that it is either a five or a red card?

12. If a single card is drawn from a standard 52-card deck, what is the probability that it is either a face card or a red card?

13. If a pair of dice is rolled, what is the probability that the total showing is either odd or greater than eight?

14. If a pair of dice is rolled, what is the probability that the total showing is either even or less than seven?

Use the following table, that we presented in Example 4, relating the amount of time consumers spend shopping on the Internet per month with their annual income, to answer Exercises 15–18.

Annual Income	10+ Hours	3–9 Hours	0–2 Hours	Totals
Above $60,000	192	176	128	496
$40,000–$60,000	160	208	144	512
Below $40,000	128	192	272	592
Totals	480	576	544	1,600

15. What is the probability that a consumer we select randomly either spends 0–2 hours per month shopping on the Internet or has an annual income below $40,000?

16. What is the probability that a consumer we select randomly either spends ten or more hours per month shopping on the Internet or has an annual income between $40,000 and $60,000?

17. What is the probability that a consumer we select randomly neither spends more than two hours per month shopping on the Internet nor has an annual income of $60,000 or less?

18. What is the probability that a consumer we select randomly neither spends more than two hours per month shopping on the Internet nor has an annual income below $40,000?

In Exercises 19–22, assume that A and B are events.

19. If $P(A \cup B) = 0.85$, $P(B) = 0.40$, and $P(A) = 0.55$, find $P(A \cap B)$.

20. If $P(A \cup B) = 0.75$, $P(B) = 0.45$, and $P(A) = 0.60$, find $P(A \cap B)$.

21. If $P(A \cup B) = 0.70$, $P(A) = 0.40$, and $P(A \cap B) = 0.25$, find $P(B)$.

22. If $P(A \cup B) = 0.60$, $P(B) = 0.45$, and $P(A \cap B) = 0.20$, find $P(A)$.

23. If we draw a card from a standard 52-card deck, what is the probability that the card is neither a heart nor a face card? (*Hint:* Draw a picture of this situation before trying to calculate the probability.)

24. If we draw a card from a standard 52-card deck, what is the probability that the card is neither red nor a queen? (*Hint:* Draw a picture of this situation before trying to calculate the probability.)

Communicating Mathematics *In Exercises 25–28, determine whether each statement is true or false for events A, B, and C. Explain your answer.*

25. $P(A) = P(A \cup B) - P(B)$

26. $P(A \cup B) - P(B) = P(A \cap B)$

27. $P(A) + P(B) - P(A \cup B) = P(A \cap B)$

28. $P(A \cup B) - P(B) = P(A)$

Probability of earning commissions. *Joanna earns both a salary and a monthly commission as a sales representative for an electronics store. The following table lists her estimates of* *the probabilities of earning various commissions next month. Use this table to calculate the probabilities in Exercises 29–32.*

Commission	Probability That This Will Happen
Less than $1,000	0.08
$1,000 to $1,249	0.11
$1,250 to $1,499	0.23
$1,500 to $1,749	0.30
$1,750 to $1,999	0.12
$2,000 to $2,249	0.05
$2,250 to $2,499	0.08
$2,500 or more	0.03

29. the probability that she will earn at least $1,000 in commissions

30. the probability that she will earn at least $1,500 in commissions

31. the probability that she will earn no more than $1,999 in commissions

32. the probability that she will earn less than $2,250 in commissions

A college administration has conducted a study of 200 randomly selected students to determine the relationship between satisfaction with academic advisement and academic success. They obtained the following information: Of the 70 students on academic probation, 32 are not satisfied with advisement; however, only 20 of the students not on academic probation are dissatisfied with advisement. Use these data to answer Exercises 33–36. In each exercise, assume that we select a student at random.

33. What is the probability that the student is not on academic probation?

34. What is the probability that the student is satisfied with advisement?

35. What is the probability that the student is on academic probation and is satisfied with advisement?

36. What is the probability that the student is not on academic probation and is satisfied with advisement?

Further Exercises

37. Communicating Mathematics Many texts that discuss probability state that if events E and F are disjoint, then $P(E \cup F) = P(E) + P(F)$. Explain why it is really not necessary to state this formula.

38. Communicating Mathematics If $P(E \cup F) = P(E) + P(F)$, what can you conclude about $P(E \cap F)$?

39. Communicating Mathematics Explain in your own words the intuition behind the formula for computing the probability of the complement of an event.

40. Communicating Mathematics Explain in your own words the intuition behind the formula for computing the probability of the union of two events.

* **41.** Communicating Mathematics If events A, B, and C are as pictured in this Venn diagram, write a formula for $P(A \cup B \cup C)$. Explain your answer.

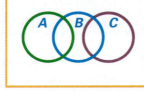

42. Communicating Mathematics If events A, B, and C are as pictured in this Venn diagram, write a formula for $P(A \cup B \cup C)$. Explain your answer.

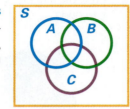

---

* Exercise numbers circled in red can be used as group exercises.

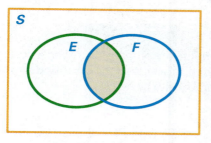

FIGURE 13.11   The intersection of two events.

13.3 CONDITIONAL PROBABILITY AND INTERSECTIONS OF EVENTS

We now move from computing the probabilities of complements and unions of events to computing the probabilities of the intersection of events. Because $E \cap F$* is a subset of both E and F (see Figure 13.11), we expect its probability to be less than the probability of both E and F. The question is, "Exactly how much less?"

In order to answer this question, we need to understand how one event can affect how we compute the probability of another event.

Conditional probability takes into account that one event occurring may change the probability of a second event.

Suppose that you and your friend Marcus cannot agree as to which video to rent and you decide to settle the matter by rolling a pair of dice. You will each pick a number and then roll the dice. The person whose total showing on the dice comes up first gets to pick the video. With your knowledge of probability, you know that the number you should pick is seven because it has the highest probability of appearing, namely $\frac{1}{6}$.

In order to begin to understand the notion of conditional probability, we will change the situation that we just described. Assume another friend, Janelle, is present who will go to either movie that you and Marcus choose. Now, however, you will have Janelle roll the dice before you pick your numbers. You are not allowed to look at the dice, but Janelle will tell you some information about the dice. Then you and Marcus will pick your numbers and finally look at the dice.

Suppose that Janelle rolls the dice and tells you that the total showing is an even number. Would you still choose a seven? Of course not because the situation has changed completely. Instead of the probability being $\frac{1}{6}$, knowing the condition that the total is even, the probability now of having a seven is 0. The point is that instead of considering all pairs in the sample space, now we must exclude those pairs such as (1, 4), (5, 6), and (4, 3) that give us an odd total.

In a similar way, suppose that you draw a card from a standard 52-card deck, put that card in your pocket, and then draw a second card. What is the probability that the second card is a king? How you answer this question depends on knowing what card is in your pocket. If the card in your pocket is a king, then there are three kings remaining in the 51 cards that are left, so the probability is

---

* Again, throughout this section, S is a sample space and E and F are events.

$\frac{3}{51}$. If the card in your pocket is not a king, then the probability of the second card being a king is $\frac{4}{51}$. Why? This discussion leads us to the formal definition of conditional probability.

> ### DEFINITION
>
> When we compute the probability of event F assuming that the event E has already occurred, this is called the **conditional probability** of F, given E. We denote this probability by $P(F \mid E)$.

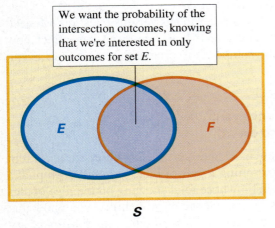

We want the probability of the intersection outcomes, knowing that we're interested in only outcomes for set E.

FIGURE 13.12 To compute the probability of F given E, we compare the outcomes in $E \cap F$ with the outcomes in E.

We read $P(F \mid E)$ as "the probability of F given that E has occurred," or in a quicker way, "the probability of F given E." The Venn diagram in Figure 13.12 will help you remember the meaning of conditional probability.

We drew E with a heavy line in Figure 13.12 to emphasize that when we assume that E has occurred, we can then think of the outcomes outside of E as being discarded from the discussion. In computing conditional probability, you will find it useful to consider the sample space to be E and the event as being $E \cap F$, rather than F. We will first state a special rule for computing conditional probability when the outcomes are equally likely (all have the same probability of occurring). We will state the more general conditional probability rule later.

> **Special Rule for Computing $P(F \mid E)$ by Counting**
>
> If E and F are events in a sample space with equally likely outcomes, then $P(F \mid E) = \frac{n(E \cap F)}{n(E)}$.

EXAMPLE 1   Computing Conditional Probability by Counting

Assume that we roll two dice and the total showing is even. What is the probability that the total showing is ten?

SOLUTION: This sample space has 36 equally likely outcomes. We will let E be the event "we roll an even total" and let T be the event "we roll a ten." There are 18 pairs that give an even total; therefore, $n(E) = 18$. Of these outcomes, the pairs (4, 6), (5, 5), and (6, 4) will give a total of ten, thus, $n(E \cap T) = 3$. Using the special rule for computing conditional probability, we get,

$$P(T \mid E) = \frac{n(E \cap T)}{n(E)} = \frac{3}{18} = \frac{1}{6}.$$

Notice how the probability of rolling a ten has increased when we know that the total showing is even.

PROBLEM SOLVING

As we have emphasized often, the order in which we do things is important in mathematics. In Example 1, $P(T \mid E)$ does not mean the same thing as $P(E \mid T)$. To understand the difference, state what $P(E \mid T)$ represents and then compute $P(E \mid T)$.

It is important to remember that the special rule for computing conditional probability only works when the outcomes in the sample space are *equally likely*. Sometimes you may be solving a problem where that is not the case; or, you may have a situation where it is not possible to count the outcomes. In such cases, we need a rule for computing $P(F \mid E)$ that is based on probability rather than counting.

> **General Rule for Computing $P(F \mid E)$.**
>
> If E and F are events in a sample space, then $P(F \mid E) = \frac{P(E \cap F)}{P(E)}$.

We still can use Figure 13.12 to remember this rule; however, now instead of comparing the number of outcomes in $E \cap F$ with the number of outcomes in E, we compare the probability of $E \cap F$ with the probability of E.

EXAMPLE 2 Using the General Rule for Computing Conditional Probability

The state bureau of labor statistics conducted a survey of college graduates comparing starting salaries to majors. The survey results are listed in Table 13.8.

Major	$20,000 and Below	$20,001 to $25,000	$25,001 to $30,000	$30,001 to $35,000	Above $35,000	Totals (%)
Liberal arts	6*	10	9	1	1	27
Science	2	4	10	2	2	20
Social sciences	3	6	7	1	1	18
Health fields	1	1	8	3	1	14
Technology	0	2	7	8	4	21
Totals (%)	12	23	41	15	9	100

* These numbers are percentages.

TABLE 13.8 Survey comparing starting salaries to major in college.

If we select a graduate who was offered between $30,001 and $35,000, what is the probability that the student has a degree in the health fields?

SOLUTION: Each entry in Table 13.8 is the probability of an event. For example, the 8 percent that we highlighted is the probability of selecting a graduate in technology who earns between $30,001 and $35,000, inclusive. The 14 percent that we highlighted tells us the probability of selecting a graduate who majored in the health fields.

Let R be the event "the graduate received a starting salary of $30,001 to $35,000, and let H be the event "the student has a degree in the health fields." Then we wish to find $P(H \mid R)$.

Since we want the probability of H given R, we can, in effect, ignore all the outcomes that do not correspond to a starting salary of $30,001 to $35,000. We darken the columns we wish to ignore in Table 13.9.

Major	$20,000 or Below	$20,001 to $25,000	$25,001 to $30,000	$30,001 to $35,000	Above $35,000	Totals (%)
Liberal arts	6	10	9	1	1	27
Science	2	4	10	2	2	20
Social sciences	3	6	7	1	1	18
Health fields	1	1	8	3	1	14
Technology	0	2	7	8	4	21
Totals (%)	12	23	41	15	9	100

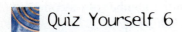

Quiz Yourself 6

Assume that we select a graduate who received more than $35,000 as a starting salary. Use Table 13.8 to find the probability that the graduate has a degree in technology.

TABLE 13.9 The columns we wish to ignore are green.

In order to use the general rule for computing conditional probability, we first need to know $P(R)$ and $P(H \cap R)$. We find $P(R)$ by adding the percentages in the column labeled $30,001 to $35,000 to get 15 percent or 0.15. Also, $P(H \cap R)$ is the 3 percent or 0.03 found in the intersection of the row labeled "Health Fields" and the column labeled $30,001 to $35,000. Thus

$$P(H \mid R) = \frac{P(H \cap R)}{P(R)} = \frac{0.03}{0.15} = 0.20.$$

Therefore, if a graduate receives an offer between $30,001 and $35,000, then the probability that the person is in the health fields is 0.20, or 20 percent.

■

We use conditional probability to find the probability of the intersection of two events.

The general rule for computing conditional probability states that $P(F \mid E) = \frac{P(E \cap F)}{P(E)}$. If we multiply both sides of this equation by the expression $P(E)$, we get the rule for computing the probability of the intersection of two events.

> **Rule for Computing the Probability of the Intersection of Events**
>
> If E and F are two events, then
>
> $$P(E \cap F) = P(E) \cdot P(F \mid E).$$

EXAMPLE 3 **Finding the Probability of the Intersection of Two Events**

Suppose that we draw two cards without replacement from a standard 52-card deck. What is the probability that both cards are kings?

SOLUTION: We can think of this event as the intersection of the two events A, "we draw a king on the first card," and B, "we draw a king on the second card." By the rule we just stated we need to calculate

$$P(A \cap B) = P(A) \cdot P(B|A).$$

Finding the probability of A is easy because we can select any one of the four kings from the 52 cards in the deck. Therefore, $P(A) = \frac{4}{52}$.

Let us think carefully about the meaning of $P(B|A)$. In words, we would read this as "the probability that event B occurs, given the fact that event A has already occurred." That is, we are assuming that we drew a king on the first card and have not replaced it in the deck. When we draw the second card, there are only three kings among the remaining 51 cards, so the probability of drawing the second king is therefore $P(B|A) = \frac{3}{51}$.

Thus the probability that we get kings on both the first and second draws is

$$P(A \cap B) = P(A) \cdot P(B|A) = \frac{4}{52} \cdot \frac{3}{51} = \frac{1}{13} \cdot \frac{1}{17} = \frac{1}{221} \approx 0.0045. \qquad \text{}$$

Quiz Yourself 7

Suppose that we draw two cards without replacement from a standard 52-card deck. What is the probability that both cards are face cards?

SOME GOOD ADVICE

A common mistake that you can make when computing conditional probability is to use the formula $P(A \cap B) = P(A) \cdot P(B)$. That is, you may forget to take into account that event A has occurred. Notice that if we had used this incorrect formula in Example 3, the probability of drawing a second king would have been $\frac{4}{52}$, not $\frac{3}{51}$. In effect, we would have computed the second probability as though the first card had been returned to the deck.

Trees help you visualize probability computations.

Recall that the Three Way Principle in Section 1.1 tells you that drawing a diagram is a good problem solving technique. You will find that it is often very helpful to draw a probability tree to help you understand a conditional probability problem.

Using Trees to Calculate Probabilities

We can represent an experiment that happens in stages with a tree whose branches represent the outcomes of the experiment. We calculate the probability of an outcome by multiplying the probabilities found along the branch representing that outcome.

In some states, in order to get a driver's license, a person must first pass a written test and then pass a driving test. If a candidate for a license fails the driving test a fixed number of times, then the person must start all over again with the written test. Example 4 shows how we can use conditional probability to decide how many times a candidate should be allowed to retake the driving test.

EXAMPLE 4   Using a Tree to Compute Probabilities

A candidate for a driver's license is currently allowed to take a driving test three times before having to retake a preliminary written test. Some feel that candidates should only be given two attempts at the driving test before retaking the written test. Others feel that such a regulation is unduly harsh to those who now pass on their third attempt.

From past history, 60 percent pass the driving test the first time, 75 percent pass on their second try, and only 30 percent who take it the third time are able to pass.

a) What is the probability that a candidate will first fail the first time and then pass the test on the second try?

b) What is the probability that a candidate will fail twice and pass the test on the third try?

SOLUTION: (a) We will solve this problem by first drawing a simple tree diagram in Figure 13.13(a), which we will expand later to solve part (b).

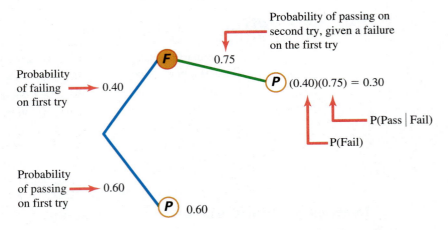

FIGURE 13.13   (a) Tree showing probability of passing a driving test in two tries.

The top branch in Figure 13.13(a), passing through F and then P, shows that the probability of failing on the first attempt and then passing on the second attempt will be

$$P(Fail\ and\ then\ a\ pass) = P(Fail)P(Pass\,|\,Fail) = (0.40)(0.75) = 0.30.$$

This means that the probability is 0.30 that the person will pass the test in two attempts.

b) Figure 13.13(b) expands the tree in Figure 13.13(a) to show the probability of passing the test in one, two, or three attempts.

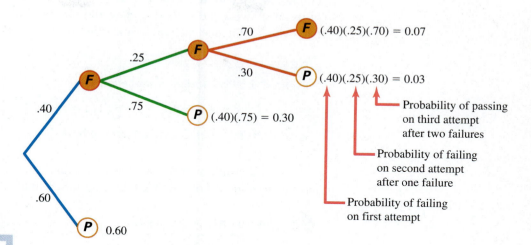

FIGURE 13.13   (b) Tree showing probabilities of passing driving test in 1, 2, or 3 tries.

The branch that passes through F, then F again, and finally through P shows that the probability of passing a test on the third try is $(0.40)(0.25)(0.30) = 0.03$. Because the probability is only 0.03 that a candidate will pass the test on the third attempt, we see that reducing the number of attempts to pass the test would not affect many candidates.

Notice that in Example 4(a), we could think of the product $(0.40)(0.75)$ as the probability of the events

A: "the candidate fails on the first try"

and

B: "the candidate passes on the second try."

The event $A \cap B$ is the event "the candidate fails on the first try and passes on the second try." By looking at the tree, we found that

$$P(A \cap B) = P(A) \cdot P(B \mid A) = (0.40)(0.75).$$

Independent events have no effect on each other's probabilities.

As we have seen, knowing that a first event has occurred may affect the way we calculate the probability of a second event. Such was the case in Example 3 when we computed the probability of drawing a second king, given that a first king was drawn and not returned to the deck.

Quiz Yourself 8

A box contains two red balls and two green balls. Balls will be drawn from the box without replacement until a green ball is selected.

a) Draw a tree to represent this situation.

b) What is the probability that a green ball is selected on the third try?

However, sometimes the occurrence of the first event has no effect whatsoever on the probability of the second event. This would have been the case in Example 3 if after we drew the first king, we then returned it to the deck. The probability of drawing the second king would remain at $\frac{4}{52}$. So you see that sometimes two events influence each other and other times they do not.

DEFINITIONS

Events E and F are **independent** events if

$$P(F \mid E) = P(F).$$

If $P(F \mid E) \neq P(F)$, then E and F are **dependent**.

This definition says that if E and F are independent, then knowing that E has occurred does not influence the way we compute the probability of F.

 PROBLEM SOLVING

Recall that the Analogies Principle in Section 1.1 tells you that mathematical terminology, symbolism, and its equations are often based on real-life ideas. Although at times it may seem difficult, if you work hard to make the connection between the intuitive ideas and the mathematical formalism, you will be rewarded by your increased understanding of mathematics. If you understand how the everyday usage of the terms *independent* and *dependent* corresponds to their mathematical definitions, it will help you to remember their meaning.

EXAMPLE 5   **Determining Whether Events Are Independent or Dependent**

Assume we roll a red and a green die. Are the events F, "a five shows on the red die" and G, "the total showing on the dice is greater than ten," independent or dependent?

SOLUTION: To answer this question we must determine whether $P(G \mid F)$ and $P(G)$ are the same or different. There are three outcomes—(5, 6), (6, 5), and (6, 6)—that give a total greater than ten, so $P(G) = \frac{3}{36} = \frac{1}{12}$.

Now,

$$F = \{(5, 1), (5, 2), (5, 3), (5, 4), (5, 5), (5, 6)\}$$

and

$$G \cap F = \{(5, 6)\},$$

so

$$P(G \mid F) = \frac{P(G \cap F)}{P(F)} = \frac{1/36}{6/36} = \frac{1}{6}.$$

Since $P(G \mid F) \neq P(G)$, the events are dependent.

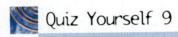 Quiz Yourself 9

The situation is the same as in Example 5. Are the events F, "a five shows on the red die," and O, "an odd total shows on the dice," dependent or independent?

EXAMPLE 6   **Selecting a Dormitory Room**

Mariella is taking part in a lottery for a room in one of the new dormitories at her college. She is guaranteed a space, but she will have to draw a card randomly to determine exactly which room she will have. Each card has the name of one dormitory, X, Y, or Z, and also has a two-person room number or an apartment number. Thirty percent of the available spaces are in X, fifty percent in Y, and twenty percent of the spaces are in Z.

Half of the available spaces in X are in rooms, forty percent of Y's spaces are in rooms, and thirty percent of the spaces in Z are in rooms.

a) Draw a probability tree to describe this situation.

b) Given that Mariella selects a card for dormitory Y, what is the probability that she will be assigned to an apartment?

c) What is the probability that Mariella will be assigned an apartment in one of the three dormitories?

SOLUTION: a) In drawing the tree, we will think of Mariella's assignment happening in two stages. First she is assigned a dormitory and then she is assigned either a room or an apartment. We show the tree in Figure 13.14, which begins with three branches corresponding to the dormitories X, Y, and Z. Each of these three branches has two further branches representing the assignment of either a room or an apartment.

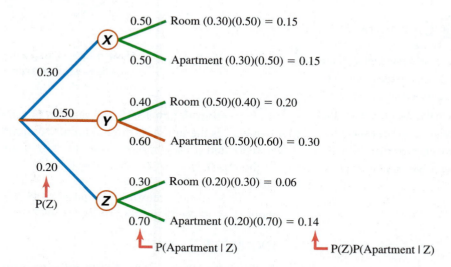

FIGURE 13.14   Probability tree for Mariella's room assignment.

We wrote various probabilities along the tree's branches in Figure 13.14. If you look carefully at the lower branch, you see that the 0.20 is the probability that Mariella will be assigned to dormitory Z; thus $P(Z) = 0.20$. The 0.70 is a conditional probability. Assuming the condition that she is assigned to dormitory Z, then there is a 0.70 chance that she will be assigned to an apartment. Symbolically, this means that $P(Apartment|Z) = 0.70$.

The product $(0.20)(0.70) = 0.14$ is the probability that Mariella will be assigned to dormitory Z and also to an apartment.* We could have written this numeric equation more formally as

$$P(Z \cap Apartment) = P(Z) \cdot P(Apartment \,|\, Z),$$

which is the formula we gave you for computing the probability of the intersection of two events. In Quiz Yourself 10, we will ask you to interpret some more of these probabilities.

b) We can answer this question easily by looking at the branch that is highlighted in red in Figure 13.14. That branch shows that after Mariella has been assigned to dormitory Y, there is then a 0.60 chance that she will get an apartment. That is, $P(Apartment \,|\, Y) = 0.60$.

c) Realize that if Mariella is assigned to an apartment, then she is in *exactly one* of dormitories X, Y, or Z. Figure 13.15 shows how to find the probability that she is assigned an apartment.

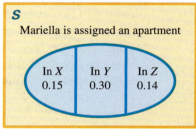

Quiz Yourself 10

a) Explain the meaning of the 0.30 on the top branch of the tree in Figure 13.14.

b) What is the probability that Mariella will be assigned a room, given that she is assigned to dormitory Y?

FIGURE 13.15 Probabilities that Mariella is assigned an apartment in dormitories X, Y, or Z.

From Figure 13.15, we see that the event "Mariella is assigned an apartment" is the union of three mutually disjoint subevents with probabilities 0.15, 0.30, and 0.14. Therefore, the probability that Mariella is assigned an apartment is $0.15 + 0.30 + 0.14 = 0.59$.

We will now return to the issue of HIV testing for engaged couples that we mentioned at the beginning of the chapter.

EXAMPLE 7 HIV Testing†

In the early 1990s, public health officials estimated that the incidence of the HIV virus in the general population was about 0.5 percent. Also, there was a test that when given to people who had the HIV virus would correctly identify the virus 95 percent of the time. This test also gave a false positive about 5 percent of the time. That is, the test results would be positive even though the person did not have HIV.

---

* It may seem that we have gotten away from thinking of events as subsets of the sample space, but we have not. The "Z" represents the set of cards that have dormitory Z written on them, and "Apartment" represents the set of all cards that have apartment written on them.

† For a more extensive discussion of this example, see Richard Isaac, *The Pleasures of Probability* (Springer-Verlag, 1995), Chapter 4.

Highlight: *Expert Systems*

Whether playing chess, doing mathematics, making soup, or practicing law, what makes an expert different from the rest of us? Is it the amount of knowledge or the way an expert reasons about the knowledge that is more important? Does an expert organize knowledge differently than the non-expert? In order to answer such questions, scientists developed complex computer programs called *expert systems*. These scientists learned that organizing knowledge effectively is important in understanding what makes a person an expert. Once information is organized, expert systems rely heavily on conditional probability to draw conclusions from the information.

The "grandfather of expert systems"* is an ingenious program called Mycin, which was developed at Stanford University to diagnose blood infections. Some other well-known expert systems are Hearsay, which understands spoken language; Prospector, which is able to predict the location of valuable mineral resources; and Internist, which can diagnose about 75 percent of internal diseases. At present, over one thousand expert systems are available and, with ongoing research, in this area, expert systems continue to be a major application of probability theory.

a) Of the people who test positive for HIV, what percentage of them do we expect to actually have the virus?

b) What is the value of requiring this test of all couples who apply for a marriage license?

SOLUTION: a) Let T be the event "the person tests positive for HIV" and let V be the event "the person has the virus." We are asking, then, if we are given that the person tests positive, what is the probability that the person has the virus? That is, we are asking for $P(V \mid T)$. The tree in Figure 13.16 will give us some insight into this problem.

Recall that $P(V \mid T) = \frac{P(V \cap T)}{P(T)}$. Branches 1 and 3 in Figure 13.16 correspond to event T, the HIV test is positive. This means that $P(T) = (.005)(.95) + (.995)(.05)$. The event $V \cap T$ corresponds to branch 1. Therefore $P(V \cap T) = (.005)(.95)$. Therefore,

$$P(V \mid T) = \frac{P(V \cap T)}{P(T)} = \frac{(.005)(.95)}{(.005)(.95) + (.995)(.05)} = 0.087.$$

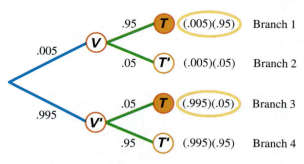

FIGURE 13.16 Tree showing probabilities for HIV testing.

This means that we would expect that less than 9 percent of the people who have positive test results would actually have the HIV virus.

b) It is not clear that this testing would be desirable because a relatively small number of new cases of HIV would be discovered. On the other hand, a great deal of distress might result from the expected large percentage of false positives.

---

* See Frank Puppe, *Systematic Introduction to Expert Systems* (Springer-Verlag, 1993), pp. 3–8.

Exercises 13.3

In many of these exercises, you may find it helpful to draw a tree diagram before computing the probabilities.

Communicating Mathematics *In Exercises 1–4, assume that we are rolling two fair dice. First compute* P(F) *and then* P(F|E). *Explain why you would expect the probability of* F *to change as it did when we added the condition that* E *had occurred.*

1. E—an odd total shows on the dice

   F—the total is seven

2. E—an even total shows on the dice

   F—the total is four

3. E—a three shows on at least one of the dice

   F—the total is less than five

4. E—a two shows on at least one of the dice

   F—the total is greater than five

Playing a store's discount game. *In order to stimulate business, a department store has introduced a game called "Register Roulette." To play, a customer selects a first purchase and a second purchase to qualify for a discount. The customer then reaches into a container and randomly selects two game tokens. There are 50 green tokens that qualify a purchase for a 10 percent discount, 10 blue tokens that result in a 20 percent discount, and 2 red tokens that give a 50 percent discount.*

    In Exercises 5–8 assume that you are randomly selecting two of these tokens, without replacement. Compute P(F|E).

5. E—you select a red token first

   F—the second token is green

6. E—you select a red token first

   F—the second token is red

7. E—you select a non-red token first

   F—the second token is red

8. E—you select a non-red token first

   F—the second token is non-red

In Exercises 9–12, we are drawing a single card from a standard 52-card deck. Find each probability.

9. $P(\text{heart}|\text{red})$

10. $P(\text{king}|\text{face card})$

11. $P(\text{seven}|\text{non–face card})$

12. $P(\text{even-numbered card}|\text{non–face card})$

In Exercises 13–16, two fair dice are rolled. Compute P(E|F) *and* P(F|E). *Explain why the two probabilities are not the same.*

13. E—a total greater than nine shows on the dice

    F—the total is odd

14. E—a total greater than nine shows on the dice

    F—the total is even

15. E—a three shows on at least one of the dice

    F—the total is odd

16. E—a four shows on at least one of the dice

    F—the total is greater than seven

Probability and starting salaries. *The following table from Example 2 relates starting salaries of college graduates and their majors. In Exercises 17–20, find* P(F|E).

Major	$20,000 and Below	$20,001 to $25,000	$25,001 to $30,000	$30,001 to $35,000	Above $35,000	Totals (%)
Liberal arts	6*	10	9	1	1	27
Science	2	4	10	2	2	20
Social sciences	3	6	7	1	1	18
Health fields	1	1	8	3	1	14
Technology	0	2	7	8	4	21
Totals (%)	12	23	41	15	9	100

* These numbers are percentages.

17. *E*—a person majored in social science

F—a person received a starting salary between $20,001 and $25,000

18. *E*—a person majored in science

F—a person received a starting salary between $25,001 and $30,000

19. *E*—a person received a starting salary between $30,001 and $35,000

F—a person majored in technology

20. *E*—a person received a starting salary between $20,001 and $25,000

F—a person majored in liberal arts

Selecting a dormitory room. *Exercises 21–24 refer to the tree diagram that we drew in Figure 13.14 of Example 6.*

21. What is the meaning of the number 0.40 on the top middle branch of Figure 13.14?

22. What is $P(Room \cap Y)$?

23. What is $P(Apartment|X)$?

24. If Mariella gets an apartment, what is the probability that she is in dorm *Z*?

Testing a cold medication. *Imagine that you are taking part in a study to test a new cold medicine. Although you don't know exactly what drug you are taking, the probability that it is drug A is 10 percent, that it is drug B is 20 percent, and that it is drug C, 70 percent. From past clinical trials, the probabilities that these drugs will improve your condition are: A (30 percent), B (60 percent), and C (70 percent).*

25. Draw a tree to illustrate this drug trial situation.

26. What is the probability that you will improve given that you are taking drug B?

27. What is the probability that you will improve?

28. If you improve, what is the probability that you are taking drug B?

29. Communicating Mathematics   In what ways are the special and general rules for computing conditional probability similar?

30. Communicating Mathematics   In what ways are the special and general rules for computing conditional probability different?

31. Communicating Mathematics   If *A* and *B* are events, in general, how does $P(B|A)$ compare with $P(A \cap B)$? Which number is generally smaller? Why?

32. Communicating Mathematics   If *A* and *B* are events, can $P(B|A) = P(B)$? When?

Probability and drawing cards. *In Exercises 33–38, assume that we draw two cards from a standard 52-card deck. Find the desired probabilities.*

  a) First assume that the cards are drawn without replacement.

  b) Next assume that the cards are drawn with replacement.

33. the probability that we draw two jacks

34. the probability that we draw two hearts

35. the probability that we draw a face card followed by a non–face card

36. the probability that we draw a heart followed by a spade

37. the probability that we draw a jack and a king

38. the probability that we draw a heart and a spade

39. We are drawing two cards without replacement from a standard 52-card deck. Find the probability that we draw at least one face card. (*Hint:* Consider the complement.)

40. We are drawing two cards with replacement from a standard 52-card deck. Find the probability that we draw at least one heart. (*Hint:* Consider the complement.)

41. We draw three cards without replacement from a standard 52-card deck. Find the probability of drawing three face cards.

42. We draw three cards without replacement from a standard 52-card deck. Find the probability of drawing three spades.

43. We draw three cards without replacement from a standard 52-card deck. Find the probability of drawing exactly two hearts.

44. We draw three cards without replacement from a standard 52-card deck. Find the probability of drawing exactly two kings.

45. We roll a pair of dice three times. Find the probability that a total of five is rolled each time.

46. We roll a pair of dice three times. Find the probability that a total of five is rolled exactly twice.

Probability and rolling dice. *In Exercises 47–50, we are rolling a pair of fair dice. For each pair of events, determine whether* E *and* F *are dependent.*

47. *E*—the total showing is greater than nine

F—the total showing is even

48. *E*—at least one two shows on the dice

    F—the total showing is less than six

49. *E*—a three shows on the first die

    F—the total showing is even

50. *E*—the total showing is odd

    F—the total showing is greater than four

51. **Probability and exam questions.** Assume that either Professor Ansah or Professor Brunich has constructed the comprehensive exam that you must pass for graduation. Since each professor has extremely different views, it would be useful for you to know who has written the exam questions so that you can slant your answers accordingly. Assume that there is a 60 percent chance that Ansah wrote the exam. Ansah asks a question about international relations 30 percent of the time and Brunich asks a similar question 75 percent of the time. If there is a question on the exam regarding international relations, what is the probability that Ansah wrote the exam?

52. **Probability and exam questions.** Assume now that a third professor, Professor Ubaru, writes the exam 20 percent of the time, Brunich 30 percent of the time, and Ansah the rest. Ubaru asks a question about international relations 40 percent of the time, Brunich 35 percent of the time, and Ansah 25 percent of the time. If there is an international relations question on the exam, what is the probability that Brunich did not write the exam?

Product reliability. *You wish to purchase a DVD drive for your laptop computer. Assume that 65 percent of the drives are made outside the United States. Of the U.S.–made drives, 4 percent are defective; of the foreign-made drives, 6 percent are defective. Determine each probability rounded to three decimal places.*

53. The probability that the drive you purchase is U.S.–made and is not defective.

54. The probability that the drive you purchase is foreign-made and is defective.

55. If your drive is defective, the probability that it is foreign-made.

56. If your drive is defective, the probability that it is made in the United States.

HIV testing. *In Exercises 57–58, do computations similar to those in Example 7 using this revised information. Assume that the incidence of the HIV virus in a particular population is 6 percent and that the test correctly identifies the virus 90 percent of the time. Assume that false positives occur 8 percent of the time.*

57. If a person tests positive for the virus, what is the probability that the person actually has the virus?

58. If a person does not test positive for the virus, what is the probability that the person does not have the virus?

Further Exercises

59. Communicating Mathematics   Explain how the formal definition of independent events corresponds to your intuitive understanding of the word *independent* in English.

60. Communicating Mathematics   Explain how the formal definition of dependent events corresponds to your intuitive understanding of the word *dependent* in English.

61. Communicating Mathematics   Explain what is wrong with the formula $P(E \cap F) = P(E) \cdot P(F)$. Give an example to show that this formula is sometimes incorrect.

62. Communicating Mathematics   Can the formula $P(E \cap F) = P(E) \cdot P(F)$ ever be used? Explain. Give

an example to show that this formula may sometimes be correct.

The birthday problem. *A surprising result that appears in many elementary discussions on probability is called the birthday problem. The question simply stated is this: "If we poll a certain number of people, what is the probability that at least two of those people were born on the same day of the year?" For example, it may be that in the survey two people were both born on March 29.*

In order to solve this problem, we will use the formula for computing the probability of a complement of an event. It is clear that

P(*duplication of some birthdays*) = 1 − P(*no duplications*)

To illustrate this, assume that we have three people. In order to have no duplications, the second person must have a birthday that is different from that of the first person, and the third person must have a birthday that is different from the first two. The probability that the second birthday is different from the first is $\frac{364}{365}$. The probability that the third person has a birthday different from the first two is $\frac{363}{365}$.

Therefore, for three people,

P(*duplication of birthdays*)

$$= 1 - P(\textit{no duplication of birthdays})$$
$$= 1 - \left(\frac{364}{365}\right)\left(\frac{363}{365}\right) = 0.0082.$$

In Exercises 63–64, we will look at several other cases:

* **63.** Assume that we have ten people. Find the probability that at least two of the ten people were born on the same day of the year.

64. Repeat Exercise 63 for 20 people.

65. Find the smallest number of people such that the probability of two of them being born on the same day of the year is greater than 0.50.

66. Conduct an experiment by surveying groups of various sizes† to see how your surveys conform to the predicted probabilities. If five people are working in a group, each person can survey four groups of 20, 30, 40, and so on to generate a reasonable amount of data.

13.4 EXPECTED VALUE

In this section, we use probability to evaluate and compare alternatives in order to make decisions. For example, consider the following situation. Matt has been working as a Web page designer through a temporary employment agency and is trying to decide whether to continue working for the agency or take a steady job as a computer graphics designer. He has recorded the hours he worked each month in the past and will use that information to predict what will happen in the future.

We can use probability in situations such as this to estimate what we can *expect* to happen. Simple calculations not only can help you compare jobs, but can also predict long-term returns for businesses, estimate the profit that an insurance company will make on a policy, or estimate a casino's earnings. We show you how to do this in Example 1.

---

* Exercise numbers circled in red can be used as group exercises.
† Instead of surveying people, you can write the numbers 1 to 365 on pieces of paper or cardboard and draw these from a box *with replacement*. Be sure to shake the container well before each draw. In this way you can quickly simulate interviews of groups of 20, 30, or 40 people, with the purpose of determining how close your experimental results conform to the predicted probabilities.

EXAMPLE 1 **Estimating the Number of Hours a Part-Time Employee Will Work**

Table 13.10 shows the hours Matt has worked per week as a Web page designer for the past 36 weeks.

Hours Worked per Week	Number of Weeks This Occurred	Probability of Working This Number of Hours in a Given Week
29	1	$\frac{1}{36}$
30	3	$\frac{3}{36}$
34	6	$\frac{6}{36}$
37	4	$\frac{4}{36}$
40	17	$\frac{17}{36}$
42	4	$\frac{4}{36}$
44	1	$\frac{1}{36}$

TABLE 13.10 Probabilities of Matt working a certain number of hours per week as a Web page designer.

Use this information to predict how many hours Matt can expect to work as a Web page designer in future weeks.

SOLUTION: We will use this information to assign probabilities to the number of hours worked per week shown in the left-most column of Table 13.10. For example, since Matt worked 29 hours only one week out of the 36, we assign the probability $\frac{1}{36}$ to the chance of him working 29 hours in a given week. Similarly, the probability of working 40 hours is $\frac{17}{36}$. In order to predict how many hours Matt can expect to work per week, we will multiply each of the values 29, 30, 34, . . . , 44 by its probability and then add these products as follows.

$$\left(\frac{1}{36} \cdot 29\right) + \left(\frac{3}{36} \cdot 30\right) + \left(\frac{6}{36} \cdot 34\right) + \left(\frac{4}{36} \cdot 37\right) + \left(\frac{17}{36} \cdot 40\right)$$

$$+ \left(\frac{4}{36} \cdot 42\right) + \left(\frac{1}{36} \cdot 44\right) = \frac{1,363}{36} = 37.86$$

According to this result, Matt can expect to work about 38 hours per week as a Web page designer. Knowing this may help him decide which job he should pursue. ◎

The value 37.86 we found in Example 1 is called the expected value of the number of hours worked per week. We can now give the formal definition of this concept.

DEFINITION

Assume that an experiment has outcomes numbered 1 to n with probabilities P_1, P_2, P_3, . . . , P_n. Assume that each outcome has a numerical value associated with it and these are labeled V_1, V_2, V_3, . . . , V_n. The **expected value** of the experiment is

$$(P_1 \cdot V_1) + (P_2 \cdot V_2) + (P_3 \cdot V_3) + \cdots + (P_n \cdot V_n).$$

In Example 1, the probabilities P_1, P_2, P_3, and so on, were $\frac{1}{36}$, $\frac{3}{36}$, $\frac{6}{36}$, ... The values V_1, V_2, V_3, and so on, were 29, 30, 34, ...

 SOME GOOD ADVICE

Pay careful attention to what notation tells you to do in performing a calculation. In calculating expected value, you are told to *first* multiply the probability of each outcome by its value and *then* add these products together.

 Quiz Yourself 11*

Change the last line in Table 13.10 to read

Hours Worked per Week	Number of Weeks This Occurred
44	5

and keep the rest of the table the same. Of course you will have to recalculate the probabilities in column three. Redo the calculations in Example 1 to find the expected number of hours per week that Matt will work.

EXAMPLE 2   **Computing Expected Value When Flipping Coins**

What is the number of heads we can expect when we flip four fair coins?

SOLUTION: Recall that there are sixteen ways to flip four coins. We will consider the outcomes for this experiment to be the different numbers of heads that could arise. Of course, these outcomes are not equally likely, as we indicate in Table 13.11. If you don't see this at first, you could draw a tree to show the 16 possible ways that four coins can be flipped. You would find that one of the 16 branches corresponds to no heads, four of the 16 branches would represent flipping exactly one head, and six of the 16 branches would represent flipping exactly two heads, and so on.

We calculate the expected number of heads by first multiplying each outcome by its probability and then adding these products as follows:

$$\left(\frac{1}{16} \cdot 0\right) + \left(\frac{4}{16} \cdot 1\right) + \left(\frac{6}{16} \cdot 2\right) + \left(\frac{4}{16} \cdot 3\right) + \left(\frac{1}{16} \cdot 4\right) = \frac{32}{16} = 2$$

Number of Heads	Probability
0	$\frac{1}{16}$
1	$\frac{4}{16}$
2	$\frac{6}{16}$
3	$\frac{4}{16}$
4	$\frac{1}{16}$

TABLE 13.11   Probabilities of obtaining a number of heads when flipping four coins.

---

* Quiz Yourself answers begin on page 849.

Thus we can expect to flip two heads when we flip four coins, which corresponds to our intuition.

We can use the notion of expected value to predict the likelihood of winning (or more likely losing) at games of chance such as blackjack, roulette, and even lotteries.

EXAMPLE 3   The Expected Value of a Roulette Wheel

Although there are many ways to bet on the 38 numbers of a roulette wheel,* one simple betting scheme is to place a bet, let's say $1, on a single number. In this case, the casino pays you $35 (you also keep your $1 bet) if your number comes up and otherwise you lose the $1. What is the expected value of this bet?

SOLUTION: We can think of this betting scheme as an experiment with two outcomes:

1. Your number comes up and the value to you is $+$35.
2. Your number doesn't come up and the value to you is $-$1.

Since there are 38 equally likely numbers that can occur, the probability of the first outcome is $\frac{1}{38}$ and the probability of the second is $\frac{37}{38}$. The expected value of this bet is therefore

$$\left(\frac{1}{38}\cdot(35)\right) + \left(\frac{37}{38}\cdot(-1)\right) = \frac{35-37}{38} = \frac{-2}{38} = -\frac{1}{19} = -0.0526.$$

This amount means that, on the average, the casino expects you to lose slightly more than 5 cents for every dollar you bet.

The roulette wheel in Example 3 is an example of an unfair game. If a game has an expected value of 0, then the game is called *fair.* A game in which the expected value is not 0 is an *unfair game.* In order for a casino or a state that runs a lottery to make a profit, the game must be favored against the player.

EXAMPLE 4   The Expected Value of a Lottery

It costs $1 to buy a ticket to play a state's daily number. The player chooses a three-digit number between 000 and 999, inclusive, and if the number is selected that day then the player wins $500 (this means the profit is $500 − $1 = $499). The expected value of this game is

$$\left(\frac{1}{1,000}\cdot(499)\right) + \left(\frac{999}{1,000}\cdot(-1)\right) = \frac{499-999}{1,000} = -0.50.$$

What should the price of a ticket be in order for this game to be fair?

---

* See Example 8 in Section 13.1 for a description of a roulette wheel.

Historical Highlight: *The History of Lotteries*

Lotteries have existed since ancient times. The Roman emperor Nero gave slaves or villas as door prizes to guests attending his banquets, and Augustus Caesar used public lotteries to raise funds to repair Rome.

The first public lottery paying money prizes began in Florence, Italy, in the early 1500s; when Italy became consolidated in 1870, this lottery evolved into the Italian National Lottery. In this lottery five numbers are drawn from 1 to 90. A winner who guesses all five numbers is paid at a ratio of 1,000,000 to 1. The number of possible ways to choose these five numbers is $C(90, 5) = 43,949,268$. Thus, as with most lotteries, these odds make the lottery a very good bet for the state and a poor one for the ordinary citizen.

Lotteries also played an important role in the early history of the United States. In 1612, King James I used lotteries to finance the Virginia Company to send colonists to the New World. Benjamin Franklin obtained money to buy cannons to defend Philadelphia, and George Washington built roads through the Cumberland mountains by raising money through another lottery he conducted. In fact, in 1776, the Continental Congress used a lottery to raise $10 million to finance the American Revolution.

SOLUTION: Let us call the price of the ticket x. Then if you win, your profit will be $500 - x$; if you lose, your loss will be x. With this in mind, we recalculate the expected value to get

$$\left(\frac{1}{1,000} \cdot (500 - x)\right) + \left(\frac{999}{1,000} \cdot (-x)\right) = \left(\frac{(500 - x) - 999x}{1,000}\right) = \left(\frac{500 - 1,000x}{1,000}\right).$$

We want the game to be fair, so we set this expected value equal to zero and solve for x. This gives us $\frac{500 - 1,000x}{1,000} = 0$, which simplifies to $500 - 1,000x = 0$. Adding $1,000x$ to both sides and dividing by 1,000, we get $x = \frac{1}{2} = 0.50$. Therefore 50 cents would be a fair price of a ticket to play this lottery. Of course, such a lottery would make no money for the state, which is why most states charge $1 to play the game.

Calculating expected value can help a student decide what is the best strategy for answering questions on standardized tests such as the GMATs.

EXAMPLE 5 Expected Value and Standardized Tests

A student is taking a standardized test consisting of multiple-choice questions, each of which has five choices. The test taker earns one point for each correct answer; $\frac{1}{3}$ point is subtracted for each incorrect answer. Questions left blank neither receive nor lose points.

a) Find the expected value of randomly guessing an answer to a question. Interpret the meaning of this result for the student.

b) If the student can eliminate one of the choices, is it wise to guess in this situation?

SOLUTION: a) Since there are five choices, the student has a probability of $\frac{1}{5}$ of guessing the correct result, and the value of this is $+1$ point. There is a $\frac{4}{5}$ probability of an incorrect guess, with an associated value of $-\frac{1}{3}$ point. The expected value is therefore

$$\left(\frac{1}{5}\cdot 1\right) + \left(\frac{4}{5}\cdot\left(-\frac{1}{3}\right)\right) = \frac{1}{5} + \frac{-4}{15} = \frac{3}{15} - \frac{4}{15} = -\frac{1}{15}.$$

Thus the student will be penalized for guessing and should not do so.

b) If the student eliminates one of the choices and chooses randomly from the remaining four choices, the probability of being correct is $\frac{1}{4}$ with a value of $+1$ point; the probability of being incorrect is $\frac{3}{4}$ with a value of $-\frac{1}{3}$. The expected value is now

$$\left(\frac{1}{4}\cdot 1\right) + \left(\frac{3}{4}\cdot\left(-\frac{1}{3}\right)\right) = \frac{1}{4} + \frac{-1}{4} = 0.$$

The student now neither benefits nor is penalized by guessing.

> **Quiz Yourself 12**
>
> Calculate the expected value as in Example 5(b), but now assume that the student can eliminate two of the choices. Interpret this result.

Life insurance companies earn large profits using calculations similar to those in Example 6.

EXAMPLE 6   Expected Value and Life Insurance

Based on mortality tables, the probability of a 20-year-old male living to age 21 is 0.99. If a $1,000 one-year term life insurance policy on a 20-year-old male costs $25, what is its expected value?

SOLUTION: There are two outcomes to this experiment: the 20-year-old male (1) lives or (2) dies. With outcome (1), he loses the cost of the policy, which is $25. With outcome (2) his beneficiary gains $1,000 - $25 = $975.
Therefore the expected value is

$$((0.99)\cdot(-25)) + ((0.01)\cdot 975) = -24.75 + 9.75 = -15.$$

This result means that a purchaser can expect to lose $15 for each policy purchased.

Exercises 13.4

In Exercises 1–2, you are playing a game in which a single die is rolled. Calculate your expected value for each game. Is the game fair? (Assume that there is no cost to play the game.)

1. If an odd number comes up, you win the number of dollars showing on the die. If an even number comes up, you lose the number of dollars showing on the die.

2. You are playing a game in which a single die is rolled. If a four or five comes up, you win $2; otherwise, you lose $1.

In Exercises 3–4, you pay $1 to play a game in which a pair of fair dice are rolled. Calculate your expected value for the game. (Remember to subtract the cost of playing the game from your winnings.) Is the game fair?

3. If a six, seven, or eight comes up, you win $5; if a two or twelve comes up, you win $3; otherwise, you lose the dollar you paid to play the game.

4. If a total less than five comes up, you win $5; if a total greater than nine comes up, you win $2; otherwise, you lose the dollar you paid to play the game.

In Exercises 5–6, a card is drawn from a standard 52-card deck. Calculate your expected value for each game. You pay $5 to play the game, which must be subtracted from your winnings.

5. If a heart is drawn, you win $10; otherwise, you lose your $5.

6. If a face card is drawn, you win $20; otherwise, you lose your $5.

In Exercises 7–8, you are playing a game in which you flip three fair coins. It costs $1 to play the game, which must be subtracted from your winnings. Calculate the expected value for the game.

7. If all coins show the same (all heads or all tails), you win $3; otherwise, you lose your dollar.

8. If you have two or three heads, you win $3; otherwise, you lose your dollar.

In Exercises 9–12, calculate the price to play the game that would make the game fair.

9. the game in Exercise 3

10. the game in Exercise 5

11. the game in Exercise 7

12. the game in Exercise 8

In Exercises 13–16, first calculate the expected value of the lottery. Determine whether the lottery is a fair game. If the game is not fair, determine a price for playing the game that would make it fair.

13. The Daily Number lottery costs $1 to play. You must pick three digits in order from 0 to 9 and duplicates are allowed. If you win, the prize is $600.

14. The Big Four lottery costs $1 to play. You must pick four digits in order from 0 to 9 and duplicates are allowed. If you win, the prize is $2,000.

15. Five hundred chances are sold at $5 apiece for a raffle. There is a grand prize of $500, two second prizes of $250, and five third prizes of $100.

16. One thousand chances are sold at $2 apiece for a raffle. There is a grand prize of $300, two second prizes of $100, and five third prizes of $25.

In Exercises 17–20, we describe several ways to bet on a roulette wheel. Calculate the expected value of each bet. We show a portion of a layout for betting on roulette in the accompanying diagram. When we say that a bet pays "k to one," we mean that if a player wins, the player wins k dollars as well as keeping his or her bet. When the player loses, he or she loses $1. Recall that there are 38 numbers on a roulette wheel.

17. A player can "bet on a line" by placing a chip at location A in the figure. By placing the chip at A, the player is betting on 1, 2, 3, 0, and 00. This bet pays 6 to 1.

18. A player can "bet on a square" by placing a chip at the intersection of two lines, as at location D. By placing a chip at D, the player is betting on 2, 3, 5, and 6. This bet pays 8 to 1.

19. A player can "bet on a street" by placing a chip on the table at location B. The player is now betting on 7, 8, and 9. This bet pays 8 to 1.

20. Another way to "bet on a line" is to place a chip at location C. By placing a chip at location C, the player is betting on 7, 8, 9, 10, 11, and 12. This bet pays 5 to 1.

In Exercises 21–24, a student is taking a standardized test consisting of several multiple-choice questions. One point is awarded for each correct answer. Questions left blank neither receive nor lose points.

21. If there are four options for each question and the student is penalized $\frac{1}{4}$ point for each wrong answer, is it in the student's best interest to guess? Explain.

22. If there are three options for each question and the student is penalized $\frac{1}{3}$ point for each wrong answer, is it in the student's best interest to guess? Explain.

23. If there are five options for each question and the student is penalized $\frac{1}{2}$ point for each wrong answer, how many options must the student be able to rule out before the expected value of guessing is zero?

24. If there are four options for each question and the student is penalized $\frac{1}{2}$ point for each wrong answer, how many options must the student be able to rule out before the expected value of guessing is zero?

25. Assume that the probability of a 25-year-old male living to age 26, based on mortality tables, is 0.98. If a $1,000 one-year term life insurance policy on a 25-year-old male costs $27.50, what is its expected value?

26. Assume that the probability of a 22-year-old female living to age 23, based on mortality tables, is 0.995. If a $1,000 one-year term life insurance policy on a 22-year-old female costs $20.50, what is its expected value?

27. Your insurance company has a policy to insure personal property. Assume you have a laptop computer worth $2,200, and there is a 2 percent chance that the laptop will be lost or stolen during the next year. What would be a fair premium for the insurance? (We are assuming that the insurance company earns no profit.)

28. Assume that you have a used car worth $6,500 and you wish to insure it for full replacement value if it is stolen. If there is a 1 percent chance that the car will be stolen, what would be a fair premium for this insurance? (We are assuming that the insurance company earns no profit.)

29. A company estimates that it has a 60 percent chance of being successful in bidding on a $50,000 contract. If it costs $5,000 in consultant fees to prepare the bid, what is the expected gain or loss for the company if it decides to bid on this contract?

30. In Exercise 29, suppose that the company believes that it has a 40 percent chance to obtain a contract for $35,000. If it will cost $2,000 to prepare the bid, what is the expected gain or loss for the company if it decides to bid on this contract?

31. Kathy is planning to buy a franchise from Home Deco to sell decorations for the home. The table below shows average weekly profits, rounded to the nearest hundred, for a number of the current franchises. If she were to buy a franchise, what would her expected weekly profit be?

Average Weekly Profit	Number Who Earned This
$100	4
$200	8
$300	13
$400	21
$500	3
$600	1

32. For the past several years, the Metrodelphia Fire Department has been keeping track of the number of fire hydrants that have been opened illegally daily during heat waves. These data (rounded to the nearest ten) are given in the table below. Use this information to calculate how many hydrants the department should expect to be opened per day during the upcoming heat wave.

Hydrants Opened	Days
20	13
30	11
40	15
50	11
60	9
70	1

Further Exercises

33. Nell's Bagels & Stuff, a local coffee shop, sells coffee, bagels, magazines, and newspapers. Nell has gathered information for the past 20 days regarding the demand for bagels. We list this information in the following table.

Demand for Bagels Sold	150	140	130	120
Number of Days with These Sales	3	6	5	6

Nell wishes to use expected value to compute her best strategy for ordering bagels for the next week. She intends to order the same number each  day and must order in multiples of ten; therefore she will order either 120, 130, 140, or 150 bagels. She buys the bagels for 65 cents each and sells them for 90 cents.

a) What is Nell's expected daily profit if she orders 130 bagels per day? *Hint:* First compute the profit Nell will earn if she can sell 150, 140, 130, and 120 bagels.

b) What is Nell's expected daily profit if she orders 140 bagels per day?

34. Mike sells the *Town Crier*, a local paper, at his newsstand. Over the past two weeks he has sold the following number of copies.

Number of Copies Sold	90	85	80	75
Number of Days with These Sales	2	3	4	1

Each copy of the paper costs him 40 cents, and he sells it for 60 cents. Assume that these data will be consistent in the future.

a) What is Mike's expected daily profit if he orders 80 copies per day? *Hint:* First compute the profit Mike will earn if he can sell 90, 85, 80, and 75 papers.

b) What is Mike's expected daily profit if he orders 85 copies per day?

35. Communicating Mathematics   In playing a lottery a person might buy several chances in order to improve the likelihood (probability) of winning. Discuss whether this is the case. Does buying several chances change your expected value of the game? Explain.

*36. Communicating Mathematics   Consider the game of Monopoly. Assume your playing piece is at a particular position in a game in progress. There will no doubt be houses or hotels on some properties, money in Free Parking, and so on. Discuss several such situations with the point of determining the expected value of your next move.

37. Communicating Mathematics   A lottery in which you must choose 6 numbers correctly from 40 possible is called a $\frac{40}{6}$ lottery. In general, an $\frac{m}{n}$ lottery is one in which you must correctly choose n numbers from m possible numbers. Investigate what kind of lotteries there are in your state. What is the probability of winning such a lottery?

CHAPTER SUMMARY

SECTION 13.1   **experiment, outcomes, sample space**
An observation of a random phenomenon is called an experiment. The different possible observations of the experiment are called outcomes. The set of all possible outcomes is called the sample space of the experiment.

event
An event is a subset of the sample space.

probability of an outcome, probability of an event
The probability of an outcome in a sample space is a number between 0 and 1, inclusive. The probability of an event is defined as the sum of the probabilities of the outcomes that make up the event.

---

* Exercise numbers circled in red can be used as group exercises.

empirical assignment of probabilities
When an experiment is repeated several times, we define the probability of event E as

$$P(E) = \frac{\text{the number of times } E \text{ occurs}}{\text{the number of times the experiment is performed}}.$$

This ratio is called the relative frequency of E.

theoretical assignment of probabilities
If S is a sample space with all *equally likely outcomes*, then each outcome has a probability equal to $\dfrac{1}{\text{number of outcomes in } S} = \dfrac{1}{n(S)}$. For an event E in this sample space, $P(E) = \dfrac{n(E)}{n(S)}$.

basic properties of probability
If S is a sample space for some experiment and E is an event in S:

1. $0 \le P(E) \le 1$
2. $P(\varnothing) = 0$
3. $P(S) = 1$

odds
If the outcomes in the sample space are *equally likely*, the odds against event E are the number of outcomes against E occurring compared to the number of outcomes in favor of E occurring. That is, $\dfrac{n(E')}{n(E)}$. An alternate formula for computing odds using probability is $\dfrac{P(E')}{P(E)}$.

SECTION 13.2

the probability of a complement of an event
If E is an event, then $P(E) = 1 - P(E')$.

the probability of a union of two events
If E and F are events, then $P(E \cup F) = P(E) + P(F) - P(E \cap F)$. If E and F are mutually exclusive events, the preceding formula simplifies to $P(E \cup F) = P(E) + P(F)$.

SECTION 13.3

conditional probability
If we compute the probability of event F assuming that event E has already occurred, this is called the conditional probability of F given E. We denote this probability by $P(F|E)$.

computing conditional probability
If E and F are events in a sample space with equally likely outcomes, then $P(F|E) = \dfrac{n(E \cap F)}{n(E)}$. In general, $P(F|E) = \dfrac{P(E \cap F)}{P(E)}$.

computing the probability of the intersection of events
If E and F are events, then $P(E \cap F) = P(E) \cdot P(F|E)$.

independent, dependent events
Events E and F are independent if $P(F|E) = P(F)$. Otherwise, they are dependent.

SECTION 13.4   **expected value**

If an experiment has n outcomes with probabilities $P_1, P_2, P_3, \ldots, P_n$, and $V_1, V_2, V_3, \ldots, V_n$ are values associated with the outcomes, then the expected value of the experiment is $(P_1 \cdot V_1) + (P_2 \cdot V_2) + (P_3 \cdot V_3) + \cdots + (P_n \cdot V_n)$.

CHAPTER TEST

SECTION 13.1

1. Describe each event as a set of outcomes.

   a) When three coins are flipped, we obtain exactly two heads.

   b) When two dice are rolled, we obtain a total of eight.

2. If a single card is selected from a standard 52-card deck, what is the probability that a red face card is selected?

3. Explain the difference between empirical and theoretical probability. Give an example of each type of probability.

4. In pea plants, purple flower color dominates white. With snapdragons, however, a pure red flowering plant crossed with pure white produces a pink flowering plant. If we begin by crossing pure red and pure white snapdragons, what is the probability of white flowers in second-generation plants?

5. a) If the odds against E are 7 to 3, what is $P(E)$?

   b) If $P(E) = 0.55$, what are the odds against E?

SECTION 13.2

6. a) State the formula for computing the probability of the complement of an event.

   b) Draw a diagram to explain this formula.

   c) In what sort of situations would you be likely to use this formula?

7. If a single card is drawn from a standard 52-card deck, what is the probability that we obtain either a face card or a red card? Draw a diagram to illustrate this situation.

SECTION 13.3

8. Explain in your own words what we mean by conditional probability.

9. If a pair of fair dice is rolled, what is the probability that we roll a total of five if we are given that the total is less than nine?

10. Assume that two cards are drawn without replacement from a standard 52-card deck.

   a) What is the probability that two hearts are drawn?

   b) What is the probability that a queen then an ace is drawn?

11. A pair of fair dice are rolled. Are events E and F independent?

        E—an odd total is obtained
        F—the total showing is less than six

12. Assume that the incidence of the HIV virus in a particular population is 4 percent and that the test correctly identifies the virus 90 percent of the time. Assume that false positives occur 6 percent of the time. If a person tests positive for the virus, what is the probability that the person actually has the virus?

SECTION 13.4

13. A card is drawn from a standard 52-card deck. If a face card is drawn, you win $15; otherwise, you lose $4. Calculate the expected value for the game.

14. You are playing a game in which four fair coins are flipped. If all coins show the same (all heads or all tails), you win $5. Calculate the price to play the game that would make the game fair.

Of Further Interest: BINOMIAL EXPERIMENTS

Have you ever taken a quiz for which you were not prepared? Perhaps it was a ten-question true-false quiz, or maybe a five-question multiple-choice quiz. What would your chances of being successful on these types of tests be by purely guessing? Maybe you have bought a box of cereal containing a collectable figure. How many purchases must you make before you can reasonably expect to have the complete set of figures? In an apparently different situation, a pharmaceutical company makes a claim regarding the effectiveness of a new vaccine. How might we test the company's claim for its reliability?

Binomial trials have only two outcomes.

All these questions are related to an important class of experiments in probability theory that deserve special consideration. These experiments are similar in that they all have two outcomes—one of the outcomes is referred to as "success" and the other as "failure." Such experiments are called *binomial* trials*. We list several of these in Table 13.12. What we have decided to call a success or a failure is arbitrary. The point is that one outcome is considered a success and the other a failure.

Experiment	Success	Failure
Flip one coin	Head	Tail
Roll two dice	Roll a total of seven	Roll a total other than a seven
Test a computer chip	Chip functions properly	Chip is defective
Inoculate a person with a flu vaccine	Person does not get the flu	Person gets the flu
Buy a box of cereal	Obtain a new collectable figure	Obtain a collectable figure that you already have
Guess at a multiple-choice question	Guess correct answer	Guess incorrectly

TABLE 13.12   Examples of binomial trials.

Properties of a Binomial Experiment

A sequence of binomial trials is called a **binomial experiment** and has the following properties.

1. The experiment is performed for a fixed number of trials.
2. The experiment has only two outcomes, "success" and "failure."
3. The probability of success is the same from trial to trial.
4. The trials are independent of each other.

---

* These are also often called Bernoulli trials, named after the brilliant seventeenth-century Swiss mathematician Jakob Bernoulli.

EXAMPLE 1   **Experiments That Are Not Binomial Experiments**

Explain why each experiment is not a binomial experiment.

a) Roll a single die until a six comes up.

b) A student takes a course and earns either an A, B, C, D, or F.

c) Select four cards without replacement and observe whether a heart is drawn.

SOLUTION: a) The number of trials is not fixed. We do not know how many times the die will have to be rolled before a six comes up.

b) There are more than two outcomes.

c) Since the cards are not being replaced, the probability of a heart will change from draw to draw.

The probabilities of binomial experiments have the same pattern.

In Example 2, we see a common pattern that occurs in computing binomial probabilities.

EXAMPLE 2   **A Binomial Experiment of Rolling a Pair of Dice Three Times**

What is the probability of rolling a total of seven exactly once if we roll a pair of dice three times?

SOLUTION: Since we can think of this experiment as occurring in three stages (roll dice first time, roll dice second time, roll dice third time), we can represent it by a tree diagram, as in Figure 13.17. We have placed the probability of each outcome of a stage of the experiment along the appropriate branch.

We have highlighted the branches in Figure 13.17 to show the three ways to roll exactly one seven in three rolls of the dice. If we consider rolling a seven a success and rolling a non-seven a failure, then we see that the probability of a success on the first roll and failure on rolls two and three is $(\frac{1}{6})(\frac{5}{6})(\frac{5}{6})$. Similarly, we could get success on the second roll and failure on the other rolls with a probability of $(\frac{5}{6})(\frac{1}{6})(\frac{5}{6})$. Finally, if the success occurs on the third roll, the probability is $(\frac{5}{6})(\frac{5}{6})(\frac{1}{6})$. Thus the probability of rolling exactly one seven on three rolls of the dice is $(\frac{1}{6})(\frac{5}{6})(\frac{5}{6}) + (\frac{5}{6})(\frac{1}{6})(\frac{5}{6}) + (\frac{5}{6})(\frac{5}{6})(\frac{1}{6}) = \frac{75}{216} \approx 0.3472$.

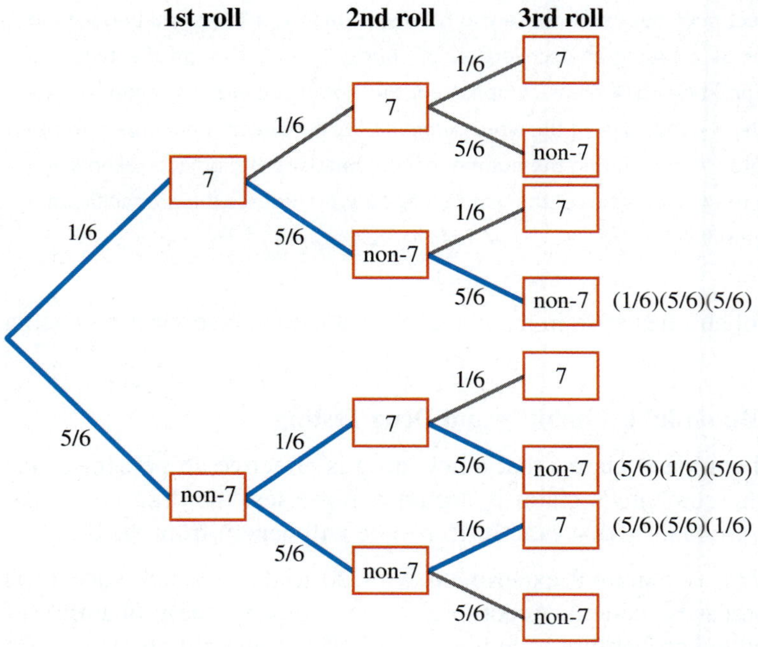

In Example 2, we can write each one of the three probabilities in the form $(\frac{1}{6})^1 (\frac{5}{6})^2$. We view this abstractly as

FIGURE 13.17   Rolling three dice can be thought of as a binomial experiment—in each roll we get a seven or a non-seven.

(probability of success)$^{(\text{number of successes})}$ ×
(probability of failure)$^{(\text{number of failures})}$.

Suppose now that we want to compute the probability of obtaining exactly two sevens when a pair of dice is rolled five times. We will compute this probability in stages.

First we have to choose which two of the five rolls correspond to sevens. From Chapter 12, recall that the number of ways to choose two objects from a set of five is $C(5, 2) = \frac{5!}{2!\,3!} = 10$.

Second, by doing computations similar to Example 2, we see that each of these ten outcomes has a probability of $(\frac{1}{6})^2 (\frac{5}{6})^3$. Thus the probability of rolling exactly two sevens in five rolls of a pair of dice is

$$C(5, 2)\left(\frac{1}{6}\right)^2\left(\frac{5}{6}\right)^3 = 10 \cdot \frac{1}{36} \cdot \frac{125}{216} = \frac{1,250}{7,776} \approx 0.1608.$$

We will now state this calculation in a general form.

Formula for Computing Binomial Probabilities

In a binomial experiment with n trials, if the probability of success in each trial is p, then the probability of exactly k successes is given by

$$C(n, k)\,(p)^k\,(1 - p)^{(n - k)}.$$

We will denote this probability by $B(n, k; p)$.

SOME GOOD ADVICE

If you connect new ideas with those you have seen before, it helps you understand mathematics as an integrated collection of concepts. Even though the binomial probability notation may seem complicated, the notation helps us remember its meaning. The B reminds us of the word *binomial*. We have seen the n, k pattern before in Chapter 12 in counting the number of combinations of n objects taken k at a time. Finally the p reminds us of the word *probability*, in this case the probability of a success in a binomial trial.

Binomial probabilities arise in many real-life situations involving the testing of products.

EXAMPLE 3   Binomial Probability and Drug Testing

A pharmaceutical company claims that a new drug is effective 75 percent of the time. Assuming the company's claim is accurate, if we test the drug on 20 patients, what is the probability that exactly 15 people will benefit from the drug?

SOLUTION: This is a binomial experiment with 20 trials in which success is that the drug benefits the patient. According to the company, the probability of success is 0.75 and the probability of failure is 0.25. The probability of exactly 15 successes is therefore $B(20, 15; 0.75)$.

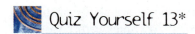

Quiz Yourself 13*

If we roll a single die eight times, what is the probability of getting exactly two threes?

Using the binomial probability formula,† we get

$$B(20, 15; 0.75) = C(20, 15)(0.75)^{15} (1 - 0.75)^{(20 - 15)}$$
$$= 15{,}504(0.75)^{15} (0.25)^5 \approx 0.2023.$$

We will investigate further how to test the pharmaceutical company's claim in the exercises.

EXAMPLE 4   Binomial Probability and Guessing on Exams

Maria has fallen behind in her anthropology class and her instructor just announced a pop quiz consisting of ten true-false questions. If she must get six or more correct to pass the quiz, what is the probability of passing the quiz by guessing?

SOLUTION: To pass the quiz, Maria must get either six, seven, eight, nine, or ten correct. The probability of this is

$$B(10, 6; \tfrac{1}{2}) + B(10, 7; \tfrac{1}{2}) + B(10, 8; \tfrac{1}{2}) + B(10, 9; \tfrac{1}{2}) + B(10, 10; \tfrac{1}{2})$$
$$= 0.2051 + 0.1172 + 0.0439 + 0.0098 + 0.0010 = 0.377.$$

Thus the probability that she will pass the quiz by randomly guessing is almost 40 percent!

We conclude this section with another interesting application involving binomial trials. Manufacturers often package sports cards in small groups so that a collector will have to buy numerous packages in order to obtain a complete set. A collector might ask, "If I know the number of items in the complete set, how many purchases should I expect to make to get a complete set?" Before we answer the collector's question, we need to know more about binomial trials.

Suppose that we are making a sequence of binomial trials where success has probability p and that we repeat the trial until we obtain a success. How many trials do we expect to perform before we get a success? If the probability of success is small, then our intuition tells us that we can expect to perform many trials before we get a success. On the other hand, if the probability of success is large, then we should not have to perform many trials before obtaining a success. In fact, the following can be proved (we will not prove it here):

> **The Number of Binomial Trials We Can Expect before a Success**
>
> If we repeat a binomial trial in which success has probability p, then the number of trials we can expect to perform before we get a success is $\frac{1}{p}$.

---

† Throughout this section, we will round these binomial probabilities to four decimal places. If you carry more decimal places, you will sometimes get answers that are slightly different than ours.

This is saying, for example, that if in a binomial trial the probability of success is $\frac{1}{100}$, then we should expect to perform the experiment 100 times before we get a success. We are now ready to answer a simplified version of the collector's question.

EXAMPLE 5   **How Many Packages Should You Expect to Buy to Obtain a Complete Set of Collectibles?***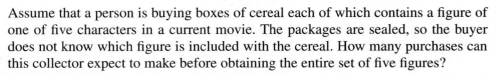

Assume that a person is buying boxes of cereal each of which contains a figure of one of five characters in a current movie. The packages are sealed, so the buyer does not know which figure is included with the cereal. How many purchases can this collector expect to make before obtaining the entire set of five figures?

SOLUTION: We will call the total number of purchases to obtain a complete set T. Then

T = the number of purchases to obtain a first new figure +
the number of purchases to obtain a second new figure +
the number of purchases to obtain a third new figure +
the number of purchases to obtain a fourth new figure +
the number of purchases to obtain a fifth new figure.

The number of purchases required to obtain a first new figure is one, since the very first purchase will give us a new figure.

At this point we have one figure; before we open our next purchase, there are four possible new figures and one old figure. We can think of buying a box of cereal now as a binomial experiment in which the probability of getting a new figure (success) is $\frac{4}{5}$ and the probability of failure is $\frac{1}{5}$. The number of trials we can expect before we get a new figure is therefore $\frac{1}{4/5} = \frac{5}{4}$.

Once we have two different figures, we change our computations. Before we open the next box of cereal, there are possibly one of two old figures in the box or one of three new ones. Buying a box of cereal is now a binomial trial with probability of success $\frac{3}{5}$. Thus we can expect to obtain the third new figure in $\frac{1}{3/5} = \frac{5}{3}$ trials.

Similarly, the expected number of purchases before we get the fourth figure is $\frac{5}{2}$ and the expected number of purchases before we get the fifth figure is $\frac{5}{1}$.

The total number of purchases T we can expect to make before we have a complete set is therefore

$$T = 1 + \frac{5}{4} + \frac{5}{3} + \frac{5}{2} + \frac{5}{1} = 11.42.$$

It is likely that the collector will obtain the complete set of five figures by the twelfth purchase.

_____

* For a more extensive discussion of the collector's problem, see Richard Isaac, *The Pleasures of Probability* (Springer-Verlag, 1995), Chapter 8.

Exercises

In Exercises 1–6, determine whether each experiment is a binomial experiment. If not, explain what property of a binomial experiment fails.

1. Select four cards with replacement from a standard 52-card deck and observe whether a king is drawn.

2. Roll a pair of dice five times and observe whether doubles are obtained.

3. Three coins are flipped until three heads are obtained.

4. A person buys five chances on the Daily Number lottery.

5. A student takes a ten-question true-false quiz.

6. Twenty people are injected with a new cold vaccine.

7. If we roll a pair of dice three times, what is the probability of rolling a total of five exactly once?

8. If we roll a pair of dice three times, what is the probability of rolling a total of eight exactly twice?

In Exercises 9–10, explain the meaning of each expression and compute its value.

9. $B(5, 3; \frac{1}{4})$

10. $B(6, 2; \frac{1}{2})$

In Exercises 11–12, explain the error in using the notation to indicate a binomial probability.

11. $B(3, 5; \frac{1}{3})$

12. $B(4, 3; 2)$

13. If you are given a twelve-question true-false quiz, what is the probability that you will answer *exactly* nine of the questions correctly by guessing?

14. If you are given an eight-question multiple-choice quiz, what is the probability that you will answer *exactly* six of the questions correctly by guessing? Assume that each question has four options.

15. If you are given a twelve-question true-false quiz, what is the probability that you will answer *at least* nine of the questions correctly by guessing?

16. If you are given a six-question multiple-choice quiz, what is the probability that you will answer *at least* four of the questions correctly by guessing? Assume that each question has five options.

17. A heart surgeon has developed a new procedure that she believes can correct a life-threatening medical condition. If the success rate for this procedure is 80 percent, and the procedure is tried on ten patients, what is the probability that at least eight of them will show improvement?

18. You are planning a vacation at the beach and the weather forecast is that there is a 30 percent chance of rain for each of the next three days. What is the probability that you will have rain less than two days?

19. A baseball player currently has a .250 batting average. If we ignore all other factors, what is the probability that if the player comes to bat five times he gets exactly two hits in today's game?

20. A WNBA basketball player has a 50 percent foul shooting average. If she attempts eight foul shots today, what is the probability that she makes exactly four foul shots?

21. A hospital has an urgent need for three units of type A+ blood. Assume that approximately 30 percent of the population has this type of blood. If there are twelve people waiting to donate one unit of blood, what is the probability that the hospital will be able to meet its need? (*Hint:* Subtract the probability that fewer than three people have A+ blood from 1.)

22. A cellular phone communications network has redundancy built into it in the sense that, if several of the components fail, the network may still be able to function properly. If a network has fifteen components, each of which is 95 percent reliable, what is the probability that the network will fail if the network requires at least twelve of the components working in order to function properly? (*Hint:* Compute the probability that the network will not fail and subtract from 1.)

In Exercises 23–24, fill in the missing item to make the equation true. Give an intuitive reason why both sides of the equation must be equal.

23. $B(5, 3; \frac{1}{4}) = B(5, 2; ?)$

24. $B(10, 3; 0.4) = B(10, ?; 0.6)$

25. Assume that a child is buying a fast-food meal that also contains a (random) figure of one of six characters in a popular movie. How many purchases can she expect to make before she collects the entire set of six figures?

26. Assume that a child is buying packages of candy that also contain a (random) picture card of one of ten players on the NBA all-star team. How many purchases can he expect to make before he collects the entire set of ten pictures?

Further Exercises

27. Assume that during the winter, 50 percent of the population will come down with the common cold. Suppose a company claims that it has developed a cold vaccine to lower the infection rate to 25 percent. Ten people are vaccinated and we agree that we will accept the claim of the company if four or fewer catch a cold. What is the probability that we are accepting a worthless claim? (*Hint:* If the vaccine is worthless, the probability of catching a cold has not changed.)

28. Continuing the situation in Exercise 27, suppose we decide on a more strict requirement in order to accept the claim that the cold vaccine is effective. We agree that now we will accept the company's claim provided that two or fewer people of the ten who were inoculated catch a cold. What is the probability that we reject a worthwhile vaccine?

CHAPTER 14

Descriptive Statistics:*
What a Data Set Tells Us

You may not realize it, but you encounter statistics dozens of times each day. The expiration date of your breakfast cereal was determined from statistics on the shelf life of its ingredients. Research scientists used statistics to estimate the mileage that you can expect to get from a set of tires and your likelihood of surviving a head-on collision. Your insurance company uses statistics to set the rates on your insurance policies. Highway engineers use statistics on traffic flow to set the timing of traffic lights. The length of time you have to wait in line at a fast-food restaurant or for a ride at an amusement park are also based on statistical distributions.

In order to understand a large amount of data in such cases, we do two things. First, we organize and present the data so that we can observe patterns and spot trends. Second, we summarize the data numerically in order to understand relationships among the data better. In this chapter you will learn how to present data effectively and also how to calculate numbers that help describe the information contained in a set of data.

Like many people, you may have learned about statistics the hard way when a product broke down just after its warranty expired. It almost seemed as though the manufacturer knew the exact moment the product would fail. In fact, manufacturers do need to know a product's life expectancy—at least statistically—in order to be able to assure consumers that the product will last for a reasonable length of time. In Section 14.4 you will study the normal

---

* For further resources on statistics, see www.aw.com/pirnot.

distribution, which describes many real-life sets of data. This knowledge of the normal distribution will enable you to understand the mathematics behind writing an effective warranty.

14.1   ORGANIZING AND VISUALIZING DATA*

We can define **statistics** as the area of mathematics in which we are interested in gathering, analyzing, presenting, and making predictions from numerical information called **data.** Suppose that you were to see a study stating that today 12.5 percent fewer college students smoke than did so ten years ago. It would be ideal if the researchers had gathered their information by actually contacting every student mentioned in the study. This entire set of students would be called the **population.** Of course, in such a situation, it would be impractical to contact every college student, so instead, a statistician contacts a subset of the population, called a **sample.** It is very important that the sample be typical of the population as a whole. The following example illustrates the danger of choosing a sample that does not represent the whole population.

In the 1980s, Shere Hite, author of *The Hite Report,* a book on sexuality, distributed 100,000 questionnaires to various women's groups to learn about women's sexual habits. From 4,500 replies, she found that over 70 percent of women who were married for more than 5 years had had an affair. At about the same time, the advice columnist, Dear Abby, conducted a survey of her readers. When the 200,000 responses were tabulated, she found that only 15 percent of the women reported that they had been unfaithful. How can both surveys be true? Which survey should we believe? As a matter of fact, there are good reasons why we should not trust either survey.†

We will describe a sample as **biased** if it does not accurately reflect the population as a whole with regard to the data that we are gathering. Bias often occurs if we use poor sampling techniques. There are many ways in which bias can creep into a sample. It could occur because of the way in which we decide how to choose the people to participate in the survey. This is called **selection bias.** As an example, if we were to do a phone survey in the middle of a weekday afternoon, we would probably get an overrepresentation of retirees and stay-at-home parents

---

* Many of the calculations that we do in this chapter can be done easily using computers and graphing calculators. For tutorials on technology, see www.aw.com/pirnot.
† Smith, Tom W. "Speaking Out: Hite vs. Abby in Methodological Messes." *The American Association of Public Opinion Research News,* (spring) 1988.

in our sample. Call-in surveys conducted by local news shows and radio stations are also prone to selection bias.

It might seem we would get better, more reliable information if we were to walk around a town and ask people "randomly" to take part in our survey. We put the word "randomly" in quotes, because studies have found that selection bias can occur using this method because interviewers tend to choose people who are better dressed and who look cooperative, thus skewing the sample.

Another issue that can affect the reliability of a survey is the way we ask the questions, which is called **leading-question bias.** For example, in a 1992 Roper poll conducted for the American Jewish Committee on the Holocaust, people were asked, "Does it seem possible or does it seem impossible to you that the Nazi extermination of the Jews never happened?" The use of double negatives in this question caused confusion in the way people responded to the survey. When the question was worded this way, 22% of those surveyed said that it was possible that the Holocaust did not occur. A new survey was conducted in which the question was rephrased, "Does it seem possible to you that the Nazi extermination of the Jews never happened, or do you feel certain that it happened?" In the new survey, only 1% of those surveyed stated that it was possible that the Holocaust never occurred.

We have only touched upon the notion of how important it is that statistical conclusions are based upon reliable, nonbiased data. There are whole books written on sampling theory. For now, we just want you, as an educated consumer of technical information, to be aware that just because a study says something, as the song says, "It ain't necessarily so!"

We will now turn our focus on what to do once we have obtained reliable data.

When we gather information about a population, we often end up with a large collection of numbers. Unless we can organize the data in a meaningful way, it is nearly impossible to interpret these facts. For example, if you glance at the financial section of a newspaper, you will find several pages containing thousands of numerical facts about the daily performance of various stocks in the stock market. This amount of detail about the market's activity seems overwhelming; it is difficult to understand the general pattern of changes in stock prices from those lists of numbers. On the other hand, if you were to tune in to the evening news, the commentator might summarize this set of data by saying, "The Dow Jones lost 99.59 points today, closing at 9,251.7. Losers outnumbered winners by three to one." In order for any large amount of data to be comprehensible, we must organize it and present it so that we can see patterns, trends, and relationships.

> **DEFINITIONS**
>
> We refer to a collection of numerical information as **data** or a **distribution**. A set of data listed with their frequencies is called a **frequency distribution**.

Sometimes we wish to show the percent of the time that each item occurs in a frequency distribution. In this case, we call the distribution a **relative frequency distribution**.

A frequency table is one method used to organize data.

We often present a frequency distribution as a **frequency table**. In a frequency table, we list the values in one column and the frequencies of the values in another column, as we show in Example 1. We can also present a relative frequency distribution in table form.

Evaluation	Frequency
E	4
A	7
V	8
B	4
P	2

TABLE 14.1 Frequency table summarizing viewer evaluations of a police drama.

EXAMPLE 1   Using Tables to Summarize TV Program Evaluations

A television network has asked 25 viewers to evaluate a new police drama. The possible evaluations are

(E)xcellent, (A)bove average, a(V)erage, (B)elow average, (P)oor.

After the show, the 25 evaluations were as follows:

A, V, V, B, P, E, A, E, V, V, A, E, P, B, V, V, A, A, A, E, B, V, A, B, V

Construct a frequency table and a relative frequency table for this list of evaluations.

SOLUTION: If we count the number of E's, A's, and so on in the list, we get the results shown in Table 14.1.

By organizing the data in this table, we can see the distribution of favorable and unfavorable evaluations more quickly. Notice that the sum of the frequencies in Table 14.1 is 25, which is the number of viewers asked to evaluate the program.

Evaluation	Relative Frequency
E	0.16
A	0.28
V	0.32
B	0.16
P	0.08

TABLE 14.2 Relative frequency table summarizing viewer evaluations of a police drama.

We construct a relative frequency distribution for this data by dividing each frequency in Table 14.1 by 25. For example, because there are 4 Es, the relative frequency of the score E is $\frac{4}{25} = 0.16$. Table 14.2 shows the relative frequency distribution for the set of evaluations.

The sum of the relative frequencies in Table 14.2 is 1; however, in other examples, the sum of the relative frequencies may not be exactly 1 due to rounding.   ◎

 Quiz Yourself 1*

Construct a frequency table and a relative frequency table for the following distribution:

1, 2, 7, 2, 6, 5, 2, 7, 8, 8,
1, 3, 10, 7, 9, 1, 7, 3, 5, 2

If there are many different values in a data set, we may group the data values into classes to make the information more understandable. Although there is no hard and fast rule, generally using 8 to 12 classes will give a good presentation of the data. We illustrate how to group data in Example 2.

---

* Quiz Yourself answers begin on page 849.

Historical Highlight: *Presidential Polls*

On July 12, 1936, George Gallup warned that the current polling methods used by newspapers and magazines would incorrectly predict the next president of the United States. Gallup claimed that the *Literary Digest*, using flawed sampling methods, would predict that Alfred Landon, the Republican governor of Kansas, would defeat President Franklin Roosevelt for reelection by a margin of 56 percent to 44 percent. Gallup claimed that, in fact, the results of the election would be exactly the reverse of what the *Digest* predicted.

Gallup's claim seemed even more far-fetched because he was using a sample of only 50,000, whereas the *Digest* intended to survey ten million. The editor of the *Literary Digest* was outraged that Gallup had been so bold as to predict the results of the *Digest*'s own poll weeks before it had even started. He claimed that the *Digest*'s "flawed" methods would correctly predict the election, as they had 100 percent of the time in the past.

However, after Roosevelt won, by the margin Gallup predicted, it became evident that correct sampling methods were more important than the size of the sample. The *Digest* had used telephone directories, magazine subscription lists, and membership lists of clubs and organizations to compile its sample. Because those in higher economic classes tended to have phones, buy magazines, and belong to clubs and professional organizations, the *Digest*'s sample suffered from extreme selection bias and thus greatly overrepresented Republicans.

After the *Digest*'s fiasco, pollsters developed a method called *quota sampling* in which they tried to have the sample reflect the makeup of the population. The idea was to have the same percentages of men, women, Catholics, Jews, blacks, whites, and so on in the sample as there were in the population. Quota sampling seemed like a good idea; however, when pollsters used this approach to predict the 1948 presidential election, disaster struck again. Crosley, Gallup, and Roper, the major polling organizations of the time, all predicted that Thomas Dewey would defeat the incumbent Harry S. Truman by a margin of roughly 50 percent to 45 percent.

After Truman's victory, researchers realized that they had to reexamine sampling techniques. One major reason that the 1948 polls failed was that the polling stopped too soon and missed a late trend toward Truman. Also, those conducting the polls had too much latitude in choosing exactly whom they would interview. Selection bias crept into the sampling because even though quotas were met, the interviewers chose whom to interview within those categories.

EXAMPLE 2   Grouping Data Values into Classes

Suppose 40 health care workers take an AIDS awareness test and earn the following scores:

79, 62, 87, 84, 53, 76, 67, 73, 82, 68,
82, 79, 61, 51, 66, 77, 78, 66, 86, 70,
76, 64, 87, 82, 61, 59, 77, 88, 80, 58,
56, 64, 83, 71, 74, 79, 67, 79, 84, 68

Construct a frequency table and a relative frequency table for these data.

SOLUTION: Because there are so many different scores in this list, the frequency of each score will be very small; constructing a frequency table as we did

Range of Scores on AIDS Awareness Test	Frequency	Relative Frequency
50–54	2	0.05
55–59	3	0.075
60–64	5	0.125
65–69	6	0.15
70–74	4	0.10
75–79	9	0.225
80–84	7	0.175
85–89	4	0.10

TABLE 14.3 Frequency table and relative frequency table for scores on AIDS awareness test.

in Example 1 would not give us any useful information.

We must decide how to group the scores before making a table. The smallest score is 51 and the largest is 88. The difference, $88 - 51 = 37$, suggests that if we take a range of 40 and divide it into equal parts, we might get a reasonable grouping of the data. We will group the data into classes, each containing five values. The first class contains numbers from 50 to 54, the second contains numbers from 55 to 59, and so on. Counting the numbers in each class gives us the frequencies in the second column of Table 14.3.

To find the relative frequencies, we divide each count in the first column by 40, which is the total number of scores. For example, in the row labeled 55–59, we divide 3 by 40 to get 0.075 in the third column.

Table 14.3 helps us to see patterns and trends in the data. For example, a large number of workers have test scores below 70, which might mean that these workers are not effective in treating AIDS patients and may require extra training.

We use bar graphs to represent frequency distributions graphically.

The saying, "A picture is worth a thousand words," certainly applies when working with large sets of data. By presenting data graphically, we can observe patterns more easily. A **bar graph** is one way to visualize a frequency distribution. In drawing a bar graph, we specify the classes on the horizontal axis and the frequencies on the vertical axis. We draw bars (rectangles) above each class; each bar has a height equal to the frequency of the class. It is customary to draw the bars with the same width and with a slight separation between the bars. If we are graphing a relative frequency distribution, then the heights of the bars correspond to the size of the relative frequencies.

EXAMPLE 3   Drawing a Bar Graph of the Viewer Evaluation Data

a) Draw a bar graph of the frequency distribution of TV viewers' responses summarized in Table 14.1 in Example 1.

b) Draw a bar graph of the relative frequency distribution of the TV viewers' responses summarized in Table 14.2 in Example 1.

SOLUTION: a) Because the largest value is 8, we chose to label the vertical axis from 0 to 8. Next we drew five bars of heights 4, 7, 8, 4, and 2 to indicate the frequencies of the evaluations E, A, V, B, and P. The resulting graph is shown in Figure 14.1(a).

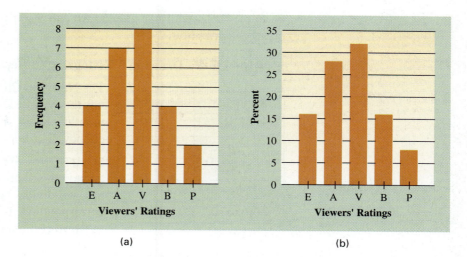

FIGURE 14.1   (a) Bar graph of frequency distribution of viewers' ratings. (b) Bar graph of relative frequency distribution of viewers' ratings.

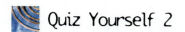

Quiz Yourself 2

Draw a bar graph representing the relative frequencies that you found in Quiz Yourself 1.

b) In Figure 14.1(b), we labeled the vertical axis from 0 to 35 because the largest relative frequency was 0.32, or 32 percent. Although both bar graphs have the same shape, it is usually better to draw a bar graph of relative frequency distributions if you are trying to compare two different data sets.   ⊙

If we are comparing two data sets of different sizes, graphing the relative frequencies, rather than the actual values in the data sets, allows us to compare the distributions. In this case, instead of drawing two separate bar graphs, we could show both distributions on a single graph, using, say, red for the bars in the first distribution and green for the bars in the second.

Until now, the data we have been organizing and graphing could not take on fractional values. By this we mean that in Example 1, a viewer could evaluate the TV show as above average or excellent, but could not give a rating between those two. Similarly, in Example 2, a score on the AIDS awareness test could be 78 or 79, but a score between these two numbers, such as 78.56, was not possible. A variable quantity that cannot take on arbitrary values is called *discrete*. Other quantities, called *continuous* variables, can take on arbitrary values. Weight is an example of a continuous variable. We may say that a person weighs 150 pounds; however, with a more accurate scale, we may find that the person actually weighs 150.3 or perhaps 150.314 pounds.

We use a special type of bar graph called a **histogram** to graph a frequency distribution when we are dealing with a continuous variable quantity. We also may use a histogram when the variable quantity is not continuous, but has a very large number of different possible values. Money is an example of such a quantity.

As with a bar graph, we specify classes for a histogram. With a histogram, however, we do not allow any spaces between the bars above each class. If a data value falls on the boundary between two data classes, then you must make it clear as to whether you are counting that value in the class to the right or to the left of the data value. We show how to draw a histogram for a frequency distribution in

Example 4. As with bar graphs, we can also draw a histogram for a relative frequency distribution.

EXAMPLE 4   Drawing a Histogram to Represent Weight Loss Data

Pounds Lost	Frequency
0 to 10	14
10 + to 20	23
20 + to 30	17
30 + to 40	8
40 + to 50	3

The New You Clinic has the following data regarding the weight lost by its clients over the past six months. Draw a histogram for the relative frequency distribution for this data.

SOLUTION: We first must find the relative frequency distribution. Because there are 65 data values, we divide each frequency by 65 to obtain the corresponding relative frequency distribution in the third column of Table 14.4.

Pounds Lost	Frequency	Relative Frequency
0 to 10	14	0.215
10 + to 20	23	0.354
20 + to 30	17	0.262
30 + to 40	8	0.123
40 + to 50	3	0.046

TABLE 14.4   Frequency and relative frequency distributions of weight loss at the New You Clinic.

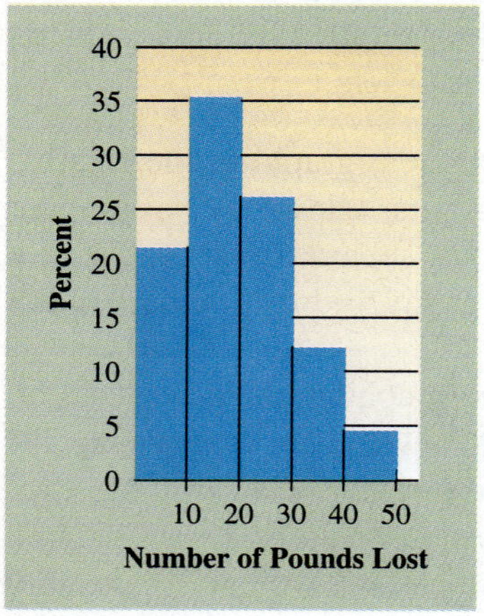

FIGURE 14.2   Histogram of weight loss at the New You Clinic.

We now draw this histogram exactly like a bar graph, as shown in Figure 14.2, except that we do not allow spaces between the bars. Also, we label the endpoints of the class intervals on the horizontal axis.

When we look at the histogram in Figure 14.2, we can see that the majority of the clients lost between 10 and 30 pounds. There is really no strict rule as to how to construct a histogram. It is up to you to decide whether to group the data and how large each data class should be; however, it is customary to have data classes all of the same size.

EXAMPLE 5   Determining Information from a Graph

Figure 14.3 shows the number of Atlantic hurricanes* over a period of years. Use this bar graph to answer the following questions.

a) What was the smallest number of hurricanes in a year during this period? What was the largest?

b) What number of hurricanes per year occurred most frequently?

---

* These data are from the Colorado State Tropical Prediction Center.

c) How many years were the hurricanes counted?

d) In what percentage of the years were there more than ten hurricanes?

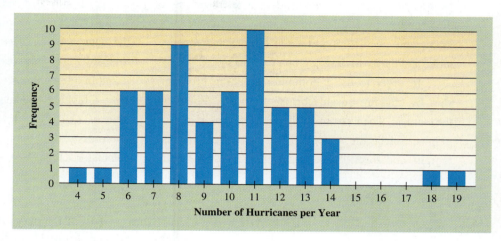

FIGURE 14.3   Number of hurricanes per year.

SOLUTION: a) The smallest number of hurricanes in any year during this time period was 4. The largest number was 19.

b) The number of hurricanes per year that occurred most frequently corresponds to the tallest bar in Figure 14.3, which appears over the number 11. Therefore, eleven hurricanes occurred in ten different years.

c) To find the total number of years for which these data were gathered, we add the heights of all of the bars to get

$$1 + 1 + 6 + 6 + 9 + 4 + 6 + 10 + 5 + 5 + 3 + 1 + 1 = 58 \text{ years.}$$

d) First we count the number of years in which there were more than ten hurricanes. If we add the heights of the bars above these values, we get $10 + 5 + 5 + 3 + 1 + 1 = 25$. Because there are 58 years of data, we calculate $\frac{25}{58} = 0.431$, which is approximately 43 percent.

A stem-and-leaf display is another way to display data.

A **stem-and-leaf display** is an effective way to present two sets of data "side by side" for analysis. This technique is used in a field of study called *exploratory data analysis*. This field was developed by John Tukey, a mathematician who worked at Princeton University and Bell Labs, the prestigious research division of AT&T.

It is common for sports fans to argue about which of two athletes was greater. Troy Aikman or John Elway? Venus Williams or Steffi Graf? The 1975 Cincinnati Reds or the 1998 Yankees? Magic Johnson or Michael Jordan? One way to answer such a question is to compare the data on the two athletes. We explain how to construct a stem-and-leaf display to analyze such data in Example 6.

Highlight: *Using Technology to Graph Data**

In organizing a large amount of data, it is tedious to draw graphs by hand. Many computer software programs and calculators allow you to enter a frequency table, and they will draw a graph for you.

In order to use a computer graphing program, start by clicking on an icon (picture) of a graph on the screen. Then, from a list of different types of graphs, select the type you wish (bar graph, histogram, and so on). Next, a series of pictures of different variations of the type of graph appears on the screen. After a few more simple mouse clicks, type in the data and the graph appears on the screen. With this software, you can draw a clear and accurate graph in a few minutes.

Graphing calculators also draw graphs very quickly. Once you have entered the frequency table, you press several buttons and the graph appears on the calculator screen.

EXAMPLE 6   Using Stem-and-Leaf Displays to Compare Babe Ruth's and Hank Aaron's Home Run Records

The following are the number of home runs hit by Babe Ruth during the seasons he played for the New York Yankees† and the number of home runs hit by Hank Aaron during his best fifteen years.

Ruth: 54, 59, 35, 41, 46, 25, 47, 60, 54, 46, 49, 46, 41, 34, 22
Aaron: 44, 30, 39, 40, 34, 45, 44, 32, 44, 39, 44, 38, 47, 34, 40

Compare these home run records using a stem-and-leaf display.

```
2 |  2 5
3 |  4 5
4 |  1 1 6 6 6 7 9
5 |  4 4 9
6 |  0
```

FIGURE 14.4   Stem-and-leaf display of Babe Ruth's home run record.

```
3 |  0 2 4 4 8 9 9
4 |  0 0 4 4 4 4 5 7
```

FIGURE 14.5   Stem-and-leaf display of Hank Aaron's home run record.

SOLUTION: We first examine Ruth's data. In constructing a stem-and-leaf display, we view each number as having two parts. The left digit is considered the stem and the right digit the leaf. For example, 54 has a stem of 5 and a leaf of 4. The stems for Ruth's data are 2, 3, 4, 5, and 6. We first list the stems in numerical order and draw a vertical bar to their right. Next, we write the leaves corresponding to each stem to the right of the stem and its vertical bar. We also list the leaves in increasing order away from the stem. Figure 14.4 shows the stem-and-leaf display for Babe Ruth's home run record.

From Figure 14.4 we see that Babe Ruth hit "40-something" home runs more frequently than any other category. In particular, there were seven seasons in which he hit between 40 and 49 home runs, while there were only three seasons when he hit between 50 and 59 home runs.

We display Hank Aaron's home run record in Figure 14.5.

By placing these two displays side by side, we can compare the data. To do this we use a common stem and place the leaves for one player to the right of the stem and the leaves of the other player to the left of the stem. Leaves are placed in increasing order away from the stem. Figure 14.6 shows the combined stem-and-leaf

---

* For tutorials on using technology to graph data, see www.aw.com/pirnot.
† *Total Baseball*, 3rd ed. (Macmillan, 1993).

Historical Highlight: *Florence Nightingale**

It may surprise you to see a reference to the legendary nurse Florence Nightingale in a discussion of statistics. Although she is most well known for her compassion toward the sick and her work in improving hospital sanitation, she was also trained in mathematics. Her parents gave her a solid classical education, but they were reluctant at first to allow her to study mathematics because they felt it was wasteful and would be of no use to her when she became a married woman. Because of her persistence, Nightingale was eventually permitted to have mathematics tutors; she had the good fortune to study with James Sylvester, one of the most eminent British mathematicians of the nineteenth century. Before she entered nursing, she worked as a mathematics tutor in arithmetic, algebra, and geometry.

In 1854, the British Secretary of War recruited Nightingale to serve as a nurse in the Crimean War at the military hospital in Uskudar (now part of Istanbul, Turkey). She was so disturbed by the high mortality rate that she developed methods for systematically gathering data on her patients. After analyzing the data, she presented her findings to her superiors in graphs that she invented, called polar-area charts. Using her statistical methods, she was able to persuade the military and political leaders to permit her to carry out her hospital reforms. As a result of improved hospital sanitation, deaths decreased, and some credit her with saving the British army during the Crimean War.

Nightingale was recognized for her revolutionary work in developing a statistical approach to medicine when she was elected to the Statistical Society of England in 1858. She also helped the U.S. government as an advisor on military health during the American Civil War and in 1874 she became an honorary member of the American Statistical Association. The famous statistician Karl Pearson described Florence Nightingale as a prophetess in the development of applied statistics.

Aaron			Ruth
	2	2	5
9 9 8 4 4 2 0	3	4	5
7 5 4 4 4 4 0 0	4	1 1 6 6 6 7 9
	5	4 4 9
	6	0

FIGURE 14.6 Combined stem-and-leaf display of Ruth's and Aaron's home run records.

Quiz Yourself 3

Use a stem-and-leaf display to represent the following collection of scores.

92, 68, 77, 98, 88,
75, 82, 62, 84, 67,
62, 91, 82, 73, 66,
81, 63, 90, 83, 71

display for Ruth and Aaron. Some call such a display a back-to-back stem-and-leaf display.

A supporter of Aaron might argue that he was more consistent, never hitting fewer than 30 home runs during his best years. A supporter of Ruth might point out that Aaron never hit 50 home runs, let alone 60. Although Figure 14.6 may not resolve the dispute as to who is the better home run hitter, it does help us to see the patterns in the data more clearly.

Since the numbers in Example 6 were two-digit numbers, we used single digits for the stems. If we had to represent a number such as 325, we could use either a stem of 32 and a leaf of 5 or a stem of 3 and a leaf of 25, depending on which way presented the data more clearly.

Organizing and displaying a collection of data is generally not our final goal; we usually want a concise numerical description of a set of data. In Section 14.2 we will discuss how to analyze data once we have organized it.

---

* This note is based on the biography of Florence Nightingale by Cynthia Audain, located at the Biographies of Women Mathematicians Web site (*http://www.agnesscott.edu/lriddle/women/women.htm*) at Agnes Scott College, Atlanta, Georgia. Also, we used "Mathematical Education in the Life of Florence Nightingale," by Sally Lipsey, *The Newsletter of the Association for Women in Mathematics*, Vol. 23, No. 4 (July/August 1993), 11–12.

Exercises 14.1

In Exercises 1–2, construct a frequency table, a relative frequency table, and a bar graph for the data given.

1. The number of hours of flight time for 20 amateur pilots last month:

  7, 8, 6, 5, 7, 10, 2, 7, 9, 5, 8, 8, 10, 9, 6, 5, 10, 7, 9, 8

2. The number of passengers per car on a high-speed railroad system:

  38, 39, 38, 37, 40, 38, 38, 37, 40, 38, 38, 39, 38, 37, 40, 37, 40, 38, 38, 37, 40, 38, 39, 38, 37, 40, 38, 38, 39, 38

In Exercises 3–4, construct a bar graph for the data given in each frequency table.

3. This table contains the EPA mileage ratings for 38 domestic cars.

MPG	19	20	21	22	23	24	25	26	27	28	29	30
Frequency	2	5	3	2	1	8	1	2	3	2	7	2

4. This table contains the weight loss by 30 patrons of a weight-loss clinic after one month.

Weight Loss in Pounds	0	1	2	3	4	5	6	7	8	9	10
Frequency	2	3	3	0	1	6	1	4	3	2	5

In Exercises 5–6, draw a bar graph for the relative frequency table for the data given.

5. The following are the ages of 60 people who have volunteered to work for Habitat for Humanity.

  21, 23, 27, 22, 23, 29, 28, 24, 24, 25, 27, 22, 26, 26, 23,
  23, 26, 28, 25, 24, 28, 27, 24, 23, 29, 28, 24, 22, 27, 26,
  22, 24, 26, 21, 24, 28, 24, 25, 22, 25, 27, 21, 23, 26, 23,
  23, 27, 27, 23, 21, 22, 27, 26, 23, 25, 29, 24, 27, 27, 26

6. The following are the ages of 40 ExecuCorps volunteers who are advising young entrepreneurs.

  51, 56, 57, 52, 53, 59, 58, 52, 54, 55,
  53, 56, 58, 55, 54, 58, 57, 52, 53, 59,
  52, 54, 53, 51, 54, 58, 54, 55, 52, 55,
  52, 57, 57, 53, 51, 52, 57, 56, 53, 55

7. Communicating Mathematics   What do you see as an advantage in grouping data? A disadvantage?

8. Communicating Mathematics   How does grouping data affect the way a bar graph looks?

In Exercises 9–12, group the data as indicated and draw a histogram.

9. The following are the heights (in inches) of the players in four Western Conference teams of the WNBA for the 2002 season.

  Los Angeles: 77, 71, 77, 69, 73, 77, 71, 68, 73, 69, 72, 70
  Houston: 77, 71, 77, 69, 73, 77, 71, 68, 73, 69, 72, 70
  Seattle: 71, 69, 76, 77, 72, 72, 70, 75, 68, 69, 68, 73, 76
  Phoenix: 70, 75, 75, 71, 73, 66, 73, 71, 71, 80, 76, 70, 78

  Use classes of width 2, starting at 65.5.

10. The following are the scores on a 100-point language aptitude test given to 60 people who are applying for the Peace Corps.

83, 71, 92, 87, 56, 64, 41, 95, 88, 91, 78, 73, 81, 79, 59,
73, 81, 93, 84, 66, 74, 51, 85, 78, 81, 98, 63, 91, 89, 64,
74, 61, 92, 77, 86, 79, 63, 91, 86, 91, 58, 83, 81, 77, 89,
83, 61, 83, 94, 76, 78, 61, 84, 88, 87, 68, 83, 71, 85, 64

Use classes of width 10, starting at 40.5.

11. The following table contains the average price (in dollars) of one gallon of unleaded regular gasoline for 1999 to 2001. (*Source*: Bureau of Labor Statistics)

	1999	2000	2001
Jan.	0.972	1.301	1.472
Feb.	0.955	1.369	1.484
Mar.	0.991	1.541	1.447
Apr.	1.177	1.506	1.564
May	1.178	1.498	1.729
Jun.	1.148	1.617	1.640
Jul.	1.189	1.593	1.482
Aug.	1.255	1.510	1.427
Sep.	1.280	1.582	1.531
Oct.	1.274	1.559	1.362
Nov.	1.264	1.555	1.263
Dec.	1.298	1.489	1.131

Use classes of width 5 cents, starting at 0.95.

12. The following data are the number of reports of mishandled baggage per 1,000 passengers for 10 U.S. airlines during three months of 2002. (*Source*: U.S. Department of Transportation)

February	1.97	3.23	3.78	3.21	3.65	3.52	3.28	4.25	4.60	9.70
March	3.59	3.32	3.37	3.55	4.20	3.74	3.40	5.35	7.35	12.38
April	2.52	2.82	2.94	3.05	3.17	3.42	3.66	4.18	4.71	9.97

Use classes of width 2, starting at 1.

13. The manager at the local espresso bar counted the customers once each hour over a busy weekend. We summarize the results she obtained in the following bar graph; use it to answer the following questions.

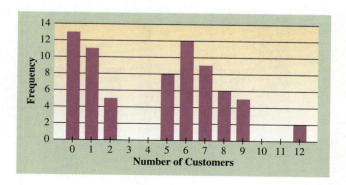

a) What was the smallest number of customers in the bar, and how often did it occur?

b) What was the largest number of customers in the bar, and how often did it occur?

c) What was the most frequently occurring nonzero customer count?

d) For how many hours were the customers counted?

e) For what fractional part of the total number of hours were there more than six customers in the bar?

14. Redo Exercise 13 using the following bar graph.

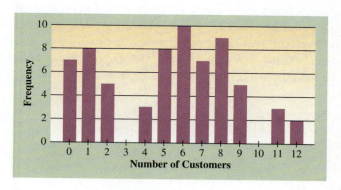

To estimate how much money to budget for snow removal, a town has kept records for the past 20 years of the number of days it has snowed each winter. The bar graph below summarizes this information. Use this graph to answer Exercises 15–16.

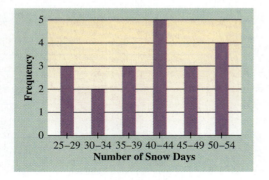
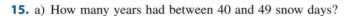

15. a) How many years had between 40 and 49 snow days?

   b) How many years had fewer than 30 snow days?

   c) How many years had at least 40 snow days?

   d) How many years had fewer than 50 snow days?

16. a) How many years had between 25 and 29 snow days?

   b) How many years had fewer than 40 snow days?

   c) How many years had at least 45 snow days?

   d) How many years had more than 34 snow days?

Comparing wage data. *The following bar graphs compare women's and men's hourly wages in 2001. Use these graphs to answer Exercises 17–22. Because you are estimating your answers by looking at the graphs, your answers may not agree exactly with ours. (Source: U.S. Department of Labor, as reported in* The World Almanac (2002))

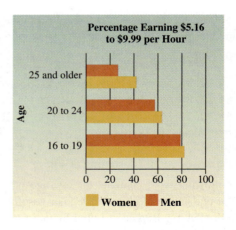

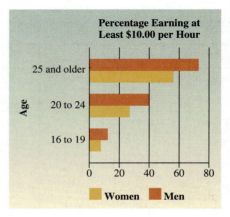

17. What percent of women ages 20 to 24 earned $5.15 or less per hour?

18. What percent of men age 25 and older earned between $5.16 and $9.99 per hour?

19. In what age and wage category do men seem to have the biggest advantage (in terms of a percentage difference) over women?

20. In what age and wage category do women seem to have the biggest advantage (in terms of a percentage difference) over men?

21. In what age and wage category do women and men seem to be most equal?

22. What general conclusions can you draw from these data?

In Exercises 23–24, represent the two sets of data on a single stem-and-leaf display.

23. A: 29, 32, 34, 43, 47, 43, 22, 38, 42, 39,
        37, 33, 42, 18, 22, 39, 21, 26, 18, 43

B: 32, 38, 22, 39, 21, 26, 28, 16, 13, 20,
21, 29, 22, 24, 33, 47, 23, 22, 18, 33

24. X: 29, 42, 34, 44, 47, 43, 22, 38, 42, 59, 41, 16,
47, 43, 42, 18, 22, 49, 21, 26, 18, 45, 24, 40

Y: 32, 48, 22, 59, 21, 26, 28, 16, 14, 20, 17, 45,
21, 29, 22, 24, 34, 47, 23, 22, 18, 45, 21, 16

25. Comparing training programs. A firm gives sales training to its newly hired employees. To determine how effective the training is, the firm compared the monthly sales of a group that has completed the training with a group that has not. The following numbers indicate thousands of dollars of sales last month. Represent the two sets of data on a single stem-and-leaf display. Does the orientation program seem to be succeeding?

No Training: 19, 22, 34, 23, 27, 43, 42, 28,
32, 29, 41, 26, 28, 26, 43, 40

With Training: 29, 21, 39, 44, 41, 36, 37, 29,
43, 45, 28, 32, 28, 33, 36, 32

26. Comparing weight loss programs. A hospital is testing two weight loss programs to determine which program is more effective. The following data represent the amount of weight lost by comparable clients for a year in each program. Represent the two given sets of data on a single stem-and-leaf display. Which program seems to be more effective?

Program A: 19, 32, 27, 34, 33, 36, 47, 32, 25,
52, 29, 26, 37, 28, 26, 43, 31, 40

Program B: 29, 21, 39, 44, 41, 36, 37, 26, 26,
43, 45, 28, 32, 28, 33, 36, 53, 39

27. If you have access to a computer with graphing capability to draw bar graphs and histograms, try to reproduce some of the graphs in this section with that computer.

Further Exercises*

Use the following table† for Exercises 28–29.

‡**28.** Organize the SAT data and present the results graphically.

State	Average SAT Score	Average Teacher Salary (in $thousands)
Connecticut	901	44
Delaware	903	35
Kansas	1040	30
Kentucky	994	29
Louisiana	993	26
Maryland	908	38
Massachusetts	900	36
Michigan	968	38
Mississippi	996	24
Minnesota	987	27
New Hampshire	928	31
New Jersey	891	38

State	Average SAT Score	Average Teacher Salary (in $thousands)
New Mexico	1007	26
New York	882	42
North Carolina	841	29
North Dakota	1069	24
Oregon	923	32
Pennsylvania	883	36
Rhode Island	883	38
South Dakota	1061	22
Tennessee	1008	28
Texas	874	28
Virginia	895	32
Washington	923	33

* You will find many of these exercises less tedious if you use some sort of technology to solve them. (See www.aw.com/pirnot.)
† These data are from the *Statistical Abstract of the United States*. The data have been rounded in some cases to simplify the computations.
‡ Exercise numbers circled in red can be used as group exercises.

29. Draw a single graph comparing the SAT data with the data on teachers' salaries.

In Exercises 30 and 31, use the data to construct a histogram that conveys the information as well as you can. Explain how you decided on the class width and the first class. Mention any problems that the data presented to you in making an attractive looking histogram.

30. According to *Variety Magazine* (World Almanac, 2002), the all-time top grossing movies through 2000 were as follows:

Movie	Gross (millions $)
Titanic	601
Star Wars (Episode IV)	461
Star Wars (Episode I)	431
ET	400
Jurassic Park	357
Forrest Gump	330
The Lion King	313
Return of the Jedi	309
Independence Day	306
The Sixth Sense	294
The Empire Strikes Back	290
Home Alone	286
Dr. Seuss' How the Grinch Stole Christmas	260
Jaws	260
Batman	251
Men in Black	250
Toy Story 2	246
Raiders of the Lost Ark	242
Twister	242
Ghostbusters	239
Beverly Hills Cop	235
Castaway	229
The Lost World	229
Mrs. Doubtfire	219
Ghost	218

31. According to Nielsen Media Research (World Almanac, 2002), the favorite prime-time shows in 2000–2001 and their average audience were: (We have not listed programs multiple times such as Who Wants to Be a Millionaire which appeared several times a week.)

Program	Average Audience (%)
Becker	10.9
CSI	11.7
Cursed	10.9
The District	8.7
ER	15.2
Everybody Loves Raymond	12.6
Frasier	10.7
Friends	12.9
JAG	9.2
Judging Amy	9.9
Just Shoot Me	10.5
Law and Order	12.3
Law and Order (SVU)	9.4
Who Wants to Be a Millionaire	13.7
Monday Night Football	12.7
NYPD Blue	9.7
The Practice	12.0
Sixty Minutes	11.1
Survivor II	17.4
Temptation Island	10.7
Touched by an Angel	9.8
The West Wing	11.6
Will and Grace	11.4
Weber Show	9.2

32. Communicating Mathematics What do you think a bar graph (with grouped data) of the number of credits completed by members of your class would look like? Why?

33. Communicating Mathematics What do you think a bar graph (with grouped data) of the day of the year that the members of your class were born (Jan 1 = 1, Jan 2 = 2, and so on) would look like? Why?

34. How might you present three sets of data in the same graph?

35. The following table is an example of a double-stem display. Without having this table explained to you, try to interpret its meaning and list the data items in the table.

(201–250)	2	34	45	49	
(251–300)	2	57	68	77	82
(301–350)	3	23	45		
(351–400)	3	62	73	78	
(401–450)	4	12	34		
(451–500)	4	82	89	93	

14.2 MEASURES OF CENTRAL TENDENCY

Although graphs are effective ways to make data more understandable, statisticians also often wish to summarize data numerically. There are two types of numerical information about a data set that are important. The first type, which we will discuss in this section, describes the center of the data set. The second type of number, which we will discuss in the next section, describes how the values in a set of data are spread out.

In this section, we will learn to compute various **measures of central tendency**, which are numbers that will help us understand the center of a set of data. The four measures of central tendency that we will discuss are the mean, the median, the mode, and quartiles. Except for the mode, each of these numbers will give us some sense of the center of the data set.

The mean is the most common notion of average.

When considering changing your long-distance phone company, you may want to know the average number of minutes that you have been calling long distance over the past six months. In order to calculate this average, you would add the number of long-distance minutes for the six months and divide by six. When we calculate an "average" this way, we are calculating the mean.

To calculate the mean and other numbers in statistics, we have to add lists of numbers. We use the Greek letter Σ (capital sigma) to indicate a sum. If we have n data values, $x_1, x_2, x_3, \ldots, x_n$, then we will represent the sum of these data values by Σx. For example, we will write the sum of the data values 7, 2, 9, 4, 10 by $\Sigma x = 7 + 2 + 9 + 4 + 10$.

Computing the Mean

If a data set contains n data values, the **mean** $\bar{x}$ of the data set is

$$\bar{x} = \frac{\Sigma x}{n}.$$

We represent the mean of a *sample* of a population by $\bar{x}$ (read as "*x* bar") and we will use the Greek letter μ (lowercase mu) to represent the mean of the whole *population*.

> Unless we state otherwise, we will assume that data sets in this chapter are samples rather than populations.

EXAMPLE 1   Finding the Mean Number of Accidents

The National Motor Corporation has been studying its safety record at its Mexican factory and has found that the number of accidents each year over the past five years was 25, 23, 27, 22, and 26. Find the mean annual number of accidents for this five-year period.

SOLUTION: To calculate the mean, we add the number of accidents and divide by 5, as follows:

$$\bar{x} = \frac{25 + 23 + 27 + 22 + 26}{5} = \frac{123}{5} = 24.6$$

This tells us that over the past five years, the factory in Mexico has averaged between 24 and 25 accidents per year.

As you saw in Example 1, the five data values balance on either side of the mean, as shown in Figure 14.7. Notice that just like children on a seesaw, a few numbers further from the mean (the balancing point) will balance with more numbers that are close to the mean.

We often use the mean to compare data to see trends. In Example 1 we can ask whether the five-year mean of 24.6 accidents per year is good or bad. We do not know unless we compare it to some other data. For example, suppose that National Motor also has a Seattle, Washington, factory of roughly the same size and workload, but that factory had a mean of 31.8 accidents per year over the past five years. This difference in means shows that perhaps a problem exists at the Seattle factory, and management must take steps to correct it.

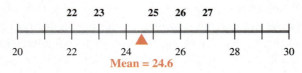

FIGURE 14.7   A set of data values balances about its mean.

In Example 1, each data value occurred once. However, often in a data set, some values occur several times, in which case we use a frequency table to compute the mean.

EXAMPLE 2   Computing the Mean of a Frequency Distribution of Water Temperatures

The Environmental Protection Agency suspects that hot water discharged from a nuclear power plant is responsible for a recent fish kill. In order to investigate this

---

Quiz Yourself 4*

Find each of the following for the data set 9, 12, 22, 6, 5, 15, 12, 25.

a) Σx b) n c) $\bar{x}$

Temperature (°F), x	Frequency, f
52	4
53	6
54	3
55	8
56	4
57	5

TABLE 14.5   Frequency table of water temperatures.

problem, the agency has recorded the water temperature at a point downstream from the plant for the last 30 days. We summarize this information in Table 14.5. What is the mean temperature for this distribution?

SOLUTION: We can simplify our calculations by remembering that repeated addition is the same as multiplication. For example, since 52 occurs four times in the distribution, we write $52 + 52 + 52 + 52$ as $52 \cdot 4$. The third column of Table 14.6 contains the products of the raw scores and their frequencies.

The sum of the frequencies, Σf, is the same

Temperature (°F), x	Frequency, f	Product, $x \cdot f$
52	4	208
53	6	318
54	3	162
55	8	440
56	4	224
57	5	285
Totals	$\Sigma f = 30$	$\Sigma(x \cdot f) = 1{,}637$

TABLE 14.6   The number of scores and the sum of the scores in the distribution of water temperatures.

as n, the total number of scores in the distribution, and $\Sigma(x \cdot f)$ is the sum of all the scores in the distribution. Therefore, the mean of the distribution is $\frac{\Sigma(x \cdot f)}{\Sigma f} = \frac{1{,}637}{30} \approx 54.6$ degrees. With this information, a biologist may determine that the water is not too hot and decide to look for another reason for the fish kill.

As you saw in Example 2, when you compute the mean of a frequency distribution, you must multiply the data values by their frequencies before adding them.

Computing the Mean of a Frequency Distribution

We use a frequency table to compute the mean of a data set as follows:

1. Write all products $x \cdot f$ of the scores times their frequencies in a new column of the table.

2. Represent the sum of the products you calculated in step 1 by $\Sigma(x \cdot f)$.

3. Denote the sum of the frequencies by Σf.

4. The mean is then $\dfrac{\Sigma(x \cdot f)}{\Sigma f}$.

SOME GOOD ADVICE

A common mistake students sometimes make when computing the mean of a frequency distribution is to divide by the number of entries in the first column of the frequency table. This is the number of different data values but it does not count repeated values. You must instead be sure to divide by the sum of the frequencies.

Quiz Yourself 5

A paramedic service kept track of the number of calls per day that it received over a two-week period. Use the information in the following frequency table to find the mean number of daily calls over this period.

Number of Calls, x	4	5	6	7	8	9	10
Frequency, f	1	4	1	2	3	2	1

One of the drawbacks in using the mean to represent the average value in a data set is that one or two extreme scores can have a strong influence on the mean.

EXAMPLE 3   The Effect of Extreme Scores on the Mean

Table 14.7 lists some of the top earnings in the entertainment industry.

a) What is the mean of the earnings of the entertainers on this list?

b) Is this mean an accurate measure of the "average" earnings for these entertainers?

SOLUTION: a) Summing the salaries and dividing by 10 gives the mean of $\frac{718}{10} = \$71.8$ million.

Name	Earnings (in $millions)
George Lucas	200
Oprah Winfrey	150
U2	69
Dave Matthews Band	50
Adam Sandler	47
Madonna	43
'N Sync	42
Ben Affleck	40
Britney Spears	40
Jennifer Lopez	37

TABLE 14.7 Earnings of some top entertainers.

b) To answer our second question, let us compare the mean to the values in the data set. We see that the earnings of eight entertainers are below the mean, whereas only two are above it. Therefore, the mean in this example does not give an accurate sense of what is "average" in this set of data, since it was unduly influenced by atypical scores such as George Lucas' earnings of $200 million and Oprah Winfrey's earnings of $150 million.

Extreme scores in a data set, such as George Lucas' and Oprah Winfrey's earnings, are called *outliers*, and you must decide what to do with them when you analyze data. In Example 3, you may feel that it is best to discard these two extreme values and calculate the mean based on the remaining eight earnings values. Or you may decide that it is better to describe the data using some other measure than the mean.

Highlight: *Using Spreadsheets in Statistics**

Spreadsheets can lighten our work when making statistical calculations. Spreadsheet (a) calculates the sum of the values in cells B1 through B5 (shaded) and places this sum in cell B7. Then cell B8 calculates the mean by dividing the contents of cell B7 by 5. We are using the cells in column A for labels in this spreadsheet. SUM is one of the many built-in functions that is available in spreadsheets. If we place the numbers 12.4, 34.6, 17.5, 67.2, and 83.4 in cells B1 through B5, the spreadsheet looks like (b).

In order to calculate a different mean, we enter different numbers in cells B1 through B5 and the spreadsheet automatically recalculates the sum and mean of the scores.

	A	B
1	Score 1	cell B1
2	Score 2	cell B2
3	Score 3	cell B3
4	Score 4	cell B4
5	Score 5	cell B5
6		
7	Sum of Scores	SUM (B1..B5)
8	Mean	B7/5

(a)

	A	B
1	Score 1	12.4
2	Score 2	34.6
3	Score 3	17.5
4	Score 4	67.2
5	Score 5	83.4
6		
7	Sum of Scores	215.1
8	Mean	43.02

(b)

The median tells us the middle of a set of data.

A measure that describes the middle of a data set is called the median.

Computing the Median

If we arrange a set of numbers in increasing (or decreasing) order, the **median** is the middle value in the list of numbers. However, there are two cases to consider.

1. If there is an odd number of numbers, then the median is the number in the middle position.

2. If there is an even number of numbers, then the median is the average of the two middle numbers.

One reason that we often use the median to describe the middle of a set of data is that the median is not affected by a few outliers.

---

* For tutorials on using spreadsheets and graphing calculators to do statistical calculations, see www.aw.com/pirnot.

T. Roosevelt	42
Taft	51
Wilson	56
Harding	55
Coolidge	51
Hoover	54
F. D. Roosevelt	51
Truman	60
Eisenhower	61
Kennedy	43
L. Johnson	55
Nixon	56
Ford	61
Carter	52
Reagan	69
Bush	64
Clinton	46

TABLE 14.8 Ages of U.S. presidents at inauguration.

SOME GOOD ADVICE

A common error made when finding the median is forgetting to arrange the scores in increasing numerical order.

During a presidential debate in 1980, when Ronald Reagan was running against Jimmy Carter, Reagan was asked whether his "advanced" age would be a drawback. In contrast, in 1992, when Bill Clinton was campaigning for president, many political observers had the opposite opinion that Clinton was too young and inexperienced to be president. How did Clinton's age compare with that of earlier presidents?

EXAMPLE 4 Finding the Median of Presidents' Ages

Table 14.8 lists the ages at inauguration of the presidents who assumed office between 1901 and 1993. Find the median age for this distribution to see how Bill Clinton's age at inauguration compares with that of other presidents.

SOLUTION: The distribution of ages, when arranged in increasing order, is

42, 43, 46, 51, 51, 51, 52, 54, 55, 55, 56, 56, 60, 61, 61, 64, 69.

Since there are 17 scores in this distribution, the middle score is the ninth score, which is 55. Because Clinton's age when he took office is well below the median, we agree with the opinion that he was young when he took office.

PROBLEM SOLVING

The Three-Way Principle in Section 1.1 says that making associations with everyday words helps you understand the meaning of mathematical ideas. Recalling that the strip between a divided highway is called the median will help you remember that the median is the middle number in a distribution.

Often consumer protection agencies test products to see whether they match the weight or volume stated on the package. In Example 5, we use the median to evaluate the accuracy of packaging information.

EXAMPLE 5 Using a Frequency Table to Find the Median

Acting on a tip from a dissatisfied customer, an agent of the consumer protection agency purchased 50 quarts of a particular brand of milk at various supermarkets to see whether they contained 32 ounces of milk. The results of this survey are reported in Table 14.9. What is the median for this distribution?

SOLUTION: Notice that because the 50 scores in Table 14.9 are in increasing order, the two middle scores are in positions 25 and 26. Counting the frequencies, we see that 29 ounces is in position 25 and 30 ounces is in position 26. Therefore,

Number of Ounces, x	Frequency, f
27	8
28	5
29	12
30	16
31	9
	$\Sigma f = 50$

TABLE 14.9   Number of ounces of milk in 50 quart cartons.

the median for this distribution is $\frac{29 + 30}{2} = 29.5$. Because the median is far below the advertised 32 ounces, the variation is probably not due to randomness in filling the containers. Perhaps the company packaging this milk should be fined, because it may be cheating consumers by systematically providing less milk than advertised.

Quiz Yourself 6

Find the median for the following frequency distribution.

x	72	86	91	95	100
Frequency, f	2	4	3	7	3

The five-number summary describes the position of data items in a data set.

> **DEFINITION   The Five-Number Summary**
>
> The median divides a data set into two halves. The set of numbers below the median is called the **lower half** and the set of numbers above the median is called the **upper half**. The median of the lower half is called the **first quartile** and is indicated by Q_1; the **third quartile**, denoted by Q_3, is the median of the upper half. The **five-number summary** of a set of data values consists of the following
>
> minimum value, Q_1, median, Q_3, maximum value.

The five-number summary is an effective way to describe a set of data, as we see in Example 6.

EXAMPLE 6   Finding the Five-Number Summary of Presidents' Ages

Consider the list of ages of the presidents from Example 4:

42, 43, 46, 51, 51, 51, 52, 54, 55, 55, 56, 56, 60, 61, 61, 64, 69.

Find the following for this data set.

a) the lower and upper halves

b) the first and third quartiles

c) the five-number summary

SOLUTION: a) The median is in the ninth position of this list. Therefore, the lower half is the set of ages in positions 1 to 8 and the upper half is the set of ages in positions 10 to 17.

lower half: 42, 43, 46, 51, 51, 51, 52, 54
upper half: 55, 56, 56, 60, 61, 61, 64, 69

b) The first quartile is the median of the lower half and is

$$Q_1 = \frac{51 + 51}{2} = 51.$$

The third quartile is

$$Q_3 = \frac{60 + 61}{2} = 60.5.$$

c) The five-number summary for this data set is 42, 51, 55, 60.5, 69.

We represent the five-number summary by a graph called a **box-and-whisker plot**. The five-number summary for the president's ages in Example 6 is 42, 51, 55, 60.5, 69. We show the box-and-whisker plot for this summary in Figure 14.8.

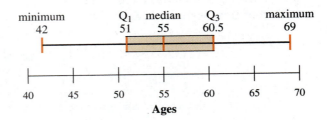

FIGURE 14.8 Box-and-whisker plot of ages of some U.S. presidents at inauguration.

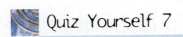

Quiz Yourself 7

a) Find the five-number summary of the entertainers' salaries listed in Example 3.

b) Represent the five-number summary in part (a) by a box-and-whisker plot.

The horizontal axis in Figure 14.8 represents the ages of the presidents and includes the smallest and largest ages. Next, we draw a box above the axis from the first quartile to the third quartile to show where the ages between Q_1 and Q_3 lie. We draw a vertical line through the box to show the median. Last, the lines extending to the left and right of the box—the "whiskers"— show the extreme values in the distribution. From this graph we can see quickly where the extreme scores, the median, and the middle 50 percent of the data lie.

The mode is the most frequent score in a data set.

We often want to know which data item occurs the most frequently in a data set. For example, a fashion designer might want to know what is the most common dress size, or an automobile manufacturer might be interested in the most common height of U.S. drivers. The mode is a measure that is easy to find to describe the most prevalent item in a data set.

Computing the Mode

The **mode** of a set of data is the data item that occurs most frequently. If two items occur most frequently, then each is a mode. If more than two scores occur most frequently, we will say that there is no mode.*

EXAMPLE 7 Finding the Mode of a Data Set

Find the mode for each data set.

a) 5, 5, 68, 69, 70

b) 3, 3, 3, 2, 1, 4, 4, 9, 9, 9

c) 98, 99, 100, 101, 102

d) 2, 3, 4, 2, 3, 4, 5

SOLUTION: a) The mode is 5 because that is the most frequently occurring item.

b) There are two modes, namely 3 and 9.

In parts (c) and (d) there is no mode because more than two items occur most often.

For a set of data, the mean, median, and mode are usually not the same. Therefore, when presenting a summary of data, you may wish to emphasize one measure over another.

EXAMPLE 8 Which Measure of Location Is the Best

Assume that you are negotiating the contract for your union at Magnum Industrial Corporation. To prepare for the next negotiation session, you have gathered annual wage data and found that three workers earn $30,000, five workers earn $32,000, three workers earn $44,000, and one worker earns $50,000. In your negotiations, which measure of location should you emphasize?

SOLUTION: You want to make the wages seem as low as possible; therefore, you should choose the smallest measure of central tendency. We list the frequency distribution of wages in Table 14.10.

We see in Table 14.10 that the mode is $32,000. Because there are twelve salaries, we look at positions 6 and 7. Both contain $32,000; therefore, the median salary is also $32,000.

We use Table 14.10 to compute the mean as follows:

Salary (in $thousands), x	30	32	44	50
Frequency, f	3	5	3	1

TABLE 14.10   Wage distribution at Magnum Industrial Corporation.

$$\bar{x} = \frac{\Sigma(x \cdot f)}{\Sigma f} = \frac{30 \cdot 3 + 32 \cdot 5 + 44 \cdot 3 + 50 \cdot 1}{12} = \frac{432}{12} = 36.$$

The mean is therefore $36,000.

It is to the union's benefit to present the median (or the mode) of $32,000. No doubt management will claim that the mean of $36,000 is the average salary.

---

* Although some authors allow for three or more modes, we will not do so.

Highlight: *Brink's Inc. v. City of New York*

In May 1978, Brink's Inc. was contracted to collect money from approximately 70,000 parking meters owned by the City of New York. Acting on a tip that some of the collected money was being stolen, city investigators planted specially treated coins in meters to determine whether they would be turned in. When meters were emptied and the contents given to the city, it was found that indeed theft was occurring; in the spring of 1980, several employees of Brink's Inc. were arrested and convicted. The city decided to sue Brink's for negligence to recover the stolen money.

Since it was not known precisely how much was stolen, lawyers for the city used statistics regarding the average amount collected before and after the firing of Brink's in presenting their case. The jury in turn used this information in determining the damages. (*Brink's Inc. v. City of New York*, 717 F.2d 700 (2d Cir. 1983))

In describing data, which measure—mean, median, or mode—is most appropriate? This is not an easy question to answer. The three measures are often different, and it is really up to you to decide how you wish to summarize the data.

If you are considering the mean, remember that one or two extreme values in a distribution have an undue influence on the mean. This is why, for example, in scoring ice skating at the Olympics the highest and lowest scores are discarded. If a data set has a few outliers, you may decide that it is best to discard them before computing the mean.

If the scores in a distribution are symmetrically balanced on either side of the mean, then the mean, the median, and the mode (assuming there is only one mode) will all be the same. However, some distributions are highly asymmetric, so there may be many small scores on one side of the mean and a few large scores on the other. In such a case the mean is not the best measure of the center of the data. For this reason, when we read about the average family income or the average cost of a new home, the median is often used rather than the mean.

The mode emphasizes what scores occur most frequently. For example, if a shoe store sells mostly size nines and size elevens, then it would not make sense in a report to say that the average size of the shoes sold was size ten.

In addition to describing the center of a data set, it is also important to understand how data values are spread out. In the next section, we will compute other numbers that measure the spread of a data set and give us a better understanding of a distribution.

Exercises 14.2

Find the mean, median, and mode for the following distributions.

1. 4, 6, 8, 3, 9, 11, 4, 7, 5
2. 8, 9, 4, 2, 10, 5, 5, 3, 3
3. 4, 6, 4, 6, 7, 9, 3, 9, 10, 11
4. 12, 11, 7, 9, 8, 6, 4, 5, 10, 1
5. 7, 3, 1, 5, 8, 6, 2, 5, 9, 4
6. 12, 4, 4, 8, 4, 7, 9, 8, 7, 7
7. 7, 8, 6, 5, 7, 10, 2, 7, 9, 5, 8, 8, 10, 9, 6, 5, 10, 7, 9, 8
8. 8, 8, 6, 6, 7, 10, 4, 7, 9, 5, 9, 8, 10, 10, 6, 5, 10, 8, 9, 8

In Exercises 9–15, find the mean, median, and mode for each data set. In each case, explain which measure you believe best describes a typical value in the distribution.

9. **Presidential vetoes.** The following is a summary of the number of vetoes (including pocket vetoes) for U.S. presidents from Franklin D. Roosevelt to Bill Clinton. (*Source: The World Almanac, 2002*)

President	Number of Vetoes
F. D. Roosevelt	635
Truman	250
Eisenhower	181
Kennedy	21
L. Johnson	30
Nixon	43
Ford	66
Carter	31
Reagan	78
Bush	44
Clinton	37

10. **Ages of Supreme Court justices.** At the start of the 2001–2002 term of the Supreme Court, the nine justices and their years of birth were Rehnquist (1924), Stevens (1920), O'Connor (1930), Scalia (1936), Kennedy (1936), Souter (1939), Thomas, (1948), Ginsburg (1933), and Breyer (1938). (*Source: The World Almanac, 2002*) For data, use the ages of the justices on December 31, 2001.

11. **Ozone pollution.** Ozone is formed when sunlight acts on nitrogen oxides. The number of millions of short tons of nitrogen oxides emitted into the air each year from 1987 to 1999 are given in the accompanying table. (*Source: U.S. Environmental Protection Agency*)

Year	Millions of Tons of Nitrogen Oxides
1987	23
1988	25
1989	24
1990	24
1991	24
1992	24
1993	24
1994	25
1995	25
1996	26
1997	26
1998	26
1999	25

12. **Baseball records.** A minor league baseball team has had the following yearly numbers of wins over the past 20 years:

38, 38, 36, 27, 21, 26, 39, 37, 23, 27, 36, 26, 26, 37, 30, 28, 37, 27, 29, 27

13. **The Super Bowl.** Many have criticized the Super Bowl as uninteresting because the games have not been close contests. The following table lists the results of the Super Bowl from 1967 to 2001. Use as your data the margin of victory in these games. Do these data support the claim that the Super Bowl is uninteresting?

Year	Winner	Loser
1967	Green Bay 35	Kansas City 10
1968	Green Bay 33	Oakland 14

continued

continued

Year	Winner	Loser
1969	New York Jets 16	Baltimore 7
1970	Kansas City 23	Minnesota 7
1971	Baltimore 16	Dallas 13
1972	Dallas 24	Miami 3
1973	Miami 14	Washington 7
1974	Miami 24	Minnesota 7
1975	Pittsburgh 16	Minnesota 6
1976	Pittsburgh 21	Dallas 17
1977	Oakland 32	Minnesota 14
1978	Dallas 27	Denver 10
1979	Pittsburgh 35	Dallas 31
1980	Pittsburgh 31	Los Angeles 19
1981	Oakland 27	Philadelphia 10
1982	San Francisco 26	Cincinnati 21
1983	Washington 27	Miami 17
1984	Los Angeles 38	Washington 9
1985	San Francisco 38	Miami 16
1986	Chicago 46	New England 10
1987	New York Giants 39	Denver 20
1988	Washington 42	Denver 10
1989	San Francisco 20	Cincinnati 16
1990	San Francisco 55	Denver 10
1991	New York Giants 20	Buffalo 19
1992	Washington 37	Buffalo 24
1993	Dallas 52	Buffalo 17
1994	Dallas 30	Buffalo 13
1995	San Francisco 49	San Diego 26
1996	Dallas 27	Pittsburgh 17
1997	Green Bay 35	New England 21
1998	Denver 31	Green Bay 24
1999	Denver 34	Atlanta 19
2000	St. Louis 23	Tennessee 16
2001	Baltimore 34	New York Giants 7

14. According to the U.S. Census Bureau, in 2000, the ten wealthiest communities in the United States were:

Community	Per Capita Income (thousands $)
Rancho Santa Fe, CA	113
Atherton, CA	112
Palm Beach, FL	109
Bloomfield Hills, MI	105
Belle Meade, TN	105
Woodside, CA	105
Indian River Shores, FL	103
North Hills, NY	101
Cherry Hills Village, CO	100
Portola Valley, CA	100

15. The following table shows the cumulative number of deaths due to AIDS from 1995 to 2000.

Year	Cumulative Number of Deaths (thousands)
1995	347
1996	385
1997	406
1998	423
1999	439
2000	448

(For this exercise, use the number of AIDS deaths per year for the years 1996–2000 as your data.)

Find the mean, median, and mode for the following frequency distributions.

16.

x	f
3	2
4	5
6	3
7	1
8	4
10	3
11	2

17.

x	f
2	3
5	2
7	3
8	2
9	1
10	5
11	2

18.

x	f
5	1
8	3
11	6
12	2
14	5
15	4
18	3

19.

x	f
3	5
6	2
8	8
11	3
15	6
17	2
23	1

20. Gina had an 84 and an 86 on her first two tests and thought that she did well enough on her final exam to keep her B average. However, when she got her grade she received a D. She checked with her instructor and learned that he had made a mistake by transposing the digits when he recorded her final exam grade. If her incorrect average (mean) for the class was 69, what was her correct final exam score?

21. The mean number of hours that Raphael worked over the past four weeks is 38.75. If he works 42 hours this week, what will the mean number of hours be that he will have worked over the five-week period?

22. Mileage ratings. The EPA has determined the number of miles per gallon (MPG) for 58 foreign cars, as given in the following frequency table. Find the mean, median, and mode for these data.

MPG	19	20	21	22	23	24	25	26	27	28	29	30
Frequency	2	5	3	5	4	8	8	6	3	5	7	2

23. Mileage ratings. The Rolls-Royce and Jaguar XJ12 have ratings of 11 MPG. Suppose this new score with its frequency of 2 is included in the table in Exercise 22. What are the new mean, median, and mode of the distribution?

24. Accident statistics. The top ten states in terms of the number of deaths by accidents and adverse effects in a recent year are listed in the following table. Calculate the mean, median, and mode for this data set and discuss which you feel is the most appropriate to represent the "average."

State	Number
California	10,172
Texas	6,366
Florida	5,170
New York	5,061
Pennsylvania	4,125
Illinois	3,984
Ohio	3,678
Michigan	3,162
North Carolina	2,914
Georgia	2,874

25. Sports. Statements similar to the following two statements are somewhat common in sports discussions.

> "I read that the Golden Bears averaged 7 yards per carry in Saturday's game."
> "I saw all of our team's basketball games and I would say that Gunner averages 21 points a game."

Which measure of location do you think is being used in each statement?

26. Consumer costs. What measure of location do you think is being used in each of the following statements?

> "The average cost of a new house is $123,000."
> "The typical new car is in the $14,000–$18,000 range."

27. Exam scores. Assume that in your History of Film class you have earned test scores of 78, 82, 56, and 72, and only one test remains.

a) If you need a mean score of 70 to earn a C, then what must you obtain on the final test?

b) If you need a mean score of 80 to earn a B, what must you obtain on the final test?

28. **Exam scores.** Assume that in your Ancient History class you have earned test scores of 74, 81, 56, and 70, and only one test remains.

   a) If you need a mean score of 70 to earn a C, then what must you obtain on the final test?

   b) If you need a mean score of 80 to earn a B, what must you obtain on the final test?

29. Communicating Mathematics   Give a real-life example of when the mean would be an appropriate measure of the average in a distribution. Explain why it would be appropriate.

30. Communicating Mathematics   Give a real-life example of when the median would be an appropriate measure of the average in a distribution. Explain why it would be appropriate.

31. Communicating Mathematics   Give a real-life example of when the mode would be an appropriate measure of the average in a distribution. Explain why it would be appropriate.

32. Give an example of two distributions which have the same mean, median, and mode. In one distribution, the data values should be tightly bunched together; in the other, the data values should be spread apart.

Credit card companies often compute the average daily balance* on your account as follows: If you start on the first of a 31-day month with a balance due of $100, and then on the 5th you charge another $50 and on the 27th you charge another $20, they would say that on days 1, 2, 3, and 4 you owed $100; on days 5, 6, 7, . . . , 25, and 26 you owed $150; and on days 27, 28, 29, 30, and 31 you owed $170. Therefore, to calculate the average daily balance, we would compute

$$\frac{4 \cdot 100 + 22 \cdot 150 + 5 \cdot 170}{31} = \frac{4,550}{31} = \$146.77$$

Use this method to answer Exercises 33–34.

33. **Credit card balance.** Assume that in a 31-day month you begin with a $50 balance due on your credit card, then charge an item for $75 on the 10th of the month and

a $120 item on the 25th of the month. What is your average daily balance on your credit card for this month?

34. **Credit card balance.** Assume that in a 31-day month you begin with an $80 balance due on your credit card, then charge an item for $60 on the 5th of the month and a $100 item on the 20th of the month. What is your average daily balance on your credit card for this month?

In Exercises 35–38, (a) give the five-number summary for the distribution, and (b) draw a box-and-whisker plot.

35. 11, 23, 25, 17, 26, 31, 45, 18, 41, 26, 31, 33, 48, 44, 53

36. 21, 24, 15, 45, 18, 31, 26, 41, 23, 18, 44, 27, 36, 21, 43

37. 31, 25, 41, 33, 28, 34, 37, 41, 33, 29, 49, 32, 38, 45, 30

38. 13, 24, 27, 45, 32, 29, 28, 39, 25, 21, 37, 36, 42, 34, 49

39. Draw box-and-whisker plots for the distributions of the number of home runs hit by Babe Ruth and Hank Aaron in Example 6 in Section 14.1.

40. Communicating Mathematics   In considering the display of Ruth's and Aaron's home runs, which do you feel conveys the information better—the stem-and-leaf displays or the box-and-whisker plots?

A college placement office has made a comparative study of the starting salaries for graduates in various majors. Use the following box-and-whisker plots, which describe the starting salaries for business, education, engineering, and liberal arts, to answer Exercises 41–46.

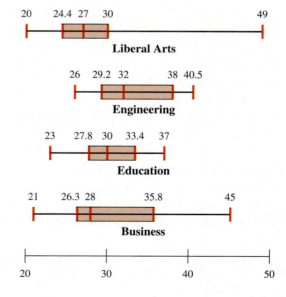

41. Salaries for college graduates. Which major, in general, receives the worst salary offers?

42. Salaries for college graduates. If a student wishes to receive $32,000 or more for a starting salary, which major offers the best chance of achieving that?

43. Salaries for college graduates. Does education or business have the higher median salary? Does this imply that salaries in education are generally higher than those in business? Discuss.

44. Salaries for college graduates. What is the significance of the long whisker in the liberal arts plot?

45. Salaries for college graduates. What is the significance of the short whiskers in the engineering plot?

46. Salaries for college graduates. Try to describe specific situations that might lead a student to a misinterpretation of the information contained in these plots.

Further Exercises

47. Population. The following table lists the percentage of the population that was over 65 in five countries in a recent year. Explain why it is not reasonable to compute the mean for this distribution using only the information in this table. (*Source: The World Almanac, 1999*)

Country	Percent over 65
Sweden	18
Norway	16
United Kingdom	16
Belgium	15
Denmark	15

48. Life expectancy. The following table lists the average life expectancy in five countries in a recent year. Explain why it is not reasonable to compute the median for this distribution using only the information in this table. (*Source: The World Almanac, 1999*)

Country	Life Expectancy
United States	72 years
Canada	73 years
Brazil	62 years
Guatemala	55 years
Venezuela	67 years

***49.** Communicating Mathematics Suppose there are nine scores from 1 to 10 in a distribution. If possible, give an example of each of the following. If you think that it is not possible to give the requested distribution, try to explain why you believe this.

a) a distribution in which the mean, median, and mode are each 5

b) a distribution in which the median is 3 and the mean is 5

c) a distribution in which the median is 5 and the mean is less than 5

50. Communicating Mathematics Suppose there are nine scores from 1 to 10 in a distribution. If possible, give an example of each of the following. If you think that it is not possible to give the requested distribution, try to explain why you believe this.

a) a distribution in which the mean is 5, the median is 4, and the mode is 2

b) a distribution in which the mean is 3 and the median is 5

c) a distribution in which the mean is 5 and the median is less than 5

If possible, give examples of data sets (sets of numbers) that have the following properties.

51. The mean, median, and mode are all the same.

52. The mean, median, and mode are all different.

53. The mean is larger than the median.

54. The median is larger than the mean.

55. Communicating Mathematics Suppose that distributions X and Y have means $\bar{x}$ and $\bar{y}$, respectively. If we combine X and Y into a single distribution, is the mean of this new distribution $\bar{x} + \bar{y}$? Explain.

56. Communicating Mathematics Suppose that distributions X and Y have medians a and b, respectively. If we combine X and Y into a single distribution, is the median of this new distribution $a + b$? Explain.

---

* Exercise numbers circled in red can be used as group exercises.

14.3   MEASURES OF DISPERSION

We summarize data numerically so that we can draw conclusions from them to make proper decisions. Although the mean and median describe the center of a data set, knowing only these measures can lead us to a wrong conclusion. For example, the following two distributions each have a median and mean of 25, so with just that information, and not knowing the data values, you may think that the distributions are similar.

$$X: 24, 25, 25, 25, 25, 26 \qquad Y: 1, 2, 3, 47, 48, 49$$

However, you can see that X and Y differ in that the values in Y are much more spread out than are the values in X.

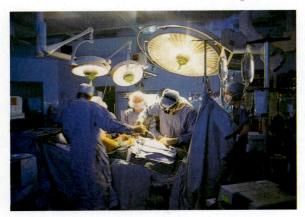

Although these two distributions have the same mean and median, the difference in spreads of the distributions may influence our view of the data sets. For instance, suppose that a hospital's cardiology unit is evaluating two types of pacemaker batteries. Periodically surgeons must operate on heart patients to replace batteries that are in danger of running down. Should the cardiologists use battery A, which lasts for a mean time of 45,000 hours (slightly over five years), or battery B, whose mean time is 46,000 hours?

On the surface, it would seem that the doctors should choose battery B. However, suppose we tell you that in testing both batteries, all of A's times were within 500 hours of the mean, but B's times were only within 2,500 hours of the mean. That is, some of B's times could be as low as $46,000 - 2,500 = 43,500$, whereas A's times were never below 44,500. Based on this information, the cardiologists should choose battery A, because the spread of the times for battery B presents a danger to the patients if it is used beyond 43,500 hours.

■

The range is a crude measure of the spread of a data set.

It is clear from the discussion of pacemaker batteries that a numerical measure of the dispersion, or spread, of a data set can provide useful information. One simple way to describe the spread of a set of data is to subtract the smallest data value from the largest.

> **DEFINITION**
>
> The **range** of a data set is the difference between the largest and smallest data values in the set.

EXAMPLE 1   Investigating Longevity

A person who is investigating his family's longevity has compiled a list of the ages at which seven family members died.

Historical Highlight: *The Origins of Statistics*

Ring a ring of roses,
A pocket full of posies,
Asha Asha, We all fall down.*

Some feel it was King Henry VII's fear of the dreaded Black Plague that prompted him in 1532 to begin registering all deaths occurring in England. This practice evolved into the publishing of weekly Bills of Mortality stating the number and causes of deaths. John Graunt, a merchant, noticed patterns in these reports regarding the percentage of deaths due to accident, suicide, and disease and concluded that social phenomena do not occur in a random fashion. In 1662 he published his observations in a paper titled "Natural and Political Observations . . . Made upon the Bills of Mortality." King Charles II was so impressed by Graunt's work

that he nominated him as one of the original members of the Royal Society of London even though he had few academic credentials.

Edmund Halley, the noted English astronomer, continued the work of Graunt and others in a paper titled "An Estimate of the Degrees of the Mortality of Mankind, Drawn from Curious Tables of the Births and Funerals at the City of Breslaw; with an Attempt to Ascertain the Price of Annuities upon Lives." Halley studied the mathematics of life expectancy and helped lay the foundations for an area of mathematics called actuarial science. Insurance companies use actuarial science to determine premiums; pension plans use it to determine how they must operate to remain financially sound.

Relative	Age at Death
Grandmother	81
Aunt	83
Grandfather	84
Cousin	23
Grandfather	79
Grandmother	91
Uncle	85

Find the range of this set of ages.

SOLUTION: The largest data value is 91 and the smallest is 23; therefore, the range is $91 - 23 = 68$.

In Example 1, the range is not very meaningful because of the outlier 23. It is probably best to drop this age from the list, in which case the range would then be $91 - 79 = 12$. Because of this problem with outliers, the range is not a very effective measure of the spread of a data set.

---

* Some analyze this rhyme as being descriptive of the symptoms and consequences of the Black Plague. If so, as statistics, the rhyme owes its origins to that horrible time in European history.

The standard deviation is a reliable measure of dispersion.

A better way to find the spread of data is to use the standard deviation, a measure based on calculating the distance of each data value from the mean.

> **DEFINITION**
>
> If x is a data value in a set whose mean is $\bar{x}$, then $x - \bar{x}$ is called x's *deviation from the mean*.

Data Value, x	Deviation from Mean, $x - \bar{x}$
16	−1
14	−3
12	−5
21	4
22	5

TABLE 14.11 Deviations from the mean for data values in distribution A.

To illustrate the deviation from the mean, consider distribution A: 16, 14, 12, 21, 22, whose mean is 17. We list the deviation of each data value from the mean in Table 14.11.

It might seem reasonable to measure the spread in A by averaging the deviations from the mean in Table 14.11. However, this will not work. As you see in Figure 14.9, for scores above the mean that have positive deviations from the mean, there must be scores below the mean that have negative deviations from the mean. When we add the positive and negative deviations, they cancel each other, giving a total of zero.

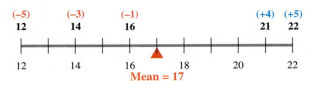

FIGURE 14.9 Values having positive and negative deviations from the mean must balance about the mean.

To avoid this cancellation, we square each of these deviations, as shown in Table 14.12.

Data Value, x	Deviation from Mean, $x - \bar{x}$	Deviation Squared, $(x - \bar{x})^2$
16	−1	1
14	−3	9
12	−5	25
21	4	16
22	5	25

TABLE 14.12 The squares of deviations from the mean for distribution A.

If we now average* these squared deviations, we get

$$\frac{1 + 9 + 25 + 16 + 25}{5 - 1} = \frac{76}{4} = 19.$$

---

* When doing this calculation for a *sample* (as opposed to a *population*), statisticians divide by $n - 1$ instead of n. There are technical reasons for doing this that are beyond the scope of this text. In doing this calculation for a population, we would divide by n rather than $n - 1$.

Clearly this number is too large to represent the spread of scores from the mean. In order to compensate for the fact that we squared the deviations in doing our calculations, we take the square root of this number to get $\sqrt{19} \approx 4.36$. This number is a more reasonable measurement of the way scores vary from the mean. The quantity that we just calculated is called the standard deviation.

> **DEFINITION**
>
> We denote the **standard deviation** of a *sample* of n data values by s,* which is defined as follows:
>
> $$s = \sqrt{\frac{\Sigma(x - \bar{x})^2}{n - 1}}.$$

We compute the standard deviation in four steps.

> **Computing the Standard Deviation**
>
> To compute the standard deviation for a sample consisting of n data values, do the following.
> 1. Compute the mean of the data set; call it $\bar{x}$.
> 2. Find $(x - \bar{x})^2$ for each score x in the data set.
> 3. Add the squares found in step 2 and divide this sum by $n - 1$; that is, find
>
> $$\frac{\Sigma(x - \bar{x})^2}{n - 1},$$
>
> which is called the variance.
> 4. Compute the square root of the number found in step 3.

EXAMPLE 2 Calculating the Standard Deviation

A company has hired six interns. After four months their work records show the following number of work days missed for each worker:

$$0, 2, 1, 4, 2, 3$$

Find the standard deviation of this data set.

SOLUTION: The mean of this data set is 2. We calculate the squares of the deviations of the data values from the mean in the following table.

Number of Days Missed	Square of Deviation from the Mean
0	4
2	0
1	1

continued

---

* We represent the standard deviation of a population by σ (instead of s), which we calculate using the formula

$$\sigma = \sqrt{\frac{\Sigma(x - \mu)^2}{n}}.$$

continued

Number of Days Missed	Square of Deviation from the Mean
4	4
2	0
3	1
	$\Sigma(x - 2)^2 = 10$

Quiz Yourself 8*

Find the standard deviation of the following sample of data values.

3, 4, 5, 6, 4, 2, 0, 8, 4

There are six scores in the distribution, so the standard deviation is

$$s = \sqrt{\frac{\Sigma(x - 2)^2}{6 - 1}} = \sqrt{\frac{10}{5}} \approx 1.41.$$

This relatively small standard deviation shows that the data values are not spread too far from the mean of 2.

In Example 3 we calculate the standard deviation of a distribution using a frequency table

EXAMPLE 3 **Using a Frequency Table to Compute the Standard Deviation**

Anya is considering investing in WebSoft, a software company that specializes in Internet applications. She intends to compute the standard deviation of its recent price changes to measure how steady its stock price has been recently. The following are the closing prices for the stock for the past 20 trading sessions:

37, 39, 39, 40, 40, 38, 38, 39, 40, 41,
41, 39, 41, 42, 42, 44, 39, 40, 40, 41

What is the standard deviation for this data set?

SOLUTION: In Table 14.13 we see that the sum of the 20 closing prices is 800, so the mean is $\frac{800}{20} = 40$. We have added columns to the table for the deviations from the mean, the squares of these deviations, and so on.

When calculating the standard deviation, we must multiply the squared deviation of each price from the mean by its frequency. We list these products in the column labeled Product, $(x - 40)^2 \cdot f$, and show the sum of these products at the bottom of the column.

We can now calculate the standard deviation. Remember that Σf is the number of data values which we have been representing by n. The standard deviation is therefore

$$s = \sqrt{\frac{\Sigma(x - 40)^2 \cdot f}{n - 1}} = \sqrt{\frac{50}{19}} \approx 1.62.$$

---

* Quiz Yourself answers begin on page 849.

Closing Price, x	Frequency, f	Product, $x \cdot f$	Deviation, $(x - 40)$	Deviation Squared, $(x - 40)^2$	Product, $(x - 40)^2 \cdot f$
37	1	37	-3	9	9
38	2	76	-2	4	8
39	5	195	-1	1	5
40	5	200	0	0	0
41	4	164	1	1	4
42	2	84	2	4	8
44	1	44	4	16	16
	$\Sigma f = 20$	$\Sigma(x \cdot f) = 800$			$\Sigma(x - 40)^2 \cdot f = 50$

TABLE 14.13 Computations necessary to find the standard deviation of WebSoft closing prices.

This relatively small standard deviation indicates that the closing prices for the WebSoft stock have not been varying very much lately. If Anya is a cautious investor, she will find the stability of WebSoft's prices appealing.

We summarize the method we use for finding the standard deviation in the following formula.

> **Formula for Computing the Sample Standard Deviation for a Frequency Distribution**
>
> We calculate the standard deviation, s, of a sample that is given as a frequency distribution as follows:
>
> $$s = \sqrt{\frac{\Sigma(x - \bar{x})^2 \cdot f}{n - 1}},$$
>
> where $\bar{x}$ is the mean of the distribution, f is the frequency of data value x, and $n = \Sigma f$, the number of data values in the distribution.

 Quiz Yourself 9

The investor in Example 3 also gathered the closing stock prices for WebRanger, another Internet software company. We list these stock prices in Table 14.14. The mean closing price is 42. Find the standard deviation for this distribution.

Closing Price, x	36	37	38	39	40	41	42	43	44	45
Frequency, f	2	0	0	3	1	3	0	1	5	5

TABLE 14.14 WebRanger closing stock prices for 20 days.

Comparing the standard deviation of 1.62 for WebSoft with the standard deviation of 3.01 for WebRanger in Quiz Yourself 9, we see that WebRanger's stock is much more volatile than WebSoft's stock.

We use the coefficient of variation to compare the standard deviations of different data sets.

In Example 3 we used the standard deviation with one set of data. If you use the standard deviation to compare two sets of data, the data must be similar and have comparable means. The following situation shows why.

Suppose we are comparing the body weights of two groups of people and find the standard deviation to be three pounds for the first group and ten pounds for the second group. Can we state that there is more uniformity in the first group than the second? Before answering this question, here is more information. The first group consists of preschool children, whereas the second is a group of National Football League linemen. Clearly, the standard deviation of 3 is more significant, *relatively speaking*, for a group of preschoolers with a mean weight of 30 pounds than the standard deviation of 10 for a group of football players with a mean weight of 300 pounds. In order to use the standard deviation effectively to compare different sets of data, we must first make the numbers comparable. We do this by finding the coefficient of variation.

> **DEFINITION**
>
> For a set of data with mean $\bar{x}$ and standard deviation s, we define the **coefficient of variation**, denoted by CV, as
>
> $$CV = \frac{s}{\bar{x}} \cdot 100\%.$$

Note that the coefficient of variation compares the standard deviation to the mean and is expressed as a percentage.

The coefficient of variation for the group of preschool children is

$$CV = \frac{3}{30} \cdot 100\% = 10\%.$$

In contrast, the coefficient of variation for the group of NFL football players is

$$CV = \frac{10}{300} \cdot 100\% = 3.3\%.$$

Highlight: *Using Computer Algebra Systems in Statistics**

Computer algebra systems can perform many statistical calculations. Here we show how a computer algebra system calculates standard deviation. The text to the left is the command you would type in; the bold text to the right is what the computer algebra system would produce for you.

First we enter the data.

> data:=[4,6,5,7,8];   **data:=[4, 6, 5, 7, 8]**

Next we ask the system to compute the mean.

> mean(data);   **6**

We now ask for the standard deviation.

> standarddeviation(data);   **1.58114**

With such computer algebra systems available, more than ever before, it is important to have the ability to analyze situations, think critically, and set up problems for solution rather than simply dwelling on the calculational aspects of mathematics.

Since the coefficient of variation is larger for the group of preschoolers, we conclude that there is more variation, *relatively speaking*, in their weights than in the weights of the football players.

EXAMPLE 4   Using the Coefficient of Variation to Compare Data

Use the coefficient of variation to determine whether the women's 100-meter race or the men's marathon (26 miles and 385 yards) has had more consistent times over the five Olympics listed in Table 14.15.

	Women's 100 Meters	Men's Marathon
2000	10.75 sec	2 h, 10 m, 11 sec (7,811 sec)
1996	10.94 sec	2 h, 12 m, 36 sec (7,956 sec)
1992	10.82 sec	2 h, 13 m, 23 sec (8,003 sec)
1988	10.54 sec	2 h, 10 m, 32 sec (7,832 sec)
1984	10.97 sec	2 h, 09 m, 21 sec (7,761 sec)

TABLE 14.15   Olympic times in women's 100-meter race and men's marathon.

SOLUTION: Using a calculator, we found that the mean time for the 100-meter race is 10.804 and the standard deviation is approximately 0.17. For the marathon, the mean is 7,872.6 and the standard deviation is approximately 102.3.

Thus the coefficient of variation for the women's 100-meter race is

$$\frac{0.17}{10.804} \cdot 100\% \approx 1.57 \text{ percent.}$$

---

* For tutorials on using technology in statistics, see www.aw.com/pirnot.

For the men's marathon, the coefficient of variation is

$$\frac{102.3}{7{,}872.6} \cdot 100\% \approx 1.30 \text{ percent.}$$

Surprisingly, using the coefficient of variation as a measure, there is less variation in the times for the marathon than for the 100-meter race!! ⊚

Exercises 14.3

If you wish, use a calculator or computer software to solve the following exercises. Some of your answers may differ slightly from ours due to differences in the way we round off our calculations.

Find the range, mean, and standard deviation for the following data set.

1. 19, 18, 20, 19, 21, 20, 22, 18, 18, 17

2. 89, 72, 100, 87, 65, 98, 77, 92

3. 5, 7, 9, 4, 6, 8, 7, 10   4. 8, 4, 7, 6, 5, 5, 4, 9

5. 2, 12, 3, 11, 14, 5, 8, 9   6. 21, 3, 5, 11, 7, 1, 9, 7

7. 18, 3, 8, 7, 7, 9, 13, 7

8. 22, 18, 15, 21, 21, 15, 19, 13

9. 3, 3, 3, 3, 3, 3, 3   10. 4, 6, 4, 6, 4, 6, 4, 6

11. The following frequency table shows the number of fires per month in a city. Complete the table entries to find the mean and standard deviaion for this distribution.

12. The frequency table at the top of page 815 summarizes the number of job offers made to graduates of a network administrator certification program. Complete the table entries to find the mean and standard deviation for this distribution.

Find the mean and standard deviation for the following frequency distributions.

13.

x	f
2	2
3	4
4	2
5	0
6	4

14.

x	f
8	3
9	2
10	1
11	2
12	3

Number, x	Frequency, f	Product, $x \cdot f$	Deviation, $(x - \bar{x})$	Deviation², $(x - \bar{x})^2$	Product, $(x - \bar{x})^2 \cdot f$
2	1				
3	1				
4	0				
5	2				
6	3				
7	4				
8	4				
9	3				
10	2				
	$\Sigma f =$	$\Sigma x \cdot f =$			$\Sigma (x - x)^2 \cdot f =$

Number, x	Frequency, f	Product, $x \cdot f$	Deviation, $(x - \bar{x})$	Deviation², $(x - \bar{x})^2$	Product, $(x - \bar{x})^2 \cdot f$
4	2				
5	0				
6	3				
7	5				
8	4				
9	5				
10	1				
	$\Sigma f =$	$\Sigma x \cdot f =$			$\Sigma (x - \bar{x})^2 \cdot f =$

15.

x	f
3	1
4	1
5	0
6	2
7	5
8	3
9	3

16.

x	f
12	3
13	0
14	5
15	0
16	1
17	2
18	3

17. Summarizing test score data. The following are the scores of 20 people who took a paramedics licensing test. Find the mean and standard deviation for these data.

Score	72	73	78	84	86	93
Frequency	7	1	3	6	2	1

18. Summarizing age data. The following are the ages of sixteen World War II veterans who are attending

a reunion commemorating the Normandy invasion. Find the mean and the standard deviation for these data.

Age	74	75	76	78	85	87
Frequency	4	2	4	2	3	1

19. Summarizing vital statistics. The following is the 2002 roster of the Los Angeles Sparks in the WNBA. Find the mean and standard deviation of these data.

Player	Weight
Marlies Askamp	198
Latasha Byears	206
Erika Desouza	190
Tamecka Dixon	148
Vedrana Grgin-Fonseca	160
Lisa Leslie	170
Mwadi Mabika	165
Nicky McCrimmon	125
DeLisha Milton	172
Rosalind Ross	160
Nikki Teasley	165
Sophia Witherspoon	145

20. Summarizing tax data. The following table lists the state income tax for a person earning $50,000 per year for several states. Find the mean and standard deviation for these data. (*Source: The World Almanac*, 1999)

State	Income Tax (percent)
Arizona	3.24
Colorado	5.00
Hawaii	10.00
Kansas	6.25
Massachusetts	5.95
New York	6.85
Pennsylvania	2.80
Virginia	5.75

21. The manager at the local espresso bar counted the customers once each hour over a busy weekend. We summarize the results she obtained in the bar graph below. Find the mean and standard deviation of these data.

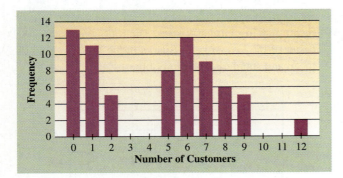

22. Find the mean and standard deviation for the distribution given by the following bar graph.

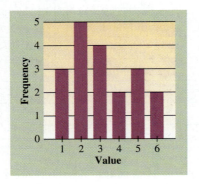

Try to answer each of the following as true or false without doing any computations. Explain your answers in words or give appropriate examples or counterexamples to support your answers.

23. **Communicating Mathematics** The standard deviation of the set of numbers $-2, 2, -2, 2, -2, 2, -2, 2$ is zero.

24. **Communicating Mathematics** If the standard deviation of a set of data is zero, then all the numbers in the data set are the same.

25. **Communicating Mathematics** The distribution 5, 6, 7, 8, 9 and 25, 26, 27, 28, 29 have the same standard deviations.

26. **Communicating Mathematics** The more numbers there are in a distribution, the larger the standard deviation.

27. **Family incomes.** The following table gives the annual incomes for eight families, in thousands of dollars. Find the number of standard deviations family H's income is from the mean.

Family	A	B	C	D	E	F	G	H
Annual Income (in $thousands)	47	48	50	49	51	47	49	51

28. **Family incomes.** The following table gives the annual incomes for eight families, in thousands of dollars. Find the number of standard deviations family A's income is from the mean.

Family	A	B	C	D	E	F	G	H
Annual Income (in $thousands)	49	51	52	51	51	50	52	52

In Exercises 29–30, you are given a distribution of averages in a literature class. The professor in this class assigns grades as follows:*

A: averages that are at least 1.5 standard deviations above the mean

B: averages that are between 0.5 standard deviations above the mean, and 1.5 standard deviations above the mean

C: averages that are between 0.5 standard deviations below the mean, and 0.5 standard deviations above the mean

---

* If an average falls on the border of two grades, the professor will assign the higher grade.

D: *averages that are between 1.5 standard deviations below the mean, and 0.5 standard deviations below the mean*

F: *averages that are more than 1.5 standard deviations below the mean*

29. Assigning grades. The averages in the class are 80, 76, 81, 84, 79, 80, 90, 75, 75, and 80. What grade does the person earning the 76 get?

30. Assigning grades. The grades in the class are 72, 71, 73, 70, 71, 79, 65, 73, 74, and 72. What grade does the person earning the 74 get?

In Exercises 31–34, we present information on the performance of various stocks during a given month. Stocks with greater coefficients of variation are considered more volatile.

31. Comparing stocks. Suppose for a given month that the mean daily closing price for Magnum Industrial common stock was 123.76 and the standard deviation was 12.3. For GrandCore Incorporated stock, the mean daily closing price was 78.6 with a standard deviation of 7.2. Which stock was more volatile?

32. Comparing stocks. Suppose for a given month that the mean daily closing price for Home Video, Inc., was 37.4 and the standard deviation was 4.1. For VideoMagic, the mean daily closing price was 18.6 with a standard deviation of 3.2. Which stock was more volatile?

33. Comparing stocks. The Dow Jones Industrial Average (DJIA) measures the stock prices of a large group of stocks. If, for a given week, the DJIA had a mean daily closing price of 11,261.12 with a standard deviation of 72.17 and WebMaster stock had a mean closing price of 37.6 with a standard deviation of 1.7, did WebMaster have more or less volatility than the DJIA during that week?

34. Comparing stocks. If, for a given week, the DJIA had a mean daily closing price of 10,834.8 with a standard deviation of 144.5 and WebRanger stock had a mean closing price of 123.6 with a standard deviation of 1.2, did WebRanger have more or less volatility than the DJIA during that week?

35. Price stability. In the following table, we list the monthly price (in dollars) of a pound of coffee and a gal-

lon of unleaded gasoline for 2001. Find the standard deviation of each set of data. (*Source:* Bureau of Labor Statistics)

	Coffee $/lb.	Gasoline $/gallon
Jan.	3.224	1.472
Feb.	3.217	1.484
Mar.	3.205	1.447
Apr.	3.128	1.564
May	3.097	1.729
Jun.	3.156	1.640
Jul.	3.097	1.482
Aug.	3.046	1.427
Sep.	3.025	1.531
Oct.	3.015	1.362
Nov.	2.988	1.263
Dec.	2.913	1.131

36. Comparing price stability. Consider the table given in Exercise 35. Use the coefficient of variation to determine whether coffee prices or gasoline prices were more stable in 2001.

37. Comparing family incomes. Consider the tables given in Exercises 27 and 28, respectively. Suppose that the eight families are the same in both tables and the incomes are for two different years that are two years apart. Which of the families had the greatest improvement in annual income over the two-year period with respect to the other families? Explain how you arrived at your conclusion.

38. Comparing family incomes. Suppose that the mean family income in the United States was $28,000 with a standard deviation of $1,000 in year X. Also, the mean family income was $31,000 with a standard deviation of $2,250 three years later, in year $X + 3$. If a family earned $30,000 during year X and $34,000 three years later, then in which of these two years did the family have a better annual income in relation to the rest of the population? Explain how you arrived at your answer.

Further Exercises

*39. Communicating Mathematics   a) Flip four coins ten times and record the number of heads that appear each time. Find the mean and the standard deviation of the number of heads you obtain.

b) Repeat part (a), but this time flip the four coins 20 times.

c) Repeat part (a), but this time flip the four coins 30 times.

d) What conclusion, if any, can you make about the mean and standard deviation as you increase the number of times you flip the four coins?

40. Communicating Mathematics   a) Pick a data set consisting of any five numbers. Compute the mean and the standard deviation of this data set. (Consider the data set to be a sample.)

b) Add 20 to each number in the data set you created in part (a) and compute the mean and standard deviation for this new data set.

c) Subtract 5 from each number in the data set you created in part (a) and compute the mean and standard deviation for this new data set.

d) What conclusion can you make about changes in the mean and the standard deviation when the same number is added to or subtracted from each score in a data set?

e) Use the conclusion reached in part (d) to simplify the calculation of the mean and standard deviation for the distribution 598, 597, 599, 596, 600, 601, 602, 603.

41. Communicating Mathematics   a) Pick any five numbers and compute the mean and standard deviation for this data set.

b) Multiply each number in the data set you created in part (a) by 4 and compute the mean and standard deviation for this new data set.

c) Multiply each number in the data set you created in part (a) by 9 and compute the mean and standard deviation for this new data set.

d) What conclusion can you make about changes in the mean and standard deviation when each score in a data set is multiplied by the same number?

e) The mean is 4 and the standard deviation is 2.2 for the data set 3, 4, 7, 1, 5. Consider your conclusion from part (d) and compute the mean and standard deviation for the data set 15, 20, 35, 5, 25.

42. Communicating Mathematics   Consider the two distributions given by the bar graphs in Exercises 21 and 22. Can you simply look at the graphs to decide which of the two distributions has the greater standard deviation? If so, describe what you look for in the graphs on which to base your decision.

43. Data on the Internet. Do a search on the Internet to locate data in an area that is of interest to you. For example, you can go to the *New York Times* Web site and find the number of weeks that the top ten fiction books have been on the bestseller list. Or, you can locate statistics from the Web site of your favorite sports team. Other good sources of data are the Bureau of Labor Statistics and the U.S. Department of Transportation Web sites. After you have located a site, choose a data set and find its mean and standard deviation.

44. Comparative cost of living. According to an Internet service that compares the cost of living in one location with the cost of living in another, a salary of $100,000 in Manhattan is comparable to $38,000 in Allentown, Pennsylvania. What statistical methods do you think this service uses in determining this information?

---

* Exercise numbers circled in red can be used as group exercises.

THE NORMAL DISTRIBUTION

When describing a set of data, statisticians take into account the shape of the graph of the data. One particular shape models so many different types of data that we study it separately from all of the others. In order to understand what we mean by the shape of a distribution,* consider the histograms in Figure 14.10.

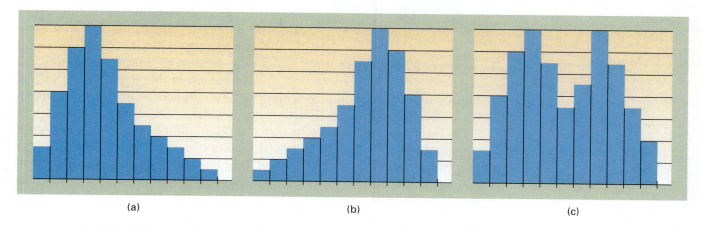

(a)   (b)   (c)

FIGURE 14.10   Shapes of distributions. (a) Distribution is skewed to the right. (b) Distribution is skewed to the left. (c) Bimodal distribution.

In Figure 14.10(a) the distribution has most of the data values to the left, and then the number of data values drops off quickly, forming a "tail" to the right. Such a distribution is said to be *skewed to the right*. Similarly, the distribution in Figure 14.10(b) is *skewed to the left*. Notice in Figure 14.10(c) that the distribution has two data values that are not next to each other that occur more frequently than the others. A distribution whose graph has two peaks, such as in Figure 14.10(c), is called *bimodal*.

■

Many different types of data sets are normal distributions.

The most common distribution in statistics is called the **normal distribution**. Normal distributions are important in statistics because many real-life data sets follow the pattern shown in the histogram in Figure 14.11 on the following page. Distributions of such diverse data sets as SAT scores, heights of people, the number of miles before an automobile tire wears out, the size of hamburgers at a fast-food restaurant, and the number of hours of use before a certain brand of DVD player breaks down are all examples of normal distributions.

---

* Recall that a distribution is just another name for a set of data.

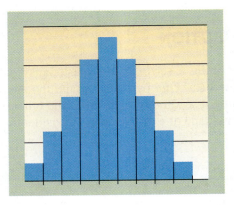

FIGURE 14.11 A normal distribution.

There are several patterns in Figure 14.11 that are common to all normal distributions. First, if we were to take a wire and attach it to the tops of the bars in the histogram and then smooth out the wire to make a curve, the curve would have a bell-shaped appearance, as shown in Figure 14.12. This is why a normal distribution is often called a *bell-shaped curve*.

Second, the mean, median, and mode of a normal distribution are the same. Third, the curve is symmetric with respect to the mean. This means that if you see some pattern in the graph on one side of the mean, then the mean acts as a mirror to give a reflection of that same pattern on the other side of the mean. Fourth, the area under a normal curve equals 1.

In Figure 14.13, we have marked the mean and two points on the normal curve called inflection points. An *inflection point* is a point on the curve where the curve changes from being curved upward to being curved downward, or vice versa. For a normal curve, inflection points are located one standard deviation from the mean. Also, because a normal curve is symmetric with respect to the mean, one-half, or 50 percent, of the area under the curve is located on each side of the mean.

In normal distributions, approximately 68 percent of the data values occur within one standard deviation of the mean, 95 percent of the data values lie within two standard deviations of the mean, and 99.7 percent of the data values lie within three standard deviations of the mean. We refer to these facts as the *68-95-99.7 rule*. Figure 14.13 illustrates the 68-95-99.7 rule. In discussing normal distributions, we usually assume that we are dealing with an *entire population* rather than a *sample*, so in Figure 14.13 we represent the mean by μ and the standard deviation by σ (rather than $\overline{x}$ and s).

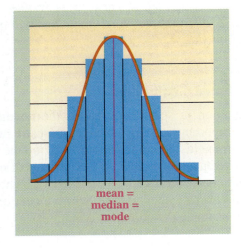

FIGURE 14.12 Smoothing a histogram of a normal distribution into a normal curve.

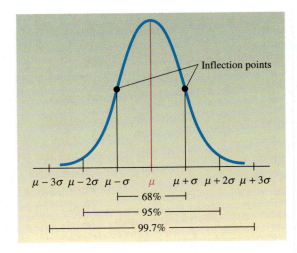

FIGURE 14.13 The 68-95-99.7 rule for a normal distribution.

We summarize the properties of a normal distribution.

> **Properties of a Normal Distribution**
> 1. A normal curve is bell-shaped.
> 2. The highest point on the curve is at the mean of the distribution.
> 3. The mean, median, and mode of the distribution are the same.
> 4. The curve is symmetric with respect to its mean.
> 5. The total area under the curve is 1.
> 6. Roughly 68 percent of the data values are within one standard deviation from the mean, 95 percent of the data values are within two standard deviations from the mean, and 99.7 percent of the data values are within three standard deviations from the mean.*

We can use the 68-95-99.7 rule to estimate how many values we expect to fall within one, two, or three standard deviations of the mean of a normal distribution.

EXAMPLE 1 The Normal Distribution and Intelligence Tests

Suppose that the distribution of scores of 1,000 students who take a standardized intelligence test is a normal distribution. If the distribution's mean is 450 and its standard deviation is 25,

a) how many scores do we expect to fall between 425 and 475?

b) how many scores do we expect to fall above 500?

SOLUTION: a) As you can see in Figure 14.14, the scores 425 and 475 are one standard deviation below and above the mean respectively.

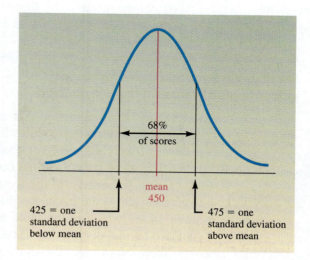

FIGURE 14.14 68 percent of the scores lie within one standard deviation of the mean.

---

* To keep this discussion simple, we are approximating these percentages. Shortly, we will use a table to improve slightly our estimate of the percentage of scores within one and two standard deviations from the mean.

From the 68-95-99.7 rule, we know that 68 percent, or about 0.68 of the scores, lie within one standard deviation of the mean. Since we have 1,000 scores, we can expect that about $0.68 \times 1,000 = 680$ scores are in the range 425 to 475.

b) Recall in Figure 14.15, that 95 percent of the scores in a normal distribution fall between two standard deviations below the mean and two standard deviations above the mean.

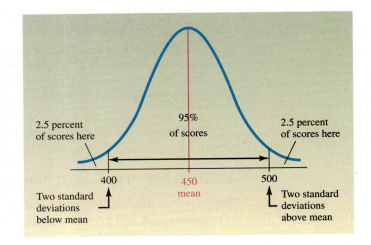

FIGURE 14.15   5 percent of the scores lie more than two standard deviations from the mean.

Quiz Yourself 10

Use the information given in Example 1 to determine the following.

a) How many scores do we expect to fall between 400 and 500?

b) How many scores do we expect to fall below 400?

This means that 5 percent, or 0.05, of the scores lie more than two standard deviations above or below the mean. Thus, we can expect to have $0.05/2 = 0.025$ of the scores to be above 500. Again multiplying by 1,000, we can expect that $0.025 \times 1,000 = 25$ scores to be above 500.

Areas under a normal curve represent percentages of values in the distribution.

In Example 1, we estimated how many values were within one standard deviation of the mean. It is natural to ask whether we can predict how many values lie other distances from the mean. For example, knowing that the weights of women between 18 and 25 form a normal distribution, a clothing manufacturer may want to know what percentage of the female population is between 1.3 and 2.5 standard deviations above the mean weight for women. It is possible to find this by using Table 14.16, which is the table of areas under the standard normal curve. The standard normal distribution has a mean of 0 and a standard deviation of 1.

Table 14.16 gives the area under this curve between the mean and a number called a z-score. A *z*-**score** represents the number of standard deviations a data value is from the mean. In Example 1, we found that for a normal distribution with a mean of 450 and a standard deviation of 25, the value 500 was two standard deviations above the mean. Another way of saying this is that the value 500 corresponds to a z-score of 2.

z	A	z	A	z	A	z	A	z	A	z	A
.00	.000	.56	.212	1.12	.369	1.68	.454	2.24	.488	2.80	.497
.01	.004	.57	.216	1.13	.371	1.69	.455	2.25	.488	2.81	.498
.02	.008	.58	.219	1.14	.373	1.70	.455	2.26	.488	2.82	.498
.03	.012	.59	.222	1.15	.375	1.71	.456	2.27	.488	2.83	.498
.04	.016	.60	.226	1.16	.377	1.72	.457	2.28	.489	2.84	.498
.05	.020	.61	.229	1.17	.379	1.73	.458	2.29	.489	2.85	.498
.06	.024	.62	.232	1.18	.381	1.74	.459	2.30	.489	2.86	.498
.07	.028	.63	.236	1.19	.383	1.75	.460	2.31	.490	2.87	.498
.08	.032	.64	.239	1.20	.385	1.76	.461	2.32	.490	2.88	.498
.09	.036	.65	.242	1.21	.387	1.77	.462	2.33	.490	2.89	.498
.10	.040	.66	.245	1.22	.389	1.78	.463	2.34	.490	2.90	.498
.11	.044	.67	.249	1.23	.391	1.79	.463	2.35	.491	2.91	.498
.12	.048	.68	.252	1.24	.393	1.80	.464	2.36	.491	2.92	.498
.13	.052	.69	.255	1.25	.394	1.81	.465	2.37	.491	2.93	.498
.14	.056	.70	.258	1.26	.396	1.82	.466	2.38	.491	2.94	.498
.15	.060	.71	.261	1.27	.398	1.83	.466	2.39	.492	2.95	.498
.16	.064	.72	.264	1.28	.400	1.84	.467	2.40	.492	2.96	.499
.17	.068	.73	.267	1.29	.402	1.85	.468	2.41	.492	2.97	.499
.18	.071	.74	.270	1.30	.403	1.86	.469	2.42	.492	2.98	.499
.19	.075	.75	.273	1.31	.405	1.87	.469	2.43	.493	2.99	.499
.20	.079	.76	.276	1.32	.407	1.88	.470	2.44	.493	3.00	.499
.21	.083	.77	.279	1.33	.408	1.89	.471	2.45	.493	3.01	.499
.22	.087	.78	.282	1.34	.410	1.90	.471	2.46	.493	3.02	.499
.23	.091	.79	.285	1.35	.412	1.91	.472	2.47	.493	3.03	.499
.24	.095	.80	.288	1.36	.413	1.92	.473	2.48	.493	3.04	.499
.25	.099	.81	.291	1.37	.415	1.93	.473	2.49	.494	3.05	.499
.26	.103	.82	.294	1.38	.416	1.94	.474	2.50	.494	3.06	.499
.27	.106	.83	.297	1.39	.418	1.95	.474	2.51	.494	3.07	.499
.28	.110	.84	.300	1.40	.419	1.96	.475	2.52	.494	3.08	.499
.29	.114	.85	.302	1.41	.421	1.97	.476	2.53	.494	3.09	.499
.30	.118	.86	.305	1.42	.422	1.98	.476	2.54	.495	3.10	.499
.31	.122	.87	.308	1.43	.424	1.99	.477	2.55	.495	3.11	.499
.32	.126	.88	.311	1.44	.425	2.00	.477	2.56	.495	3.12	.499
.33	.129	.89	.313	1.45	.427	2.01	.478	2.57	.495	3.13	.499
.34	.133	.90	.316	1.46	.428	2.02	.478	2.58	.495	3.14	.499
.35	.137	.91	.319	1.47	.429	2.03	.479	2.59	.495	3.15	.499
.36	.141	.92	.321	1.48	.431	2.04	.479	2.60	.495	3.16	.499
.37	.144	.93	.324	1.49	.432	2.05	.480	2.61	.496	3.17	.499
.38	.148	.94	.326	1.50	.433	2.06	.480	2.62	.496	3.18	.499
.39	.152	.95	.329	1.51	.435	2.07	.481	2.63	.496	3.19	.499
.40	.155	.96	.332	1.52	.436	2.08	.481	2.64	.496	3.20	.499
.41	.159	.97	.334	1.53	.437	2.09	.482	2.65	.496	3.21	.499
.42	.163	.98	.337	1.54	.438	2.10	.482	2.66	.496	3.22	.499
.43	.166	.99	.339	1.55	.439	2.11	.483	2.67	.496	3.23	.499
.44	.170	1.00	.341	1.56	.441	2.12	.483	2.68	.496	3.24	.499
.45	.174	1.01	.344	1.57	.442	2.13	.483	2.69	.496	3.25	.499
.46	.177	1.02	.346	1.58	.443	2.14	.484	2.70	.497	3.26	.499
.47	.181	1.03	.349	1.59	.444	2.15	.484	2.71	.497	3.27	.500
.48	.184	1.04	.351	1.60	.445	2.16	.485	2.72	.497	3.28	.500
.49	.188	1.05	.353	1.61	.446	2.17	.485	2.73	.497	3.29	.500
.50	.192	1.06	.355	1.62	.447	2.18	.485	2.74	.497	3.30	.500
.51	.195	1.07	.358	1.63	.449	2.19	.486	2.75	.497	3.31	.500
.52	.199	1.08	.360	1.64	.450	2.20	.486	2.76	.497	3.32	.500
.53	.202	1.09	.362	1.65	.451	2.21	.487	2.77	.497	3.33	.500
.54	.205	1.10	.364	1.66	.452	2.22	.487	2.78	.497		
.55	.209	1.11	.367	1.67	.453	2.23	.487	2.79	.497		

TABLE 14.16 Standard normal distribution.

Notice that Table 14.16 gives areas only for positive z-scores—that is, for data that lie above the mean. We use the symmetry of a normal curve to find areas corresponding to negative z-scores.

Example 2 shows how to use Table 14.16. Because the total area under a normal curve is 1, we can interpret areas under the standard normal curve as percentages of the data values in the distribution. Also note that because the mean is 0 and the standard deviation for the standard normal curve is 1, the value of a z-score is also the same as the number of standard deviations that z-score is from the mean. After you have practiced working with the standard normal distribution, we will show you how to use the techniques that you have learned to solve problems involving real-life data.

PROBLEM SOLVING

The Three-Way Principle in Section 1.1 suggests that you can often understand a problem graphically. This is certainly true in solving problems involving the normal curve. If you draw a good picture of the situation, it often becomes more clear to you as to what to do to set up and solve the problem.

EXAMPLE 2   Finding Areas under the Standard Normal Curve

Use Table 14.16 to find the percentage of the data (area under the curve) that lie in the following regions for a standard normal distribution.

a) Between $z = 0$ and $z = 1.3$

b) Between $z = 1.5$ and $z = 2.1$

c) Between $z = 0$ and $z = -1.83$

SOLUTION: a) The area under the curve between $z = 0$ and $z = 1.3$ is shown in Figure 14.16. We find this area by looking in Table 14.16 for the z-score 1.30. We see that A is 0.403 when $z = 1.30$. Thus, we expect 0.403, or 40.3 percent, of the data to fall between 0 and 1.3 standard deviations above the mean.

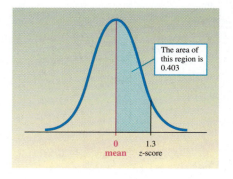

FIGURE 14.16   Area under the standard normal curve between $z = 0$ and $z = 1.3$.

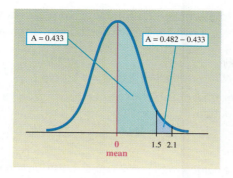

FIGURE 14.17   Finding the area under the standard normal curve between $z = 1.5$ and $z = 2.1$.

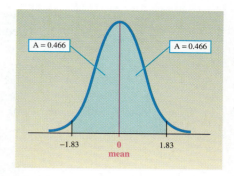

FIGURE 14.18   Finding the area under the standard normal curve between $z = -1.83$ and $z = 0$.

 Quiz Yourself 11

Use Table 14.16 to find the following areas under the standard normal curve.

a) between $z = 0$ and $z = 1.45$

b) between $z = 1.23$ and $z = 1.85$

c) between $z = 0$ and $z = -1.35$

b) Figure 14.17 shows the area we want. Areas in Table 14.16 correspond to regions from $z = 0$ to the given z-score. Therefore, in order to find the area under the curve between $z = 1.5$ and $z = 2.1$, we must first find the area from $z = 0$ to $z = 2.1$ and then subtract the area from $z = 0$ to $z = 1.5$. From Table 14.16 we see that when $z = 2.1$, $A = 0.482$, and when $z = 1.5$, $A = 0.433$. Therefore the area we seek is $0.482 - 0.433 = 0.049$. This means that in the standard normal distribution, 4.9 percent of the data lie between 1.5 and 2.1 standard deviations above the mean.

c) Due to the symmetry of the normal distribution, the area between $z = 0$ and $z = -1.83$ is the same as the area between $z = 0$ and $z = 1.83$ (see Figure 14.18). From Table 14.16 we see that when $z = 1.83$, $A = 0.466$. Therefore, 46.6 percent of the data values lie between 0 and -1.83.

SOME GOOD ADVICE

A common mistake in solving a problem such as Example 2(b) is to subtract 1.5 from 2.1 to get 0.6 and then wrongly use this for the z-score in Table 14.16. If you think about the graph of the standard normal curve, clearly the area between $z = 0$ and $z = 0.6$ is not the same as the area between $z = 1.5$ and $z = 2.1$.

We convert values in a nonstandard normal distribution to z-scores.

A real-life normal distribution, such as the set of all weights of women between 18 and 25, may have a mean of 120 pounds and a standard deviation of 25 pounds. Such a distribution will have the properties that we stated earlier for a normal distribution, but because the distribution does not have a mean of 0 and a standard deviation of 1, we cannot use Table 14.16 directly, as we did in Example 2. We *can* use Table 14.16, however, if we first convert nonstandard values, called *raw scores*, to *z*-scores. The following formula shows the desired relationship between values in a nonstandard normal distribution and *z*-scores.

> **Formula for Converting Raw Scores to z-Scores**
>
> Assume a normal distribution has a mean of μ and a standard deviation of σ. We use the equation
>
> $$z = \frac{x - \mu}{\sigma}$$
>
> to convert a value *x* in the nonstandard distribution to a *z*-score.

EXAMPLE 3   Converting Raw Scores to *z*-Scores

Suppose the mean of a normal distribution is 20 and its standard deviation is 3.

a) Find the *z*-score corresponding to the raw score 25.

b) Find the *z*-score corresponding to the raw score 16.

SOLUTION: a) Figure 14.19 gives us a picture of this situation.

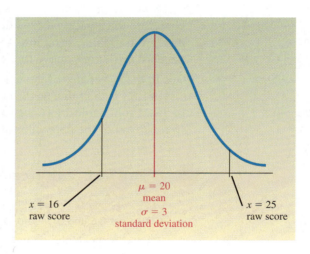

FIGURE 14.19   Normal distribution with mean of 20 and standard deviation of 3.

> ![Quiz icon] **Quiz Yourself 12**
>
> Suppose a normal distribution has a mean of 50 and a standard deviation of 7. Convert each raw score to a *z*-score.
>
> a) 59   b) 50   c) 38

We will use the conversion formula $z = \frac{x - \mu}{\sigma}$, with the raw score $x = 25$, the mean $\mu = 20$, and the standard deviation $\sigma = 3$. Making the substitutions, we have, $z = \frac{25 - 20}{3} = \frac{5}{3} = 1.67$. We can interpret this result as telling us that in this distribution, 25 is 1.67 standard deviations above the mean.

b) We will use Figure 14.19 and the conversion formula again, but this time with $x = 16$, $\mu = 20$, and $\sigma = 3$. Therefore, the corresponding *z*-score is $z = \frac{16 - 20}{3} = \frac{-4}{3} = -1.33$. This tells us that 16 is 1.33 standard deviations below the mean.   ◎

The Three-Way Principle in Section 1.1 suggests that you can remember the formula for converting raw scores to z-scores by interpreting the formula geometrically. In Example 3, we had a normal distribution with a mean of 20 and standard deviation 5. In order to make the mean of 20 in the nonstandard distribution correspond to the mean of 0 in the standard normal distribution, you can think of "sliding" the distribution to the left 20 units. That is why we compute an expression of the form $x - 20$ in the numerator of the z-score formula. Because the standard deviation was 5, we divided $x - 20$ by 5 to "narrow" the distribution so that it would have a standard deviation of 1 and therefore allow us to use Table 14.16.

EXAMPLE 4   Interpreting the Significance of an Exam Score

Suppose that to qualify for a management training program offered by your employer you must score in the top 10 percent of those employees who take a standardized test. Assume that the distribution of scores is normal and you received a score of 72 on the test, which had a mean of 65 and a standard deviation of 4. What percentage of those who took this test had a score below yours?

SOLUTION: We will find the number of standard deviations your score is above the mean and then determine what percentage of scores fall below this number. First we calculate the z-score, which corresponds to your 72:

$$z = \frac{72 - 65}{4} = \frac{7}{4} = 1.75.$$

Now, looking in Table 14.16 for $z = 1.75$, we find that A $= 0.460$. Therefore, 46 percent of the scores fall between the mean and your score (see Figure 14.20).

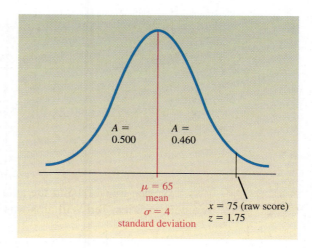

FIGURE 14.20   Normal distribution with mean of 65 and standard deviation of 4.

But, we must not forget the scores that fall below the mean. Because a normal curve is symmetric, another 50 percent of the scores fall below the mean. So, there are 50 percent + 46 percent = 96 percent of the scores below yours. Congratulations! You qualify for the program. ◎

We can use the ideas we have been discussing about the normal curve to compare data.

EXAMPLE 5   Using *z*-Scores to Compare Data

Consider the following information:

Ty Cobb hit .420 in 1911.
Ted Williams hit .406 in 1941.
George Brett hit .390 in 1980.

In the 1910s, the mean batting average was .266 and the standard deviation was .0371.
In the 1940s, the mean batting average was .267 and the standard deviation was .0326.
In the 1970s, the mean batting average was .261 and the standard deviation was .0317.

Assume that the batting averages were normally distributed in each of these three decades. Use *z*-scores to determine which of the three batters was ranked the highest in relationship to his contemporaries.

SOLUTION: We will convert each of the three batting averages to a corresponding *z*-score for the distribution of batting averages for the decade in which the athlete played.

In the 1910s, Ty Cobb's average of .420 corresponded to a *z*-score of

$$\frac{.420 - .266}{.0371} = \frac{.154}{.0371} = 4.1509.$$

In the 1940s, Ted Williams's average of .406 corresponded to a *z*-score of

$$\frac{.406 - .267}{.0326} = \frac{.139}{.0326} = 4.2638.$$

In the 1970s, George Brett's average of .390 corresponded to a *z*-score of

$$\frac{.390 - .261}{.0317} = \frac{.129}{.0317} = 4.0694.$$

We see that in using the standard deviation to compare each hitter with his contemporaries, Ted Williams was ranked as the best hitter. ◎

We conclude our discussion of normal distributions by explaining how manufacturers can use information regarding the life of their products as a basis for warranties.

EXAMPLE 6   Using the Normal Distribution to Write Warranties

In order to increase sales, a manufacturer plans to offer a warranty on a new model of DVD player. In testing the DVD player, quality-control engineers found that the player has a mean time to failure of 3,000 hours with a standard deviation of 500 hours. Assume that the typical purchaser will play the DVD player for four hours per day. If the manufacturer does not want more than 5 percent of the players to be returned as defective within the warranty period, how long should the warranty period be to guarantee this?

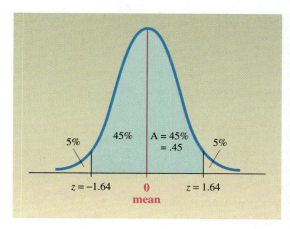

FIGURE 14.21   Finding the area under the standard normal curve for a negative z-score.

SOLUTION: In solving this problem, we first work with the standard normal distribution and then convert the answer to fit our nonstandard normal distribution. Figure 14.21 gives a picture of the situation for the standard normal curve.

In Figure 14.21 we see that we need to find a z-score such that at least 95 percent of the entire area is *beyond* this point. Observe that this score is left of the mean and is negative. Therefore, we use symmetry and find the z-score such that 95 percent of the entire area is *below* this score. But Table 14.16 gives the areas only for the upper half of the distribution. However, this is not a problem, because we know that 50 percent of the entire area lies below the mean. Therefore, our problem reduces to finding a z-score greater than 0 such that 45 percent of the area lies between the mean and that z-score.

In Table 14.16, we see that if $A = 0.450$, the corresponding z-score is 1.64. This means that 95 percent of the area underneath the standard normal curve falls below $z = 1.64$. By symmetry, we also can say that 95 percent of the values lie above -1.64. We must now interpret this z-score in terms of original normal distribution of raw scores.

Recall the equation that relates values in a distribution to z-scores, namely,

$$z = \frac{x - \mu}{\sigma} \tag{1}$$

Previously we used this equation to convert raw scores to z-scores. We now use the equation in answering the reverse question. Specifically, what is x when $\mu = 3000$, $\sigma = 500$, and $z = -1.64$? Substituting these numbers in equation (1), we get

$$-1.64 = \frac{x - 3000}{500}. \tag{2}$$

We solve for x in this equation by multiplying both sides by 500:

$$-1.64 \cdot (500) = \frac{x - 3000}{500} \cdot (500).$$

Simplifying the equation, we get

$$-820 = x - 3000.$$

Last, we add 3,000 to both sides to obtain $x = 2,180$ hours. We expect 95 percent of the breakdowns to occur beyond 2,180 hours.

Since we know that owners use the DVD player about four hours per day, we divide 2,180 by 4 to get $\frac{2,180}{4} = 545$ days. This is equivalent to $\frac{545}{31} = 17.58$ months if we use 31 days per month. Therefore, if the manufacturer wants to have no more than 5 percent of the breakdowns occur within the warranty period, the warranty should be for roughly eighteen months.

Exercises 14.4

Assume that all distributions in this exercise set are normal.

Assume that the distribution in Exercises 1–6 has a mean of 10 and a standard deviation of 2. Use the 68-95-99.7 rule to find the percentage of values in the distribution that we describe.

1. between 10 and 12

2. between 12 and 14

3. above 14

4. below 8

5. above 12

6. below 10

Assume that the distribution in Exercises 7–12 has a mean of 12 and a standard deviation of 3. Use the 68-95-99.7 rule to find the percentage of values in the distribution that we describe.

7. below 9

8. between 12 and 15

9. above 6

10. below 12

11. between 15 and 18

12. above 18

Due to random variations in the operation of an automatic coffee machine, not every cup is filled with the same amount of coffee. Assume that the mean amount of coffee dispensed is 8 ounces and the standard deviation is 0.5 ounces. Use Figure 14.13 to solve Exercises 13–14.

13. Analyzing vending machines.

    a) What percentage of the cups should have at least 8 ounces of coffee?

    b) What percentage of the cups should have less than 7.5 ounces of coffee?

14. Analyzing vending machines.

    a) What percentage of the cups should have at least 8.5 ounces of coffee?

    b) What percentage of the cups should have less than 8 ounces of coffee?

A machine fills bags of candy. Due to slight irregularities in the operation of the machine, not every bag gets exactly the same number of pieces. Assume that the number of pieces per bag has a mean of 200 with a standard deviation of 2. Use Figure 14.13 to solve Exercises 15–16.

15. Analyzing vending machines.

    a) How many of the bags do we expect to have between 200 and 202 pieces in them?

    b) How many of the bags do we expect to have at least 202 pieces in them?

16. Analyzing vending machines.

    a) How many of the bags do we expect to have between 198 and 200 pieces in them?

    b) How many of the bags do we expect to have at least 196 pieces in them?

Use Table 14.16 to find the percentage of area under the standard normal curve that is specified.

17. Between $z = 0$ and $z = 1.23$.

18. Between $z = 0$ and $z = 2.06$.

19. Between $z = 0.89$ and $z = 1.43$.

20. Between $z = 0.76$ and $z = 1.52$.

21. Between $z = 0$ and $z = -0.75$.

22. Between $z = 0$ and $z = -1.35$.

23. Between $z = 1.25$ and $z = 1.95$

24. Between $z = 0.37$ and $z = 1.23$

25. Between $z = -0.38$ and $z = -0.76$

26. Between $z = -1.55$ and $z = -2.13$

27. Above $z = 1.45$

28. Below $z = -0.64$

29. Below $z = -1.40$

30. Above $z = 0.78$

31. Below $z = 1.33$

32. Above $z = -0.46$

33. Above $z = -0.84$

34. Below $z = 1.22$

35. Find a z-score such that 10 percent of the area under the standard normal curve is above that score.

36. Find a z-score such that 20 percent of the area under the standard normal curve is above that score.

37. Find a z-score such that 12 percent of the area under the standard normal curve is below that score.

38. Find a z-score such that 24 percent of the area under the standard normal curve is below that score.

39. Find a z-score such that 60 percent of the area is below that score.

40. Find a z-score such that 90 percent of the area is below that score.

41. Find a z-score such that 75 percent of the area is above that score.

42. Find a z-score such that 60 percent of the area is above that score.

43. Communicating Mathematics If a raw score corresponds to a z-score of 1.75, what does that tell you about that score in relationship to the mean of the distribution?

44. Communicating Mathematics If a raw score corresponds to a z-score of -0.85, what does that tell you about that score in relationship to the mean of the distribution?

45. Communicating Mathematics Explain why, if you wish to compute the area under a normal curve between $z = 1.3$ and $z = 2.0$, it is not correct to subtract $2.0 - 1.3 = 0.7$ and then use the z-score of 0.7 to find the area.

46. Communicating Mathematics Without looking at Table 14.16, can you determine whether the area under the standard normal curve between $z = 0.5$ and $z = 1.0$ is greater than or less than the area between $z = 1.5$ and $z = 2.0$? Explain.

In Exercises 47–52, we give you a mean, a standard deviation, and a raw score. Find the corresponding z-score.

47. mean 80, standard deviation 5, $x = 87$

48. mean 100, standard deviation 15, $x = 117$

49. mean 21, standard deviation 4, $x = 14$

50. mean 52, standard deviation 7.5, $x = 61$

51. mean 38, standard deviation 10.3, $x = 48$

52. mean 8, standard deviation 2.4, $x = 6.2$

In Exercises 53–58, we give you a mean, a standard deviation, and a z-score. Find the corresponding raw score.

53. mean 60, standard deviation 5, $z = 0.84$

54. mean 20, standard deviation 6, $z = 1.32$

55. mean 35, standard deviation 3, $z = -0.45$

56. mean 62, standard deviation 7.5, $z = -1.40$

57. mean 28, standard deviation 2.25, $z = 1.64$

58. mean 8, standard deviation 3.5, $z = -1.25$

59. **Product reliability.** Suppose that for a particular brand of television set, the distribution of picture tube failures has a mean of 120 months with a standard deviation of 12 months. With respect to this distribution, 140 months corresponds to what z-score?

60. **Product reliability.** Redo Exercise 59 using 130 months instead of 140.

61. **Distribution of heights.** Assume that in the NBA the distribution of heights has a mean of 6 feet, 8 inches, and a standard deviation of 3 inches. If there are 324 players in the league, how many players do you expect to be over seven feet tall?

62. **Distribution of heights.** Redo Exercise 61, but now we want to know how many players you would expect to be over 6 feet, 6 inches.

63. **Distribution of heart rates.** Assume that the distribution of 21-year-old women's heart rates at rest is 68 beats per minute with a standard deviation of 4 beats per minute. If 200 women are examined, how many would you expect to have a heart rate of less than 70?

64. **Distribution of heart rates.** Redo Exercise 63, but now we ask how many you would expect to have a heart rate less than 75.

In Exercises 65–66, we assume that the mean length of a human pregnancy is 268 days.

65. **Birth statistics.** If 95 percent of all human pregnancies last between 250 and 286 days, what is the standard deviation of the distribution of lengths of human pregnancies? What percentage of human pregnancies do we expect to last at least 275 days?

66. **Birth statistics.** Babies born before the 37th week (that is, before day 252 of a pregnancy) are considered premature. What percentage of births do we expect to be premature? (See Exercise 65.)

67. **Analyzing Internet use.** An Internet provider has analyzed the amount of time that its customers are online per session. It found that the distribution of connect times has a mean of 37 minutes and a standard deviation of 11 minutes. For this distribution, find the raw score that corresponds to a z-score of 1.5.

68. **Analyzing customer service.** The distribution of times that customers spend waiting in line in a supermarket has a mean of 3.6 minutes and a standard deviation of 1.2 minutes. For this distribution, find the raw score that corresponds to a z-score of -1.3.

In Exercises 69–70, use the information and methods of Example 5 to make the indicated comparisons.

69. **Comparing athletes.** In 1949, Jackie Robinson hit .342 for the Brooklyn Dodgers; in 1973, Rod Carew hit .350 for the Minnesota Twins. Using z-scores to compare each batter with his contemporaries, determine which batting average was more impressive.

70. **Comparing athletes.** In 1940, Joe DiMaggio hit .352 for the New York Yankees; in 1975, Bill Madlock hit .354 for the Pittsburgh Pirates. Using z-scores to compare each batter with his contemporaries, determine which batting average was more impressive.

The nth percentile in a distribution is a number x such that n percent of the values in that distribution fall below x. For example, 20 percent of the values in the distribution fall below the 20th percentile.

71. What z-score in the standard normal distribution is the 80th percentile?

72. What z-score in the standard normal distribution is the 30th percentile?

73. If a distribution has a mean of 40 and a standard deviation of 4, what is the 75th percentile?

74. If a distribution has a mean of 80 and a standard deviation of 6, what is the 35th percentile?

75. **Finding percentiles.** If you score a 75 on a commercial pilot's exam that has a mean score of 68 and a standard deviation of 4, to what percentile does your score correspond?

76. **Finding percentiles.** If you know that the mean salary for your profession is $53,000 with a standard deviation of $2,500, to what percentile does your salary of $57,000 correspond?

77. **Writing a warranty.** A manufacturer plans to provide a warranty on a new 3D interactive video game. In testing, the manufacturer has found that the set of failure times for the game is normally distributed with a mean time of 2,000 hours and a standard deviation of 800 hours. Assume that the typical purchaser will play the game for two hours per day. If the manufacturer does not want more than 4 percent of the games returned as defective within the warranty period, how long should the warranty period be to guarantee this? (Assume that there are 31 days in each month.)

78. **Writing a warranty.** Redo Exercise 77, but now assume that the mean of the distribution of failure times is 2,500 hours, the standard deviation is 600 hours, and the manufacturer wants no more than 2.5 percent of the games returned during the warranty period.

79. **Analyzing investments.** A certain mutual fund over the last fifteen years has a mean yearly return of 7.8 percent with a standard deviation of 1.3 percent.

   a) If you had invested in this fund over the past fifteen years, in how many of these years would you expect to earn at least 9 percent on your investment?

   b) In how many years would you expect to earn less than 6 percent?

80. **Analyzing investments.** A certain bond fund over the last twelve years has a mean yearly return of 5.9 percent with a standard deviation of 1.8 percent.

   a) If you had invested in this fund over the past twelve years, in how many of these years would you expect to earn at least 8 percent on your investment?

   b) In how many years would you expect to earn less than 4 percent?

81. Analyzing the SATs. Assume that the math SAT scores are normally distributed with a mean of 500 and a standard deviation of 100. If you score 480 on this exam, what percentage of those taking the test scored below you?

82. Analyzing the SATs. Assume that the verbal SAT scores are normally distributed with a mean of 500 and a standard deviation of 100. If you score 520 on this exam, what percentage of those taking the test scored below you?

Further Exercises

83. Communicating Mathematics   Explain how you would estimate the mean and the standard deviation of the two normal distributions in the given figure.

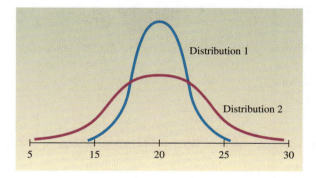

84. Communicating Mathematics   Explain how you would estimate the mean and the standard deviation of the two normal distributions in the given figure.

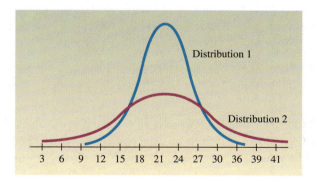

85. Communicating Mathematics   If you had a large collection of data, how might you go about determining whether the distribution was normal?

86. Communicating Mathematics   Discuss whether you feel that the method we used to compare Ty Cobb, Ted Williams, and George Brett would be a valid way to compare the great African-American athletes who played in the Negro leagues with the great white athletes of the same era.

87. Communicating Mathematics   Gather data about several sports figures that you wish to compare. Use the methods of this chapter to show that one of these athletes is superior to the other.

CHAPTER SUMMARY

SECTION 14.1

data, distribution, frequency distribution, frequency table, relative frequency distribution

We refer to a collection of numerical information as data. A set of data listed with their frequencies is called a frequency distribution. We often present a frequency distribution in a frequency table. A relative frequency distribution shows the percentage of the time that each value occurs in a frequency distribution.

bar graph, histogram

In drawing a bar graph, we specify the classes on the horizontal axis and frequencies on the vertical axis. We draw bars (rectangles) above each class having a height equal to the frequency of the class. A histogram is a type of bar graph used to graph a frequency distribution when we are dealing with a continuous variable quantity.

stem-and-leaf display

A stem-and-leaf display is an effective way to present two sets of data "side by side" for analysis.

SECTION 14.2

measures of central tendency

The mean, the median, the mode, and quartiles are measures of central tendency.

mean of a data set

The mean of a data set is $\frac{\Sigma x}{n}$, where Σx is the sum of the data values and n is the number of scores. We indicate the mean of a sample by $\bar{x}$ and the mean of a population by μ.

mean of a frequency distribution

We compute the mean of a frequency distribution using the formula

$$\frac{\Sigma(x \cdot f)}{\Sigma f},$$

where $x \cdot f$ represents the product of a data value x and its frequency f and Σf is the sum of the frequencies of the data values in the distribution.

median of a data set

To find the median, first arrange the data in order. If there is an odd number of values, the median is the value in the middle position. If there is an even number of values, the median is the mean of the two middle values.

lower half, upper half, first quartile, third quartile

If a data set is arranged in increasing order, the set of values below the median is called the lower half. The set of values above the median is called the upper half. The first quartile, Q_1, is the median of the lower half. The third quartile, Q_3, is the median of the upper half.

five-number summary

The five-number summary of a data set is the minimum value, the first quartile, the median, the third quartile, and the maximum value.

constructing a box-and-whisker plot

Draw a horizontal axis with a scale including the extreme values of the data. Draw a box from Q_1 to Q_3 and draw a vertical line in the box indicating the median. Draw whiskers (lines) from Q_3 to the largest score and from Q_1 to the smallest score.

mode

The mode of a data set is the most frequently occurring value. If two values are most frequent, then we say that there are two modes. We do not allow more than two modes.

SECTION 14.3

range

The range is the difference between the largest and smallest values in a data set.

standard deviation

The standard deviation of a sample, indicated by s, is defined as

$$s = \sqrt{\frac{\Sigma(x - \bar{x})^2}{n - 1}},$$

where $\bar{x}$ is the mean of the sample and n is the number of data values. For a population, we use the formula

$$\sigma = \sqrt{\frac{\Sigma(x - \mu)^2}{n}}.$$

standard deviation of a frequency distribution

If f represents the frequency of data value x in the distribution, then the standard deviation, s, is given by the formula

$$s = \sqrt{\frac{\Sigma(x - \bar{x})^2 \cdot f}{n - 1}}.$$

The number $n = \Sigma f$ is the number of data values in the distribution.

coefficient of variation of a distribution

The coefficient of variation expresses the standard deviation of a sample as a percentage of the mean and is calculated as follows:

$$CV = \frac{s}{\bar{x}} \cdot 100\%$$

SECTION 14.4

properties of a normal distribution

1. A normal curve is bell-shaped.
2. The highest point on the curve is at the mean of the distribution.
3. The mean, median, and mode of the distribution are the same.
4. The curve is symmetric with respect to its mean.
5. The total area under the curve is 1.

6. Roughly 68 percent of the data values are within one standard deviation from the mean, 95 percent of the data values are within two standard deviations from the mean, and 99.7 percent are within three standard deviations from the mean.

z-score

A z-score represents the number of standard deviations a data value is from the mean of a data set. We can use a table to relate positive z-scores to the amount of area under the standard normal curve between the mean and the given z-scores.

converting data values to z-scores

If a normal distribution has a mean of μ and a standard deviation of σ, the equation $z = \frac{x - \mu}{\sigma}$ converts a data value x to a z-score.

CHAPTER TEST

SECTION 14.1

The following data set represents the number of automobile accidents that have occurred in a city each day over the past month.

8, 8, 6, 6, 7, 10, 4, 6, 5, 10, 8, 9, 8, 7, 8, 8,
6, 9, 5, 9, 8, 10, 8, 8, 6, 10, 6, 5, 10, 8, 9

1. Construct a frequency table for these data.

2. Construct a relative frequency table for these data.

3. Display the relative frequency table in Question 2 as a bar graph.

4. An operator of an amusement park ride counted the number of people on the ride at ten-minute intervals for a certain period of time. The results he obtained are summarized in the accompanying bar graph. Use the graph to answer the following questions.

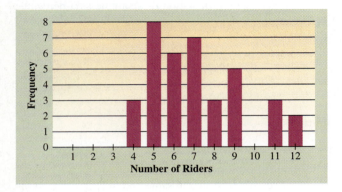

a) What was the smallest number of people on the ride and how often did it occur?

b) What was the most frequently occurring rider count?

c) For how many ten-minute intervals were the riders counted?

5. A hospital is testing two weight loss programs to determine which program is more effective. The following data represent the amount of weight lost in a year by patients in each program. Represent the two given sets of data on a single stem-and-leaf display. Which program seems to be more effective?

Program A: 19, 32, 27, 34, 33, 36, 47, 32, 25,
52, 29, 26, 37, 28, 26, 43, 31, 40
Program B: 29, 21, 39, 44, 41, 36, 37, 26, 26,
43, 45, 28, 32, 28, 33, 36, 53, 39

SECTION 14.2

6. Find the mean, median, and mode of the following distribution:

10, 7, 6, 12, 6, 10, 8, 7, 6, 7, 4, 9, 13, 10, 7, 13, 7

7. Briefly describe what information is conveyed by the mean, median, and mode.

8. Calculate the five-number summary for the following distribution:

10, 8, 6, 12, 6, 3, 11, 7, 6, 17, 4, 9, 13, 20, 7

9. Construct a box-and-whisker plot for the distribution given in Question 4.

SECTION 14.3

10. Calculate the standard deviation for the following distribution:

$$4, 6, 7, 3, 5, 6, 4, 5$$

11. Explain what the standard deviation tells you about a distribution.

SECTION 14.4

12. State several basic properties of the normal distribution.

13. If a distribution has a mean of 80 and a standard deviation of 7, to what z-score does the raw score 85 correspond?

14. If a distribution has a mean of 60 and a standard deviation of 5, the z-score 1.35 corresponds to what raw score?

15. Suppose you earned an 82 on a history exam that had a mean score of 78 and a standard deviation of 3, and you also earned an 84 on an anthropology exam that had a mean of 79 and a standard deviation of 4. On which exam did you do better in relationship to the rest of the class?

16. Explain how the manufacturer of the DVD player in Example 6 of Section 14.4 wrote a guarantee so that no more than 5 percent of the DVD players would be returned for defects.

Of Further Interest: LINEAR CORRELATION

How can we tell whether two sets of data are related? For example, is there a relationship between:

- the amount of dietary fat eaten and the number of pounds a person is overweight?
- time spent at tutoring sessions and grades?
- the amount of alcohol consumed and blood pressure?

In each of these cases we seek some connection between the first quantity and the second. We say there is a **correlation** between two variables if one of them is related to the other in some way—for example, "Is there a correlation between the amount of dietary fat and number of pounds a person is overweight?" Although there are many different types of correlation, later in this section we will discuss one particular type called *linear correlation*.

In order to determine whether there is a correlation between two variables, we obtain pairs of data, called *data points*, relating the first variable to the second. For example, we might examine 100 patients and relate their dietary fat to the number of pounds they are overweight. In order to understand such data, we plot data points in a graph called a **scatterplot**.

EXAMPLE 1   Constructing a Scatterplot

An instructor wants to know whether there is a correlation between the number of times students attended tutoring sessions during the semester and their grades on a 50-point examination. The instructor has collected data for ten students, as shown in Table 14.17. Represent these data points by a scatterplot and interpret the graph.

SOLUTION: We plot the points (18, 42), (6, 31), (16, 46), and so on in Figure 14.22.

The scatterplot shows that as the number of tutoring sessions increases, the grades also generally increase. Based on this scatterplot, the instructor might conclude that there is some relationship between the number of study sessions students attend and their grades.

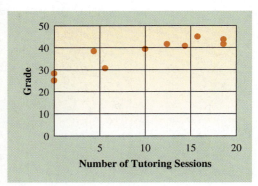

FIGURE 14.22   Scatterplot of data in Table 14.17.

In Example 1, we must be careful not to conclude that there is a cause-and-effect relationship between the number of tutoring sessions attended and the

Number of tutoring sessions, x	18	6	16	14	0	4	0	10	12	18
Exam Grade, y	42	31	46	41	25	38	28	39	42	44

TABLE 14.17   Tutoring sessions attended and exam grades.

students' grades. There are probably many factors that affect the students' grades, such as class attendance, motivation, and the amount of time spent doing home-work. Statisticians often say, "Correlation does not imply causation." Based on the scatterplot, there seems to be some kind of relationship between the two vari-ables; however, at this point we cannot say more than this.

 Quiz Yourself 13*

Draw a scatterplot of the data shown in Table 14.18 and interpret the graph.

x	5	16	14	0	10	20	3	2
y	39	20	29	41	30	15	42	44

TABLE 14.18   Paired data for variables x and y.

We will now discuss one type of correlation called linear correlation. **Lin-ear correlation** exists between two variables if, when we graph them in a scatter-plot, the points in the graph tend to lie in a straight line. Although this definition may seem vague, there is a number called the *linear correlation coefficient* that allows us to compute to what degree the points of a scatterplot lie along a straight line. The derivation of this number is beyond the scope of this text, but, we can explain its meaning and show you how to use it.

Formula for Computing the Linear Correlation Coefficient

If we have pairs of data for two variables x and y, the linear correlation coefficient† for these data, denoted by r, is given by the formula

$$r = \frac{n\Sigma xy - (\Sigma x)(\Sigma y)}{\sqrt{n(\Sigma x^2) - (\Sigma x)^2}\sqrt{n(\Sigma y^2) - (\Sigma y)^2}}. \tag{1}$$

The number n is the number of pairs of data we have for x and y.

We will first show you how to compute the linear correlation coefficient and we will then explain how to interpret its meaning.

EXAMPLE 2   Computing the Linear Correlation Coefficient

Compute the linear correlation coefficient for the data given in Table 14.17.

SOLUTION: We will first describe each expression in equation (1) and then compute these expressions in Table 14.19.

---

* Quiz Yourself answers begin on page 849.
† In this section we will assume that we are dealing with samples. In this case, the linear correlation coefficient is sometimes called the sample linear correlation coefficient.

Σx is the sum of the first coordinates of the data pairs.
Σy is the sum of the second coordinates of the data pairs.
Σx^2 is the sum of the squares of the first coordinates of the data pairs.
Σy^2 is the sum of the squares of the second coordinates of the data pairs.
Σxy is the sum of the products of the first and second coordinates of the data pairs.

x	y	x^2	y^2	xy
18	42	324	1,764	756
6	31	36	961	186
16	46	256	2,116	736
14	41	196	1,681	574
0	25	0	625	0
4	38	16	1,444	152
0	28	0	784	0
10	39	100	1,521	390
12	42	144	1,764	504
18	44	324	1,936	792
$\Sigma x = 98$	$\Sigma y = 376$	$\Sigma x^2 = 1,396$	$\Sigma y^2 = 14,596$	$\Sigma xy = 4,090$

TABLE 14.19 Computations to calculate the correlation coefficient.

The linear correlation coefficient is therefore

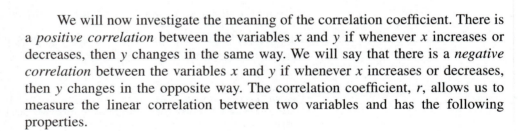

$$r = \frac{n\Sigma xy - (\Sigma x)(\Sigma y)}{\sqrt{n(\Sigma x^2) - (\Sigma x)^2}\,\sqrt{(n(\Sigma y^2) - (\Sigma y)^2}}$$

$$= \frac{10 \cdot 4{,}090 - (98)(376)}{\sqrt{10 \cdot (1{,}396) - 98^2}\,\sqrt{10 \cdot (14{,}596) - 376^2}}$$

$$= \frac{4{,}052}{\sqrt{4{,}356}\,\sqrt{4{,}584}} \approx 0.9068$$

Quiz Yourself 14

Compute the linear correlation coefficient for the data shown in Table 14.18.

We will now investigate the meaning of the correlation coefficient. There is a *positive correlation* between the variables x and y if whenever x increases or decreases, then y changes in the same way. We will say that there is a *negative correlation* between the variables x and y if whenever x increases or decreases, then y changes in the opposite way. The correlation coefficient, r, allows us to measure the linear correlation between two variables and has the following properties.

> **Properties of the Linear Correlation Coefficient**
>
> 1. *r* is a number between −1 and 1, inclusive. If *r* = 1 or −1, then the scatterplot lies on a straight line.
>
> 2. If *r* is positive, there is positive correlation between the variables. If *r* is negative, there is negative correlation between the variables.
>
> 3. If *r* is close to 1, then there is a significant positive linear correlation between the variables and the scatterplot is close to lying along a straight line that rises from left to right.
>
> 4. If *r* is close to −1, then there is a significant negative linear correlation between the variables and the scatterplot is close to lying along a straight line that falls from left to right.
>
> 5. If *r* is close to 0, then there is little linear correlation between the variables.

We show the relationship between scatterplots and the linear correlation coefficient in Figure 14.23.

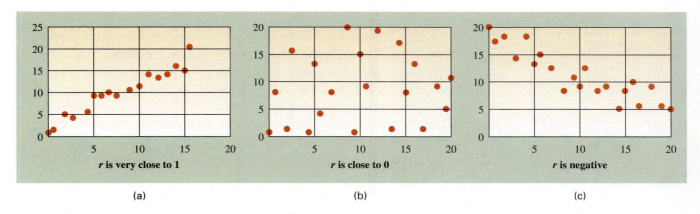

| (a) | (b) | (c) |

FIGURE 14.23   The value of *r* is related to the shape of the scatterplot.

If *r* is close to 1 or −1, then there is significant positive or negative linear correlation. But how close to 1 or −1 does *r* have to be in order for us to conclude that there is significant linear correlation? Table 14.20 contains a list of critical values that we can use to decide whether there is significant linear correlation between two variables. We use this table as follows.

1. Compute the linear correlation coefficient *r* for *n* pairs of data.

2. Go to line *n* in Table 14.20.

3. If the absolute value* of *r* exceeds the number in the column labeled α = .05 on line *n*, then there is less than a .05 (5 percent) chance that the variables do

_____

* The absolute value of a nonnegative number is the number itself. The absolute value of a negative number is its opposite. For example, the absolute value of 5 is 5, and the absolute value of −13 is 13. The absolute value of 0 is 0.

Highlight: *Using Technology to Compute the Linear Correlation Coefficient**

When you are working with a large set of data, a statistical calculator will speed your work. The following screens are taken from a graphing calculator. Screen (a) shows the data for the car weights and gas mileage from Example 3. Screen (b) shows the command (we chose 4) we enter in order to instruct the calculator to compute the linear correlation coefficient. Screen (c) shows the value of *r*, which is the same as we computed in Example 3.

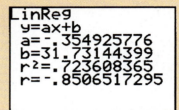

Data has been entered	**Select command**
(a)	(b)

Calculator computes linear correlation coefficient

(c)

n	α = .05	α = .01
4	.950	.999
5	.878	.959
6	.811	.917
7	.754	.875
8	.707	.834
9	.666	.798
10	.632	.765
11	.602	.735
12	.576	.708
13	.553	.684
14	.532	.661
15	.514	.641
16	.497	.623
17	.482	.606
18	.468	.590
19	.456	.575
20	.444	.561

TABLE 14.20   Critical values for the linear correlation coefficient.

not have significant linear correlation. That is, we can be 95 percent confident that there is significant linear correlation between the variables.

4. If the absolute value of *r* exceeds the number in the column labeled α = .01 on line *n*, then there is less than a .01 (1 percent) chance that the variables do not have significant linear correlation. In other words, we can be 99 percent confident that there is significant linear correlation between the variables.

We will now use the linear correlation coefficient to determine how confident we are that there is a linear relationship between two variables.

EXAMPLE 3   **Determining Correlation between Car Weight and Mileage**

Let us suppose that you are considering buying a used car. After talking to several car dealers, you are convinced that there is a relationship between the weight of the car and gas mileage. In order to check this, you gather data† regarding gas mileage of various weight cars, summarized in Table 14.21, to see whether there is a linear correlation between weight and gas mileage for these data.

Weight (in hundreds of pounds)	29	28	31	24	25	30	24	28	32	26
City MPG	21	22	22	23	23	21	24	21	20	22

TABLE 14.21   Car weight versus mileage.

Find the correlation coefficient for these data and determine whether there is linear correlation at the 5 percent or 1 percent level.

---

* See www.aw.com/pirnot.
† This set of data is based on actual mileage ratings.

SOLUTION: We will represent the weight by x and the mileage by y and by doing calculations similar to those in Example 2, we find that $\Sigma x = 277$, $\Sigma y = 219$, $\Sigma x^2 = 7{,}747$, $\Sigma y^2 = 4{,}809$, and $\Sigma xy = 6{,}040$. Substituting these values in the formula for the correlation coefficient, we get $r = -0.8507$. This number is close to -1, so we expect that there is significant negative linear correlation for this sample of data.

Because there are ten pairs of data, we will use line 10 of Table 14.20 to determine how confident we can be that there is a significant linear correlation. The absolute value of r is 0.85, which exceeds .765 in line 10 of Table 14.20. Therefore we can be 99 percent confident that there is significant negative linear correlation between the variables of car weight and gas mileage.

Although the computations in Example 3 show that we can be very sure that there is a significant linear correlation between the variables of car weight and gas mileage, remember that you cannot assume that there is a cause-and-effect relationship between these variables.

We use linear regression to find the line of best fit.

In Example 2, you saw that there was a strong positive linear correlation between the number of times that students attended tutoring sessions and their scores on an examination. We will now show you how to find the line that best models the data that we used in Example 2. This line is called **the line of best fit.*** Although the line of best fit usually does not pass through many of the data points, it is the line which minimizes the sum of all the vertical distances from the data points to the line, as we show in Figure 14.24.

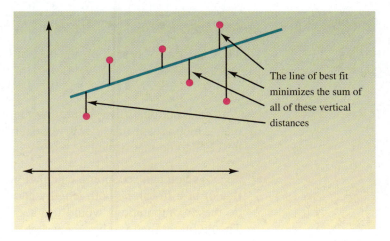

The line of best fit minimizes the sum of all of these vertical distances

FIGURE 14.24 The line of best fit is the best linear approximation for a set of data points.

Recall that (except for vertical lines) we can always write a linear equation in the form $y = mx + b$, where m is the slope of the line, and b is the

---

* The line of best fit is also called the regression line or the least squares line.

y-intercept.* Notice in the next definition, that the calculations we do to find the slope and the *y*-intercept of the line of best fit are similar to the calculations that you did in finding the linear regression coefficient.

DEFINITION

The **line of best fit** for a set of data points of the form (x, y) is of the form $y = mx + b$, where

$$m = \frac{n\Sigma xy - (\Sigma x)(\Sigma y)}{n(\Sigma x^2) - (\Sigma x)^2} \text{ and } b = \frac{\Sigma y - m(\Sigma x)}{n}.$$

We will use this definition to find the line of best fit for the data points in Example 2.

EXAMPLE 4   Finding the Line of Best Fit for a Set of Data Points

Find the line of best fit for the set of data points we used in Example 2.

SOLUTION: Recall in Example 2, that $n = 10$, $\Sigma xy = 4{,}090$, $\Sigma x = 98$, $\Sigma y = 376$, and $\Sigma x^2 = 1{,}396$. Substituting these values in the formula for the slope of the line, we get

$$m = \frac{n\Sigma xy - (\Sigma x)(\Sigma y)}{n(\Sigma x^2) - (\Sigma x)^2} = \frac{10 \cdot 4{,}090 - 98 \cdot 376}{10 \cdot 1{,}396 - (98)^2}$$
$$= \frac{40{,}900 - 36{,}848}{13{,}960 - 9{,}604} = \frac{4{,}052}{4{,}356} \approx 0.9302$$

The *y*-intercept is given by,

$$b = \frac{\Sigma y - m(\Sigma x)}{n} = \frac{376 - (0.9302)98}{10} = \frac{376 - 91.1596}{10} \approx \frac{284.84}{10} = 28.48.$$

Thus, the line of best fit for the data of Example 2 is

$$y = 0.93x + 28.46.$$

In Example 2, we found that the linear correlation coefficient was close to one, which meant that it was reasonable for us to model the data by a line. The best line to do that is the line that we just found.   ◎

As you can see, doing calculations to find the regression coefficient and the line of best fit can be both lengthy and tedious. For that reason, most people who do these computations would use a graphing calculator or a computer program. The highlight that we showed earlier to do the correlation computations in Example 3 regarding the gas mileage data, also shows the slope and the *y*-intercept of the line of best fit. However, the TI-83 calls the slope *a* instead of *m*. As you can see from the highlight, the line of best fit for the car mileage data is

$$y = -0.355x + 31.73.$$

---

* If you are a little rusty on this, see Section 6.1.

Highlight: *Health Scares*

If you eat apples, drink coffee, or use a cellular phone, you can breathe a little easier. According to an article in the *New York Times*,* the American Council on Science and Health consider these to be three of the greatest unfounded health scares of the last 50 years. Often, scientists may believe that there is a correlation between some product and a disease only to find after further studies that the connection does not exist.

In 1989, Alar, a chemical used for ripening apples, was thought to cause childhood cancer. These claims were based on a study in which a by-product of Alar caused tumors in mice. Later studies by the National Cancer Institute and the Environmental Protection Agency showed that Alar produced cancer in mice only when it was given at a dosage of over 100,000 times what a child would consume by eating apples or drinking apple juice. Nonetheless, this scare cost the apple industry about $375 million.

In 1981, Harvard researchers found that drinking two cups of coffee a day doubled the risk of pancreatic cancer, which is almost always fatal. Other researchers found no connection between pancreatic cancer and coffee. Five years later, the Harvard team was unable to reproduce its original results. The council stated, "This brief scare illustrates the danger of putting too much credence in a single study without analyzing any possible biases or confounding factors."

A more recent scare involves the use of cellular phones. During the 1990s, several cancer patients sued cell phone manufacturers, claiming that the electromagnetic field of the cell phones caused their illness. The cell phone industry paid for studies to determine whether cell phones were a health risk. This research was not able to find evidence that cell phones are a cause of cancer.†

These examples point out that we must look at reported studies very carefully and decide whether the group performing the study seems reliable.

Notice that the slope of the line is negative, which is consistent with the fact that we found negative linear correlation in Example 3.

Exercises

In Exercises 1–2, state what kind of correlation, if any, each scatterplot indicates.

1.

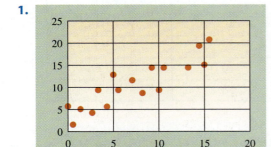

2.

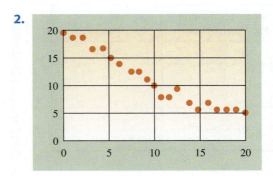

---

* Jane E. Brody, "Health Scares That Weren't So Scary," *New York Times*, August 18, 1998.
† However, more recent studies show that there may be a correlation between the use of cell phones and certain types of cancer.

For Exercises 3–6, do the following.

a) *Plot the xy pairs in a scatterplot.*

b) *Estimate the linear correlation coefficient between the two data sets.*

c) *Calculate the linear correlation coefficient to see how accurate your guess was.*

3. (3, 5), (7, 8), (4, 6), (6, 7)

4. (4, 5), (6, 8), (5, 3), (5, 6)

5. (11, 5), (15, 8), (12, 3), (12, 6)

6. (3, 12), (7, 10), (4, 8), (6, 0)

In Exercises 7–10, use Table 4 and the linear correlation coefficient you found in the given exercise to determine whether we can be 95 percent or 99 percent confident that there is significant linear correlation between x and y.

7. Exercise 3   **8.** Exercise 4

9. Exercise 5   **10.** Exercise 6

In Exercises 11–14, find the line of best fit for the data in the specified exercise.

11. Exercise 3   **12.** Exercise 4

13. Exercise 5   **14.** Exercise 6

In Exercises 15–18, determine the linear correlation coefficient for the data and also determine whether we can be 95 percent or 99 percent confident that there is significant linear correlation between the two variables.

15. The following table lists the number of years of education past high school and the annual income of five people.

Years of Education Past High School	Annual Income (in $thousands)
0	23
1	22
2	27
2	28
5	35

16. The accompanying table lists the average SAT score and teachers' salaries for certain states. (*Source: Statistical Abstract of the United States*)

State	SAT Score	Teacher Salary (in $thousands)
Connecticut	901	44
Delaware	903	35
Kansas	1040	30
Kentucky	994	29
Louisiana	993	26

17. As a car begins to accelerate, the gas mileage is poor. As the speed increases, the gas mileage continues to increase. The gas mileage increases for a while, then as the speed increases further, the mileage begins to decrease. The accompanying table of data illustrates this.

Speed in Miles per Hour	Mileage in Miles per Gallon
30	26
40	31
50	33
60	31
70	26

18. The accompanying table is based on data from the National Center for Health Statistics (*http://www.cdc.gov/nchs/*).

Years of Education	Percentage Who Are Overweight
8	62.7
10	61.1
12	65.5
16	57.6
18	53.7

In Exercises 19–22, find the line of best fit for the data in the specified exercise.

19. Exercise 15   **20.** Exercise 16

21. Exercise 17   **22.** Exercise 18

Further Exercises

23. Communicating Mathematics   a) Describe a situation in which you would expect to have a significant positive correlation between two data sets.

b) Describe a situation in which you would expect to have a significant negative correlation between two data sets.

c) Describe a situation in which you would expect to have no correlation between two data sets.

24. What is the linear correlation coefficient for a set of xy pairs if each y score is the double of the corresponding x score? We are looking at pairs of the sort (12, 24), (10, 20), (23, 46), and so on.

25. What is the linear correlation coefficient for a set of xy pairs if in each pair, x and y are related by the equation $y = 3x + 5$? We are looking at pairs such as (4, 17), (10, 35), (21, 68), and so on.

ANSWERS TO Quiz Yourself PROBLEMS

1.

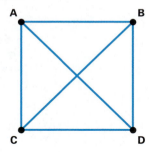

2. a) *h* and *l*
b) *p, d,* and *q*

3.

Line 4:	Fail	Fail	Pass
Line 7:	Fail	Pass	Pass

4. a)

1	6	15	20	15	6	1
1	7	21	35	35	21	7	1

b) 1 and 100

5. The eight we already have are:

   none, *s, b, m, s + b, s + m, b + m, s + b + m.*

If we add real estate to each of these we get eight new ways—namely *r, s + r, b + r, m + r, s + b + r, s + m + r, b + m + r, s + b + m + r.*

6. 20 pennies, 40 nickels, and 18 quarters

7. 45

8. Almost any choice of numbers for *a, b,* and *c* will work. For example, let

$a = 3, b = 4,$ and $c = 5.$ Then $\dfrac{a + b}{a + c} = \dfrac{3 + 4}{3 + 5} = \dfrac{7}{8}.$
However, $\dfrac{b}{c} = \dfrac{4}{5}.$

9. a) We are first adding two numbers and then taking the square root of the sum.
b) We are first taking the square roots of two numbers and then forming the sum of these square roots.
c) To "first add and then take the square root" is not the same as to "first take the square roots and then add."

10. a) There is an extra line in the symbol on the right.
b) The symbol on the left has a pair of braces around the ∅ symbol.
c) The symbol on the left is rounded; the symbol on the right is pointed.
d) The symbol on the right is the number 0; the symbol on the left is the number 0 with braces around it.

11. a) intersection of streets
b) happening at the same time

12. a) 635,000   b) 724,000

13. a) $25 \times 400 = 10,000$ b) $250 \div 50 = 5$

14. a) {Monday, Tuesday, . . . , Sunday}
b) {*x : x* is a natural number less than or equal to 60}

15. a) well defined   b) not well defined

16. a) true   b) true   c) false

17. a) 10   b) 2   c) 50

18. a) true   b) false

19. a) true   b) false

20. ∅, {1}, {2}, {3}, {1, 2}, {1, 3}, {2, 3}, {1, 2, 3}

21. a) $2^5 = 32$ b) $2^{26} = 67,108,864$

22. $M \cup B = \{m, a, t, h, e, i, c, s, b, u, y\},$
$M \cap B = \{a, t, e\}$

23. a) $C' = \{c, i, k, o, q\}$
b) $B - A = \{2, 7, 8\}; A - B = \{1, 5\}$

24. a) {1, 3, 4, 6, 8, 9, 10}
b) {3, 4, 6, 10}
c) {1, 4, 8, 9, 10}
d) {1, 3, 4, 6, 8, 9, 10}

25. a) $A' \cap B' \cap C$ b) $(A \cap B \cap C') \cup (A' \cap B \cap C')$

26. 57

27. There are several correct answers; one would be

Reagan	Clinton	Carter
↕	↕	↕
Gore	Mondale	Bush

28. There are many correct answers for this, one is:

1	2	3	· · ·	*n*	· · ·
↕	↕	↕		↕	
2	3	4	· · ·	*n* + 1	· · ·

29. a) 6, 5/2, 4/3   b) 2/6

1. a) $31 + 7$ b) $5 + 41$ **2.** 66

3. a) simple   b) compound   c) compound

4. a) $\sim c \vee \sim v$ b) $(\sim c) \wedge v$

5. a) $\sim f \rightarrow \sim q$ b) $f \leftrightarrow q$

6. a) Some professors are not friendly.
b) All dogs do not bite. (No dogs bite.)

7. a) true   b) false   c) false   d) true

8.

		2	1	5	3	4		
p	q	$(p$	$\wedge$	$\sim q)$	$\vee$	$(\sim p$	$\vee$	$q)$
T	T	T	F	F	**T**	F	T	T
T	F	T	T	T	**T**	F	F	F
F	T	F	F	F	**T**	T	T	T
F	F	F	F	T	**T**	T	T	F

9.

		2		1		
p	q	$\sim$	$(p$	$\vee$	$q)$
T	T	**F**	T	T	T
T	F	**F**	T	T	F
F	T	**F**	F	T	T
F	F	**T**	F	F	F

	1	3	2
	$(\sim p)$	$\wedge$	$(\sim q)$
	F	F	F
	F	F	T
	T	F	F
	T	T	T

The statements have identical truth tables and so are logically equivalent.

10. It is not the case that: the car is more than five years old or has been driven over 50,000 miles.

11.

		2	1	6	3	5	4	
p	q	$(p$	$\wedge$	$\sim q)$	$\rightarrow$	$(\sim p$	$\vee$	$q)$
T	T	T	F	F	**T**	F	T	T
T	F	T	T	T	**F**	F	F	F
F	T	F	F	F	**T**	T	T	T
F	F	F	F	T	**T**	T	T	F

12. Inverse: "If the Federal Reserve does not raise the prime rate, then interest rates will not increase."
Contrapositive: "If interest rates will not increase, then the Federal Reserve does not raise the prime rate.

13. a) If you travel to Europe, then you have updated your immunization.
b) If you increase your cardiovascular fitness, then you exercise three times a week.

14. The argument is invalid.

15. fallacy of the inverse; invalid

16.

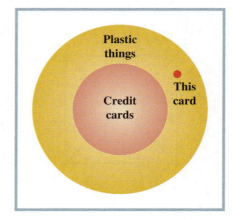

invalid

17. valid

18. Answers may vary; however, here are some possibilities:
a) Some B's are not A's.
b) Some A's are not C's.
c) Some B's are not C's.

19. d) The truth value of "*Gone with the Wind* is not a great movie" is 0.27.
e) "You find mathematics uninteresting" has a truth value of 0.3.

20. a) 0.45   b) 0.82

21. a) 0.45   b) 0.55

22. a) 0.80   b) 0.55   c) 0.80

23. a) $(\sim i) \vee c = 0.95$ b) $(\sim i) \vee c = 0.75$

CHAPTER **3**

1. The graph is connected. A, B, D, E, and F are even. C and G are odd. Edge CG is a bridge.

2. a) Can be traced; C, B, A, C, D, B, A, D.

b) Cannot be traced; it has four odd vertices.

3. a) CABCFEGDFIH

b) No. Because C and H are both odd vertices, we must begin at one and end at the other.

4. A, B, C, and D are the odd vertices in the graph. If we duplicate edge CD and also all edges along the indicated path from A to B, the new graph will be Eulerian.

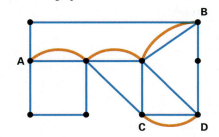

5.

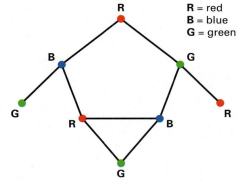

R = red
B = blue
G = green

6. 6! = 720

7. ABDCA has the minimal weight of 11. (Of course, the reverse circuit ACDBA also has the same weight.)

8. MAPNCM; $1,200

9. a) yes; ABC and AFBC   b) no   c) 4

10.

	To			
	A	**B**	**C**	**D**
A	0	1	0	0
B	0	0	0	0
C	1	2	0	0
D	1	2	1	0

From

11.

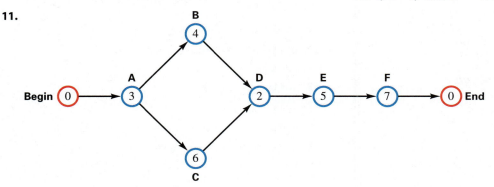

12. a) Begin, B, D, E, G
  b) Begin, B, D, E, G, End
  c) To begin on the tenth day of the project
  d) fifteen days

CHAPTER **4**

1. a) 21,346
  b) 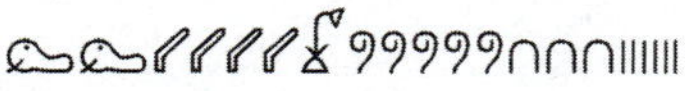 𓋹𓋹�/////𓎡99999∩∩∩𓏤𓏤𓏤𓏤𓏤𓏤

2. a)  𓎡9∩∩∩∩∩∩∩𓏤
  b)  𓎡𓎡999999𓏤𓏤𓏤𓏤

3. a) 1 + 8 + 16   b) 43 + 344 + 688 = 1075

4. a) We cannot subtract L; we can only subtract I from V and X
  b) 548
  c) 9 × 100 = 900; 5 × 1,000 = 5,000

5. a) 12 × 60² + 23 × 60 + 32 = 44,612
  b) ▼▼ ▼▼▼▼▼▼ ◀▼▼▼

6.

	3	2	8	
1	0 / 9	0 / 6	2 / 4	3
2	2 / 7	1 / 8	7 / 2	9
	7	9	2	

Product is 12,792.

7. 142

8. 1, 2, 10_3, 11_3, 12_3, 20_3, 21_3, 22_3, 100_3, 101_3

9. 423   **10.** 11_5 and 10_5 **11.** a) $1,243_6$ b) 154_6

12. 13_5 and 31_5 **13.** a) $3,133_5$ b) $3,402_8$

14. a) 3 b) 6   **15.** a) false b) true

16. a) 5 b) 0, 3, 6   **17.** 33

CHAPTER **5**

1. a) prime   b) composite   c) composite   d) prime

2. a) yes   b) yes   c) no   d) yes   e) yes

f) no   g) yes   h) yes

3. a)

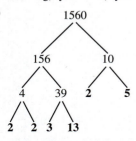

1560

156   10

4   39   2   5

2   2   3   13

b) $1560 = 2 \cdot 2 \cdot 2 \cdot 3 \cdot 13 \cdot 2 \cdot 5 = 2^3 \cdot 3 \cdot 5 \cdot 13$

4. 42   **5.** a) $+10$ b) -5 c) -10 d) $+22$

6. a) $(+5) - (+12) = (+5) + (-12) = -7$

b) $(-3) + (+9) = +6$

7. a) $+24$ b) -56

8. a) -4 b) $+5$ c) $+13$ d) -5

9. a) Equal   b) Not equal

10. $\dfrac{3}{4}$ **11.** $\dfrac{29}{24}$ and $\dfrac{17}{48}$ **12.** a) $\dfrac{14}{5}$ b) $\dfrac{10}{9}$

13. a) $\dfrac{5}{22}$ b) $\dfrac{12}{7}$ **14.** a) $5\dfrac{3}{4}$ b) $\dfrac{13}{5}$

15. $0.\overline{63}$

16. a) $\dfrac{2{,}548}{10{,}000} = \dfrac{637}{2{,}500}$ b) $\dfrac{6}{11}$

17. a) $3\sqrt{5}$ b) $\dfrac{\sqrt{5}}{4}$

18. $\dfrac{\sqrt{15}}{5}$ **19.** a) $11\sqrt{5}$ b) 0   **20.** $3^6 = 729$

21. a) 5^{12} b) $(-2)^8 = 256$

22. a) $\dfrac{1}{36}$ b) 64   c) 1

23. a) $3^8 \cdot 3^{-6} = 3^{8+(-6)} = 3^2 = 9$

b) $(2^{-2})^{-1} = 2^{(-2)\cdot(-1)} = 2^2 = 4$

24. a) 5.372841×10^4 b) 0.00245   **25.** 50   **26.** 440

27. 354,294   **28.** 6,560   **29.** 144 and 233

CHAPTER **6**

1. $c = 40 + 0.15m$

2. $W = \dfrac{P - 2L}{2}$

3. x-intercept, $(4\frac{1}{2}, 0)$; y-intercept, $(0, 6)$

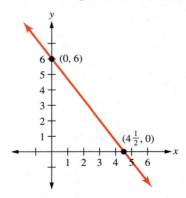

4. $\dfrac{5}{4}$

5. $y = -4x + 5$

6. $y = \dfrac{-5}{9}x + \dfrac{10}{3}$

7. $x = 4$ and $x = -11$

8.

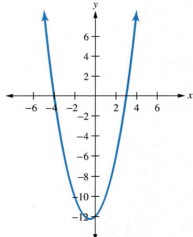

9. $3,327.50

10. slightly over 119 million

11. 0.6096, or 60.96 percent

12. 4.5 hours

13. 104,000

14. 133

15. 496 mg

16. a) $\frac{1}{2}$; unstable   b) -6; stable

CHAPTER **7**

1. (3, 1)   **2.** (− 4, 5)

3. a) There are no solutions.

b) The system represents a pair of distinct parallel lines.

4. a) yes   b) no   c) no

5.

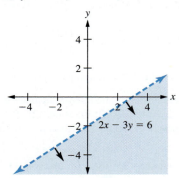

6.

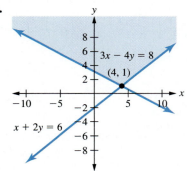

CHAPTER **8**

1. a) a line   b) line segment *CD*   c) a ray

d) a half line

2. a) vertical angles   b) a right angle

c) supplementary angles   d) an obtuse angle

e) an acute angle

3. a) 104°   b) 76°

4. 5 inches

5. 360°

6. 135°

7. 126 square units

8. 18 square feet

9. approximately 17.3 inches

10. circumference ≈ 50.24; area ≈ 200.96

11. volume = 144 cubic centimeters; surface area = 180 square centimeters

12. volume ≈ 785 cubic centimeters; surface area ≈ 471 square centimeters

13. volume ≈ 83.73 cubic yards; surface area ≈ 80.42 square yards

14. $\frac{4}{3}\pi(6)^3 \approx 904$ cubic centimeters

15. a) six liters   b) one meter   c) 9 kilograms

16. a) 1,000   b) 100

17. a) 56,300   b) 0.04850

18. 2,145 feet

19. 133.76 ounces

20.

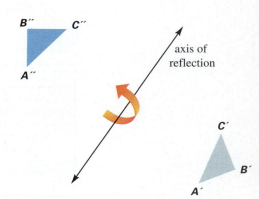

21.

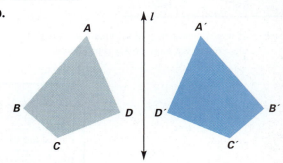

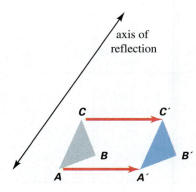

22. Each vertex is the vertex of two octagons and one square. The interior angle of each octagon is 135° and the interior angle of the square is 90°. The tessellation is possible because 135° + 135° + 90° = 360°.

23. 256

24. 243

25. $\frac{64}{27}$

26. $\frac{27}{64}$

27. 1.5

CHAPTER **9**

1. On a 12-member board, Naxxon gets 6 representatives, Aroco 4, and Eurobile 2.

2. a) 284.61   b) 284.613579

3.

Spending will increase	32.0
Spending will decrease	24.7
Spending will stay the same	22.8
Unsure	20.5
Total	100.0%

4. a) The average constituency of the electricians is 140.
b) The electricians are more poorly represented.

5. 64

6. $\dfrac{2{,}351 - 2{,}342}{2{,}342} = \dfrac{9}{2{,}342} \approx 0.004$

7. a) 0.216   b) 0.175
c) State B should get the additional representative.

8. For Iowa the Huntington–Hill number is $\dfrac{(2.8)^2}{6 \times 7} = 0.187$; for

Nebraska the Huntington–Hill number is $\dfrac{(1.6)^2}{3 \times 4} = 0.213$.

Therefore, Nebraska should get the additional representative.

9. The Huntington–Hill numbers for the three companies are as follows: Naxxon, 368.2; Aroco, 228.2; Eurobile, 128. Because 368.2 is the largest, Naxxon gets the sixth representative.

10. The Huntington–Hill numbers are 184.1 for Naxxon, 114.1 for Aroco, and 128.0 for Eurobile. Therefore Naxxon gets the eighth representative.

11. a) 1,500,000   b) A: 2   B: 2.67   C: 3.33

12. a) 600   b) 4.9   c) 4

13. a) 480   b) 2.6   c) 3

14. a) 1,500   b) 2.79   c) 3

15. a) discrete   b) continuous
c) discrete   d) continuous

16.

	Rosa $(\frac{1}{4})$	Juan $(\frac{1}{4})$	Carlos $(\frac{1}{4})$	Luis $(\frac{1}{4})$
Bid on house	$250,000	$240,000	$260,000	$200,000
Fair share of estate	$62,500	$60,000	$65,000	$50,000
Item obtained with highest bid			House	

17.

	Rosa $(\frac{1}{4})$	Juan $(\frac{1}{4})$	Carlos $(\frac{1}{4})$	Luis $(\frac{1}{4})$
Pays to estate (+) or receives from estate (−)	$62,500 (−)	$60,000 (−)	$195,000 (+)	$50,000 (−)
Division of estate balance ($22,500)	$5,625 (−)	$5,625 (−)	$5,625 (−)	$5,625 (−)
Summary of cash	Receives $68,125	Receives $65,625	Pays $189,375	Receives $55,625

CHAPTER **10**

1. A wins with 78 points.

2. C wins.

3. N wins with 2 points. B gets 1 point and T gets none.

4. The winner is A with 67 points. Because C has the majority of first-place votes, the majority criterion is not satisfied.

5.

1st	A	A	A	B	B	C	C
2nd	B	B	B	C	C	B	B
3rd	C	C	C	A	A	A	A

6. Using the pairwise comparison method, the winner is W. However, if X and Y are removed, then Z wins the election. The independence-of-irrelevant-alternatives criterion is not satisfied.

7. a) 4 (Let's call them A, B, C and D.)

   b) 5

   c) {A, B}, {A, C}, {A, B, C}, {A, B, D}, {A, C, D}, {B, C, D}, {A, B, C, D}

8. a)

	Winning Coalitions	Critical Members
1	{A, B}	A, B
2	{A, C}	A, C
3	{A, B, C}	A
4	{A, B, D}	A, B
5	{A, C, D}	A, C
6	{B, C, D}	B, C, D
7	{A, B, C, D}	none

   b) A, $\frac{5}{12}$;   B, $\frac{3}{12}$;   C, $\frac{3}{12}$;   D, $\frac{1}{12}$

9. $5! = 120$

CHAPTER **11**

1. a) 0.1745   b) 0.0005   c) 245%   d) 2.5%

2. 24%

3. a) 9   b) 75   c) 30%

4. $2,000

5. $4,440.73

6. $1,126.49

7. 1.511391594

8. 23 months

9. $36.30

10. $289.25; $5.06

11.

Day	Balance	Number of Days × Balance
1, 2	$240	$2 \times 240 = 480$
3, 4, 5, 6, 7, 8, 9, 10	$264	$8 \times 264 = 2,112$
11, 12, 13, 14, 15, 16, 17, 18, 19, 20, 21, 22	$204	$12 \times 204 = 2,448$
23, 24, 25, 26, 27, 28, 29, 30	$216	$8 \times 216 = 1,728$

Average daily balance $= \frac{6,768}{30} = \$225.60$.

12. $\dfrac{x^4 - 1}{x - 1}$

13. $1,907.40

14. $95.47

15. 240 months, or 20 years

16. $253.94

17.

4	$798.37	$698.27	$100.10	$119,603.07

18. $9,493.49

19. a) $17.76   b) about 11 percent

20. 13 percent

CHAPTER **12**

1. $7 \times 6 = 42$ 2. 16   3. 24

4. $2 \times 8 \times 3 \times 2 = 96$ different cars

5. 2,116,800

6. a) $P(8, 3)$ is the number of ways we can select three different objects from a set of eight and arrange them in a straight line.

   b) 336

7. 792

8. a) 1   7   21   35   35   21   7   1

   b) 35

9. $C(3, 0)$ $C(3, 1)$ $C(3, 2)$ $C(3, 3)$

10. $13 \times C(4, 2) = 13 \times 6 = 78$

CHAPTER **13**

1. a) {(1, 5), (2, 4), (3, 3), (4, 2), (5, 1)}

   b) {bbb, bbg, bgb, gbb}

2. 0.67

3. a) $\dfrac{5}{36}$ b) $\dfrac{C(12, 2)}{C(52, 2)} = \dfrac{66}{1{,}326} = \dfrac{11}{221}$

4. $1 - \dfrac{3}{36} = \dfrac{33}{36} = \dfrac{11}{12}$

5. 0.45

6. $\dfrac{0.04}{0.09} = 0.44$

7. $\dfrac{12}{52} \cdot \dfrac{11}{51} \approx 0.05$

8. a)

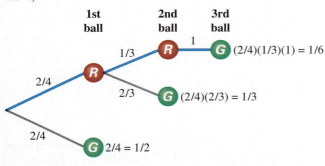

   b) $\frac{1}{6}$

9. Since $P(O) = \frac{1}{2} = P(O\,|\,F)$, the events are independent.

10. a) This is the probability that she will be assigned to dormitory X.

    b) 0.40

11. 38.48

12. $+\frac{1}{9}$; it is now to the student's advantage to guess.

13. $B(8, 2; \frac{1}{6}) = 0.2605$

CHAPTER **14**

1.

Score	Frequency	Relative Frequency
1	3	0.15
2	4	0.20
3	2	0.10
4	0	0.00
5	2	0.10
6	1	0.05
7	4	0.20
8	2	0.10
9	1	0.05
10	1	0.05

2.

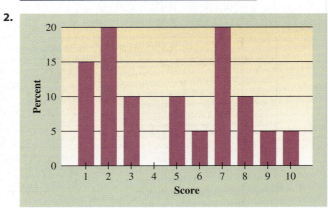

3.

6	2 2 3 6 7 8
7	1 3 5 7
8	1 2 2 3 4 8
9	0 1 2 8

4. a) $\Sigma x = 106$ b) $n = 8$ c) $\bar{x} = 13.25$

5. The mean number of calls is $\dfrac{\Sigma(x \cdot f)}{\Sigma f} = \dfrac{96}{14} \approx 6.86$.

6. 95

7. a) 37, 40, 45, 69, 200

b)

minimum   median
37   45
Q₁   Q₃   maximum
40   69   200

0   25   50   75   100   125   150   175   200

Salaries (in $millions per year)

8. 2.29

9. $s = \sqrt{\dfrac{\Sigma(x - 42)^2 \cdot f}{n - 1}} = \sqrt{\dfrac{172}{19}} \approx 3.01$

10. a) 950 b) 25

11. a) 0.427 b) $0.468 - 0.391 = 0.077$ c) 0.412

12. a) 1.29 b) 0 c) -1.71

13.

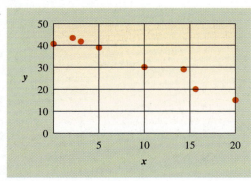

We see in this graph that as x increases, y generally decreases.

14. -0.9718

ANSWERS TO Exercises

CHAPTER 1

Section 1.1

1.

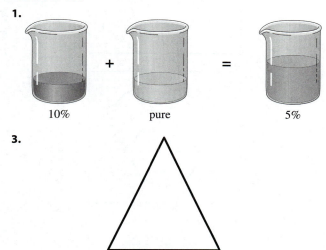

10%   +   pure   =   5%

3.

5. length is *l* and width is *w*   **7.** *M, C, R, J,* and *S*

9. *c* for amount invested in collectibles; *p* for amount invested in precious metals

11. *HH, HT, TH, TT*   **13.** *CS, CM, CT, SM, ST, MT*

15. 35, 42, 49, 56, 63   **17.** *bf, cd, ce, cf, cg*

19. Possible answer: Three people A, B, C can be lined up in 6 ways; four people can be lined up in 24 ways.

21. Possible answer: With three letters there are 9 ways; with five letters there are 25 ways.

23. Possible answer: For only air conditioning (*A*) and CD player (*C*) there are four possibilities: neither, A, C, AC. With three options there are eight possibilities.

25. 107, 214   **27.** 9 in the store and 6 giving lessons

29. Emmit 240, Barry 211, Ricky 291

31. $3,500 at 8% and $4,500 at 6%   **33.** 3 students

35. false; September has 30 days

37. false; almost any numbers give a counterexample

39. false; A is the grandfather of C

41. false; $-9 < -5$, but $(-9)^2 = 81$, which is not less than $(-5)^2 = 25$

43. false   **45.** not the same   **47.** the same

49. 5 is a number; {5} is a number with braces around it

51. one is uppercase, the other is lowercase

53. { } are different from ( )

Section 1.2

1. 37,300   **3.** 8,300   **5.** $20 + 40 + 190 + 40 = 290$

7. $35 - 15 = 20$ **9.** $5 \times 16 = 80$ **11.** $18/3 = 6$

13. $0.1 \times 800 = 80$ **15.** $9\% \times 1,000 = 0.09 \times 1,000 = 90$

17. $4 \times 5 \times 6 = 120$

19. $325/50 = 6.5$ more hours, 7:30 PM.

21. $\$80 \times 0.15 = \$12$

23. $3 \times \$3 + 4 \times \$1.50 + \$3.00 = \$18$

25. It seems safe. Alicia probably weighs less than 200 pounds, so that leaves $2,300 - 200 = 2,100$ pounds for the 21 students. They probably do not weigh 100 pounds each.

27. $\$40,000 \times 4\% = \$40,000 \times 0.04 = \$1,600$

29. Her total expenses are about $100/month; $100/7 \approx 14$, $12 \times \$14 = \$168$

31. $66,810   **33.** $18,663   **35.** 29 percent

37. modems; 39 percent   **39.** $701.44 billion

41. $219.2 billion   **43.** 212,799   **45.** 18.5 percent

47. If you divide the grassy area into rectangles, you get 10,354 square feet. They need slightly over two bags of fertilizer.

Section 1.3

1. {10, 11, 12, 13, 14, 15}

3. {17, 18, 19, 20, 21, 22, 23, 24, 25}

5. {4, 8, 12, 16, 20, 24, 28}

7. {Sunday, Monday, Tuesday, Wednesday, Thursday, Friday, Saturday}

9. $\varnothing$

11. {Ronald Reagan, George H.W. Bush, Bill Clinton}

13. $\{x : x$ is a multiple of 3 between 3 and 12 inclusive$\}$

15. $\{y : y$ is a letter in the word *music*$\}$

17. $\{-1, -2, -3, \ldots\}$ **19.** $\varnothing$

21. {101, 102, 103, . . .}

23. $\{x : x$ is an even natural number between 1 and 101$\}$

25. $\{x : x$ is a humanities elective$\}$

27. {History012, History223, Geography115, Anthropology111}

29. well defined   **31.** not well defined

33. not well defined   **35.** well defined

37. $\notin$ **39.** $\in$ **41.** $\in$ **43.** $\in$ **45.** $\notin$ **47.** $\in$

49. 6   **51.** 0   **53.** 4   **55.** 2   **57.** finite

59. infinite   **61.** finite   **63.** finite   **65.** 4.5

67. Sony   **69.** George W. Bush   **71.** Sunday

Section 1.4

1. equal   **3.** not equal   **5.** equal   **7.** equal

9. equal   **11.** true   **13.** false   **15.** false   **17.** true

19. false   **21.** equivalent   **23.** equivalent

25. not equivalent   **27.** equivalent

29. not equivalent; 2004 is a leap year

31. {1,2}, {1,3}, {2,3}

33. {1,2,3}, {1,2,4}, {1,3,4}, {2,3,4}

35. Equal sets have exactly the same elements. Equivalent sets have the same number of elements, but the elements in the two sets do not have to be the same.

37. The line in the notation $\subseteq$ allows for the *possibility* that the two sets being compared can be equal.

39. T **41.** L

43. N is a subset of all the sets listed; W is a subset of W, I, Q, and R; I is a subset of I, Q, and R; Q is a subset of Q and R; R is a subset of R.

45. 32; 31   **47.** 25 is not a power of 2   **49.** 5

51. At the first branching in the tree, the "yes" or "no" indicates that in forming a subset of {1,2}, we will either take the 1 or omit it. The second branchings in the tree indicate whether or not we are going to take the 2 as a member of the subset that we are forming. The tree shows all possible ways that we can decide to take or not to take 1 and 2 in forming a subset of {1,2}. The top branch corresponds to the subset {1,2}, the second branch corresponds to the subset {1}, and so on.

53. The fifth line counts the number of subsets of sizes 0, 1, 2, 3, 4, and 5 of a five-element set.

55. 84   **57.** The total in the nth row is 2^n.

59. 7   **61.** If $A \subseteq B$ and $B \subseteq C$, then $A \subseteq C$.

Section 1.5

1. {1, 3, 5}   **3.** {1, 2, 3, 4, 5, 6, 7, 8}   **5.** {1, 3, 5, 7, 9}

7. {1, 2, 3, 4, 5, 6, 7, 8, 9, 10}   **9.** {1, 3, 5, 7}   **11.** {9}

13. These will use the computer for education and business.

15. These will use the computer for all three.

17. These will use the computer for education and business, but not home management.

19. {potato chip, bread, pizza}   **21.** {apple, fish, banana}

23. {fish}   **25.** {d, g}   **27.** {c, d, f, g}

29. {b, d, g, h}   **31.** {d}   **33.** {m, mc, hc}

35. {m, mc, bc, c, hc}

37.

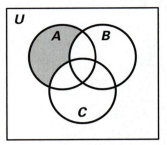

39.

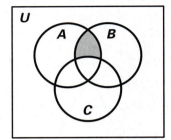

41.

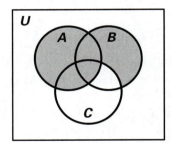

43.

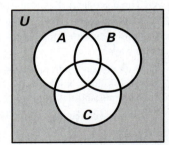

45. $B - A$ **47.** $(A \cup B)'$ **49.** $A \cap B \cap C$

51. $(A \cup C) - B$ **53.** A **55.** B **57.** 30   **59.** 28

61. 20   **63.** 27   **65.** false   **67.** true   **69.** true

71. false   **73.** true   **75.** true   **77.** true

79. $A \cap B = B \cap A$; true

81. $(A \cap B) \cap C = A \cap (B \cap C)$; true

Section 1.6

1. 2, 3 **3.** 3 **5.** 2 **7.** 2, 3, 5, 6 **9.** 5, 6

11. 4, 7 **13.** 2 **15.** 7 **17.** 6

19. $n(A) = 18, n(B) = 15, n(C) = 14$

21. $n(A) = 5, n(B) = 14, n(C) = 9$

23. 59 **25.** 35 **27.** 30, 49 **29.** 37

31. a) 44; b) 27; c) 8 **33.** a) 106; b) 15; c) 40

35. 117 **37.** 20 **39.** Answers will vary.

41.

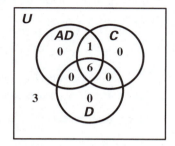

Chapter 1 Test

1. *ab, ac, ad, ba, bc, bd, ca, cb, cd, da, db, dc*

2. Possible answer: You might try two appetizers, three entrées, and two desserts, in which case you get twelve meals.

3. Eight hours as stockperson; twelve hours as ski instructor

4. false

6. a) 46,000; b) 28,000

7. a) $210 - 60 = 150$; b) $6 \times 15 = 90$

8. $150/50 = 3$ hours; 7:00 PM

9. a) $36,755; b) $12,322

10. a) $\{x : x \text{ is an odd counting number less than 10}\}$ b) $\{h, a, p, i, n, e, s\}$ c) $\{y : y \text{ is a letter of the alphabet}\}$ d) $\varnothing$

11. a) yes; they contain the same elements

b) no; the second set contains elements which the first set does not

c) yes; they contain exactly the same elements

d) no; $\{\varnothing\}$ is not empty

12. a) true; every element of the left-hand set is contained in the right-hand set

b) false; since the sets are equal, one cannot be a proper subset of the other

c) false; *e* is an element of the left-hand set, but not of the right-hand set

d) true; we cannot find an element of the empty set that fails to be in $\{1, 3, 5\}$

e) false; since the sets are equal, one cannot be a proper subset of the other

13. a) equivalent b) equivalent c) equivalent

14. a) $\varnothing, \{a\}, \{b\}, \{c\}, \{a, b\}, \{a, c\}, \{b, c\}, \{a, b, c\}$
b) $2^{10} = 1,024$

15. a) $\{1, 3, 7\}$ b) $\{1, 2, 3, 4, 5, 6, 7, 8\}$ c) $\{1, 8\}$
d) $\{2, 4, 6\}$

16. a) $\{3, 5, 7\}$ b) $\{1, 5\}$

17. a)

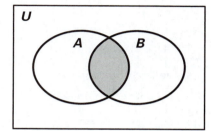

b)

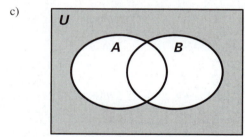

c)

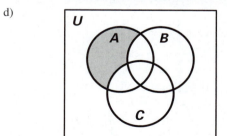

d)

18. $A' \cup B'$

19. a) closure, commutativity, associativity, identity

b) closure, commutativity, associativity, identity

c) union distributes over intersection; intersection distributes over union

20. a) 10 b) 24   **21.** 57, 14

Of Further Interest

1.

1	2	3	4	5	$\cdots$	n	$\cdots$
$\updownarrow$	$\updownarrow$	$\updownarrow$	$\updownarrow$	$\updownarrow$		$\updownarrow$	
4	8	12	16	20	$\cdots$	$4n$	$\cdots$

2.

1	2	3	4	5	$\cdots$	n	$\cdots$
$\updownarrow$	$\updownarrow$	$\updownarrow$	$\updownarrow$	$\updownarrow$		$\updownarrow$	
5	10	15	20	25	$\cdots$	$5n$	$\cdots$

3.

1	2	3	4	5	$\cdots$	n	$\cdots$
$\updownarrow$	$\updownarrow$	$\updownarrow$	$\updownarrow$	$\updownarrow$		$\updownarrow$	
8	11	14	17	20	$\cdots$	$3n+5$	$\cdots$

4.

1	2	3	4	5	$\cdots$	n	$\cdots$
$\updownarrow$	$\updownarrow$	$\updownarrow$	$\updownarrow$	$\updownarrow$		$\updownarrow$	
7	11	15	19	23	$\cdots$	$4n+3$	$\cdots$

5.

1	2	3	4	5	$\cdots$	n	$\cdots$
$\updownarrow$	$\updownarrow$	$\updownarrow$	$\updownarrow$	$\updownarrow$		$\updownarrow$	
2	4	8	16	32	$\cdots$	2^n	$\cdots$

6.

1	2	3	4	5	$\cdots$	n	$\cdots$
$\updownarrow$	$\updownarrow$	$\updownarrow$	$\updownarrow$	$\updownarrow$		$\updownarrow$	
3	9	27	81	243	$\cdots$	3^n	$\cdots$

7.

1	2	3	4	5	$\cdots$	n	$\cdots$
$\updownarrow$	$\updownarrow$	$\updownarrow$	$\updownarrow$	$\updownarrow$		$\updownarrow$	
1	1/2	1/3	1/4	1/5	$\cdots$	$1/n$	$\cdots$

8.

1	2	3	4	5	$\cdots$	n	$\cdots$
$\updownarrow$	$\updownarrow$	$\updownarrow$	$\updownarrow$	$\updownarrow$		$\updownarrow$	
1/2	2/3	3/4	4/5	5/6	$\cdots$	$n/(n+1)$	$\cdots$

9.

1	2	3	4	5	$\cdots$	n	$\cdots$
$\updownarrow$	$\updownarrow$	$\updownarrow$	$\updownarrow$	$\updownarrow$		$\updownarrow$	
3	7	9	12	15	$\cdots$	$3n$	$\cdots$

10.

1	2	3	4	5	$\cdots$	n	$\cdots$
$\updownarrow$	$\updownarrow$	$\updownarrow$	$\updownarrow$	$\updownarrow$		$\updownarrow$	
5	7	9	11	13	$\cdots$	$2n+3$	$\cdots$

11.

1	2	3	4	5	$\cdots$	n	$\cdots$
$\updownarrow$	$\updownarrow$	$\updownarrow$	$\updownarrow$	$\updownarrow$		$\updownarrow$	
1	4	7	10	13	$\cdots$	$3n-2$	$\cdots$

12.

1	2	3	4	5	$\cdots$	n	$\cdots$
$\updownarrow$	$\updownarrow$	$\updownarrow$	$\updownarrow$	$\updownarrow$		$\updownarrow$	
9	13	17	21	25	$\cdots$	$4n+5$	$\cdots$

13. Match $\{2, 4, 6, 8, 10, \ldots\}$ with $\{4, 6, 8, 10, 12, \ldots\}$; in general, match $2n$ with $2n + 2$

14. Match $\{5, 10, 15, 20, 25, \ldots\}$ with $\{10, 15, 20, 25, 30, \ldots\}$; in general, match $5n$ with $5n + 5$

15. Match $\{7, 10, 13, 16, 19, \ldots\}$ with $\{10, 13, 16, 19, 22, \ldots\}$; in general, match $3n + 4$ with $3n + 7$

16. Match $\{6, 9, 12, 15, 18, \ldots\}$ with $\{9, 12, 15, 18, 21, \ldots\}$; in general, match $3n + 3$ with $3n + 6$

17. Match $\{2, 4, 8, 16, 32, \ldots\}$ with $\{4, 8, 16, 32, 64, \ldots\}$; in general, match 2^n with 2^{n+1}

18. Match $\{3, 9, 27, 81, 243, \ldots\}$ with $\{9, 27, 81, 243, 729, \ldots\}$; in general, match 3^n with 3^{n+1}

19. Match $\{1, 1/2, 1/3, 1/4, 1/5, \ldots\}$ with $\{1/2, 1/3, 1/4, 1/5, 1/6, \ldots\}$; in general, match $\dfrac{1}{n}$ with $\dfrac{1}{n+1}$

20. Match $\{1/2, 2/3, 3/4, 4/5, 5/6, \ldots\}$ with $\{2/3, 3/4, 4/5, 5/6, 6/7, \ldots\}$; in general, match $\dfrac{n}{n+1}$ with $\dfrac{n+1}{n+2}$

21. Match $\{1/2, 1/4, 1/6, 1/8, 1/10, \ldots\}$ with $\{1/4, 1/6, 1/8, 1/10, 1/12, \ldots\}$; in general, match $\dfrac{1}{2n}$ with $\dfrac{1}{2n+2}$

22. Match $\{1/2, 1/4, 1/8, 1/16, 1/32, \ldots\}$ with $\{1/4, 1/8, 1/16, 1/32, 1/64, \ldots\}$; in general, match $\dfrac{1}{2^n}$ with $\dfrac{1}{2^{n+1}}$

23. 6   **24.** 3/4   **25.** 25   **26.** 17   **29.** 6   **30.** 24

31. If we take the union of $\{1\}$, which has cardinal number 1 and $\{2, 3, 4, \ldots\}$, which has cardinal number $\aleph_0$, we get $\{1, 2, 3, 4, \ldots\}$ which has cardinal number $\aleph_0$.

32. Take the union of $\{1, 3, 5, \ldots\}$ and $\{2, 4, 6, \ldots\}$, both of which have cardinal number $\aleph_0$ to get $\{1, 2, 3, 4, 5, 6, \ldots\}$ which has cardinal number $\aleph_0$.

CHAPTER 2

Section 2.1

1. inductive   **3.** deductive   **5.** inductive

7. deductive   **9.** inductive

11. Inductive reasoning is the process of drawing a general conclusion by observing a pattern in specific instances. In deductive reasoning, we use accepted facts and general principles to arrive at a specific conclusion.

13. answers will vary   **15.** 16   **17.** 96

19. 1/64   **21.** 21

23.

25.

27. Hint: Think of prime numbers

	X	X		X		X				X		X		

29. 4   **31.** 4   **33.** 5 + 11   **35.** 7 + 13   **37.** 55

39. 100   **41.** 30

43. The total of all the numbers in the square is 1 + 2 + 3 + . . . + 16 = 136, so the numbers in each of the four rows is 136/4 = 34. The same is true for the columns and diagonals. From this you can deduce the missing numbers.

7	6	12	9
10	11	5	8
13	16	2	3
4	1	15	14

45. The algebraic interpretation of the steps in the trick are as follows:

a) Call the number n

b) $3n$

c) $3n + 9$

d) $\dfrac{3n + 9}{3} = \dfrac{3n}{3} + \dfrac{9}{3} = n + 3$

e) $n + 3 - n = 3$

In this trick, you will always get the number 3.

47. The algebraic interpretation of the steps in the trick are as follows:

a) Call the number n

b) $8n$

c) $8n + 12$

d) $\dfrac{8n + 12}{4} = \dfrac{8n}{4} + \dfrac{12}{4} = 2n + 3$

e) $2n + 3 - 3 = 2n$

In this trick, you will always get a result that is twice the number that you started with.

49.

51. 20   **53.** a) 60   b) 210   **55.** 50

59. Each new point needs to be connected to all of the previous points. So, if we have already connected four points, we need four new line segments to join a new point to the existing four.

61. If you expand the expression as follows

$$(2n + 5)50 + 1754 - 1986 = 100n + 250 + 1754 - 1986$$
$$= 100n + 2004 - 1986$$
$$= 100n + 18,$$

you see that the 1754 + 250 gives a multiple of 100 plus the current year (2004). If you then subtract the year that you were born, you get your age. If you have already had your birthday, you need to add the extra year, which is why we would then add 1755.

Section 2.2

1. statement   **3.** not a statement   **5.** not a statement

7. statement   **9.** not a statement   **11.** compound

13. compound   **15.** simple   **17.** compound

19. compound   **21.** $r \vee \sim s$ **23.** $(\sim r \wedge t) \rightarrow \sim s$

25. $\sim s \leftrightarrow t$

27. The radial tires are included or the sunroof is not extra.

29. It is not true that: the sunroof is extra and the radial tires are included.

31. If the radial tires are included, then the sunroof is extra or the power windows are not optional.

33.–49. We have not provided answers for these exercises. These statements are complex and sophisticated. You will probably find that you, your classmates, and your instructor do not always agree as to which connectives are present. Nevertheless, it is interesting to try to determine the form of these statements.

51. There exist snakes that are not poisonous.

53. All personal items are covered by this insurance policy.

55. All scientists believe that it is not true that an asteroid collision led to the extinction of the dinosaurs.

57. Some modern art is not difficult to understand.

59.

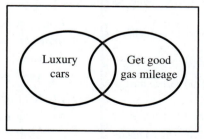

Some luxury cars do not get good gas mileage.

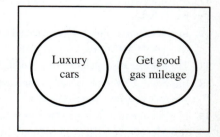

It is not true that some luxury cars get good gas mileage.

61. $(p \vee q) \wedge r$ **63.** $(\sim p \vee q) \wedge r$

65.

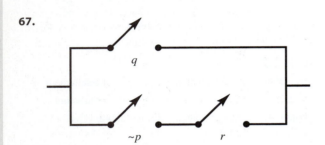

67.

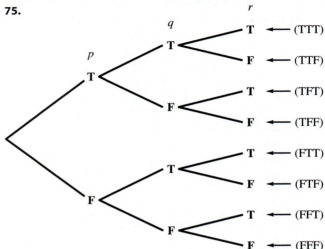

Section 2.3

1. F   **3.** T   **5.** F   **7.** T

9. $\overset{3}{\sim}(\overset{2}{p} \vee \overset{1}{\sim q})$ **11.** $\overset{4}{p} \overset{3}{\wedge} \sim(\overset{2}{p} \vee \overset{1}{\sim q})$

13. T   **15.** F   **17.** F   **19.** T   **21.** 32

23. yes   **25.** no

27. We want the truth table to cover every possibility for truth values for the three variables. There are eight different ways to assign truth values to the three variables.

In Exercises 29–40, we will give the final column of the truth table. We assume that the columns in the truth table are labeled p, q, r, and so on in the usual way.

29. FTFF   **31.** TTFT   **33.** FFFT   **35.** FFTTFFFF

37. TTTTTFFF   **39.**   FFFFTFFF   **41.** exclusive or

43. inclusive or   **45.** Bill is not tall or Bill is not thin.

47. Christian will not apply for a loan and he will not apply for work study.

49. Ken does not qualify for a rebate and he does not qualify for a reduced interest rate.

51. The number x is equal to 5 or s is odd.

53. It is not true that: the earned income tax did reduce the tax you owe and gave you a refund.

55. It is not true that: you are single or the head of a household.

57. We wanted to be able to do straightforward logic computations to decide whether or not two statements were saying the same thing rather than relying on our intuition.

59. yes   **61.** no   **63.** yes   **65.** yes

67. $p \vee (q \vee r)$ **69.** $p \vee \sim q$

71. $p \wedge (q \wedge r)$ is logically equivalent to $(p \wedge q) \wedge r$.

73. $p \vee (q \wedge r)$ is logically equivalent to $(p \vee q) \wedge (p \vee r)$.

75.

	p	q	r	
			T	(TTT)
		T	F	(TTF)
	T		T	(TFT)
		F	F	(TFF)
			T	(FTT)
		T	F	(FTF)
	F		T	(FFT)
		F	F	(FFF)

77. Let the subsets of $\{p, q, r\}$ indicate which of the variables should be true. For example, the subset $\{p, r\}$, specifies that p and r should be true and that q is false. This corresponds to the line labeled TFT in the truth table.

79. The form $\sim(\sim p \vee \sim q)$ is logically equivalent to $p \wedge q$.

81. The stroke connective is logically equivalent to $\sim(p \wedge q)$.

83.

p	q	$(p \mid p)$	$\mid$	$(q \mid q)$
T	T	F	T	F
T	F	F	T	T
F	T	T	T	F
F	F	T	F	T

Section 2.4

1. true   **3.** true   **5.** true   **7.** false

In Exercises 9–20, we assume that the columns of the truth tables are labeled p, q, r, and so on in the usual way.

9. FTTT   **11.** FTFF   **13.** FFTTFTFT

15. TTTTTTTT   **17.** TTTTTFFT   **19.** TTTT

21. If it pours, then it rains.

23. If you do not buy the all-weather radial tires, then they will not last for 80,000 miles.

25. If its sides are not all equal in length, then a geometric figure is not an equilateral triangle.

27. If x does not evenly divide 6, then x does not evenly divide 9.

29. If you can be claimed by someone else as a dependent, then your gross income is not over $2,250.

31. If you decrease the amount being withheld from your pay, then the amount you overpaid is large.

33. converse: $q \rightarrow (\sim p)$; inverse: $p \rightarrow (\sim q)$; contrapositive: $\sim q \rightarrow p$

35. converse: $\sim(q \wedge r) \rightarrow (\sim p)$; inverse: $p \rightarrow (q \wedge r)$; contrapositive: $(q \wedge r) \rightarrow p$

37. equivalent   **39.** not equivalent

41. If I finish my workout, then I'll take a break.

43. If you qualify for this deduction, then you complete Form 3093.

45. If you receive a free cell phone, then you sign up before March 1.

47. If you remain accident-free for three years, then you get a reduction on your auto insurance.

53. $\sim p \vee q$ **55.** Jamie has not been a member for ten years.

57. $r \rightarrow (p \wedge q)$

63. You do not buy the all-weather radial tires or they will last for 80,000 miles.

65. $p \vee (r \wedge (p \wedge q))$ is logically equivalent to p.

67.

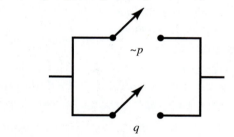

69.

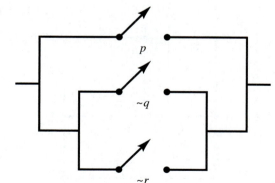

Section 2.5

1. law of detachment; valid

3. fallacy of the converse; invalid

5. disjunctive syllogism; valid

7. fallacy of the inverse; invalid

9. law of syllogism; valid

11. law of contraposition; valid

13. fallacy of the inverse; invalid

15. law of contraposition; valid

17. disjunctive syllogism; valid

21. valid   **23.** invalid   **25.** invalid   **27.** invalid

29. valid   **31.** invalid   **33.** invalid   **35.** valid

37. valid   **39.** invalid

45. If $a \wedge b$ is true, then b is true. If b is true and $b \rightarrow c$ is true, by the law of detachment, then c is true. The contrapositive of $d \rightarrow \sim c$ is $c \rightarrow \sim d$. If both c and $c \rightarrow \sim d$ are true, then $\sim d$ is true.

Section 2.6

1. valid   **3.** invalid   **5.** invalid   **7.** invalid

9. valid   **11.** invalid   **13.** invalid   **15.** valid

17. Some taxes should be abolished.

19. Some teams that wear red uniforms do not play in a domed stadium.

Chapter 2 Test

1. a) inductive   b) deductive

2. Inductive reasoning is the process of drawing a general conclusion by observing a pattern in specific instances. In deductive reasoning, we use accepted facts and general principles to arrive at a specific conclusion.

3. a) 27   b) 47

4.

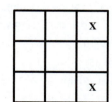

5. $48 = 37 + 11$ (there are other possibilities)

6. The following sequence of algebraic expressions explains each step of the trick.

a) n b) $8n$ c) $8n + 12$ d) $\dfrac{8n}{4} + \dfrac{12}{4} = 2n + 3$

e) $2n + 3 - 3 = 2n$

The result you get is twice your original number.

7. (a) not a statement   (b) statement   (c) not a statement

8. (a) $\sim b \vee s$ (b) $\sim(b \wedge \sim s)$

9. (a) It is not true that: Antonio is fluent in Spanish and he has not lived in Spain for a semester.

(b) Antonio is not fluent in Spanish or he has not lived in Spain for a semester.

10. (a) There is a writer who is not passionate.

(b) All graduates do not receive several job offers.

11. (a) true   (b) false   (c) true

12. (a) 8   (b) 32

13. (a) FFTF   (b) FFFFFTFF

14. (a) You do not qualify for the extended warranty and you do not qualify for the free maintenance contract.

(b) I will sign the lease and I will accept the housing agreement.

15. (a) logically equivalent   (b) not logically equivalent

16. (a) true   (b) false   (c) true

17. (a) TTTF   (b) TFTFFFTT

18. converse: If we cannot recover the data, then the disk drive has been damaged.
inverse: If the disk drive has not been damaged, then we can recover the data.
contrapositive: If we can recover the data, then the disk drive has not been damaged.

19. (a) If you qualify for the tax rebate, then you earned less than $6,500 last year.

(b) If you are an astronaut, then you have a pilot's license.

20. (a) fallacy of the inverse   (b) law of contraposition

21. invalid   **22.** valid   **23.** invalid

Of Further Interest

9. 0.05   **10.** 0.85   **11.** 0.25   **12.** 0.60

13. 0.85   **14.** 0.75   **15.** 0.15   **16.** 0.15

17. 0.27   **18.** 0.36   **19.** 0.64   **20.** 0.27

21. 0.29   **22.** 0.73   **23.** 0.71   **24.** 0.64

25. marketing trainee (0.50)

26. stay in apartment (0.65)

27. attend Good Old State (0.60)

28. invest in C (0.65)

CHAPTER **3**

Section 3.1

1. connected; odd: A, B; even: C, D

3. connected; odd: A, B, E, F; even: C, D

5. not connected; all vertices are even

7. connected; odd: A, B, E, F; even: C, D

9.

11. not possible

13. 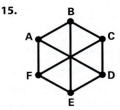   **15.**

17. Each edge connects two vertices, so it contributes two to the total of all the degrees. The total of all the degrees of all the edges will be twice the number of edges.

19. yes   **21.** no; four odd vertices

23. no; not connected   **25.** no; four odd vertices

27. If a vertex is in the middle of the tracing of a graph, every time we enter the vertex by one edge, we must leave the vertex by another edge. If the vertex were odd, eventually we would enter by one edge, but be unable to leave.

In Exercises 29–34, we put spaces in the sequences of vertices to make them more readable.

29. ABFHI GCBEF GECDA

31. duplicate edges: GH, GI, FG; ABCDE FDIGF GIBHG HA

33. duplicate edges: CF, DE, EI; ABDEB CEFCF IEIHE DHGDA

35. no; if we represent this map by a graph, it has four odd vertices

In Exercises 37–38, after duplicating edges as indicated, follow edges in numerical order.

37.
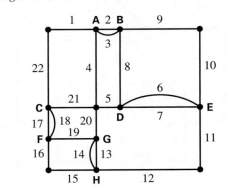

39.

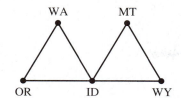

41.

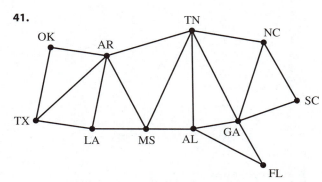

43. The trip is possible because the graph in Exercise 39 has no odd vertices.

45. The trip is not possible because the graph in Exercise 41 has more than two odd vertices.

47. not possible; in the graph model, B will be an odd vertex which is neither a beginning nor ending vertex in the attempt to trace the graph

49. Model this with a graph in which two vertices are joined if the ambassadors from the two countries are unfriendly with each other. The resulting graph can be colored with three colors, which tells us that at most three tables are required.

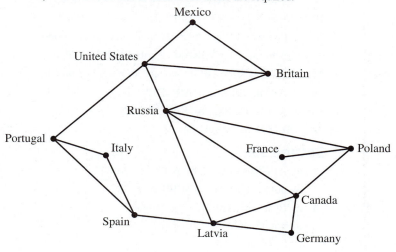

51. Model the committees with vertices. Two vertices are joined
if they have common members. Color the graph and then
committees that are colored the same can meet at the same
time. This graph requires three colors, so three meeting times
are needed.

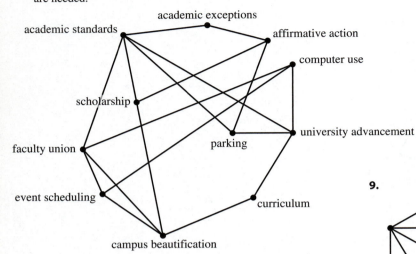

53.

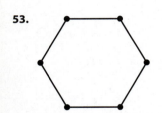

55.–56. A graph containing the following cannot be colored
with three colors.

57. One sequence is C, F, E, C#, D, A, E♭, B.

Section 3.2

1. a) ADBCEA b) EBCDAE

3. a) ADCBFEA b) EADCBFE

5. ABCDEA, ABCEDA, ABDCEA, ABECDA **7.** 6! = 720

9.

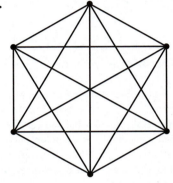

11. a) 20 b) 23 **13.** ACBDA has weight 14

15. ACBDA has weight 99 **17.** ABDECA

19. ADCFEBA **21.** ABEDCA **23.** ADBECFA

29. a) (We won't count reversals of circuits) 720/2 = 360

  b) 6 hours c) PNBRAMCP d) PNBCMRAP

31. 181,440/1,000 = 181.44 seconds, or roughly 3 minutes

33. a) ABMDHEFA b) ABMDHEFA

35. a) XFEBACDPX b) XFEBACDPX

37. ABEDCA has weight 19

Section 3.3

1. a) ABCE, ABCDE b) ABC c) ABCE

  d) not possible e) ABCEBA

3. a) BCE b) ABCFG c) ACFGCE

  d) not possible e) ABCFGC

5.

From

To

	A	B	C	D	E
A	1	1	1	0	0
B	1	1	1	1	1
C	1	1	0	2	2
D	1	2	0	1	1
E	2	1	1	1	1

7.

From

To

	A	B	C	D	E	F	G
A	0	1	2	0	1	1	0
B	0	0	1	0	1	1	0
C	0	0	0	0	1	1	1
D	0	0	1	0	2	1	0
E	0	0	0	0	0	0	0
F	0	0	1	0	0	0	1
G	0	0	1	0	1	1	0

9. Any one of the five neighbors.

11. chief financial officer, sales manager, production manager, president, marketing director

13. I and B are in a block of vertices that do not have any directed edges leaving them.

15. The graph would have to have a directed edge or path from F to I and from F to J. We cannot do this by changing the direction of only one edge.

17.

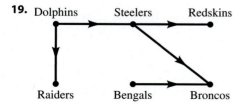

19.

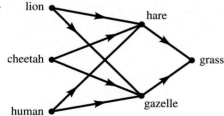

21. Quigley (5), Thums (4), Ross (3), Stickle (1), Pratt (0)

23. C (8), D (7), A (6), E (3), B (2)

Chapter 3 Test

1. a) 12 edges b) F, G, H, and I are odd; the rest are even
   c) yes d) FH, GH, and HI are bridges

2. The objects are represented by vertices. Any two objects that are related are joined by an edge in the graph.

3. Graph (a) can be traced, because it has only two odd vertices. Graph (b) cannot be traced, because it has more than two odd vertices.

4. There are many answers to this. One circuit is ABCDB EDFGH FKHIJ KCA.

5.

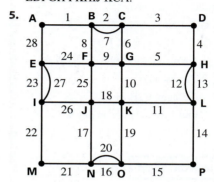

Duplicate edges BC, EI, HL, and NO. The resulting graph will have all even edges, which can be traced using Fleury's algorithm. This tracing will give the indicated route.

6.

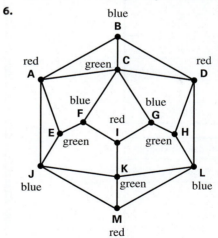

7. ABCDEA, ABCEDA, ABDCEA, ABDECA, ABECDA, ABEDCA

8. ABDECA has weight 13

9. ACEDBA

10. ABDECA

11. a) CDEB b) not possible c) HGEBF
   d) FGEB, FHGEB

12. Directed graphs are used when the relationship modeled may apply in one direction. For example, object X is related to Y, but Y may not be related to X.

13. Pham (8), Robinson (6), Jackson (5), Stein (4), Vaccaro (3), Bartkowski (2)

Of Further Interest

1. a) Begin, A, B, E, I   b) Begin, A, B, E   c) day 14
   d) day 16   e) 17 days   f) Begin, A, B, E, I, J, End

2. a) Begin, A, B, E, H   b) Begin, A, B, E, G
   c) day 17   d) day 20   e) 22 days
   f) Begin, A, B, E, G, I, End

3. a) Begin, B, E, F, G   b) Begin, B, E, F, H
   c) Begin, B, E, F, G, I, End   d) to begin on day 6
   e) to begin on day 9   f) 14 days

4. a) Begin, C, E, G   b) Begin, C, E, H
   c) Begin, C, E, H, I, End   d) to begin on day 9
   e) to begin on day 9   f) 17 days

5.

Task	Day Task Can Begin
A	1
B	1
C	3
D	3
E	6
F	8
G	8
H	12

6.

Task	Day Task Can Begin
A	1
B	1
C	2
D	3
E	7
F	7
G	6
H	14

7.

Task	Day Task Can Begin
A	1
B	1
C	1
D	3
E	4
F	5
G	7
H	7
I	11
J	11

8.

Task	Day Task Can Begin
A	1
B	1
C	1
D	6
E	7
F	11
G	4
H	11
I	14
J	17

9.

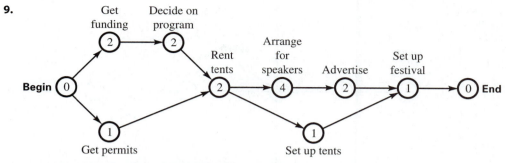

Task	Week Task Can Begin
Get funding	1
Get permits	1
Decide on program	3
Rent tents	5
Arrange for speakers, etc.	7
Advertise	11
Set up tents, etc.	7
Set up festival	13

10.

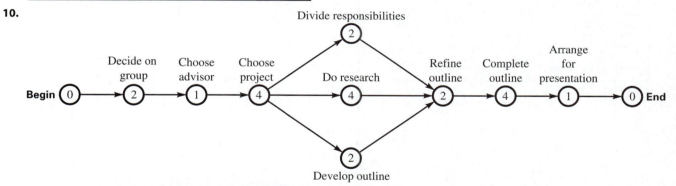

Task	Week Task Can Begin
Decide on group	1
Choose advisor	3
Choose project	4
Divide responsibilities	8
Do research	8
Develop outline	8
Refine outline	12
Complete project	14
Arrange for presentation	18

11.

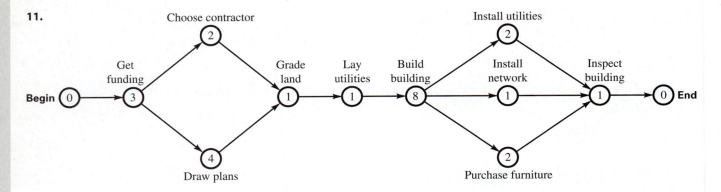

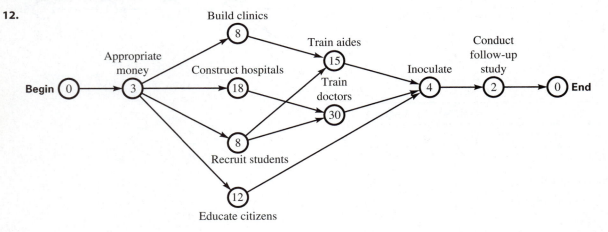

Task	Month Task Can Begin
Get funding	1
Choose contractor	4
Draw plans	4
Grade land	8
Lay utilities	9
Build building	10
Install utilities	18
Install network	18
Purchase furniture	18
Inspect building	20

12.

Task	Month Task Can Begin
Appropriate money	1
Build clinics	4
Construct hospitals	4
Educate citizens	4
Recruit students	4
Train doctors	22
Train aides	12
Inoculate	52
Conduct follow-up study	56

13.

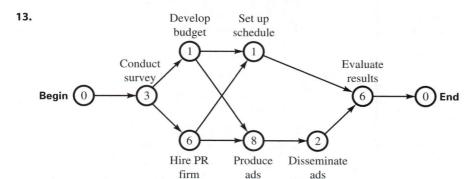

Task	Month Task Can Begin
Conduct survey	1
Develop budget	4
Hire PR firm	4
Set up production schedule	10
Produce ads	10
Disseminate ads	18
Evaluate results	20

CHAPTER **4**

Section 4.1

1. 324   **3.** 21,324   **5.** 2,121,210

7. 99∩∩∩∩||||| **9.** ⌓⌓⌓99∩∩∩∩|||||

11. ⌓⌓⌓⌐⌐⌐⌐⌐⌐⌐99∩

13. ⌐9∩∩∩∩∩||| **15.** ⍦⍦⌓⌓⌐⌐⌐⌐9∩

17. ⌐∩∩||||||| **19.** 99∩∩∩∩∩∩∩∩|||||

21. ⌐⌐⌐⌐⌐⌐9∩∩∩∩||||||| **23.** ∩∩∩∩∩∩∩∩|||||||

25. ⌓⌓⌓⌐⌐⌐⌐⌐⌐⌐⌐99999∩∩

27. 602   **29.** 1,284   **31.** 4,235   **33.** 59   **35.** 564

37. 1963   **39.** 7,544   **41.** 50,262   **43.** 501,420

45. XLIV   **47.** CCLXXVIII   **49.** CDXLIV

51. $\overline{\text{IV}}$DCCXCV   **53.** $\overline{\text{LXXXIX}}$CDXXIII

55. MCMXXXIX   **57.** MCMXCIV   **59.** 436

61. 5,067   **63.** 9,999   **65.** 六十八   **67.** 四百九十七

69. 二十八百三十五   **71.** 九十八百四十六

73. 一千五百二十三 ; 一千二十七

75. No. A number tells us "how many," whereas a numeral is the symbol or symbols that we use to represent the number.

77. 78,124

79. The Roman system had the subtraction principle, which allowed people to write numbers like 4 and 9 more efficiently. They also had a multiplication principle. For example, $|\text{V}|$ represents $5 \times 100 = 500$.

81. If the Chinese chose to omit some power of 10, they simply did not use that symbol in the numeral. If they chose to include some power of 10, they wrote a symbol to tell how many times to include that power of 10 in the number.

Section 4.2

1. 32   **3.** 28   **5.** 731   **7.** 8,187

9.

11.

13.

15.

17. five hundred   **19.** five thousand

21. $2 \times 10^4 + 5 \times 10^3 + 3 \times 10^2 + 8 \times 10^1 + 9 \times 10^0$

23. $2 \times 10^5 + 7 \times 10^4 + 8 \times 10^3 + 0 \times 10^2 + 6 \times 10^1 + 3 \times 10^0$

25. $1 \times 10^6 + 2 \times 10^5 + 0 \times 10^4 + 0 \times 10^3 + 0 \times 10^2 + 4 \times 10^1 + 5 \times 10^0$

27. 5,368   **29.** 370,082   **31.** 3,070,502   **33.** 3,288

35. 6,238   **37.** 142   **39.** 3,655   **41.** 940

43. 20,148   **45.** 136,245   **47.** 1,884   **49.** 3,936

51. 5,544

53. In writing , it could be difficult to tell whether there is one space or two spaces between the two groups of symbols. If there is one space, the number is $1 \times 60 + 5$. If there are two spaces, the number is $1 \times 60^2 + 5$.

55. We need to be able to indicate which powers of the base (say ten) are missing from the numeral.

57. It was based on 60 and also could use the same symbols in different positions to represent different powers of the base, 60.

59. 回三六   **61.** 五〇六七   **63.** ᏂΟΟᏂ

65. The ancient Chinese system requires a separate symbol for each power of 10, e.g., one billion, ten billion, and so on. The Hindu–Arabic system uses different powers of 10, 10^9, 10^{10}, and so on. The positioning of symbols in the numeral indicates these powers of 10.

67. 199,260

69. Using Napier's rods is identical to using a galley to multiply a one-digit number by another number, say 3×546.

71.

$685 \times 47 = 32,195$

73. 18,733   **75.** 290,173

Section 4.3

1. $23_5, 30_5$ **3.** $455_6, 501_6$ **5.** $1010_2, 1100_2$

7. $66_8, 70_8$ **9.** FD_{16}, FF_{16} **11.** 117   **13.** 184

15. 39   **17.** 117   **19.** 183   **21.** 452   **23.** 756

25. 3,336   **27.** 2314_5 **29.** 12302_6 **31.** 1100111_2

33. 1011110_2 **35.** 6513_8 **37.** 6115_8 **39.** $AE8_{16}$

41. DEA_{16} **43.** 4143_5 **45.** 10310_5 **47.** $2E53_{16}$

49. 110000_2 **51.** 2041_5 **53.** 648_{16} **55.** 100110_2

57. 2043_5 **59.** 24041_5 **61.** $114_5 R11_5$ **63.** $44_5 R24_5$

65. 1355_8; $2ED_{16}$ **67.** 1075_8; $23D_{16}$ **69.** 1751_8; $3E9_{16}$

71. 10100110_2 **73.** 101000111110_2

75. CANDY   **77.** LOVE

79. 1, 2, 3, 4, 5, 6, 7, 8, 9, A, B, 10_{12}, 11_{12}, 12_{12}, 13_{12}, 14_{12}, 15_{12}, 16_{12}, 17_{12}, 18_{12}, 19_{12}, $1A_{12}$, $1B_{12}$, 20_{12}, 21_{12}

81. $\oplus$, $\neq$, \$, @, %, $\oplus\mp$, $\oplus\oplus$, $\oplus\neq$, $\oplus$\$, $\oplus$@, $\oplus$%, $\neq\mp$, $\neq\oplus$, $\neq\neq$, $\neq$\$, $\neq$@, $\neq$%, \$$\mp$, \$$\oplus$, \$$\neq$

83. %$\mp$

85. The base b is larger because it requires fewer places to represent the number than when using base c.

87. In a base 8 system, you would need 64 addition and 64 multiplication facts. In a base 16 system, you would need 256 addition and 256 multiplication facts.

89. 60; 5

Chapter 4 Test

1. 1,232,210   **2.** ⟨ᒣ𐎛ꓥꓥꓥꓥꓥꓥꓥIIIIIIIIII   **3.** 1,961

4. MMMMDCCXCV   **5.** 7,593

6. No. A number tells us "how many," whereas a numeral is the symbol or symbols that we use to represent the number.

7. The Roman system had the subtraction principle, which allowed people to write certain numbers more efficiently. They also had a multiplication principle.

8. ▼▼▼ ⟨ᒣ▼▼ ⟨▼▼   **9.** 1,564   **10.** 4,738

11. We need to be able to ndicate which powers of the base (say ten) are missing from the numeral.

12. 97; 2,877   **13.** 6513_8 **14.** 110000_2

15. $134_5\,R20_5$ **16.** 1342_8; $2E2_{16}$

Of Further Interest

1. 7   **2.** 3   **3.** 4   **4.** 5   **5.** 3   **6.** 1   **7.** true

8. false   **9.** true   **10.** false   **11.** false   **12.** true

13. 8   **14.** 10   **15.** 5, 12, 19, 26, 33, . . .

16. 8, 20, 32, 44, 56, . . .   **17.** 6, 16, 26, 36, 46, . . .

18. 3, 12, 21, 30, 39, . . .   **19.** 5   **20.** 3   **21.** 2

22. 3   **23.** 7   **24.** 7   **25.** 4   **26.** 6   **27.** 0

28. 6   **29.** 3   **30.** 13   **31.** 10   **32.** 9   **33.** 6

34. 3   **35.** 6   **36.** 8   **37.** 3   **38.** 4   **39.** 4

40. 7   **41.** 3, 8   **42.** 2, 6   **43.** 24   **44.** 7

45. 42   **46.** 23   **47.** 28   **48.** 42

49. a) 7   b) 3, 7   **50.** 4   **51.** 1982   **52.** 1959

53. 5   **54.** 4   **55.** 9   **56.** 7

CHAPTER **5**

Section 5.1

1. true   **3.** false   **5.** false   **7.** true

9. 4 is a factor of 48; 48 is a multiple of 4

11. 7 is a factor of 28; 7 divides 28

13. 9 divides 72; 72 is a multiple of 9

15. 53, 59, 61, 67, 71, 73, 79, 83, 89, 97

17. 9   **19.** 12   **21.** 21×11 **23.** prime

25. prime   **27.** 7×17 **29.** 2, 3, 5, 6, 10   **31.** 3, 5, 9

33. 30   **35.** 20   **37.** 4   **39.** 20   **41.** $2^2 \times 5 \times 7^2$

43. $2^2 \times 3^2 \times 5^2 \times 11$ **45.** $3^3 \times 23$ **47.** 11×29

49. 4; 120   **51.** 14; 280   **53.** 72; 864   **55.** 21; 3969

57. After we factor a and b, we form the product of the smallest powers of the common primes that divide each number.

59. 28   **61.** 45   **63.** 108   **65.** 360   **67.** 140

69. 6   **71.** 3 feet by 3 feet   **73.** 432

75. Four will always divide a multiple of 100, so all we have to do to check if four divides a number is to determine if four divides the number formed by the last two digits. So, four automatically divides 36,800. Therefore we only have to check if four divides 24, which it does.

77. 59 and 61; 71 and 73; 101 and 103

79. No. Both 4 and 6 divide 12, but $4 \times 6 = 24$ which does not divide 12.

81. If the numbers are a, b, and c, you could first find the GCD of a and b, call it g, and then find the GCD of g and c.

83. Fifteen will divide a number if both three and five divide the number.

85. It means that if we express a natural number as a product of prime numbers, we cannot do it in any other way, which is what the Fundamental Theorem of Arithmetic tells us.

Section 5.2

1. $+3$ **3.** $+10$ **5.** $+9$ **7.** $+19$ **9.** $+23$

11. -13 **13.** -65 **15.** $+50$

17. For the next four days you spend six dollars, so in four days, your net change in finances will be $-\$24$. Thus $(+4)(-6) = -24$.

19. For the past three days you have gained seven dollars, so three days ago, you had \$21 less. Thus $(-3)(+7) = -21$.

21. $+35$ **23.** -56 **25.** $+48$ **27.** -38

29. -63 **31.** $+72$

33. $+16 = (-2) \cdot c$; $c = -8$;

35. $-14 = (-2) \cdot c$; $c = +7$;

37. $-40 = (+8) \cdot c$; $c = -5$;

39. $+12 = (+3) \cdot c$; $c = +4$;

41. -3 **43.** $+15$ **45.** $+5$ **47.** -6 **49.** $+6$

51. -4 **53.** -26 **55.** -12 **57.** $+16$

59. -50 **61.** 3   **63.** not possible   **65.** -5

67. not possible   **69.** true

71. false; $(-5) + (+8) = +3$ **73.** false; $\dfrac{-6}{-3} = +2$

75. true   **77.** 57,260 feet   **79.** 7,201 feet

81. 265 degrees   **83.** 66

85. 1,919 years

87. 37 degrees

89. No; the sums of the entries in row one and row three are not the same.

91. If $\dfrac{8}{0} = x$, then $8 = 0 \cdot x$ which is not possible.

93. The sum of the nine entries is 9×37.

95. The reasoning is similar to Exercise 94, except now there is no center number. So, put a point, C, in the middle of the 4 by 4 square and notice that any two numbers that are symmetric with respect to C have the same total. There are eight such pairs in a 4 by 4 box, therefore, add two numbers in the box that are symmetric with respect to the center, C, and multiply this total by eight to get the total of all 16 numbers.

97. $5 + x = 3$

Section 5.3

1. equal   **3.** not equal   **5.** not equal   **7.** equal

9. not equal

11. $\dfrac{3}{7}$ **13.** $-\dfrac{1}{3}$ **15.** $-\dfrac{11}{18}$ **17.** $\dfrac{9}{14}$ **19.** $\dfrac{13}{14}$

21. $\dfrac{7}{6}$ **23.** $-\dfrac{1}{3}$ **25.** $\dfrac{5}{48}$ **27.** $\dfrac{43}{72}$ **29.** $\dfrac{59}{24}$

31. $-\dfrac{1}{24}$ **33.** $\dfrac{1}{3}$ **35.** $\dfrac{1}{3}$ **37.** $-\dfrac{21}{5}$

39. $-\dfrac{7}{24}$ **41.** $-\dfrac{3}{8}$ **43.** $-\dfrac{33}{4}$ **45.** 2

47. $\dfrac{62}{81}$ **49.** $-\dfrac{49}{10}$ **51.** $\dfrac{79}{5} = 15\dfrac{4}{5}$

53. $\dfrac{29}{3} = 9\dfrac{2}{3}$ **55.** $\dfrac{95}{12} = 7\dfrac{11}{12}$ **57.** $6\dfrac{3}{4}$

59. $8\dfrac{1}{15}$ **61.** $60\dfrac{16}{17}$ **63.** $\dfrac{11}{4}$ **65.** $\dfrac{55}{6}$

67. $\dfrac{35}{3}$ **69.** 0.75 **71.** 0.1875 **73.** $5.\overline{3}$ **75.** $0.\overline{81}$

77. $0.\overline{307692}$ **79.** $\dfrac{16}{25}$ **81.** $\dfrac{209}{250}$ **83.** $\dfrac{469}{200}$

85. $\dfrac{61}{5}$ **87.** $\dfrac{4}{9}$ **89.** $\dfrac{7}{37}$ **91.** $\dfrac{7}{22}$ **93.** $\dfrac{5}{13}$

95. $\dfrac{1}{4}$ **97.** $\dfrac{5}{48}$ **99.** 24; no   **101.** $3\dfrac{3}{4}$

103. $1\dfrac{3}{8}$

105. The small tube costs 49.5 cents per ounce; the large tube costs 50.5 cents per ounce; so the smaller tube is the better buy.

107. He can get $3\dfrac{5}{9}$ strips from a 12-foot strip and $4\dfrac{4}{9}$ strips from a 15-foot strip. Thus there is more waste per strip from the 12-foot strips.

109. $2\dfrac{23}{32}$ feet   **111.** $29\dfrac{1}{4}$

113. If we assume that $\dfrac{x+5}{y+5} = \dfrac{x}{y}$, then we can cross multiply to get $(x+5)y = (y+5)x$. Thus $xy + 5y = yx + 5x$. This means that $5y = 5x$. Therefore, $y = x$.

115. $\dfrac{4}{10} = \dfrac{2}{5}$ **117.** $\dfrac{4}{15}$ **119.** 6   **121.** $\dfrac{10}{9}$

Section 5.4

1. rational   **3.** irrational   **5.** rational   **7.** irrational

9. $\sqrt{9} = 3$ **11.** $3\sqrt{2}$ **13.** $2\sqrt{3}$ **15.** $5\sqrt{3}$

17. $4\sqrt{3}$ **19.** $3\sqrt{21}$ **21.** $11\sqrt{5}$ **23.** not possible

25. $8\sqrt{5}$ **27.** not possible   **29.** not possible   **31.** 6

33. $6\sqrt{5}$ **35.** $14\sqrt{3}$ **37.** 2   **39.** $\dfrac{4}{3}$ **41.** $\dfrac{2\sqrt{3}}{3}$

43. $\dfrac{3\sqrt{5}}{5}$ **45.** $2\sqrt{6}$ **47.** $\dfrac{5\sqrt{22}}{11}$ **49.** $\dfrac{4\sqrt{3}}{3}$

51. $\dfrac{\sqrt{5}}{2}$

53. False. $\sqrt{2}$ is irrational; 1.414215362 is a rational, decimal approximation of $\sqrt{2}$.

55. False. Counterexample: $\sqrt{25} = 5$ which is rational.

57. True. $\sqrt{3}\sqrt{5} = \sqrt{15}$, and 15 is not a perfect square.

59. True. If a is rational and b is irrational, and $a + b = c$, then if c is rational we could say $b = c - a$. But $c - a$ is the difference of two rational numbers and therefore is rational. This would contradict the fact that b is irrational.

61. 13 pounds   **63.** 144.22 pounds

65. 0.435; 0.43512112111211112 . . .

67. 0.45785; 0.45785121121112111112 . . .

69. 0.6; 0.612112111211112 . . .

71. 0.12113; 0.1211312121121112111112 . . .

73. a,d,b,c   **75.** a,c,d,b   **77.** distributive

79. commutative(addition)   **81.** associative(addition)

83. identity element for addition   **85.** commutative(addition)

87. a) we can cancel common factors from the numerator and denominator   b) square both sides of a)   c) multiply through equation by b^2 d) 2 divides the left side of c) so 2 divides the right side of c)   e) if a were odd then a^2 would be odd, but it is not   f) a is even   g) substitute $2k$ for a in c)   h) square both sides of g)   i) divide both sides of h) by 2   j) 2 divides b^2 k) if b were odd then b^2 would be odd

l) We assumed that $\dfrac{a}{b}$ was reduced, so a and b cannot both be even.

89. 8.42   **91.** 10.29   **93.** 2

95. These numbers are possible lengths of the three sides of a right triangle. If a right triangle has legs of length 3 and 4 and hypotenuse of length 5, then $3^2 + 4^2 = 5^2$

Section 5.5

1. 32   **3.** -16 **5.** -9 **7.** 9   **9.** 1   **11.** 0

13. 729   **15.** 117,649   **17.** $\dfrac{1}{25}$ **19.** $\dfrac{1}{729}$

21. -3 **23.** 25   **25.** 36   **27.** $\dfrac{1}{27}$ **29.** 1

31. 11   **33.** $-\dfrac{1}{64}$ **35.** 8

37. In evaluating -2^4, we raise 2 to the fourth power to get 16 and then negate that result to get -16. In evaluating $(-2)^4$, we first negate 2 and then raise that negative number to the fourth power to get positive 16.

39. $a^2 a^3 = aaaaa = a^5$, so in this case, we add exponents.

41. $\dfrac{a^8}{a^6} = \dfrac{aaaaaaaa}{aaaaaa} = a^2$, so in this case, we subtract exponents.

43. 4.356×10^6 **45.** 7.83×10^2 **47.** 2.4×10^{-3}

49. 3.824×10^5 **51.** 4.0×10^{-1} **53.** 8.0×10^{-3}

55. 32,500   **57.** 0.00178   **59.** 63   **61.** 0.0627

63. 0.00000045   **65.** 1,000,000   **67.** 2.381×10^7

69. 8.4×10^2 **71.** 6.0×10^{11} **73.** 3.6×10^2

75. 2.6064×10^4 **77.** 4.0×10^{-5} **79.** 3.36×10^{-1}

81. 2.6311×10^{-6} **83.** 8.076×10^{10} **85.** 1.564×10^{13}

87. 4.0×10^{-7} **89.** 6.6×10^{-7} **91.** 1.7×10^3

93. 2.215×10^1 **95.** 1.011×10^3 **97.** 3.17×10^1

99. a) 5.902×10^4;   b) 5.902×10^7 c) The 130 pound person weighs as much as 59 million mosquitoes.

101. 7.86×10^1 **103.** 3.72×10^3 **105.** 3.87×10^1

107. 2.17×10^4; 6 hours

109. Scientific notation allows us to compute easily with very large or very small numbers.

111. $\left(\dfrac{a}{b}\right)^n = \dfrac{a^n}{b^n}$

Chapter 5 Test

1. 71, 73, 79, 83, 89   **2.** $13 < \sqrt{180} < 14$

3. 191 is prime; $441 = 3^2 7^2$

4. 3, 4, 6, and 9 divide the number

5. GCD = 66; LCM = 1980

6. To calculate the GCD, multiply the *smallest* powers of any primes that are common to both numbers. To calculate the LCM, multiply the *largest* powers of all primes that occur in either number.

7. a) 9   b) 11   c) 72   d) -16

8. The number -4 represents four days ago. The $+3$ means that you have been gaining 3 dollars per day. So, four days ago you had 12 fewer dollars. Thus $(-4)(+3) = -12$.

9. If $\dfrac{-24}{-8} = c$, then $-24 = -8 \cdot c$. Thus, $c = +3$.

10. 20

11. If $\dfrac{5}{0} = c$, then $5 = 0 \cdot c$, which is impossible.

12. No; $28 \cdot 35 \neq 65 \cdot 14$ **13.** a) $\dfrac{5}{27}$ b) $\dfrac{18}{37}$ **14.** $6\dfrac{5}{8}$

15. $14\dfrac{1}{2}$ **16.** a) $\dfrac{3}{8}$ b) $\dfrac{7}{11}$

17. $3\dfrac{3}{4}$ cups of peppers and $4\dfrac{7}{8}$ cups of tomatoes

19. Rational numbers have repeating expansions and irrational numbers have nonrepeating expansions.

20. a) $6\sqrt{3}$ b) $\sqrt{3}$ **21.** $\dfrac{8}{\frac{4}{2}}$ does not equal $\dfrac{\frac{8}{4}}{2}$

22. a) 81   b) $\dfrac{1}{256}$ c) 216   d) $\dfrac{1}{64}$

23. In evaluating -2^4, we first raise 2 to the fourth power and then take its opposite to get -16. In evaluating $(-2)^4$, we raise -2 to the fourth power to get 16.

24. 4.5×10^{-4}; 1.23×10^6 **25.** 1,325,000; 0.0000863

26. 2.4×10^{-6} **27.** 6.036×10^1

Of Further Interest

1. arithmetic; 17; 20   **2.** arithmetic; -5; -9

3. geometric; 648; 1,944   **4.** arithmetic; 36; 44

5. geometric; $\dfrac{1}{16}$; $\dfrac{1}{32}$ **6.** geometric; 0.00001; 0.000001

7. arithmetic; -10; -15 **8.** arithmetic; 56; 69

9. geometric; 32; 64   **10.** geometric; $\dfrac{1}{81}$; $\dfrac{1}{243}$

11. arithmetic; 3.5; 4.0   **12.** geometric; 1; -1

13. 35; 220   **14.** 59; 315   **15.** 86; 660   **16.** 78; 792

17. 10.5; 115   **18.** 5.5; 46.75   **19.** 436   **20.** 3,043

21. 59,049; 88,573   **22.** 768; 1,533

23. $\dfrac{1}{64}$; $\dfrac{127}{64} = 1.984375$ **24.** $-1,024$; -682

25. 0.00002; 2.22222   **26.** 500,000,000; 555,555,555

27. 144   **28.** 10,946   **29.** 377   **30.** 75,025

31. $36,810   **32.** 66   **33.** $1,475.11

34. At the 11th stage, more than 8.5 billion letters would have to be sent.

35. 4.10 feet

36. After 9 bets, you would have lost $1,533 and can no longer continue doubling your bets.

37. $\dfrac{(n - k + 1)(a_k + a_n)}{2}$

38. a) This is the sum of the terms a_1 to a_n in a geometric sequence.

     b) Use distributivity to factor out a.

     c) If you multiply $(1 - r)(1 + r + r^2 + \ldots + r^{n-1}) = 1 - r^n$ and then divide both sides of the equation by $(1 - r)$, you get the result in c).

     d) Multiply numerator and denominator by -1 to get the final result.

39. 195.45 million   **40.** 1,392.26 million

41. $F_1 + F_2 + F_3 + \ldots + F_n = F_{n+2} - 1$

42. $F_n{}^2 + F_{n+1}{}^2 = F_{2n+1}$

CHAPTER **6**

Section 6.1

1. 5   **3.** $-\frac{4}{5}$ **5.** 40   **7.** -12 **9.** $\frac{16}{7}$ **11.** 25

13. $w = \frac{P - 2l}{2}$ **15.** $\mu = x - \sigma z$ **17.** $r = \frac{A - P}{Pt}$

19. $x = \dfrac{-3y + 6}{2}$ **21.** $l = \dfrac{V}{wh}$ **23.** $h = \dfrac{S - 2\pi r^2}{2\pi r}$

25. 168   **27.** 32

29. 7; If Shaun sells more than seven systems, Best Deal is better for him.

31. 70; If Cassandra uses more than 70 minutes, then Cingleton is cheaper.

33. $n = 0.7715g - 27$; $281.60

35. $c = 1.072b + 80$; $2,438.40   **37.** $10l + 25t = 200$

39. $22l + 13t = 341$ **41.** $35t + 55d = 14,500$

43. $9e + 15c = 342$

47. intercepts: $(4, 0), (0, 6)$

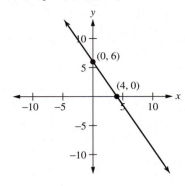

49.

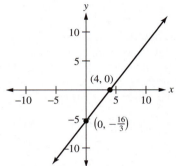

51. intercepts: $(9, 0), (0, 6)$

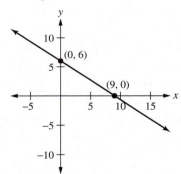

53.

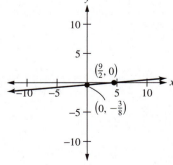

55. intercepts: (8, 0), (0, −0.4)

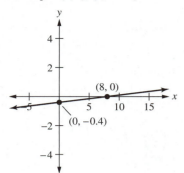

57.

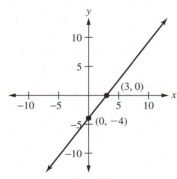

59. $\frac{3}{4}$　**61.** $\frac{-4}{5}$　**63.** $\frac{5}{2}$　**65.** does not exist

67. *a, d*　**69.** *f*　**71.** The rise is zero.

73. *y*-intercept = −3; slope = 4

75. *y*-intercept = −3; slope = −5

77. $y = -23.85x + 262.35$; the slope, −23.85, is the rate at which the DVD player is being paid off; the *y*-intercept, 262.35, is the amount owed on the player

79. a) By the sixth week, Save-U is the cheaper plan.

81. after 160 tokens　**83.** 5.246 percent

85. $y = 108d$; $d = \dfrac{y}{108}$　**87.** $d = 406.7e$　**89.** 17.75 feet

Section 6.2

1. $y = 3x - 5$　**3.** $y = 4x + 14$　**5.** $y = -2x + 11$

7. $y = -5x - 31$　**9.** $y = 2x - 1$　**11.** $y = \frac{1}{4}x + \frac{31}{4}$

13. $y = \frac{-6}{19}x - \frac{10}{19}$　**15.** $y = \frac{7}{2}x + 13$

17. a) $y = 0.3x + 78.8$　b) 81.8

19. a) $y = 25x + 1{,}984$　b) 2,259

21. a) $y = 157x + 3{,}506$　b) 5,076

23. a) $\frac{-112}{3}t$　b) $v = \frac{-112}{3}t + 637$　c) 301

25. $y = 6.8x + 87$; 121,000

27. $y = -0.0157x + 8.5$; 2441

29. a) 1.2　b) $l = 1.2t + 131.5$　c) 139.9 million

31.

	1998 (Year 0)	1999 (Year 1)	2000 (Year 2)	2001 (Year 3)
Actual data	131.5	133.5	135.2	135.1
Values predicted by your model	131.5	132.7	133.9	135.1
Values predicted by line of best fit	131.95	133.2	134.45	135.7

Section 6.3

1. $x = 2, x = 8$ **3.** $x = 1, x = \frac{3}{2}$

5. $-3, \frac{2}{3}$ **7.** $-\frac{3}{5}, 4$

9. opening down; vertex $(3, 1)$; $(2, 0), (4, 0)$; $(0, -8)$
(see graph 9)

Graph 9

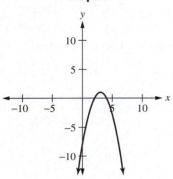

11. opening down; vertex $(1, 9)$; $(-\frac{1}{2}, 0), (\frac{5}{2}, 0)$; $(0, 5)$
(see graph 11)

Graph 11

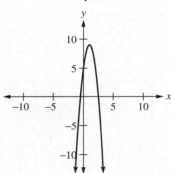

13. opening up; vertex $(\frac{1}{2}, -3)$; $\left(\dfrac{1 + \sqrt{3}}{2}, 0\right)\left(\dfrac{1 - \sqrt{3}}{2}, 0\right)$;
$(0, -2)$ (see graph 13)

Graph 13

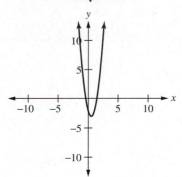

15. opening up; vertex $(\frac{-7}{6}, \frac{-121}{12})$ $(-3, 0), (\frac{2}{3}, 0), (0, -6)$; (see
graph 15)

Graph 15

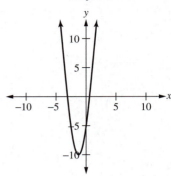

17. opening down; vertex $(\frac{7}{2}, \frac{1}{4})$ $(3, 0), (4, 0), (0, -12)$; (see
graph 17)

Graph 17

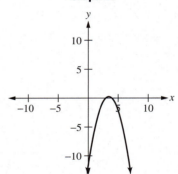

19. b) 8.125

21. b) at the end of the fourth month, the cumulative
loss is $410   c) by the end of the eighth month
d) $790   e) $350

23. by the fourth week

25. a) see graph 25   b) t (time) cannot be negative
c) 3.16 seconds

Graph 25

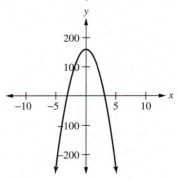

27. a) see graph 27 b) *t* (time) cannot be negative
c) 3.125 seconds d) 6.25 seconds

Graph 27

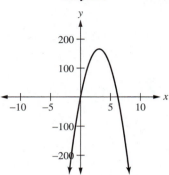

29. The crate begins falling at a point above the ground and picks up speed as it falls. A linear equation would not be appropriate, because the rate of change of the distance of the crate above the ground is not constant.

31. a) see graph 31 b) *t* cannot be negative
c) the runner is running more slowly when *t* is close to 0
d) 10.45 seconds

Graph 31

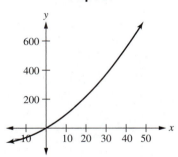

35. no; when we solve a quadratic equation to find the *x*-intercepts, we can have at most two solutions.

Section 6.4

1. $1,050 **3.** $4,100 **5.** $6,381.40 **7.** $4,665.60

9. $12,667.70 **11.** $22,666.02 **13.** 1,188.2 million

15. 18.9 million **17.** 1,377.4 million **19.** 1.86

21. 2.10 **23.** 0.51 **25.** 1.57 **27.** 14.21 years

29. 2059 **31.** 1,224.4 million **33.** 198 years

35. 0.3669 **37.** 0.7291 **39.** 35.2 percent

41. 30.9 percent **43.** 63.2 percent **45.** 56.4 percent

47. 307 mg **49.** after eight hours

51. The population will exceed 300 by the end of the fourth year; fishing can begin toward the end of the fourth year.

53. 200

55. An exponential model is appropriate when the rate of growth is proportional to the amount present. Bacteria and money grow exponentially.

Section 6.5

1. 4 **3.** 62.5 **5.** 18 **7.** 36 **9.** 2.6 mg

11. 62.5 mph **13.** 3 bags **15.** $437.76 **17.** $1,408

19. 12,000 **21.** 5 or 6

23. 435.12; The total number of students is less than the number on the dean's list. He was not consistent in the way he set up the two ratios in the proportion.

25. In $m:n$, we are comparing *m* objects with *n* objects. In $n:(m + n)$, we are comparing the *n* objects with the total of all the $m + n$ objects.

27. 65 **29.** 1 **31.** 100 **33.** $\dfrac{1}{16}$ **35.** 25

37. 6,860 **39.** $\dfrac{4}{3}$ **41.** 192 **43.** $3.38

45. $13\dfrac{1}{3}$ pounds **47.** 5.66 feet

49. 2,230 **51.** 6.4 pounds per square inch

53. directly **55.** strength doubles **57.** strength triples

Chapter 6 Test

1. a) 4 b) $-\frac{1}{2}$

2. $\frac{A - P}{Pt} = r$

3. a) $s = 200 + 10(h - 40)$ b) 260

4. $5.80c + 11.35d = 160.40$

5.

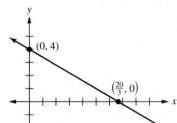

6. $\frac{3}{4}$ **7.** line 1 is *a*; line 2 is *d*; line 3 is *b*; line 4 is *c*

8. after the fourth month

9. when there is a constant rate of change between the two variables

10. $y = \frac{5}{3}x - 1$ **11.** $y = 0.0021x + 0.062$; 14.6 percent

12. a technique by which we find a linear equation that best models a set of data

13. a) opens down   b) $\left(-\frac{5}{2}, \frac{9}{2}\right)$

c) $(-1, 0), (-4, 0)$ d) $(0, -8)$

e)

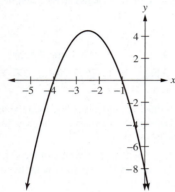

14. a technique by which we find a quadratic equation that best models a set of data

15. week 8

16. \$12,641.72

17. It will double by the fifteenth year.

18. 0.5075

19. In a logistic model we adjust the rate of growth as the population grows. The larger the population, the smaller the rate of growth.

20. a) $\dfrac{25}{4} = 6.25$ b) $\dfrac{75}{2}$ **21.** 6.25 gallons

22. 825   **23.** $\dfrac{7}{3}$ **24.** 1,408

25. There is no change in the strength of the beam.

Of Further Interest

1. 9   **2.** 17   **3.** 82   **4.** 2

5. 18.2464   **6.** -0.0416

7. $A_{n+1} = 1.05A_n$, $n = 0, 1, 2, \ldots$ where $A_0 = 1,000$; \$1,102.50

8. $A_{n+1} = 1.06A_n$, $n = 0, 1, 2, \ldots$ where $A_0 = 1,500$; \$1,786.52

9. $P_{n+1} = [1 + (0.08)(1 - P_n)]P_n$, $n = 0, 1, 2, \ldots$ where $P_0 = 0.30$; 33.4 percent

10. $P_{n+1} = [1 + (0.12)(1 - P_n)]P_n$, $n = 0, 1, 2, \ldots$ where $P_0 = 0.20$; 24 percent

11. $D_{n+1} = 0.60D_n + 250$, $n = 0, 1, 2, \ldots$ where $D_0 = 0$; 490 mg

12. $D_{n+1} = 0.25D_n + 1000$, $n = 0, 1, 2, \ldots$ where $D_0 = 0$; 1312.5 mg

13. $a = -3$; unstable

14. $a = \frac{5}{3}$; unstable

15. $a = \frac{16}{3}$; stable

16. $a = -\frac{20}{9}$; stable

17. 900 years

18. 4,300 years

19. 333

20. 515

CHAPTER **7**

Section 7.1

1.

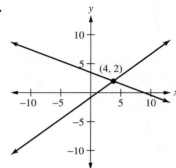

3.

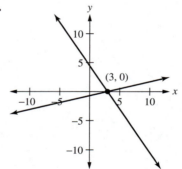

5.

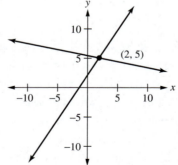

7.

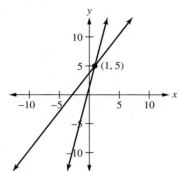

9.

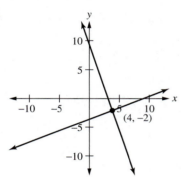

11.

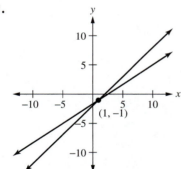

13. $(1, 4)$ **15.** $(4, 5)$

17. $(-6, -2)$ **19.** no solution

21. infinite number of solutions

23. no solution   **25.** $\left(\dfrac{1}{4}, \dfrac{1}{2}\right)$ **27.** $\left(\dfrac{1}{5}, \dfrac{2}{5}\right)$

29. a) There are no solutions.

b) There are an infinite number of solutions.

c) There is one, unique solution.

31. 116 wins, 46 losses

33. 4 bagels, 5 ounces of cream cheese

35. At 7 systems, he makes $540 at either store. After that, he earns more at Best Deal.

37. 21 wins, 11 losses

39. Hartsfield—80 million, O'Hare—72 million

41. $11 per hour   **43.** $34   **45.** $(-1, 3)$ **47.** $(-1, 4)$

49. no solution   **51.** infinite number of solutions

55. $(4, 0, 1)$ **57.** $(5, 2, 1)$

Section 7.2

1. a, d   **3.** c, d

9.

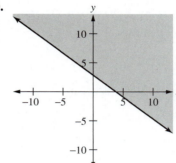

11.

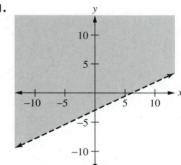

13.

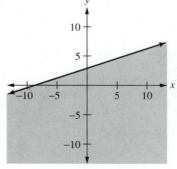

15.

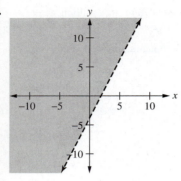

17.

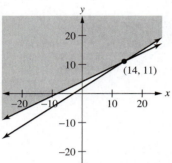

(14, 11)

19.

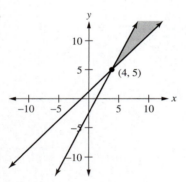

(4, 5)

21.

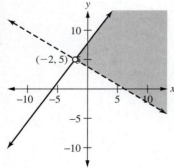

(−2, 5)

23.

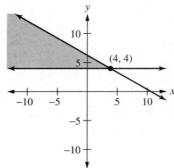

(4, 4)

25.

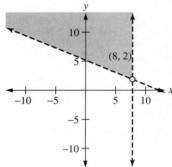

(8, 2)

27.

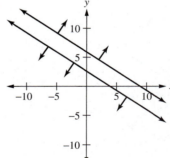

29.

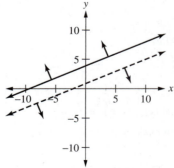

31.

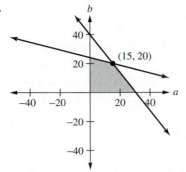

33.

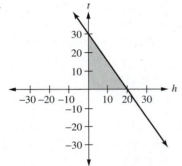

35.

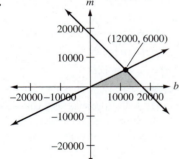

37.

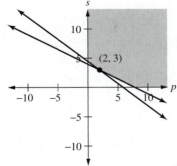

39.

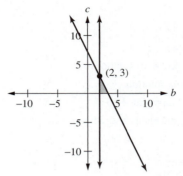

41.

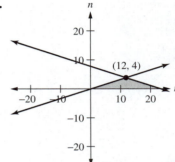

43.

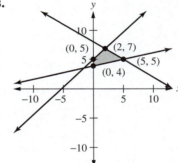

45.

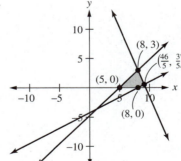

47. $3y \le 43 - 5x$
$11y \ge 27 - 2x$
$3y \ge 1 + 2x$

49. $3y \le 43 - 5x$
$11y \ge 27 - 2x$
$3y \le 1 + 2x$

Chapter 7 Test

1. $(3, -5)$

2. a) Case 2; no solutions

b) Case 3; infinite number of solutions

c) Case 1; one, unique solution

3. United States—70; Opponent—53

4. 12 hours at the fast food restaurant; 5 hours as a personal trainer

5.

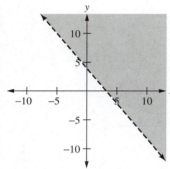

6.

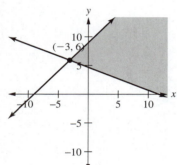

7.

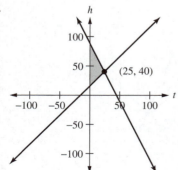

Of Further Interest

1. $(0, 0), (6, 0), (5, 4), (4, 6), (0, 8)$

2. $(0, 0), (5, 0), (8, 3), (5, 5), (0, 4)$

3. $(0, 0), \left(\frac{22}{3}, 0\right), (6, 2), \left(\frac{8}{3}, \frac{16}{3}\right), (0, 6)$

4. $(2, 4), \left(\frac{17}{2}, 3\right), \left(5, \frac{13}{2}\right)$

5. Maximum is 36 at $(4, 6)$ **6.** Maximum is 8 at $(5, 3)$

7. Minimum is -3 at $(0, 3)$ **8.** Minimum is 0 at $(0, 0)$

9. Minimum is 4 at $(2, 3)$ **10.** Maximum is 48 at $(6, 2)$

11. Maximum is 34 at $(2, 2)$ **12.** Maximum is 14 at $(8, 6)$

13. Minimum is 16 at $\left(\frac{5}{2}, 3\right)$ **14.** Minimum is 51 at $(9, 4)$

15. 15 Athens, 20 Barcelona

16. \$2,250 in pharmaceuticals, \$750 in communications

17. 5 hats, 10 T-shirts   **18.** 17 plain, 12 graphics

19. 2 PowerUp, 3 StressTabs   **20.** 12 plates, 16 cups

21. 80 lizards, 120 frogs   **22.** 20 of each type of apartment

CHAPTER **8**

Section 8.1

1. 7 and 9   **3.** 7 and 3   **5.** 10   **7.** 9 and 10

9. true   **11.** false   **13.** true   **15.** false   **17.** false

19. e, d **21.** c, g **25.** $60°, 150°$

27. no complementary angle; supplementary angle is $60°$

29. $38.8°, 128.8°$

31. $m\angle a = 144°, m\angle b = 36°, m\angle c = 144°$

33. $m\angle a = 45°, m\angle b = 135°, m\angle c = 45°$

35. $m\angle a = 52°, m\angle b = 90°, m\angle c = 128°$

37. Yes, if the lines are perpendicular.

39. No more than two could be obtuse. Because verticals are equal, if there were a third obtuse angle, then the fourth would also have to be obtuse, giving an angle sum of more than 360 degrees.

41. $90 - x$ **43.** $3(90 - x)$ **45.** $30°$ **47.** $75°$

49. 12   **51.** arc AB has length 6 feet

53. $m\angle ACB = 120°$

55. circumference $= 1,200$ millimeters

57. $14.4°$

61. Yes, if the angles are both 45-degree angles.

63.

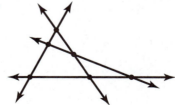

65. 180

Section 8.2

1. true   **3.** false   **5.** true   **7.** false   **9.** false

11. polygon   **13.** not a polygon; not made of line segments

15. No. A scalene triangle has no sides equal.

17. Both have four equal sides. A rhombus does not have right angles.

19. $m\angle A = 30°, m\angle B = 60°, m\angle C = 90°$

21. $m\angle A = 45°, m\angle B = 55°, m\angle C = 80°$

23. Divide the hexagon into four triangles. The angle sum is $4 \times 180° = 720°$.

25. 162°   **27.** 18

29. $m \angle E = 40°$, length of EH is 7

31. $m \angle I = 35°$, length of HI is 84

33. 192.5 feet   **35.** 54°   **37.** possible

39. Not possible; the angle sum would be greater than 180 degrees.

43. The measure of the interior angles gets larger. For very large n, the interior angles have measures close to 180°.

45. Construct the level (see figure) so that lengths AB and BC are equal and so that $\angle EDB = \angle BED$. Mark a vertical line at M, the midpoint of DE. The level is resting on level ground when the string lines up with the mark made at point M.

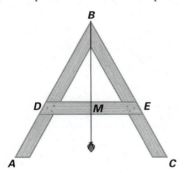

47. a) Triangles are rigid. The scaffolding containing triangles will not shift.

b) The rectangles in this scaffolding can deform into nonrectangular parallelograms, causing the scaffolding to collapse.

Section 8.3

1. 160 ft²   **3.** 140 in²   **5.** 108 cm²   **7.** 72 yd²

9. 78.5 cm²   **11.** 50.24 m²   **13.** 100 cm²   **15.** 42 m²

17. 6.88 m²   **19.** 0.86 m²   **21.** 30 sq in.   **23.** 54 sq in.

25. 12 sq yd   **27.** 72 sq yd   **29.** 7 in.²   **31.** 3.5 in.²

33. 42.15 cm²   **35.** 59.81 m²   **37.** 6.62 cm   **39.** 13 m

41. 6.93 yd   **43.** 57.5 in.²   **45.** 66.33 ft²

47. 127.28 ft   **49.** 2 m   **51.** area   **53.** perimeter

55. 86.13 ft²   **57.** 4 ft   **59.** false   **61.** 1,000.96 m²

63. a) 22.61 ft

b) 203.47 ft²

65. The area of WXYZ is one half of the area of ABCD.

67. 50 feet by 50 feet   **69.** about 21,195 ft²

71. $(2r)^2 - \pi r^2 = (4 - \pi)r^2$

Section 8.4

1. a) 148 cm²   b) 120 cm³

3. a) 80.48 in.²   b) 75.36 in.³

5. a) 408.2 ft²   b) 628 ft³

7. a) 5,024 cm²   b) 33,493.33 cm³

13. 75 ft³   **15.** 180 in.³   **17.** 18.33 m³   **19.** 75.36 in.³

21. 91   **23.** 1,728   **25.** about 3.7 yd³

27. the rectangular cake

29. a) 2,198 in.³   b) 2,260.8 in.³

31. a) 188.4 in.³   b) 200.96 in.³

33–34. In both cases we get a larger increase by increasing the radius, because in the formulas for volume of the cylinder and the cone, we square the radius.

35. Let d be the diameter of the can. The height of the can is $3d$; the circumference of the can is $\pi d \approx 3.14d$, which is larger.

37. radius ≈ 0.68 ft   **39.** 84 ft³   **41.** 2,164 mi

43. The volume is four times as large, because in the formula for the volume of a cone, the radius is squared.

45. The smaller cubes have more surface area. For example, compare the surface area of 1 3-inch cube with the surface area of 27 1-inch cubes.

47. 6.08 in.

49. $\pi r \sqrt{r^2 + h^2} - \pi\left(\dfrac{r}{2}\right)\sqrt{\left(\dfrac{r}{2}\right)^2 + \left(\dfrac{h}{2}\right)^2} = \dfrac{3}{4}\,\pi r \sqrt{r^2 + h^2}$

Section 8.5

1. 1,135 dg   **3.** 4,572 mm   **5.** 15.24 cm

7. 0.02159 dam   **9.** 1.77 dl   **11.** 24,000 dl

13. 34.5 hm   **15.** 8,500 dg   **17.** 1,800 g

19. 4.5 m   **21.** 350 dl

23. Because dekagrams are much larger than milligrams, we would need fewer dekagrams. Therefore, we would move the decimal point to the left.

25. The prefix "centi" reminds us of 100 and "deci" reminds us of ten. Therefore, "centi" means hundredths and "deci" means tenths.

27. b; A pencil is about 7 inches long or about 17.78 cm

29. c; Six kilograms weighs about 13.2 pounds.

31. c; A bottle of wine might be a little less than a quart (or a liter).

33. b; The dog might be about 2 feet tall or about $\frac{2}{3}$ of a meter.

35. Don't give him an inch.

37. It is first down and ten yards to go.

39. 3.22 pounds   **41.** 102.20 liters

43. 365.76 centimeters   **45.** 10.57 quarts

47. 56.21 yards   **49.** 21.34 kilograms   **51.** 1.12 pounds

53. 10.57 pounds   **55.** 38.28 yards   **57.** 49.56 tons

59. 0.91 liters   **61.** 36.58 dekameters

63. a) 160 cubic meters   b) 160,000 liters

    c) 160,000 kilograms

65. one hektare = 10,000 square meters

67. 90   **69.** 135.93   **71.** \$1.25 per pound   **73.** 60 m

75. 12.75 kilometers per liter   **77.** 65°C   **79.** 140°F

81. 68°F   **83.** 45°C

Section 8.6

1.–4.

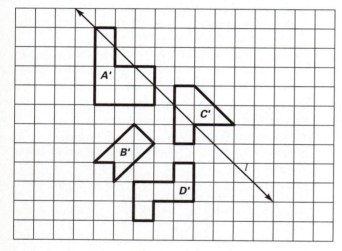

5.–6.

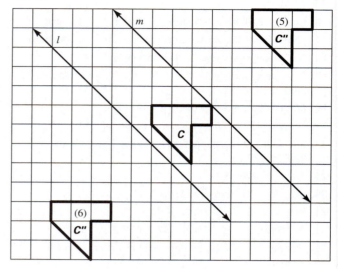

7.–8.

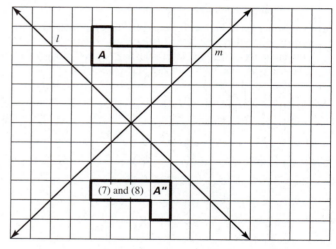

9. yes; the effect is the same as if we performed a translation

11.–12.

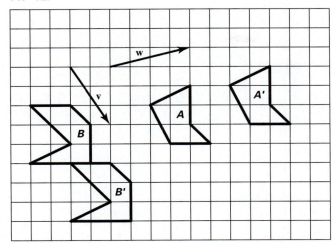

13.

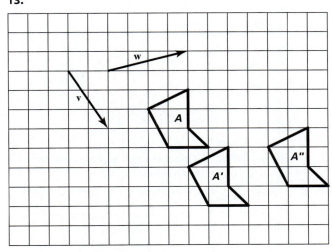

15.

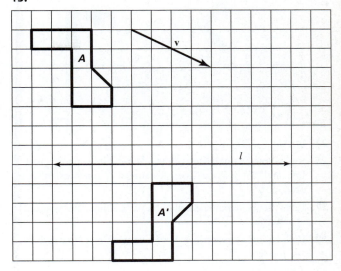

17.

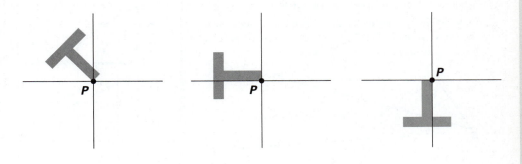

19.

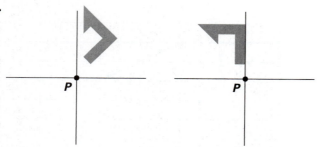

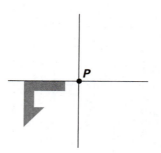

21. (c), (d)   **23.** (a), (e)

25. Reflectional symmetries: about a vertical line, a horizontal line, and two diagonal lines. Rotational symmetries: 90°, 180°, 270°.

27. Reflectional symmetries: about a vertical line. No rotational symmetries.

29. Reflectional symmetries: about a vertical line. No rotational symmetries.

31. 1,800°, 150°

33. Using the interior angles of a regular pentagon, we cannot obtain an angle sum of 360° around a point.

35.

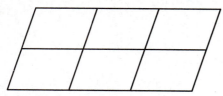

37. At the points where the vertices of the equilateral triangles and regular hexagons meet, the angle sum is 120° + 60° + 120° + 60° = 360°.

39. At the points where the vertices of the equilateral triangle, the squares, and the regular hexagon meet, the angle sum is 60° + 90° + 120° + 90° = 360°.

41. Because the sum of the interior angles of a convex quadrilateral is 360°, you can always arrange four copies of the quadrilateral around a point to make an angle sum of 360°.

43. In the figure we constructed the perpendicular bisectors of segments AA' and CC'. The point where these perpendicular bisectors meet is the center of rotation.

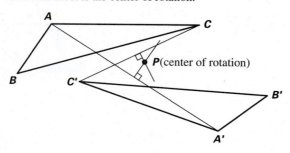

Chapter 8 Test

1. a) *b, g*

   b) *e, d*

2. $m\angle a = 42°$, $m\angle b = 138°$, $m\angle c = 48°$

3. 160°

4. $m\angle E = 45°$, length of $EH = \frac{20}{3}$, length of $GF = 4$

5. a) 80 cm² b) 20 in.²

6. a) 12 m² b) 41.12 ft²

7. a) 15.2 cm² b) 3.04 cm

8. 1,235.5 m²

9. a) 37.68 in.³ b) 330 cm³

10. 72

11. The volume will be four times as large, because in the formula for the volume of a cone, we square the radius.

12. a) 3.5 meters b) 43,150 centigrams c) 38,600 deciliters

13. 56.21 yards

14. 2,219.07 quarts

15.

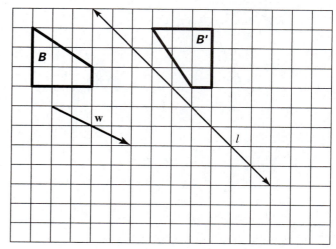

16. a) *c, e* b) *d*

17. Reflectional symmetries: about a vertical and horizontal line. Rotational symmetries: 180°.

18.

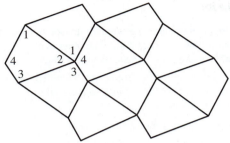

Of Further Interest

1. $4^5 = 1,024$ **2.** $(\frac{4}{3})^5 \approx 4.21$ **3.** $(\frac{4}{3})^{10} \approx 17.76$

4. false

5. a)

b) $(\frac{5}{3})^5$

6. a)

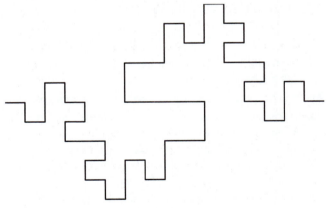

b) 2^5

7.

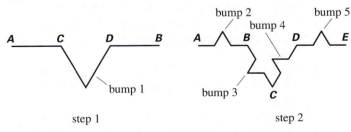

step 1   step 2

8. The dimension is the same as the Koch curve, 1.26.

9. 1.29 **10.** 1.79 **11.** 1.46 **12.** 1.5

13.

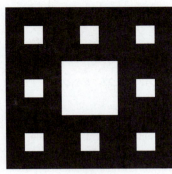

step 2

14.

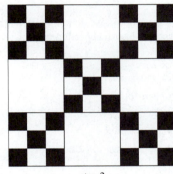

step 2

15. 5^n **16.** 7^n **17.** $(\frac{3}{4})^{10} \approx 0.06; (\frac{3}{4})^n$

18. $(\frac{8}{9})^{10} \approx 0.31; (\frac{8}{9})^n$

CHAPTER 9

Section 9.1

1. a) integer part: 24; fractional part: 0.098

b) integer part: 5; fractional part: 0.02

3. California: 4; Arizona: 4; Nevada: 3

5. painters: 14; sculptors: 8; weavers: 9

7. Alabama: 7; Mississippi: 5; Louisiana: 7

11. c) paradox occurs when assigning fourteenth seat; business

13. a) 353.7   (b) 353.785

15. 12.6, 34.2, 25.4, 11.7, 16.1

17. 34.79, 11.20, 23.90, 12.03, 2.99, 15.09

19. 12.5, 31.1, 36.0, 15.7, 4.7

21. 101.453, 96.001, 33.980, 67.133

23. 59,136   **25.** A; 75; 0.045   **27.** 15; 0.016

29. 160,000; 0.364

31. a) 0.181   b) 0.270

   c) Giving the extra representative to Naxxon results in a smaller unfairness.

33. An Alabama paradox occurs when a legislature is increased in size, but with no change in any state's population, a state loses a representative.

35. After any representative is assigned, we do not go back and reconsider its assignment in a later reapportionment.

Section 9.2

1. a) 0.162   b) 0.434   c) musicians

3. a) 0.012   b) 0.254   c) red line

5. New York   **7.** Colorado   **9.** 41,223.2   **11.** 1,470.9

13. 0.315   **15.** 0.259   **17.** Illinois

19. New Hampshire   **21.** musicians   **23.** yellow

25. artists   **27.** yes

29. The apportionment principle allows us to assign representatives one at a time, at each stage, giving the next representative to the state that most deserves it. We make the assignments so as to minimize the relative unfairness of the apportionment.

31. With the Huntington–Hill method, after we assign representatives, we do not reassign the representatives if the size of the legislature changes.

33. 5.7 million

Section 9.3

1. row 1, carpenters   **3.** 104.2, 162.0, 160.2, plumbers

5. row 1, Carney   **7.** 264.5, 181.5, 308.2, Carney

9. MTDMTMTMTD

11. a) Alsace, 3; Bradford, 5; Cambria, 2

   b) BACBABCBAB

13. OR, UT, ID, OR, UT, OR, OR, ID, UT, OR

15. KS, AR, NE, KS, AR, NE, KS, AR, KS, AR, NE

17. B, B, H; B, 5; H, 4; S, 2

19. region 1, 4, region 2, 1, region 3, 2

21. one class is high-impact aerobics, the other is step exercise

25. 8

Section 9.4

1. 12,500; 2.16; 2   **3.** 20,000; 4.6; 4   **5.** 8,000; 4.6; 5

7. 4,800; 11.7; 12   **9.** 28,000; 2.65; 3

11. 8,500; 2.35; 2

13. The Hamilton method uses the standard divisor and the apportionment follows the quota rule. The other methods use modified divisors and may violate the quota rule.

15. Because the Jefferson method rounds the modified quotas down, we often need larger modified quotas and hence smaller modified divisors.

17. She needs to try a larger modified divisor.

19. 13,545.45; CA 4.134, NV 3.027, AZ 3.839

21. CA 4, NV 3, AZ 4

23. 33.2; performers 6.416, food workers 8.223, maintenance 5.361

25. performers 6, food workers 8, maintenance 6

27. 5.15; electricians 4.854, plumbers 3.495, painters 5.631, carpenters 6.019

29. electricians 5, plumbers 4, painters 5, carpenters 6

31. 4.263; fiction 7.037, poetry 4.692, technical 3.988, media 3.284

33. fiction 7, poetry 5, technical 4, media 3

35. 23.154; region 1 4.621, region 2 2.807, region 3 1.857, region 4 2.548, region 5 1.166

37. region 1 4, region 2 3, region 3 2, region 4 3, region 5 1

39. 16.667; high impact 3.36, low impact 1.74, jazzercise 0.660, step exercise 0.240

41. high impact 2, low impact 2, jazzercise 1, step exercise 1

43. Yes, because A's population has increased faster than B's, but A loses a representative to B in the reapportionment.

45. Yes, because the number of passengers per week at Bakerstown has increased faster than at Columbia City, but Bakerstown loses a security guard to Columbia City.

47. Yes; when C's representatives are added to the commission, B loses a representative to A.

49. No; when D's representatives are added to the commission, there is no change in A's, B's, or C's apportionment.

51. D's standard quota is 87.961; however, using the Jefferson method, D would receive 89 representatives, which is greater than its upper quota.

53. D's standard quota is 240.642; however, using the Adams method, D would receive 239 representatives, which is less than its lower quota.

55. smallest 926; largest 940

57. smallest 1,058; largest 1,066

Chapter 9 Test

2. Revolutionary War, 3; Civil War, 3; World War I, 2; World War II, 3

4. a) B; 26   b) 0.139

5. 11.1, 27.3, 36.9, 3.0, 21.7

6. a) SC, 0.292; MD, 0.233   b) South Carolina

8. GRBYGRBGRGBRY

9. three more classes of tae-bo and one class of karate

10. high-impact aerobics, 0; low-impact aerobics, 5; jazzercise, 1; step exercise, 2

11. high-impact aerobics, 1; low-impact aerobics, 4; jazzercise, 1; step exercise, 2

12. high-impact aerobics, 1; low-impact aerobics, 4; jazzercise, 1; step exercise, 2

13. No, because the apportionments are the same in both cases.

14. The new-states paradox occurs when a new state is added, and its share of seats is added to the legislature causing a change in the allocation of seats previously given to another state.

15. Jefferson's, Adams', and Webster's

Of Further Interest

1. a) discrete   b) continuous   c) discrete

2. a) discrete   b) discrete   c) continuous

3. a) $55,000   b) $42,500   c) Car

4. a) $13,500   b) $17,500   d) Book

5. a) $8,000   b) $16,800   d) Copyright

6. a) $42,000   b) $58,500   c) pizza stand

7. a) $21,000   b) $24,000   c) $9,000
d) $7,000   e) $8,000

8. a) $150,000   b) $150,000   c) $140,000   d) $165,000
e) $37,500   f) $37,500   g) $35,000   h) $41,250

9.

	Greg ($\frac{1}{2}$)	Dharma ($\frac{1}{2}$)
Pays to estate ($+$) or receives from estate ($-$)	$55,000 ($+$)	$42,500 ($-$)
Division of estate balance ($12,500)	$6,250 ($-$)	$6,250 ($-$)
Summary of cash	Pays $48,750	Receives $48,750

10.

	Karl ($\frac{1}{2}$)	Friedrich ($\frac{1}{2}$)
Pays to estate ($+$) or receives from estate ($-$)	$13,500 ($-$)	$17,500 ($+$)
Division of estate balance ($4,000)	$2,000 ($-$)	$2,000 ($-$)
Summary of cash	Receives $15,500	Pays $15,500

11.

	Alex (40%)	Marianne (60%)
Pays to pool ($+$) or receives from pool ($-$)	$8,000 ($-$)	$11,200 ($+$)
Division of pool balance ($3,200)	$1,280 ($-$)	$1,920 ($-$)
Summary of cash	Receives $9,280	Pays $9,280

12.

	Matt (35%)	Tony (65%)
Pays to pool ($+$) or receives from pool ($-$)	$78,000 ($+$)	$58,500 ($-$)
Division of pool balance ($19,500)	$6,825 ($-$)	$12,675 ($-$)
Summary of cash	Pays $71,175	Receives $71,175

13.

	Ed ($\frac{1}{3}$)	Al ($\frac{1}{3}$)	Jerry ($\frac{1}{3}$)
Items obtained with highest bid	Painting		Statue
Pays to estate (+) or receives from estate (−)	$4,000 (+)	$7,000 (−)	$7,000 (+)
Division of estate balance ($4,000)	$1,333.33 (−)	$1,333.33 (−)	$1,333.33 (−)
Summary of cash	Pays $2,666.67	Receives $8,333.33	Pays $5,666.67

14.

	Franz ($\frac{1}{4}$)	Ida ($\frac{1}{4}$)	Bill ($\frac{1}{4}$)	Monica ($\frac{1}{4}$)
Stock obtained with highest bid	Pharmaceutical		Oil	Computer
Pays to pool (+) or receives from pool (−)	$2,500 (+)	$37,500 (−)	$10,000 (+)	$48,750 (+)
Division of pool balance ($23,750)	$5,937.50 (−)	$5,937.50 (−)	$5,937.50 (−)	$5,937.50 (−)
Summary of cash	Receives $3,437.50	Receives $43,437.50	Pays $4,062.50	Pays $42,812.50

15. Betty receives the books, Dennis receives the ring and desk; Dennis pays $9,125 to Betty.

16. Matt receives D, Dani A, Christian B and C; Christian pays $161,666.67; Matt receives $138,333.33; Dani receives $23,333.33.

CHAPTER **10**

Section 10.1

1. no; Edelson   **3.** A (33 votes)   **5.** D

7. C (23 votes)   **9.** M   **11.** T (15 votes)   **13.** G

15. L (17 votes)   **17.** L

19. The Borda count method is being used, with three points for first, two for second, and one for third; (a) 98   (b) 137   (c) 67   (d) 58

21. The first eliminated is in last place, the second eliminated is in second-last place, and so on.

23. A, D, C, S   **25.** D, A, C, S

27. T, G, E, P, F   **29.** G, T, E, P, F

31. With only two candidates, whoever has more votes (assuming no tie) will have a majority of the votes.

33. There is no third candidate to eliminate, so whoever wins is the candidate with the majority of first-place votes.

35. 200   **37.** 110   **39.** 6   **41.** 4

Section 10.2

1. B

5. C wins; B defeats everyone else head to head.

7. B wins; if we remove A, then C defeats B 105 to 104.

9. C wins; however, A defeats all other options head to head.

11. B wins; No; no one has a majority of first place votes.

13. There are many correct answers; one is (a) 5;   (b) 4.

19. There are many correct answers; one is (a) 30.

Section 10.3

1. a) 5   b) each voter has 1 vote
   c) no dictator   d) each voter has veto power

3. a) 11   b) A, 10 votes; B, 3; C, 4; D, 5
   c) no dictator   d) no voter has veto power

5. a) 15   b) A, 1 vote; B, 2; C, 3; D, 3; E, 4
   c) no dictator
   d) no voter has veto power (note: no resolutions can be passed)

7. a) 12   b) A, 1 vote; B, 3; C, 5; D, 7
   c) no dictator   d) C and D have veto power

9. a) 25   b) A, 4 votes; B, 4; C, 6; D, 7; E, 9
   c) no dictator   d) C, D, and E have veto power

11. a) 51   b) A, 20 votes; B, 20; C, 20; D, 10; E, 10

   c) no dictator   d) no voter has veto power

13. {C, D}, {A, C, D}, {B, C, D}, {A, B, C, D}

15. {A, C, D, E}, {B, C, D, E}, {A, B, C, D, E}

17. C and D are critical in all coalitions.

19. All are critical in {A, C, D, E} and {B, C, D, E}; C, D, and E are critical in {A, B, C, D, E}.

21.

Coalition	Weight	
{P}	5	
{T}	4	
{S}	2	
{P, T}	9	Winning
{P, S}	7	Winning
{T, S}	6	Winning
{P, T, S}	11	Winning

23.

Coalition	Weight	
{A}	3	
{F}	4	
{T}	3	
{N}	2	
{A, F}	7	
{A, T}	6	
{A, N}	5	
{F, T}	7	
{F, N}	6	
{T, N}	5	
{A, F, T}	10	Winning
{A, F, N}	9	Winning
{A, T, N}	8	Winning
{F, T, N}	9	Winning
{A, F, T, N}	12	Winning

25. All voters are critical in {P, T}, {P, S}, and {T, S}; none are critical in {P, T, S}

27. All voters are critical in all coalitions, with the exception of {A, F, T, N}, which has no critical voters.

29. Each voter has index $\frac{1}{5}$.

31. A and B have index $\frac{1}{11}$; all others have index $\frac{3}{11}$.

33. A and B have index 0; C and D have index $\frac{1}{2}$.

35. K and C have index $\frac{13}{36}$; all others have index $\frac{1}{18}$.

37. All have index $\frac{1}{6}$.

39. All have index $\frac{1}{5}$.

41. A has index 1.

Chapter 10 Test

1. no; Myers   **2.** A   **3.** social justice   **4.** B

5. D wins; no; R has the majority of first-place votes.

6. R wins; yes; R beats all others head to head.

7. B wins; no; if A is removed, then C wins with 80 points.

8. A wins; yes; if either B or C is removed, then A still wins the election.

9. quota, 17; A, 1 vote; B, 5; C, 7; D, 8; no dictator; B, C, and D have veto power

10. Assume that A has 2 votes, B has 3, C has 5, and D has 7. Winning coalitions: {C, D}, {A, B, D}, {A, C, D}, {B, C, D}, {A, B, C, D}

11. A and B have index $\frac{1}{10}$, C has index $\frac{3}{10}$, and D has index $\frac{1}{2}$.

12. a) Each has index $\frac{1}{4}$.   b) A is a dictator with index 1.

Of Further Interest

1. (X, Y, Z), (X, Z, Y), (Y, X, Z), (Y, Z, X), (Z, X, Y), (Z, Y, X)

2. (P, Q, R), (P, R, Q), (Q, P, R), (Q, R, P), (R, P, Q), (R, Q, P)

3. There are 24 permutations: (W, X, Y, Z), (W, X, Z, Y), (W, Y, X, Z), . . . , (Z, Y, X, W)

4. There are 24 permutations: (P, Q, R, S), (P, Q, S, R), (P, R, S, Q), . . . , (S, R, Q, P)

5. 720   **6.** 40,320   **7.** 479,001,600

8. 6,227,020,800

9.

Coalition	Weight after 1st, 2nd, and 3rd Voter Is Added	Pivotal Voter
a) (A, B, C)	3, 5, 7	B
b) (A, C, B)	3, 5, 7	C
c) (B, A, C)	2, 5, 7	A
d) (B, C, A)	2, 4, 7	A
e) (C, A, B)	2, 5, 7	A
f) (C, B, A)	2, 4, 7	A

10.

Coalition	Weight after 1st, 2nd, and 3rd Voter Is Added	Pivotal Voter
a) (A, B, C)	4, 7, 9	B
b) (A, C, B)	4, 6, 9	B
c) (B, A, C)	3, 7, 9	A
d) (B, C, A)	3, 5, 9	A
e) (C, A, B)	2, 6, 9	B
f) (C, B, A)	2, 5, 9	A

11. A and B have index $\frac{1}{2}$; C has index 0.

12. Each has index $\frac{1}{3}$.

13. A and B have index $\frac{5}{12}$; C and D have index $\frac{1}{12}$.

14. A, B, and C have index $\frac{1}{3}$; D has index 0.

15. A has index $\frac{1}{2}$; the others have index $\frac{1}{6}$.

16. Each has index $\frac{1}{4}$.

17. Each has index $\frac{1}{5}$.

18. Each has index $\frac{1}{12}$.

19. A has index 1; the others have index 0.

20. We cannot compute the Shapley-Shubik index for any person.

21. A has index $\frac{3}{4}$; the others have index $\frac{1}{12}$.

22. Each has index $\frac{1}{3}$.

23. Each has index $\frac{1}{4}$.

24. K and C have index $\frac{2}{5}$; the associates all have index $\frac{1}{20}$.

CHAPTER 11

Section 11.1

1. 0.78 **3.** 0.08 **5.** 0.2735 **7.** 0.0035 **9.** 43%

11. 36.5% **13.** 145% **15.** 0.2% **17.** 75%

19. 31.25% **21.** 250% **23.** 1.6% **29.** 98

31. 15% **33.** 350 **35.** 19.6 **37.** 17.5% **39.** 128

41. 19.4% **43.** $290,000 **45.** 27.8% **47.** 54.3%

49. 23.9% **51.** 30.2% **53.** 249 million

55. 1,061 million **57.** 8%

59. She divided by 16,065 instead of 14,875.

61. $37,800 **63.** $680 **65.** 25% **67.** 11.5%

69. $12,800 **71.** 24 mpg **73.** $10,794.52

75. No. The price after the reduction will be less than the original price.

77. No. It is the same as an increase by 32%.

Section 11.2

5. $I = \$240$ **7.** $r = 5$ percent **9.** $A = \$3,100$

11. $P = \$1,500$ **13.** $r = 6$ percent

15. (a) $4,896; (b) $1,296 **17.** $66.67 **19.** $5,000

21. 12.5 percent **23.** $4,725 **25.** $2,060

27. 75 percent

29.

	I (Interest)	A (Value of Account)
End of year 1	$160	$2,160
End of year 2	$172.80	$2,332.80
End of year 3	$186.62	$2,519.42

31. 1.5 percent **33.** 12/365 percent **35.** $6,381.41

37. $4,686.64 **39.** $23,457.76 **41.** $4,885.48

43. 4.95 percent compounded quarterly

45. $12,278.88 **47.** 2.096 **49.** 14.207 **51.** 2.1544

53. 1.7783 **55.** 4.56 percent **57.** 9.65 percent

59. 10.34 **61.** 31.49

63. (a) 4.74 percent; (b) 4.67 percent **65.** 10.07 years

67. $8.40 **69.** $76,689.31 **71.** $1,712.15

73. 7.76 percent **75.** 6.14 percent

77. 10.5170863 percent **79.** 10.5170918 percent

Section 11.3

1. $46.50 **3.** $37.40 **5.** $243.20; $63.4

7. $3,019.68; $197.91 **9.** 288 percent

11. 384 percent

13. (a) $313.50 (b) down to $256.50 (c) $3,982.46

15. (a) $536.67 (b) down to $429.34 (c) $6,614.91

17. $4.38 **19.** $7.96 **21.** $9.67 **23.** $5.37

25. $4.95 **27.** $6.77 **29.** $5.03

31. add-on method, $87.50; credit card, $82.50 **33.** $61.44

Section 11.4

1. $\dfrac{x^5 - 1}{x - 1}$ **3.** $\dfrac{x^8 - 1}{x - 1}$ **5.** $922.80 **7.** $803.01

9. $1,221.21 **11.** $23,247.07 **13.** $12,148.68

15. $23,008.28 **17.** $85,785.11 **19.** $2,435.99

21. $6,286.36 **23.** $121,417.91 **25.** $162.14

27. $193.75 **29.** $102.78 **31.** $98.31

37. 2.7268 **39.** 1.1073 **41.** 2.3219

43. 1.9527 **45.** 42.62 **47.** 30.91 **49.** 22.43

51. 61 months **53.** 47 months **55.** $141.33

57. $197,395.14

Section 11.5

1. $126.82   **3.** $193.44   **5.** $306.64   **7.** $112.37

9. $68.58

11.

Payment Number	Amount of Payment	Interest Payment	Applied to Principal	Balance
				$5,000.00
1	$126.82	$41.67	$85.15	$4,914.85
2	$126.82	$40.96	$85.86	$4,828.99
3	$126.82	$40.24	$86.58	$4,742.41

13.

Payment Number	Amount of Payment	Interest Payment	Applied to Principal	Balance
				$12,500.00
1	$306.64	$85.94	$220.70	$12,279.30
2	$306.64	$84.42	$222.22	$12,057.08
3	$306.64	$82.89	$223.75	$11,833.33

15. (a)

Payment Number	Amount of Payment	Interest Payment	Applied to Principal	Balance
				$100,000.00
1	$665.31	$583.33	$81.98	$99,918.02
2	$665.31	$582.86	$82.45	$99,835.57
3	$665.31	$582.37	$82.94	$99,752.63

(b)

Payment Number	Amount of Payment	Interest Payment	Applied to Principal	Balance
				$100,000.00
1	$765.31	$583.33	$181.98	$ 99,818.02
2	$765.31	$582.27	$183.04	$ 99,634.98
3	$765.31	$581.20	$184.11	$ 99,450.87

(c) $1.76

17. (a) $276.46   (b) $1,770.08

19. (a) $118.54   (b) $233.72

21. (a) $57.50   (b) $12.50

23. (a) $115.07   (b) $151.05   **25.** $986.13

27. The present value of the lottery winnings is $425,678.19; this is slightly better than the lump sum of $425,000.

29. $13,593.02

31. The present value of the retirement plan is $38,900.73; this is slightly worse than the lump sum of $40,000.

33. $5,416.17   **35.** $107,389.08   **37.** $4,159.37

39. (a) $240.46   (b) $88.08   **41.** (a) $786.70   (b) $8,955

43. (a) $971.60   (b) $118.08

Chapter 11 Test

1. 12.45%   **2.** 0.01365   **3.** 68.75%   **4.** 85%

5. $1,545.60   **6.** $1,770   **7.** 10 percent

8. $6,744.25   **9.** $5,496.33   **10.** 131 months

11. 3.13 percent   **12.** $326.70; $45.75   **13.** $4.57

14. $34,543.31   **15.** $50.85   **16.** 2.8928

17. 30 months   **18.** $126.82

19.

Payment Number	Amount of Payment	Interest Payment	Applied to Principal	Balance
				$100,000.00
1	$836.45	$666.67	$169.78	$ 99,830.22
2	$836.45	$665.53	$170.92	$ 99,659.30

20. The present value is $490,907.37; the lump sum is the better deal.

21. $8,374.15

Of Further Interest

1. 12 percent   **2.** 19.2 percent   **3.** $15

4. $28   **5.** $13   **6.** $22.50   **7.** 13 percent

8. 12 percent   **9.** 14 percent   **10.** 11 percent

11. 15 percent   **12.** 13 percent   **13.** 11 percent

14. 13 percent   **15.** 15 percent   **16.** 14 percent

17. 15 percent   **18.** 16 percent   **19.** (b) is better

20. (b) is better   **21.** (b) is better   **22.** (b) is better

25. over 16 percent   **26.** over 16 percent

CHAPTER **12**

Section 12.1

1. AB, AC, AD, AE, BC, BD, BE, CD, CE, DE

3. AB, AC, AD, AE, BA, BC, BD, BE, CA, CB, CD, CE, DA, DB, DC, DE, EA, EB, EC, ED

7. 24   **9.** 32   **11.** 16   **13.** 4   **15.** 6   **17.** 30

19. 120   **21.** 16   **23.** 9   **25.** 16

27. (1, 4), (2, 3), (3, 2), (4, 1)

29. (1, 1), (2, 2), (3, 3), (4, 4), (5, 5), (6, 6)

31. (1, 1), (1, 2), (2, 1), (1, 3), (2, 2), (3, 1), (1, 4), (2, 3), (3, 2), (4, 1)

35. 120   **37.** 36   **39.** 24   **41.** 12   **43.** 55

45. 27   **47.** 18

49. Follow the branches in the tree below to make the schedules.

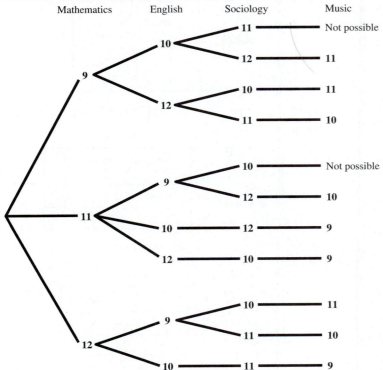

Mathematics English Sociology Music

51. RRBG, RBRG, BRRG

53. RGBR, RBGR, GRBR, GBRR, BRGR, BGRR

Section 12.2

1. 42 **3.** 336 **5.** 480 **7.** 1,560 **9.** 64

11. 96 **13.** 17,576

15. $12 \times 11 \times 10 \times \cdots \times 2 \times 1 = 479,001,600$

17. 9,000 **19.** 1,120 **21.** 2,500 **23.** 243

25. a) 256 b) over $8\frac{1}{2}$ hours **27.** 6

29. We can use a tree diagram to visualize the total number of possibilities that we count when we use the Fundamental Counting Principle.

31. 1,058,400 **33.** 720 **35.** 72

Section 12.3

1. 24 **3.** 6 **5.** 720 **7.** 120 **9.** 30 **11.** 120

13. 336 **15.** 1,814,400 **17.** 56 **19.** 45

21. In a permutation, order is important; in a combination, it is not.

23. $P(8, 8)$ **25.** $C(17, 3)$ **27.** $P(5, 5)$ **29.** $C(10, 3)$

31. $C(9, 5)$ **33.** $C(17, 8)$ **35.** 177,100 **37.** 1,865

39. $P(15, 5) = 360,360$

41. $P(15, 5) \times P(15, 5) \times P(15, 4) \times P(15, 5) \times P(15, 5) =$ a VERY large number

43. $C(6, 2) \times C(8, 3) = 15 \times 56 = 840$

45. 57,750 **47.** 60 **49.** 5,940 **51.** 3,300 **53.** 70

55. $C(13, 2) \times C(13, 3) = 78 \times 286 = 22,308$

57. 1 8 28 56 70 56 28 8 1 **59.** 21

61. the second entry in the 18th row

63. the sixth entry in the 20th row

65. Because of the special conditions on the numbers, we cannot use all of the digits at each position in the number.

67. This is the number of ways to choose no elements from a five-element set. The set we are choosing is the empty set, which can be chosen in only one way.

69. This is the number of ways to choose a four-element set from five elements—this can be done in five ways, depending on which one element we omit.

71. The number of ways to do something must be an integer.

75.  **77.** 6 **79.** $C(n - 1, r - 1)$

81. $C(n - 1, r - 1) + C(n - 1, r) = C(n, r)$

83. The number of ways that we can choose an r-element set from n elements is the same as the number of ways that we can choose $n - r$ elements from n elements; for example, the number of ways that we can choose two elements from five elements is the same as the number of ways that we can choose three elements from five elements.

Chapter 12 Test

1. PQ, PR, PS, QP, QR, QS, RP, RQ, RS, SP, SQ, SR

2. 64   **3.** 240   **4.** 2,184

5. 288   **6.** 90   **7.** 240

8. In a permutation, order is important; in a combination, it is not.

9. 40,320   **10.** 680   **11.** 1,404,000

12. $C(n, r) = \frac{P(n, r)}{r!}$ **13.** 21

14. the second entry in row 18

Of Further Interest

1. 120   **2.** 80   **3.** 20   **4.** 84   **5.** 10   **6.** 6

7. three oranges

8. three bells

9. 4

10. a) 4   b) 10   c) excluding royal flushes, $40 - 4 = 36$

11. a) 4   b) 1,287   c) $5,148 - 40 = 5,108$

12. a) 10   b) 1,024   c) 10,240   d) 10,200

13. 123,552   **14.** 1,098,240   **15.** 1,302,540

CHAPTER **13**

Section 13.1

1. {HHH, HHT, HTH, HTT, THH, THT, TTH, TTT}

3. There are sixteen pairs: {(1, 1), (1, 2), (1, 3), (1, 4), (2, 1), . . . , (4, 3), (4, 4)}

5. {(b, f), (b, r), (i, f), (i, r), (h, f), (h, r)}

7. {(i, i), (i, d), (i, s), (d, i), (d, d), (d, s), (s, i), (s, d), (s, s)}

9. {(1, 6), (2, 5), (3, 4), (4, 3), (5, 2), (6, 1)}

11. {hhh, hht, hth, thh}

13. {(b, f), (i, f), (h, f)}

15. {(i, d), (i, s), (d, i), (s, i)}

17. {(r, b), (r, y), (b, r), (y, r)}

19. {(r, r, b), (r, r, y), (r, b, r), (r, y, r), (b, r, r), (y, r, r)}

21. An outcome is an element in a sample space; an event is a subset of a sample space.

23. Rolling a total less than 13 when rolling two dice.

25. The probability that the child is a carrier is $\frac{1}{2}$.

		Second Parent	
		s	n
First Parent	s	ss	sn
	s	ss	sn

27. $\frac{8}{20} = \frac{2}{5}$ **29.** $\frac{30}{100} = \frac{3}{10}$ **31.** 3 to 2   **33.** 7 to 3

37. 0.49   **39.** 0.26

41. a)

		First-Generation Plant	
		r	r
First-Generation Plant	w	wr	wr
	w	wr	wr

b) All flowers will be pink. $P(\text{pink}) = 1$; all other probabilities are 0.

43. a)

		Second Parent	
		N	c
First Parent	N	NN	Nc
	c	cN	cc

b) $P(\text{disease}) = \frac{1}{4}$

45. 0.43   **47.** $\frac{4}{36} = \frac{1}{9}$ **49.** $\frac{13}{52} = \frac{1}{4}$ **51.** $\frac{6}{16} = \frac{3}{8}$

53. $\frac{22}{36} = \frac{11}{18}$ **55.** $\frac{2}{36} = \frac{1}{18}$ **57.** 0.000495   **59.** 0.002

61. 0.0178   **63.** 0.7747   **65.** $\frac{1}{336}$ **67.** 0.000027

69. $\frac{2}{7}$ **71.** $\frac{5}{12}$ **73.** 3 to 7; 7 to 3

Section 13.2

1. 0.985   **3.** $\frac{999}{1,000}$ **5.** $\frac{25}{36}$ **7.** $\frac{31}{32}$ **9.** 0.916

11. $\frac{7}{13}$ **13.** $\frac{11}{18}$ **15.** 0.54   **17.** 0.08   **19.** 0.10

21. 0.55   **23.** $\frac{15}{26}$ **25.** false   **27.** true   **29.** 0.92

31. 0.84   **33.** 0.65   **35.** 0.19

37. If E and F are disjoint then $P(E \cap F) = 0$.

41. $P(A) + P(B) + P(C) - P(A \cap B) - P(B \cap C)$

Section 13.3

1. $\frac{1}{6}; \frac{1}{3}$ **3.** $\frac{1}{6}; \frac{2}{11}$ **5.** $\frac{50}{61}$ **7.** $\frac{2}{61}$ **9.** $\frac{1}{2}$ **11.** $\frac{1}{10}$

13. $\frac{1}{9}; \frac{1}{3}$ **15.** $\frac{1}{3}; \frac{6}{11}$ **17.** $\frac{1}{3}$ **19.** $\frac{8}{15}$

21. Forty percent of the available spaces in dorm Y are rooms.

23. 0.50

25.

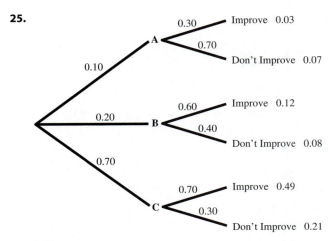

27. 0.64

31. Generally $P(B|A)$ is larger. If you consider the equation $P(B|A)P(A) = P(B \cap A)$, you see that we have to multiply $P(B|A)$ by $P(A)$, which is usually less than one to get $P(B \cap A)$.

33. a) $\frac{1}{221}$ b) $\frac{1}{169}$ **35.** a) $\frac{40}{221}$ b) $\frac{30}{169}$

37. a) $\frac{8}{663}$ b) $\frac{2}{169}$ **39.** $\frac{7}{17}$ **41.** $\frac{11}{1,105}$ **43.** $\frac{117}{850}$

45. $\frac{1}{729}$ **47.** dependent **49.** independent

51. 0.375 **53.** 0.336 **55.** 0.736 **57.** 0.418

63. 0.117 **65.** 23

Section 13.4

1. −0.5; no **3.** $1.39; no **5.** −$2.50 **7.** −$0.25

9. $2.39 **11.** $0.75 **13.** −$0.40; $0.60

15. −$2.00; $3.00 **17.** −$0.08 **19.** −$0.29

21. Student should guess; expected value is $\frac{1}{16}$.

23. two options **25.** −$7.50 **27.** $44 **29.** $25,000

31. $328 **33.** $29.80; $27.35

Chapter 13 Test

1. a) {HHT, HTH, THH}

  b) {(2, 6), (3, 5), (4, 4), (5, 3), (6, 2)}

2. $\frac{3}{26}$ **4.** $\frac{1}{4}$ **5.** a) $\frac{3}{10}$ b) $\frac{9}{11}$

6. a) $P(E') = 1 - P(E)$ **7.** $\frac{8}{13}$ **9.** $\frac{2}{13}$

10. a) $\frac{1}{17}$ b) $\frac{4}{663}$

11. no **12.** 0.385 **13.** $0.38 **14.** $0.63

Of Further Interest

1. yes **2.** yes

3. no; the number of trials is not fixed **4.** yes **5.** yes

6. yes **7.** 0.2634 **8.** 0.0498 **9.** 0.0879

10. 0.2344 **11.** k is larger than n **12.** p is greater than 1

13. 0.0537 **14.** 0.0038 **15.** 0.0730 **16.** 0.0170

17. 0.6778 **18.** 0.7840 **19.** 0.2637 **20.** 0.2734

21. 0.7472 **22.** 0.0055 **23.** $\frac{3}{4}$ **24.** 7 **25.** 14.7

26. 29.29 **27.** 0.3770 **28.** 0.4744

CHAPTER **14**

Section 14.1

1.

x	Frequency	Relative Frequency
2	1	0.05
3	0	0.00
4	0	0.00
5	3	0.15
6	2	0.10
7	4	0.20
8	4	0.20
9	3	0.15
10	3	0.15

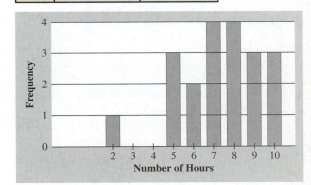

3.

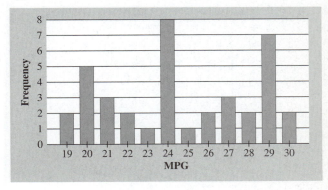

5.

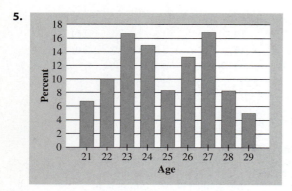

9.

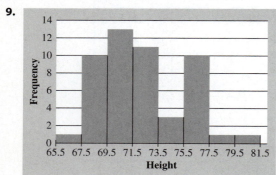

11.

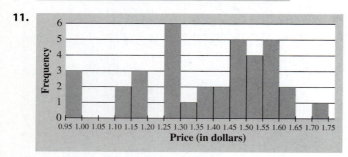

13. a) 0 occurred thirteen times   b) 12 occurred two times

c) 6   d) 71   e) $\frac{22}{71}$

15. a) 8   b) 3   c) 12   d) 16   **17.** 10.2 percent

19. 25 and older, earning at least $10.00 per hour

21. 16 to 19, earning at $5.15 or less per hour

23.

	A			B
	8 8	1		3 6 8
	9 6 2 2 1	2		0 1 1 2 2 2 3 4 6 8 9
	9 9 8 7 4 3 2	3		2 3 3 8 9
	7 3 3 3 2 2	4		7

25.

No Training				With Training
	9	1		
	9 8 8 7 6 6 3 2	2		1 8 8 9 9
	4 2	3		2 2 3 6 6 7 9
	3 3 2 1 0	4		1 3 4 5

35. 234, 245, 249, 257, 268, 277, 282, 323, 345, 362, 373, 378, 412, 434, 482, 489, 493

Section 14.2

1. mean, 6.3; median, 6; mode, 4

3. mean, 6.9; median, 6.5; no mode

5. mean, 5; median, 5; mode, 5

7. mean, 7.3; median, 7.5; mode, 7, 8

9. mean, 128.7; median, 44; no mode

11. mean, 24.7; median, 25; mode, 24

13. mean, 16.3; median, 16; mode, 17

15. mean, 20.2; median, 17; no mode

17. mean, 7.4; median, 8; mode, 10

19. mean, 10.04; median, 8; mode, 8   **21.** 39.4

23. mean, 24.28; median, 25; mode, 24, 25

25. mean; mean   **27.** a) 62   b) 112   **33.** $130.32

35. a) min, 11; Q_1, 23; median, 31; Q_3, 44; max, 53

b)

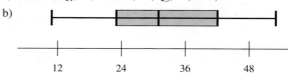

37. a) min, 25; Q_1, 30; median, 33; Q_3, 41; max, 49

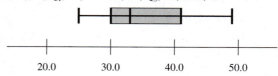

39. (Ruth)

(Aaron)

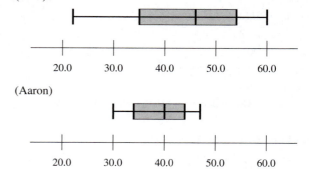

41. liberal arts   **43.** education; no

45. The maximum and minimum salaries are not too far from the middle 50 percent of the salaries.

47. The countries no doubt have different-sized populations; if you think in terms of a frequency table, the countries with large populations contribute more than the smaller countries.

Section 14.3

1. 5; 19.2; 1.55

3. range, 6; mean, 7; standard deviation, 2

5. range, 12; mean, 8; standard deviation, 4.34

7. range, 15; mean, 9; standard deviation, 4.57

9. range, 0; mean, 3; standard deviation, 0   **11.** 7; 2.13

13. mean, 4; standard deviation, 1.60

15. mean, 7; standard deviation, 1.73

17. 79; 6.59   **19.** 167; 22.88   **21.** 4.41; 3.38

23. false   **25.** true

27. Family H's income is 1.25 standard deviations above the mean

29. D

31. *CV* for Magnum is 9.9%; *CV* for GCI is 9.2%; Magnum is more volatile

33. DJIA has a *CV* of 0.6%; WebMaster has a *CV* of 4.5%; WebMaster is more volatile

35. coffee, 0.098; gasoline, 0.159

37. Family G went from the mean income to 0.94 standard deviations above the mean.

Section 14.4

1. 34%   **3.** 2.5%   **5.** 16%   **7.** 16%   **9.** 97.5%

11. 13.5%   **13.** a) 50%   b) 16%

15. a) 34%   b) 16%   **17.** 39.1%   **19.** 11.1%

21. 27.3%   **23.** 8%   **25.** 12.8%   **27.** 7.3%

29. 8.1%   **31.** 90.8%   **33.** 80%   **35.** 1.28

37. −1.175   **39.** 0.2525   **41.** −0.673

43. The score is almost two standard deviations above the mean.

45. By looking at the graph of the normal curve, we see that there is clearly more area under the curve between 0.0 and 0.7 than there is between 1.3 and 2.0.

47. 1.4   **49.** −1.75   **51.** 0.97   **53.** 64.2

55. 33.65   **57.** 31.69   **59.** 1.67

61. $0.092 \times 324 = 29.808$ (we expect about 29 or 30)

63. $69.2\% \times 200 = 138.4$

65. 9 days; 21.8%   **67.** 53.5

69. Robinson was 2.3 standard deviations above the mean batting average for the 1940s and Carew was 2.81 standard deviations above the mean for the 1970s. Carew was more dominant.

71. 0.84   **73.** 42.68   **75.** 96th   **77.** about 10 months

79. a) 2.685 (2 or 3 years)   b) 1.26 (1 or 2 years)

81. 42.1%

83. mean is 20 in both distributions; standard deviation is about 2 or 3 in distribution 1; standard deviation is about 4 or 5 in distribution 2

Chapter 14 Test

1.

Number of Accidents	Frequency
4	1
5	3
6	6
7	2
8	10
9	4
10	5

2.

Number of Accidents	Relative Frequency
4	0.03
5	0.10
6	0.19
7	0.06
8	0.32
9	0.13
10	0.16

3.

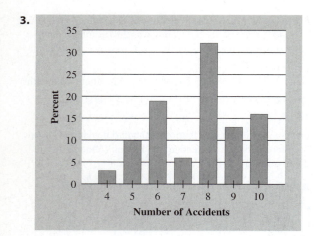

4. a) four riders; three times   b) five riders   c) 37

5.

```
        A            |   |        B
              9      | 1 |
        9 8 7 6 6 5  | 2 | 1 6 6 8 8 9
      7 6 4 3 2 2 1  | 3 | 2 3 6 6 7 9 9
            7 3 0    | 4 | 1 3 4 5
                2    | 5 | 3
```

The programs are much the same in their effectiveness. B might be a little better.

6. 8.35; 7; 7   **8.** 3, 6, 8, 12, 20

9.

10. 1.31   **13.** 0.71   **14.** 66.75

15. the history exam

Of Further Interest

1. positive correlation

2. significant negative correlation

3. $r = 0.99$

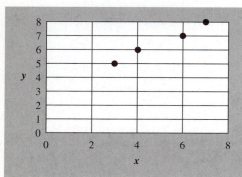

4. $r = 0.59$

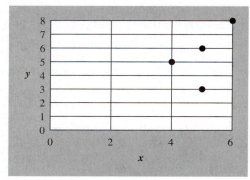

5. $r = 0.74$

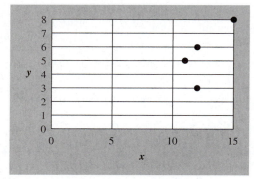

6. $r = -0.42$

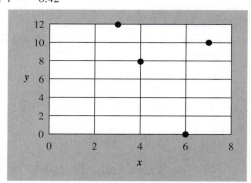

7. we can be 95% confident, but not 99% confident

8. neither   **9.** neither   **10.** neither

11. $y = 0.7x + 3$ **12.** $y = 1.5x - 2$

13. $y = 0.89x - 5.61$ **14.** $y = -1.2x + 13.5$

15. 0.96; we can be 99% confident that there is positive linear correlation

16. -0.78; neither   **17.** 0.00; neither   **18.** -0.80; neither

19. $y = 2.64x + 21.71$ **20.** $y = -0.09x + 119.59$

21. $y = 0 \cdot x + 29.4$ **22.** $y = -0.88x + 71.45$

24. 1   **25.** 1

INDEX